Lecture Notes in Computer Science 2601

Edited by G. Goos, J. Hartmanis, and J. van Leeuwen

Springer
Berlin
Heidelberg
New York
Barcelona
Hong Kong
London
Milan
Paris
Tokyo

Marco Ajmone Marsan Giorgio Corazza
Marco Listanti Aldo Roveri (Eds.)

Quality of Service in Multiservice IP Networks

Second International Workshop, QoS-IP 2003
Milano, Italy, February 24-26, 2003
Proceedings

Series Editors

Volume Editors

Marco Ajmone Marsan
Politecnico di Torino, Dipartimento di Elettronica
Corso Duca degli Abruzzi 24, 10129 Torino, Italy
E-mail: ajmone@polito.it

Giorgio Corazza
Università di Bologna, DEIS
Viale Risorgimento 2, 40136 Bologna, Italy
E-mail: gcorazza@deis.unibo.it

Marco Listanti
Aldo Roveri
Università di Roma "La Sapienza", Dipartimento di INFOCOM
Via Eudossiana, 18, 00184 Roma, Italy
E-mail: {marco/roveri}@infocom.ing.uniroma1.it

Cataloging-in-Publication Data applied for

A catalog record for this book is available from the Library of Congress

Bibliographic information published by Die Deutsche Bibliothek
Die Deutsche Bibliothek lists this publication in the Deutsche Nationalbibliografie; detailed bibliographic data is available in the Internet at <http://dnb.ddb.de>.

CR Subject Classification (1998): C.2, D.2, H.4.3, K.6

ISSN 0302-9743
ISBN 3-540-00604-4 Springer-Verlag Berlin Heidelberg New York

Springer-Verlag Berlin Heidelberg New York
a member of BertelsmannSpringer Science+Business Media GmbH

http://www.springer.de

Typesetting: Camera-ready by author, data conversion by Boller Mediendesign
Printed on acid-free paper SPIN 10872718 06/3142 5 4 3 2 1 0

Preface

IP has become the dominant paradigm for all sorts of networking environments, from traditional wired networks to innovative networks exploiting WDM links and/or wireless accesses. Key to the success of IP is its flexibility that allows the integration within one network of a number of systems which at lower layers may have different characteristics, as well as the transport over a common infrastructure of the traffic flows generated by a variety of applications (web browsing, e-mail, telephony, audio and video distribution, multimedia multicasting, financial transactions, etc.) whose performance requirements may be extremely different. This situation has generated a great interest in the development of techniques for the provision of quality-of-service guarantees in IP networks offering a variety of services through a range of different user interfaces (wired and wireless, of different nature).

In 2001 and 2002, the Italian Ministry of Education, Universities and Research funded four research programmes on these topics, named IPPO (IP packet optical networks), NEBULA IP (techniques for end-to-end quality-of-service control in multidomain IP networks), PLANET-IP (planning IP networks), and RAMON (reconfigurable access module for mobile computing applications).

At the end of their activity, these four programmes organized in Milan, Italy in February 2003 the *Second International Workshop on QoS in Multiservice IP Networks* (QoS-IP 2003), for the presentation of high-quality recent research results on QoS in IP networks, and for the dissemination of the most relevant research results obtained within the four programmes.

This volume of the LNCS series contains the proceedings of QoS-IP 2003, including 1 invited paper as well as 53 papers selected from an open call. These very high quality papers provide a clear view of the state of the art of the research in the field of quality-of-service provisioning in multiservice IP networks, and we hope that these proceedings will be a valuable reference for years to come.

This volume benefitted from the hard work of many people: the researchers of IPPO, NEBULA, PLANET-IP, and RAMON, the paper authors, the reviewers, and the LNCS staff, Alfred Hofmann in particular. We wish to thank all of them for their cooperation. Special thanks are due to Luigi Fratta and Gian Paolo Rossi for taking care of the conference organization in Milan, and to Michela Meo, Claudio Casetti, and Maurizio Munafò for their help in the preparation of the technical program of QoS-IP 2003.

December 2002

Marco Ajmone Marsan
Giorgio Corazza
Marco Listanti
Aldo Roveri

Organization

Program Committee

Program Committee Cochairs

Marco Ajmone Marsan (Politecnico di Torino)
Giorgio Corazza (Università di Bologna)
Marco Listanti (Università di Roma, La Sapienza)
Aldo Roveri (Università di Roma, La Sapienza)

Program Committee Members

Anthony Acampora
Ian F. Akyildiz
Mohammed Atiquzzaman
Tulin Atmaca
Andrea Baiocchi
Mario Baldi
Albert Banchs
Roberto Battiti
Giuseppe Bianchi
Nicola Blefari-Melazzi
Chris Blondia
Franco Callegati
Pietro Camarda
Andrew T. Campbell
Augusto Casaca
Olga Casals
Claudio Casetti
Carla-Fabiana Chiasserini
Franco Davoli
Edmundo A. de Souza e Silva
Christophe Diot
Lars Dittman
Josè Enrìquez-Gabeiras
Serge Fdida
Enrica Filippi
Viktoria Fodor
Andrea Fumagalli
Rossano Gaeta
Richard Gail
Giorgio Gallassi
Stefano Giordano
Weibo Gong
Enrico Gregori
Ibrahim Habib
Mounir Hamdi
Boudewijn Haverkort
Sundar Iyer
Bijan Jabbari
Laszlo Jereb
Denis A. Khotimsky
Daniel Kofman
Udo Krieger
Santosh Krishnan
Anurag Kumar
T.V. Lakshman
Jean-Yves Le Boudec
Luciano Lenzini
Emilio Leonardi
Christoph Lindemann
Alfio Lombardo
Saverio Mascolo
Marco Mellia
Michela Meo
Sandor Molnar
Masayuki Murata
Michele Pagano
Sergio Palazzo
Krzysztof Pawlikowski
George C. Polyzos
Ramon Puigjaner

Carla Raffaelli
Roberto Sabella
Stefano Salsano
Fausto Saluta
Andreas Schrader
Matteo Sereno
Dimitris Serpanos
Rajeev Shorey
Aleksandra Smiljanic
Josep Solè-Pareta
Tatsuya Suda
Guillaume Urvoy-Keller
Yutaka Takahashi
Tetsuya Takine
Asser Tantawi
Phuoc Tran-Gia
Sergio Treves
Adam Wolisz

Organizing Committee

Organizing Committee Cochairs

Luigi Fratta (Politecnico di Milano)
Gian Paolo Rossi (Università di Milano)

Organizing Committee Members

Antonio Capone (Politecnico di Milano)
Fiorella Moneta (Politecnico di Milano)
Elena Pagani (Università di Milano)

Sponsoring Institution

Ericsson Lab Italy

Table of Contents

Analytical Models I

QoS Routing

Measurements and Experimental Results

QoS Below IP

End-to-End QoS in IP Networks

QoS Multicast

Analytical Models II

Optical Networks

Reconfigurable Protocols and Networks

Provision of Multimedia Services

QoS in Multidomain Networks

Invited Paper

Congestion and Admission Control

Architectures and Protocols for QoS Provision

On Evaluating Loss Performance Deviation: A Simple Tool and Its Practical Implications*

Ying Xu and Roch Guérin

Department of Electrical and Systems Engineering, University of Pennsylvania
(yingx,guerin)@ee.upenn.edu

Abstract. The focus of this paper is on developing and evaluating a practical methodology for determining if and when different types of traffic can be safely multiplexed within the same service class. The use of class rather than individual service guarantees offers many advantages in terms of scalability, but raises the concern that not all users within a class see the same performance. Understanding when and why a user will experience performance that differs significantly from that of other users in its class is, therefore, of importance. Our approach relies on an analytical model developed under a number of simplifying assumptions, which we test using several real traffic traces corresponding to different types of users. This testing is carried out primarily by means of simulation, to allow a comprehensive coverage of different configurations. Our findings establish that although the simplistic model does not accurately predict the absolute performance that individual users experience, it is quite successful and robust when it comes to identifying situations that can give rise to substantial performance deviations within a service class. As a result, it provides a simple and practical tool for rapidly characterizing real traffic profiles that can be safely multiplexed.

1 Introduction

To design a successful QoS service model, it is often necessary to trade-off *both* the performance guarantees that the service model can provide *and* the associated operational and management complexity so that the QoS guarantees needed by end users can be provided while the service model is kept simple enough to meet deployment constraints. Most recently, there has been a renewed interest in investigating aggregate QoS solutions, due to the fact that they can greatly improve scalability, and therefore reduce the associated operational and management complexity. In a typical aggregate QoS service model, individual flows are multiplexed into a common service class and treated as a single stream when it comes to QoS guarantees, which eliminates the need for per-flow based state and processing. Such an approach is reflected in recent proposals such as the IETF Diff-Serv model [3]. By coarsening the different levels of QoS that the network offers into a small number of service classes, the Diff-Serv model is expected to

* This work was supported in part through NSF grant ITR00-85930.

M. Ajmone Marsan et al. (Eds.): QoS-IP 2003, LNCS 2601, pp. 1–18, 2003.

scale well. However, this gain in scalability through aggregation comes at the cost of losing awareness of the exact level of performance that an individual user experiences. More specifically, because performance is now monitored and enforced only at the aggregate class level, it is not clear whether performance guarantees extend to all individual users in the service class. In particular, it is possible that a given user experiences poor performance without this being noticed at the class level.

This issue was explored in an earlier work [9], which focused on the loss probability as the relevant metric. [9] developed a number of *explicit* models to evaluate the individual loss probabilities and the overall loss probability, in which user traffic are represented by either Markov ON-OFF sources or periodic sources. Using these analytical models, the deviation between individual loss probabilities and the overall loss probability was evaluated for a broad range of configurations. The major conclusion was that there are indeed cases where significant performance deviations can exist, and that user traffic parameters can have a major influence on whether this is the case or not.

In this paper, we go beyond the work of [9] by developing a simple methodology capable of evaluating loss performance deviations in realistic settings. Specifically, we concentrate on the ability to predict performance deviations when aggregating *real* traffic sources such as voice and video, which may not be accurately captured by the simplified source models introduced in [9]. In contrast to the existing body of work on call admission control, which is aimed at dynamically deciding whether or not to accept a flow-level request, our methodology is aimed at providing *off-line* decisions about whether or not to assign different types of traffic streams to the same service class, given their source statistics. Our investigation starts with mapping real traffic sources onto "equivalent" Markovian sources. This mapping is, however, limited to matching first order statistics. Our rationale for using such an approach is, as is discussed later in the paper, based on our expectation that performance deviations are less sensitive to source modeling inaccuracies than absolute performance measures. The efficacy of this simple approach is then evaluated by means of simulations for a broad range of scenarios where multiple real-time traffic sources, i.e., voice and/or video sources, are aggregated, which we believe can also help gain a better understanding of whether and when an aggregate service model is suitable for supporting real-time applications. The reliability and robustness of the approach are further analyzed in Section 6, where we use worst case analysis to explicitly evaluate its intrinsic limitations in predicting performance deviations using only first order statistics.

The rest of the paper is structured as follows: Section 2 describes our model and methodology for loss deviation prediction. Sections 3 and 4 report experimental results for two different boundary configurations, while results for the intermediate configuration are reported in Section 5. In Section 6, we conduct a worst case analysis of the performance of our solution. We concentrate on the main results, and most of the derivations are relegated to a technical report [10]. Finally, Section 7 summarizes the findings and implications of the paper.

2 Model and Methodology

The system we consider consists of multiple users that belong to a common service class, each of which is connected through its access link to an aggregation point, which is modeled as a single server, finite buffer, FIFO queue. In this section, we review the different configurations that we consider and the performance measures that we use to evaluate loss performance deviation. The voice and video traffic sources and their characteristics are then described together with the method we use for mapping them to sources that our analytical model can deal with. The simulation environment that we use is also briefly presented at the end of the section. Due to space constraints, we omit detailed descriptions of the analytical models used, as they were presented in [9].

2.1 System Parameters and Loss Deviation Evaluation

System Parameters. The first parameter we vary is the number of users aggregated in the same service class, i.e., the number of traffic sources multiplexed in the FIFO queue. In particular, we focus on two cases: a two-source configuration and a "many-source" one, which correspond to two different boundary conditions, i.e., an environment where few large bandwidth connections share resources, and one where many "small" flows are multiplexed into the same queue. Another system parameter we consider is the size of the FIFO queue into which flows are multiplexed. In our simulations, we consider both a system with a relative small buffer size and another system with a relative large buffer size. In most cases, the qualitative behavior of the two systems is similar, and thus typically only results corresponding to one of them are reported.

Evaluation Methodology. In a system with N users, let P_L denote the overall loss probability, P_L^i the loss probability of user i, and P_{max}^N the maximum loss probability experienced by an individual user (without loss of generality, it is usually assumed to be user N) among all the users. We first evaluate the *minimum* amount of bandwidth C needed to ensure an overall loss probability $P_L \leq P_{max}$, where P_{max} corresponds to the target loss probability of the service class. We then compute the *maximum* loss probability ratio $PLR = P_{max}^N / P_L$. In addition, we also evaluate, the *minimum* percentage by which the bandwidth allocated to the service class needs to be increased to ensure that $P_{max}^N \leq P_{max}$. In other words, if C_N denotes the minimum amount of bandwidth needed so that $P_{max}^N \leq P_{max}$, we evaluate the quantity $(C_N - C)/C$.

The (maximum) loss probability ratio reflects the magnitude of the loss deviation when service guarantees are only provided at the aggregate level, and is therefore the metric we use to assess whether traffic aggregation is safe or not. Conversely, the additional bandwidth needed measures the cost of eliminating deviations for a given traffic mix. In addition to these two metrics, we also introduce a "consistency test" that measures the ability of our model to successfully predict scenarios where severe performance deviations will arise. The test relies on a "threshold" region, $[PLR_{min}, PLR_{max}]$, that is used to separate risky traffic mixes, i.e., mixes that give rise to large deviations, from safe ones. Specifically, if the loss probability ratio is less than PLR_{min}, the traffic mix is considered "definitely safe", if it is greater than PLR_{max}, the traffic mix is deemed "definitely dangerous", and if it falls between those two values, i.e., inside the threshold region, the traffic mix is flagged as potentially risky and requiring additional investigations. Ideally, the threshold region should be as narrow as possible, but depending on the accuracy of the model, too narrow a region may either yield a large number of false alarms or fail to properly identify harmful configurations. A wider threshold region reduces those problems at the cost of some imprecision in terms of classifying certain scenarios. We believe that introducing such a "grey" region provides greater flexibility, and can help minimize the impact of inaccuracies inherent in our evaluation methodology. Our hope is that a reasonably narrow threshold region is sufficient to flag the various traffic mixes for which the model might provide inaccurate answers.

2.2 Real-Time Traffic Traces and Mapping Methodology

Source Traces. The video traces we use to conduct our study are obtained from an online video trace library [2]. We chose this public library because of the diversity of the traces it provides. The library contains traces of 26 different hour-long video sequences, each encoded using two different standards (the MPEG4 and the H.263 standard), and under three different quality indices (low, medium and high). For each recorded trace, statistics reflecting its characteristics are also available. We refer to [4] for a detailed description of the trace collection procedure and the methodology used to extract the trace statistics.

Table 1. Video Trace Statistics

Trace Name	Mean Frame Size (Bytes)	Mean Bit Rate (Kb/sec)	Frame Size (Peak/Mean)	Frame Size (Covariance)
Jurassic (low)	768.6	153.7	10.6	1.4
Jurassic (high)	3831.5	766.3	4.4	0.6
Lecture (low)	211.7	42.3	16.2	2.3

In our study, we rely mainly on 3 different sequences encoded using the MPEG4 codec with a frame rate of 25 frames/sec. These include the low quality and high quality traces of the movie Jurassic Park I and the low quality trace of a lecture video. The high quality trace of Jurassic Park I (Jurassic high) represents a video source that requires high transmission rate due to its quality requirement. The low quality trace of Jurassic Park I (Jurassic low) represents a lower rate source encoded from the same movie scene. At last, due to its scene characteristics, the low quality lecture trace (Lecture low) is selected because it has the lowest mean rate among all the traces in the video library. Several important statistics of the three video sources are given in Table 1.

Table 2. Voice Trace Statistics

Trace Name	Mean Spurt Length (ms)	Mean Gap Length (ms)	Peak Rate (Kb/sec)	Utilization
Voice	326	442	64	0.425

We also use voice traces from [7] in our investigation. In particular, the voice source that we choose corresponds to Fig. 4b in [7] and was generated using the NeVoT Silence Detector(SD) [8]. We refer to [7] for a complete description of the codec configuration and trace collection procedure. In Table 2 important statistics of the voice source are given.

Converting to Markov ON-OFF Sources. Due to the nature of voice and video traffic generation, both the voice traffic and the video traffic inherit built-in ON-OFF patterns, which facilitates the usage of a Markov ON-OFF source to model them. A Markov ON-OFF source can be represented by a 3-tuple $< R, b, \rho >$, where R is the transmission (peak) rate when the source is active, b is the average duration of an active or ON period, and ρ represents the fraction of time that the source is active, or its utilization. Our task, is thus to capture a real traffic source using the above 3-tuple, based on its characteristics reflected by statistics given in Table 1 and 2.

The raw video traffic consists of variable size frames that arrive at regular time intervals (40ms for the traces we use). In our study, we assume that a *whole* frame arrives instantaneously, so that the access link serves as a shaper, bounding the maximum rate of the output traffic by its own transmission speed. Assuming a fluid model, the output traffic can be thought of as a distorted version of the input traffic, where the frame transmission (ON period) extends over a time interval that is a function of the frame size and the input link speed. If the link speed is too slow to transmit a frame in 40ms, consecutive frames will be concatenated and form an extended burst. In the experiments that we conduct, we deliberately eliminate such frame concatenation by setting the link speed larger than the peak frame-level bit rate [1]. The shaped traffic emanated by the access link can thus be modeled by an equivalent ON-OFF source, with:

$$R = R_{link}; \quad b = \frac{F_{avg}}{R_{link}}; \quad \rho = \frac{b}{40 \cdot 10^{-3}} \ . \tag{1}$$

where R_{link} is the access link speed and F_{avg} is the average frame size of the video trace.

We do not expect that such a simple source model will fully capture all the complex traffic characteristics of a video source. Specifically, although the mean of the burst duration is accurately captured, its distribution does not have to

[1] We found that if the access link speed is less than the peak frame-level bit rate, then several extremely long bursts may occur and the delay at the access link could be quite large.

be exponential. This may affect the accuracy of estimating the absolute loss probability. However, we believe this is still a reasonable approach based on the experience obtained from [9], where we saw that deviations were to a large extent a function of the first order statistics of a source, e.g., the peak rate, the utilization, and the average burst duration, among which the first two parameters usually played a dominant role. Furthermore, as discussed in Section 6, there are actually many cases for which the differences between the model's prediction and the actual deviation can be shown to be small, and most important, independent of the burst duration distribution.

As with video sources, the traffic of voice sources can be divided into "ON" and "OFF" periods, corresponding to the talk spurt and silence period in human speech. Again, while those periods need not always be exponentially distributed, our expectation is that this will only minimally impact our ability to predict loss probability ratios when mixing voice and other traffic sources. Hence, the mapping of voice sources onto ON-OFF sources is carried out using the values shown in Table 2, i.e., $R = 64$Kb/sec; $b = 326$ms and $\rho = 0.425$. Note that since voice traffic is reasonably smooth, i.e., during a burst period the voice source generates multiple packets of 80 Bytes with an equal spacing of 10ms, the peak rate of the equivalent ON-OFF voice source is fixed at 64Kb instead of equal to the link capacity[2]. As we shall see later, this smoothness enables the voice source to avoid major performance degradation when aggregated with video sources, in spite of its small traffic contribution.

Once video and voice sources have been mapped onto Markov ON-OFF sources, we can apply the analytical models developed in [9] to evaluate both individual and overall loss probabilities. Specifically, in the cases where only two sources are aggregated, we use Equations (1) and (2) in [9] to calculate both the individual and the overall loss probabilities. As in the many-source case, Equations (6) and (7) of [9] are applied.

2.3 Simulation Environment and Experimental Configuration

In order to assess the accuracy of our simplified model, we rely on simulations to evaluate the true loss probabilities and loss probability ratios. Moreover, in order to examine the cost associated with the aggregation process, we also conduct simulations to evaluate the amount of additional bandwidth needed. We use the NS2[1] simulator, as it can accurately model packet level behaviors. In all of our simulations, the packet size of the voice traffic is fixed at 80Bytes while the packet size of the video traffic is fixed at 100Bytes, so that a video frame larger than 100Bytes is fragmented into multiple packets. Throughout the paper, the threshold region is set as $[3, 5]$ and the target loss probability P_{max}, when fixed, is set to 10^{-4}. For simplicity, most of our experiments consider only two different types of sources, where one of the two types of sources is selected and has its parameters varied so that it experiences the larger loss probability. The parameters of the other type of sources are always fixed.

[2] The peak rate will be equal to the access link speed only when it is less than 64Kb/sec, which is lower than the values assumed in this paper.

3 Loss Deviation Prediction when Aggregating Two Sources

This section is devoted to scenarios with only two sources aggregated. This basic configuration, despite or rather because of its simplicity, provides useful insight in capturing the complex behavior of loss deviation. From [9], we know that a user experiences losses that differ significantly from the overall losses, if it both contributes significantly to the onset of congestion when active, and does it in a way that is undetected by resource allocation mechanism. When applied to two (few) sources, this rule indicates that the utilization of an individual source plays a dominant role in determining any loss deviation it experiences. Specifically, when multiplexing two sources, if one of them has a much smaller utilization than the other but a comparable peak rate, its total traffic contribution will be minor and unlikely to trigger the allocation of significant resources, yet its impact on congestion, and therefore losses, when active can be substantial. This observation was quantified in [9] using analytical models, and as we will see in this paper, also apply to the real-life traffic sources we consider. More important, we will establish that the simple methodology we have developed, is indeed capable of accurately capturing this qualitative behavior.

3.1 Aggregating Two Video Sources

In this sub-section, we consider scenarios consisting of aggregating two video sources. The first case involves multiplexing Jurassic low and Jurassic high, which corresponds to two video streams of similar characteristics but different quality and, therefore, bit rate. A second scenario considers the case of two video streams, Lecture low and Jurassic high, which differ in both characteristics and quality (bit rate). In each case, we increase the access speed of the lower bit rate sequence so that it has a larger peak rate and lower utilization than the higher bit rate sequence, and thus is expected to experience a larger loss probability. The buffer size is set to 5000 Bytes and to 2000 Bytes[3] in the first and second scenarios, respectively.

The loss probability ratio of each scenario is shown in Figure 1. From Figure 1(a), we see that when Jurassic low and Jurassic high are multiplexed, performance deviations can arise, as the loss probability ratio approaches a value of about 6. More important, the simple model we propose can be seen to accurately capture the trend and magnitude of this deviation. The relatively limited range of the loss probability ratio (it never exceeds 6) can be explained by the fact that although the two sources differ in utilization, the low bit rate source, Jurassic low, is not a negligible contributor to the overall traffic. This in turn limits the extent to which the performance experienced by the Jurassic low stream can deviate for the overall performance. In general, the impact of the relative traffic contribution of the two sources being multiplexed can be captured through an

[3] We also explore different buffer size (200Bytes) but didn't find significantly different results from the data here.

upper bound on the loss probability ratio that only involves the mean rate of the two sources. Specifically, in a configuration with N sources being multiplexed, where, without loss of generality, the N-th source is the one experiencing the larger loss probability, the loss probability ratio satisfies:

$$PLR = \frac{r_L^N/r_S^N}{r_L/r_S} \leq \frac{r_L^N/r_S^N}{r_L^N/r_S} = \frac{1}{r_S^N/r_S} \ . \tag{2}$$

where r_L^N and r_S^N correspond to the long term loss speed, i.e., the amount of data lost per unit of time, and transmission speed of source N, and r_L and r_S correspond to the overall long term loss speed and transmission speed, respectively. Equation (2) states that the loss probability ratio PLR can never exceed a value that is inversely proportional to the fraction of traffic contributed by the worst performer in a service class. The smaller the fraction this user contributes, the larger the maximum loss probability ratio it may experience. In this example, Jurassic low is the source that loses more and from Table 1, we see that $r_S^2 = 153.7$Kb/sec, which is about 16.7% of the total traffic, resulting in a loss probability ratio satisfying $PLR \leq 6$. As can be seen from Figure 1(a), as the access link speed of Jurassic low increases, the loss probability ratio will indeed approach (and never exceed) this value. Equation 2 indicates that when multiplexing two (or a small number of) sources, the more likely candidate for experiencing significant loss deviation is the source with the smallest bit rate. Furthermore, Equation 2, also shows that in the case of two sources, the higher bit rate source can never experience a loss probability of more than twice the overall loss probability.

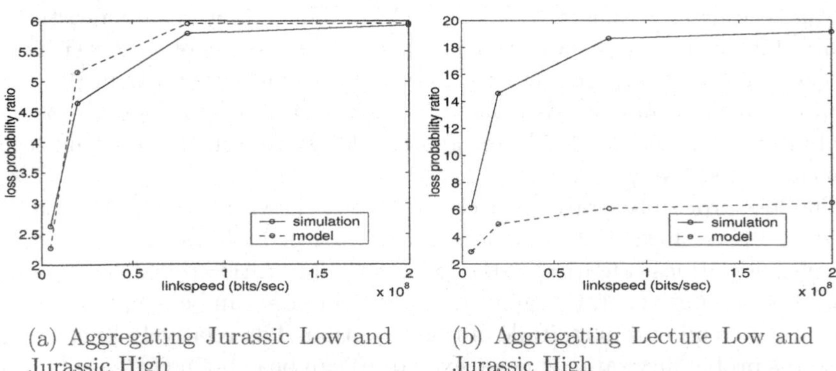

(a) Aggregating Jurassic Low and Jurassic High

(b) Aggregating Lecture Low and Jurassic High

Fig. 1. Loss Probability Ratio When Aggregating Video Sources

Figure 1(b) displays the loss probability ratio when multiplexing Lecture low and Jurassic high, for which a maximum value even greater than that of the first scenario can be attained. In particular, we see that the Lecture low source can experience losses that are up to 20 times larger than the overall losses. This is primarily due to the lower (relative) rate of Lecture low, which only contributes about 5.2% of the total traffic. From equation (2), this implies that the loss probability ratio should satisfy $PLR \leq 19.1$. From Figure 1(b), we see that as the link speed increases, the loss probability ratio indeed approaches this upper bound. The loss probability ratio predicted by the model is, however, not as accurate as in the case of Figure 1(a), as it consistently underestimates the actual loss probability ratio. This is because high order statistics of the traffic generated by both sources still influence the magnitude of the loss probability ratio. This impact notwithstanding, the model still generates a value that in most cases is sufficiently high to at least trigger a warning regarding to the potential danger of multiplexing these two sources. In particular, except for the lowest link speed, the loss probability ratio predicted by the model is always within or above the selected threshold region of [3, 5].

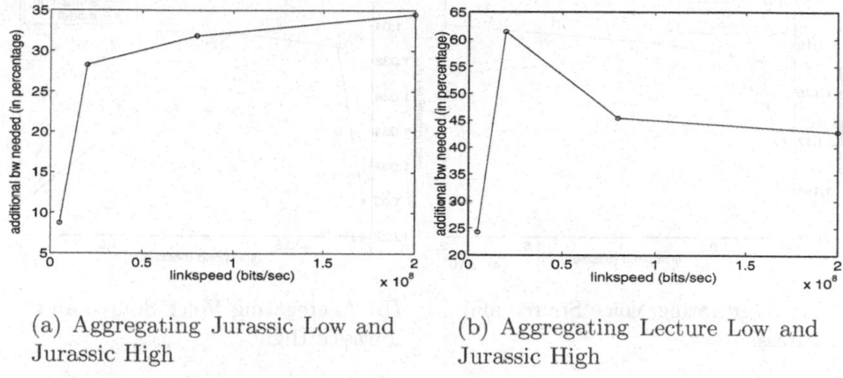

(a) Aggregating Jurassic Low and Jurassic High

(b) Aggregating Lecture Low and Jurassic High

Fig. 2. Additional Bandwidth Needed When Aggregating Video Sources

Figure 2 provides a different view on the differences in performance between the two sources. It shows the additional bandwidth (in percentage) needed in order to ensure the desired loss probability for even the poorer performer of the two sources. As can be seen, this amount can be quite large (30% for the more benign case of Jurassic low and Jurassic high, and up to 60% for the more severe case of Lecture low and Jurassic high). This means that the observed differences in performance cannot be easily fixed through standard over-provisioning, and identifying such potentially dangerous scenarios is, therefore, important.

3.2 Aggregating One Video Source and One Voice Source

In this section, we study scenarios where one video source and one voice source are multiplexed. From Table 1 and Table 2, we can see that the voice source has a mean transmission rate that is much lower than that of the video source, for both the Jurassic low and Jurassic high sources. When used in equation (2) this translates into a relatively large upper bound for PLR, which could, therefore, indicate a potentially dangerous traffic mix. This turns out not to be the case because of the reasonably smooth nature of the voice source, which makes it mostly immune to performance deterioration when multiplexed with either the Jurassic low or the Jurassic high video sources. Instead, it is the video source that experiences somewhat poorer performance. The magnitude of this performance deviation is, however, limited, because the video source is the major contributor to the overall transmission rate (equation (2) gives an upper bound of 1.04 and 1.18 for the PLR for Jurassic low and Jurassic high, respectively). As a result, our main concern in these scenarios is whether the model will generate a false alarm, i.e., falsely predict that mixing a voice and a video sources is potentially dangerous.

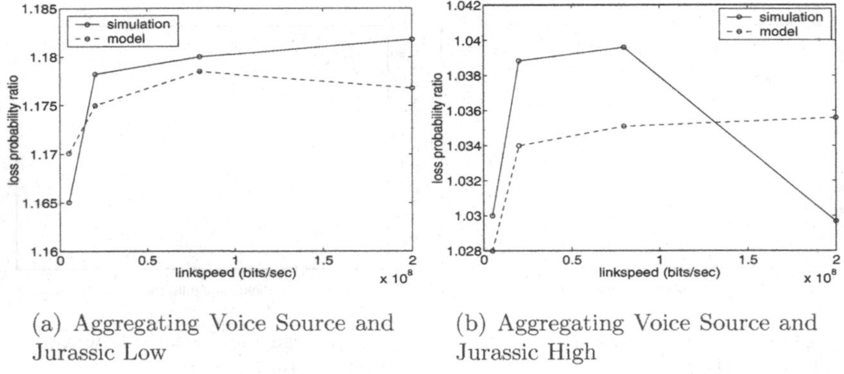

(a) Aggregating Voice Source and Jurassic Low

(b) Aggregating Voice Source and Jurassic High

Fig. 3. Loss Probability Ratio When Aggregating One Video and One Voice

The corresponding results are shown in Figure 3. From the graphs, we see that for both Jurassic low and Jurassic high, the model tracks the actual loss probability ratio reasonably accurately. More important, it will not generate any false alarm and will consistently identify the traffic mixes as safe. As an aside, the amount of additional bandwidth needed to ensure that both the voice and video sources experience the desired loss probability target can also be found to be small (less than 3%), which further confirms that mixing a voice and a video source should be fine in practice.

4 Loss Deviation Prediction when Aggregating Many Sources

The previous section dealt with the special case of only two sources, which while interesting in its own right, is clearly not the only or even most common scenario one would expect to encounter in practice. In this section, we consider another set of possibly more realistic scenarios consisting of mixing many voice and video sources. As before, our goal is to determine whether the model is capable of accurately distinguishing between safe and unsafe traffic mixes. For sake of simplicity and in order to limit the number of combinations to consider, we limit ourselves to multiplexing two different types of sources.

According to the discussion of Section 3, for one or more sources to experience substantially worse losses than other sources, the peak rate of the source should typically be high while its contribution to the total rate should be small enough to ensure that the losses experienced by itself only have a small impact on the overall loss probability. The latter typically implies a combination of a low average transmission rate and a small number of its peers. We therefore concentrate on such scenarios and evaluate the model's ability to properly identify cases when severe loss deviations can occur.

4.1 A First Many-Source Example

In this first example, the dominant traffic contributors consist of 500 voice sources that are multiplexed with a *single*, Jurassic low video source[4]. The loss performance deviation is examined for a system where the buffer size is set to 5000 Bytes and for configurations where the access link speed of the video source increases from 5Mbps to 200Mbps, while the access link speed of the voice sources is fixed at 5Mbps. Given that the Jurassic low source has a reasonably high peak rate but only contributes a small amount to the total bit rate, we expect this scenario to be a prime candidate for generating significant differences in the loss probabilities that the voice and video sources experience. The resulting loss probability ratio is shown in Figure 4(a).

As seen in Figure 4(a), as the video source's access link speed grows, so does the loss probability ratio, and this evolution is accurately captured by the model even if it slightly over-estimates the actual value. This over-estimation does, however, not affect the model's ability to correctly identify cases where significant deviations occur. The figure also shows that for the configurations considered, the range of possible deviations is substantially larger than that of the two-source scenarios of Section 3. This is not unexpected, as the traffic contributed by the video source is much smaller (only 1.12%) in comparison to the total traffic volume than was the case in the two-source scenarios. This very small contribution to the overall traffic translates into an upper bound of 89.5, when using Equation 2, and we can again see that this value is essentially achieved at high link speeds.

As with the two-source case, we also investigate by simulation the amount of additional bandwidth needed to ensure that all sources experience at least the

[4] We have conducted another set of experiments in which 500 voice sources and a single Jurassic high video source are aggregated. Since the results are similar to those reported in this section, both qualitatively and quantitatively, we omit reporting it here.

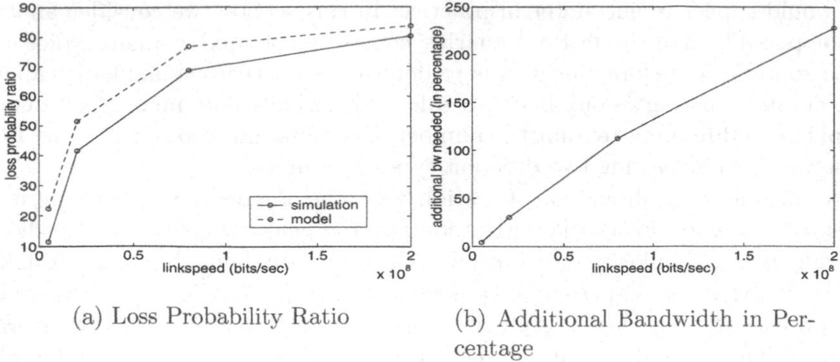

(a) Loss Probability Ratio

(b) Additional Bandwidth in Percentage

Fig. 4. Aggregating 500 voice sources and one video source (Jurassic low)

target performance of 10^{-4}. The results are shown in Figure 4(b), where we can again see that over-allocation is not an option in the cases where substantial performance deviations exist. In particular, when the link speed is high, more than twice the allocated bandwidth is required to guarantee adequate performance for the video sources.

Based on the earlier discussion, significant performance deviation arises in the many-source case only when sources have both a high peak rate (high enough to trigger the onset of congestion when the source becomes active) and represent a small contribution to the overall bit rate (their higher losses do not contribute much to the overall loss probability). The one video source, many voice sources scenario is a perfect illustration of such an "unsafe" configuration. In contrast, a scenario with a small number of voice sources and many video sources would not result in any significant deviations and would be very close to what was observed in the previous section for a one voice source and one video source configuration. In the next section, we investigate the behavior of the system when it transits from an "unsafe" region to a "safe" one.

4.2 Transiting from Unsafe to Safe Traffic Mixes

In this section, we consider a number of scenarios consisting of a mixture of voice and video sources, and where the fraction of traffic contributed by each type of sources varies. Specifically, we start with a configuration consisting of 500 voice sources and one video source (Jurassic low). The access link speed of voice sources is taken to be 5 Mbps, while it is set to 200 Mbps for the video source. We then increase the number of video sources while keeping the total traffic rate fixed. As a result, the number of voice sources decreases and the total traffic contribution of video sources increases, which affects the possibility for performance deviations. In particular, as the traffic contribution from video sources becomes dominant, performance deviations are all but eliminated.

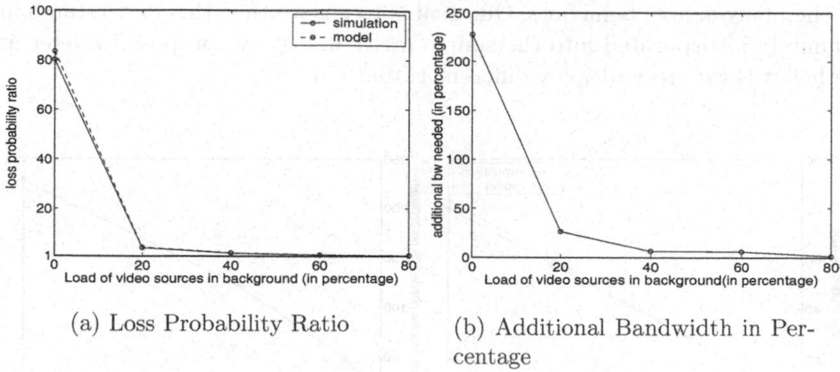

(a) Loss Probability Ratio

(b) Additional Bandwidth in Percentage

Fig. 5. A Variable Traffic Mix Scenario

The results are shown in Figure 5(a), which illustrates that as the traffic contribution from the video sources first increases, the loss probability ratio experiences a sharp initial drop. This is followed by a slower decrease towards a level where performance deviations are essentially negligible. Specifically, when video sources contribute about 20% of the total traffic, the loss probability ratio has dropped to 4.6, down from 80 for a single video source. The loss probability ratio drops further to 2.3 for a video traffic contribution of 40%, and to 1.7 for a video traffic contribution of 60%. As illustrated in Figure 5(b), this general behavior is mirrored in the amount of additional bandwidth needed to eliminate the performance penalty incurred by video sources.

More important from the perspective of assessing the ability of our model to accurately predict whether it is safe or not to mix traffic, the model consistently tracks the actual loss probability ratios across the range of scenarios. From a practical standpoint, this section provides an important guideline for determining when it is safe to mix traffic sources that differ significantly in both their peak and mean rates. Specifically, although there are a number of configurations for which it is safe to mix such sources, transiting from unsafe to safe configurations is not a well demarcated process. As a result, a generally safe practice is to altogether avoid multiplexing sources that differ too much in both their peak and mean rates.

5 Intermediate Configurations

As is seen in Section 3 and Section 4, the behavior of loss performance deviation differs significantly between the two-source and the many-source cases. Specifically, in the two-source case, source utilization alone can result in significant performance deviations, while in the many-source case both peak rate and utilization are involved. In this section, we study a few intermediate configurations that can help shed some light on the transition between the two-source and the many-source behaviors. Our goal is to ensure that this understanding is adequately incorporated into the simple methodology we propose to determine whether it is safe to multiplex different traffic sources.

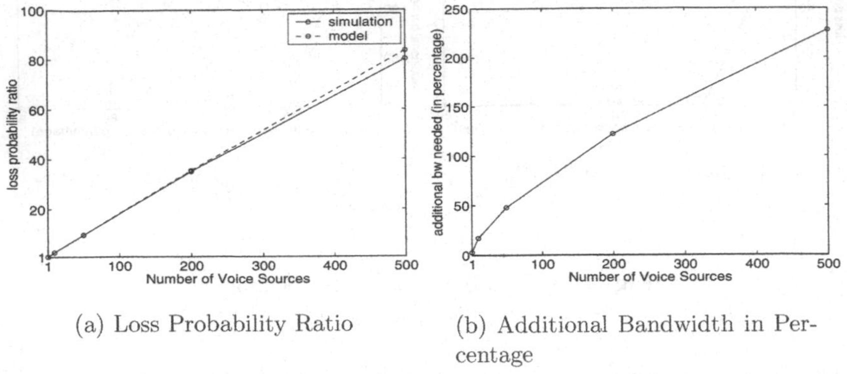

(a) Loss Probability Ratio

(b) Additional Bandwidth in Percentage

Fig. 6. Increasing the Number of Voice Sources

The configuration we use for our investigation consists of multiplexing a variable (from 1 to 500) number of voice sources with one video source (Jurassic low). This allows us to explore the evolution from a safe, two-source traffic mix to an unsafe, many-source mix. The loss probability ratios and the additional bandwidth required (in percentage) are plotted for this range of scenarios in Figure 6(a) and Figure 6(b) respectively.

As can be seen from the figure, the model is quite accurate in predicting the loss probability ratio across all scenarios. In addition, the transition from safe to unsafe regions is shown to be progressive, roughly following a linear function, but with a fairly steep slope. Specifically, mixing the video source with just 50 voice sources already yields a loss probability ratio close to 10, and it would take 50% more bandwidth to fix the problem. This is quite significant, especially in light of the fact that the traffic contribution of the video source is about one tenth of the total traffic, i.e., not an insignificant amount. This again highlights the fact that it is better to err on the side of caution when considering mixing traffic sources with rather disparate peak and mean rates.

6 Discussions of Prediction Errors

One of the main assumptions behind the approach used in this paper, is that while the underlying analytical model may be too simplistic to accurately compute *absolute loss probabilities,* it is capable of robust predictions when it comes to *loss probability ratios.* The obvious question that this assumption raises is "why would two wrongs make a right?" This section is an attempt at answering part of this question.

6.1 Prediction Error in the Two-Source Case

The error in computing loss probabilities and, therefore, the loss probability ratio, originates from mapping the actual distribution of the burst duration into an exponential distribution. Specifically, the peak rate and the utilization (mean rate) of a source can usually be captured reasonably accurately, and the burst duration is the main parameter whose estimation in problematic. As a result, modeling a traffic source using a Markov ON-OFF source involves selecting a triplet of the form $< R, b^{'}, \rho >$, where $b^{'}$ is an *equivalent* burst duration chosen to yield the same loss probability as the one experienced by the actual source. Because the choice of $b^{'}$ is very much source dependent (see [5] for details on a method for deriving estimates for $b^{'}$) and although its value is often different from the average burst duration b, we nevertheless make the assumption that $b^{'} = b$ in our evaluation. This therefore represents the major source of error, when it comes to computing loss probabilities and the loss probability ratio. Our goal in this section is thus to provide a worst case analysis of the magnitude of this error.

Consider a system consisting of two sources multiplexed onto the same link and where, without loss of generality, source 2 is the one that contributes the smallest amount of traffic. Denote PLR_{est} as the loss probability ratio obtained from the approach proposed in this paper, and let E_{PLR} be the *worst case* possible difference between PLR_{est} and the actual loss probability ratio PLR. The determination of E_{PLR} is based on evaluating the *maximum* possible value of $|PLR_{est} - PLR|$ as a function of the system parameters. Specifically, we first derive bounds (intervals) for both PLR and PLR_{est}, and then derive the expression of E_{PLR} based on the maximum distance between the boudaries of these intervals. Due to space constrain, we refer to [10](Appendix I) for details of the derivation, and only present the main results here. In [10], it is found that one can distinguish three main regions for which different bounds can be derived for PLR (and PLR_{est}):

- Case 1: $R_1 \leq C, R_2 \leq C$
- Case 2: $R_1(R_2) < C < R_2(R_1)$
- Case 3: $C \leq R_1, C \leq R_2$

where Case 2 can be further divided into two sub-regions depending on whether R_1 is larger than R_2 or not.

In addition, because the allocated bandwidth C in actual system might be different from that estimated by the model (more precisely, it should be denoted as C_{est}) [5], the total number of regions need to be considered is actually nine (all possible combinations). In [10], the expression of E_{PLR} is derived for all the nine cases, and the results are given in Table 3:

From the table, we can infer that in many cases the value of E_{PLR} will be reasonably small. The only cases where a potentially large estimation error can

[5] This is because the capacity allocated in the model is based on the assumption that the ON period is exponentially distributed with mean b, while the actual capacity allocated incorporate the effect of the real source statistics.

Table 3. Worst Case Prediction Error: E_{PLR}

Model Prediction / Actual Situation	Case 1	Case 2	Case 3
Case 1	0	$\begin{cases} <1 & (r_S^1/R_1<r_S^2/R_2,\ R_1>R_2) \\ \max[2, \frac{1}{r_S^2/r_S} * \frac{1}{1+R_1/R_2}]-1 & (r_S^1/R_1>r_S^2/R_2,\ R_1>R_2) \\ \frac{1}{r_S^2/r_S} * (\frac{R_1/R_2}{1+R_1/R_2}) & (R_1<R_2) \end{cases}$	$\frac{1}{r_S^2/r_S}-1$
Case 2	$\begin{cases} <1 & (r_S^1/R_1<r_S^2/R_2,\ R_1>R_2) \\ \max[2, \frac{1}{r_S^2/r_S} * \frac{1}{1+R_1/R_2}]-1 & (r_S^1/R_1>r_S^2/R_2,\ R_1>R_2) \\ \frac{1}{r_S^2/r_S} * (\frac{R_1/R_2}{1+R_1/R_2}) & (R_1<R_2) \end{cases}$	$\begin{cases} <1 & (r_S^1/R_1<r_S^2/R_2,\ R_1>R_2) \\ \max[2, \frac{1}{r_S^2/r_S} * \frac{1}{1+R_1/R_2}]-1 & (r_S^1/R_1>r_S^2/R_2,\ R_1>R_2) \\ \frac{1}{r_S^2/r_S} * (\frac{R_1/R_2}{1+R_1/R_2}) & (R_1<R_2) \end{cases}$	$\frac{1}{r_S^2/r_S}-1$
Case 3	$\frac{1}{r_S^2/r_S}-1$	$\frac{1}{r_S^2/r_S}-1$	$\frac{1}{r_S^2/r_S}-1$

occur is when the allocated capacity C for either the model or the actual system is smaller than the peak rates of the two sources. In all other configurations, one can find in [10] that the maximum error remains relatively small.

For example, when both the model and the actual system correspond to Case 1, one can establish that $E_{PLR} = 0$. This is because the loss probability ratio of the two sources satisfies $P_L^2/P_L^1 = \rho_1/\rho_2$, which is independent of the burst duration, and hence of errors in estimating its statistics. Similarly, when neither system falls into Case 3, it is possible to show $E_{PLR} \leq 1$ if the smaller traffic contributor, source 2, satisfies $\rho_1 < \rho_2$ and $R_1 > R_2$. In this case, one can easily verify that $P_L^1 > P_L^2$, and since it is then the major traffic contributor (source 1) that loses more, $PLR = P_L^1/P_L$. This implies that PLR can never exceed r_S/r_S^1, which is less than 2. This then yields $E_{PLR} \leq 1$. In other scenarios, as long as neither system falls into Case 3, it is possible to show that E_{PLR} will be a decreasing function of both the fraction of traffic contributed by source 2 and the difference between R_1 and R_2. This is reflected in the expressions $\max\left\{1, \frac{1}{r_S^2/r_S} \cdot \frac{1}{1+R_1/R_2} - 1\right\}$ and $\frac{1}{r_S^2/r_S} \cdot \frac{R_1/R_2}{1+R_1/R_2}$ for the cases $R_1 > R_2$ and $R_1 < R_2$, respectively. In those cases, E_{PLR} will be large only when both the value of r_S^2/r_S is small (the traffic contribution of source 2 is much smaller than that of source 1) and R_1/R_2 is close to 1.

The only other cases where we commit it is possible for E_{PLR} to be large is, as mentioned earlier, when either the modeled or the actual system correspond to Case 3. In this case, either PLR or PLR_{est} can be anywhere in the range 1 to r_S/r_S^2, so that $E_{PLR} = r_S/r_S^2 - 1$, which can be arbitrarily large if the traffic contribution of source 2 is very small. The question is then to determine how common this configuration is. A brief and arguably limited review of the many different configurations we tested found that this scenario occurred only when multiplexing video sources with rather different rates, e.g., Lecture low and Jurassic high. In those cases, the modeled system fell (erroneously) in Case

3 while the actual system is in Case 2. This indeed translated into substantial differences between the loss probability ratio predicted by the model and the value of PLR obtained in the simulations, as is seen in Figure 1. However, it is worth noting that in spite of those differences, the tests performed using the model to determine if it was safe or not to multiplex the sources, still provided a correct answer in most cases. In other words, the methodology based on the average burst duration still successfully captured the trend in loss probability ratio and thus, was able to correctly identify the potential danger of mixing the traffic sources.

6.2 Prediction Error in the Many-Source Case

In the many-source case, the equations used to predict the loss probability ratio, i.e., Equation (6) and (7) of [9], are derived assuming a bufferless model. For such a system, the loss probability ratio is only affected by the peak rate and the utilization [6]. This is because data losses occur *only* when the arrival rate is greater than the link capacity, and thus are determined only by their difference. As a result, the loss probability ratio does not depend on the statistics of burst duration. Instead, the model's accuracy depends only on how good a fit a bufferless model is for the real system. This is a function of both the buffer size and the relative burstiness of the source(s).

When the number of sources is large, the ability of a buffer, even a large one, to accommodate many large bursts is very limited, so that a bufferless model is often quite accurate. From the simulations of Section 4 and Section 5, we have indeed seen that a buffer size of 5000 Bytes gives results that are barely distinguishable from those of a bufferless system, even in cases where the number of sources is moderate. In other simulations that are not included in the paper due to lack of space, similar observations were made for systems with much larger buffers, i.e., 100 KBytes. In those cases, although there were instances where the absolute prediction was not entirely accurate, it always remained sufficient to clearly identify when it was dangerous to multiplex sources of different types. As a result, we believe that the performance of an approach based on a bufferless model should remain robust across a wide range of buffer sizes, as long as the number of sources being multiplexed is large enough.

7 Conclusions

This paper is concerned with the problem of deciding when it is possible to multiplex different traffic sources into a common service class, when both resource allocation and performance monitoring is done at the class level. Our goal is to develop a reasonably simple methodology for identifying traffic mixes that can generate significant performance deviations. The methodology we propose is based on a set of analytical tools developed in [9], and which are used together with a simple method for mapping real traffic sources onto source models that the analysis is capable of handling.

The effectiveness of the methodology is evaluated through simulations by exploring its prediction ability across a number of scenarios involving real traffic sources. The paper establishes that the methodology, though simple, is quite successful and robust in identifying traffic mixes that can result in severe loss performance deviations. In particular, it reliably identifies cases where large deviations would occur and mostly avoids false alarms. In addition, the region for which it gives somewhat inconclusive results, its so-called threshold region, is relatively narrow. Furthermore, the paper also shows that in most cases where large performance deviations are observed, the amount of additional bandwidth needed to provide all users in the service class with the desired performance, can often be quite large. This indicates that simply over-provisioning the bandwidth allocated to the service class cannot easily eliminate loss performance deviations, when present. As a result, providing a simple yet accurate method for identifying potentially dangerous traffic mixes is important when contemplating deploying class-based services.

8 Acknowledgement

The authors would like to acknowledge Wenyu Jiang from Columbia University for providing the voice traffic traces, and Frank Fitzek from the Technical University of Berlin, Germany, for providing the video traffic traces.

References

[1] The network simulator – ns-2. http://www.isi.edu/nsnam/ns/.

[2] Video trace library. http://www-tkn.ee.tu-berlin.de/research/trace/trace.html.

[3] S. Blake, D. Black, M. Carlson, E. Davies, Z. Wang, and W. Weiss. An architecture for differentiated services. Request For Comments (Proposed Standard) RFC 2475, IETF, December 1998.

[4] F. Frank and R. Martin. MPEG-4 and H.263 video traces for network performance evaluation. *IEEE Network Magazine*, 15(6):40–54, November 2001.

[5] L. Gün. An approximation method for capturing complex traffic behavior in high speed networks. *Performance Evaluation*, 19(1):5–23, January 1994.

[6] U. Mocci J. W. Roberts and J. Virtamo, editors. *Broadband Network teletraffic - Final Report of Action COST 242*, volume 1155. Springer-Verlag, 1996.

[7] W. Jiang and H. Schulzrinne. Analysis of on-off patterns in voip and their effect on voice traffic aggregation. In *The 9th IEEE International Conference on Computer Communication Networks*, 2000.

[8] H. Schulzrinne. Voice communication across the Internet: a network voice terminal. Technical Report 92–50, Department of Computer Science, University of Massachusetts, Amherst, MA, USA, 1992.

[9] Y. Xu and R. Guérin. Individual QoS versus aggregate QoS: A loss performance study. In *Proceedings of the IEEE Infocom*, NYC, USA, June 2002.

[10] Y. Xu and R. Guérin. On evaluating loss performance deviation: A simple tool and its practical implications. Technical report, University of Pennsylvania, September 2002. (Available at http://einstein.seas.upenn.edu/mnlab/publications.html).

Extending the Network Calculus Pay Bursts Only Once Principle to Aggregate Scheduling

Markus Fidler

Department of Computer Science, Aachen University
Ahornstr. 55, 52074 Aachen, Germany
fidler@i4.informatik.rwth-aachen.de

Abstract. The Differentiated Services framework allows to provide scalable network Quality of Service by aggregate scheduling. Services, like a Premium class, can be defined to offer a bounded end-to-end delay. For such services, the methodology of Network Calculus has been applied successfully in Integrated Services networks to derive upper bounds on the delay of individual flows. Recent extensions allow an application of Network Calculus even to aggregate scheduling networks. Nevertheless, computations are significantly complicated due to the multiplexing and de-multiplexing of micro-flows to aggregates. Here problems concerning the tightness of delay bounds may be encountered.

A phenomenon called Pay Bursts Only Once is known to give a closer upper estimate on the delay, when an end-to-end service curve is derived prior to delay computations. Doing so accounts for bursts of the flow of interest only once end-to-end instead of at each link independently. This principle also holds in aggregate scheduling networks. However, it can be extended in that bursts of interfering flows are paid only once, too. In this paper we show the existence of such a complementing Pay Bursts Only Once phenomenon for interfering flows. We derive the end-to-end service curve for a flow of interest in an arbitrary aggregate scheduling feed forward network for rate-latency service curves, and leaky bucket constraint arrival curves, which conforms to both of the above principles. We give simulation results to show the utility of the derived forms.

1 Introduction

The Differentiated Services (DS) architecture [2] is the most recent approach of the Internet Engineering Task Force (IETF) towards network Quality of Service (QoS). DS addresses the scalability problems of the former Integrated Services approach by an aggregation of micro-flows to a small number of different traffic classes, for which service differentiation is provided. Packets are identified by simple markings that indicate the respective class. In the core of the network, routers do not need to determine to which flow a packet belongs, only which aggregate behavior has to be applied. Edge routers mark packets and indicate whether they are within profile or, if they are out of profile, in which case they might even be discarded by a dropper at the edge router. A particular marking on

M. Ajmone Marsan et al. (Eds.): QoS-IP 2003, LNCS 2601, pp. 19–34, 2003.

a packet indicates a so-called Per Hop Behavior (PHB) that has to be applied for forwarding of the packet. Currently, the Expedited Forwarding (EF) PHB [9], and the Assured Forwarding (AF) PHB group are specified. The EF PHB is intended for building a service that offers low loss, low delay, and low delay jitter, namely a Premium service. The specification of the EF PHB was recently redefined to allow for a more exact and quantifiable definition [5]. Especially the derivation of delay bounds is of interest, when providing a Premium service. In [4] such bounds are derived for a general topology and a maximum load. However, these bounds can be improved, when additional information concerning the current load, and the special topology of a certain DS domain is available.

In [15] a central resource management for DS domains called a Bandwidth Broker (BB) is presented. A BB is a middleware service which controls and facilitates the dynamic access to network services of a particular administrative domain [10]. The task of a BB in a DS domain is to perform a careful admission control, and to set up the appropriate configuration of the domain's edge routers, whereas the configuration of core routers is intended to remain static to allow for scalability. While doing so, the BB knows about all requests for capacity of certain QoS classes. Besides it can easily learn about the DS domains topology, either statically, or by implementing a listener for the domains routing protocol. Thus, a BB can have access to all information that is required, to apply the mathematical methodology of Network Calculus [3,13], in order to base its admission control on delay boundaries that are derived for the current load, and the special topology of the administrated domain [19].

In this paper we address the derivation of end-to-end delay guarantees based on Network Calculus. We derive a closed form solution for the end-to-end delay in feed forward First In First Out (FIFO) networks, for links that have a rate-latency property, and flows that are leaky bucket constraint. In particular this form accounts for bursts of interfering flows only once, and thus implements a principle for aggregate scheduling that is similar to the Network Calculus Pay Bursts Only Once principle [13]. The derived boundaries can be applied as a decision criterion to perform the admission control of a DS domain by a BB. The remainder of this paper is organized as follows: In Section 2 the required background on Network Calculus, and the notation that is applied in the sequel are given. Section 3 introduces two examples, which show how bursts of interfering flows worsen the derived delay bounds and, which prove the existence of a counterpart to the Pay Bursts Only Once phenomenon for interfering flows in aggregate scheduling networks. The first example can be satisfactorily solved by direct application of current Network Calculus, whereas to our knowledge for the second example a tight solution is missing in current literature. This missing piece is addressed in Section 4, where a tight closed form solution for arbitrary feed forward networks with FIFO rate-latency service curves and leaky bucket constraint arrival curves is derived. In Section 5 we describe the implementation of the admission control in a DS BB that is based on worst-case delay bounds. Numerical results on the performance gain that is achieved by applying the previously derived terms are given. Section 6 concludes the paper.

2 Network Calculus Background and Notation

Network Calculus is a theory of deterministic queuing systems that is based on the early work on the calculus for network delay in [6,7], and on the work on Generalized Processor Sharing (GPS) presented in [16,17]. Further extensions, and a comprehensive overview on current Network Calculus are given in [12,13], and from the perspective of filtering theory in [3]. Here only a few concepts are covered briefly, to give the required background, and to introduce the notation that is mainly taken over from [13]. In addition since networks consisting of several links n that are used by a flow of interest, and a number of interfering flows m are investigated, upper indices j indicate links, and lower indices i indicate flows in the sequel.

The scheduler on an outgoing link can be characterized by the concept of a service curve, denoted by $\beta(t)$. A special characteristic of a service curve is the rate-latency type that is given by $\beta_{R,T}(t) = R \cdot [t - T]^+$ with a rate R and a latency T. The term $[x]^+$ is equal to x, if $x \geq 0$, and zero otherwise. Service curves of the rate-latency type are implemented for example by Priority Queuing (PQ), or Weighted Fair Queuing (WFQ). The latency of a PQ scheduler is given in [5] for variable length packet networks with a Maximum Transmission Unit (MTU) according to $T = \text{MTU}/R$. Nevertheless, routers can implement additional non-preemptive layer 2 queues on their outgoing interfaces for a smooth operation, which can add further delay to a layer 3 QoS implementation [20]. Thus $T = (l_2 + 1) \cdot \text{MTU}/R$ might have to be applied, whereby l_2 gives the layer 2 queuing capacity in units of the MTU.

Flows are defined either by their arrival functions denoted by $F(t)$, or by their arrival curves $\alpha(t)$, whereas $\alpha(t_2 - t_1) \geq F(t_2) - F(t_1)$ for all $t_2 \geq t_1$. In DS networks, a typical characteristic for incoming flows can be given by the leaky bucket constraint $\alpha_{r,b}(t) = b + r \cdot t$ that is also known as sigma-rho leaky bucket in [3]. Usually the ingress router of a DS domain meters incoming flows against a leaky bucket algorithm, and either shapes, or drops non-conforming traffic, which justifies the application of leaky bucket constraint arrival curves.

If a link j is traversed by a flow i, the arrival function of the output flow F_i^{j+1}, which is the input arrival function for an existing, or an imaginary subsequent link $j+1$, can be given according to (1) for $t \geq s \geq 0$ [12].

$$F_i^{j+1}(t) \geq F_i^j(t - s) + \beta^j(s) \tag{1}$$

From (1) the term in (2) follows. The operator $\otimes$ denotes the convolution under the min-plus algebra that is applied by Network Calculus.

$$F_i^{j+1}(t) \geq (F_i^j \otimes \beta^j)(t) = \inf_{t \geq s \geq 0} [F_i^j(t - s) + \beta^j(s)] \tag{2}$$

Further on, the output flow is upper constrained by an arrival curve α_i^{j+1} that is given according to (3), with $\oslash$ denoting the min-plus de-convolution.

$$\alpha_i^{j+1}(t) = (\alpha_i^j \oslash \beta^j)(t) = \sup_{s \geq 0} [\alpha_i^j(t + s) - \beta^j(s)] \tag{3}$$

If the path of a flow i consists of two or more links, the formulation of the concatenation of links can be derived based on (4).

$$\begin{array}{rl} F_i^{j+2}(u) - F_i^{j+1}(u-(t-s)) & \geq \beta^{j+1}(t-s) \\ F_i^{j+1}(u-(t-s)) - F_i^{j}(u-t) \geq & \beta^{j}(s) \\ \hline F_i^{j+2}(u) \qquad -F_i^{j}(u-t) & \geq \beta^{j+1}(t-s)+\beta^{j}(s) \end{array} \tag{4}$$

The end-to-end service curve is then given in (5), which covers the case of two links, whereas the direct application of (5) also holds for n links.

$$\beta^{j+1,j}(t) = (\beta^{j+1} \otimes \beta^{j})(t) = \inf_{t \geq s \geq 0} [\beta^{j+1}(t-s) + \beta^{j}(s)] \tag{5}$$

The maximal virtual delay d for a system that offers a service curve of $\beta(t)$ with an input flow that is constraint by $\alpha(t)$, is given as the supremum of the horizontal deviation according to (6).

$$d \leq \sup_{s \geq 0} [\inf[\tau \geq 0 : \alpha(s) \leq \beta(s+\tau)]] \tag{6}$$

For the special case of service curves of the rate-latency type, and sigma-rho leaky bucket constraint arrival curves, simplified solutions exist for the above equations. The arrival curve of the output flow according to (3) is given in (7) for this case, whereas the burst size of the output flow $b_i + r_i \cdot T^j$ is equal to the maximum backlog at the scheduler of the outgoing link.

$$\alpha_i^{j+1}(t) = b_i + r_i \cdot T^j + r_i \cdot t \tag{7}$$

The concatenation of two rate-latency service curves can be reduced to (8).

$$\beta^{j+1,j}(t) = \min[R^{j+1}, R^j] \cdot [t - (T^{j+1} + T^j)]^+ \tag{8}$$

Finally, (9) gives the worst case delay for the combination of a rate-latency service curve, and a leaky bucket constraint arrival curve.

$$d \leq T^j + b_i / R^j \tag{9}$$

If a flow i traverses two links j and $j+1$, two options for the derivation of the end-to-end delay exist. For simplicity, the service curves are assumed here to be of the rate-latency type, and the arrival curve is chosen to be leaky bucket constraint. At first the input arrival curve of flow i at link $j+1$ can be computed as in (7). Then the virtual end-to-end delay is derived to be the sum of the virtual delays at link j and $j+1$ according to (9). Doing so results in $d \leq T^j + b_i/R^j + T^{j+1} + (b_i + r_i \cdot T^j)/R^{j+1}$. The second option is to derive the end-to-end service curve as is done in (8), and compute the delay according to (9) afterwards, resulting in $d \leq T^j + T^{j+1} + b_i / \min[R^{j+1}, R^j]$. Obviously, the second form gives a closer bound, since it accounts for the burst size b_i only once. This property is known as the Pay Bursts Only Once phenomenon from [13].

Until now only networks that perform a per-flow based scheduling, for example Integrated Services networks, have been considered. The aggregation or the multiplexing of flows can be given by the addition of the arrival functions $F_{1,2}(t) = F_1(t) + F_2(t)$, or arrival curves $\alpha_{1,2}(t) = \alpha_1(t) + \alpha_2(t)$. For aggregate scheduling networks with FIFO service curves, families of per-flow service curves $\beta_\theta(t)$ according to (10) with an arbitrary parameter $\theta \geq 0$ are derived in [8,13]. $\beta_\theta^j(t)$ gives a family of service curves for a flow 1 that is scheduled in an aggregate manner in conjunction with a flow 2 on a link j. $1_{t>\theta}$ is zero for $t \leq \theta$.

$$\beta_\theta^j(t) = [\beta^j(t) - \alpha_2(t-\theta)]^+ 1_{t>\theta} \tag{10}$$

The parameter θ has to be set to zero in case of blind multiplexing. In FIFO networks for $u = \sup[v : F_{1,2}^j(v) \leq F_{1,2}^{j+1}(t)]$, the conditions $F^j(u) \leq F^{j+1}(t)$, and $F^j(u^+) \geq F^{j+1}(t)$ with $u^+ = u + \epsilon$, $\epsilon > 0$ hold for the sum of both flows, and for the individual flows, too. These additional constraints allow to derive (10) for an arbitrary $\theta > 0$. The complete derivation is missed out here. It can be found in [13]. From (10) it cannot be concluded that $\sup_\theta[\beta_\theta^j]$, or $\inf_\theta[\beta_\theta^j]$ is a service curve, but for the output flow $\alpha_1^{j+1}(t) = \inf_{\theta \geq 0}[(\alpha_1^j \oslash \beta_\theta^j)(t)]$ can be given [13].

For the special case of rate-latency service curves, and leaky bucket constraint arrival curves, (11) can be derived from (10) to be a service curve for flow 1 for $r_1 + r_2 < R^j$.

$$\beta^j(t) = (R^j - r_2) \cdot [t - (T^j + b_2/R^j)]^+ \tag{11}$$

The methodology that is shown up to here already allows to implement the algorithmic derivation of delay bounds in a DS domain by a BB, with the restriction that the domain has to be a feed forward network. In non feed forward networks analytical methods like time stopping [3,13] can be applied. Nevertheless, for an algorithmic implementation, it is simpler to prevent from aggregate cycles, and transform the domains topology into a feed forward network, to allow for the direct application of Network Calculus. Such a conversion of the network topology can be made by means of breaking up loops by forbidding certain links, for example by turn prohibition as presented in [21] for networks that consist of bidirectional links. Doing so, the BB can inductively derive the arrival curves of micro-flows at each outgoing link, then compute the service curve of each link from the point of view of each micro-flow, and concatenate these to end-to-end service curves, to give upper bounds on the worst case delay for individual flows.

3 Extended Pay Bursts Only Once Principle

Though Section 2 gives the required background to derive per micro-flow based end-to-end delay bounds in aggregate scheduling networks, these bounds are likely to be unnecessarily loose. Figure 1 gives a motivating example of a simple network consisting of two outgoing links that are traversed by two flows. This example is given to introduce a similar concept to the Pay Bursts Only Once principle [13] for aggregate scheduling. Flow 1 is the flow of interest, for which the end-to-end delay needs to be derived. According to the Pay Bursts Only Once principle, at first the end-to-end service curve for flow 1 has to be derived, and then the delay is computed, instead of computing the delay at each link

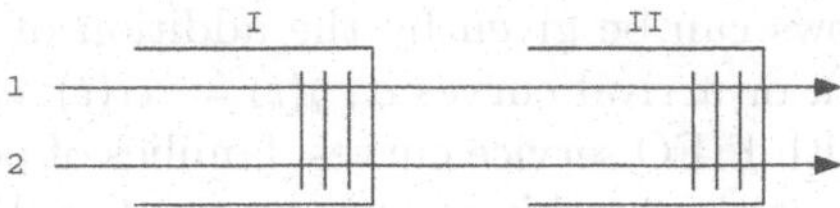

Fig. 1. Two Flows Share Two Consecutive Links

independently, and summing these delays up. A similar decision has to be made, when deriving the service curve for flow 1 by subtracting the arrival curve of flow 2 from the aggregate service curves. Either this subtraction is made at each link independently before concatenating service curves, or the subtraction is made after service curves have been concatenated. To illustrate the difference the two possible derivations of the end-to-end service curve for flow 1 are given here for rate-latency service curves, and leaky bucket constraint arrival curves.

For the first option, which is already drafted at the end of Section 2, the service curve of link I from the point of view of flow 2 can be derived to be of the rate-latency type with a rate $R^{\mathrm{I}} - r_1$ and a latency $T^{\mathrm{I}} + b_1/R^{\mathrm{I}}$. Then, the arrival curve of flow 2 at link II can be given to be leaky bucket constraint with the rate r_2 and the burst size $b_2 + r_2 \cdot (T^{\mathrm{I}} + b_1/R^{\mathrm{I}})$. The service curves at link I and II from the point of view of flow 1 can be given to be rate-latency service curves with the rates $R^{\mathrm{I}} - r_2$, respective $R^{\mathrm{II}} - r_2$, and the latencies $T^{\mathrm{I}} + b_2/R^{\mathrm{I}}$, respective $T^{\mathrm{II}} + (b_2 + r_2 \cdot (T^{\mathrm{I}} + b_1/R^{\mathrm{I}}))/R^{\mathrm{II}}$. The concatenation of these two service curves yields the rate-latency service curve for flow 1 given in (12).

$$\begin{aligned}\beta^{\mathrm{I,II}}(t) = \min&[R^{\mathrm{I}} - r_2, R^{\mathrm{II}} - r_2] \\ &\cdot [t - (T^{\mathrm{I}} + b_2/R^{\mathrm{I}}) - (T^{\mathrm{II}} + (b_2 + r_2 \cdot (T^{\mathrm{I}} + b_1/R^{\mathrm{I}}))/R^{\mathrm{II}})]^+\end{aligned} \tag{12}$$

The second option requires that the service curves of link I and II are convoluted prior to subtraction of the flow 2 arrival curve [19]. The concatenation yields a service curve of the rate-latency type with a rate of $\min[R^{\mathrm{I}}, R^{\mathrm{II}}]$ and a latency of $T^{\mathrm{I}} + T^{\mathrm{II}}$. After subtracting the arrival curve of flow 2 from the concatenation of the service curves of link I, and II, (13) can be given for flow 1.

$$\beta^{\mathrm{I,II}}(t) = (\min[R^{\mathrm{I}}, R^{\mathrm{II}}] - r_2) \cdot [t - (T^{\mathrm{I}} + T^{\mathrm{II}} + b_2/\min[R^{\mathrm{I}}, R^{\mathrm{II}}])]^+ \tag{13}$$

Obviously, the form in (13) offers a lower latency than the one in (12), which accounts for the burst size of the interfering flow 2 twice, once with the size b_2 at link I, and then with $b_2 + r_2 \cdot (T^{\mathrm{I}} + b_1/R^{\mathrm{I}})$ at link II. The increase of the burst size of flow 2 at link II is due to the aggregate scheduling of flow 2 with flow 1 at link I. Thus, based on (12), and (13) a counterpart to the Pay Burst Only Once phenomenon has been shown for interfering flows in aggregate scheduling networks.

However, most problems of interest cannot be solved as simple as the one in Figure 1. One such example is shown in Figure 2. Flow 2 is the flow of interest, for which the end-to-end service curve needs to be derived. Links I and II can

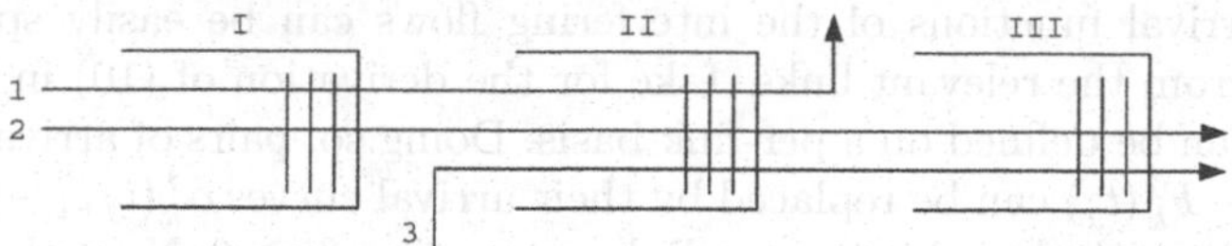

Fig. 2. Three Flows Share Three Consecutive Links

be concatenated after the arrival curve of flow 3 is subtracted from the service curve of link II. Then the arrival curve of flow 1 can be subtracted from the concatenation of the service curves of link I and II. Unfortunately the direct application of the Network Calculus terms given in Section 2 then requires that the arrival curve of flow 3 at link III is subtracted from the service curve of link III independently, which violates the extended Pay Bursts Only Once principle. An alternative derivation is possible, if the arrival curve of flow 1 is subtracted from the service curve of link II before links II and III are concatenated. Doing so does unfortunately encounter the same problem.

4 Closed Form Solution for Feed Forward Networks

Consider an end-to-end path of a flow of interest i in an arbitrary aggregate scheduling feed forward network with FIFO service curve elements. Further on, assume that a concatenation of consecutive links offers a FIFO characteristic, too. The path consists of n links that are indexed by j in ascending order from the source to the destination of the flow of interest i. These links are given in the set $\mathbb{J}_i$. Further on, the links of the path of flow i are used in an aggregate manner by an additional number of flows m that are indexed by k. The paths of the flows k are given by the sets $\mathbb{J}_k$. For each link j a set $\mathbb{K}^j$ is defined to hold all other flows k that traverse the link, not including flow i. Note that flows may exist that share a part of the path of the flow of interest i, then follow a different path, and afterwards again share a part of the path of flow i. These flows are split up in advance into as many individual flows as such multiplexing points exist. Thus, all resulting flows do have only one multiplexing point with flow i. In [13] the term Route Interference Number (RIN) is defined to hold the quantity of such multiplexing points. Then, the set $\mathbb{K}_i = \bigcup_{j \in \mathbb{J}_i} \mathbb{K}^j$ is defined to hold all m flows that use the complete path or some part of the path of flow i. Further on, the set $\mathbb{J}_{i,k} = \mathbb{J}_i \bigcap \mathbb{J}_k$ holds all links of a sub-path that are used by both flow i and a flow k. Based on these definitions, on the terms in (4), and on the rules for multiplexing, (14) can be given for time pairs $t_{j+1} - t_j \geq 0$ for all $j \in \mathbb{J}_i$. The time indices are chosen here to match link indices, since arrival functions are observed at these time instances at the belonging links.

$$F_i^{n+1}(t_{n+1}) - F_i^1(t_1) \geq \sum_{j \in \mathbb{J}_i} \beta^j(t_{j+1} - t_j) - \sum_{k \in \mathbb{K}_i} \sum_{j \in \mathbb{J}_{i,k}} (F_k^{j+1}(t_{j+1}) - F_k^j(t_j)) \quad (14)$$

Now, the arrival functions of the interfering flows can be easily split up and subtracted from the relevant links. Like for the derivation of (10) in [13], FIFO conditions can be defined on a per-link basis. Doing so, pairs of arrival functions $F_k^{j+1}(t_{j+1}) - F_k^j(t_j)$ can be replaced by their arrival curves $\alpha_k^j(t_{j+1} - t_j - \theta^j)$ for $t_{j+1} - t_j > \theta^j$, with the arbitrary per link parameters $\theta^j \geq 0$. Nevertheless, doing so results in paying the bursts of all interfering flows at each link independently.

However, FIFO conditions can in addition to per link be derived per sub-path $\mathbb{J}_{i,k}$. With $\sum_{j \in \mathbb{J}_{i,k}} (F_k^{j+1}(t_{j+1}) - F_k^j(t_j)) = F_k^{j_{\max}+1}(t_{j_{\max}+1}) - F_k^{j_{\min}}(t_{j_{\min}})$, whereby $j_{\max} = \max[j \in \mathbb{J}_{i,k}]$, and $j_{\min} = \min[j \in \mathbb{J}_{i,k}]$, (15) can be derived for $t_{j_{\max}+1} - t_{j_{\min}} > \theta_k$, based on (10).

$$F_i^{n+1}(t_{n+1}) - F_i^1(t_1) \geq \sum_{j \in \mathbb{J}_i} \beta^j(t_{j+1} - t_j) - \sum_{k \in \mathbb{K}_i} \alpha_k^{j_{\min}}(t_{j_{\max}+1} - t_{j_{\min}} - \theta_k) \quad (15)$$

To motivate the step from (14) to (15), the FIFO conditions that are applied, are shown exemplarily for the small network in Figure 2. The derivation, is a direct application of the one given in [13] for the form in (10). Define an u_1, and u_3 for flow 1, respective flow 3 according to (16), and (17).

$$u_1 = \sup\{v : F_1^{\mathrm{I}}(v) + F_2^{\mathrm{I}}(v) \leq F_1^{\mathrm{III}}(t_3) + F_2^{\mathrm{III}}(t_3)\} \quad (16)$$

$$u_3 = \sup\{v : F_2^{\mathrm{II}}(v) + F_3^{\mathrm{II}}(v) \leq F_2^{\mathrm{IV}}(t_4) + F_3^{\mathrm{IV}}(t_4)\} \quad (17)$$

Now, from (16), and with $u_1^+ = \inf\{v : v > u_1\}$, $t_1 \leq u_1 \leq t_3$, $F_1^{\mathrm{I}}(u_1) + F_2^{\mathrm{I}}(u_1) \leq F_1^{\mathrm{III}}(t_3) + F_2^{\mathrm{III}}(t_3)$, and $F_1^{\mathrm{I}}(u_1^+) + F_2^{\mathrm{I}}(u_1^+) \geq F_1^{\mathrm{III}}(t_3) + F_2^{\mathrm{III}}(t_3)$ can be given. Due to the per sub-path FIFO conditions, the above terms also hold for the individual flows 1, and 2, for example $F_1^{\mathrm{I}}(u_1) \leq F_1^{\mathrm{III}}(t_3)$, and $F_1^{\mathrm{I}}(u_1^+) \geq F_1^{\mathrm{III}}(t_3)$. Similar conditions can be derived for flows 2, and 3 from (17). The following term in (18) can be set up based on (4).

$$\begin{aligned} F_2^{\mathrm{IV}}(t_4) - F_2^{\mathrm{I}}(t_1) \geq\ & \beta^{\mathrm{III}}(t_4 - t_3) + \beta^{\mathrm{II}}(t_3 - t_2) + \beta^{\mathrm{I}}(t_2 - t_1) \\ & - (F_3^{\mathrm{IV}}(t_4) - F_3^{\mathrm{II}}(t_2)) - (F_1^{\mathrm{III}}(t_3) - F_1^{\mathrm{I}}(t_1)) \end{aligned} \quad (18)$$

With $F_1^{\mathrm{I}}(u_1^+) \geq F_1^{\mathrm{III}}(t_3)$, and $F_3^{\mathrm{II}}(u_3^+) \geq F_3^{\mathrm{IV}}(t_4)$, (19) can be derived.

$$\begin{aligned} F_2^{\mathrm{IV}}(t_4) - F_2^{\mathrm{I}}(t_1) \geq\ & \beta^{\mathrm{III}}(t_4 - t_3) + \beta^{\mathrm{II}}(t_3 - t_2) + \beta^{\mathrm{I}}(t_2 - t_1) \\ & - (F_3^{\mathrm{II}}(u_3^+) - F_3^{\mathrm{II}}(t_2)) - (F_1^{\mathrm{I}}(u_1^+) - F_1^{\mathrm{I}}(t_1)) \end{aligned} \quad (19)$$

Choose two arbitrary parameters $\theta_1 \geq 0$, and $\theta_3 \geq 0$. If $u_1 < t_3 - \theta_1$, $F_1^{\mathrm{I}}(u_1^+) \leq F_1^{\mathrm{I}}(t_3 - \theta_1)$ holds. Further on, $t_3 - t_1 > \theta_1$ can be given in this case, since $u_1 \geq t_1$. With a similar condition for flow 3, the form in (20) is derived.

$$\begin{aligned} F_2^{\mathrm{IV}}(t_4) - F_2^{\mathrm{I}}(t_1) \geq\ & \beta^{\mathrm{III}}(t_4 - t_3) + \beta^{\mathrm{II}}(t_3 - t_2) + \beta^{\mathrm{I}}(t_2 - t_1) \\ & - \alpha_3(t_4 - t_2 - \theta_3) - \alpha_1(t_3 - t_1 - \theta_1) \end{aligned} \quad (20)$$

Else, for $u_1 \geq t_3 - \theta_1$, the FIFO condition $F_2^{\mathrm{III}}(t_3) \geq F_2^{\mathrm{I}}(u_1)$ is applied. Further on, the term $F_2^{\mathrm{III}}(t_3) - F_2^{\mathrm{I}}(t_1) \geq \beta^{\mathrm{I,II}}(t_3 - t_1)$ can be set up, with $\beta^{\mathrm{I,II}}$ denoting the service curve for flow 2 that is offered by link I, and II. Substitution of t_1 by u_1, yields $F_2^{\mathrm{III}}(t_3) \geq F_2^{\mathrm{I}}(u_1) + \beta^{\mathrm{I,II}}(t_3 - u_1)$. Hence, $\beta^{\mathrm{I,II}}(t_3 - t_1) = 0$ is a trivial service curve for $u_1 \geq t_3 - \theta_1$. Similar forms can be derived for $u_3 \geq t_4 - \theta_3$.

In the following the form in (15) is solved for the simple case of rate-latency service curves, and leaky bucket constraint arrival curves.

Proposition 1 (End-to-End Service Curve) *The end-to-end service curve for a flow of interest i in an aggregate scheduling feed forward network with FIFO service curve elements of the rate-latency type $\beta_{R,T}$, and leaky bucket constrained arrival curves $\alpha_{r,b}$ is again of the rate-latency type, and given according to (21).*

$$\beta_i(t) = \min_{j \in \mathbb{J}_i} \Big[R^j - \sum_{k \in \mathbb{K}^j} r_k \Big] \cdot \Big[t - \sum_{j \in \mathbb{J}_i} T^j - \sum_{k \in \mathbb{K}_i} \frac{b_k^{j_{\min}}}{\min_{j \in \mathbb{J}_{i,k}} [R^j]} \Big]^+ \quad (21)$$

The form in (21) gives an intuitive result. The end-to-end service curve for flow i has a rate R, which is the minimum of the remaining rates at the traversed links, after subtracting the rates of interfering flows that share the individual links. The latency T is given as the sum of all latencies along the path, plus the burst size of interfering flows at their multiplexing points indicated by $j_{\min}$, divided by the minimum rate along the common sub-path with the flow of interest i. Thus, bursts of interfering flows account only once with the maximal latency that can be evoked by such bursts along the shared sub-paths.

Proof 1 In (22) the form in (15) is given for rate-latency service curves, and leaky bucket constraint arrival curves for $t_{j_{\max}+1} - t_{j_{\min}} > \theta_k$.

$$\begin{aligned} F_i^{n+1}(t_{n+1}) - F_i^1(t_1) \geq & \sum_{j \in \mathbb{J}_i} (R^j \cdot [t_{j+1} - t_j - T^j]^+) \\ & - \sum_{k \in \mathbb{K}_i} (b_k^{j_{\min}} + r_k \cdot (t_{j_{\max}+1} - t_{j_{\min}} - \theta_k)) \end{aligned} \quad (22)$$

Here, we apply a definition of the per flow θ_k by per link θ^j according to (23). Now the failure of any of the conditions $t_{j_{\max}+1} - t_{j_{\min}} > \theta_k$ requires the failure of at least one of the conditions $t_{j+1} - t_j > \theta^j$ for any j with $j_{\max} \geq j \geq j_{\min}$.

$$\theta_k = \sum_{j \in \mathbb{J}_{i,k}} \theta^j \quad (23)$$

Reformulation of (22) yields (24) for $t_{j+1} - t_j > \theta_j$, with $j \in \mathbb{J}_{i,k}$.

$$\begin{aligned} F_i^{n+1}(t_{n+1}) - F_i^1(t_1) \geq & \sum_{j \in \mathbb{J}_i} (R^j \cdot [t_{j+1} - t_j - T^j]^+) \\ & - \sum_{k \in \mathbb{K}_i} \Big(b_k^{j_{\min}} + r_k \cdot \sum_{j \in \mathbb{J}_{i,k}} (t_{j+1} - t_j - \theta^j) \Big) \end{aligned} \quad (24)$$

Next, the fact that (10) gives a service curve for any setting of the parameter θ with $\theta \geq 0$ is used. Thus, the per-flow θ_k can be set arbitrarily with $\theta_k \geq$

0, whereas this condition can according to (23) be fulfilled by any $\theta^j \geq 0$. Hence, (24) gives a service curve for any $\theta^j \geq 0$. We define an initial setting of the parameters θ^j, for which an end-to-end service curve for flow i is derived for all $t_{j+1} - t_j > \theta^j$. Then, for the special cases in which $t_{j+1} - t_j > \theta^j$ does not hold for one or several links j, θ^j is redefined. We show for these cases by means of the redefined θ^j that the end-to-end service curve that is derived before still holds true. Thus, we prove that this service curve is valid for all $t_{j+1} - t_j \geq 0$. The definition of different settings of the parameters θ^j is not generally allowed to derive a service curve. As already stated in Section 2, it cannot be concluded that for example $\inf_\theta \beta_\theta(t)$ is a service curve [13]. Nevertheless, there is a difference between doing so, and the derivation that is shown in the following. Here, a service curve β_{θ^j} is derived for fixed θ^j for all $t_{j+1} - t_j > \theta^j$. This service curve is not modified later on, but only proven to hold for any $t_{j+1} - t_j \geq 0$ by applying different settings of the θ^j. The initially applied setting of the θ^j is given in (25). The θ^j are defined to be the latency of the scheduler on the outgoing link j plus the burst size of interfering flows k, if j is the multiplexing point of the flow k with flow i, divided by the minimum rate along the common sub-path of flow k and i.

$$\theta^j = T^j + \sum_{k \in \mathbb{K}^j | j = j_{\min}} \frac{b_k^{j_{\min}}}{\min_{j' \in \mathbb{J}_{i,k}} [R^{j'}]} \tag{25}$$

With (25) the term in (24) can be rewritten according to (26) for all $t_{j+1} - t_j > \theta^j$.

$$\begin{aligned} F_i^{n+1}(t_{n+1}) - F_i^1(t_1) \geq & \sum_{j \in \mathbb{J}_i} \left(R^j \cdot [t_{j+1} - t_j - T^j]^+ \right) \\ & - \sum_{k \in \mathbb{K}_i} \left(b_k^{j_{\min}} + r_k \cdot \sum_{j \in \mathbb{J}_{i,k}} \left(t_{j+1} - t_j - T^j - \sum_{k' \in \mathbb{K}^j | j = j_{\min}} \frac{b_{k'}^{j_{\min}}}{\min_{j' \in \mathbb{J}_{i,k'}} [R^{j'}]} \right) \right) \end{aligned} \tag{26}$$

Some reordering, while scaling up the subtrahends by adding $[\ldots]^+$ conditions, and a replacement of $\sum_{k \in \mathbb{K}_i} \sum_{j \in \mathbb{J}_{i,k}}$ by $\sum_{j \in \mathbb{J}_i} \sum_{k \in \mathbb{K}^j}$ yields (27).

$$\begin{aligned} F_i^{n+1}(t_{n+1}) - F_i^1(t_1) \geq & \sum_{j \in \mathbb{J}_i} \left(\left(R^j - \sum_{k \in \mathbb{K}^j} r_k \right) \cdot [t_{j+1} - t_j - T^j]^+ \right) \\ & - \sum_{k \in \mathbb{K}_i} \left(b_k^{j_{\min}} - r_k \cdot \sum_{j \in \mathbb{J}_{i,k}} \sum_{k' \in \mathbb{K}^j | j = j_{\min}} \frac{b_{k'}^{j_{\min}}}{\min_{j' \in \mathbb{J}_{i,k'}} [R^{j'}]} \right) \end{aligned} \tag{27}$$

With the replacement of $\sum_{k \in \mathbb{K}_i} \sum_{j \in \mathbb{J}_{i,k}}$ by $\sum_{j \in \mathbb{J}_i} \sum_{k \in \mathbb{K}^j}$, (28) can be derived.

$$\begin{aligned} F_i^{n+1}(t_{n+1}) - F_i^1(t_1) \geq & \sum_{j \in \mathbb{J}_i} \left(\left(R^j - \sum_{k \in \mathbb{K}^j} r_k \right) \cdot [t_{j+1} - t_j - T^j]^+ \right) \\ & - \sum_{k \in \mathbb{K}_i} b_k^{j_{\min}} + \sum_{j \in \mathbb{J}_i} \sum_{k \in \mathbb{K}^j} \left(r_k \cdot \sum_{k' \in \mathbb{K}^j | j = j_{\min}} \frac{b_{k'}^{j_{\min}}}{\min_{j' \in \mathbb{J}_{i,k'}} [R^{j'}]} \right) \end{aligned} \tag{28}$$

Further on $\sum_{k\in\mathbb{K}_i} b_k^{j_{\min}} = \sum_{j\in\mathbb{J}_i}\sum_{k'\in\mathbb{K}^j|j=j_{\min}} b_{k'}^{j_{\min}}$ yields (29).

$$F_i^{n+1}(t_{n+1}) - F_i^1(t_1) \geq \sum_{j\in\mathbb{J}_i}\Big(\big(R^j - \sum_{k\in\mathbb{K}^j} r_k\big)\cdot[t_{j+1} - t_j - T^j]^+\Big)$$
$$-\sum_{j\in\mathbb{J}_i}\Big(\sum_{k'\in\mathbb{K}_{j|j=j_{\min}}} b_{k'}^{j_{\min}} - \sum_{k\in\mathbb{K}^j} r_k\cdot\sum_{k'\in\mathbb{K}^j|j=j_{\min}}\frac{b_{k'}^{j_{\min}}}{\min_{j'\in\mathbb{J}_{i,k'}}[R^{j'}]}\Big) \quad (29)$$

Applying the common denominator, while scaling up the subtrahend, and with $j \in \mathbb{J}_{i,k'}$ and thereby $R^j \geq \min_{j'\in\mathbb{J}_{i,k'}}[R^{j'}]$ with $k' \in \mathbb{K}^j$, (30) can be derived.

$$F_i^{n+1}(t_{n+1}) - F_i^1(t_1) \geq \sum_{j\in\mathbb{J}_i}\Big(\big(R^j - \sum_{k\in\mathbb{K}^j} r_k\big)\cdot[t_{j+1} - t_j - T^j]^+\Big)$$
$$-\sum_{j\in\mathbb{J}_i}\Big(\big(R^j - \sum_{k\in\mathbb{K}^j} r_k\big)\cdot\sum_{k'\in\mathbb{K}^j|j=j_{\min}}\frac{b_{k'}^{j_{\min}}}{\min_{j'\in\mathbb{J}_{i,k'}}[R^{j'}]}\Big) \quad (30)$$

Then, (30) can be reformulated according to (31), still for $t_{j+1} - t_j > \theta_j$.

$$F_i^{n+1}(t_{n+1}) - F_i^1(t_1) \geq \sum_{j\in\mathbb{J}_i}\Big(\big(R^j - \sum_{k\in\mathbb{K}^j} r_k\big)$$
$$\cdot\Big([t_{j+1} - t_j - T^j]^+ - \sum_{k'\in\mathbb{K}^j|j=j_{\min}}\frac{b_{k'}^{j_{\min}}}{\min_{j'\in\mathbb{J}_{i,k'}}[R^{j'}]}\Big)\Big) \quad (31)$$

The $\inf_{(t_{j+1}-t_j>\theta_j)|j\in\mathbb{J}_i}$ of (31) can be derived to be the form that is given in (21). Thus, the service curve in (21) is approved for all $t_{j+1} - t_j > \theta^j$ with the parameter settings of θ^j according to (25).

Now, if some of the conditions $t_{j+1} - t_j > \theta^j$ fail, the θ^j that are given in (25) can be redefined according to (32), based on the arbitrary parameters $\delta_k^j \geq 0$ with $\sum_{j\in\mathbb{J}_{i,k}} \delta_k^j = 1$.

$$\theta^j = T^j + \sum_{k\in\mathbb{K}_j}\frac{\delta_k^j\cdot b_k^{j_{\min}}}{\min_{j'\in\mathbb{J}_{i,k}}[R^{j'}]} \quad (32)$$

The burst size of interfering flows $b_k^{j_{\min}}$ is arbitrarily accounted for by θ^j in (25), whereas, if an interfering flow k traverses more than one link, the burst size $b_k^{j_{\min}}$ could be part of any θ^j, with $j \in \mathbb{J}_{i,k}$. For such redefined θ^j, it can be shown that the same derivation as above holds, resulting in (33). Again, applying the $\inf_{(t_{j+1}-t_j>\theta_j)|j\in\mathbb{J}_i}$ leads to the same form (21), as shown for (31) before.

$$F_i^{n+1}(t_{n+1}) - F_i^1(t_1) \geq \sum_{j\in\mathbb{J}_i}\Big(\big(R^j - \sum_{k\in\mathbb{K}^j} r_k\big)$$
$$\cdot\Big([t_{j+1} - t_j - T^j]^+ - \sum_{k'\in\mathbb{K}_j}\frac{\delta_{k'}^j\cdot b_{k'}^{j_{\min}}}{\min_{j'\in\mathbb{J}_{i,k'}}[R^{j'}]}\Big)\Big) \quad (33)$$

However, there can be pairs of $t_{j+1} - t_j \geq 0$, for which no setting of the parameters δ_k^j according to (32) allows a redefinition of θ^j, for which $t_{j+1} - t_j > \theta^j$ for all $j \in \mathbb{J}_i$ can be achieved. A condition $t_{j+1} - t_j > \theta_j$ can fail for $\delta_k^j = 0$ for all $k \in \mathbb{K}_j$, if $t_{j+1} - t_j \leq T^j$, without violating any of the per-flow conditions $t_{j_{\max}+1} - t_{j_{\min}} > \theta_k$. In this case the terms that are related to link j in (33) are nullified immediately by the $[\ldots]^+$ condition. Nevertheless, if some of the per flow conditions $t_{j_{\max}+1} - t_{j_{\min}} > \theta_k$ are violated for a number of flows $k \in \mathbb{L}_i$, the service curves of the sub-paths $\bigcup_{k \in \mathbb{L}_i} \mathbb{J}_{i,k}$ have according to the derivation of (10) to be set to zero. However, setting the service curves of sub-paths to zero is the same as setting the service curves of all links along these paths to zero. Regarding (33), it can be seen that any links for which the service curve is set to zero possibly increase the resulting rate of the service curve, compared to (21), whereas the resulting maximum latency is not influenced. This holds true for any θ^j according to (32), respective for any δ_k^j with $\sum_{j \in \mathbb{J}_{i,k}} \delta_k^j = 1$. Thus, also the case of $\delta_k^j = 0$ for all $k \in \mathbb{K}_i \backslash \{\mathbb{L}_i\}$, and $j \in \bigcup_{k \in \mathbb{L}_i} \mathbb{J}_{i,k}$ is covered. The latter setting ensures that bursts of flows that share part of the sub-paths that are set to zero, but that also traverse further links, are accounted for at these links. Then by scaling down the term $\min_{j \in \mathbb{J}_i \backslash \{\mathbb{J}_{i,k|k \in \mathbb{L}_i}\}} [R^j - \sum_{k \in \mathbb{K}^j} r_k]$ to $\min_{j \in \mathbb{J}_i} [R^j - \sum_{k \in \mathbb{K}^j} r_k]$, (21) is also a service curve for cases in which $t_{j+1} - t_j \leq \theta^j$, and finally holds for any $t_{j+1} - t_j \geq 0$ with $j \in \mathbb{J}_i$. □

Finally, (34) gives a tight end-to-end service curve for flow 2 in Figure 2. As intended, the bursts of the interfering flows 1, and 3 are accounted for only once, with their initial burst size at the multiplexing, or route interference point.

$$\begin{aligned}\beta_2(t) = \min[R^{\mathrm{I}} - r_1, R^{\mathrm{II}} - r_1 - r_3, R^{\mathrm{III}} - r_3] \\ \cdot \left[t - T^{\mathrm{I}} - T^{\mathrm{II}} - T^{\mathrm{III}} - \frac{b_1}{\min[R^{\mathrm{I}}, R^{\mathrm{II}}]} - \frac{b_3}{\min[R^{\mathrm{II}}, R^{\mathrm{III}}]}\right]^+\end{aligned} \tag{34}$$

5 Numerical Results

In this section we give numerical results on the derivation of edge-to-edge delay bounds in a DS domain, which can be efficiently applied for the definition of so-called Per Domain Behaviors (PDBs) [14]. We compare the options of applying either the Extended Pay Burst Only Once principle, the Pay Bursts Only Once principle, or none of the two principles.

We implemented an admission control for an application as a BB in a DS domain. The BB currently knows about the topology of its domain statically, whereas a routing protocol listener can be added. Requests for Premium capacity are sent via socket communication to the BB. The requests consist of a start, and an end time to allow for both immediate, and advance reservation, a Committed Information Rate (CIR), a Committed Burst Size (CBS), and a target maximum delay. Whenever the BB receives a new request, it computes the edge-to-edge delay for all requests that are active during the period of time of the new request, as described in Section 2. If none of the target maximum per-flow delays is violated, the new request is accepted, which otherwise is rejected.

For performance evaluation we implemented a simulator that generates such Premium resource requests. Sources and sinks are chosen uniformly from a predefined set. Start, and end times are modelled as negative exponentially distributed with a mean λ, respective μ, that is a mean of $\rho = \mu/\lambda$ requests are active concurrently. This modelling has been found to be appropriate for user sessions, for example File Transfer Protocol (FTP) sessions in [18]. The target delay, CIR, and CBS are used as uniformly distributed parameters for the simulations.

The topology that is used is shown in Figure 3. It consists of the level one, and level two nodes of the German Research Network (DFN) [1]. The level one nodes are core nodes. End systems are connected to the level two nodes that are edge nodes. In detail, we connect up to five sources and sinks to each of the level two nodes. Links are either Synchronous Transfer Mode (STM) 4, STM 16, or STM 64 connections. The link transmission delay is assumed to be 2 ms. Shortest Path First (SPF) routing is applied to minimize the number of hops along the paths. Further on, Turn Prohibition (TP) [21] is used, to ensure the required feed forward property of the network. Figure 3 shows how loops are broken within the level one mesh of the DFN topology by the TP algorithm. The nodes have been processed by TP in the order of their numbering. For example the turn $(8; 7; 9)$ that is the turn from node 8 via node 7 to node 9 is prohibited, whereas for instance the turn $(8; 7; 6)$, or simply the use of the link $(8; 7)$ is permitted. For the DFN topology the SPF TP algorithm does only increase the length of one of the paths by one hop compared to SPF routing. Further on, the TP algorithm can be configured to prohibit turns that include links with a comparably low capacity with priority [21], as is shown in Figure 3, and applied by our BB. The Premium service is implemented based on PQ. Thus, service curves are of the rate-latency type with a latency set to the time it takes to transmit 3 MTU of 9.6 kB, to account for non-preemptive scheduling, due to packetization, and a router internal buffer for 2 Jumbo frames [20].

The performance measure that we apply is the ratio of accepted requests divided by the overall number of requests, and as an alternative the distribution of the derived delay bounds. Simulations have been run, until the 0.95 confidence interval of the acceptance ratio was smaller than 0.01. Initial-data deletion [11] has been applied to capture only the steady-state behavior, and the replication and deletion approach for means, that is for example shown in [11], was used. The results are given for different settings of the requested maximum delay, CBS, and CIR, and for a varying load $\rho = \mu/\lambda$ in Figure 4, and 5. In addition Figure 6 shows the fraction of flows for which a delay bound that is smaller than the delay given on the abscissa was derived.

As one of our main results we find a significant performance gain, when accounting for the Extended Pay Bursts Only Once phenomenon. The Pay Bursts Only Once principle alone allows to derive noticeable tighter delay bounds, based on edge-to-edge service curves, compared to an incremental delay computation, as described already in Section 2. The advantage can further on be seen in terms of the acceptance ratio in Figure 4, and 5, whereas the importance of the Pay Bursts Only Once phenomenon increases, if a larger CBS is used. In addition

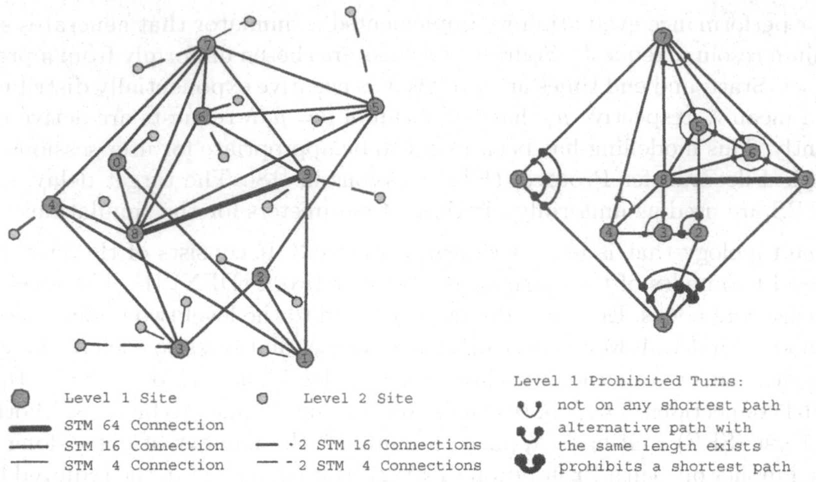

Fig. 3. DFN Topology and Example Level 1 Prohibited Turns

the Extended Pay Bursts Only Once principle is of major relevance, if the load on the network is high, and if large bursts are permitted. In case of a high load, flows are likely to share common sub-paths, and the interference of such flows with each other is much more accurately described by the Extended Pay Bursts Only Once principle. Thus, the form presented in (21), allows to derive tighter delay bounds, and thus to increase the acceptance ratio. In particular, as Figure 6 shows, the delay bound for the 99-percentile of the flows is reduced from above 112 ms to 81 ms in case of the Pay Bursts Only Once principle, and than to 59 ms in case of the extended principle for aggregate scheduling. For the 95-percentile, 85 ms, 63 ms, and 49 ms can be given. Further on, the load, up to which an acceptance ratio of for example 0.95 can be achieved, can be multiplied, when accounting for the Extended Pay Bursts Only Once phenomenon.

6 Conclusions

In this paper we have shown that a counterpart to the Pay Bursts Only Once phenomenon exists for interfering flows in aggregate scheduling networks. We then have derived a closed form solution for the end-to-end per-flow service curve in arbitrary feed forward aggregate scheduling networks, where links are of the rate-latency type, and flows are sigma-rho leaky bucket constraint. Our solution accounts for the known Pay Bursts Only Once principle, and extends it to aggregate scheduling in that bursts of interfering flows are paid only once, too. Thus, our form allows to give significantly closer bounds on the delay, while the intuitive form reduces computational complexity, if for example applied as a decision criterion for a Differentiated Services admission control in a Bandwidth Broker. A significant performance gain has been shown by simulation results.

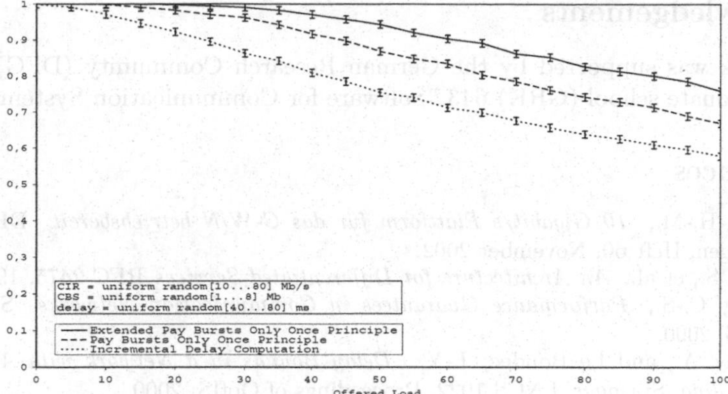

Fig. 4. Acceptance Ratio versus Load

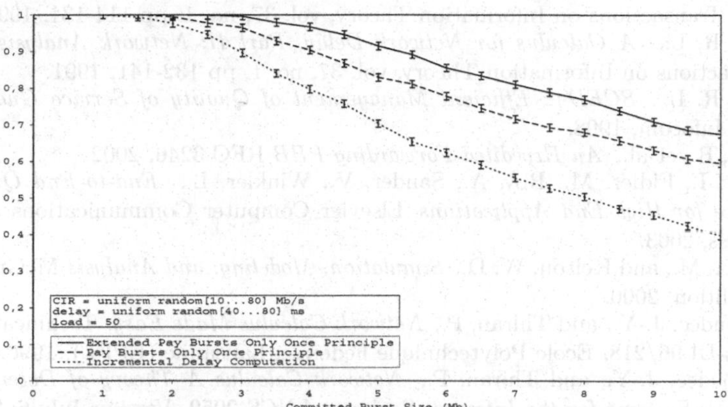

Fig. 5. Acceptance Ratio versus Committed Burst Size

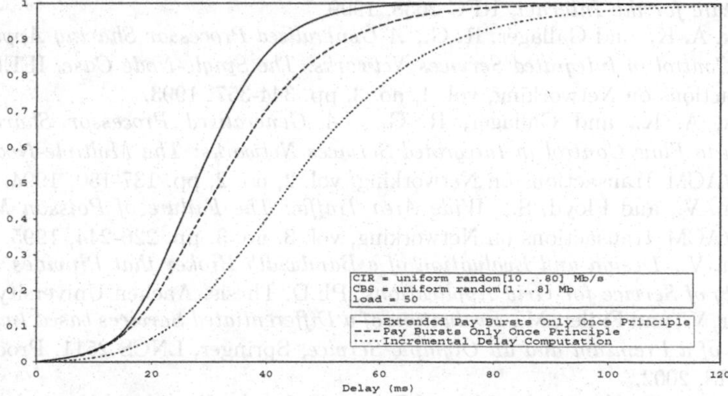

Fig. 6. Fraction of Flows with a smaller Delay Bound

Acknowledgements

This work was supported by the German Research Community (DFG) under grant graduate school (GRK) 643 "Software for Communication Systems".

References

1. Adler, H.-M., *10 Gigabit/s Plattform für das G-WiN betriebsbereit*, DFN Mitteilungen, Heft 60, November 2002.
2. Blake, S., et al., *An Architecture for Differentiated Services* RFC 2475, 1998.
3. Chang, C.-S., *Performance Guarantees in Communication Networks*, Springer, TNCS, 2000.
4. Charny, A., and Le Boudec, J.-Y., *Delay Bounds in a Network with Aggregate Scheduling*, Springer, LNCS 1922, Proceedings of QofIS, 2000.
5. Charny, A., et al., *Supplemental Information for the New Definition of EF PHB (Expedited Forwarding Per-Hop-Behavior)* RFC 3247, 2002.
6. Cruz, R. L., *A Calculus for Network Delay, Part I: Network Elements in Isolation*, IEEE Transactions on Information Theory, vol. 37, no. 1, pp 114-131, 1991.
7. Cruz, R. L., *A Calculus for Network Delay, Part II: Network Analysis*, IEEE Transactions on Information Theory, vol. 37, no. 1, pp 132-141, 1991.
8. Cruz, R. L., *SCED+: Efficient Management of Quality of Service Guarantees*, IEEE Infocom, 1998.
9. Davie, B., et al., *An Expedited Forwarding PHB* RFC 3246, 2002.
10. Foster, I., Fidler, M., Roy, A., Sander, V., Winkler, L., *End-to-End Quality of Service for High-End Applications* Elsevier Computer Communications Journal, in press, 2003.
11. Law, A. M., and Kelton, W. D., *Simulation, Modeling, and Analysis* McGraw-Hill, 3rd edition, 2000.
12. Le Boudec, J.-Y., and Thiran, P., *Network Calculus Made Easy*, Technical Report EPFL-DI 96/218. Ecole Polytechnique Federale, Lausanne (EPFL), 1996.
13. Le Boudec, J.-Y., and Thiran, P., *Network Calculus A Theory of Deterministic Queuing Systems for the Internet*, Springer, LNCS 2050, Version July 6, 2002.
14. Nichols, K., and Carpenter, B., *Definition of Differentiated Services Per Domain Behaviors and Rules for their Specification*, RFC 3086, 2001.
15. Nichols, K., Jacobson, V., and Zhang, L., *A Two-bit Differentiated Services Architecture for the Internet*, RFC 2638, 1999.
16. Pareck, A. K., and Gallager, R. G., *A Generalized Processor Sharing Approach to Flow Control in Integrated Services Networks: The Single-Node Case*, IEEE/ACM Transactions on Networking, vol. 1, no. 3, pp. 344-357, 1993.
17. Pareck, A. K., and Gallager, R. G., *A Generalized Processor Sharing Approach to Flow Control in Integrated Services Networks: The Multiple-Node Case*, IEEE/ACM Transactions on Networking, vol. 2, no. 2, pp. 137-150, 1994.
18. Paxson, V., and Floyd, S., *Wide Area Traffic: The Failure of Poisson Modeling* IEEE/ACM Transactions on Networking, vol. 3, no. 3, pp. 226-244, 1995.
19. Sander, V., *Design and Evaluation of a Bandwidth Broker that Provides Network Quality of Service for Grid Applications*, Ph.D. Thesis, Aachen University, 2002.
20. Sander, V., and Fidler, M., *Evaluation of a Differentiated Services based Implementation of a Premium and an Olympic Service*, Springer, LNCS 2511, Proceedings of QofIS, 2002.
21. Starobinski, D., Karpovsky, M., and Zakrevski, L., *Application of Network Calculus to General Topologies using Turn-Prohibition*, Proceedings of IEEE Infocom, 2002.

Modelling the Arrival Process for Packet Audio

Ingemar Kaj[1] and Ian Marsh[2]

[1] Dept. of Mathematics, Uppsala University, Sweden
ikaj@math.uu.se
[2] Ian Marsh, CNA Lab, SICS, Sweden
ianm@sics.se

Abstract. Packets in an audio stream can be distorted relative to one another during the traversal of a packet switched network. This distortion can be mainly attributed to queues in routers between the source and the destination. The queues can consist of packets either from our own flow, or from other flows. The contribution of this work is a Markov model for the time delay variation of packet audio in this scenario. Our model is extensible, and show this by including sender silence suppression and packet loss into the model. By comparing the model to wide area traffic traces we show the possibility to generate an audio arrival process similar to those created by real conditions. This is done by comparing the probability density functions of our model to the real captured data.

Keywords: Packet delay, VoIP, Markov chain, Steady state

1 Introduction

Modelling the arrival process for audio packets that have passed through a series of routers is the problem we will address. Figure 1 illustrates this situation: Packets containing audio samples are sent at a constant rate from a sender, shown in step one. The

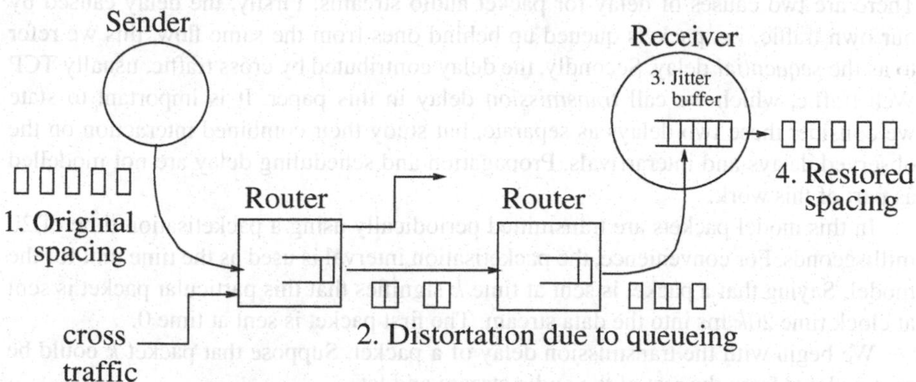

Fig. 1. The networks effect on packet audio spacing

M. Ajmone Marsan et al. (Eds.): QoS-IP 2003, LNCS 2601, pp. 35–49, 2003.

spacing between packets is compressed and elongated relative to each other. This is due to the buffering in intermediate routers and mixing with cross-traffic, shown in step two. In order to replay the packets with their original spacing, a buffer is introduced at the receiver, commonly referred to as a *jitter buffer* shown in step three. The objective of the buffer is to absorb the variance in the inter-packet spacing introduced by the delays due to cross traffic, and (potentially) its own data. In step four, using information coded into the header of each packet, the packets are replayed with their original timing restored.

The motivation for this work derives from the inability of using known arrival processes to approximate the packet arrival process at the receiver. Using a known arrival process, even a complex one, is not always realistic as the model does not include characteristics that real audio streams experience. For example the use of silence suppression or the delay/jitter contribution of cross traffic. One alternative is to use real traffic traces. Although they produce accurate and representative arrival processes, they are inherently static and do not offer much in the way of flexibility. For example, observing the affect of different packet sizes without re-running the experiments. When testing the performance of jitter buffer playout algorithms, for example, this inflexibility is undesirable. Thus, an important contribution of this paper is to address the deficiencies of these approaches by *combining* the advantages of both a model of the process, with data from real traces.

This paper presents in a descriptive manner, a packet delay model, based on the main assumption that packets are subjected to independent transmission delays. It is intended that readers not completely familiar with Markovian theory can follow the description. We assume no prior knowledge of the model as it is built from first principles starting in section 2. We give results for the mean arrival and interarrival times of audio packets in this section. We add silence suppression to the model in section 3 and packet loss in the next section, 4. Real data is incorporated in section 5, related work follows in section 6 and we customarily round off with some conclusions in section 7.

2 The Packet Delay Model

There are two causes of delay for packet audio streams. Firstly, the delay caused by our own traffic, i.e. packets queued up behind ones from the same flow, this we refer to as the *sequential* delay. Secondly, the delay contributed by cross traffic, usually TCP Web traffic, which we call *transmission* delay in this paper. It is important to state we consider these two delays as separate, but study their combined interaction on the observed delays and interarrivals. Propagation and scheduling delay are not modelled as part of this work.

In this model packets are transmitted periodically using a packetisation time of 20 milliseconds. For convenience, the packetisation interval is used as the time unit for the model. Saying that a packet is sent at time k signifies that this particular packet is sent at clock time $20k$ ms into the data stream. The first packet is sent at time 0.

We begin with the transmission delay of a packet. Suppose that packet k could be sent isolated from the rest of the audio stream and let

$$Y_k = \text{transmission delay of packet } k \text{ (no. of 20 ms periods).}$$

To see the impact of the sequential delay, let

$$T_k = \text{the arrival time of packet } k \text{ at the jitter buffer}, \quad k \geq 1.$$

The model used in this paper is shown in Figure 2. The figure shows packets being trans-

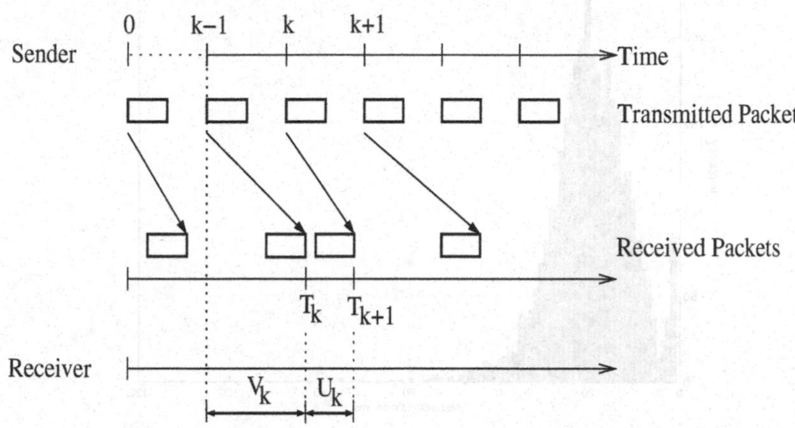

Fig. 2. T_k arrival times before playout, V_k observed delays, U_k observed interarrival times

mitted from a sender at regular intervals. They traverse the network, where as stated, their original spacing is distorted. Packet k arrives at time T_k at the receiver. The difference in time between when it departed and arrived we call the *observed delay*, which we denote

$$V_k = \text{arrival time} - \text{departure time} = T_k - k + 1 \quad k \geq 1.$$

The time when the next packet (numbered $k+1$) arrives is T_{k+1} and so the *observed interarrival times* are obtained as the differences between T_{k+1} and T_k, denoted

$$U_k = T_{k+1} - T_k.$$

A packet k, sent at time $k-1$, requires time Y_k to propagate through the network and arrives therefore at $T_k = k - 1 + Y_k$, as long as it is not delayed further by other audio packets (which we call sequential delays). It may however catch up to audio packets transmitted earlier ($1 \to k-1$). This packet is forced to wait before being stored in the playout buffer. This shows that the actual arrival times satisfy:

$$\begin{aligned} T_1 &= Y_1 \\ T_k &= \max(T_{k-1},\, k - 1 + Y_k), \quad k \geq 2. \end{aligned} \tag{1}$$

Since T_{k-1} and Y_k are independent, we conclude from the relation above (1) that T_k forms a transient Markov chain. Moreover, the interarrival times satisfy

$$U_k = T_k - T_{k-1} = \max(0,\, k - 1 + Y_k - T_{k-1}) \quad k \geq 2. \tag{2}$$

The arrival times (T_k), interarrival times (U_k) and observed delays (V_k) can be easily observed from traffic traces. As an example, Figure 3 shows the histogram for an

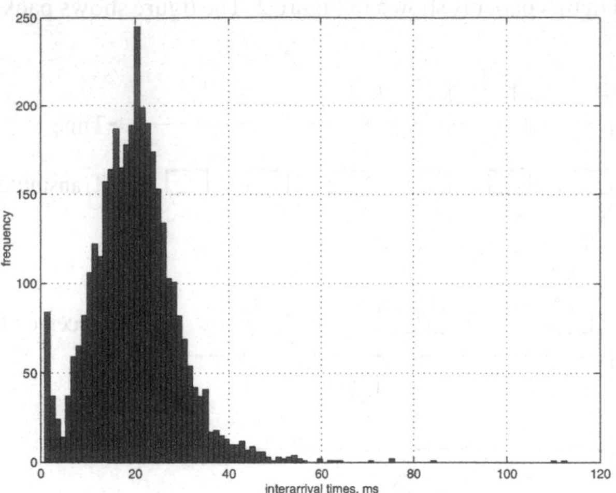

Fig. 3. Histogram of the interarrival times (U_k)

empirical sequence of interarrival times. The data is from a recording of a Voice over IP session between Argentina and Sweden, more details of the traffic traces are given in section 5.1. The transmission delay sequence (Y_k) should be on the other hand considered as non-observable. The approach in this study is to consider (Y_k) having a general (unknown) distribution and investigate the resulting properties of the observed delay (V_k) and interarrival times (U_k). Since the latter sequences can be empirically observed, this leads to the question to whether the transmission delay distribution can be reconstructed using statistical inference. In this direction we will indicate some methods that could be used to compare the theoretical results with the gathered empirical data.

To carry out the study, we assume from this point the sequence (Y_k) is independent and identically distributed, with distribution function

$$F(x) = P(Y_k \leq x), \quad k \geq 1,$$

and finite mean transmission delay $\nu = \int_0^\infty (1 - F(x))\,dx < \infty$. For the data in our study, typical values of ν are 20-40, i.e. 400-800 ms. We consider these assumptions justified for the purpose of studying a reference model, obviously it would be desirable to allow dependence over time.

2.1 Mean Arrival and Interarrival Times

It is intuitively clear that in the long run $E(U_k) \approx 1$ as on average packets arrive with 20 ms spacing, which we will now verify for the model. The representation (1) for T_k can be written

$$T_k = \max(Y_1, 1+Y_2, \ldots, k-1+Y_k) \quad k \geq 1,$$

which gives the alternative representation

$$T_k = \max(Y_1, 1+T'_{k-1}), \quad k \geq 2 \tag{3}$$

where on the right side

$$T'_{k-1} = \max(Y_2, 1+Y_3, \ldots, k-2+Y_k)$$

has the same marginal distribution as T_{k-1} but is independent of Y_1. From (3) follows that we can write $\{T_k > t\}$ as a union of two disjoint events, as

$$\{T_k > t\} = \{1+T'_{k-1} > t\} \cup \{Y_1 > t, 1+T'_{k-1} \leq t\}.$$

Hence, using the independence of T'_{k-1} and Y_1,

$$\begin{aligned} P(T_k > t) &= P(1+T'_{k-1} > t) + P(Y_1 > t, 1+T'_{k-1} \leq t) \\ &= P(1+T_{k-1} > t) + P(Y_1 > t)P(1+T_{k-1} \leq t) \end{aligned}$$

and so

$$\begin{aligned} E(T_k) &= \int_0^\infty P(T_k > t)\,dt \\ &= E(1+T_{k-1}) + \int_1^\infty P(Y_1 > t)P(T_{k-1} \leq t-1)\,dt. \end{aligned} \tag{4}$$

Therefore

$$E(U_k) = 1 + \int_1^\infty P(Y_1 > t)P(T_{k-1} \leq t-1)\,dt \to 1, \quad k \to \infty \tag{5}$$

(since $\nu = \int_0^\infty P(Y_1 > t)\,dt < \infty$ and $T_k \to \infty$, the dominated convergence theorem applies forcing the integral to vanish in the limit).

A further consequence of (4) is obtained by iteration,

$$E(T_k) = k - 1 + E(Y_1) + \int_1^\infty P(Y_1 > t) \sum_{i=1}^{k-1} P(T_i \leq t-1)\,dt.$$

If we introduce

$N(t)$ = the number of arriving packets in the time interval $(0, t]$,

so that $\{N(t) \geq n\} = \{T_n \leq t\}$, this can be written

$$E(V_k) = E(Y_1) + \int_1^\infty P(Y_1 > t) \sum_{i=1}^{k-1} P(N(t-1) \geq i)\,dt, \tag{6}$$

which, as $k \to \infty$, gives an asymptotic representation for the average observed delay as

$$E(V_k) \to \nu + \int_1^\infty P(Y_1 > t) E(N(t-1)))\,dt. \tag{7}$$

2.2 Steady State Distributions

By (1),

$$P(T_k \le x) = \prod_{i=1}^{k} P(i + Y_i \le x + 1) = \prod_{i=0}^{k-1} F(x - i),$$

and therefore the sequence (V_k), which we defined by $V_k = T_k - k + 1, k \ge 1$, satisfies

$$P(V_k \le x) = \prod_{i=0}^{k-1} F(x + k - 1 - i) = \prod_{i=0}^{k-1} F(x + i) \quad x \ge 0.$$

This shows that (V_k) is a Markov chain with state space the positive real line and asymptotic distribution given by

$$P(V_\infty \le x) = \prod_{i=0}^{\infty} F(x + i) \quad x \ge 0. \tag{8}$$

Furthermore, for $x \ge 0$

$$\begin{aligned} P(U_k \ge x) &= P(k - 1 + Y_k - T_{k-1} \ge x) = P(V_{k-1} \le Y_k + 1 - x) \\ &= \int_0^\infty P(V_{k-1} \le y + 1 - x)\, dF(y), \end{aligned}$$

where in the step of conditioning over Y_k we use the independence of Y_k and V_{k-1}. Therefore the sequence (U_k) has the asymptotic distribution

$$P(U_\infty \le x) = 1 - \int_0^\infty \prod_{i=1}^{\infty} F(y - x + i)\, dF(y) \quad x \ge 0, \tag{9}$$

in particular a point mass in zero of size

$$P(U_\infty = 0) = 1 - \int_0^\infty \prod_{i=1}^{\infty} F(y + i)\, dF(y). \tag{10}$$

This distribution has the property that $E(U_\infty) = 1$ for any given distribution F of Y with $\nu = E(Y) < \infty$. In fact, this follows from 5 under a slightly stronger assumption on Y (uniform integrability), but can also be verified directly by integrating (9). Figure 4 shows numeric approximations of the (non-normalised) density function $\frac{d}{dx}P(U_\infty \le x)$ of (9) for three choices of F. All three distributions show a characteristic peak close to time 1 corresponding to the bulk of packets arriving with more or less correct spacing of 20 ms. A fraction of the probability mass is fixed at $x = 0$ in accordance with (10), but not shown explicitly in the figure. These features of the density functions can be compared with the shape of the histogram in Figure 3 with its peak at the 20 ms spacing. Also, close to the origin is a small peak which corresponds to packets arriving back-to-back usually arriving as a burst, probably due to a delayed packet ahead of them. In Figure 4 the density function with the highest peak close to 1 time unit is a Gaussian distribution with arbitrarily selected parameters mean 5 and variance 0.2. Of the two exponential distributions, the one with the higher variance (Exp(3)) has a lower peak and more mass at zero compared with an exponential with smaller variance (Exp(2)).

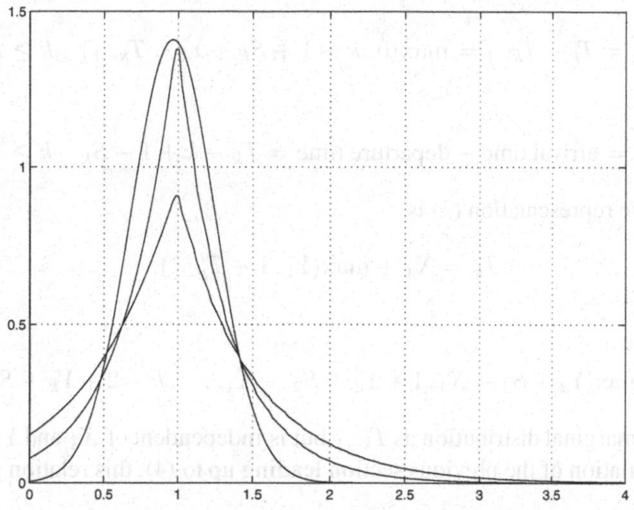

Fig. 4. Density functions of U for N(5,0.2), Exp(2) and Exp(3)

3 Silence Suppression Mechanism

In this section we incorporate an additional source of random delays due to silence suppression into the model. Silence suppression is employed at the sender so as not to transmit packets when there is no speech activity. During a normal conversation this accounts for about half of the total number of packets, considerably reducing the load on the network. Assign to packet number k the quantity

$$X_k = \text{duration of silent period between packets } k-1 \text{ and } k.$$

A silent period is the time interval during which the silence suppression mechanism is in effect. We assume that the silence suppression intervals are independent of $(Y_k)_{k\geq 1}$ and are given by a sequence of independent random variables $X_1, X_2, \ldots$, such that

$$G(x) = P(X_k \leq x), \quad 1-\alpha = G(0) = P(X_k = 0) > 0, \quad \mu = E(X_k) < \infty.$$

The (small) probability $\alpha = P(X_k > 0)$ represents the case where silence suppression is activated just after packet $k-1$ is transmitted from the sender. Note that

$$S_k = \sum_{i=1}^{k} X_i = \text{total time of silence suppression affecting packet } k,$$

which implies that the delivery of packet k from the sending unit now starts at time $k-1+S_k$. The representation (1) takes the form

$$T_1 = S_1 + Y_1, \quad T_k = \max(T_{k-1}, k-1+S_k+Y_k), \quad k \geq 2, \tag{11}$$

hence

$$U_k = T_k - T_{k-1} = \max(0, k-1+S_k+Y_k-T_{k-1}) \quad k \geq 2. \tag{12}$$

Similarly,

$$V_k = \text{arrival time} - \text{departure time} = T_k - k + 1 - S_k \quad k \geq 1.$$

The alternative representation (3) is

$$T_k = X_1 + \max(Y_1, 1 + T'_{k-1}), \tag{13}$$

where

$$T'_{k-1} = \max(Y_2 + S_2 - X_1, 1 + Y_2 + S_2 - X_1, \ldots, k-2+Y_k+S_k-X_1)$$

has the same marginal distribution as T_{k-1} but is independent of X_1 and Y_1. In analogy with the calculation of the previous section leading up to (4), this relation gives

$$E(T_k) = E(X_1 + 1 + T_{k-1}) + \int_1^\infty P(X_1 + Y_1 > t, X_1 + T'_{k-1} \leq t-1)\, dt. \tag{14}$$

Exchanging the operations of integration and expectation shows that the last integral can be written

$$E\left[\int_{1+X_1}^\infty 1\{Y_1 > t - X_1, T'_{k-1} > t - X_1 - 1\}\, dt\right]$$

where we have also used that the integrand vanishes on the set $\{t \leq 1 + X_1\}$. Apply the change-of-variables $t \to t - X_1$ to get $E\left[\int_1^\infty 1\{Y_1 > t, T'_{k-1} > t-1\}\, dt\right]$. Then shift integration and expectation again to obtain from (14) the relations

$$E(T_k) = 1 + E(X_1) + E(T_{k-1}) + \int_0^\infty P(Y_1 > t)P(T_{k-1} \leq t-1)\, dt$$

and

$$E(U_k) = 1 + E(X_1) + \int_1^\infty P(Y_1 > t)P(T_{k-1} \leq t-1)\, dt.$$

Hence with silence suppression, as $k \to \infty$,

$$E(U_k) \to 1 + \mu, \quad E(V_k) \to \nu + \int_1^\infty P(Y_1 > t)E(N(t-1))\, dt, \tag{15}$$

using the same arguments as in the simpler case of the previous section.

4 Including Packet Loss in the Model

We return to the original model without silence suppression but consider instead the effect of lost packets. Suppose that each IP packet is subject to loss with probability p, independently of other packet losses and of the transmission delays. Lost packets are unaccounted for at the receiver and hence, in this section, the sequence (T_k) records arrival times of non-lost packets only. To keep track of their delivery times from the sender introduce

$$K_k = \text{number of attempts required between successful packets } k-1 \text{ and } k, \quad k \geq 1,$$

which gives a sequence $(K_k)_{k\geq 1}$ of independent, identically distributed random variables with the geometric distribution

$$P(K_k = j) = (1-p)p^j, \quad j \geq 0.$$

Moreover,

$$L_k = K_1 + \ldots + K_k = \text{number of attempts required for } k \text{ successful packets}$$

is a sequence of random variables with a negative binomial distribution. The arrival times of packets are now given by

$$T_1 = K_1 - 1 + Y_{K_1}, \quad T_k = \max(T_{k-1}, L_k - 1 + Y_{L_k}), \quad k \geq 2.$$

Due to the independence we may re-index the sequence of Y_{L_k}'s to obtain

$$T_1 = K_1 - 1 + Y_1, \quad T_k = \max(T_{k-1}, L_k - 1 + Y_k), \quad k \geq 2.$$

and thus

$$T_k = K_1 - 1 + \max(Y_1, 1 + T'_{k-1}), \quad k \geq 2 \tag{16}$$

with K_1, Y_1 and T'_{k-1} all independent, and again T_{k-1} and T'_{k-1} identically distributed. This is the same relation as (13) with X_1 replaced by $K_1 - 1$ and hence, as in (15), $E(U_k) \to 1 + E(K_1 - 1) = \frac{1}{1-p}$, $k \to \infty$, which provides a simple method to estimate packet loss based on observed interarrival times. Similarly, combining silence suppression and packet loss,

$$E(U_k) \to 1 + E(X) + E(K_1 - 1) = \mu + \frac{1}{1-p}, \quad k \to \infty, \tag{17}$$

5 Incorporating Real Data

5.1 Trace Data

We give a brief description of the experiments we performed in order to obtain estimates for the parameters in the model. Pulse Code Modulated (PCM) packet audio streams were sent from a site in Buenos Aires, Argentina to Stockholm, Sweden over a number of weeks[1]. The streams are sent with a 64kbits/sec rate in 160 byte payloads. This implies the packets leave the sender with a inter-packet spacing of 20 ms. The remote site is approximately 12,000 kilometres, 25 Internet hops and four time zones from

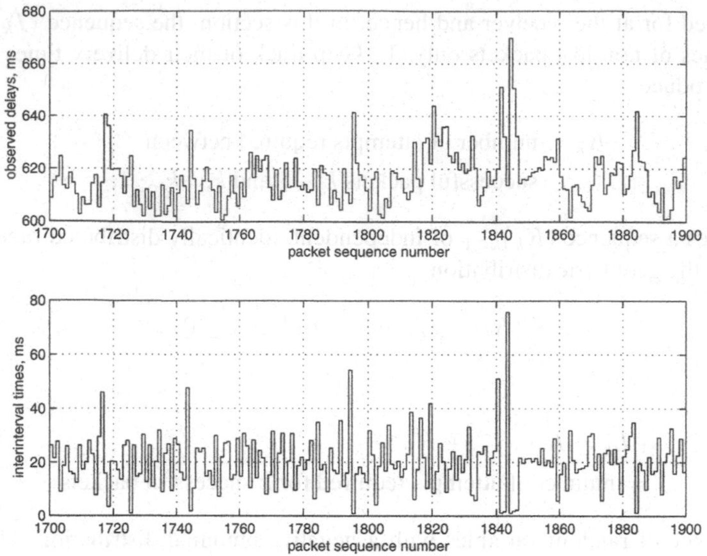

Fig. 5. Four second audio packet traces: a)delays b)interarrival times

our receiver. The tool is capable of silence suppression, in which packets are not sent when the speaker is silent. Without silence suppression, 3563 packets are sent during 70 seconds and with suppression 2064 are sent. We record the absolute times the packets leave the sender and the absolute arrival times at the receiver. This gives an observed sequence

$$v_k = \text{arrival time no } k - \text{departure time no } k$$

of the Markov chain (V_k). In particular, the sample mean $\bar{v}$ is an estimate of the one-way delay. Similarly,

$$u_k = \text{arrival time no } k - \text{arrival time no } (k-1)$$

is a sample of the interarrival time sequence (U_k).

A typical sequence of trace data used in this study *without* silence suppression is shown in Figure 5, which shows (v_k) and (u_k) for a small sequence of 200 packets $(1700 \leq k < 1900)$, corresponding to four seconds of audio. To further illustrate such trace data, Figure 6 shows a histogram of the delays (v_k) and Figure 3 a histogram for the interarrival times (u_k). It can be noted that large values of interarrival times are sometimes followed by very small ones, manifesting that a severely delayed packet forces subsequent packets to arrive back-to-back. The fraction of packets arriving in this manner corresponds to the height of the leftmost peak in the histogram of Figure 3.

Returning to traces with silence suppression, Figure 7 gives the statistics of the recorded voice signal used. The upper part shows a histogram of the talkspurts and the lower part the corresponding histogram for the non-zero part of the distribution G of

[1] Available from http://www.sics.se/~ianm/COST263/cost263.html

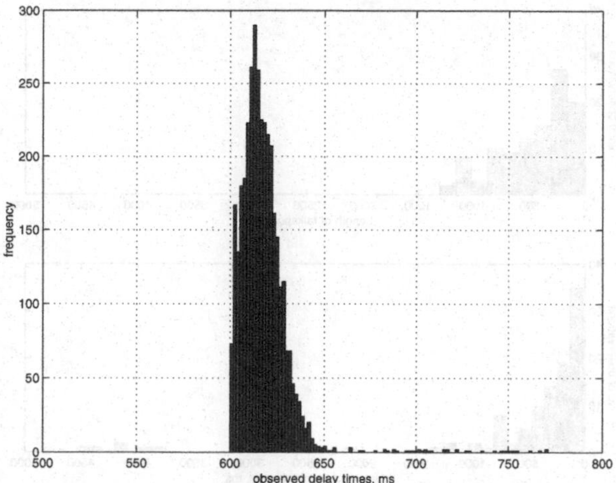

Fig. 6. Histogram of the observed delays (V_k)

the silence intervals X discussed in section 3. The probability $\alpha = P(X = 0)$ and the expected value $\mu = E(X)$ were estimated to

$$\alpha^* = 0.0456 \qquad \mu^* = 25.7171.$$

5.2 Numerical Estimates

In this section we indicate a few simple numerical techniques that give parameter estimates based on trace data. In principle such methods based on the model presented here can be used for systematic studies of the delays and losses and for comparison of traces sampled in different environments.

Considering first the case of no silence suppression, it was pointed out in section 4 that given an observed realization $(u_k)_{k=1}^n$ of (U_k), a point estimate of the packet loss probability p is obtained from (17) (with $\mu = 0$), using

$$p^* = 1 - \frac{20}{\bar{u}}, \qquad \bar{u} = \frac{1}{n}\sum_{k=1}^{n} u_k \text{ ms.}$$

Our measurements gave consistently $\bar{u} \approx 20.002 - 20.005$ ms, indicating loss probabilities of the order 10^{-4}.

Next we look at an experiment where the pre-recorded voice is transmitted at seven different times using silence suppression, and the interarrival times measured at the receiver during each transmission. Table 1 shows the expected silence interval $E(X)$ and the estimated μ from the trace files. The obtained estimates indicate a systematic bias of the order 0.5 milliseconds in the mean values of the silence suppression intervals.

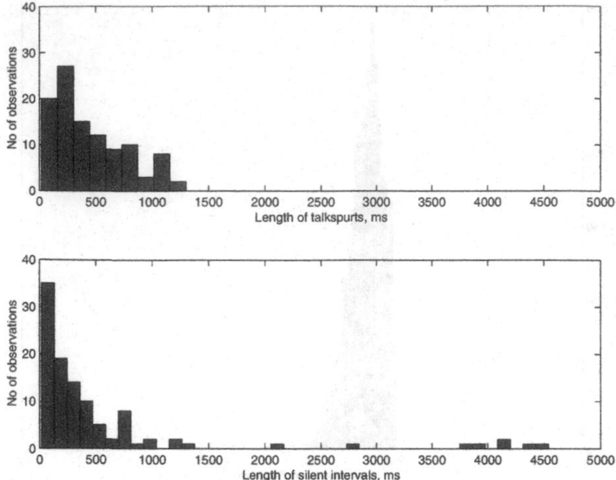

Fig. 7. Lengths of talkspurts and silence periods

Table 1. Silence Interval Parameters

Trace	**E(X)**	**E(X)**-μ^*
trace 1	25.7492	0.0321
trace 2	26.2204	0.4639
trace 3	26.2284	0.5113
trace 4	26.2164	0.4993
trace 5	26.2186	0.5015
trace 6	26.2124	0.4953
trace 7	26.2209	0.5038

Packet losses do not seem to explain fully the observed deviation. A more comprehensive statistical analysis might reveal the source of this slight mismatch. For the present preliminary investigation we find the numerical estimates satisfactory.

We now consider the problem of estimating the distribution F of packet delays Y given a fixed length sample observation (v_k) of the Markov chain (V_k) for observed delays. One method for this can be based on the steady state analysis in section 2.2. Indeed, rewriting (8) as the simple relation

$$P(V_\infty \leq x) = F(x) \prod_{i=1}^{\infty} F(x+i) = F(x)\, P(V_\infty \leq x+1)$$

shows that if we let $\bar{F}_V$ denote an empirical distribution function of V, then we obtain an estimate $\bar{F}$ of F by taking

$$\bar{F}(x) = \frac{\bar{F}_V(x)}{\bar{F}_V(x+1)} \qquad x \geq 0, \tag{18}$$

where we recall that the variable x is measured in units of 20 ms intervals. An application of this numerical algorithm to the trace data of the previous figures (5 and 6) yields an estimated density function for Y as in Figure 8. The numerical scheme is sensitive

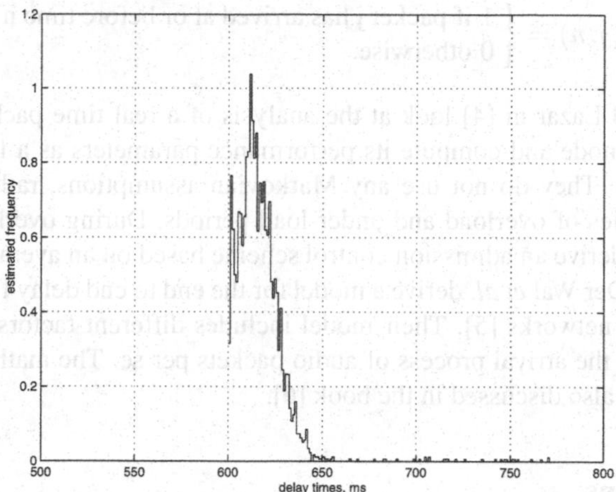

Fig. 8. Estimated density of Y

for small changes in the data, so it is difficult to draw conclusions on the finer details of the distribution of F. As expected the graph is very similar to that of the observed delays, Figure 6, but with certain differences due to the Markovian dependence structure in the sequence (V_k) as opposed to the independence in (Y_k). The main difference is the shift towards smaller values for Y in comparison to those of V. This corresponds to the inequality $\bar{F}(x) \geq \bar{F}_V(x)$ valid for all x, which is obvious from (18).

6 Related Work

Many researchers have looked at the needs in terms of buffer size for packet streams characterised by Markov (semi or modulated) behaviour especially in the case of multiplexed sources. Their goal was to derive the waiting time of packets spent in the buffer shown as probability density function of the waiting times. Relatively few, however, have looked at the arrival process using a stage of buffers and identifying embedded Markov chains from a single source. Additionally we concentrate on this scenario, including both streams with and without silence suppression. Additionally as far as we know, no-one has used real trace data to enhance and verify their models to the level we show.

Some early analytical work on the buffer size requirements for packetised voice is summarised by Gopal *et al.* [1]. One often cited piece of work is Barberis [2]. As part of this work he assumes the delays experienced by packets of the same talkspurt are i.i.d according to an exponential distribution $p(t) = \lambda e^{-\lambda T}$ where $1/\lambda$ is the average transmission delay and standard deviation. M.K. Mehmet Ali *et al.* in their work of buffer requirements [3] model the arrival process as a Bernoulli trial with probability $[1 - F(j, n - j + 1)]$ of the event "no arrival yet" at each interval up to its arrival. The outcome of the trial is represented by the random variable $k(j, n)$:

$$k(j,n) = \begin{cases} 1 \text{ if packet j has arrived at or before time n} \\ 0 \text{ otherwise.} \end{cases}$$

Ferrandiz and Lazar in [4] look at the analysis of a real time packet session over a single channel node and compute its performance parameters as a function of their model primitives. They do not use any Markovian assumptions, rather an approach which uses a series of overload and under-load periods. During overload packets are discarded. They derive an admission control scheme based on an average of the packet arrival rate. Van Der Wal *et al.* derive a model for the end to end delay for voice packets in large scale IP networks [5]. Their model includes different factors contributing to the delay but not the arrival process of audio packets per se. The mathematical model described here is also discussed in the book [6].

7 Conclusions

We have addressed the problem of modelling the arrival process of a single packet audio stream. The model can be used to produce packet audio streams with characteristics, at least, quite similar to the particular traces we have obtained. The model is suitable for generating streams both with and without silence suppression applied at the source, in addition the case where packets are lost has been included.

The work can be generally applied to research where modelling arriving packet audio streams needs to be performed. A natural next step is to use the arrival model presented here for evaluation of jitter buffer performance, such as investigating waiting times and possible packet loss in the jitter buffer. We observed from our model that the interarrival times are negatively correlated (as mentioned Section 5.1). This will have an impact on the dynamics and performance of a jitter buffer. With an accurate model, based on real data measurements, a realistic traffic generator can be written. In separate work we have gathered nearly 25,000 VoIP traces from ten globally dispersed sites which we can utilise for 'parameterising' the model, depending on the desired scenario.

References

[1] Prabandham M. Gopal, J. W. Wong, and J. C. Majithia, "Analysis of playout strategies for voice transmission using packet switching techniques," *Performance Evaluation*, vol. 4, no. 1, pp. 11–18, Feb. 1984.

[2] Giulio Barberis, "Buffer sizing of a packet-voice receiver," *IEEE Transactions on Communications*, vol. COM-29, no. 2, pp. 152–156, Feb. 1981.

[3] Mehmet M. K. Ali and C. M. Woodside, "Analysis of re-assembly buffer requirements in a packet voice network," in *Proceedings of the Conference on Computer Communications (IEEE Infocom)*, Bal Harbour (Miami), Florida, Apr. 1986, IEEE, pp. 233–238.

[4] Josep M. Ferrandiz and Aurel A. Lazar, "Modeling and admission control of real time packet traffic," Technical Report Technical Report Number 119-88-47, Center for Telecommunications Research, Columbia University, New York 10027, 1988.

[5] Walm van der Wal, Mandjes Michel, and Rob Kooij, "End-to-end delay models for interactive services on a large-scale IP network," in *IFIP*, June 1999.

[6] Ingemar Kaj, *Stochastic Modeling for Broadband Communications Systems*, Society for Industrial and Applied Mathematics, Philadelphia, PA, USA, 2002, Prepared with LaTeX.

On Packet Delays and Segmentation Methods in IP Based UMTS Radio Access Networks

Gábor Tóth and Csaba Antal

Traffic Analysis and Network Performance Laboratory, Ericsson Research
Laborc u. 1, Budapest, H-1037, Hungary
{gabor.toth,csaba.antal}@eth.ericsson.se

Abstract. In UMTS radio access networks (UTRAN), the delay requirement - bandwidth product of real-time classes is small, therefore large packets need to be segmented. Furthermore, low priority segmented packets may also have real-time requirements. Assuming that UTRAN traffic is characterized as the superposition of periodic sources with random offset, we propose analytic results and we show simulation results for the delay distribution of real-time classes in a priority system. We then compare segmentation methods considering the delay requirements of all real-time classes.

1 Introduction

UMTS Terrestrial Radio Access Networks (UTRAN) [1] will be based on packet switched technologies to utilize the gain of multiplexing services with various QoS requirements. First UTRAN networks will be deployed with ATM infrastructure, IP based UTRAN networks will be introduced in later releases. As opposed to generic multiservice networks, most traffic types in UTRAN have real-time requirements. It is well-known that voice traffic, which is expected to be the dominant traffic type at first UMTS systems, has inherent real-time requirements. In addition to that, applications used at mobile terminals also have real-time requirements [2] in the access network due to the timing requirements of the Radio Link Control (RLC) protocol and the support of soft handover. The end-to-end delay requirement of real-time data packets, however, is less stringent (15-40 ms) than that of voice (5-7 ms) [3]. Other traffic types that do not span to the air interface are synchronization of base stations, which has the most stringent delay requirement, and the non-real-time operation and maintenance and non-UTRAN traffic in IP based UTRAN.

RLC protocol provides reliable transmission between mobile terminals and the corresponding radio network controller via a retransmission mechanism to overcome packet losses at the air interface. Therefore, it tolerates packet losses in the transport network. Accordingly, both voice and data applications have *statistical delay requirements* in the radio access network.

A special property of real-time traffic types over the transport network is periodicity; that is, mobile terminals and RNCs emit RLC-blocks periodically.

M. Ajmone Marsan et al. (Eds.): QoS-IP 2003, LNCS 2601, pp. 50–63, 2003.

The initial phase of periodic flows is random, hence the traffic can be modeled as a superposition of independent periodic streams with a random initial offset. In the general case, periodic streams are modulated with an on-off process due to Discontinuous Transmission (DTX) for voice sources and higher layer protocols for data applications. Packet size of real-time flows is heterogeneous. Synchronization and voice packet are small (30-50 bytes), while the size of data packets might be large. For example if the transmission time interval of data packets, i.e. period length, is 20 ms and the bitrate of the bearer is 384 kbps then the packet size is larger than 1000 bytes. The exact packet size including all protocol overheads depends on the applied technologies.

The above characteristics of UTRAN networks indicate that to assure optimal operation for real-time traffic with the most stringent delay requirements, small packets (such as voice, some signalling and synchronization) should be differentiated from real-time data traffic in routers and switches. It is also apparent that real-time data traffic has to have preference over maintanence and best effort non-UTRAN traffic. Therefore, in IP based UTRAN, which will be based on the DiffServ concept [4], multiple real-time classes should be distinguished in the routers.

Furthermore, even if voice has absolute priority over real-time data traffic, the delay of voice packets is significantly increases due to data packets (a 1000 byte long data packet may increase the voice delay by 4 ms at an E1 link). To minimize this effect, large data packets are segmented. In ATM based UTRAN ATM does this task, while in IP UTRAN layer 2 segmentation is applied based on the PPP Multilink Protocol (MP) [5]. If multiple QoS classes are needed at layer 2 then multi-class extension of MP [6] should also be supported.

Segmentation of real-time data packets, however, has an effect on the delay of data packets themselves. Therefore, all real-time classes should be included in the delay analysis of the system. The aim of this paper is to derive new analytical results that allow the evaluation of the effect of segment size and segmentation methods on the delay of all real-time classes. Therefore, we analyze the packet delay distribution in a two-class strict-priority system, where traffic sources generate packets periodically with random initial offset and packets are segmented at the entry of the queues. Thus, results presented in this paper are based on a constant bitrate source model. Approximation for the delay distribution of on-off sources is also possible based on the presented results if the methods presented in [2] are applied.

The rest of the paper is structured as follows. In Section 2, we propose analytic results for the exact packet delay distribution for high priority packets in a priority system. We derive then the low priority packet delay distribution for the case of a single low priority source in Section 3. In Section 4, we verify the analytic results by simulations. We propose new segmentation methods in Section 5 and we evaluate their performance in Section 6. Finally, we draw the conclusions in Section 7.

2 Analytic Evaluation of High Priority Packet Delay

2.1 Prior Work

The delay distribution of high priority constant bit-rate traffic has been analyzed since the idea of transmitting real-time traffic over packet switched networks emerged. The N*D/D/1 system, which means that deterministic periodic sources are served by a single deterministic server, received dedicated attention at the beginning of the 90s. The exact virtual waiting time distribution was expressed first in [7] and in [8]. These results were extended to more general cases and a closed form approximation was also proposed in several papers including [9], [10] and [11]. For a detailed review of the work on N*D/D/1 systems see [12] and [13].

Denoting the number of voice sources by N_H and assuming that they generate a b_H long packet in each T long time period, the CDF of the queuing delay of a voice packet can be expressed based on equation 3.2.2 in [12] as

$$F_{D_q}(t) = 1 - \sum_{\frac{tC}{b} < n \leq \hat{N_H}} \binom{\hat{N_H}}{n} h_n^n (1-h_n)^{\hat{N_H}-n-1}(1-h_{\hat{N_H}}) \tag{1}$$

where the server rate is C, $\hat{N_H} = N_H - 1$ is the number of voice sources excluding the observed one, $h_i = \rho(i) - \frac{t}{T}$ and $\rho(x)$ is the system utilization with x voice sources: $\rho(x) = \frac{x b_H}{CT}$.

The queuing delay does not include the deterministic transmission delay $\frac{b_H}{C}$. Thus, the PDF and CDF of the total packet delay is the PDF and CDF of the queuing delay shifted by $\frac{b_H}{C}$ along the time axis, $F_{D_t}(t) = F_{D_q}(t - \frac{b_H}{C})$.

The effect of low priority packets on the delay of high priority voice packets was also analyzed in recent studies. In [14], the exact delay distribution of voice packets was derived in a strict priority system based on an N*D/D/1 system with vacations. Analytic results, however, are valid only for the cases when the low priority buffer is saturated. Voice delay distribution at strict priority and weighted round robin scheduling was also studied in [15]. The waiting time distribution of voice packets was calculated based on an M/D/1 model in [15] assuming that the low priority buffer is saturated. Both papers assumed that the delay of high priority voice packets is the sum of the voice delay in the preemptive strict priority system and the residual transmission time of low priority segments.

2.2 Effect of Non-saturated Low Priority Traffic

In this subsection, we express the distribution of the high priority voice delay in non-preemptive strict priority system where the low priority buffer is not saturated, which is not considered in previous studies. For the calculation of the voice delay we assume, such as [15] and [14], that the voice delay expressed in Section 2.1 increases by an independent additional term. That term is equal to the residual transmission time of actual low priority segment in the corresponding *preemptive* strict priority system at the instant when the observed high priority packet arrives (t_0). Thus, we aim to express the distribution of that residual transmission time first and then we derive the total non-preemptive voice delay by convolution.

Denote the number of low priority data sessions by N_L and assume that each injects one b_L long low priority data packet into the system in each T long time-period. Assume that long data packets are fragmented into short segments such that the total number of segments in a T long time interval is K, the number of s_i long segments during T is k_i and $\sum_i k_i = K$. Thus, the relative frequency of a segment with size of s_i is $\Pr(s_i) = \frac{k_i}{K}$, and the average segment length is $\bar{S} = \sum_i \frac{s_i k_i}{K}$.

Let A denote the event that the residual transmission time of the data segment (D_s) at t_0 is less than t, i.e. $\Pr(A) = \Pr(D_s < t)$. Let B denote the event that the server is serving a data segment at t_0, and C the event that a segment of size s_i is in the server at t_0. Now, $\Pr(A)$ can be expressed as follows

$$\Pr(A) = \Pr(B) \sum_{\forall s_i} \Pr(A|BC)\Pr(C|B) + \Pr(A|\bar{B})\Pr(\bar{B}) \tag{2}$$

In the second term of (2), $\Pr(A|\bar{B}) = 1$ if $t > 0$, otherwise it is 0. The first term can be expressed by using that $\Pr(A|BC) = \min(1, \frac{tC}{s_i})$, $\Pr(C|B) = \frac{s_i \Pr(s_i)}{\bar{S}}$, and $\Pr(B) = \frac{N_L b_L}{TC - N_H b_H}$. After substituting these formulae into (2) we get the CDF that yields the probability density function by derivation, which has the following form:

$$f_{D_s}(t) = p_0 \delta(t) + \sum_{\forall i: s_i > tC} \frac{C k_i}{TC - N_H b_H} \tag{3}$$

where $p_0 = 1 - \frac{N_L b_L}{TC - N_H b_H}$ and $\delta(x)$ is the Dirac delta function.

The PDF of the total voice delay, $f_D(t)$, then is obtained by convolving the density functions of the preemptive voice delay, $f_{D_t}(t) = \frac{d}{dt} F_{D_q}(t)$, and that of the residual transmission time of data segments, $f_{D_s}(t)$. Accomplishing the necessary calculations, the PDF is obtained:

$$f_D(t) = -p_0 f_{D_t}(t) + \sum_{\forall i: s_i < \hat{t}C} c_i \left[F_{D_t}(t) - F_{D_t}(t - \frac{s_i}{C})\right] - \sum_{s_i \geq \hat{t}C} c_i F_{D_t}(t) \tag{4}$$

where $c_i = \frac{k_i}{TC - N_H b_H}$ and p_0 is defined above.

3 Analytic Evaluation of Low Priority Delay

As we highlighted before, the delay distribution of low priority data packets is of interest in UTRAN networks as they also have statistical delay requirements. Analytic results to be presented in this section are restricted to the case when a single low priority session is present in the system. We first derive the distribution of the delay of the low priority packet in a *preemptive system*, and then we show how these results can be applied to the *non-preemptive* case.

3.1 Preemptive System

We start with the definition of the PDF of the delay, that is

$$f_D(t) = \lim_{dt \to 0} \frac{\Pr(t < D \leq t + dt)}{dt} \tag{5}$$

where D stands for the delay of low priority packets. As the investigated system is preemptive, the delay of high priority traffic is independent of the low priority traffic. Assuming that the data packet arrives at time instant 0, the PDF of data delay, $f_D(t)$, can be expressed in terms of the cumulative length of idle periods of the high priority system in $[0, t]$.

$$f_D(t) = \lim_{dt \to 0} \frac{\Pr\left(Idle^{high}(0,t) < \frac{b_L}{C} \leq Idle^{high}(0,t+dt)\right)}{dt} \tag{6}$$

where $Idle^{high}(t_1, t_2)$ is the total length of the idle periods of the high priority system between t_1 and t_2. Equation 6 can be further transformed to (7):

$$f_D(t) = \lim_{dt \to 0} \frac{\Pr\left(V_t = 0, \frac{b_L}{C} - dt \leq Idle^{high}(0,t) < \frac{b_L}{C}\right)}{dt} \tag{7}$$

where V_t represents the high priority system content at time t. The probability in the numerator can be split into two parts depending on whether the voice queue is empty or not at the arrival of the data packet.

$$P_C = \Pr\left(V_0 \neq 0, V_t = 0, Idle^{high}(0,t) \in \left[\frac{b_L}{C} - dt, \frac{b_L}{C}\right)\right) \tag{8}$$

$$P_D = \Pr\left(V_0 = 0, V_t = 0, Idle^{high}(0,t) \in \left[\frac{b_L}{C} - dt, \frac{b_L}{C}\right)\right) \tag{9}$$

Thus, the sum of a continuous and a discrete distribution yields the density function of the data delay. We will handle these two parts separately, first P_C and P_D are calculated and then the limits.

Continuous part: First, we consider the case when the voice queue is not empty at the moment of the arrival of the data packet. In this case, P_C can be calculated if it is conditioned on the number of voice packets arrived during the service of the data packet, and summed over all possible arrivals. In Fig.1 a situation is shown when the number of voice arrivals in (t_0, t_d) is m, i.e. $\nu(t_0, t_d) = m$, where t_d represents the departure of the data packet, and t_0 signifies the first time instant when the voice queue gets empty after the arrival of the data packet.

In such a case, $t_d = t_0 + \frac{mb_H + b_L}{C}$ and $V_{t_d} = V_{t_0} = 0$. The condition $t_d \in (t, t + dt)$ can stand only if the service of the data packet starts in a certain time interval, i.e. $t_0 \in (t - \frac{mb_H + b_L}{C}, t - \frac{mb_H + b_L}{C} + dt)$. If we further condition the obtained expression on the exact length of the voice busy period that finishes

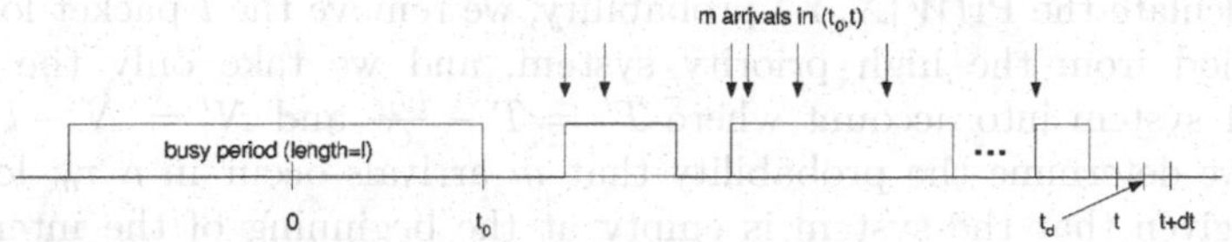

Fig. 1. Important time instants during the service of a low priority packet

at t_0, which is denoted by B_0, and sum it over all possible voice busy period lengths, we get

$$P_C = \sum_l \sum_m \Pr\left(B_0 = l, V_t = 0, t_0 \in I, \nu(t_0, t_d) = m\right) \tag{10}$$

where I represents the time interval $(t - \tau_m, t - \tau_m + dt]$ and $\tau_m = \frac{mb_H + b_L}{C}$. For the sake of easing the calculations let us define the following events:

- X: $B_0 = l > 0$
- Y: $t_0 \in (t - \tau_m, t - \tau_m + dt]$
- W: $\nu(t_0, t_d) = m$
- Z: $V_t = 0$.

Using the new notations P_C can be expressed as

$$P_C = \sum_l \sum_m \Pr(Z|X,Y,W)\Pr(W|X,Y)\Pr(Y|X)\Pr(X) \tag{11}$$

The rightmost probability is the probability that the length of a randomly chosen burst is l, multiplied with the probability that the arrival of the data packet occurs when a voice packet is being served.

$$\Pr(X) = \Pr(B_0 = l | V_0 > 0)\Pr(V_0 > 0) = \frac{lb_H \Pr(B = l)}{\mathbf{E}(B)} \frac{Nb_H}{TC} \tag{12}$$

where the distribution of busy period length $Pr(B = l)$ and the mean busy period length $E(B)$ is expressed based on [16]. The probability of event X has the form of:

$$\Pr(X) = \binom{N}{l}\left(1 - \frac{Nb_H}{TC}\right)\left(\frac{lb_H}{TC}\right)^l \left(1 - \frac{lb_H}{TC}\right)^{N-l-1} \tag{13}$$

The next probability to be calculated is the probability that the service of the data packet starts in a specific dt long time interval given that the length of the served voice busy period is l. As the arrival time of the packet is distributed evenly over the time-interval of the served voice burst, the second probability is

$$\Pr(Y|X) = \frac{C}{lb_H} dt \tag{14}$$

if $l > \frac{tC - b_L}{b_H} - m$ and 0 otherwise.

To calculate the $\Pr(W|X,Y)$ probability, we remove the l packet long voice busy period from the high priority system, and we take only the reduced N*D/D/1 system into account where $T' = T - \frac{lb_H}{C}$ and $N' = N - l$. In this system, we determine the probability that m arrivals occur in a τ_m long time interval, given that the system is empty at the beginning of the interval. We use parameters with apostrophe for the truncated system. Applying the Bayes-formula, the wanted probability can be formalized as

$$\Pr(W|X,Y) = \Pr(\nu'(t',t'+\tau_m) = m|V'_{t'} = 0) = \frac{\Pr(V'_{t'} = 0|\nu'(t',t'+\tau_m) = m)\Pr(\nu'(t',t'+\tau_m) = m)}{\Pr(V'_{t'} = 0)} \tag{15}$$

The probability $\Pr(V'_{t'} = 0|\nu'(t',t'+\tau_m) = m)$ can be divided into two parts depending on whether the virtual waiting time at $t' + \tau_m$ reaches the threshold $\hat{V} = \frac{TC - Nb_H - b_L}{C}$ or not. If the virtual waiting time exceeds $\hat{V}$ at $t' + \tau_m$ then the system can not empty until $t' + T'$, so $\Pr(V'_{t'} = 0|\nu'(t',t'+\tau_m) = m, V'_{t'+\tau_m} > \hat{V}) = 0$. That is, $\Pr(V'_{t'} = 0|\nu'(t',t'+\tau_m) = m)$ can be expressed as the product of two probabilities: $\mathrm{P}_a = \Pr(V'_{t'} = 0|\nu'(t',t'+\tau_m) = m, V'_{t'+\tau_m} < \hat{V})$ and $\mathrm{P}_b = \Pr(V'_{t'+\tau_m} < \hat{V}|\nu'(t',t'+\tau_m) = m)$.

There is some idle period in $(t' + \tau_m, t' + T')$ if $V'_{t'+\tau_m} < \hat{V}$, so it can be handled as an individual N*D/D/1, which yields $\mathrm{P}_a = \frac{TC - Nb_H - b_L}{TC - (l+m)b_H - b_L}$. For the calculation of probability P_b, the virtual waiting time distribution of a system with m users and τ_m long period can be used as an approximation. P_b is the value of this distribution at $\hat{V}$.

$$\mathrm{P}_b = 1 - \sum_{n=n_0}^{m} \binom{m}{n} \frac{\left(1 - \frac{Nb_H}{TC}\right)\left(R(N+n) - 1\right)^n R(m)^{-m}}{\left(1 - R(N+n) + R(m)\right)^{n+1-m}} \tag{16}$$

where $n_0 = \lceil \frac{TC}{b_H}(1 - R(N)) \rceil$ and $R(x) = \frac{xb_H + b_L}{TC}$. Probability P_b is independent of l, so it will be referred to as K_m in the following.

The next two unknown probabilities of (15) are $\Pr(\nu'(t',t'+\tau_m) = m) = \binom{N-l}{m}\left(\frac{mb_H + b_L}{TC - lb_H}\right)^m \left(1 - \frac{mb_H + b_L}{TC - lb_H}\right)^{N-l-m}$ and $\Pr(V'_{t'} = 0) = \frac{TC - Nb_H}{TC - lb_H}$. Therefore, the final form of $\Pr(W|X,Y)$ is

$$\Pr(W|X,Y) \approx \binom{N-l}{m} \frac{K_m \left(\frac{TC - Nb_H - b_L}{TC - Nb_H}\right)\left(\frac{mb_H + b_L}{TC - lb_H}\right)^m}{\left(\frac{TC - (m+l)b_H - b_L}{TC - lb_H}\right)^{l+m+1-N}} \tag{17}$$

To express the last probability we use that $\nu(t_0, t_d) = m$ and $t_d = t_0 + \frac{mb_H + b_L}{C}$. This is the empty probability at an arbitrarily chosen time instant in such an N*D/D/1 system where there are m sources with period length of $\frac{mb_H + b_L}{C}$: $\Pr(Z|X,Y,W) = \frac{b_L}{mb_H + b_L}$

Composing the product of the conditional probabilities calculated above and substituting $l+m$ with k we obtain

$$P_C \approx \sum_{\forall m} \sum_{k > \frac{tC-b_L}{b_H}} \frac{\binom{N}{m}\binom{N-m}{k-m} K_m R(0) R(m)^{m-1} \big(1-R(N)\big)}{R(k-m)^{m+1-k}\big(1-R(k)\big)^{k+1-N}} \frac{dt}{T} \tag{18}$$

Discrete part: The next step is to calculate the probability P_D, which is for the case when the low priority packet arrives into an empty system. It follows from the preemptive operation, that the service time in this case can only take such a value that exactly equals to $\frac{mb_H+b_L}{C}$, where m is a non-negative integer that is less than or equal to N. Therefore, the probability that the delay of the low priority packet is in a certain dt period equals to the sum (over all m) of the probabilities that m high priority packets arrive during the service time of the low priority packet. That is, the $\frac{mb_H+b_L}{C}$ value falls in the given dt long time interval.

$$P_D = \sum_{m=m_{min}}^{m_{max}} \Pr\Big(V_0 = 0, V_t = 0, \nu(0,\tau_m) = m\Big) \tag{19}$$

where $m_{min} = \frac{tC-b_L}{b_H}$ and $m_{max} = \frac{(t+dt)C-b_L}{b_H}$. Following the same train of thoughts as in the continuous case we can get the following expression for the P_D probability

$$P_D \approx \sum_{m=m_{min}}^{m_{max}} K_m \binom{N}{m} \frac{R(0)\big(1-R(N)\big)R(m)^{m-1}}{\big(1-R(m)\big)^{m+1-N}} \tag{20}$$

Applying that $\Pr(t < D \le t+dt) = P_C + P_D$, and substituting this into (5) we get the formula of

$$\begin{aligned} f_D(t) \approx \frac{1}{T} \sum_{\forall m} \sum_{k > \frac{tC-b_L}{b_H}} \frac{\binom{N}{m}\binom{N-m}{k-m} K_m R(0) R(m)^{m-1} \big(1-R(N)\big)}{R(k-m)^{m+1-k}\big(1-R(k)\big)^{k+1-N}} \\ + \sum_{m} K_m \binom{N}{m} \frac{R(0)\big(1-R(N)\big)R(m)^{m-1}}{\big(1-R(m)\big)^{m+1-N}} \delta\Big(t - \frac{mb_H+b_L}{C}\Big) \end{aligned} \tag{21}$$

3.2 Non-preemptive System with Segmentation

In this section we show how the analytic results obtained for the preemptive system can be applied to the non-preemptive case.

Fig. 2 illustrates the preemptive and non-preemptive systems with the same arrival process. In the non-preemptive case, the data packet is segmented according to the method depicted in the upper-left corner of the figure. In the preemptive case, it does not matter whether the data packet is segmented or not (if we neglect segmentation overhead), so for the sake of comparability we assumed that the packet is segmented according to the same method. In Fig.

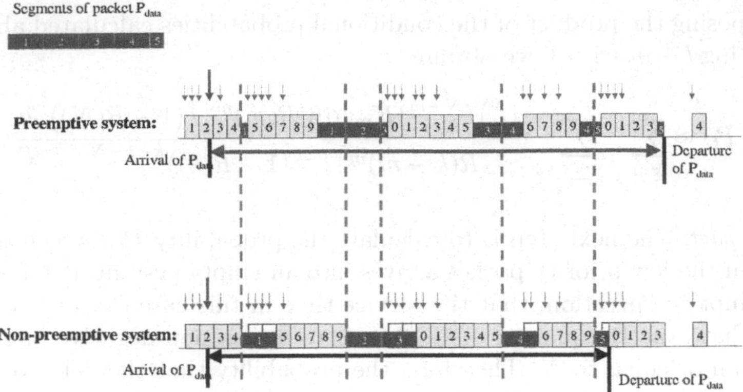

Fig. 2. Timeline of serving packets in preemptive and non-preemptive systems

2 it can be observed that the service of the data segment is started in such a time instant in the non-preemptive case when the voice queue is empty and all data segments preceding the observed one have already been served. Thus, if the stability criteria for the system are met, the start times of the segments in the non-preemptive system are the same as in the preemptive case. Since the service of a segment can not be interrupted in the non-preemptive system, the total service time of the data packet can be calculated as the sum of two components. One is the delay that a $\hat{b_L} = b_L - S_l$ long low priority packet suffers in a preemptive system, the other is the transmission time of a S_l long segment at full service rate. Thus the CDF of the delay of a single b_L long data packet in a preemptive system with segmentation is

$$F_D(t) = \hat{F}_D\left(t - \frac{S_l}{C}\right)|_{\hat{b}_L = b_L - S_l} \tag{22}$$

where S_l is the length of the last segment, $\hat{F}_D(t)$ is the CDF of the delay of a $\hat{b}_L$ long data packet in a preemptive system and C is the server rate.

4 Verification

We have carried out simulations to verify the analytic results for various scenarios. Here we present results for a router with a 1920 kbps (E1) link, which carries the multiplexed traffic of 1 data and 80 voice sources. The length of the voice and data packets are 44 and 1013 bytes, respectively, and the period is 20 ms for each sources. The comparison was done for two different segmentation sizes. The data packets are divided into two segments in both cases. In the first case, the sizes of the first and second segment are 800 and 213 bytes, in the second case, they are 506 and 507 bytes.

Fig. 3(b) and Fig. 3(a) show the complement of the normalized cumulative histograms and the cumulative density functions of the delays for both cases. In

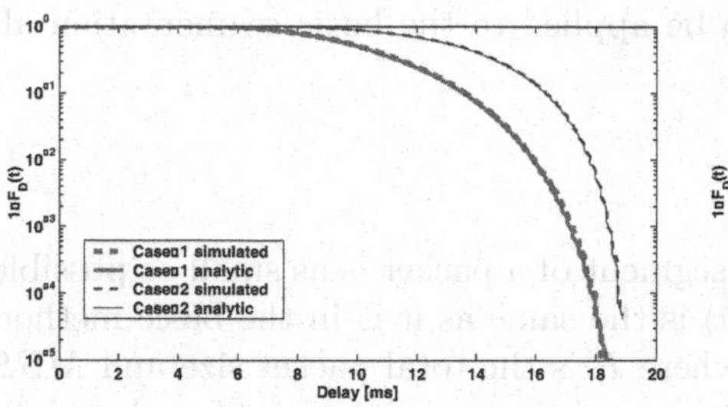

(a) Complementary distribution functions of the data delay

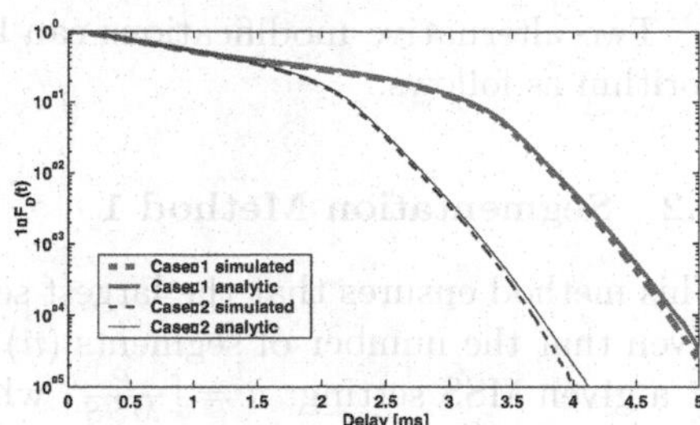

(b) Complementary distribution functions of the voice delay

Fig. 3. Verification of the analytic formulae

the figures it is apparent that the simulated and the analytical results are hardly distinguishable.

5 Segmentation Methods

In this section we investigate how the segmentation affects the delay of high and low priority packets. We also show that the simplest segmentation method, which is widely applied in current equipments, is suboptimal in this sense.

First consider the voice traffic. As it was shown in Section 2, the voice delay is the sum of two independent random variables of which only f_{D_s} depends on the segmentation of the data packet. The quantiles of the total delay are minimized via proper segmentation in practical cases if the quantiles of the segmentation-dependent component are also minimized. This can be reached by setting the size of the *largest data segment as small as possible.*

Now let us consider how segmentation effects data delay. We have shown in this paper that the delay distribution of the data traffic in a non-preemptive system is determined by the size of the last segment of the segmented data packet. How the former part of the packet is segmented does not play an important role in forming the distribution. Since the service of the last data segment in the non-preemptive system can not be interrupted once it is started, *the larger the last segment the smaller the delay* that the data packet suffers given that the total packet size is fixed.

5.1 Basic Segmentation Algorithm

The simplest method for segmenting packets cuts the packet into MSS (maximum segment size) long segments until the remaining part is shorter than the MSS. Thus, the last segment contains the remaining bytes only. This relatively simple procedure is suboptimal for both voice and data delay because the largest data segment could be smaller and the last segment could be smaller.

Two alternative modifications can be applied to the basic segmentation algorithm as follows.

5.2 Segmentation Method 1

This method ensures that the largest segment of a packet is as small as possible given that the number of segments (n) is the same as it is in the basic method at a given MSS setting: $n = \lceil \frac{P}{MSS} \rceil$ where P is the total packet size and MSS is the maximum segment size. This is to avoid the segmentation of packets into arbitrarily small pieces, which would result in too large header overhead. At these constraints voice delay is optimal when data packets are segmented such that the size of segments is set as even as possible. The algorithm produces two types of segments with sizes of S_1 and S_2 that can be calculated as $S_1 = \lceil \frac{P}{n} \rceil$ and $S_2 = \lfloor \frac{P}{n} \rfloor$. The number of the S_1-size segments is $n_1 = \mod (P, n)$ and the number of S_2-size segments is $n_2 = n - n_1$. Segments can be sent in an arbitrary sequence because one byte difference does not have a noticeable effect on data delay. Even though, the main purpose of this algorithm is to further reduce the tail distribution of the voice delay, it can be seen that this algorithm never produces shorter last segment than the basic algorithm does, thus this algorithm may also increase the performance of the data traffic.

5.3 Segmentation Method 2

The largest last segment provides the best performance for data traffic. For not to increase the delay of the voice packets, the algorithm keeps the original maximum segment size (MSS) as a parameter. The size of the segment to be sent at last is set to the largest possible value i.e. MSS, while the remaining part of the packet is segmented into $n - 1$ pieces that are equal in size to the extent possible. The method used to calculate the size of these $n - 1$ pieces of segments is identical to the one described at *Method 1.* It can be seen that this algorithm never produces more MSS sized segments than the basic algorithm does, thus the quantile of the segmentation-dependent part of the voice delay may decrease. Therefore, besides decreasing the tail distribution of the data delay, the algorithm might also increase the performance of the voice traffic.

6 Comparison of Segmentation Methods

To illustrate how the described algorithms reduce delays when two different traffic classes are carried over the same link, a simple network scenario was investigated using the analytical results obtianed in Section 2. The investigated system includes a 1920 kbps link (E1) that carries the multiplexed periodic voice and data traffic. Each high priority voice source transmits one 44-byte packet every 20 milliseconds, and each low priority data source transmits one 1013-byte long packet every 20 milliseconds, which corresponds to a 384 kbps bearer in IP UTRAN. The traffic mix includes the traffic of 80 voice sources and one data source. The header overhead introduced by the segmentation was neglected during the investigation because it does not influence the comparison.

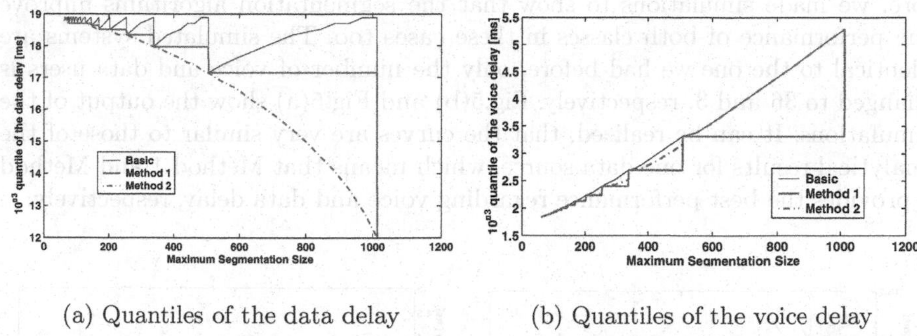

(a) Quantiles of the data delay (b) Quantiles of the voice delay

Fig. 4. 10^{-3} quantiles of the delays in the case of 80 voice and 1 data flows

Fig.4(a) shows the resulted 10^{-3} quantile of the data delay as a function of the maximum segmentation size. Fig.4(b) shows the 10^{-3} quantile of the voice delay as the function of the MSS.

The difference between the shapes of curves can be attributed to the way segment sizes depend on MSS at the studied methods. Data delay curve at the basic algorithm is monotonously increasing if the number of segments (n) does not change because then the size of the last segment decreases at increasing MSS. However, if the number of segments is decremented (due to the increased MSS) then the size of the last segment has a jump (from 0 to MSS), which results in a jump in the data delay curve as well. Segmentation Method 1 also has jumps in the data delay curve when the number of segments changes. It is constant otherwise because S_1 and S_2 depends only on n and not on MSS. At Segmentation Method 2, the size of the last segment is always equal to MSS, so the data delay curve is continuous. Regarding voice delay we can observe that the Basic Method and Method 2 have continuous curves because the size of the largest segment is equal to MSS. At Method 1, the largest segment size is $\lceil \frac{P}{n} \rceil$, so the voice delay is constant if n is constant.

It can be observed that using either Segmentation Method 1 or Segmentation Method 2, the 10^{-3} quantile of both voice and data delays are always smaller than or equal to that of the basic segmentation method. No improvement achieved at all regarding voice delay when Method 2 is used with such an MSS where the data packet is divided into two pieces. In this case, the length of the produced data segments are the same as the basic algorithm would produce, but in reverse order. Therefore, the PDFs of the segmentation-dependent part of the voice delays, and thus the PDFs of the total voice delays are the same in the two cases. It can be concluded from Fig.4(b) that the quantile of voice delay is reduced the most if Method 1 is used. Regarding data delay, however, Method 2 significantly overperforms Method 1.

We do not have analytical formulae to calculate the delay distributions for the case when more than one data source generate traffic in the system. Therefore, we made simulations to show that the segmentation algorithms improve the performance of both classes in these cases too. The simulated systems are identical to the one we had before, only the number of voice and data users is changed to 36 and 3, respectively. Fig.5(b) and Fig.5(a) show the output of the simulations. It can be realized, that the curves are very similar to those of the analytical results for one data source, which means that Method 1 and Method 2 provides the best performance regarding voice and data delay, respectively.

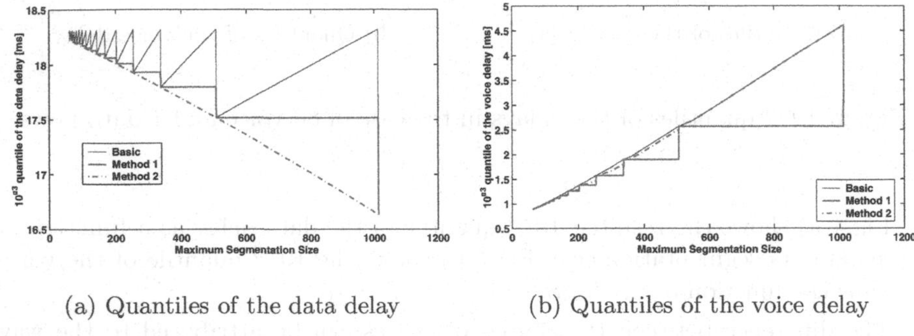

(a) Quantiles of the data delay (b) Quantiles of the voice delay

Fig. 5. 10^{-3} quantiles of the delays in the case of 36 voice and 3 data flows

7 Conclusions and Future Work

We have shown that the delay quantile of voice and real-time data packets depends significantly on the segmentation method applied for large data packets. We proposed two methods that decrease the delay of both data and voice packets compared to conventional segmentation methods. We have concluded based on analytic and simulation results that the best segmentation method regarding data delay is if the size of the data segment that is sent last has the maximum size allowed by the MSS parameter. On the other hand, voice performance is optimal when data segments are as even as possible. As applying these methods makes it possible to admit more connections from both traffic classes without violating their delay constraints, the bandwidth efficiency of small links in IP UTRAN networks can be increased. For the performance evaluation of the proposed methods, we have also derived analytic results for a two-class single server strict priority system where both high priority and low priority traffic are superpositions of periodic constant bit-rate sources with real-time requirements. We have derived the exact delay distribution of voice packets assuming that the low priority buffer is not saturated. We have also expressed the low priority delay distribution for the case when there is a single low priority session in the system. Extending the analytic results to cases of many low priority sources and for more traffic classes is in progress. The investigation of the sources modulated with an on-off process is also planned in the future.

References

1. 3GPP "UTRAN overall description," *Technical Specification*, TS 25.401.
2. Sz. Malomsoky, S. Racz, Sz. Nadas, "Connection admission control in UMTS radio access networks," *Computer Communications* - Special Issue on 3G Wireless and Beyond, Fall, 2002
3. 3GPP, "Delay budget within the access stratum," *Technical Report*, TR 25.853, V4.0.0, May 2001
4. 3GPP, "IP Transport in UTRAN," *Work Task Technical Report*, TR 25.933 V1.0.0, March 2001
5. K. Sklower, B. Lloyd, G. McGregor, D. Carr, T. Coradetti, "The multilink PPP protocol," RFC 1990, August 1996
6. C. Bormann, "The multi-class extension to multi-link PPP," RFC 2686, Sept. 1999
7. A. Bhargava, P. Humblet and H. Hluchyj, "Queuing analysis of continuous bit stream transport in packet networks," *Globecom '89*, Dallas, TX, October 1989.
8. I. Norros, J. W. Roberts, A. Simonian and J. Virtamo, "The superposition of variable bitrate sources in ATM multiplexers," *IEEE Journal on Selected Areas in Communications*, 9(3):378-387, April 1991.
9. B. Hajek, "Heavy traffic limit theorems for queues with periodic arrivals," *In ORSA/TIMS Special Interest Conference on Applied Probability in the Engineering*, Informational and Natural Sciences, pp. 44, January 1991
10. B. Sengupta, "A queue with superposition of arrival streams with an application to packet voice technology," *Performance'90*, pp. 53-60, North-Holland 1990.
11. J. Roberts and J. Virtamo, "The superposition of periodic cell arrival streams in an ATM multiplexer," *IEEE Trans. Commun.*, vol. 39, pp. 298-303
12. COST 242 Management Committee, "Methods for the performance evaluation and design of broadband multiservice networks," The COST 242 Final Report, Part III., June 1996
13. J. He, K. Sohraby, "On the Queueing Analysis of Dispersed Periodic Messages," *IEEE Infocom 2000*, March 26-30, Tel-Aviv, Israel
14. K. Iida, T. Takine, H. Sunahara, Y. Oie, "Delay analysis of CBR traffic under static-priority scheduling," *IEEE/ACM Trans. On Networking*, Vol. 9, No. 2, pp.177-185
15. M. J. Karam, F. A. Tobagi, "Analysis of the Delay and Jitter of Voice Traffic over the Internet," *IEEE Infocom 2001*, April 22-26, Anchorage, Alaska, USA
16. J.T. Virtamo, "Idle and busy period distribution of an infinite capacity N*D/D/1 queue," *14th International Teletraffic Congress - ITC 14*, Antibes Juan-les-Pins, France, June 1994.

On Performance Objective Functions for Optimising Routed Networks for Best QoS

Huan Pham, Bill Lavery

School of Information Technology
James Cook University, QLD 4811, Australia
{huan, bill}@cs.jcu.edu.au

Abstract. This paper contributes to understanding of how Quality of Service (QoS) can be optimised in Autonomous System (AS) networks using Open Shortest Path First (OSPF) routing. We assume that, although guaranteed QoS can not be provided, network wide performance can be optimised in a manner which significantly improves QoS. We particularly focus on devising a network performance objective, which is convenient for the network optimization process, while closely representing QoS experience for end users. Our results show that, firstly, network-wide performance can be significantly improved by optimizing OSPF weights. Secondly, the resulting optimal set of weights and network performance depends very much on the definition of the optimization objective function. We argue that min max utilisation is an appropriate function for heavily loaded network, while end to end delay is more appropriate optimization objective for lightly loaded condition, at which most of the backbones are operating.

1 Introduction

The Internet continues to experience remarkable growth. It carries not only non real-time traffic (such as Web, E-mail, and FTP…), but also other time sensitive traffic (such as VoIP, audio and video on demand). In the very competitive Internet Service Provider (ISP) market, even for the applications traditionally considered as non real-time traffic, ISPs that provide a "world wide wait" service will loose out to those ISPs that can provide a higher level of QoS to their customers. Typically a lower response time and a higher download speed result in a higher QoS for the end user.

To guarantee an explicit QoS (e.g. guaranteed delay, packet loss…) new technologies such as IntServ [1], DiffServ [2], MPLS [3], are required to provide the necessary "traffic policing" and "resource reservation" features. This means hardware and software upgrades for the existing AS network. For the time being, most of the ISP networks do not support these explicit QoS features, and would want to provide their best possible level of non-guaranteed QoS using "Traffic Engineering" principles.

The effectiveness of traffic engineering in IP routed networks is directly governed by the routing processes. Routing determines the paths taken by the traffic flows through a network and therefore the router settings control how much traffic travels through each link in the network. The most popular "plain IP" routing protocol is

M. Ajmone Marsan et al. (Eds.): QoS-IP 2003, LNCS 2601, pp. 64-75, 2003.

OSPF, and it routes traffic along the shortest path(s) subject to the link weight setting. As a result, traffic engineering for IP networks using OSPF reduces to adjusting the set of link weights so as to optimise some performance objective function.

Several recent publications have addressed related issues. In [4], the authors present a modeling tool incorporating the network topology, traffic demands and routing models, which lets operators change the link weights to see the traffic patterns before actually implementing it in a real network. This is a very useful tool helping network operators to manually engineer traffic. However, for large networks, manual operation can become so complex and unpredictable that network operators can not cope. It is therefore preferable that optimization be done automatically.

Fortz and Thorup [5] described a local search heuristic for optimizing OSPF weights. Their results showed that, although many claim OSPF is not flexible enough for effective traffic engineering, their heuristic could usually find a set of weights that would greatly improve the network performance. However, their work has limitations resulting from their formulation of the objective function (we call it FT00 function). Their objective function is not based on the quality of service from users' point of view, but based on an "arbitrary" convex function of link utilization.

Our work extends the work of [4] and [5]. In addition, we devise an alternative objective function that better represents the network performance, which is discussed in section 2. Then in section 3, the optimization problem is formalized and solved. Section 4 presents results for an arbitrary network and section 5 discusses their significance.

2 Optimization Objective Function

The network optimization can be viewed from customers' perspective and network operator's perspective, i.e. traffic oriented performance or resource oriented performance objectives [6]. Resource oriented performance objectives relate to the optimization of the utilisation of network assets. On the other hand, traffic oriented performance objectives relate to the improvement of the QoS provisioned to the traffic.

Often ISP networks have links with fixed capacities and fixed monthly costs independent of the link utilization. It is only when a link is so overloaded that it must be augmented that the ISP experiences increased cost. In such situations, while no links must be upgraded to higher capacities, the objective of traffic engineering should purely be based on traffic oriented performance objective. These are the QoS parameters as experienced by the customers using the network: packet loss, end-to-end packet delay, and throughput.

Unfortunately, those QoS metrics are difficult to be determined analytically for any given link capacities, and traffic loads. In addition, different applications also have different interpretation of those QoS metrics: for example, file transfers are more dependent on the throughput, interactive applications are more sensitive to the average end-to-end delay, while video/audio streaming applications are more sensitive to the packet loss rate.

On the other hand, for the sake of computational tractability, traffic engineering by optimizing OSPF weights using local search heuristics, such as the one described in

[5], requires a single performance criterion. It should only be based on the traffic load on each link and link capacities, but independent of the network topology.

It is very common to turn to maximum link utilization as the performance measure to drive the optimization process, such as proposed in [4] and [7]. However, this min max objective function prefers a solution that may involve very long routes to avoid the heaviest link, resulting in a very long delay. In addition, since it does not take into account performance of lower loaded links, such an optimization process may get stuck once the maximum utilization can not be further decreased, although other less congested parts can still be further optimized. Even if the average link utilization is used as a secondary optimizing objective, this may not help because, in some cases, we have to increase the average link utilization to help reduce network congestion. For example, see Figure 1, where all the links have a capacity of 10Mbit/s and there is a demand of 6Mbit/s between Router1 and Router5. The load balanced situation shown has average utilisation of 26.2%, and is "less congested" than would be the case where all traffic were routed via R1-R4-R5, where average utilisation would be 20% but R1-R4 would be operating at undesirably high utilisation.

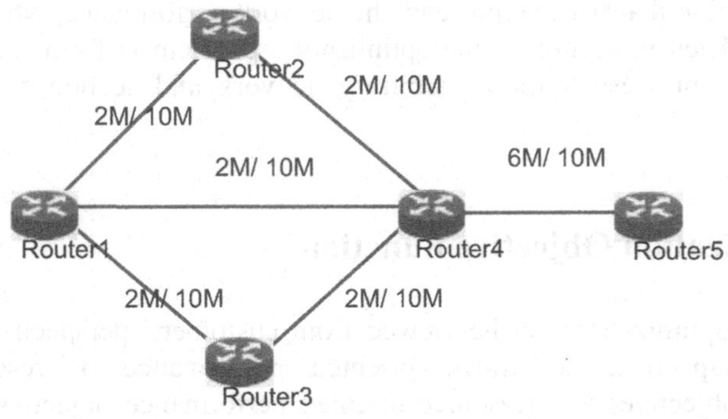

Fig. 1. Load balancing reduces congestion while increasing average utilisation

Given the drawback of using maximum link utilisation as the objective function, it would be reasonable to minimise the network-wide sum of an increasing convex function of the load and the capacity. For each link, a performance objective function, Φ_a, is to be defined and the network-wide performance, Φ, is then summed over all links, i.e.

$$\Phi = \sum_{a \in A} \Phi_a(l_a, C_a) \tag{1}$$

The convex increasing shape of the objective function, $\Phi_a(l_a, C_a)$, ensures that the more heavily loaded links have more negative influence on the overall network performance than lightly loaded links do. We next attempt to identify which are the performance parameters which most impact on end user QoS, at both low load and high load.

When Network Is Lightly Loaded

At low load, and hence at low utilisation, given the stochastic nature of Internet traffic, queues at nodes will rarely overflow, i.e. there will be negligible packet loss. In this situation, the end user's experience of performance is almost solely in terms of delay, i.e. is determined by how long packets take on average to be delivered. Thus we use an average end to end packet delay as the performance objective function.

Average queuing delay has been studied for long time in traffic queuing theory [8]. However, it assumes that the packet arrival is independent of one another and follows the Poisson distribution. It has been shown recently that for the Internet traffic this assumption is no longer valid [9, 10, 11]. These recent studies show that the Internet traffics have shown some level of self-similarity. The self-similarity level of the traffic is characterized by Hurst parameter (*H*) between [0.5 - 1].

As a result of the self-similarity characteristics, it is very difficult to devise a precise average packet delay given an average traffic demand and a link capacity. We use the approximation of the theoretical average queue length, *q*, given in [12], for an infinite buffer that is not overflowing:

$$q = \frac{\rho^{1/2(1-H)}}{(1-\rho)^{H/(1-H)}}, \tag{2}$$

where ρ is the link utilization, and *H* is the Hurst parameter. This parameter varies depending on the characteristics of the source traffic. Typically, for Internet traffic *H* is between (0.75, 0.9) [12]

The packet delay, *d*, on any link is the sum of the transmission delay and the queuing delay, which is:

$$d = \frac{1}{C} + \frac{q}{C} = \frac{1}{C} \times \left[1 + \frac{\rho^{1/2(1-H)}}{(1-\rho)^{H/(1-H)}} \right], \tag{3}$$

where *C* is the link capacity in packets per second. If we use $H = 0.75$, we have:

$$d = \frac{1}{C} \times \left[1 + \frac{\rho^2}{(1-\rho)^3} \right]. \tag{4}$$

The average end to end packet delay is:

$$\bar{d} = \frac{\sum_{a \in A} f_a \times d_a}{\sum_{i,j} r_{i,j}}, \tag{5}$$

where f_a is the traffic load in packets per second on link *a*; d_a is the average packet delay on link *a*; $r_{i,j}$ is the traffic demand between source node *i* and destination node *j*

For any given set of traffic demand, $r_{i,j}$, the sum $\sum_{s,t} r_{i,j}$ is fixed, then we can use the total packet delay as the objective function instead. The objective function can be written as follows:

$$\Phi = \sum_{a} \Phi_a \ , \tag{6}$$

with

$$\Phi_a = f_a \times d_a = \rho_a + \frac{\rho_a^{\ 3}}{(1-\rho_a)^3} \ . \tag{7}$$

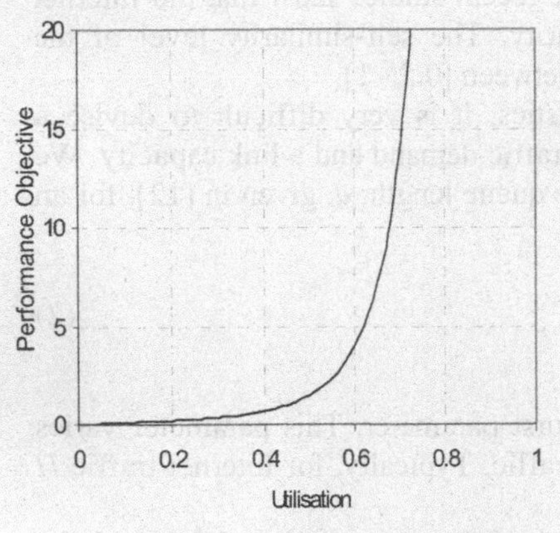

Fig. 2. Performance vs Utilisation

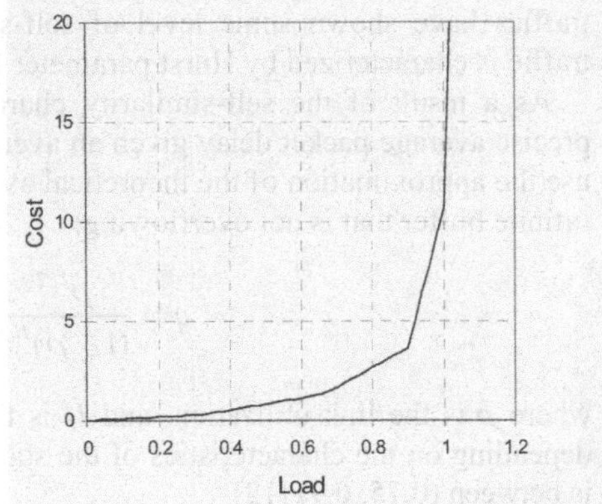

Fig. 3. FT00 "Cost" function in [5]

Figure 2 shows the relationship between the objective function, Φ_a applied for one link as a function of link utilisation ρ_a. This objective function keeps on increasing monotonically, and increases rapidly at high utilisation. Such a function conveniently behaves in a very desirable way for traffic-engineering optimization, for the network optimization will be most sensitive to the heavily utilized links.

When Traffic Load on Some Links Exceed the "Saturation" Level

When traffic demands are so big that result in some unavoidable saturated links, network capacities should be upgraded to accommodate the traffic demands, which can not be served by the current network capacities. Which links should be upgraded, and the new routing scheme for the upgraded networks is a different "resource oriented" network optimization problem, where the financial and administrative constraints play most important roles.

However, prior to any such upgrades, the ISP will still require a routing scheme that attempt to provide as good performance for end users as possible. In this congestion case, the dominant QoS is the throughput experienced by end users, or more precisely the bandwidth share allocated for each flow in progress.

To evaluate the performance of a routing scheme (given a set of link weights), the offered traffic on each link is first calculated without capacity constraints, i.e. the "link utilisation" can be greater than 100%. Assuming a fair bandwidth sharing scheme [16] between flows and the number of flows in progress proportional to the amount of offer traffic, the actual bandwidth share of each flow is inversely proportional to the offer traffic on the link, or the "link utilisation". Flows that have the lowest bandwidth share are those travelling via the link with the highest "link utilisation". As a result, the optimization objective in the overload case reduce to minimizing the maximum "link utilization"; here the "link utilization" is calculated for a given routing scheme, without link capacity constraints, and can be greater than 100%.

We can achieve this min max utilization optimization objective by shaping the objective function based on the delay (7) so that it keeps increasing steadily as the offer traffic exceed the link capacity.

If we denote ρ_{max} as the saturation utilisation level, our objective is devised as follows:

$$\Phi_a = f_a \times d_a = \rho_a + \frac{\rho_a^3}{(1-\rho_a)^3} \tag{8}$$

for $\rho_a < \rho_{max}$,

$$\begin{aligned}\Phi_a &= f_a \times d_a |_{f_a=\rho_{\max}C_a} + (f_a - \rho_{\max} C_a)\times k \\ &= \rho_{\max} + \frac{\rho_{\max}^3}{(1-\rho_{\max})^3} + (f_a - \rho_{\max} C_a)\times k\end{aligned} \tag{9}$$

for $\rho \geq \rho_{max}$,

where $k = \frac{\partial \Phi_a}{\partial f_a}\Big|_{f_a=\rho_{\max}C_a} = \frac{1}{C_a}\left(1 + \frac{3\rho_{\max}^2}{(1-\rho_{\max})^4}\right)$.

If we chose $\rho_{max} = 0.9$, we have

$$\Phi_a = \begin{cases} \rho_a + \dfrac{\rho_a^3}{(1+\rho_a)^2}, for\ \rho_a = \dfrac{f_a}{C_a} < 0.9 \\ 29301\rho_a - 25641, for\ \rho_a \geq 0.9 \end{cases} . \tag{10}$$

3 Modeling and Solving

The general network optimization or traffic-engineering problem can be formalised as following. Given a network of nodes and directed links G=(N, A), with the link capacity of C_a for link a and a set of demand $r_{i,j}$; find a set of link flows that minimise the network performance objective Φ

minimise
$$\Phi = \sum_a \Phi_a ,$$

where
$$\Phi_a = \begin{cases} \rho_a + \dfrac{\rho_a^3}{(1+\rho_a)^2}, \; for\ \rho_a = \dfrac{f_a}{C_a} < 0.9 \\ 29301\rho_a - 25641, \; for\ \rho_a \geq 0.9 \end{cases}$$

subject to
$$\sum_j f_{i,j} - \sum_j f_{j,i} = \begin{cases} r_{i,j} & if\ node\, i = source \\ -r_{i,j} & if\ node\, i = destination \\ 0 & otherwise \end{cases}$$

$$f_{i,j} \geq 0 \qquad \forall i,j \ ,$$

$f_{i,j}$ or f_a is the traffic load on link a from node i to node j ,

$\rho_a = \dfrac{f_a}{C_a}$ is the link utilisation of link a .

To be able to further compare the performance objective between networks of different sizes and capacity, we introduce a scaling factor Φ_T which is the minimum total packet transmission delay. Since there is no queuing delay in the scaling factor, packets will follow shortest paths, given the length of a link equal to the transmission time of one packet on that link, or inverse link capacity. For the same reason, Φ is always greater or equal to Φ_T, therefore after being scaled, $\Phi_* = \Phi/\Phi_T \geq 1$. The higher value of Φ_*, the more congested the network is.

For plain IP OSPF traffic engineering problem, in addition to the conditions of general traffic engineering (flow conservation), there is also other requirement that traffic from sources to destinations can not travel arbitrarily but follow shortest paths. The shortest paths from sources to destinations are calculated based on the configurable link weight parameters assigned for each link. As a result, the IP traffic engineering problem becomes the problem of finding a set of link weights so that it results in the minimum value of Φ. We will reuse the local search heuristic (using a hashing table as a guide) to determine the link weight settings, as proposed in [5].

Because of the restrictions in plain IP routing, its performance objective is always going to be greater or equal to that found by optimal general routing. Therefore, we can use the performance objective value Φ_o, found by the general optimal routing, as a lower bound to the value of Φ found by plain IP routing. Here "general optimal routing" is the idealised case where traffic streams can be routed at will; this is of course not practical for OSPF routing, but it gives a lower bound by which we can judge how good our solution Φ is.

To solve the optimal routing problem, we use Flow Deviation algorithm (FD) as proposed in [8]. The flow deviation algorithm is based on the observation that if a small amount of traffic is moved from a path with a higher incremental delay to a shorter one then the total delay will decrease. Incremental delay on a path, I_a, is the difference in the delay suffered by the flow on that link, if a small amount of flow is added to the path. When combined with (10), I_a becomes

$$I_a = \frac{d(\Phi_a)}{d(f_a)} = \begin{cases} \frac{1}{C_a}\left[1+\frac{3\rho_a^2}{(1-\rho_a)^4}\right], & for\ \rho_a \le 0.9 \\ 29301/C_a & for\ \rho_a > 0.9 \end{cases} \tag{11}$$

4 Results

To study how IP traffic engineering performs subject to different performance objective functions, we created an arbitrary network of 64 arcs and 20 nodes (figure 4). Link capacities were randomly selected from the values of 1000, 2000 or 4000 (in packets per seconds); default link weights are set to inverse link capacity as per recommendation of Cisco [14]. Traffic demands between nodes are random between 0 and a parameter *MaxDemand*, also with units of packets per second. We varied this *MaxDemand* value to see how the Traffic Engineering behaves with different traffic loads from low utilisation to overload scenarios.

We ran the local search heuristic proposed in [5] to find the best set of link weights subject to their cost function, and then subject to our network performance objective. The resulting maximum utilisation and our network performance objective are shown for different traffic loads in Figures 5 and 6.

When the Network Is Lightly Loaded (Max Demand <1000)

Given our performance objective function (Figure 2), for lightly loaded network, where all link utilisation is less than 30%, the incremental delay for each link from (11) is approximately $1/C_a$. Thus, using the FD algorithm to determine "general optimal routing", we find that the resulting traffic flows exactly match the traffic flows when OSPF networks the default link weight recommended by Cisco.

On the other hand, when the cost function FT00 proposed in [5] is used (Figure 3), the incremental cost on every link is 1. When the FD algorithm is used to determine the "general optimal routing", the resulting traffic flows match those for IP networks using least hop count routing or RIP protocol.

From the customer quality of service point of view, when the link is lightly loaded, queuing delay is negligible compared to transmission delay, and the later is proportional to inverse capacity. The packet loss rate is also very small when the utilisation is low. Clearly, the Cisco recommended default weight setting results in the least packet delay, because each packet follows the shortest transmission delay path. Therefore from QoS performance point of view, networks using Cisco recommended link weights perform better than the "optimal" least hop count routing found by [5].

This observation is also shown in Figures 5 and 6, from which we can see that the objective function as well as the maximum link utilisation of our solutions (OSPF-PL) is only 2/3 of the corresponding counterparts of OSPF-FT. This gap may be wider if the link capacities of the network differ widely.

When the Network Starts to Experience Congestion (Max Demand > 1000)

When the traffic demands increase, the default Cisco recommended weight setting results in some overloaded links. The local search heuristic using our performance objective and FT00 both can avoid the local bottlenecks. In average the maximum link utilisation of OSPF-FT using the cost function FT00 is about 10% higher than our maximum utilisation given the same unlimited weight changes. However, from the average packet delay point of view, our solutions OSPF-PL performs much better than OSPF-FT, the latter being about 50% worse than our solution.

This may result partly from the fact that FT00 has a piecewise linear shape, which gives the same penalty to links with utilisations in the same block, such as from 1/3 to 2/3, in other words such links are considered as at same level of congestion.

Compared with the optimal general routing, our OSPF-PL gives a similar performance, within about 10% in terms of delay and max utilisation if the Max Demand is less than 2000.

It is stated in [5] that the actual definition of the performance objective function is not so important to the local search heuristic, as long as it is a piece-wise convex function. In contrast, we have found that although the actual definition of the objective function does not change the way their local search heuristic operates, it does change the resulting set of weights and routes. Those routes can be very different from that found by [5]. That is, the performance objective function chosen and its shape are important in finding the solutions that is optimal from user QoS perspective.

Our objective function is based on the delay, which take into account the Hurst parameter, *H*. From equation (3) we can see that, as *H* gets close to 1, the delay function increases sharply even at very low utilisation. This means, the higher the value of H, the more sensitive to maximum utilisation the objective function is.

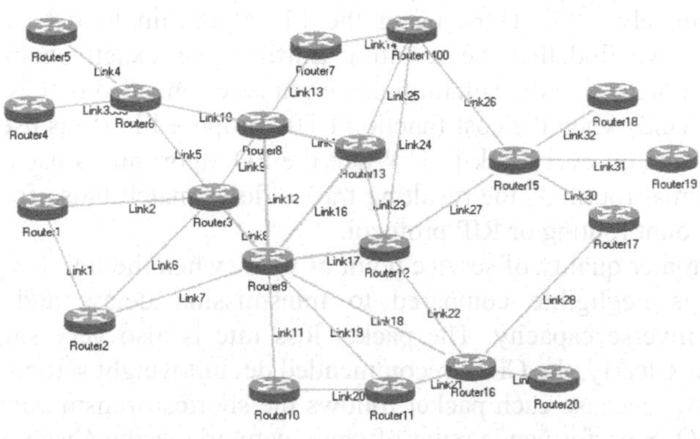

Fig. 4. Network Diagram

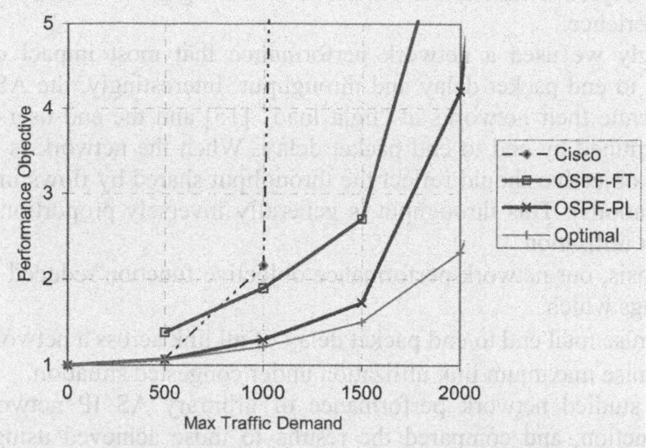

Fig. 5. Performance Objective under normal loading condition

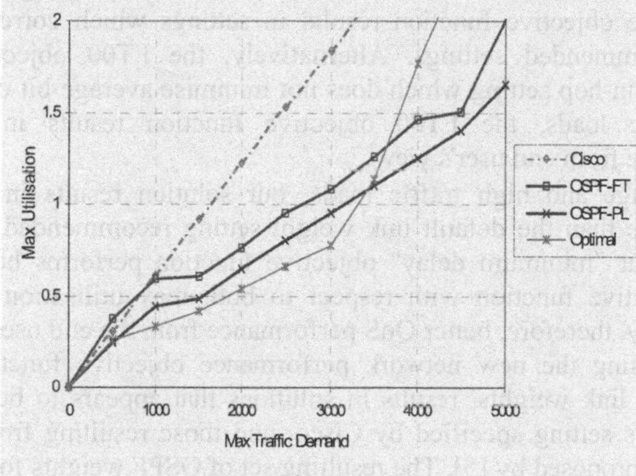

Fig. 6. Max Utilisation vs Max Traffic Demand

5 Conclusion

This study investigates the effectiveness of traffic engineering in Autonomous System IP networks using OSPF routing. The aim is to determine the optimum link weight settings that the OSPF routing algorithm relies on to direct traffic flows through networks, and in so doing to optimise network performance.

Our results indicate that we can achieve much better network performance than if the default Cisco setting is used, by optimising the OSPF link weights. However, while optimising the weight setting, the definition of an appropriate network performance objective function is crucial to achieving performance which optimizes end user experience.

Accordingly we used a network performance that most impact on users - the network end to end packet delay and throughput. Interestingly, the AS operator will typically operate their networks at "light load" [15] and the end users experience is largely determined by end to end packet delay. When the network is congested, the performance objective should reflect the throughput shared by flows in progress (end users' applications). This throughput is generally inversely proportional to the offer traffic or link utilisation.

On this basis, our network performance objective function reduced to finding link weight settings which

- minimise total end to end packet delay of all link across a network, and
- minimise maximum link utilization under congested situation.

We then studied network performance of arbitrary AS IP networks using our objective function, and compared the results to those achieved using a previously published objective function.

We found the following

- At light traffic loads, which is the "normal" AS operating scenario, our performance objective function results in settings which correspond to the Cisco recommended settings. Alternatively, the FT00 objective function results in min-hop setting which does not minimise average bit delay. Thus at light traffic loads, the FT00 objective function results in sub-optimal performance from end user's view.
- At mid range and high traffic loads, our solution results in much better performance than the default link weight setting recommended by Cisco. In addition, our "minimum delay" objective function performs better than the FT00 objective function with respect to both max utilisation and average packet delay, therefore, better QoS performance from the end user's view.

In summary, using the new network performance objective function for local search to optimise link weights, results in solutions that appears to be much better than default routers setting specified by Cisco, and those resulting from the use of objective function proposed by [5]. The resulting set of OSPF weights found using the heuristic proposed in [5] and our new objective function helps the AS operator to increase their network performance experienced by end users without upgrading their network capacity.

For further work, we will investigate the results of the optimization process to a range of networks, and their sensitivity to the Hurst parameter, H, which characterizes the self-similarity level of the Internet traffic.

Acknowledgement

We would like to thank A/Prof. Greg Allen for his comments. This work has been supported by James Cook University through an International Post-Graduate Research Scholarship.

References

1. Braden, R. et al: Integrated Services in the Internet Architecture: an Overview. IETF RFC1633, http://ietf.org/rfc/rfc1633.txt, 1994
2. Blake, S. et al: An Architecture for Differentiated Services. IETF RFC2475, http://ietf.org/rfc/rfc2475.txt, 1998
3. Rosen, E. et al: Multiprotocol Label Switching Architecture. IETF RFC3031, http://ietf.org/rfc/rfc3031.txt, 2001
4. Feldmann, A. et al.: NetScope: Traffic Engineering for IP Networks. IEEE Network, March/April 2000
5. Fortz, B., Thorup, M.: Internet traffic engineering by optimizing OSPF weights. Proc. IEEE INFOCOM, March 2000
6. Awduche, D. O.: MPLS and Traffic Engineering in IP networks. IEEE Communications Magazine, December 1999
7. Wang, Y.et al.: Internet Traffic Engineering without Full Mesh Overlaying. Proc. Infocom2001
8. Kershenbaum, A.: Telecommunications Network Design Algorithms. McGraw-Hill, 1993
9. Paxson, V. et al.: Wide Area Traffic: The failure of Poisson Modeling. Proc. ACM SIGCOMM '94, pages 257--268, 1994
10. Leland, W. et al.: On the self similar nature of Ethernet traffic. IEEE/ACM Transactions on Networking, April 1994
11. Crovella, M. et al.: Self-Similarity in World Wide Web Traffic Evidence and Possible Causes. IEEE/ACM Transactions on Networking, December 1997, pages 835-846
12. Stallings, W.: High-Speed Networks: TCP/IP and ATM Design Principles. Prentice Hall, 1998
13. Massoulie, L., Roberts, J.W.: Bandwidth sharing and admission control for elastic traffic. Telecommunications Systems, 15:185-201, 2000
14. Cisco: OSPF Design Guide. Cisco Connection Online, http://www.cisco.com/warp/public/104/1.html, 2001
15. Odlyzko, A.: Data Networks are Lightly Utilized, and Will Stay That Way. http://www.research.att.com/ ~amo/doc /network.utilization.pdf
16. Roberts, J. W., Massoulié, L.: Bandwidth Sharing and Admission Control for Elastic Traffic. Telecommunication Systems, v.15, pp.185-201, 2000

An Adaptive QoS-routing Algorithm for IP Networks Using Genetic Algorithms

Hicham Juidette[1], Noureddine Idboufker[2], and Abdelghfour Berraissoul[2]

[1] Faculty of Sciences, University Mohamed V, BP 736 Agdal Rabat, Morocco
hichamj@netcourrier.com
[2] Faculty of Sciences, University Chouaib Doukali, El Jadida, Morocco
n_idboufker@yahoo.fr

Abstract. In this paper, a genetic algorithm is designed for adaptive and optimal data routing in an internet network. The algorithm called QORGA, supports QoS-routing for the two simulated services '*Voice*' and '*Data*'. The used genetic operators permit to obtain the optimal path, while providing others feasible paths for differentiated services. This algorithm is not based on the knowledge of the full network topology by each node. That allows the optimisation of the network resources through the diminution of the exchanged accessibility information, leading to a scalable network deployment in addition to the possible use of source routing protocols in IP networks.

1 Introduction

In full expansion, Telecommunications are supported by a continual change in technologies and present a challenge for many optimisation and design concepts. Among these lasts, the QoS-based routing has to provide Quality-of-Service (QoS) guarantees to multimedia applications and an efficient use of network resources. From another part, there is an increasing interest during these last years for the use of Evolutionary computation to resolve problems in Telecommunications such as network planning deployment [1], call and data routing [2]... In this context, the problem of data routing in IP networks is considered in this paper by using Genetic Algorithms (GA's) as tools of search, optimisation and decision in order to obtain adequate routes.

This paper is presented as follow: the next section introduces the interconnection in networks and the QoS-based routing. Section 3 presents the structure of the proposed genetic-based routing algorithm and section 4 considers the simulation of differentiation of services (DiffServ). A conclusion closes the paper.

2 Networks Interconnection

The concept of interconnection is based on the implementation of a network layer to hide physical communication details of the various networks and separating the application layer from data routing problems [3]. A centralised architecture based on a single node manager (Router) is a simple solution for data routing networks. However, it

M. Ajmone Marsan et al. (Eds.): QoS-IP 2003, LNCS 2601, pp. 76-89, 2003.

becomes a non scalable architecture specially when the transported traffic has a dynamic behaviour, leading to an increase in the overflow due to the accessibility information exchange and the necessary time for routing decisions. To try to resolve these problems, one can adopt either a network segmentation into entirely independent autonomous systems that are connected by a BackBone Network, or a topology aggregation to increase the scalability.

2.1 QoS-based Routing

The majority of routing processes can be categorised into four principal types : fixed routing (FR), routing according to time (TDR), routing according to the state (SDR) and routing according to events (EDR) [5]. These processes are based on *vector distance* or *link state* protocols that use variants of Bellman-Ford and Dijsktra algorithms in order to find the path with the least cost [4]. Based on the network topology, *vector distance* and *link state* algorithms try to transfer each packet along the shortest path from the source to the destination node. In addition to a slowness convergence and the probable loops creation, a *vector distance* algorithm uses the number of hops as a single metric to make a route decision and it does not take into account QoS parameters. To overcome these limitations, the *link state* algorithm assigns a weighting coefficient to each link separating two nodes. The cost of each connection is conditioned by principal metrics like transmission delay, bandwidth, blocking probability, loss ratio, jitter…

The constraints of a QoS-based routing include the network topology, the information about network availability resources, the traffic QoS requirements and policy constraints. QoS-based routing can then help to provide better performances and to improve the network utilisation. However, it is much more complex and may consume network resources, leading to a potential routing instability. Moreover, the volume of exchanged accessibility information and the frequency of update messages represent a compromise between the accuracy of the QoS parameters and the network overload [6].

2.2 Routes Computation

In QoS-routing, the routes computation is constrained by multiple QoS parameters that are relative to classes of services and traffic profiles [7]. A routing methodology that takes into account the QoS needs for the various transported applications, must then be defined [8]. The principal difficulty is thus to define metrics to use and to establish the correlation between them. For instance, the end-to-end delay transmission depends on several factors such as bandwidth, hardware and software routers resources, the blocking probability, the loss probability and the profile of the transported traffic.

Choosing between a single or multiple metrics is decisive for a routing algorithm conception. The single metric is used when there is no correlation between the other metrics. In the case when multiple metrics are considered, the routing algorithm becomes however more complex and heavy to implement. The algorithm proposed in this paper considers a distributed routing computation with a mixed metric that takes into account different QoS parameters.

3 QORGA Algorithm

The use of Genetic Algorithms in order to solve a routing problem is justified by the fact that the search of a best route linking two nodes in a network is an optimisation problem. Besides, fundamental difficulties relative to the data routing problem can be resolved using GA's. These difficulties could be summarised by the network latency time which causes uncertainty in the observation and the determination of the network state as well as the overflow due to the exchange of routing information. GA's permit a distributed process in which each individual acted independently in order to adapt to the surrounding environment, even with less information than necessary, leading to a overflow reduction. It is also considered robust to the fast and frequent change of the network state. The proposed algorithm named 'QoS Routing Genetic Algorithm' (QORGA), permits to obtain optimal paths in IP networks. It is a QoS-Routing algorithm since it permits to obtain feasible paths taking into account QoS parameters while optimising the use of network resources.

3.1 Description of the Problem

The QORGA structure helps to build an intelligent path optimisation system that is able to adapt continually to the randomness of a IP network. The researched paths are modelled as chromosomes with different lengths. A path is a following of links connecting different nodes (Fig. 1.). A link is thus represented by a gene and this last is described by the extremity nodes and the link weight (Fig. 2.). A specific sequence of genes forms a chromosome (Fig. 3.) and a set of chromosomes forms a population. The weight accumulated over links composing a path, constitutes its evaluation. The link weight could depend on several parameters and its value is varying in time according to the network dynamic state. The weight of a link $L_{1,2}$ joining nodes 1 and 2 is noted $P_{1,2}$. In Fig. 1, nodes 2, 3 and 4 are the destination nodes corresponding to node 1. Links $L_{1,2}$, $L_{1,3}$ and $L_{1,4}$ are then feasible and have a finite weight : $P_{1,2} < \infty$, $P_{1,3} < \infty$ and $P_{1,4} < \infty$.

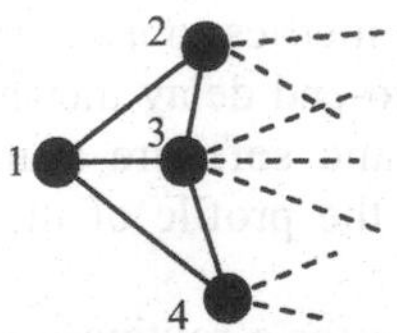

Fig. 1. Example of nodes interconnection

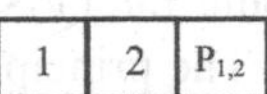

Fig. 2. A gene representing link $L_{1,2}$

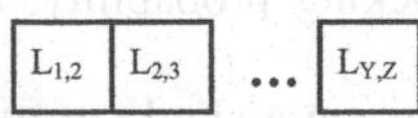

Fig. 3. A chromosome representing a path connecting nodes 1 and Z

3.2 Formation of the Initial Population

The formation of the initial population is an important stage in the development of the optimal path search algorithm. Paths forming this population must be feasible and sufficiently varied in order to guarantee a fast convergence to an optimal solution. The feasibility of a path means that this last is continuous, does not pass by a node more than once, avoiding the creation of loops, and allows to hint that a link or a path portion is browsed more than once.

During the population initialisation, constraints are imposed in order to insure the paths continuity. The arriving node of a link must then be identical to the departure node of the next link in the path. Moreover, a link $L_{A,B}$ must exist and be feasible in order to be chosen as a link composing the final path. The node B must therefore exist among the destination nodes to which lead the node A outgoing links. At the level of each node, the choice of the next link is done randomly so long as this last exists. In order to avoid loops, an infinite weight is assigned to the link that is already used in the current not yet completed path in order to be temporarily inaccessible. The final length of a path varies in reason of the random choice of links that is made in cascades from the departure node until the final node.

3.3 Genetic Operators

The genetic operators used in QORGA are the *Reproduction*, the *Crossover*, the *Mutation* that improves progressively the path by reducing if possible its cost, the *Insertion* of genes in order to insure the continuity of path, the *Deletion* of genes in order to eliminate and replace successive unfeasible links and the *Shortening* that permits to merge two feasible links into only one feasible and more optimal link that conserves the continuity of the considered path.

3.3.1 Reproduction

A mating pool is formed with couples of chromosomes chosen among the best individuals on one hand and among the remainder of the population on the other hand. The selection is processed using the lottery wheel. These couples will be subjected to ulterior genetic operations in order to generate offspring that would replace the weakest chromosomes in order to form a new population. The size of the mating pool corresponds to the *Gap Generation* [9].

3.3.2 Crossover

In [10], the crossover is done only at a node present in both chromosomes genitors. In QORGA, *Crossover* does not need necessarily this condition since it consists of exchanging parts of the chromosomes at a randomly chosen crossover site. The exchanged parts concern those that are on the right of the crossover site. The rank of this last must be inferior to the minimal length corresponding to the two genitors. The offspring could represent discontinuous paths (Fig. 4.). However, the path continuity is re-established thank to the genetic operators of *Insertion* or *Deletion*, described later.

3.3.3 Mutation

The mutation consists of two distinct operations : '*Simple Mutation*' and '*Double Mutation*'.

3.3.3.1 Simple Mutation : Mutation of a Link Node

The two successive links L_{i-1} and L_i are concerned by the '*Simple Mutation*' at a site 'i' corresponding to the arriving node of L_{i-1} and the departure one of L_i (Fig. 5.). Considering that the extremity nodes of L_{i-1} and L_i are respectively (1, 2) and (2, 3), the replacement of node 2 is done searching a node X among the destination nodes of node 1 and which must be connectable with nodes 1 and 3 in such manner that the sum of weights of the eventual new created links L_{i-1} and L_i is the most minimal among possible solutions. The two new successive links L_{i-1} and L_i must insure the feasibility of the represented path.

3.3.3.2 Double Mutation : Consecutive Mutation of the Extremity Nodes of a Link

At a chosen site, this mutation proceeds to the replacement of the extremity nodes of a link. This implies the modification of three successive links. The added weight corresponding to the links that are upstream, downstream and at the mutation site, must be the most minimal possible (Fig. 6.). This operation is important, because it alone permits in some cases the local minimisation of the path cost in modifying the two nodes of a link.

3.3.4 Insertion

Insertion operation consists in inserting a gene between two other when these lasts don't allow to insure the continuity of a chromosome (Fig. 7.). The inserted link must be feasible. This genetic operation increases the path length.

3.3.5 Deletion

After a crossover operation, the offspring could present a discontinuity case. An *Insertion* can then resolve this problem. However, a deletion operation is preferable because of the possible reduction of the chromosome length. *Deletion* consists in merging two successive links L_i and L_{i+1} presenting a discontinuity, removing genes corresponding respectively to the arriving node of L_i and the departure one of L_{i+1} and which are different (Fig. 8.). The fusion is done only if the resulting link is feasible. Otherwise, the *Insertion* operation is processed. *Deletion* decreases the path length.

3.3.6 Shortening

Shortening operation is similar to *Deletion* when considering the fusion procedure, except that it unifies two feasible and successive links in such manner that the resulting link is feasible and more optimal (Fig. 9.). *Shortening* decreases the chromosome length. Therefore and in case of links $L_{1,2}$ and $L_{2,3}$, the following condition must then be verified : $P_{1,3} < (P_{1,2}+P_{2,3})$.

Shortening, *Deletion* and *Insertion* are varied forms of mutation. The treatment computation time depends on a chromosome length which can present positive or negative influences for the ulterior solutions search.

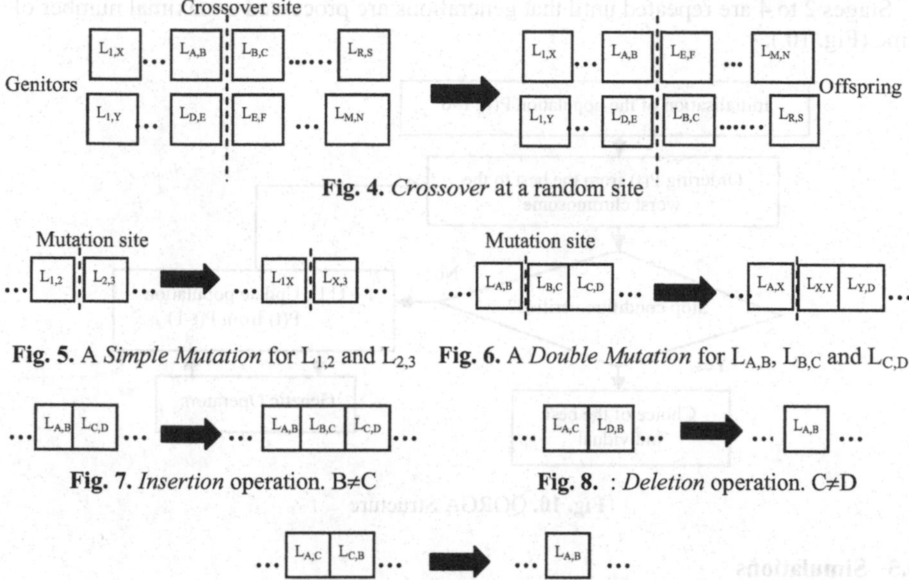

Fig. 4. *Crossover* at a random site

Fig. 5. A *Simple Mutation* for $L_{1,2}$ and $L_{2,3}$

Fig. 6. A *Double Mutation* for $L_{A,B}$, $L_{B,C}$ and $L_{C,D}$

Fig. 7. *Insertion* operation. B≠C

Fig. 8. : *Deletion* operation. C≠D

Fig. 9. *Shortening* operation

In order to make genetic operators more effective in the search of optimal solutions, genetic operators probabilities are represented by significant values. Excessive encouragement of these genetic operations is beneficial thanks to the use of knowledge and not by using the random factor. Thus, the probabilities adaptation is not used in QORGA.

3.4 Procedure of QORGA

1- Initialisation of the population.

2- The current population is organized from the best to the worst chromosome according to the used evaluation function and a mating pool is formed. The selection of chromosomes is done using the evaluation function or its adaptation, according to selected cases.

3- For each reproduced chromosomes pair and if the crossover probability allows it, the two chromosomes are crossed at a randomly chosen site. Before being evaluated, the two descendants are subjected to genetic operations of *Deletion*, *Insertion* and *Shortening* that try to reduce locally the links cost at the crossover site. This involves a total cost reduction of the searched path. Thereafter and if the mutation probability allows it, the '*Simple Mutation*' generates modifications on the two descendants at a randomly chosen mutation site. These muted chromosomes are then affected by a shortening at the mutation site before being evaluated.

4- A new population is created by replacing eventually the worse chromosomes by the generated offspring. Once the population ordered, the best chromosome undergoes genetic operations acting on all its genes (*Simple Mutation*, *Shortening* and *Double Mutation*) in order to reduce its cost.

Stages 2 to 4 are repeated until that generations are processed a maximal number of time (Fig. 10.).

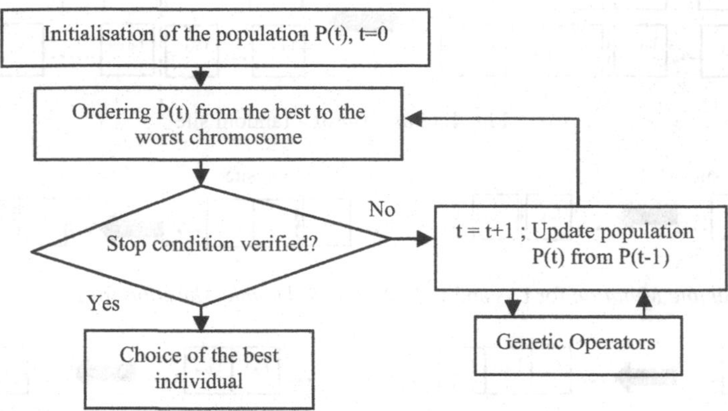

Fig. 10. QORGA Structure

3.5 Simulations

Chromosomes are evaluated by the evaluation function *fe* that is the sum of the costs of all connections constituting the searched path. The minimal value of *fe* corresponds then to the best path. A link cost can depend on a metric mixing the available rate, the transmission delay, the connection availability, the physical quality, the availability and performances of active elements (Routers). In simulations supported by QORGA, the chromosomes selection phase uses a linear adaptation of *fe* in order to exploit the diversity of different chromosomes and to avoid a rapid domination of the population by a super-individual [9]. The chance for a weak chromosome to be selected is then improved in opposition to a best chromosome whose adapted *fe* is increased and its opportunity to be selected is thus decreased. The use of a linear adaptation of *fe* involves however a number of generations larger than that necessary to find an optimal solution in case of the non-use of an adaptation. Thus and when searching a single solution, the non adaptation of *fe* is recommended because of the speed of convergence towards an optimal solution. Fig. 11 illustrates this case and represents the evolution of the minimal fitness *fe* of a population. Arrows mark positions relative to the best paths fitness's. When multiple solutions are needed, the adaptation of *fe* allows the creation of various sub-optimal solutions that could be useful as backup solutions. Fig. 12 illustrates the evolution of the minimal fitness *fe* through generations when using a linear adaptation of *fe*, according to parameters of table 1 for a network of 100 nodes represented in Fig. 15. Arrows show positions corresponding to the obtained feasible paths that are more numerous in comparison with the non adaptation case. These multiple paths constitute eventual alternative solutions.

In simulations, a node is supposed known through its coordinates, its rank in the network, the number of its outgoing connections (maximum 4 links in the simulated cases), its destination nodes as well as a temporary weight of each link. Initially, a global metric represented by random costs assigned to each connection, is considered. The network size represents the number of nodes. Optimising a path is the main purpose of the proposed algorithm. According to the used metric, an optimal path should comprise neither loops, nor the repetition of usage of a same link.

Each initial population is constituted of feasible paths, contributing to reduce considerably the processing time. Table 1 illustrates the used genetic parameters. Figures 13 and 14 show simulations results of QORGA as optimal paths obtained for networks of respectively 20 and 35 nodes. Fig. 15 shows all configurations of the possible connections in a 100 nodes network. Fig. 16 illustrates the final obtained path using QORGA that represents the best path (Shortest path) in the network represented in Fig. 15, after 15 generations with a linear adaptation of *fe* and after 9 generations without *fe* adaptation. The obtained paths for these different networks are identical to those obtained using the Dijkstra algorithm.

Moreover, QORGA allows the multi-paths possibility, necessary when other alternative solutions are needed. In this study, a linear adaptation of *fe* is used and multiple feasible paths are thus formed during a small number of generations, representing best paths according to the used metric. The adopted structure for search and decision allows to eliminate negative aspects such as loops creation or blocking at a node.

Table 1. QORGA genetic parameters

Gap Generation '*G*'	50
Mutation probability 'p_m' (*Simple Mutation*)	0,99
Crossover probability 'p_c'	1
Population size '*popsize*'	50
Number of generations	30

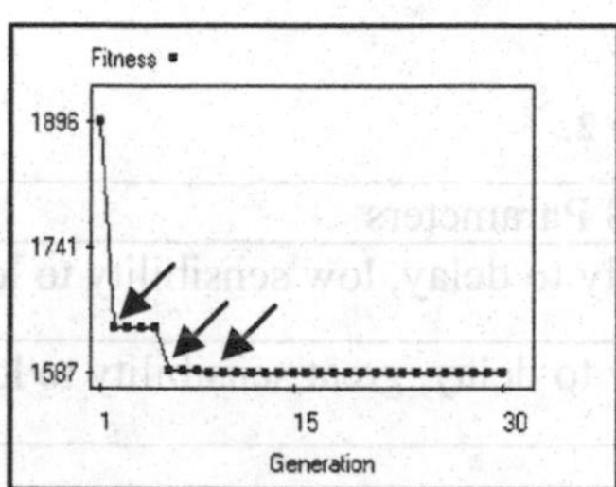

Fig. 11. Non adaptation of *fe*

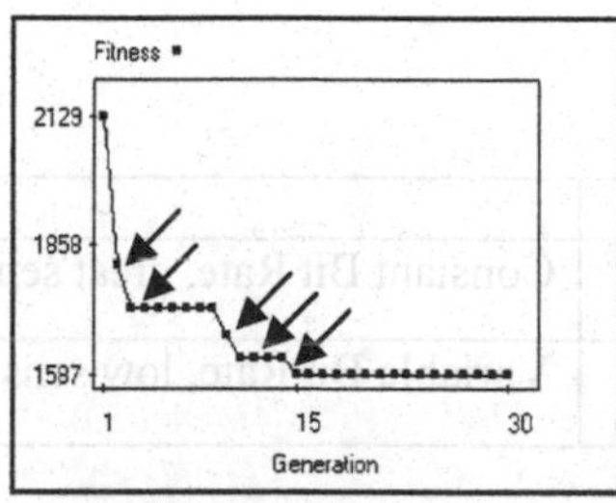

Fig. 12. Adaptation of *fe*

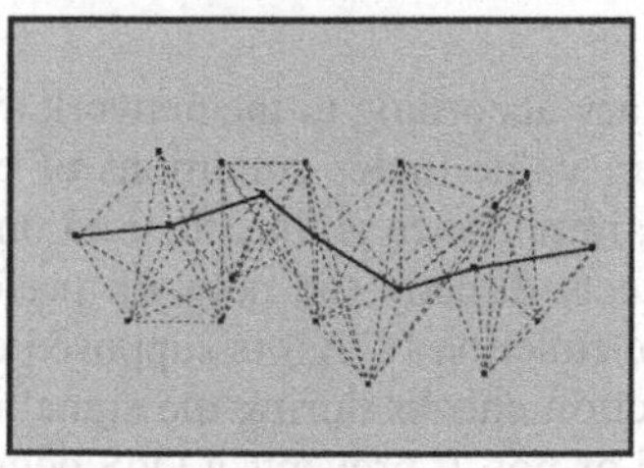

Fig. 13. Best path obtained using QORGA in a 20-nodes network

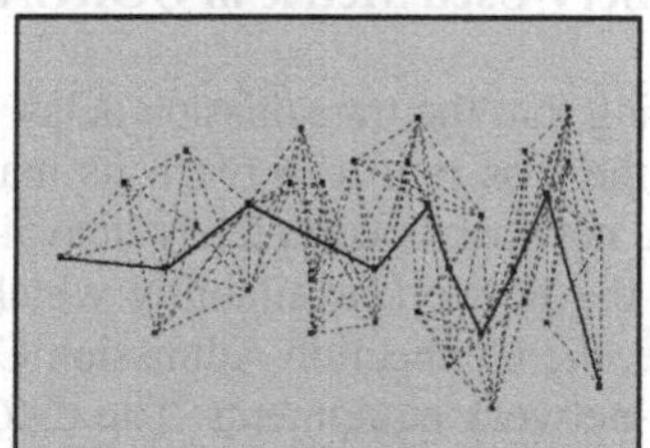

Fig. 14. Best path in a 35-nodes network

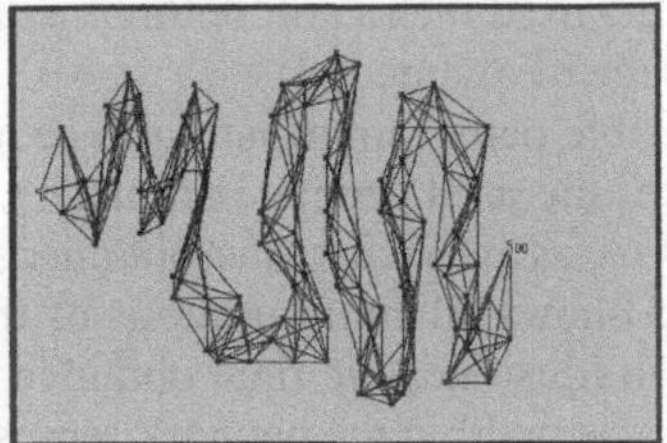

Fig. 15. A 100 nodes-network

Fig. 16. Best obtained path in the 100-nodes network

In what follows, the interest is focused on the differentiation of services (DiffServ) considering the *Voice* and *Data* services whose QoS parameters are different. The transport of these services on the same packet network, is more and more needed because of the evolution toward the Next Generation Networks (NGN).

4 Differentiation of Services (DiffServ)

In order to take into account the differentiation of services (Diffserv), two basic services, *Voice* and *Data*, are simulated. Table 2 describes QoS parameters of these services. The bandwidth problem can be solved by supposing a sufficiently large routing and transport capacities. QORGA adopted metric is based mainly on the modelling of three principal parameters which are transmission delay, network availability and loss ratio.

Table 2.

Service	QoS Parameters
Voice	Constant Bit Rate, great sensibility to delay, low sensibility to loss ratio
Data	Variable Bit Rate, low sensibility to delay, great sensibility to loss ratio

4.1 DiffServ Used Metric in QORGA

Considering that the transmission delay can vary according to the network availability and that the loss ratio can reach its maximum value under conditions of congestion where the network is completely unavailable, importance is thus focused on the modelling of the network availability which reflects directly the network performances. Therefore, the Connection Admission Control function (*CAC*) is supposed enabled in all active network equipments. The *CAC* function checks during the signalling procedure whether a connection can be established or not. It prevents a QoS degradation if the number of active connections exceeds the connections threshold number that can be handled by the network. If this last is considered with arrivals according to the Poisson law, the network load ρ and the maximal number N_{max} of calls handled by the network are as follow:

$$\rho = \frac{N_t MDC}{3600}. \tag{1}$$

$$N_{\max} = \frac{B_{\min Net}}{B_{\min traffic}}. \tag{2}$$

N_t is the number of connections attempts during the busy hour and MDC is the mean duration of a communication. N_{max} depends on both the smallest bandwidth B_{minNet} of network links and the smallest bandwidth $B_{mintraffic}$ occupied by the arriving traffic. If the network uses the *CAC* function, the blocking probability will be given by the Erlang B law [3]. Thereafter, the transmission delay DT and the loss ratio TP will depend on the network availability represented by a blocking probability P_B :

$$DT = \frac{T_P}{1-P_B}. \tag{3}$$

$$TP = \frac{P_B}{1-P_B}. \tag{4}$$

T_p is the necessary time to traverse a link and it is proportional to the distance between the link extremity nodes. Each connection will depend on the triplet (DT, P_B, TP) taking into account the QoS requirements. For implementation and simulation of the differentiation of services using QORGA, two evaluation functions fe_{Voice} and fe_{Data} concerning respectively *Voice* and *Data* services, are considered. fe_{Voice} and fe_{Data} will depend both on DT and TP: $fe_{Voice} = f(DT, TP)$; $fe_{Data} = h(DT, TP)$.

fe_{Voice} and fe_{Data} cumulate respectively the 'transmission delays' and the 'loss ratio' relative to all path links. Thus:

$$DT(L_{C,D}) = \frac{T_P(L_{C,D})}{1-P_B(D)} = \frac{T_P(C,D)}{1-P_B(D)} = DT(C,D). \tag{5}$$

$$TP(L_{C,D}) = \frac{P_B(D)}{1-P_B(D)} = TP(C,D). \tag{6}$$

The blocking probability $P_B(D)$ at a node D is simulated as follows:

$$P_B(D) = \begin{cases} 1 \quad if \quad rank(D) = 1 \\ \dfrac{E(D)}{nb_{node} S(D)} \quad if \quad 2 \le rank(D) \prec \dfrac{nb_{node}}{4} \quad and \quad \dfrac{3nb_{node}}{4} \prec rank(D) \prec nb_{node} \\ \dfrac{E(D)}{2S(D)} \quad if \quad \dfrac{nb_{node}}{4} \le rank(D) \le \dfrac{3nb_{node}}{4} \\ 0 \quad if \quad rank(D) = nb_{node} \end{cases} \tag{7}$$

nb_{node} is the number of nodes in the network, $E(D)$ and $S(D)$ represent respectively the numbers of input and output connections of a node D. The definition of $P_B(D)$ traduces a great probability of congestion in the network core and the possible relaxation of blocking in the zones between the network core and the access nodes. It is important to note that there is a great correlation between $E(D)$, $S(D)$ and P_B. This last must be the weakest possible.

The evaluation functions according to QoS constraints for each class of service can be established as follow :

$$fe_{Voice} = DT + \mu TP \,. \tag{8}$$

$$fe_{Data} = TP + \mu DT \,. \tag{9}$$

DT and *TP* are relative to all links composing the final path. μ is a low value factor used to minimise the influence of the multiplied parameter as the loss ratio for *Voice* service and the transmission delay for the *Data* service. Worth to mention that minimal values of fe_{Voice} and fe_{Data} correspond to the best paths.

4.2 Simulations

Two populations represent solutions for each service. The first population is evaluated according to fe_{Voice} and it is composed of paths generated according to the QoS constraints simulated for the *Voice* class. The second population, relative to the *Data* service, is ordered following the increasing values of fe_{Data}. Genetic parameters of Table 1 are used in simulations and a linear adaptation of the fitness's is used in the selection phases. Various generations include the successive creation of the two distinct populations and according to the same process described above (See Procedure of QORGA). From another part, the processing time can be consequently reduced, stopping the genetic process as soon as the algorithm detects that the population is dominated by a super-individual and that the average of the evaluation function does not change any more from generation to another. This is showed in Fig. 17 where the fitness's average relative to the 100-nodes network illustrated in Fig. 15, does not vary in the neighbourhood of the 15th generation for both simulated services.

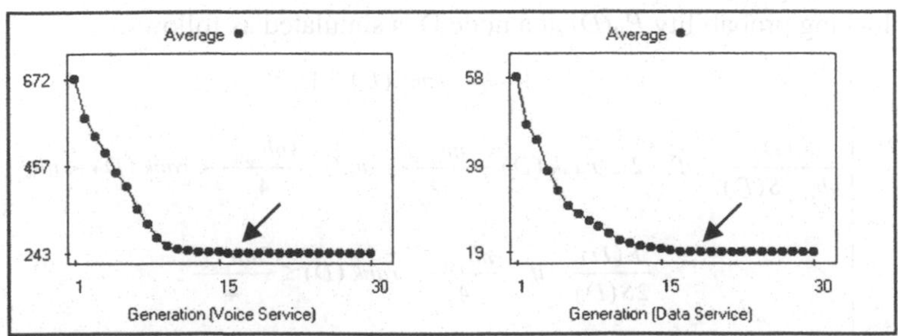

Fig. 17. Evolution of the '*Fitness Average*' for the 100-nodes network

Figures 18-a and 18-b show the obtained optimal paths simulating respectively *Voice* and *Data* services for the 100-nodes network of Fig. 15. Using QORGA allows the exploitation of the same network structure in order to obtain two different and optimal paths relative to both services The used value of μ is as follow : μ=0,02.

In a care to diversify the evaluations, more simulations processed on different networks are presented in figures 19 to 22. They concern networks of respectively 20, 35, 100 and 120 nodes. The results show that the differentiation of services is processed through the obtained optimal paths that are different following the nature of the corresponding service. Fig. 23 concerns the variations of both '*Minimal Fitness*' and '*Fitness Average*' during generations for the simulated services in the 120-nodes network of figures 22-a and 22-b. Improvements of the '*Fitness Average*' through the genetic process lead to the creation of feasible paths that are represented by the different positions of the '*Minimal Fitness*'.

Fig. 18-a. Best path obtained for the simulation of '*Voice*' service (DiffServ) in the 100-nodes network

Fig. 18-b. Best path obtained for the simulation of '*Data*' service (DiffServ) in the 100-nodes network

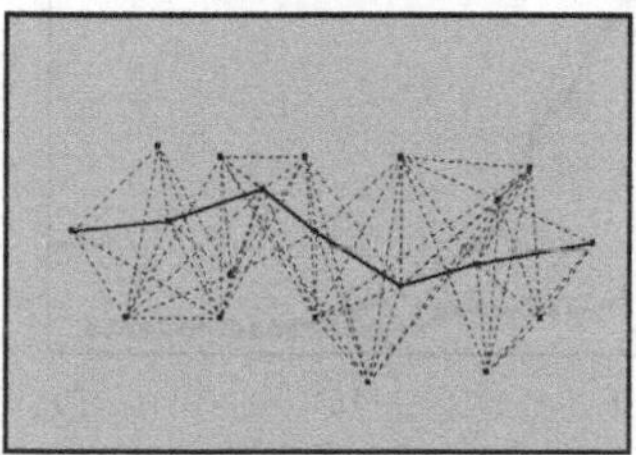

Fig. 19-a. Best path for the '*Voice*' service (DiffServ) in a 20-nodes network

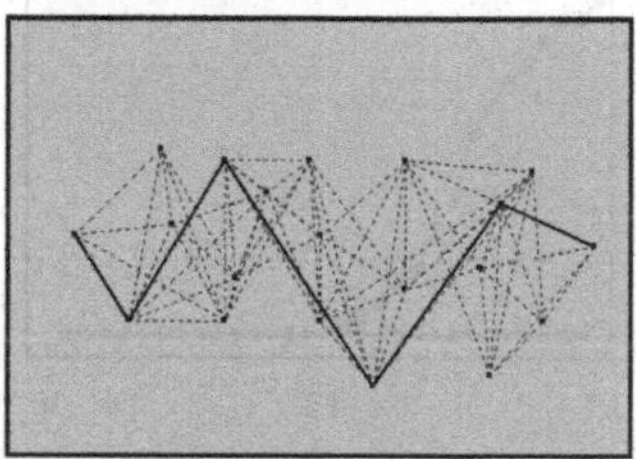

Fig. 19-b. Best path for the '*Data*' service (DiffServ) in a 20-nodes network

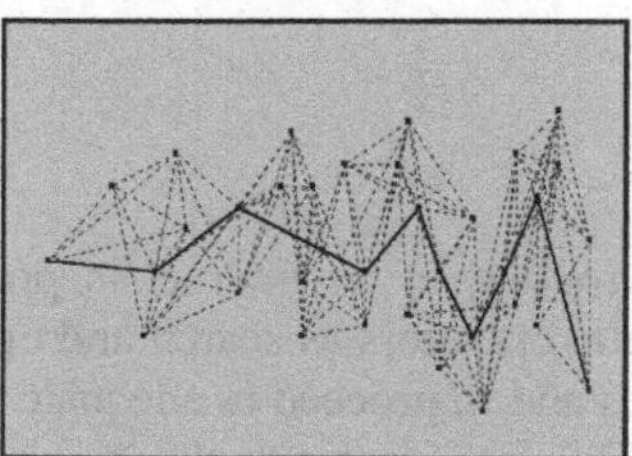

Fig. 20-a. Best path for the '*Voice*' service (DiffServ) in a 35-nodes network

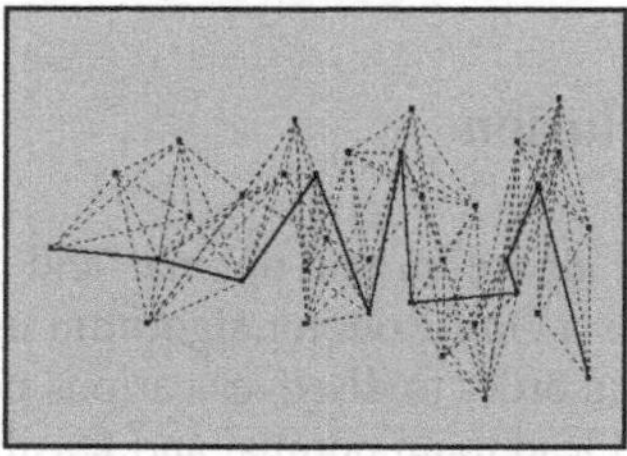

Fig. 20-b. Best path for the '*Data*' service (DiffServ) in a 35-nodes network

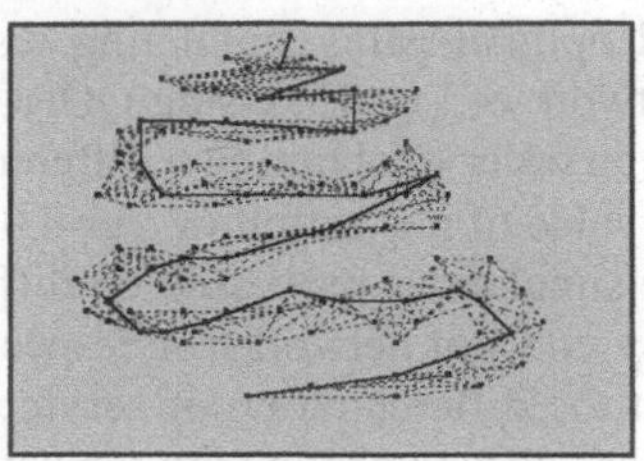

Fig. 21-a. Best path for the '*Voice*' service (DiffServ) in a 100-nodes network

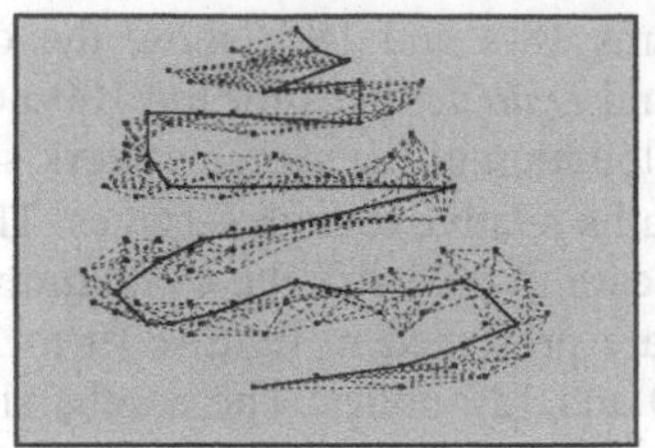

Fig. 21-b. Best path for the '*Data*' service (DiffServ) in a 100-nodes network

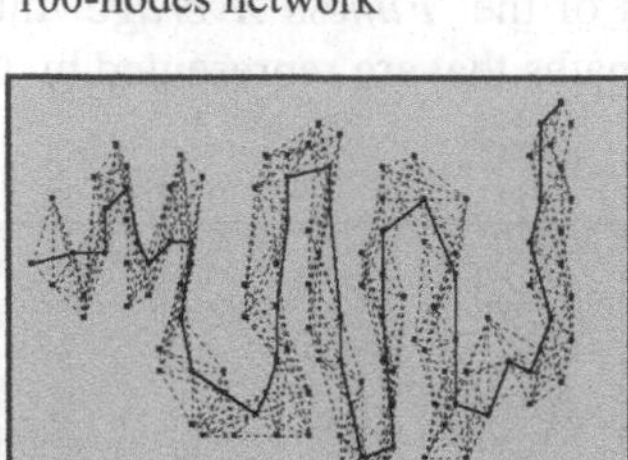

Fig. 22-a. Best path for the '*Voice*' service (DiffServ) in a 120-nodes network

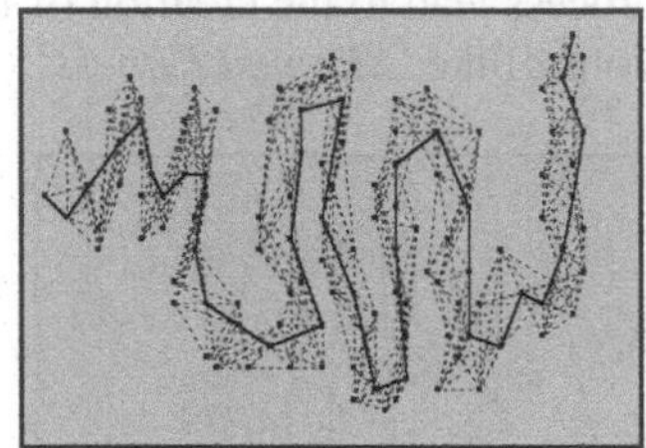

Fig. 22-b. Best path for the '*Data*' service (DiffServ) in a 120-nodes network

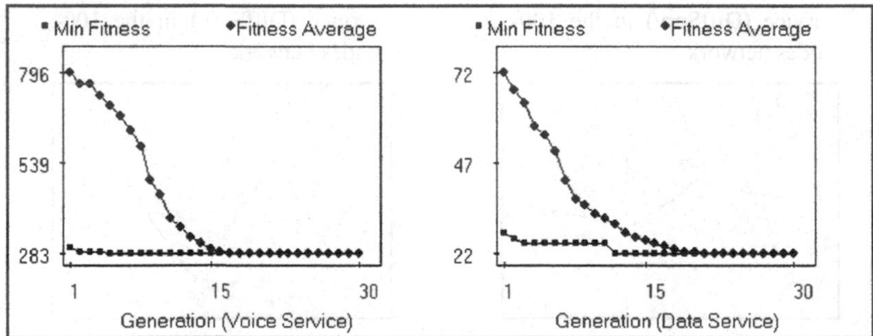

Fig. 23. Evolution of '*Minimal Fitness*' and '*Fitness Average*' for the 120-nodes network

5 Conclusion

A genetic algorithm for optimal and adaptive routing in an internet is presented. QORGA is a QoS-Routing algorithm using genetic operators as search and optimisation tools in order to discover various possibilities and to proceed to adequate choices that would lead to an optimal and feasible path. This last represents the lower cost according to the QoS requirements and does not present loops or multi-used links. During the genetic process, the algorithm is adaptive to any change affecting the weights of the network links.

Scalability is a fundamental requirement for any Internet QoS scheme. For QORGA, the computation is based only on local state information and it is not necessary that each node knows the full network topology, nevertheless necessary for the majority of traditional routing algorithms. Thus, the volume of exchanged accessibility information decreases, allowing the optimisation of network resources and taking into account the trade off between scalability and stability of the algorithm. Moreover, the possibility of alternate paths is confirmed through the availability of multiple optimal paths obtained through generations and which can be considered as backup solutions. Besides, QORGA is a QoS-based routing algorithm that supports the differentiation of services (DiffServ). Paths corresponding to distinct QoS requirements are generated while considering one population for each simulated class of service.

References

1. Webb, A., Turton, B. C. H., Brown, J. M. : Application of Genetic Algorithms to a Network Optimisation Problem. In: Proc. 6th IEE Conf. on Telecommunications. Edinburgh UK (1998) 62-66
2. Shami, S. H., Kirkwood, I. M. A., Sinclair, M. C. : Evolving simple fault-tolerant routing rules using genetic programming. Electronic Letters, Vol. 33 n°17 (1997) 1440-1441
3. Pujolle, G. : Les Réseaux. Editions Eyrolles (2000)
4. Comer, D. : TCP/IP – Architecture, Protocoles, Applications. Editions Masson, Paris (1998)
5. ITU-T E.350 : Dynamic Routing Interworking (March 2000)
6. Costa, L. H., Fdida, S., Duarte, O. C. : Un algorithme résistant au facteur d'échelle pour le routage avec trois métriques de QoS. In: Colloque Francophone sur l'ingénierie des Protocoles CIFP'2000. Toulouse (October 2000)
7. Ma, Q., Steenkiste, P. : Quality of service routing for traffic with performance guarantees. In: IFIP 5th International Workshop on Quality of Service. (May 1997) 115-126
8. ITU-T E.351 : Routing of multimedia connections across TDM-, ATM-, and IP-based networks (March 2000)
9. Goldberg, D. E. : Genetic Algorithms in Search, Optimisation and Machine Learning. Addison-Wesley, Reading MA (1989)
10. Munetomo, M. : Designing genetic algorithms for adaptive routing algorithms in the Internet. In: Proceedings GECCO'99 Workshop on Evolutionary Telecommunications: Past, Present and Future. Orlando USA (July 1999)

A QoS Routing Mechanism for Reducing the Routing Inaccuracy Effects*

Xavier Masip-Bruin, Sergio Sánchez-López,
Josep Solé-Pareta, Jordi Dominigo-Pascual

Departament d'Arquitectura de Computadors, Universitat Politècnica de Catalunya
C/Jordi Girona, 1-3 – 08034 Barcelona, Spain
{xmasip, sergio, pareta, jordid} @ac.upc.es

Abstract. In highly dynamic large IP/MPLS networks, when routing information includes not only topology information but also information to provide QoS, such as available link bandwidth, network state databases must be frequently updated. This updating process generates an important signaling overhead. Reducing this overhead implies having inaccurate routing information, which may cause both non-optimal path selection and a call-blocking increase. In order to avoid both effects in this paper we suggest a new QoS explicit routing mechanism called *BYPASS Based Routing* (BBR), which is based on bypassing those links along the selected path that potentially cannot cope with the traffic requirements. Routing algorithms derived from the proposed BBR mechanism reduce the call-blocking ratio without increasing the amount of routing control information.

1 Introduction

Emerging real time applications, such as video on demand, videoconferences or virtual reality, cannot be supported under the Internet conventional best effort model, due to both the variable delays in the queuing process and the problem of congestion. Before these applications can be deployed, the network has to be modified to support end-to-end QoS. *Multiprotocol Label Switching* (MPLS) [1] provides a fast-forwarding mechanism to route the traffic associated with each ingress-egress node pair by using labels, and can support QoS requirements if necessary. Nodes in an MPLS domain are named *Label Switching Routers* (LSR) and the path between an ingress-egress node pair is called *Label Switched Path* (LSP). In order to establish LSPs, MPLS networks use a signaling mechanism managed by the *Label Distribution Protocol* (LDP) [2], which allows the allocation and distribution of labels.

In the IP/MPLS context, the routing algorithms are extremely important. Traditionally, in IP networks, routing (OSPF, IS-IS) is done only considering network topology information, which is just updated due to either link/node failure or restoration. For

* This work was partially funded by the MCyT (Spanish Ministry of Science and Technology) under contract FEDER-TIC2002-04531-C04-02, and the CIRIT (Catalan Research Council) under contract 2001-SGR00226.

M. Ajmone Marsan et al. (Eds.): QoS-IP 2003, LNCS 2601, pp. 90-102, 2003.

QoS provision the routing algorithm must take into account more parameters than those related with topology and connectivity (QoS Routing). In the QoS Routing algorithms, the QoS attributes must be considered when selecting the path. This path selection has to be done according to the information contained in the *Traffic Engineering Database* (TED). Some examples of recently proposed QoS routing algorithms are the *Widest-Shortest Path* (WSP) [3], the *Shortest-Widest Path,* (SWP) [4], the *Minimum Interference Routing Algorithm,* (MIRA) [5], the *Profile-Based Routing,* (PBR) [6], and the *Maximum Delay-Weighted Capacity Routing Algorithm,* (MDWCRA) [7].

All the routing algorithms mentioned above rely on the accuracy of the information contained in the TEDs to optimize the path selection. However, many situations exist that lead to a certain degree of inaccuracy in the TED information. This uncertainty has negative effects in the path selection process, such as the increase of the call-blocking ratio. In order to avoid these effects, in this paper we suggest a new QoS explicit routing mechanism called *BYPASS Based Routing* (BBR), which is based on bypassing those links along the selected path that potentially cannot cope with the traffic requirements. The routing algorithms derived from the BBR mechanism reduce the call-blocking ratio without increasing the signaling overhead.

The remainder of this paper is organized as follows. In Section 2, the problem addressed in this paper is identified. Then, Section 3 presents the review of the related work and Section 4 describes our proposal. After that, Section 5 includes some performance evaluation results obtained by simulation. Finally, Section 6 concludes the paper.

2 Origins of Routing Inaccuracy and Scope of This Paper

The routing inaccuracy directly depends on the procedure used to update the TED information. Two different aspects, namely the number of nodes and links composing the network and the frequency at which the TED has to be updated, mainly affect this procedure.

In large networks the number of nodes and links able to generate updating messages must be limited. Such a limitation, clearly obtained in a hierarchical structure, as for example the PNNI [8], introduces an aggregation process that implies a loss of information. In fact, this aggregation process reduces the network to a logical network made of logical links and logical nodes in such a way that information about physical links and nodes is often lost. In this way, a certain routing inaccuracy is introduced in the link state information.

Concerning the frequency at which updating messages are sent throughout the network, in highly dynamic networks where link state changes are frequently expected, it is impractical to keep the network state databases correctly updated [9] due to the non desirable signaling overhead introduced by the large number of signaling messages needed. Such a signaling overhead is usually reduced by applying a certain triggering policy, which specifies when the nodes have to send a message to inform the network about changes in one or more of the links directly connected to them. In fact, a trade-off exists between the need of having accurate routing information and the need of reducing the number of updating messages. Three different triggering policies are described in [10], namely, *Threshold based policy*, *Equal class based policy* and *Exponential class based policy*. The first one is based on a threshold value (*tv*). Let b^i_r be the last advertised residual bandwidth of the link *i*, where the residual capacity of a link is defined as the difference between the total amount of bandwidth that this link can support and the bandwidth that is currently in use and b^i_{real} the current real residual bandwidth of that link. Then an updating message is triggered when

$$\frac{\left|b^i_r - b^i_{real}\right|}{b^i_r} > tv \ . \tag{1}$$

The other two policies are based on a link partitioning, in such a way that the total link capacity is divided into several classes. Being *Bw* a fixed bandwidth value, the *Equal class based policy* establishes its classes according to (0, Bw), (Bw, 2Bw), (2Bw, 3Bw), etc., and the *Exponential class based policy* according to (0, Bw), (Bw, (f+1)Bw), ((f+1)Bw, (f^2+f+1)Bw), etc., where *f* is a constant value. Then, an updating message is triggered when the link capacity variation implies a change of class.

This paper deals with the effects (call-blocking ratio increase and non-optimal path selection) produced in the path selection process when considering bottleneck requirements (e.g. bandwidth), and when the routing process relies on partial or inaccurate network state information. If the information contained in the network state databases is not perfectly updated, the routing algorithm could select a path unable to support the path incoming request, given that one or more links of that path could actually have less available resources than the specified by TED and that were required to set up that path. In this way, the incoming path request will be rejected in the setup process producing an increase of the call-blocking ratio. In this paper the inaccuracy introduced by the triggering policies is considered. Unfortunately this inaccuracy can be only defined for those triggering policies whose behavior can be perfectly modeled. This excludes the timer-based triggering policies of our study.

In summary, in this paper a new QoS explicit routing mechanism to improve the network performance for traffic flows with bandwidth requirements in a highly dynamic on-demand IP/MPLS environment is suggested. A simple IP/MPLS domain is considered, but in a previous work [11] a solution for computing and signaling explicit routes when two IP/MPLS domains are interconnected via an ATM backbone has been presented.

3 Review of Related Work

Several recent works exist in the literature addressing the problem of having inaccurate routing information when selecting a path. Documents [12-14] deal with finding the path that maximizes the probability of supporting the incoming traffic requirements. Based on this idea, R.Guerin and A.Orda [12] present different proposals for reducing the routing inaccuracy effects depending on the QoS constraint required by the incoming traffic. On one hand, in order to solve the problem for flows with bandwidth requirements authors suggest applying a modified version of the standard *Most Reliable Path* (MRP). On the other hand, when the objective is to compute an end-to-end delay bound, authors present two different approaches to deal with this problem, i.e. the rate-based approach and the delay-based approach and different solutions are generated for each model. The first approach has the advantage that once the delay is mathematically represented, the end-to-end delay bound only depends on the available bandwidth on each link. The second approach has the disadvantage that tractable solutions can be only applied for relatively limited cases. Nevertheless, authors introduce a simplification based on splitting the end-to-end delay guarantee to a several minor problems that extends the cases where solutions can be applied. In [13], D.Lorenz and A.Orda try to solve the problem of selecting an optimal path that guarantees a bounded delay by searching for the path most likely to satisfy this QoS requirement, namely the *problem MP (Most Probable Path)*. As in the solution presented in [12] for the delay-based approach, the complexity of this problem is reduced after splitting the end-to-end delay constraint in several local end-to-end delay constraints. How this partition is done and the optimization of this partition is analyzed as the *Optimally Partitioned MP Problem (OP-MP)*. Solutions based on using programming dynamics methods are presented to address the *problem OP-MP*. Also searching for the most likely path in [14] G.Apostopoulos et al, propose a new routing mechanism named *Safety Based Routing* (SBR), to address the routing inaccuracy effects when computing explicit paths with bandwidth constraints. The SBR is based on computing the *safety* (S), a new link attribute that is incorporated to the path selection process, which represents the effects of the routing inaccuracy in the link state reliability. The SBR can only be implemented when the performance and characteristics of the triggering policies that generate the routing inaccuracy are well known. Two algorithms inferred from the SBR mechanism are proposed in the document, the *Shortest-Safest Path* and the *Safest-Shortest Path*. The first selects the shortest path among the path that minimize the *safety* parameter and the second algorithm selects the path that minimizes the *safety* value among the shortest paths. Obtained simulation results show a lower bandwidth-blocking ratio when the *Shortest-Safest Path* (SSP) is the routing algorithm in use.

Another work was presented by S.Chen and K.Nahrstedt in [15]. Although the routing mechanism deals with the NP-complete delay-constraint least-cost routing problem, authors ensures that it can be perfectly applied to the bandwidth-constrained least-cost routing as well. Unlike other mechanisms it is not based on computing the path able to support the traffic requirements with a larger probability but rather a new multipath distributed routing scheme named *Ticket based probing* (TBP) is implemented. The TBP defines the imprecise state model by defining which information must be stored in every node, and then sends routing messages named *probes*, from the source node to the destination node to find the low cost path that fulfills the delay requirements. Obtained simulation results show that the TBP achieves high success ratio and low-cost feasible path with minor overhead.

Finally, in [16] the problem of selecting the most likely path, named *Maximum Likely Path selection* (MLPS) is implemented in an analog computer and solved by using a Hopfield Neural Network routing algorithm. This method has an important cost on complexity that can be obviate due to the analog structure in use. However, authors pointed out that this method is useful when hierarchical routing is applied and as a consequence small networks exist.

The main difference between our proposal and the existing solutions is the routing behavior when, even applying any of these new algorithms a path that really cannot cope with the traffic bandwidth requirements is selected. This situation is managed differently in the routing mechanism proposed in this paper. In fact, unlike other algorithms that reject the incoming LSP, our solution tries to jump over those links that impede the end-to-end path establishment by using a different pre-computed path.

4 BYPASS Based Routing

In this Section we describe a new QoS routing mechanism, named *BYPASS-Based Routing* (BBR) aiming to reduce the routing inaccuracy effects, that is the increase of the call-blocking ratio and the non-optimal path selection, in an IP/MPLS scenario. The BBR mechanism is based on computing more than one feasible route to reach the destination. The BBR instructs the ingress node to compute both the working route and as many paths to bypass the links (named *bypass-paths*) that potentially cannot cope with the incoming traffic requirements. Nevertheless as we discuss below, only those paths that bypass links that truly lack enough available bandwidth to support the required bandwidth are set up.

Note that the idea of the BBR mechanism is derived from the protection switching for fast rerouting discussed in [17]. However, unlike the use of protection switching for fast rerouting, in our proposal both the working and the alternative path (*bypass-path*) are computed simultaneously but not set up, they are only set up when required. The key aspects of the BBR mechanism to decide which links should be bypassed and how the *bypass-paths* are computed are the following:

Obstruct-sensitive links: A new policy must be added in order to find those links (*obstruct-sensitive links, OSLs*) that could not support the traffic requirements associated with an incoming LSP demand. This policy should guarantee that whenever a path setup message sent along the explicit route reaches a link that has not enough residual bandwidth to support the required bandwidth, this link had been defined as *OSL*.

Working path selection: Using the BBR two different routing algorithms can be initially analyzed. These algorithms are obtained from the combination of the Dijkstra algorithm and the BBR mechanism. Therefore, two different strategies may be applied:

- The *Shortest-Obstruct-Sensitive Path* (SOSP), computing the shortest path among all the paths which have the minimum number of *obstruct-sensitive links*.
- The *Obstruct-Sensitive-Shortest Path* (OSSP), computing the path that minimizes the number of *obstruct-sensitive links* among all the shortest paths.

***Bypass-paths*, calculation and utilization:** Once the working path is selected the BBR computes the *bypass-paths* needed to bypass those links in the working path defined as *OSL*. When the working path and the *bypass-paths* are computed, the working path setup process starts. Thus, a signaling message is sent along the network following the explicit route included in the setup message. When a node detects that the link by which the traffic must be flown has not enough available bandwidth to support the required bandwidth, it sends the setup signaling message along the *bypass-path* that bypasses this link. Thus, the set of bypassed links must always be defined as *OSLs* so that a feasible *bypass-path* exists. Moreover, it is important to note that the *bypass-paths* nodes are included in the setup signaling message as well, i.e. *bypass-paths* are also explicitly routed. In order to minimize the setup message size, *bypass-paths* are removed from the setup message when passing the link that bypass.

4.1 Description of the BYPASS Based Routing

Let $G(N,L,B)$ describe a defined network, where N is the set of nodes, L the set of links and B the capacity bandwidth of the links. Suppose that a set of source-destination pairs (s,d) exists, named P, and that all the LSP requests occur between elements of P. Let b_{req} be the bandwidth requested in an element $(s,d) \in P$.

Rule 1. Let $G_r(N_r,L_r,B_r)$ represent the last advertised residual graph, where N_r,L_r and B_r are respectively the remaining nodes, links and residual bandwidths at the time of path setup. Let L^{os} be the set of *OSLs* (l^{os}), where l^{os} are found depending on the triggering policy in use. Therefore,

- Threshold policy: Let b^i_r be the last advertised residual bandwidth for a link l_i. This link l_i is defined as *OSL*, l^{os}_i if

$$l_i = l^{os}_i \mid l^{os}_i \in L^{os} \Leftrightarrow b_{req} \in (b^i_r(1\text{-}tv), b^i_r(1+tv)]\,. \quad \textbf{(2a)}$$

- Exponential class policy: Let $B^i_{l\_j}$ and $B^i_{u\_j}$ be the minimum and the maximum bandwidth values allocated to class j for a link l_i. So, l_i. is an *OSL*, l^{os}_i if

$$l_i = l^{os}_i \mid l^{os}_i \in L^{os} \Leftrightarrow b_{req} \in (B^i_{l\_j}, B^i_{u\_j}]\,. \quad \textbf{(2b)}$$

Rule 2. Let L^{os} be the set of *OSLs*. Let i_j and e_j be the edge nodes of a link $l^{os}_j \in L^{os}$. Let l_k be one link adjacent to l^{os}_j. The edge nodes of the *bypass-paths* to be computed are

$$(i_j, e_j) \quad \Leftrightarrow \quad l_k \notin L^{os} \quad \textbf{(3a)}$$

or

$$(i_j, e_k) \text{ and } (i_k, e_k) \quad \Leftrightarrow \quad l_k = l^{os}_k \in L^{os}\,. \quad \textbf{(3b)}$$

In this way two or more adjacent *OSLs* could be bypassed by a single *bypass-path*.

In accordance with these rules, in Fig. 1 a brief description of the BBR mechanism is presented. Steps 4 and 5 should be in detail explained. Once a link is defined as *OSL*, the BBR mechanism computes the *bypass-path* that bypasses this link. The *bypass-paths* are computed according to de *SOSP* performance, namely, the shortest path among those paths minimizing the number of *OSLs* is chosen. Other criteria could be used to select the *bypass-paths*, such as simply apply the *OSSP* performance or to maximize the residual available bandwidth. These different approaches are left for further studies.

BYPASS BASED ROUTING (BBR)

Input: The input graph $G_r(N_r,L_r,B_r)$. The *LSP* request is between a source-destination pair *(s,d)* and the bandwidth requirement is b_{req}.

Output: A route from *s* to *d* with enough *bypass-paths* to bypass the routing inaccuracy effects in the *obstruct-sensitive links*.

Algorithm:

1. Mark those links that are defined as *obstruct-sensitive link (OSL)* according to Rule 1.
2. Depend on the algorithm to be used, *OSSP, SOSP:*
   SOSP (shortest-obstruct-sensitive path):
   - Compute the weight of a link *l* as
     $$w(l) = 1 \Leftrightarrow l \in L^{os}, \qquad w(l) = 0 \Leftrightarrow l \notin L^{os}$$
   - Apply Dijkstra's algorithm to select the paths $p \in P$ that minimize the number of *OSLs* by using *w(l)* as the cost of each link
   - If more than one path *p* exists selects the path that minimizes the number of hops

   OSSP (obstruct-sensitive-shortest path):
   - Apply Dijkstra's algorithm to select the paths $p \in P$ that minimize the number of hops by using *w(l)* =1 as the cost of each link.
   - If more than one path *p* exists selects the path that minimizes the *OSLs*.
3. Compute a *bypass-path* for all the *OSLs* included in the selected path according to Rule 2.
4. Decide which *bypass- paths* must be used in accordance with real available resources in the path setup time.
5. Route the traffic from *s* to *d* along the setup path.

Fig. 1. BYPASS Based Routing Mechanism

Regarding the BBR complexity, two main contributors exist. On one hand selecting the shortest path by using a binary-heap implementation of the Dijkstra algorithm, introduces a cost of $O(L \cdot \log N)$. On the other hand, additional cost is introduced due to the *bypass-path* computation. Assuming that the *bypass-path* cannot include a network element which is also included in the working path, G(V, E) stands for the reduced network, where $V < N$ and $E < L$. Hence, a factor of $O(E \cdot \log V)$is added in order to compute one *bypass-path*. However, since a variable number M of *bypass-paths* may be computed along a working path, the cost is $O(M(E \cdot \log V))$. Being $\hat{M}$ an upper bound of the number of computed *bypass-paths* along a working path, the complexity reduces to $O(\hat{M}(E \cdot \log V))$, i.e., effectively to $O(E \cdot \log V)$. So, the complexity is $O(L \cdot \log N)+O(E \cdot \log V)$. This expression may be finally reduced if considering that the *bypass-paths* are computed based on a reduced graph. Therefore, the complexity is $O(L \cdot \log N)$.

4.2 Example for Illustrating the BBR Behavior

Before analyzing the suggested algorithms in a large topology, we can test the BBR performance in the simple topology shown in Fig. 2, which shows the residual network topology where the number associated to each link shows the residual available units of bandwidth.

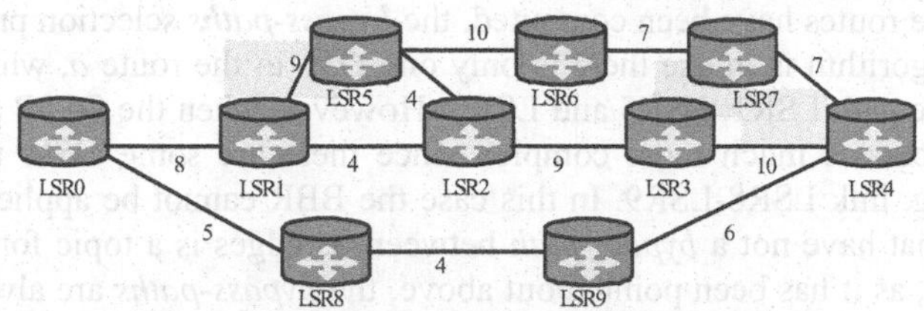

Fig. 2. Network topology used in the example

We suppose an *Exponential class* triggering policy with $f = 2$ and $Bw = 1$ (as used in [14]), in such a way that the resulting set of classes on each link are the following, (0,1], (1,3], (3,7], (7,15], etc. Moreover, we assume that an LSP incoming request is demanding b_{req} of 4 units of bandwidth between LSR0 and LSR4. In order to compare the BBR mechanism with other related work, the *Shortest-Safest Routing* algorithm presented in [14] is chosen as a sample of routing algorithm, which computes the path based on maximizing the *"probability of success"* of supporting the bandwidth requirements. Thus, the algorithms tested in this example are the SSP, WSP, SOSP, OSSP and the shortest path algorithm (SP) implemented in the OSPF routing protocol as a routing algorithm that does not consider the routing inaccuracy when selecting the path. Table 1 describes the link QoS parameters used to compute the path, where *B*, *Class* and *S* are the bandwidth, class and *safety* associated with each link. The *S* value has been computed according to the expressions found in [14]. Remind that *S* represents the probability that the requested amount of bandwidth is indeed available on a given link. Using this information the *BBR* mechanism is applied.

Table 2 shows different possible routes from LSR0 to LSR4 including the number of hops *H*, the number of *obstruct-sensitive links OSL*, the minimum last advertised residual bandwidth b_r^{min}, and the *safety* parameter *S*. As a result, different paths are selected depending on the algorithm in use, as it is shown in Table 2. Although the SOSP and the SSP algorithms select the same route, the key difference between both algorithms is the use of the *bypass-paths* when it is needed.

Table 1. Link QoS attributes

Link	B_t	*Class*	*S*	*Link*	B_t	*Class*	*S*	*Link*	B_t	*Class*	*S*
0-1	8	7,15	1	1-5	9	7,15	1	7-4	7	3,7	0,75
1-2	4	3,7	0,75	5-2	4	3,7	0,75	0-8	5	3,7	0,75
2-3	9	7,15	1	5-6	10	7,15	1	8-9	4	3,7	0,75
3-4	10	7,15	1	6-7	7	3,7	0,75	9-4	6	3,7	0,75

Table 2. Feasible routes and selected paths depending on the algorithm in use

Id	*Route (LSR)*	*H*	*OSL*	b_r^{min}	*S*
a	0-1-2-3-4	5	1	4	0.75
b	0-1-5-6-7-4	6	2	7	0.56
c	0-1-5-2-3-4	6	1	4	0.56
d	0-8-9-4	4	3	4	0.42

Alg	*Path*
SP	d
WSP	b
OSSP	d
SOSP	a
SSP	a

Once feasible routes have been computed, the *bypass-paths* selection process starts. If the SOSP algorithm is in use there is only one *OSL* in the route *a*, which can be bypassed by the path LSR1, LSR5 and LSR2. However, when the OSSP algorithm is in use, the process is much more complex since there are some *OSLs* that cannot be bypassed, e.g. link LSR8-LSR9. In this case the BBR cannot be applied. How to bypass *OSLs* that have not a *bypass-path* between its edges is a topic for further study. In this paper, as it has been pointed out above, the *bypass-paths* are always computed by minimizing the number of *OSLs*.

Finally, after computing the *bypass-paths*, a path setup message is sent along the working path. Each node checks the real available link bandwidth and depending on this value the setup message is sent through either the working or the *bypass-path*.

5 Performance Evaluation

In this section we compare by simulation the BBR algorithms introduced in this paper that is the SOSP and the OSSP algorithms, with the WSP and the SSP algorithms. We exclude the SWP due to its worse performance behavior shown in [18].

5.1 Performance Metrics

The parameters used to measure the algorithms behavior are the routing inaccuracy and the blocking ratio.

Routing Inaccuracy: This parameter represents the percentage of paths that have been incorrectly selected. It is defined as

$$\text{routing inaccuracy} = \frac{\text{number of paths incorrectly selected}}{\text{total number of requested paths}} . \tag{4}$$

A path can be incorrectly selected because of two factors. The first factor is the LSP request rejection when really there was a route with enough resources to support that demand. The second factor is the blocking of an LSP that initially was routed by the ingress node but, due to the insufficient bandwidth in an intermediate link, it is rejected.

Blocking Ratio: We use the bandwidth-blocking ratio defined as

$$\text{bandwidth blocking ratio} = \frac{\sum_{i \in \text{rej\_LSP}} \text{bandwidth}_i}{\sum_{i \in \text{tot\_LSP}} \text{bandwidth}_i} \tag{5}$$

where *rej_LSP* are the set of blocked demands and *tot_LSP* are the set of total requested LSP.

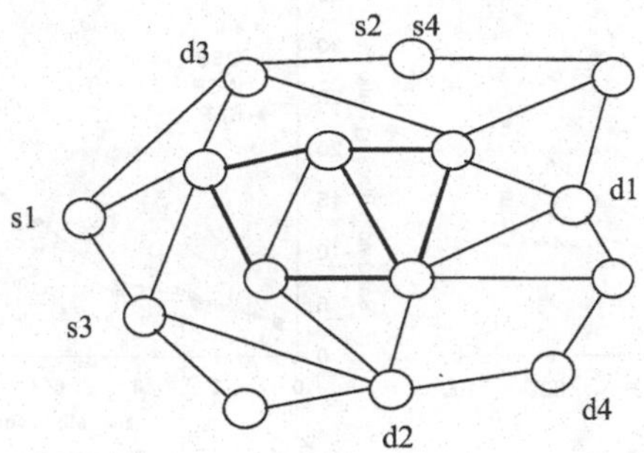

Fig. 3. Network topology used in simulations

5.2 Simulation Scenario

The simulations are performed over the network topology shown in Fig. 3, borrowed from [5], using the ns2 simulator extended with MPLS and BBR features. We use two link capacities, 622 Mb/s represented by a light line and 2.5 Gb/s represented by a dark line. The source nodes (s) and the destination nodes (d) are those shown in Fig. 3. Every simulation requests 1300 LSPs from s_i to d_i, which arrive following a Poisson distribution where the requested bandwidth is uniformly distributed between 1 Mb/s and 5 Mb/s. The holding time is randomly distributed with a mean of 60 sec. The *Threshold based triggering policy* and the *Exponential class based triggering policy* with f = 2, are implemented in our simulator.

5.3 Results

The results presented in this paper have been obtained after repeating the experiment 10 times, considering that every simulation lasts 259 sec. Fig. 4 shows the bandwidth-blocking ratio for the *Threshold* and the *Exponential class triggering* policies. The algorithms derived from the BBR mechanism (OSSP and SOSP) perform better than the WSP. In addition, while the OSSP presents similar results than the SSP, the SOSP substantially improves the SSP performance. Specifically, for the SOSP algorithm the Threshold value can be increased a 10% keeping the same bandwidth blocking ratio than the SSP.

Fig.5 represents the routing inaccuracy for both triggering policies. The SOSP algorithm presents the best behavior as well, that is, the SOSP is the algorithm that computes a lower number of incorrect routes.

Fig. 6 shows the cost of the BBR mechanism in terms of number of computed *by-pass-paths*. The figure shows that the cost is similar for both algorithms derived from the BBR mechanism. It reinforces the conclusion that the SOSP behaves better than the OSSP algorithm. The SSP and the WSP do not incur in the cost depicted in Fig. 5. Nevertheless, note that this cost is low given the benefits provided by the BBR mechanism shown in Fig. 4 and Fig. 5.

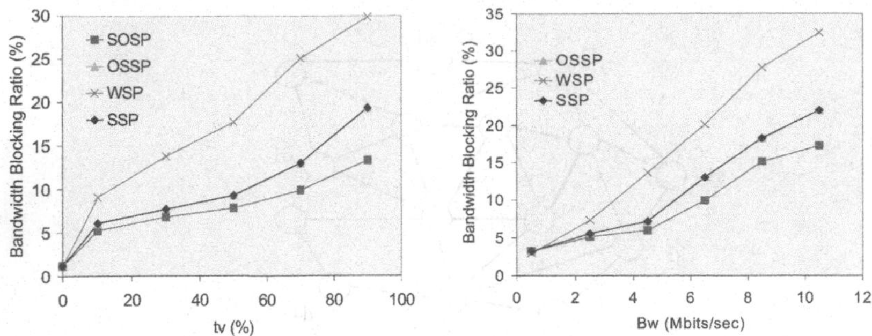

Fig. 4. Bandwidth Blocking Ratio for both Threshold and Exponential class triggering policies

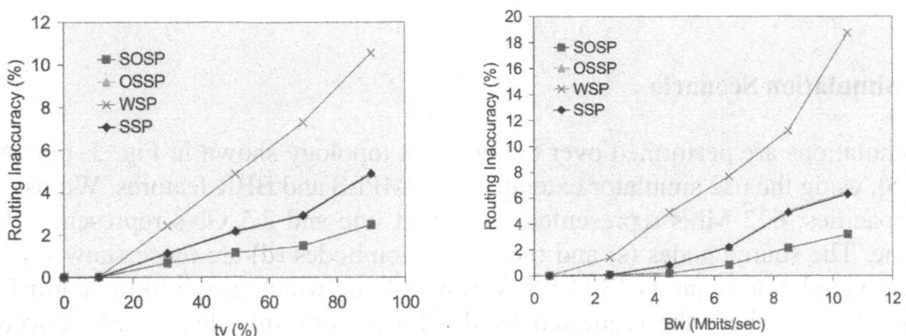

Fig. 5. Routing Inaccuracy for the Threshold and the Exponential class triggering policies

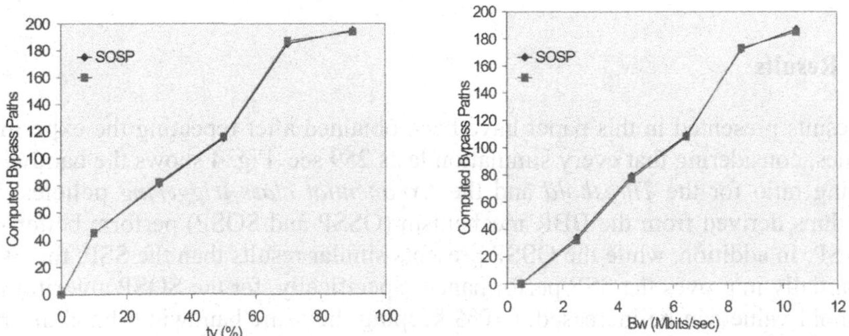

Fig. 6. Computed *bypass* paths for the Threshold and the Exponential class triggering policies

In summary, as a numeric example we take a *tv* value of 70 % and if we analyze the results provided by the BBR mechanism and those provided by the SSP algorithm, it is shown that the bandwidth blocking ratio presented by the SOSP (9.7 %) is substantially lower than that provided by the SSP (12.9 %). Regarding to the routing inaccuracy, the SOSP (1.49 %) presents a lower number of paths incorrectly selected when comparing to the SSP (2.91 %). In both cases the OSSP presents similar results than the SSP, and the WSP is that algorithm which presents the worst behavior. This is due to the fact that the WSP does not consider the routing inaccuracy when selecting the path. Finally, for a tv = 70 % the number of *bypass-paths* computed during the simulation for the BBR mechanism is close to 180 and almost the same for the SOSP and the OSSP algorithms. That means an overhead in computation of LSP about 14% but not in signaling as has been explained before.

6 Conclusions

In this paper a new QoS routing mechanism for establishing LSPs in an IP/MPLS network under inaccurate routing information has been suggested and its performance evaluated by simulation in comparison with the existing solution, the *Safety Based Routing* mechanism.

We called this new QoS routing mechanism, *BYPASS Based Routing* (BBR). The BBR minimizes the routing inaccuracy effects due to implementing a certain triggering policy to reduce the volume of updating messages. Basically, the main idea of BBR is to bypass those links that potentially are unable to support the traffic requirements associated with the incoming LSP request. These links are defined as *obstruct-sensitive links (OSL)* and a new mechanism is proposed to both define which links are to be *OSL* and find *bypass-paths* to bypass the *OSLs*.

Two algorithms are derived from combining BBR with the shortest path algorithm, namely the *Shortest-Obstruct-Sensitive* (SOSP) and the *Obstruct-Sensitive-Shortest* (OSSP). The simulation results obtained when comparing these BBR algorithms with the SSP and the WSP algorithms confirm the BBR effectiveness to improve the routing performance when the network state databases are not perfectly updated. In fact, the SOSP algorithm exhibits a lower bandwidth blocking ratio than the other tested routing algorithms, substantially improving the *Safety Based Routing* behavior. This improvement is achieved without incrementing the use of resources since the *bypass-paths* are established only when they are needed at the time of setup.

References

1. E. Rosen, A. Viswanathan, R. Callon, "Multiprotocol Label Switching Architecture", RFC3031, January 2001.
2. L.Anderson, P.Doolan, N.Feldman, A.Fredette, B.Thomas, "LDP Specification", RFC3036, Jan.2001.
3. R. Guerin, A. Orda, D. Williams, "QoS Routing Mechanisms and OSPF Extensions", In Proceedings of 2nd Global Internet Miniconference (joint with Globecom'97), Phoenix, AZ, November 1997
4. Z. Wang, J. Crowcroft, "Quality-of-Service Routing for Supporting Multimedia Applications", IEEE JSAC, 14(7): 1288-1234, September 1996
5. M. Kodialam, T.V. Lakshman, "Minimum Interference Routing with Applications to MPLS Traffic Engineering" In Proceedings of INFOCOM 2000.
6. S.Suri, M.Waldvogel, P.R.Warkhede, "Profile-Based Routing: A New Framework for MPLS Traffic Engineering".In Proceedings of QofIS 2001.
7. Y.Yang, J.K.Muppala, S.T.Chanson, "Quality of Service Routing Algorithms for Bandwidth-Delay Constrained Applications", in Proc. 9th IEEEICNP2001, Riverside, Nov. 11-14, 2001.

8. Private Network-Network Interface Specification v1.0 (PNNI), ATM Forum , March 1996
9. G. Apostolopoulos, R. Guerin and S. Kamat, " Implementation and performance measurements of QoS Routing Extensions to OSPF", in Proceedings INFOCOM, March 1999, pp. 680-688
10. G. Apostolopoulos, R. Guerin, S. Kamat, S. Tripathi, "Quality of Service Based routing: A Performance Perspective" In Proceedings of SIGCOMM 1998.
11. S.Sánchez, X.Masip, J.Domingo, J.Solé, J.López "A Path Establishment Approach in an MPLS-ATM Environment", In Proceedings of IEEE Globecom 2001.
12. R.A. Guerin, A.Orda, "QoS Routing in Networks with Inaccurate Information: Theory and Algorithms", IEEE/ACM Transactions on Networking, Vol.7, nº.3, pp. 350-364, June 1999
13. D.H.Lorenz, A.Orda, "QoS Routing in Networks with Uncertain Parameters", IEEE/ACM Transactions on Networking, Vol.6, nº.6, pp.768-778, December 1998.
14. G. Apostolopoulos, R. Guerin, S. Kamat, S.K. Tripathi, "Improving QoS Routing Performance Under Inaccurate Link State Information", In Proceedings of ITC'16, June 1999.
15. S.Chen, K.Nahrstedt, "Distributed QoS Routing with Imprecise Sdtate Information", in Proceedings of 7th IEEE International Conference of Computer, Communicati0ons and Nettworks, 1998
16. J.Levendovszky, A.Fancsali, C.Vegso, G.Retvari, "QoS Routing with Incomplete Information by Analog Computing Algorithms", in Proceedings QofIS, Coimbra, Portugal, September 2001.
17. T.M.Chen, T.H.Oh, "Reliable Services in MPLS", IEEE Communications Magazine, 1999 pp. 58-62
18. Q. Ma, P. Steenkiste, "On Path Selection for Traffic with Bandwidth Guarantees", In Proceedings of IEEE International Conference on Networks Protocols, October 1997.

Stability and Scalability Issues in Hop-by-Hop Class-Based Routing

Marília Curado, Orlando Reis, João Brito, Gonçalo Quadros, Edmundo Monteiro

Laboratory of Communications and Telematics
CISUC/DEI
University of Coimbra
Pólo II, Pinhal de Marrocos, 3030-290 Coimbra
Portugal
marilia@dei.uc.pt, {oreis, jbrito}@student.dei.uc.pt,
{quadros, edmundo}@dei.uc.pt

Abstract. An intra-domain Quality of Service (QoS) routing protocol for the Differentiated Services framework is being developed at the University of Coimbra (UC-QoSR). The main contribution of this paper is the evaluation of the scalability and stability characteristics of the protocol on an experimental test-bed. The control of protocol overhead is achieved through a hybrid approach of metrics quantification and threshold based diffusion of routing messages. The mechanisms to avoid instability are: (i) a class-pinning mechanism to control instability due to frequent path shifts; (ii) the classification of routing messages in the class of highest priority to avoid the loss of accuracy of routing information. The results show that a hop-by-hop, link-state routing protocol, like Open Shortest Path First, can be extended to efficiently support class-based QoS traffic differentiation. The evaluation shows that scalability and stability under high loads and a large number of flows is achieved on the UC-QoSR strategy.

1 Introduction

Quality of Service plays a major role in the deployment of communication system for applications with special traffic requirements, such as video-conferencing or Internet telephony. The need to support these types of traffic has motivated the communication research community to develop new approaches. Some of this work resulted in the Differentiated and Integrated Services architectures proposed by the Internet Engineering Task Force (IETF) [1, 2].

Current routing protocols used in the Internet lack characteristics for QoS provision to support emerging new services. All traffic between two endpoints is forwarded on the same path, even if there are other alternative paths with more interesting properties for the requirements of a specific flow or traffic class. Usually, the shortest path is selected, based on a single static metric, that does not reflect the availability of resources. In these situations, congestion easily occurs on the shortest path, with the corresponding degradation of traffic performance, despite the underutilization of network resources on alternative paths. This scenario has motivated the development of QoS aware routing protocols.

M. Ajmone Marsan et al. (Eds.): QoS-IP 2003, LNCS 2601, pp. 103-116, 2003.

The most significant developments on QoS routing are aimed at communication systems where traffic differentiation is done per flow, as in the Integrated Services [1]. The Differentiated Services framework does not explicitly incorporate QoS routing. It is, thus, essential to develop QoS routing protocols for networks where traffic differentiation is done per class. The Quality of Service Routing strategy of the University of Coimbra (UC-QoSR) was conceived to fulfill this purpose.

The UC-QoSR strategy selects the best path for each traffic class based on information about the congestion state of the network. This strategy extends the Open Shortest Path (OSPF) routing protocol [3] in order to select paths appropriate for all traffic classes as described in [4, 5].

A prototype of UC-QoSR was implemented over the GateD[1] platform, running on the FreeBSD operating system [4]. The description of the mechanisms introduced to allow for scalability and to avoid instability and the evaluation of its robustness are the main objectives of the present paper. The rest of the paper is organized as follows: Section 2 presents some related work; Section 3 describes the UC-QoSR strategy; test conditions and analysis of results are presented in Section 4; the main conclusions and issues to be addressed in future work are presented in Section 5.

2 Related Work

The issues concerning stability of congestion based routing have been addressed by several researchers. This problem becomes even more important in protocols for QoS routing.

The advertisement of quantified metrics, instead of the advertisement of instantaneous values, is a common approach to avoid the instability of dynamic routing protocols. The quantification can be done using a simple average [6], or using hysteresis mechanisms and thresholds [7].

Another methodology to avoid routing oscillations is to use load-balancing techniques, allowing for the utilization of multiple-paths from a source towards the same destination. A simple approach of load balancing is to use alternate paths when congestion rises, as in the algorithm Shortest Path First with Emergency Exits (SPF-EE) [8]. This strategy prevents the excessive congestion of the current path because it deviates traffic to an alternate path when congestion starts to rise, and thus avoids routing oscillations. As an alternative to shortest path algorithms, algorithms that provide multiple paths of unequal cost to the same destination were proposed by Vutukury and Garcia-Luna-Aceves [9]. The algorithm proposed by these authors finds near-optimal multiple paths for the same destination based on a delay metric.

Even tough the proposals described above permit load balancing and avoid routing oscillations, they do not take into consideration the requirements of the different types of traffic. This problem has been addressed by several proposals within the connection oriented context. Nahrstedt and Chen conceived a combination of routing and scheduling algorithms to address the coexistence of QoS and best-effort traffic flows. In their approach, traffic with QoS guarantees is deviated from paths congested with

[1] <http://www.gated.org>

best-effort traffic in order to guarantee the QoS requirements of QoS flows and to avoid resource starvation of best-effort flows. Another routing strategy that addresses inter-class resource sharing was proposed by Ma and Steenkiste [10, 11]. Their strategy comprises two algorithms: one to route best-effort traffic and the other to route QoS traffic. The routing decisions are based on a metric that enables dynamic bandwidth sharing between traffic classes, particularly, sending QoS traffic through links that are less-congested with best-effort traffic. Although these proposals achieve active load balancing, they use source routing algorithms that do not reflect the actual hop-by-hop Internet routing paradigm and thus are not able to adapt dynamically to changes in network.

Routing stability can be achieved using some other mechanisms, like route pinning and doing load sensitive routing at the flow level [12]. However these approaches are not suitable for a situation where routing is done hop-by-hop and there is not connection establishment phase.

The development of hop-by-hop QoS routing protocols for communication systems where traffic differentiation is made per class has been the subject of recent studies. Van Mieghem *et al.* evaluated the impact of using an exact QoS routing algorithm in the hop-by-hop routing context [13]. These authors showed that such an algorithm is loop-free, however it may not find the exact solution. To solve this problem, the use of active networking is proposed. An algorithm called Enhanced Bandwidth-inversion Shortest-Path (EBSP) has been proposed for hop-by-hop QoS routing in Differentiated Services networks [14]. This algorithm is based on a widest-shortest path algorithm that takes into account the hop count. The hop count is included in the cost function in order to avoid oscillations due to the increased number of flows sent over the widest-path. Although the algorithm EBSP selects the best path for Premium-class traffic, it does not consider other traffic classes.

QoS routing introduces additional burden in the network, pertaining to the processing overhead due to more complex and frequent computations and the increased routing protocol overhead. The trade-off between the cost of QoS routing and its performance was evaluated in some works [15, 16]. The results included in these references are applicable to systems where there is a flow establishment phase. Furthermore, the above references do not address QoS routing scalability in terms of number of flows and packet sizes, treating only the different rates of arrival of flows and bandwidth requirements.

Despite the relevant QoS issues addressed, the proposals for QoS routing analyzed lack the analysis of the applicability to a class-based framework and are only evaluated theoretically or by simulation. The use of a prototype approach limits the dimension of the test-bed, however it introduces processing and communication systems dynamics, being closest to a real situation.

3 UC-QoSR Strategy

In this section the main characteristics of the routing strategy UC-QoSR are briefly described. A more detailed description can be found in previous publications of the authors [4, 5]. The mechanisms used to allow for scalability and to control instability, which consist the main objectives of the present work, are presented in detail.

3.1 UC-QoSR System Model

The UC-QoSR strategy was designed for intra-domain hop-by-hop QoS routing in networks where traffic differentiation follows the class paradigm. This strategy is composed of three main components, as follows:

a) A QoS metric that represents the availability of resources in the network;
b) Traffic class requirements in terms of QoS parameters;
c) A path computation algorithm to calculate the most suitable path for each traffic class, according to the dynamic state of the network expressed by a QoS metric.

The availability of resources in the network is measured through a QoS metric that represents the congestion state of routers interfaces. This metric consists of two congestion indexes, one relative to packet delay (*DcI*) and other to packet loss (*LcI*). These indexes evaluate the impact that delay and loss at the router, will have on application performance [17]. The congestion indexes are distributed to all routers in the domain through modified OSPF routing messages (Router Link State Advertisements – R-LSA).

The UC-QoSR strategy was conceived for communication systems where traffic characterization is based on class sensitivity to delay and loss. Currently, four classes are considered with different delay and loss sensitivities.

The problem of QoS routing when using two additive or multiplicative metrics, or one additive and one multiplicative metrics is a NP-complete problem [18, 19]. Thus, since the congestion indexes are additive metrics, the selection of a path that minimizes both congestion indexes is a NP-complete problem. However, due to their nature, the indexes represent comparable measures, and can be combined in a single metric without loss of information from aggregation of different kinds of units.

The path computation algorithm uses a cost function that combines both congestion indexes, weighted according to delay (δ) and loss sensitivity (λ) of each traffic class. The cost of link l for class i ($c_{l,i}$) results from the combination of the loss congestion index of link l (l_l) and the delay congestion index of link l (l_l), according to:

$$c_{l,i} = \delta_i d_l + \lambda_i l_l \,. \tag{1}$$

The merging of the congestion indexes origins a value that represents the congestion state of the interface, as it is perceived by traffic belonging to each class. The Dijkstra algorithm is then used to compute the shortest path tree for each traffic class. The UC-QoSR strategy remains fully compatible with original OSPF because the path selection algorithm is not altered, and because the OSPF configured costs are also advertised in R-LSAs. It is thus possible to establish adjacencies among routers running UC-QoSR and OSPF.

3.2 Mechanisms for Scalability

QoS routing protocols must contribute to a significant improvement in traffic performance and network resource usage to compensate for the burden they introduce on the network. This overhead is twofold, comprising an increase in the communication load due to routing traffic and a raise in the processing capacity of routers caused by the frequency of path computations. In UC-QoSR, these overheads are controlled by a policy that controls the emission of link state updates. This policy combines metrics quantification and threshold based diffusion. A similar approach was followed by Apostolopoulos *et al.* but in flow establishment context [15].

The quantification rule is a moving average of the congestion indexes, with a variable window size (*N*). The congestion indexes are monitored every second (the lowest time granularity provided by GateD) and the samples are taken continuously. In Equation 2, $MA_d(k)$ is the moving average of *N* values of the delay congestion indexes at sample *k*. This function is used to filter the peaks of the QoS metric.

$$MA_d(k) = \sum_{i=k-N}^{k} \frac{d(i)}{N}. \tag{2}$$

The filtered values, resulting from the application of Equation 2, are then presented to the diffusion control module. In this module, the new value is compared with the one that was previously advertised, and will be diffused only if it significantly different. The decision to issue the advertisements is controlled by the value of a defined threshold.

Besides the link state update policy described above, in UC-QoSR, OSPF was modified, in order to control even further the protocol overhead and thus increase the possibility of scalability. In original OSPF, the routing messages denominated Network-LSA (N-LSA) identify the routers connected to the network and its diffusion occurs wherever R-LSAs are issued. In the UC-QoSR strategy, the emission of N-LSAs has been detached from the emission of R-LSAs, because R-LSAs are issued at a higher rate than in OSPF and the information transported in N-LSAs does not change at such a rate. Thus, in the UC-QoSR, the emission of N-LSA remains periodic and dependent on router connectivity, while the emission of R-LSA is controlled through the threshold of the diffusion module. This strategy allows for a significant reduction of routing messages in the network.

The policy to control protocol overhead described above contributes also to avoid the number of path shifts that may occur in the network. Combined with these procedures, the UC-QoSR strategy uses a mechanism named class-pinning, that controls the path shifting frequency of all traffic classes.

3.3 Mechanism of Class-Pinning

The main role of QoS routing is to dynamically select paths based on information about the state of the network. Therefore, they enable the avoidance of congested paths, contributing to the improvement of application performance. However, the dynamic selection of paths may cause routing instability and network oscillatory behavior. This will naturally degrade application performance. In face of this scenario it is necessary to achieve a compromise between the desired adaptability of the protocol and the unwanted instability [20, 21].

In this work a mechanism of class-pinning to avoid instability is proposed. This mechanism addresses the stability problem described above by controlling the instant when a traffic class shifts to a new path.

When the state of the network changes (due to an event like the start of a new flow or a traffic burst) routing messages are sent to all routers, and new paths are computed. After the calculation, traffic will shift to the less congested paths, leaving the paths currently used. The next time this process occurs, traffic will eventually go back to the original path, and, thus, instability happens.

With the class-pinning mechanism, new paths are computed upon the arrival of routing messages. However, they will be used only if they are significantly better than the path that is currently used by that class. The *Degree of Significance* (DS) parameter is used to support the pinning decision. This parameter establishes the threshold for path shift from the old to the new path.

When the routing information about the state of the network is outdated, bad routing decisions can be made, and congestion may rise. Routing information may be outdated because it is not distributed with enough frequency or because it does not arrive due to congestion. The first cause can be avoided with the appropriate tuning of threshold used in the diffusing module. In order to avoid the delay or loss of routing information, the UC-QoSR routing messages are classified in the class with higher priority. The importance of the priority treatment of routing messages was shown by Shaikh *et al.* [20].

4 Experimentation

In this section the experimentation made to evaluate the stability and scalability of UC-QoSR are presented and its results are analyzed.

4.1 Test Conditions

The test-bed used for the experiments presented in this section is depicted in Figure 1. The *endpoints* 1 to 4 are traffic sources and *endpoints* 5 to 8 are traffic destinations. Each endpoint only generates or receives traffic of a single class to avoid the influence of endpoint processing on traffic patterns. Traffic was generated and measured with the traffic analysis tool Chariot from NetIQ[2].

The routers are PCs with the FreeBSD operating system. The kernel is modified, at the IP level, to include the delay and loss metric modules and to schedule and drop packets according to class sensitivity to these parameters. The monitoring of the delay and loss congestion indexes is needed for the routing decision. The kernel is also modified to interact with the UC-QoSR protocol embedded in GateD. It keeps the routing table with paths for all traffic classes and makes packet forwarding decisions based on destination IP address and Differentiated Services Code Point (DSCP) [2].

The interfaces between endpoints and routers are configured at 100 Mbps. Interfaces between routers are configured at 10 Mbps to introduce bottlenecks. In the results presented, the moving average window size is 80 samples and the threshold that controls the diffusion of R-LSAs is 30%. These values resulted from the tuning that was done by extensive experimentation with combinations of configurations [5].

[2] <http://www.netiq.com>

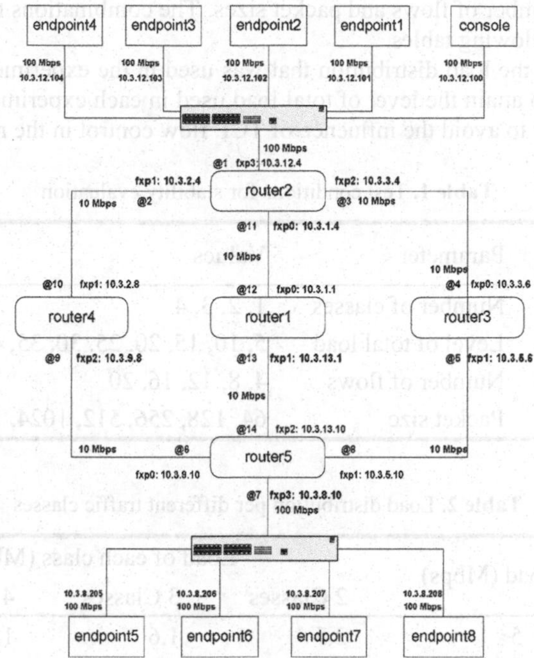

Fig. 1. Experimental test-bed

The evaluation of the UC-QoSR stability and scalability capabilities was done at protocol behavior and traffic performance levels. At the protocol level, the indicators used to measure the degree of stability and scalability of the UC-QoSR strategy are the parameters that represent the dynamics of the protocol, as follows:

a) Number of routing messages issued (Router-LSA);
b) Number of routing messages received (Router and Network[3] LSA);
c) Number of times the Shortest Path First (SPF) algorithm is applied;
d) Number of path shifts.

Each experiment was carried out for five minutes and was repeated ten times. The results present the averaged values of all tests. The inspection of protocol dynamics was done in all routers using the OSPF-Monitor tool included in GateD. The evaluation of traffic performance was made according to throughput and loss rate of all active traffic classes. These values were measured by the application Chariot. The plotted results have a degree of confidence of 95%.

4.2 Scalability Evaluation

The evaluation of the scalability of the UC-QoSR strategy was done considering different combinations of traffic characteristics namely, number of traffic classes, levels

[3] The measure used is the total number of LSAs received, that is, the number of Router and Network LSAs received.

of traffic load, number of flows and packet sizes. The combinations that were used are depicted in the following tables.

Table 2 shows the load distribution that was used in the experiments with 2, 3 and 4 traffic classes to attain the level of total load used in each experiment. The traffic of all classes is UDP to avoid the influence of TCP flow control in the results.

Table 1. Test conditions for stability evaluation

Set of tests	Parameter	Values
A	Number of classes	1, 2, 3, 4
	Level of total load	5, 10, 15, 20, 25, 30, 35, 40 Mbps
B	Number of flows	4, 8, 12, 16, 20
	Packet size	64, 128, 256, 512, 1024, 1460 byte

Table 2. Load distribution per different traffic classes

Total load (Mbps)	Load of each class (Mbps)		
	2 Classes	3 Classes	4 Classes
5	2,5	1,6	1,25
10	5	3,3	2,5
15	7,5	5	3,75
20	10	6,6	5
25	12,5	8,3	6,25
30	15	10	7,5
35	17,5	11,6	8,75
40	20	13,3	10

The results of the evaluation of the impact of the number of classes and level of load on the scalability of the UC-QoSR strategy are presented in Figures 2 and 3. The results shown are measured at *router 2*, since this is the router where exists the most significant bottleneck. These figures show that the load on the network only influences the indicators used for scalability evaluation for loads over 20 Mbps. This stems from the fact that congestion starts to occur in the routers while they adapt the paths to the traffic that is being generated. In the case where there are two traffic classes, each class is generated at 10 Mbps (Table 2). When four traffic classes are generated, each at 5 Mbps, two of those classes must share one path, since there are only three alternative paths in the test-bed. This setting clearly creates congestion, since all the capacity of the interfaces is used. However, the impact of higher loads is not significant, showing that the mechanisms used in UC-QoSR allow for scalability, controlling the rate of routing information diffusion and path calculation.

The application of the smoothing moving average filter to the QoS metric and the utilization of the threshold to control the emission of routing messages limit the number of R-LSA that the router emits. The control over the emission rate of R-LSA will naturally influence the other protocol indicators, the number of SPF and the number of path changes.

The comparison of figures 2 and 3 also shows that the impact of the number of classes is not significant in protocol dynamics. There is only a slight increase in the number of LSA received by *router 2* and, consequently, on the number of SPF calculated. This is, in fact, what was expected, since with four traffic classes all alternative paths are used. In these circumstances, all routers experience traffic changes, issuing the adequate routing messages to distribute their state.

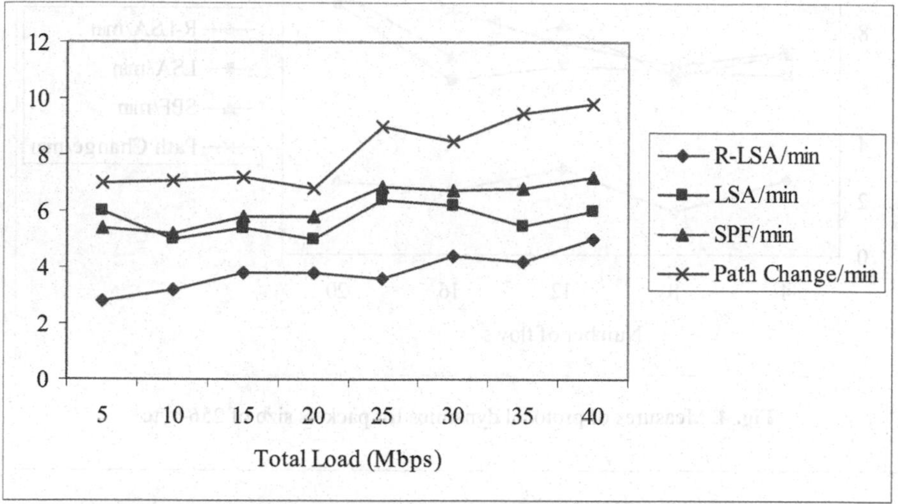

Fig. 2. Measures of protocol dynamics for different levels of traffic load (2 traffic classes)

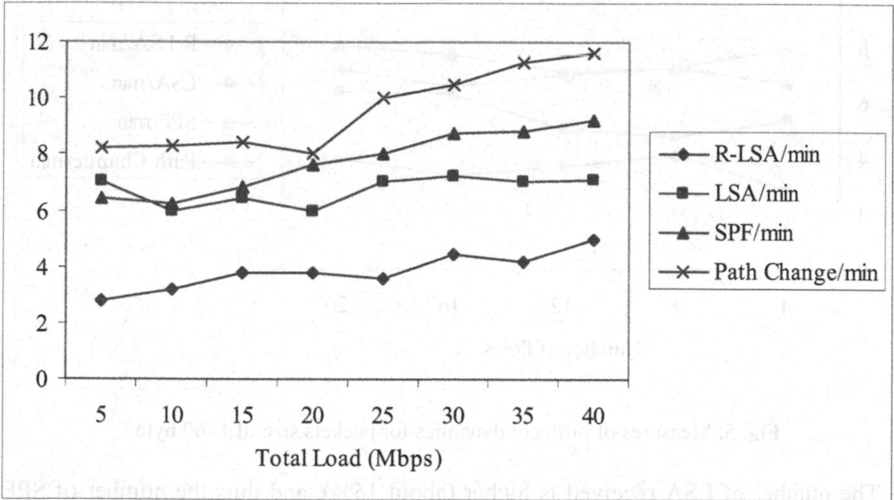

Fig. 3. Measures of protocol dynamics for different levels of traffic load (4 traffic classes)

In the evaluation of the scalability of the UC-QoSR strategy according to the number of flows and packet size, were generated four traffic classes and the traffic load was kept constant at 20 Mbps. The results are shown in Figures 4 and 5. Figure 4 shows that the indicators of protocol dynamics increase with the number of flows.

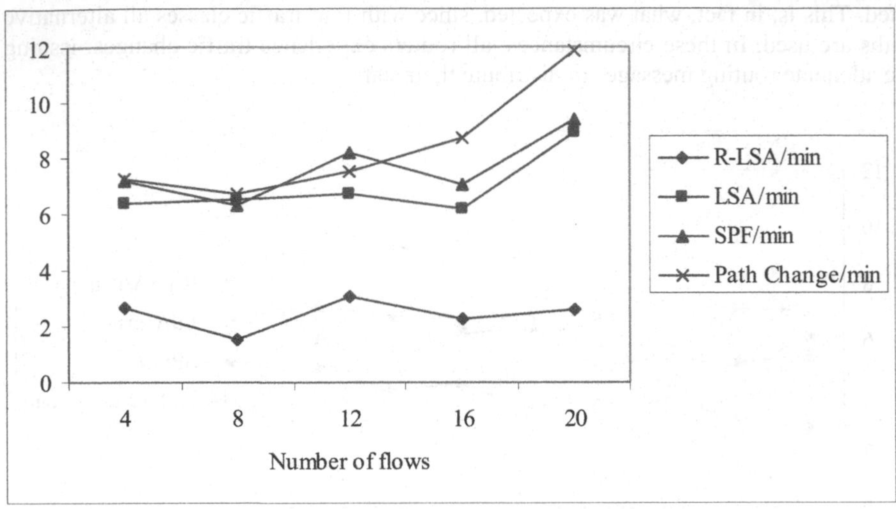

Fig. 4. Measures of protocol dynamics for packets size of 256 byte

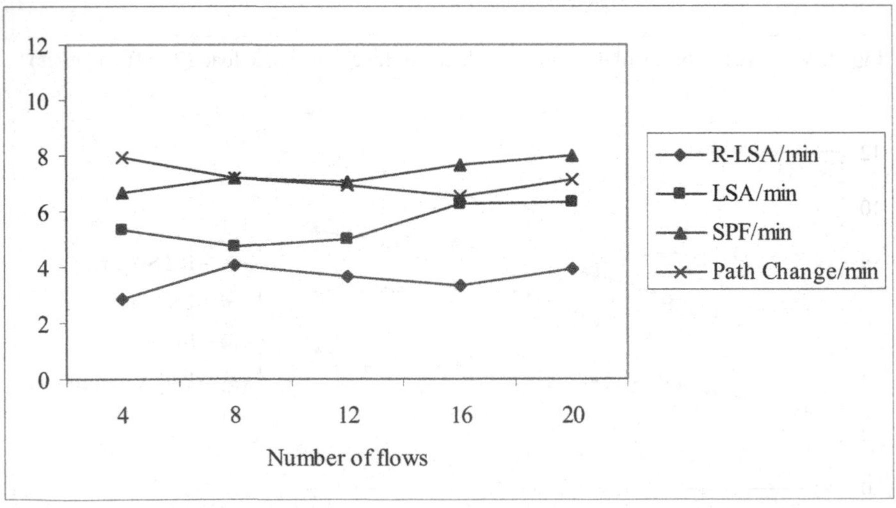

Fig. 5. Measures of protocol dynamics for packets size of 1460 byte

The number of LSA received is higher (about 15%), and thus the number of SPF also increases. The raise in the number of path shifts is due to the adaptation process of the flows in each class. For packets of 512 byte and over, the relationship between the number of flows and the indicators of protocol dynamics does not have the same nature (Figure 5). In this case, there is only an increase in the LSA received and the number of SPF. The number of path shifts is not affected by the number of flows. These results show that the UC-QoSR strategy has good scaling capabilities under the presence of an increasing number of flows, unless packets are too small. However, this problem can be overcome if faster processors are used.

Another conclusion from the observation of the results above is that the UC-QoSR strategy achieves load balancing over alternate paths. This capability is achieved through class-based routing and avoids sudden traffic fluctuations and the corresponding routing oscillations. However, this has shown insufficient to avoid oscillations in the presence of high levels of congestion, leading to the proposal of class-pinning included in the next sub-section.

4.3 Stability Evaluation

In the evaluation of the stability of the UC-QoSR strategy four traffic classes were generated. In order to evaluate the degree of stability achieved by the class-pinning mechanism, the *Degree of Significance* parameter was varied from 0% to 50%. The indicator used to evaluate stability at the protocol level was the number of path shifts as in [22].

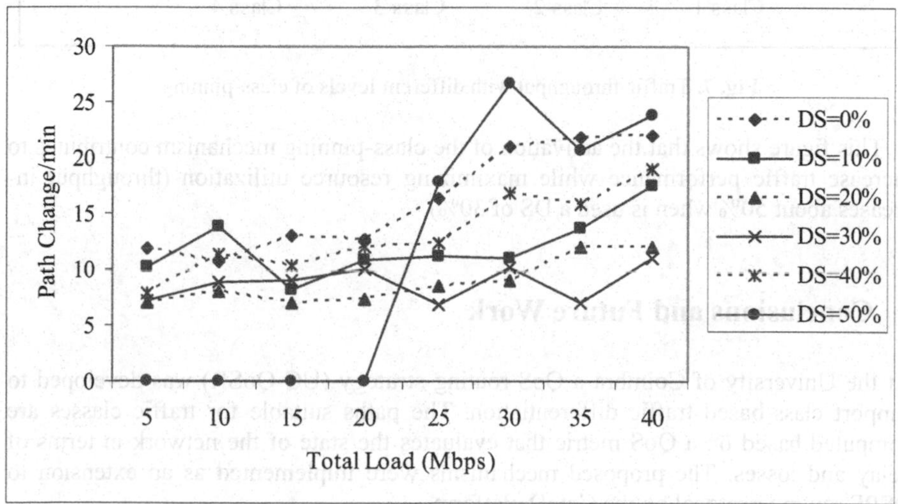

Fig. 6. Evaluation of the class-pinning mechanism

Figure 6 shows the results of the experiments concerning stability evaluation under different traffic loads. The level of the load introduced in the network has a direct impact in the number of path shifts. However, this impact is reduced for DS values of 20% and 30%, showing the effectiveness of the class-pinning mechanism. It is interesting to remark that the use of strong class-pinning, under higher load, introduces instability in the network, showing results similar to those obtained when the mechanism is not active. This fact is originated by the exaggerated postponement of path change. When the difference in path cost is sufficient to shift traffic, it is high and all traffic classes tend to shift at the same time to the same path, generating instability.

The throughput of traffic classes under two different values of DS is deployed in Figure 7. Traffic of classes 1 and 2 was generated at 6 Mbps and traffic of classes 3 and 4 at 4 Mbps.

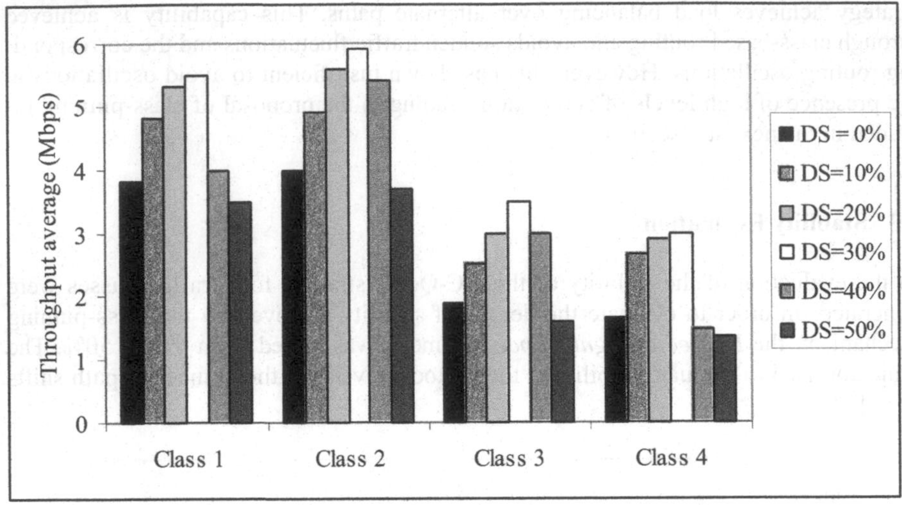

Fig. 7. Traffic throughput with different levels of class-pinning

This figure shows that the activation of the class-pinning mechanism contributes to increase traffic performance while maximizing resource utilization (throughput increases about 50% when is used a DS of 30%).

5 Conclusions and Future Work

At the University of Coimbra a QoS routing strategy (UC-QoSR) was developed to support class-based traffic differentiation. The paths suitable for traffic classes are computed based on a QoS metric that evaluates the state of the network in terms of delay and losses. The proposed mechanisms were implemented as an extension to OSPF routing protocol on the GateD platform.

Previous experiments with the UC-QoSR showed that the overhead introduced was affordable by the communication system and traffic differentiation was achieved. However some instability was noticed when the network was congested. To overcome this instability, a set of mechanism where conceived to integrate the UC-QoSR. Besides instability, scalability is other important issue in QoS routing. The focus of this paper was the proposal and evaluation of stability and scalability mechanisms to integrate the UC-QoSR strategy.

The results showed that the UC-QoSR strategy with the proposed mechanisms scales well under heavy loads. The impact of the number of flows is only significant when packets are small. The experiments also showed that the use of the class-pining mechanism significantly improves stability and traffic performance. However, for strong pinning, the behavior is worst due to the difference between paths.

Acknowledgements

This work was partially supported by the Portuguese Ministry of Science and Technology (MCT), under program POSI (Project QoS II and IPQoS) and under the PhD grant PRAXIS XXI/ BD/13723/97.

References

[1] R. Braden, D. Clark, S. Shenker, "Integrated Services in the Internet Architecture: an Overview", Request For Comments 1633, Internet Engineering Task Force, June 1994.

[2] S. Blake, D. Black, M. Carlson, E. Davies Nortel, W. Weiss, "An Architecture for Differentiated Services", Internet Engineering Task Force, Request for Comments 2475, December 1998.

[3] J. Moy, "OSPF Version 2", Internet Engineering Task Force, Request For Comments 2328, April 1998.

[4] M. Oliveira, J. Brito, B. Melo, G. Quadros, E. Monteiro, "Quality of Service Routing in the Differentiated Services Framework", Proceedings of SPIE's International Symposium on Voice, Video, and Data Communications (Internet III: Quality of Service and Future Directions), Boston, Massachusetts, USA, November 5-8, 2000.

[5] M. Oliveira, J. Brito, B. Melo, G. Quadros, E. Monteiro, "Evaluation of a Quality of Service Routing Strategy for the Differentiated Services Framework", Proceedings of the 2001 International Conference on Internet Computing (IC'2001), Monte Carlo Resort, Las Vegas, Nevada, USA, June 25-28, 2001.

[6] A. Khanna, J. Zinky, "The Revised ARPANET Routing Metric", Proceedings of SIGCOMM'89, Austin, Texas, September 19-22, 1989

[7] R. Guérin, S. Kamat, A. Orda, T. Przygienda, D. Williams, "QoS Routing Mechanisms and OSPF Extensions", Internet Engineering Task Force, Request For Comments 2676, August 1999.

[8] Z. Wang, J. Crowcroft, "Shortest Path First with Emergency Exits", Proceedings of SIGCOMM'90, Philadelphia, USA, September 1990.

[9] S. Vutukury and J.J. Garcia-Luna-Aceves, "A Simple Approximation to Minimum-Delay Routing", Proceedings of ACM SIGCOMM'99, Harvard University Science Center, Cambridge, Massachusetts, USA, 31 August-3 September, 1999.

[10] K. Nahrstedt, S. Chen, "Coexistence of QoS and Best Effort Flows - Routing and Scheduling", Proceedings of 10th IEEE Tyrrhenian International Workshop on Digital Communications: Multimedia Communications, Ischia, Italy, September, 1998.

[11] Q. Ma, P. Steenkiste, "Supporting Dynamic Inter-Class Resource Sharing: A Multi-Class QoS Routing Algorithm", Proceedings of IEEE INFOCOM'99, New York, USA, March 1999.

[12] A. Shaikh, J. Rexford, K. Shin, "Load-Sensitive Routing of Long-Lived IP Flows", Proceedings of ACM SIGCOMM'99, Harvard University Science Center, Cambridge, Massachusetts, USA, August 31-September 3, 1999.

[13] P. Van Mieghem, H. De Neve and F. Kuipers, "Hop-by-hop Quality of Service Routing", Computer Networks, vol. 37. No 3-4, pp. 407-423, 2000.

[14] J. Wang, K. Nahrstedt, "Hop-by-Hop Routing Algorithms for Premium-class Traffic in DiffServ Networks", Proceedings of IEEE INFOCOM 2002, New York, NY, June, 2002.

[15] G. Apostolopoulos, R. Guerin, S. Kamat, and S. Tripathi. "Quality of Service Based Routing: A Performance Perspective", Proceedings of SIGCOMM'98, Vancouver, British Columbia, September 1998.

[16] B. Lekovic, P. Van Mieghem, "Link State Update Policies for Quality of Service Routing", Proceedings of Eighth IEEE Symposium on Communications and Vehicular Technology in the Benelux (SCVT2001), Delft, The Netherlands, October 18, 2001.

[17] G. Quadros, A. Alves, E. Monteiro, F. Boavida, "An Approach to Support Traffic Classes in IP Networks", Proceedings of QoFIS'2000 – The First International Workshop on Quality of future Internet Services, Berlin, Germany, September 25-26, 2000.

[18] Z. Wang, J. Crowcroft, "Quality of Service Routing for Supporting Multimedia Applications", IEEEJSAC, , vol. 14. No 7, pp. 1228-1234, September 1996.

[19] H. De Neve, P. Van Mieghem, "TAMCRA: A Tunable Accuracy Multiple Constraints Routing Algorithm", Computer Communications, 2002, Vol. 23, pp. 667-679.

[20] A. Shaikh, A. Varma, L. Kalampoukas and R. Dube, "Routing stability in congested networks" Proceedings of ACM SIGCOMM'00, August 28-Sepetmber 1, Grand Hotel, Stockholm, Sweden, 2000.

[21] A. Basu, J. Riecke, "Stability Issues in OSPF Routing", Proceedings of ACM SIGCOMM'01, San Diego, California, USA, August 27-31, 2001.

Network Independent Available Bandwidth Sampling and Measurement

Manthos Kazantzidis, Dario Maggiorini, and Mario Gerla

Computer Science Department, University of California, Los Angeles, CA 90095
{kazantz,dario,gerla}@cs.ucla.edu

Abstract. Available bandwidth knowledge is very useful to network protocols. Unfortunately, available bandwidth is also very difficult to measure in packet networks, where methods to guarantee and keep track of the bandwidth (eg, weighted fair queuing scheduling) do not work well, for example the Internet. In this paper we are dealing with an available bandwidth sampling technique based on the observation of packet time dispersion in a packet train or pair. In standard techniques the available bandwidth is sampled by using a "bytes divided by dispersion" (or "bytes over time", BoT) calculation and then filtered. This method of calculating available bandwidth samples has been used in all packet dispersion related work. We propose a new sampling method of available bandwidth called ab-probe. The ab-probe method uses an intuitive model that helps understand and correct the error introduced by the BoT sample calculation. We theoretically compare the new model with the previous one, exploring their differences, observability and robustness. We argue that the model may significantly improve protocols that can use an available bandwidth measurement, in particular transport-level protocols that currently use the BoT calculation.

1 Introduction

Let us define the available bandwidth over one link as the bottleneck link bandwidth minus the used bandwidth, i.e. the un-utilized bandwidth. A path's available bandwidth is the minimum available bandwidth across all links in the end-to-end path. If the latter could be known to end hosts, it could be used to apply efficient network and flow control, for example, by means of multimedia content adaptation [8], application/web routing and dynamic server selection [1], congestion control transports [6]. At the same time it is hard to measure available bandwidth. It is a highly variable quantity and constrained to an end-to-end observation, as the Internet Protocol scalable architecture dictates. Another restriction, which we will attempt to relax, is that current "available bandwidth" techniques assume that the network is performing weighted fair queuing on its flows [7]. This is the only case where the rate observed at the receivers may be the path available bandwidth and can be calculated as the bytes over the time dispersion of two or more successive packets (from now on we will refer to this scheme as Bytes over Time or "BoT approach"). The above assumption holds in

M. Ajmone Marsan et al. (Eds.): QoS-IP 2003, LNCS 2601, pp. 117–130, 2003.

a Diff Serv environment, with well defined traffic and service classes, and with WFQ routers. The Internet today, however, cannot classify/distinguish flows, may employ a variety of queuing disciplines and currently has pre-dominantly FIFO routers. It is known that the WFQ model assumption often leads to relatively high bandwidth estimation errors especially at high loads. Unfortunately, this is exactly when available bandwidth information is most essential. We show an alternative sampling method that assumes no specific queuing discipline and therefore does not contain the error caused by the above assumption.

An "active" heuristic technique called cprobe [1] is widely used to measure available bandwidth. It was developed for server selection purposes, and it is the simplest example of active BoT sample calculation. “Passive” measurements of available bandwidth using packet dispersion techniques are also found in transport level. For example, the rate estimation proposed in TCP Westwood [6] reminds us of the packet pair bandwidth estimation technique. Likewise, Microsoft's video streaming protocol MSTFP employs bandwidth estimation concepts [8].

Packet pair techniques can measure the bottleneck link bandwidth in FIFO queuing networks by using the BoT model and a combination of techniques such as kernel density estimation and Potential Bandwidth Filtering, as implemented in nettimer and proposed in [2]. In this work a kernel density estimator is used to find the highest density point by effectively dealing with well-known problems arising from constant bin size histograms. The basic assumption, under which this estimation is correct, is that packet pair results tend to aggregate around the bottleneck link bandwidth . This work has progressed from [3] and has also produced packet tailgating [4]. The latter is a technique that uses a train of packets to measure multiple bandwidths along a path. Unfortunately, few applications are directly interested in bottleneck link bandwidth. A bottleneck link bandwidth measurement is needed however for all current available bandwidth estimation techniques.

In [5] it is shown that packet dispersion techniques converge to an Asymptotic Dispersion Rate that is lower than the capacity but is not quite the available bandwidth in the path. The method in [5] investigates the effects of cross traffic on packet dispersion in the Internet. Our work is in agreement with that study, It proposes similar models, but it uses a sample calculation method instead of the BoT method that was the basis of that work as well.

In essence, all of these schemes have in common the same probing technique, ie packet pair or train. The difference is in the filtering method. For example, when a packet dispersion based scheme is used for available bandwidth estimation, the filter may average the samples [1], discover modes [5] or discover density points [3]. In all mechanisms used so far, the samples are computed using the simple BoT formula. The main contribution of this paper is to use a different equation for computing bandwidth samples. Our equation is independent of a specific queuing discipline. It is slightly more complex than simple division, but it produces much more accurate samples, as we will show in the sequel.

In this paper, after presenting our model, we theoretically explore its robustness and explain the way it differs from the BoT model. We then develop an "active" measurement technique that uses our sample calculation in packet pairs and then in trains. Next to it, we also develop a "passive" measurement technique, again using both packet pairs and trains. We consider the passive measurement especially important as it may be used in end-to-end transports to aid in flow control. We use real MPEG and H263 video traces to provide the packets for the passive measurement. We extensively test and compare our model to the BoT model in various Internet scenarios. All the experimental results support our theoretical claims and show that our sampling technique improves upon the BoT method, especially when the network load is high.

In the next section we analyze the packet pair behavior link by link. In Section 3 we discuss the need for a more sophisticated sampling technique, present our general model for the packet train behavior related to the available bandwidth, use our model to study the BoT problems and finally develop an active and a passive measurement. In 4 we present the real network experiments, which are in keeping with the theoretical models and provide insight on how effective is the new measurement based on our model. We conclude the paper in Section 5.

2 Packet Pair Background

In this section, we qualitatively analyze the interference that an end-to-end packet pair based measurement may experience from cross traffic. Our goal is to understand the source of noise and errors in a packet pair measurement, especially on an available bandwidth measurement that we are dealing with here.

Let us focus separately on three segments of the path: 1) the path consisting of links from the source up to the bottleneck link, 2) the bottleneck link, 3) the path from after the bottleneck link to the destination. When we refer to a link we refer to the outgoing queue. Let us assume:

- $N-1$ links on the path, with the bottleneck at link k
- Pb_i, the potential bandwidth upon entering the link i, determined by the pair dispersion after link i ; Pb_0 is then the pair potential bandwidth at the sender, defined as the payload bytes over the initial time separation.
- B_i, the actual bandwidth of link i
- $Pb_0 > B_k$

Then we can analyze in details the behavior of the three segments.

The first segment of the path is interesting because it ultimately defines the potential bandwidth (Pb_k) with which the pair will enter the bottleneck link k. In order to make sure that we can measure the bottleneck link, Pb_k must be greater than the bottleneck link bandwidth. After each link the pair may decrease its potential bandwidth when $Bi < Pb_{i-1}$; the same can happen when cross traffic arrives in the time between the pair arrivals (Note that the probability of this event occurring is inversely proportional to Pb_{i-1} for a given network load). When enough cross traffic is queued already when the packet pair arrives then we may get useful time compression. It is called time compression because the time separation of the packet pair is potentially compressed (it becomes the payload of one packet over the link bandwidth). We call it "useful" because it increases the pair's potential to correctly measure large bottleneck bandwidths. Note that in the case of available bandwidth measurement this compression in not useful.

The packet pair arrives to the bottleneck link with a potential bandwidth of Pb_{k-1} and exits with Pb_k, hopefully equal to B. If Pb_{k-1} is less than B, then we already have an underestimate of B. This underestimate may be further increased by intercepting cross traffic. If Pb_{k-1} is more than B, then the exiting potential bandwidth is either B or some overestimation due to compression because of preexisting traffic.

Other events may occur after the bottleneck link. As the pair is traveling after the bottleneck link it should sustain its time separation. However, enough intercepting cross traffic (i.e. traffic that enters the node and gets queued in the middle of the packet pair interval) may cause further underestimation over a link. Traffic queued in front of it may cause time compression (if service/transmission time for the packets queued in front is more than the packet pair time separation). This is an adverse time compression that causes over-estimation of the bottleneck link bandwidth.

3 Extension of Packet Pair/Train for Available Bandwidth Sampling

3.1 Why Do We Need a New Method to Calculate the Samples?

Let us attempt to calculate what the available bandwidth over a single link should be and how this is different than the BoT calculation. We first illustrate the BoT method. Consider two packets, Packet 1 and Packet 2, each of size S, entering the bottleneck link with a potential bandwidth of Pb_{k-1} which is close to B, the bottleneck link bandwidth. The time that Packet 1 exits from the link, ie, exit from service, marks the beginning of time interval and Packet 2's exit from service marks the end of the interval, denoted as time separation d. The BoT method calculates the available bandwidth A at S/d.

Now assume that A is indeed the available bandwidth seen by those packets. Then, the consumed bandwidth is obviously $B - A$. Let us call the interval over which the available bandwidth A is measured the sampled interval. This interval can be defined, for example, by the separation of the packets, from the time the Packet 1 enters the queue until Packet 2 enters the queue, which is by definition S/Pb_{k-1}.

Since during that time the bandwidth consumed (by other connections) is $B-A$ then the number of bits that enter the link queue during the sampled interval is $(B - A) \cdot S/Pb_{k-1}$. The observed separation at the link exit, as defined in the previous paragraph, will be the transmission/service time of the intervening bits that have entered the queue after Packet 1, and the transmission/service of

Packet 2. Therefore $d = (B - A) \cdot (S/Pb_{k-1})/B + S/B$. It is obvious then, by solving for the available bandwidth A, that A is not generally S/d as the BoT calculation dictates. We elaborate on their difference in the rest of the paper and attempt to develop a new sampling method based on it.

3.2 The Ab-probe Model

In this paragraph we use an approach similar to that described in 3.1. to calculate the available bandwidth samples from a packet train of N equally sized and spaced in time packets. Assume N ($N \geq 2$) equally spaced packets each of size S bits. Assume the packets reach the bottleneck link with a potential bandwidth of Pb, that the bottleneck link bandwidth is B and the actual available bandwidth (during the train transit) at the bottleneck link is A, as before. We are calculating the available bandwidth from the perspective of an application, without counting the probing packet overhead. The sampled interval (see 3.1) is:

$$Sampled\ Interval = \frac{(N-1)S}{Pb} \tag{1}$$

If the available bandwidth during this interval is A then, using (1), the traffic (in bits) the queue should be receiving from other flows during this interval is:

$$Cross\ Traffic\ bits = \frac{(B-A)(N-1)S}{Pb} \tag{2}$$

Then the observed time dispersion (separation) between Packet 1 exiting service and Packet N exiting this link's service is the transmission time of the packet train bits and the intervening traffic bits through the link.

$$\begin{aligned} T = T_N - T_1 &= \tfrac{(N-1)S}{B} + \tfrac{(B-A)(N-1)S}{Pb \cdot B} \\ &= \tfrac{(N-1)S}{B}\left(1 + \tfrac{B-A}{Pb}\right) \end{aligned} \tag{3}$$

Solving for the available bandwidth A we get:

$$A = B - \frac{B \cdot T - (N-1)\,S}{(N-1)\frac{S}{Pb}} \tag{4}$$

Note that an available bandwidth sample, averaged over the interval sampled by the packet train, may be negative. This is intuitive and is consistent with available bandwidth definition. It may be negative when, during a transient, period more traffic enters the link than the total bandwidth can support. (More traffic than the link's capacity may enter a link for a short interval because of the queue buffer space). Averaged over longer period of times the available bandwidth with samples given by these equations will turn out to be positive. The above model can be used (i) to help understand the error when sampling available bandwidth using the BoT model in non-WFQ networks. (ii) to develop a network independent sampling method. We study these issues further.

3.3 The BoT Sampling

In this paragraph we look at the model used so far in measuring available bandwidth that uses the BoT sample calculation. Equation 4 gives the correct available bandwidth averaged over a period of time (defined by a train of packets). This equation is to be compared with the BoT calculation $(N-1) \cdot S/T$.

First, we notice that following the above logic, the BoT calculation gives a different available bandwidth result for different size trains. This effect has been in fact noticed in real testbed experiments with cprobe [1]. Figure 1 shows that applying this equation has the exact error effect shown experimentally in [1]. A train of 4.1 Mbps potential bandwidth (i.e. the potential (sender) bandwidth of the packet pairs in the train) is used over a 4 Mbps capacity link and 2 Mbps available bandwidth. The equations of the model are applied to calculate the actual available bandwidth using different values for N, the number of packets in the packet train.

Figure 2 and Fig. 3 show the relative error for the BoT calculation used over a 4 Mbps link and trains of 10 packets and 160 bits each packet. Note that the BoT error becomes exponentially larger as the network gets congested. Specifically it reaches a 100% (relative error) when the 4 Mbps link has 1.28 Mbps available bandwidth. When the network is in a very congested state, and the link has only 100 Kbps available bandwidth, out of its 4 Mbps, we have a 2000% relative error.

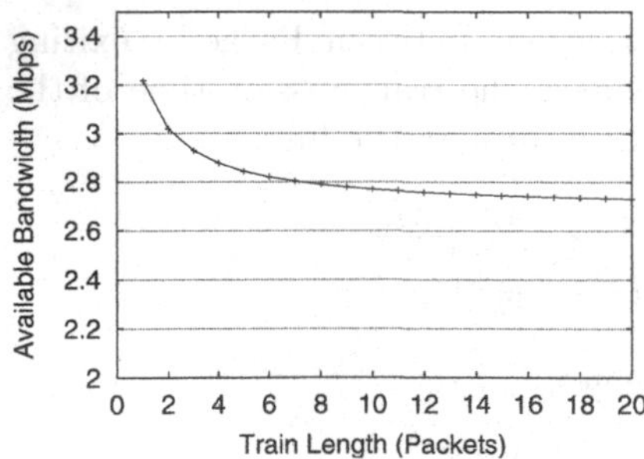

Fig. 1. BoT available bandwidth calculation varies with train size under same network conditions.

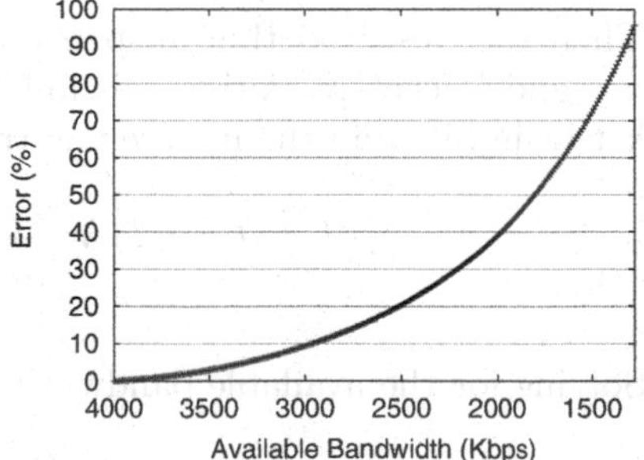

Fig. 2. The BoT relative error for a 4 Mbps link with 4 Mbps up to 1.280 Mpbs available.

3.4 Observability and Robustness of Ab-probe

The ab-probe (available bandwidth probe) model just described becomes a viable measurement method, if B and Pb can be accurately estimated. The measurement can be implemented both as an end-to-end measurement and as a network

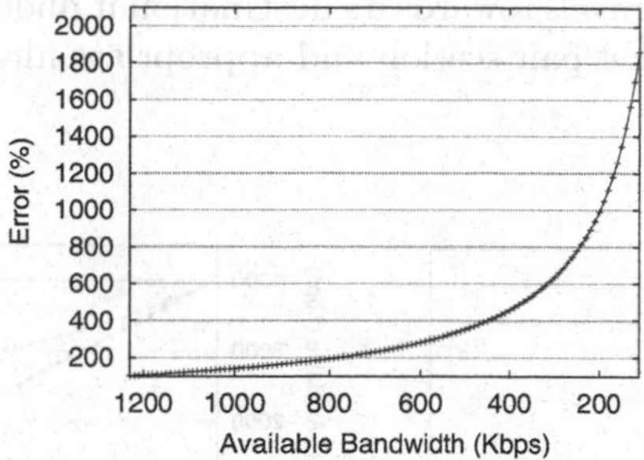

Fig. 3. The BoT relative error for a 4 Mbps link with 1.230 Mbps to 100 Kbps available.

layer measurement in network feedback architectures, in both active and passive forms.

In the network layer case, the measurement can be done link by link, thus Pb is known and B can be very easily estimated over each link [3].

In an end-to-end measurement, the bottleneck bandwidth (as opposed to available bandwidth) can be estimated accurately by proper filtering of the BoT samples as shown in the nettimer method [3]. The above method relies on the fact that with non zero probability the two packets in the pair are transmitted one immediately after the other by the bottleneck link. Thus, the kernel density estimator can identify the bottleneck link "mode" of the samples and compute the corresponding value. The "potential bandwidth" (Pb) at the entry of the bottleneck link however is only known to be between the sender bandwidth (assumed larger than the bottleneck bandwidth) and the bottleneck link bandwidth in an end-to-end measurement. This is because, when distorting events do not occur, specifically the time compression before the bottleneck link, then the sender bandwidth cannot increase. It may only decrease by entering through smaller capacity links. If we sent at a sender (potential) bandwidth of the bottleneck link capacity then with no distortions by definition the Pb will be B. In Fig. 4, we see how possible distortions in the Pb estimation affect the available bandwidth measurement. The graph is based on the equations presented by introducing errors in the Pb estimation. It shows that when the actual Pb is lower than the estimation, the available bandwidth estimation becomes negative, a 50% underestimation in AB will result from a 35% underestimation in Pb. When Pb is estimated higher than the actual Pb then the error in available bandwidth is practically bounded by a 50% overestimation. Since it is preferable to have an overestimation we sent at a sender bandwidth of slightly more than B as mentioned before. Very negative available bandwidth samples can then be filtered out and not be accounted for in the measurement. Next we show how the above can be successfully put into practice.

As the packet train travels towards its destination it undergoes the distortions we described in the packet pair section and appropriate filtering of the resulting samples should be used.

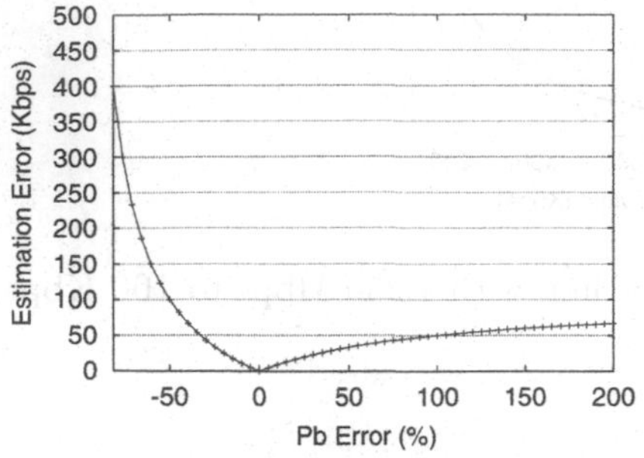

Fig. 4. How an error in *Pb* estimation affects the available bandwidth sampled with ab-probe.

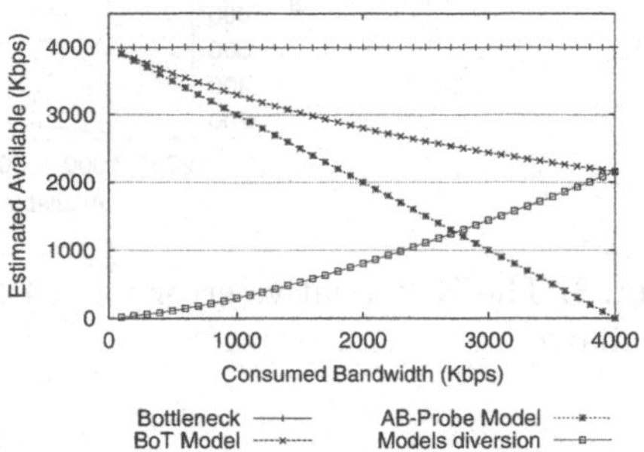

Fig. 5. Analytical Difference between BoT and ab-probe measurement.

4 Experiments

We have run real experiments using long range Internet connections and short range campus Internet connections. The experiments are performed using the active and passive measurement and testing both options of packet pairs or trains. In all cases we have found that practice agrees with theory.

The short range Internet topology is shown in Fig. 6 The two LANs are connected by a Cisco 2600 router in two 10 Mbps interfaces. On each LAN one host is performing the "probing" and another is used for injecting extra cross traffic. Normal traffic exists in the network as usual, but we inject the extra traffic so that we may observe how the measurements react to it. The long range Internet topology (Fig. 7) has the two source and destination LANs more than 20 hops apart. The sending hosts (probing and cross traffic) are switched through a Cisco Catalyst switch series 6509, located in California. The two receiving hosts are connected to a Cabletron SmartSwitch 2100, located in Italy.

The probing connection is either active or passive and uses the appropriate time-stamping in the packet headers as required for the measurement. We use application level timestamps as a measurement application or an adaptive multimedia streaming application would have to. We use the nettimer tool to compute the bottleneck link bandwidth just prior to the experiment. The active measurement uses 128 byte packets changing the inter-packet delay to achieve the required sender bandwidth. In the case of packet pairs, 40 packet pairs (80 packets) are sent at the required bandwidth following a one second idle interval. When packet trains are tested we sent 10 trains of 8 packets following a one

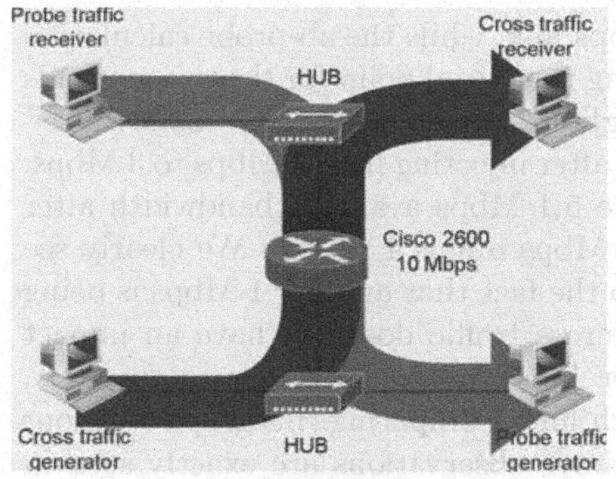

Fig. 6. The short range campus Internet experiment topology.

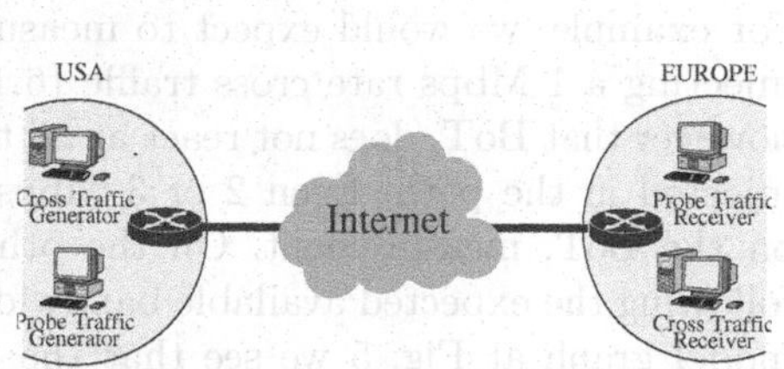

Fig. 7. The long range campus Internet experiment topology.

second idle interval. We manage to experiment with high potential bandwidth packet pairs by performing busy wait when necessary, dealing effectively with the non-real time Linux (Redhat 6.2) kernel 10 msec timer granularity problem. The passive measurement uses a real source of MPEG video. Traces are pre-captured from a Star Wars trailer clip [1]. It is simply smoothed to 200 byte packets sent uniformly over the frame time. The cross traffic is simply 512 bytes payload CBR of the reported rate in all cases. As mentioned the bottleneck link bandwidth is determined using the nettimer tool.

4.1 Active Measurement

In this paragraph we present the active measurements experiments. We perform one experiment per measurement method using the same probing traffic. All experiments are performed one after the other. These experiments prove that our model is correct in practice and it is in fact a significant improvement over BoT.

The active measurements for the short range experiment are summarized in Fig. 8. The figure depicts a graph similar to that in Fig. 5 but using real experiment results instead of equations. In this one graph we summarize results for all relevant active measurement experiments, i.e. using pairs and trains and with the BoT and the ab-probe model sample calculation. The link capacity was measured to be 9 Mbps using the Nettimer method. This is a fairly accurate measure, given that the network is a 10 Mbps E-net. The x axis is the rate of the "additional" injected cross traffic. Note that prior to our experiment there was already some traffic present in the network, over which we have no control. That is why the curves do not start at 9 Mbps at zero cross traffic. If the network had been unloaded initially, all the curves would have started from 9 Mbps at zero crossload, as clearly indicated in Fig. 5. Using the BoT sample calculation we get

[1] Star Wars trailer clip: http://www.trailersworld.com/movie.asp?movie_id=692

6.1 Mbps available bandwidth over a 9 Mbps link, while the ab-probe calculation gives approximately 4.5 Mbps. After getting the initial point for the two methods (at cross traffic equal to 0) we draw a lighter "expected available bandwidth" line. It represents the available bandwidth after injecting from 1 Mbps to 3 Mbps. For example, we would expect to measure 5.1 Mbps available bandwidth after injecting a 1 Mbps rate cross traffic (6.1 Mbps minus 1 Mbps). We clearly see however that BoT. does not react at all to the fact that another 1 Mbps is being injected in the path. Even 2 or 3 Mbps cross traffic does not have an impact on the BoT. measurement. On the other hand, ab-probe measures correctly, following the expected available bandwidth line. Comparing this graph with our model graph at Fig. 5 we see that the above observations are exactly seen in our model equations as well. Our theoretical model is very closely validated in the experiments. The ab-probe available bandwidth improvement is visible in practice as well.

In Fig. 9 we see the same graph regarding the active measurement in the long range Internet scenario. In this case the bottleneck link is measured at 4 Mbps. The available bandwidth using the BoT samples in our averaging filter is measured at approximately 2.5 Mbps when no other cross-traffic is injected intentionally. In this case too, when we inject cross traffic of 1 Mbps and 2 Mbps the BoT based measurement fails to drop appropriately, and seems only slightly affected by the extra traffic. The ab-probe method initially measures a 1.5 Mbps available bandwidth (when no cross traffic is injected intentionally). When a 1 Mbps stream is injected the ab-probe reacts to the available bandwidth drop as predicted by theory, measuring around 200 Kbps available. When a 2 Mbps stream is injected the ab-probe measures −1 Mbps. Note that as we mentioned before the available bandwidth may become negative for some short intervals. Looking at the connection loss rates for this case, we see that the probing traffic lost 80% of its packets and the cross traffic lost 33% of its packets.

Even in this highly congested, unstable condition the ab-probe manages to track the available bandwidth changes while the BoT is unable to measure or even detect the oncoming of congestion. Figure 10 shows more details regarding the long distance experiment. This graph shows the packet pair bandwidth estimation points in the above experiment in the ab-probe case on the left and in the bot case on the right. We note the many different distinct concentration points (modes) of the samples in both sample calculation methods, as has been noted by previous work.

In all, the ab-probe active measurement managed to successfully measure the available bandwidth in all cases, even in long distance (more than 20 hops) end-to-end observation, while the BoT. samples have failed.

4.2 Passive Measurement

In the passive measurement, as mentioned before, we probe bandwidth using a VBR real video source. This case is especially interesting because it shows that our measurement can be successfully piggybacked on a rate-based transport. We

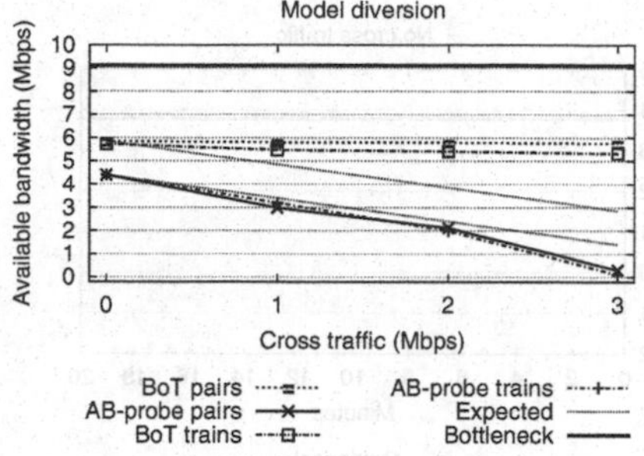

Fig. 8. Active Measurement using BOT and AB-probe in the short range Internet topology.

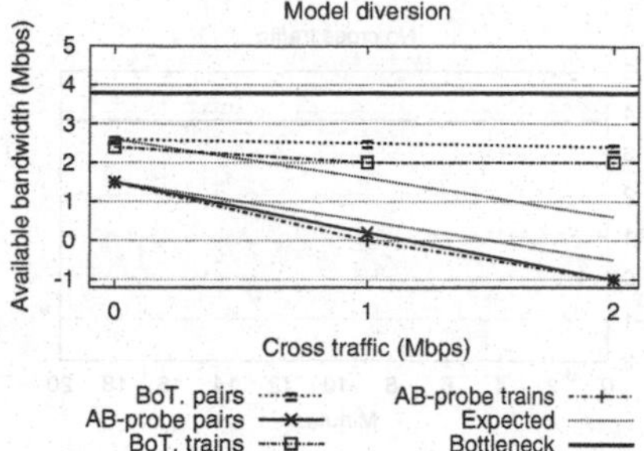

Fig. 9. Active Measurement using BOT and AB-probe in the long range Internet topology.

take advantage of the fluctuating nature of the VBR video source to probe higher bandwidths than the stream average bandwidth.

The experiments show that ab-probe is less accurate than in the active case, as expected, due to the lower and uncontrollable rate of the probing packet pairs. Figure 11 shows our usual model diversion graph for the short range Internet case where a 4 Mbps average video trace is used over a 9 Mbps reported bottleneck link bandwidth. Ab-probe is somewhat successful in reporting the extra cross traffic injected in the network whereas the BoT is practically unaffected by the additional traffic.

We used the 4 Mbps average video for the long range case. The bottleneck link reported was 4 Mbps and the loss rates of the video traffic were: 1%, 1.5% and 50%, for (*a*) no additional cross traffic, (*b*) 1 Mbps and (*c*) 2 Mbps cross traffic respectively. In this case we note: (*i*) Our probing (video) traffic consumes the bottleneck link. The 1 Mbps cross traffic connection suffered over 40% loss therefore leaving the measurements almost unaffected. (*ii*) There is significant queuing of other uncontrolled traffic after of the bottleneck link as evidenced by the fact that the BoT pairs average at higher than bottleneck point rates (many time-compressed samples). Ab-probe is more effective in dealing with compressed packets since it can average them out with the negative samples. (*iii*) Both BoT and ab-probe were able to sense the 2Mbps cross traffic. Note that this is the only satisfactory result regarding the BoT model and is reported when there is significant loss in probing and competing traffic. The BoT is able to capture this effect due to the lost packets (the size of the packets is the nominator). (*iv*) The trains fail to "sense" the 2 Mbps injected cross traffic because the lost packets cannot be accounted for correctly. Specifically, trains may become smaller when first and last packets are lost leaving large intervals without having samples.

In all, a simple averaging filtering was being used, ab-probe showed measurements close to the expected at all times while the BoT showed deficiencies similar to those found in the active measurement case.

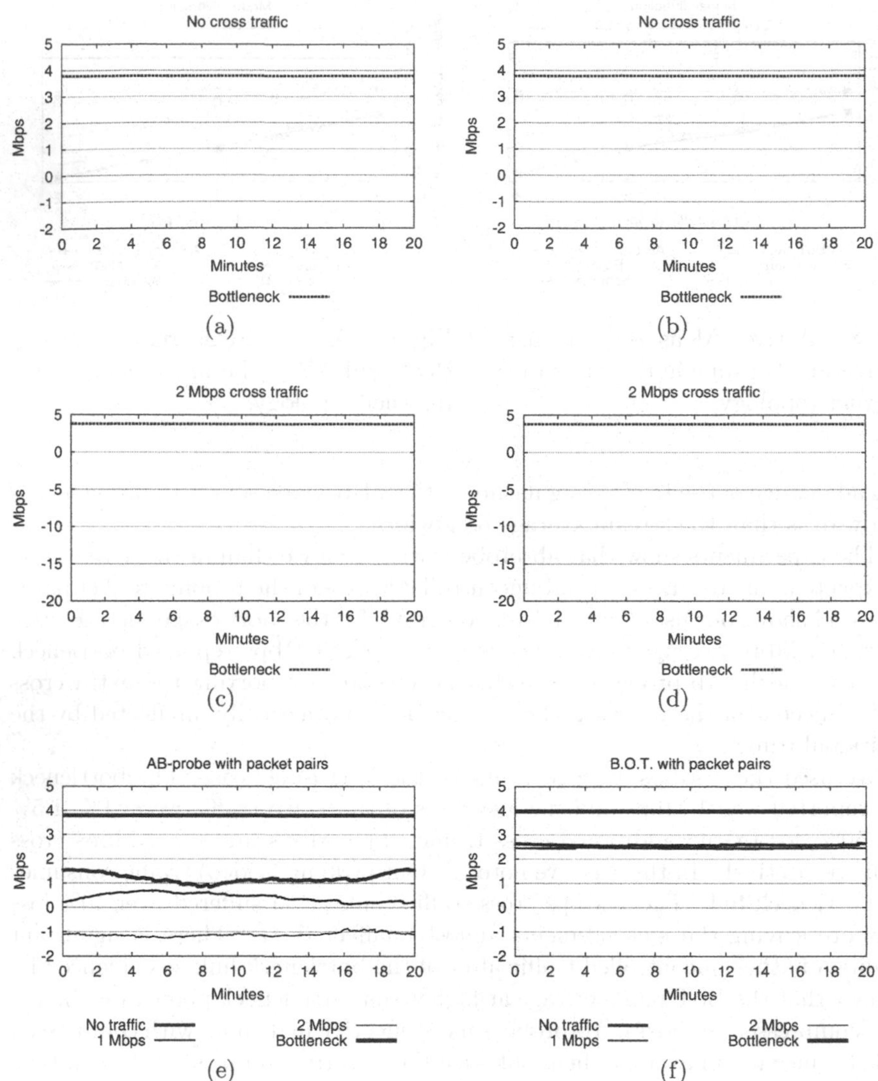

Fig. 10. Graphs from the long distance active measurement experiment. a) Ab-probe short distance packet pair samples when injected traffic is 0 Mbps. b) BoT packet pair samples when injected cross traffic is 0 Mbps. c) Ab-probe short distance packet pair samples when injected cross traffic is 3 Mbps. d) BoT packet pair samples when injected cross traffic is 3 Mbps. e) Ab-probe packet pair samples filtered. f) BoT packet pair samples filtered. The measurement is not affected by the injected traffic in all cases.

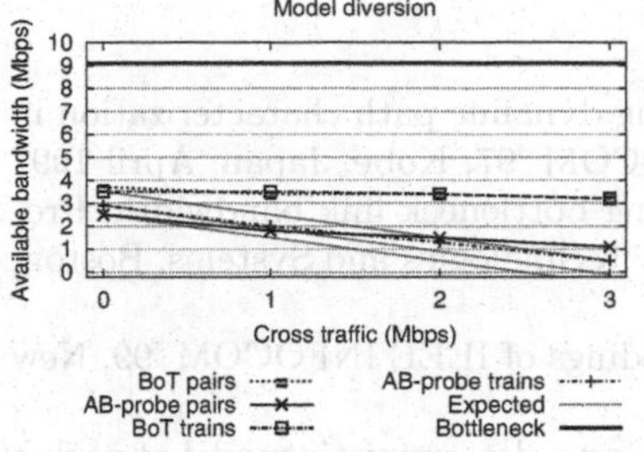

Fig. 11. Model diversion graph for pairs and trains for the short distance case.

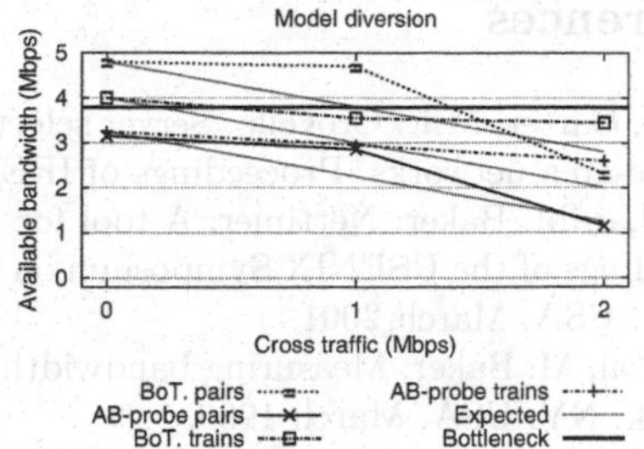

Fig. 12. Model diversion graph for pairs and trains for the long distance case.

5 Conclusions

We have presented ab-probe, a new model and measurement technique to estimate the path available bandwidth from end-to-end observation of packet dispersion. The model directs us towards a new sample calculation that is independent of the network queuing discipline and therefore can be used successfully on the Internet. Two simple measurement methods have been implemented, one active and one passive, mainly with the purpose to show that any measurement implementation based on the new model is reasonably accurate. This work can improve the performance of user sessions, or more generally any network protocol that may need available bandwidth measurement; a very useful quantity, especially for server selection applications, routing, end-to-end transports, adaptive multimedia etc.

Our motivation for this research has been the unintuitive, currently widespread method for calculating available bandwidth samples from packet dispersion used in protocols. The simple BoT calculation is theoretically correct only to the degree that the network approximates a perfect fair queuing for flows. We found that its use generally causes significant errors in non-conforming networks like the Internet, especially at high loads. We developed a model that quantifies this error and used it to produce a new sample calculation that is correct without assumptions for queuing disciplines. The model also captured and explained previously reported c-probe behavior as seen by its researchers in measurements over a real network that has been to this day unexplained. Long range Internet and short range campus experiments showed that ab-probe is accurate in practice as well and a significant improvement over BoT . The improved accuracy in available bandwidth measurements will be very beneficial to adaptive multimedia applications. It will improve network stability and will provide better congestion protection.

References

1. R.L. Carter, M.E. Crovella: Server selection using dynamic path characterization in wide-area networks. Proceedings of IEEE INFOCOM '97, Kobe, Japan, April 1997
2. K. Lai, M. Baker: Nettimer: A tool for measuring bottleneck link bandwidth. Proceedings of the USENIX Symposium on Internet Technologies and Systems, Boston, MA, USA, March 2001.
3. K. Lai, M. Baker: Measuring bandwidth. Proceedings of IEEE INFOCOM '99, New York, NY, USA, March 1999.
4. K. Lai, M. Baker: Measuring link bandwidths using a deterministic model of packet delay. Proceedings of ACM SIGCOMM 2000, Stockholm, Sweden, September 2000.
5. C. Dovrolis, P. Ramanathan, D. Moore: What do packet dispersion techniques measure? In Proceedings of IEEE INFOCOM 2001, Anchorage, AK, USA, April 2001.
6. C. Casetti, M. Gerla, S.S. Lee, S. Mascolo, M. Sanadidi: TCP with Faster Recovery. Proceedings of IEEE Military Communications Conference MILCOM 2000, Los Angeles,CA, USA, October 2000.
7. S. Keshav: Packet-pair flow control. http://www.cs.cornell.edu/skeshav/papers.html, 1994
8. Q. Zhang, Y.Q. Zhang, W. Zhu: Resource Allocation for Audio and Video Streaming over the Internet. Proceedings of International Symposium on Circuits and Systems ISCAS 2000, Geneva, Switzerland, May 2000.
9. D. Poulton, J. Oksman: Digital filters for non uniformly sampled signals. Proceedings of Nordic Signal Processing Symposium NORSIG 2000, Vildmarkshotellet Kolmarden, Sweden, June 2000.
10. V. Paxson: Measurements and Analysis of End-to-End Internet Dynamics Ph.D. Thesis, University of California, Berkeley, 1997.
11. S. Floyd, K. Fall: Promoting the use of end-to-end congestion control in the Internet. IEEE/ACM Transactions on Networking, vol. 7, no. 4, August 1999.
12. D. Bansal, H. Balakrishnan: Binomial congestion control algorithms. Proceedings of IEEE INFOCOM 2001, Anchorage, Alaska, April 2001.

Less than Best Effort: Application Scenarios and Experimental Results*

Tim Chown[1], Tiziana Ferrari[2], Simon Leinen[3], Roberto Sabatino[4], Nicolas Simar[4], and Stig Venaas[5]

[1] University of Southampton, United Kingdom, tjc@ecs.soton.ac.uk
[2] Istituto Nazionale di Fisica Nucleare, Italy, Tiziana.Ferrari@cnaf.infn.it
[3] SWITCH, Switzerland, simon@switch.ch
[4] DANTE, United Kingdom, {Roberto.Sabatino,Nicolas.Simar}@dante.org.uk
[5] UNINETT, Norway, Stig.Venaas@uninett.no

Abstract. In this article we present the work done to study the potential benefits to end users and network operators and the feasibility of deploying a Less than Best Effort (LBE) service on a wide area scale. LBE is a Per-Domain Behaviour based on the Differentiated Services Quality of Service architecture. We present a brief overview of the evolution of the case for LBE, through the IETF DiffServ WG and the Internet2 QBone project, and then describe some proposed scenarios for LBE deployment in European research networks and GÉANT, the research backbone providing interconnectivity to the European National Research and Educational Networks (NRENs). The experimental results presented demonstrate the viability and importance of Quality of Service to meet a large set of network providers and users' requirements even in presence of communication infrastructures characterized by very high-speed connections.

Keywords: Quality of Service, Differentiated Services, Less than Best Effort (LBE), Scavenger

1 Introduction

In recent years there has been a growing demand by users in the research community for a high quality of service. The most common approaches to delivering the performance that the users require are either to increase the network provision in advance of demand (a technique commonly referred to as "overprovisioning"), or to deploy some kind of "Better than Best Effort" quality of service mechanism in the network (e.g. the Premium IP service[1][2][3].) However, overprovisioning can only be applied where the funds of the network operator permit. For example, many universities and research institutes are "only" connected at capacities in the order of 155 Mbps, and many further education colleges have 2Mbit/s links.

* This work is supported by the IST projects IST-2000-26417 GN1 (GÉANT) and IST-2000-25182 DataGrid.

M. Ajmone Marsan et al. (Eds.): QoS-IP 2003, LNCS 2601, pp. 131–144, 2003.

While it is expected that Differentiated Services-based (DiffServ) Quality of Service (QoS) deployment can offer a good service for many users[4], widespread deployment can be a far from trivial exercise where multiple administrative domains are involved, as dynamic bandwidth brokering is required, and aggregate reservations have to be considered.

In this article we evaluate a different approach to quality of service based on the Less than Best Effort (LBE) Per-Domain Behaviour (PDB) similar to the one defined in [5]. We illustrate the PDB, its potential application cases and we estimate its performance in a large number of traffic and configuration scenarios. Intuitively, one might think that very few users would be willing to run applications that would receive a worse service than regular Best Effort (BE) traffic. However, we argue that in providing a traffic class that can expand to utilise the available bandwidth on a link, without any significant effect on the BE (or Premium IP) traffic, a number of new network usage scenarios can be met.

We begin by giving a service description for LBE. In this description we choose to be interoperable with the Internet2 Qbone Scavenger Service (QBSS)[6] by using the same diffserv code point (DCSP) value of 001000 [8][9].

The goal is the demonstration of the feasibility of Quality of Service in the GÉANT[10] backbone in Europe to meet a number of users and providers' requirements, and possibly, its extension to a number of European research networks in a fully inter-domain scenario.

The results obtained to date suggest that an LBE service can be deployed on GÉANT and NREN networks, and that both the NRENs and their end users can benefit.

2 The Less than Best Effort Service

Between 1999 and 2002 there have been several initiatives to design and implement services capable of offering guaranteed and predictable network QoS to end-users. The Internet2 Qbone initiative started in 1999, whilst within GÉANT work for the definition of a Premium IP service started a little later, in November 2000.

The original ideas for an LBE-like service arose from work in the IETF DiffServ working group. However, it is the work of the Scavenger team that has helped raise the profile of such a service, to the extent that adoption is now being evaluated for GÉANT and by NRENs. The main idea of LBE is that this traffic class is able to make use of unutilised bandwidth in the network, but in a way such that in cases of competition for resources, the LBE traffic will be discarded before any Best Effort (BE) or higher-class traffic.

2.1 LBE Description

The definition of the LBE service follows the basis that a given differentiated services code-point (DSCP) is used to convey the meaning that packets bearing such a DSCP value can be given a lesser service than regular best effort (BE) traffic. If congestion at a given interface is produced by LBE traffic, then congestion should be completely transparent to packets belonging to higher-priority classes like IP Premium and BE. Congestion is produced by LBE traffic if the output capacity of the interface is exceeded because of the injection of LBE packets, but both the instantaneous and average BE traffic rate can be handled by the interface without introducing BE packet loss. Performance of packets belonging to higher-priority classes cannot be protected against congestion if the amount of traffic belonging to that class or to higher-priority classes at a given output interface is sufficient to produce either short or long-term congestion, regardless of the presence of LBE packets. Protection of higher-priority classes from LBE traffic has to be supported both with and without congestion. Protection requires that packet loss, one-way delay, Instantaneous packet delay variation and throughput of streams belonging to higher-priority traffic classes should not be affected by the presence of LBE traffic, either with or without LBE congestion.

No end-to-end guarantees are provided to flows adopting the LBE service. This means that the LBE service is not parameterised, i.e. no performance metrics are needed to quantitatively describe the service. In addition, no seamless end-to-end service is provided by LBE. This implies that LBE can be supported incrementally on congested interfaces as needed without requiring any LBE service support in peering networks.

For routers not implementing the LBE queuing and drop policy, the minimum requirement is that the DSCP value chosen to indicate the LBE service is passed transparently across the network (not set to another value or cleared to zero). In routers not implementing LBE drop policies, it is expected that LBE receives the same queuing priority as BE.

While it is possible that an LBE-tagged packet may traverse all but the last hop into its target network before being dropped, that apparent "waste" of bandwidth is compensated by the resultant TCP backoff in the LBE application resulting from the packet loss.

3 LBE Application Scenarios

In this section provide an overview of example scenarios where the use of LBE may be beneficial to a user, a network provider or both.

3.1 Mirroring

Content on the Internet that is accessed by a large number of hosts, which in turn are located at many different places, is often replicated in several locations on the Internet. Users can then retrieve content from a replica or mirror site that is topologically close to them, which ensures more efficient access and imposes less resource utilisation on the server with the master copy of the content (to the extent that in some cases the master will only be accessed by replicas). In turn, this enables the provider of the content to serve a larger community of users.

One way to manage the replication of data in the network is to simply have a number of caches spread around the network, where caches have their own copies of data that has recently been accessed or are in high demand.

With asynchronous mirror updates changes at the master are propagated to the replicas at regular intervals independently of when clients access the data. Ideally the update should not penalise users accessing the data. In order to achieve this, the update will either be done at fixed times that are expected to be off-peak periods, or the master and replica might perform updates only when the load is below a certain threshold. By using LBE, updates can be done at any time of the day without penalising the user data traffic.

In this application scenario, it is the sender of the traffic (the master) that needs to classify as LBE all packets for the receiver (the replicas), but the request for the traffic comes from the replica. This leads to an important implementation issue. If the master knows the replicas, the master can distinguish between replica and user access, and classify only packets destined for replicas as LBE. This requires some detailed configuration management, and there are also cases where the master may not be able to distinguish between user and replica access.

The user may not be aware of LBE whilst the user's applications or the operating system may somehow decide whether LBE should be used. This implies that in the case of downloads, it is the receiver that knows best whether LBE should be used, whilst it is the sender that has to mark the packets (receiver denotes the party receiving the downloaded data). Because of this, we need a mechanism that allows the receiver to signal the sender that it should use LBE to send the traffic to its destination. One possibility is that the replica (or user) uses LBE to send the request to the master, in which case the sender uses LBE to send traffic if data packets from a replica or user are marked as LBE. Another approach is to signal LBE at the application level. For instance with FTP, there exists a command (SITE) that allows an FTP client to pass any string to an FTP server. Thus the client can signal LBE by passing a special agreed upon string to the server.

3.2 Production and Test Traffic

The LBE service can be effectively used for protection of high-priority traffic from low-priority traffic. For example research centres involved in experimental exchanges of large data volumes or in testing of new applications/middleware, like the GRID community, may be interested in protecting high-priority production traffic from potential congestion produced by test packets. The service is particularly interesting when the link providing access to the National Research Network (NRN) infrastructure or one or more interfaces within the local area network are subject to congestion.

3.3 Support of New Transport Protocols

The capacity of high-speed links can be efficiently used by long TCP sessions only if TCP socket sizes are properly tuned according to the bandwidth-delay product of a given flow and if the stream does not suffer from loss. In case of packet drop, the traditional congestion control and avoidance algorithms in TCP can severely limit the TCP performance. For this reason, several TCP extensions that improve the protocol efficiency and also alternative new transport protocols are under study and definition by the research community.

The LBE service could be used in production networks to protect TCP-compliant traditional traffic from the test applications using the more aggressive non-TCP compliant transmission techniques.

3.4 Traffic Management from/to Student Dormitory Networks

One of the deployment scenarios for the QBSS on Internet2 has been in student dormitory networks at university sites. The premise is that student traffic from living quarters that is passing to the Internet is generally deemed lower priority than staff and research traffic.

In such cases LBE marking can be applied on routers connecting the dormitory networks to the campus network. Some or all traffic can be so marked. Note though that where students are downloading data to their rooms the received data is not LBE-tagged unless the server side application specifically honours the DSCP seen on incoming requests. As this is unlikely, the major impact of LBE-tagging would be on data being exported from room networks, e.g. peer-to-peer file transfers, or FTP servers run in student networks.

In this scenario the LBE-tagged traffic (at the campus-dormitory border) may never leave the university network if dropped at the campus-Internet border. However, this is not always the case, and the LBE tagging may be useful on ingress to the target network (if that is the other point on the data path most likely to have congestion); thus we should argue to apply LBE rather than purely using site-specific marking and dropping.

3.5 Estimation of Available Bandwidth

It may be possible to use LBE-tagged traffic to gain some non-disruptive (to BE traffic) estimate of available bandwidth on a given link or between end points.[1] However, this area of study requires further work to prove its potential, and to do so in a way that can be shown to be non-disruptive.

Available Bandwidth has been identified as an important network performance metric in GÉANT[12], but no tools were found that were able to measure it. The use of LBE could fill this gap.

3.6 Access Capacity Management

NRNs may be interested in the support of innovative access capacity management techniques based on the LBE service that guarantee a more efficient use of capacity within the NRN backbone.

[1] Suggestion made by Sylvain Ravot of the DataGrid project in conjunction with tests between the California Institute of Technology and CERN.

Customers are traditionally connected at a limited maximum access speed, especially when technologies such as ATM or Fast/GigaEthernet are used, despite of the fact that the capacity of the physical medium available in the local loop might be much higher.

A different access capacity management can be adopted if the LBE service is supported. Best-effort traffic injected and/or received by a given customer network can be rate limited according to the traditional scheme.

However, the customer could also be allowed to inject and receive an additional amount of LBE traffic that is only limited by the physical capacity at the local loop. Forwarding of LBE traffic is not guaranteed in case of congestion, but the amount of bandwidth the customer pays for, is still guaranteed in case of LBE congestion.

The proposed access capacity management scheme stimulates the definition and use of traffic classes at different priorities within customer network and, in addition, encourages the exchange of LBE traffic, which gives the customer the possibility to achieve a much higher aggregate access link utilization.

4 Queuing Techniques for LBE Support

Different scheduling algorithms can be adopted to service the LBE queue and the higher priority queues enabled on a given output interface. In case of algorithms requiring a bandwidth share assignment to each configured queue – such as Weighted Round Robin and Weighted Fair Queuing - it is required that a very small bandwidth share be assigned to the LBE queue. The most appropriate configuration is an implementation issue that depends on the specific router platforms in use, on the number and type of QoS services enabled on a given interface and in general on the network set-up.

4.1 Packet Scheduling in the GÉANT Network

The experimental results collected on the GÉANT infrastructure are based on the WRR scheduling algorithm, where each queue is characterized by two fundamental parameters: the queue weight and the queue priority. The queue weight ensures the queue is provided a given minimum amount of bandwidth that is proportional to the weight. As long as this minimum has not been served, the queue is said to have a "positive credit". Once this minimum amount is reached, the queue has a "negative credit". A queue can have either a "high" or a "low" priority. A queue having a "high" priority will be served before any queue having a "low" priority.

The leftover bandwidth (from the positive credited queues) is fairly shared between all the "high priority negative credit" queues until these ones become empty independently of the queue weight. If the high priority negative credit queues are empty and if there is still some available bandwidth that can be allocated to packets, the "low priority negative credit" queues will equally share it. The credits are decreased immediately when a packet is sent.

The support of the three different PHBs supported by GÉANT requires that packets are associated to output-queues according to their DSCP. In particular, DSCP 46 is used for the Premium IP queue, DSCP 8 for the LBE queue while DSCP 48 and 56 identify traffic for Network Control and are associated to a dedicated queue. Any other DSCP is mapped to the BE queue.

4.2 WRR Configurations

The various WRR configurations used during the LBE test session clarify the importance of the queue weight and priority configuration for proper isolation between different traffic classes, and in particular for the protection of BE and Premium IP packets from LBE congestion.

Queue Weight Configuration. A weight of one percent was initially allocated to the LBE queue. During the test described in section 5.3, when BE and LBE had the same priority, it was noticed that the BE traffic was not protected enough in terms of end-to-end TCP throughput, which tended to decrease in case of congestion when a relatively high BE load is produced by the SmartBits. According to our expectations, non-significant drop of BE TCP throughput should have been experienced, since the SmartBits reported no BE packet loss.

One possible explanation of the throughput drop was a too small difference in weight between the BE and the LBE queues[2]. For this reason, the LBE weight was then reduced to zero. By default, the WRR serves one byte per round out of the "zero-weighted queues". The service rate of these queues can increase if some bandwidth is left by the "non-zero weighted" queues, i.e. when "non-zero weighted queues" are empty.

Experimental results of similar tests conducted with queue weight 0 and 1, indicate that performance of each class – BE, Premium IP and LBE – are equivalent in the two cases.

Queue Priority Configuration. Priority configuration is particularly important since it defines the queue service order when multiple queues have negative credit. In this way, when both BE and LBE queues have a negative credit, it is always the BE queue which is served first until it's empty. Two different priorities were assigned to the BE and LBE queues: high and low respectively.

A 0% weight was configured for the LBE queue in the previous phase and was kept in the final configuration. A weight of 1% could have been chosen and the results would have been similar. The most important point of configuration is the "low" priority attributed to the LBE queue while the other queues have a "high" priority. With this configuration, no significant drop of BE TCP throughput was observed during congestion.

[2] The parameter that is actually responsible of such throughput decrease is still unknown.

5 Experimental Results of Tests Performed on GÉANT

The support of the Less than Best Effort (LBE) quality of service was enabled on a subset of GÉANT routers in order to carry out preliminary test activities, whose goals are manifold.

We have tested the co-existence of three traffic classes (LBE, BE and Premium IP) in different traffic load scenarios and have analysed their performance in terms of the following network metrics: packet loss, throughput, one-way delay and instantaneous packet delay variation (IPDV). Only UDP traffic was used for the LBE class. The performance analysis of TCP LBE flows is subject of future work.

Three different router configurations have been adopted during the test sessions for the tuning of two scheduling configuration parameters: the *queue weight* assigned to the BE and LBE and the *queue priority*. While both the (small) weights assigned correctly protect BE traffic from LBE congestion, the queue priority proved to be critical for the minimisation of packet reordering, as explained in [16][3].

5.1 Test Equipment and Network Infrastructure

A subset of the GÉANT infrastructure consisting of a set of STM-16 and STM-64 links and of the relevant terminating routers has been used to test the class of service mechanisms needed to support the LBE class. Different traffic scenarios have been generated by combining different transport protocols – UDP and TCP – and a variety of traffic loads and streams for each of the three above-mentioned classes of service. Performance has been analysed both with and without congestion. Medium-term congestion for a maximum continuous time span of 10 sec was produced to test traffic isolation between different classes.

The equipment involved in the tests includes as a set of Juniper router M160s, three SUN workstations – located in the French, Spanish and Italian network Points of Presence (PoPs) respectively – and two SmartBits 600s by Spirent – located in Frankfurt and Milan, as illustrated in Figure 1. The SmartBits is a network device specialised in traffic generation; the SmartFlows application version 1.50 was adopted to drive the device. Two of the interfaces available on the SmartBits located in the Frankfurt PoP were used: one STM-16 interface connecting it to the M160 `de1.de.geant.net` and a FastEthernet interface connecting it to `de2.de.geant.net`. In the latter case connectivity was provided via a Giga/FastEthernet switch connected to router DE2 by a GigaEthernet interface. On the other hand, for the SmartBits in Milan just the STM-16 interface connecting it to the M160 in the Italian PoP was used for traffic generation.

The two SmartBits are used to generate large amounts of constant bit rate UDP test traffic (BE, EF and LBE), to produce congestion when needed, and

[3] The comparison of UDP and TCP performance of LBE flows with the performance experienced by such flows in presence of a single class of service is out of scope in this article.

to collect accurate one-way delay measurements thanks to the high accuracy of the SmartBits clock (10 nsec).

While the two SmartBits 600s only managed UDP traffic, the three workstations, running Solaris 2.8, were used to evaluate Best Effort TCP performance with and without LBE congestion. *Netperf* [13] was the main tool used for TCP throughput measurement. All of them are connected to the GÉANT infrastructure by GigaEthernet interfaces.

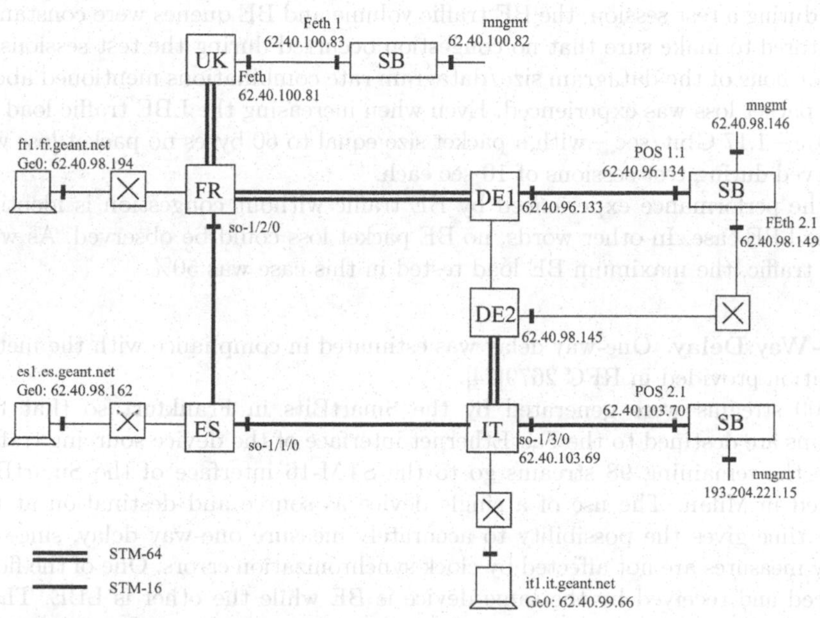

Fig. 1. LBE test infrastructure

Sections 5.2 and 5.3 of this paper illustrate the preliminary results of IP Premium, BE and LBE performance without and with congestion respectively. Tests of different complexity were carried out by gradually increasing the number of traffic classes in use.

Unless differently specified, in what follows the test traffic is sourced by the SmartBits, it is based on UDP and the size of the datagram is 512 bytes. In addition, the SmartBits traffic load will be expressed as a percentage of the capacity of the STM-16 interface connecting the SmartBits in Frankfurt – used as main test traffic source – to the network infrastructure. In addition, the total test traffic generated was the sum of the traffic sent by the SmartBits generators with the production traffic exchanged on the test data path (currently between 50 Mbps and 100 Mbps) and additional test TCP traffic of approximately 210Mbps, as reported by the test workstations.

5.2 LBE and BE Performance Measurement without Congestion

Performance measurement for each traffic class was fundamental to estimate that basic functionality of the production infrastructure in case of no congestion.

Packet Loss and Throughput. A variety of LBE traffic loads between 10% and 20% for different LBE UDP datagram sizes: {128, 256, 384, 512, 640, 768, 896, 1024, 1152, 1280, 1408} bytes were generated by a single constant bit rate LBE stream. Since the background BE production traffic can potentially greatly vary during a test session, the BE traffic volume and BE queues were constantly monitored to make sure that no congestion occurred during the test sessions.

For none of the datagram size/datagram rate combinations mentioned above LBE packet loss was experienced. Even when increasing the LBE traffic load up to 50% – 1.17 Gbit/sec – with a packet size equal to 60 bytes no packet loss was observed during test sessions of 10 sec each.

The performance experienced by BE traffic without congestion is identical to the LBE case. In other words, no BE packet loss could be observed. As with LBE traffic, the maximum BE load tested in this case was 50%.

One-Way Delay. One-way delay was estimated in compliance with the metric definition provided in RFC 2679[14].

100 streams were generated by the SmartBits in Frankfurt so that two streams are destined to the FastEthernet interface of the device sourcing traffic, while the remaining 98 streams go to the STM-16 interface of the SmartBits located in Milan. The use of a single device as source and destination at the same time gives the possibility to accurately measure one-way delay, since latency measures are not affected by clock synchronization errors. One of the flows sourced and received by the same device is BE while the other is LBE. These two reference streams were used for one-way delay measurement and were run concurrently so that a direct performance comparison can be drawn between the two classes. A fraction of the remaining flows received by the SmartBits in Italy is LBE while the remaining part is BE, so that 25% of the overall traffic load is BE while the remaining part is LBE.

Figure 2 plots the minimum, average and maximum one-way delay experienced by the two reference streams. It can be noticed that in case of no congestion one-way delay is extremely stable: the difference between minimum and maximum is almost negligible for all the traffic loads tested and for both BE and LBE traffic. The graph also shows that the maximum one-way delay experienced by BE packets is slightly higher than in case of LBE packets. This could be related to the fact that BE test traffic shares its queue with production BE traffic, which is primarily TCP and characterized by a variety of packet sized. However, this phenomenon is the subject of future analysis.

Instantaneous Packet Delay Variation. In case of no congestion, also the instantaneous packet delay variation (IPDV)[15] experienced by one BE stream

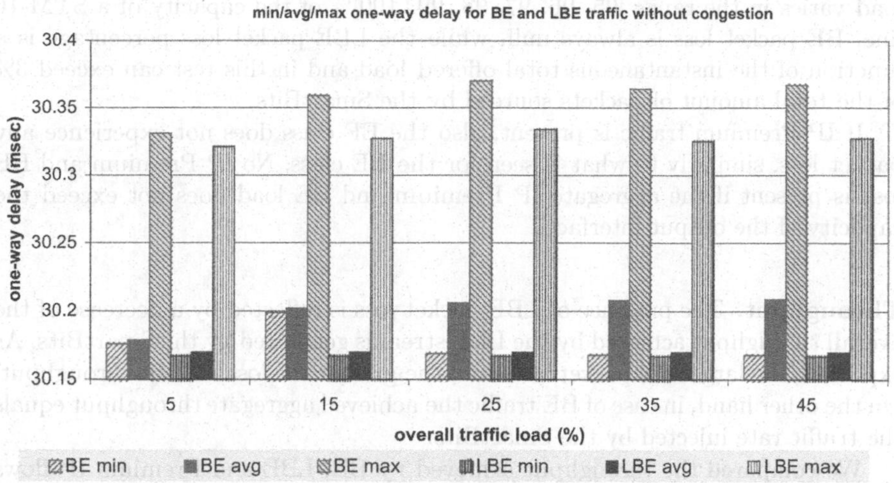

Fig. 2. One-way delay for BE and LBE flows and different traffic loads without congestion

and one LBE stream, where the two flows are run concurrently and produce one half of the overall load, is comparable for the two classes when the traffic volume varies in the range: 10, 50%.

Results show that for both services IPDV is almost negligible: the maximum IPDV recorded during the test session was experienced by the BE stream and was approximately equal to only 11 μsec. In this case, IPDV performance was measured by injecting SmartBits traffic from DE to IT. For a packet sample of 100 consecutive packets, for each traffic class IPDV was computed by calculating the absolute value of the difference of the one-way delay experienced by two consecutive packets. Even if the clocks of the sending and receiving device were not synchronised, the clock offset did not affect the IPDV measurement, since IPDV is a relative metric. The clock skew of the two devices is negligible during each test session, given the short duration equal to 10 sec.

5.3 LBE, BE and Premium IP Performance in Case of Congestion

Packet Loss. Independently of the weight and priority values assigned to the BE and LBE queues, no packet loss has ever been experienced by BE traffic in case of congestion produced by LBE packets. If both the BE and LBE traffic class are active at the same time and the output interface capacity is exceeded – but the BE offered load is less than the available output capacity – no BE packet loss is ever reported both by the BE queue statistics of the congested interface and by the per-flow packet loss statistics provided by the SmartBits.

For example, when injecting four UDP streams – three LBE and one BE flow – so that the BE offered load is 25% of the overall test traffic, if the aggregate load varies in the range [95, 96, 97, 98, 99, 100]% of the capacity of a STM-16 line, BE packet loss is always null, while the LBE packet loss percentage is a function of the instantaneous total offered load and in this test can exceed 3% of the total amount of packets sourced by the SmartBits.

If IP Premium traffic is present, also the EF class does not experience any packet loss, similarly to what is seen for the BE class. No IP Premium and BE loss is present if the aggregate IP Premium and BE load does not exceed the capacity of the output interface.

Throughput. The presence of LBE packet loss is reflected by a decrease of the overall throughput achieved by the LBE streams generated by the SmartBits. As expected, the larger the packet loss rate, the greater the loss in LBE throughput. On the other hand, in case of BE traffic the achieved aggregate throughput equals the traffic rate injected by the SmartBits.

We compared the throughput achieved by BE, LBE and Premium IP flows injecting traffic at same output rate. The test was run by sourcing 10 streams: seven of them are LBE, three are BE and one is Premium IP. The aggregate load was increased from 50 to 100%. It can be seen that the throughput of three flows, one of each class of service, is comparable only in case of no congestion. As soon as packet loss occurs (for an overall traffic load of 100%), only the LBE stream is affected.

One-Way Delay. One-way delay measurement in case of congestion was based on the same traffic pattern described in the previous section for one-way delay measurement in case of no congestion, i.e. with 100 streams of which 2 reference streams (one BE and one LBE) are sourced and received by the same SmartBits. In this case the overall amount of test traffic produced by the SmartBits is higher and varies in the range: 95, 100%.

While no effect on one-way delay could be observed in case of no congestion for different traffic loads, a different behaviour is shown in case of congestion.

LBE traffic experiences an increase of both average and maximum one-way delay when congestion starts, while the minimum latency is constant. Also BE traffic experiences a slight increase in one-way delay, but in case of BE traffic the increase is negligible. The maximum difference between minimum and maximum one-way latency for LBE traffic – experienced with 100% of overall traffic – is 1.865 msec, while the maximum difference for LBE traffic, which was observed in similar traffic load conditions, is only 0.537 msec.

More information about one-way delay experiments can be found in [16].

Instantaneous Packet Delay Variation. IPDV was tested with the same traffic profile used in case of no congestion, i.e. with two individual flows: a BE and a LBE flow generated by the SmartBits in Germany and received by the SmartBits in Italy. In this case, the overall traffic load is higher and varies in the range: 95, 100%.

The maximum LBE IPDV tends to increase with congestion, i.e. for high traffic loads, while the minimum and average IPDV are not dependent on either traffic load or congestion. We computed the IPDV frequency distribution over a sample of 100 consecutive LBE packets. We observed that the two distributions with congestion (100% of the capacity of a STM-16 interface) and without congestion (95% of the same capacity) are equivalent.

6 Conclusions and Future Work

In this article we have proposed a service description for a new LBE service whose primary function is to be able to make use of available bandwidth, but in such a way as to always defer to BE (or better) traffic where congestion occurs (LBE packets are always dropped before BE packets). We have run a first extensive set of tests to evaluate this proposed LBE service. Results are encouraging in that they largely validate the feasibility of operating an LBE service on the GÉANT backbone. The LBE traffic in the configuration tested does not adversely affect the regular BE or Premium IP traffic. As concluded in section 5.3, no significant influence on BE TCP and UDP performance metrics was observed during congestion.

In particular, if the amount of available bandwidth in a given router is not exceeded by BE and Premium IP traffic, no packet loss for these two classes is introduced by the addition of LBE traffic. In addition, congestion seems to cause an increase in BE average one-way delay that we think is almost negligible. The one-way delay and IPDV profile of LBE traffic itself is not particularly affected by congestion either.

The tests performed to date focus on the network backbone GÉANT. The results are also applicable to NREN networks. However, we note that in the context of GÉANT most network congestion is occurring at the edges of the network; thus implementation at the edge (university access links) is also important, especially where a university may be receiving (or sending) LBE traffic and giving it equal treatment to BE traffic at the campus edge router

It is also important to note that it is the sender of the data that must mark the traffic with the LBE DSCP. The general question of LBE signalling, and of voluntary against enforced use of LBE, is open for further investigation.

Having identified a number of scenarios to which LBE can be applied, the logical next step is to progress with the implementation of LBE on the GÉANT backbone, to promote its adoption within the NRENs (or at least DSCP transparency – the routers do not reset the LBE DSCP when observed) and between NREN sites and Internet2(Abilene) network.

To help in this process, we will also run a further set of experiments, where we evaluate the use of LBE where the intermediate network may offer only DCSP transparency, and the queuing and drop policy is applied at the network edge (the university) where congestion occurs. We have shown how the LBE service presented here is interoperable with Scavenger by use of a common DSCP value. We also plan to identify LBE trial applications for use with other networks (e.g.

in Japan) and to promote an interoperable service in those networks. Finally, we observe that the implementation of LBE is IP-independent. This article only considers IPv4 networks. We hope to be able to run an LBE service over IPv6 in collaboration with the 6NET project[17].

7 Acknowledgments

We thank Spirent for having given us the possibility to use SmartBits 600s on a loan basis and the SmartFlow 1.50 software, and for their support. We are also grateful to Juniper Networks for the continuous technical support provided during our test sessions. We thank the TF-NGN LBE interest group[18] especially for the collaboration in the definition of the set of DiffServ QoS application scenarios described in this paper. We are also extremely grateful to the colleagues from the DataGrid[11] Work Package 7 for the collaboration provided for the definition of the LBE test programme and their participation to the tests reported in this article.

References

1. *Specification and Implementation Plan for a Premium IP service,* GÉANT Deliverable 9.1 with two addenda
2. *Premium IP service,* `http://axgarr.dir.garr.it/~cmp/tf-ngn/IPPremium.html`
3. A. Charny et al., *An Expedited Forwarding PHB,* RFC 3246, March 2002
4. S. Blake, D. Black, M. Carlson, E. Davies, Z. Wang, *An Architecture for Differentiated Services,* RFC 2475, Dec 1998.
5. Bless, R. et al. *A Lower Effort Per-Domain Behavior for Differentiated Services,* Internet-Draft, November 2002, `draft-bless-diffserv-pdb-le-01.txt` (work in progress)
6. The QBone Initiative, *QBone Scavenger Service*
7. *QBone Premium IP service,* `http://qbone.internet2.edu/premium/`
8. K. Nichols et al., *Definition of the Differentiated Services Field (DS Field) in the IPv4 and IPv6 Headers,* RFC 2474, December 1998.
9. D. Black et al., *Per Hop Behavior Identification Codes,* RFC 3140, June 2001.
10. *GÉANT,* `http://www.dante.net/geant/`
11. *The DataGrid project,* `http://eu-datagrid.web.cern.ch/`
12. *Testing of Traffic Measurement Tools,* GÉANT Deliverable 9.4
13. *The Netperf Home Page,* `http://www.netperf.org/`
14. G. Almes et al., *One-way Delay Metric for IPPM,* RFC 2679, September 1999.
15. C. Demichelis, P. Chimento, *IP Packet Delay Variation Metric for IP Performance Metrics (IPPM),* RFC 3393, November 2002
16. *Experiments with LBE Class of Service,* GÉANT Deliverable 9.9
17. *The 6NET Project Home Page,* `http://www.6net.org/`
18. *TF-NGN LBE WG home page,* `http://www.cnaf.infn.it/~ferrari/tfngn/lbe/`

TStat: TCP STatistic and Analysis Tool*

Marco Mellia[1], Andrea Carpani[2], and Renato Lo Cigno[1]

[1] Politecnico di Torino - Dipartimento di Elettronica
C.so Duca degli Abruzzi - Torino - 10129 - Italy
{mellia,locigno}@mail.tlc.polito.it
[2] Vitaminic S.p.A
Via Cervino, 50 - Torino - 10155 - Italy
ancarpan@vitaminic.net

Abstract. Internet traffic analysis is one of the core topics of research in the evolution and planning of the next generation integrated networks. In spite of this fact, standard, open source tools for the collection and, most of all, the elaboration of traffic data are very few and far from comprehensive and easy to use. This paper present a new tool named `Tstat`, that allows the collection of traffic data and is able to reconstruct the traffic characteristics at several different logical level, from the packet level up to the application level. `Tstat` is made available to the scientific and industrial community [1] and, as of today, offer more than 80 different types of measurements, starting from classical traffic volume measurements, up to sophisticated analysis regarding the round trip times measured by TCP or the loss probability and correlation of each flow.

One of the key characteristics of `Tstat` is that, though it is capable of analyzing the traffic at the flow level, it is an entirely passive tool, that does not alter in any way the traffic pattern at the network interface where its data collection module is installed.

1 Measuring Traffic in the Internet

Measurements are the primary instrument to understand physical laws and to verify models predictions. Measurements also provide the final response concerning the validation of complex systems. Thus it should not surprise that traffic measurements are the primary source of information in networking engineering. Dimensioning and planning of telecommunication networks are based on traffic models built upon traffic measurements. While the problem of traffic characterization in traditional, circuit switched, telephone networks, was successfully solved decades ago, the definition of satisfactory traffic models for integrated networks like the Internet still puzzles researchers and engineers. The failure of the classic traffic assumptions,[2, 3], only enhance the importance of traffic measurements and, most of all, of traffic stochastic analysis and modeling.

* This work was supported by the Italian Ministry for University and Scientific Research through the PLANET-IP Project.

M. Ajmone Marsan et al. (Eds.): QoS-IP 2003, LNCS 2601, pp. 145–157, 2003.

The humongous size assumed by the Internet, continuously challenges data collection and, most of all, the organization and analysis of the enormous amount of collected data. First of all, data traffic must be characterized to a higher level of detail than voice traffic, since the diversities in the services offered and carried by the network is much wider in data networks, where different applications coexist and interact together. This inherently identifies different levels of analysis, that requires the collections of data at the session, flow, and packet level, while circuit switched traffic is satisfactorily characterized by the connection level alone. Second, the client-server communication paradigm implies that the traffic behavior does have meaning only when the forward and backward traffic are jointly analyzed. This problem makes measuring inherently difficult; it can be solved if measurements are taken on the network edge, where the outgoing and incoming flows are necessarily coupled, but it can prove impossible in the backbone, where the peering contracts among providers often disjoint the forward and backward routes [4]. Notice that the presence of purely connectionless services in the Internet, makes the notion of connection itself becomes quite fuzzy. Finally, the complexity and layered structure of the TCP/IP protocol suite, requires the joint analysis of traffic at least at three different layers (IP, TCP/UDP, Application/User layer) in order to have a picture of the traffic clear enough to allow the interpretation of data.

Measurement based works are widespread in the literature, but they are often hard to compare, given the different perspective and the different technique used. Early works like [5, 6, 7, 2, 8] examined traffic data to characterize the network, protocols or the user behavior. After the birth of the Web, lots of effort has been devoted to study caching and content delivery architecture, analyzing traces at the application level (typically log files of web or proxy servers) [9, 10, 11]. In [12], the authors use large traces collected at the university campus at Berkeley to characterize the HTTP protocol. In [13] the authors start from data collected from a large Olympic server in 1996 to understand TCP behavior, like loss recovery efficiency and ACK compression. In [14], authors analyzed more than 23 millions of HTTP connections, and derived a model for the connection interarrival time. More recently, the authors of [15] analyze and derive models for the Web traffic, starting from the TCP/IP header analysis.

None of the above works, however, share a common methodology of data collection or analysis. Even the architectural point where the data is collected is different, since the data is quite often collected in end-hosts and not directly from the network, rebuilding the status of TCP connections, independently from the application level.

Measuring tools can be broadly classified either as *passive* tools, which analyze traces without injecting traffic into the network, or as *active* tools, that derive performance indices injecting traffic into the network. Some example of active tools are the classic `ping` or `traceroute` utilities, or the recently introduced `TBIT` [16]. While these tools allow to perform detailed measurements in a controlled scenario, their use on large scale is hampered by the intrusive nature of the approach. On the other hand, passive tools allow to infer network dynamic without interfering with the network itself. Many of these freely available tools are based on the `libpcap` [17] library developed with the `tcpdump` tool [18], that allow different protocol level analysis. For example, `tcpanaly` is a tool for inferring a TCP implementation's behavior by inspecting the TCP connection packet trace. Another interesting tool is `tcptrace` [19], which is able to rebuild a TCP connection status from traces, matching data segments and ACKs. For each connection, it keeps track of elapsed time, bytes/segments sent and received, retransmissions, round trip times, window advertisements, throughput, etc. At IP level, `ntop` [20] is able to collect statistics, enabling users to track relevant network activities including traffic characterization, network utilization, network protocol usage.

Indeed, none of the above tools offer means for statistical data analysis and post-processing and, to the best of our knowledge, there aren't any such tools available to the research community. The main contribution of the work we present here is thus the development of a novel, expandable tool-set for traffic measurements, consisting of several parts. This paper presents two different parts of the tool-set, that have a major role in the field of Internet traffic measurements:

- A new engine, called `Tstat` and described in Sect. 2, for gathering and, most of all elaborating, Internet measurements, which is freely available to the community;
- A public database where the analysis performed on collected traces (anyone can upload his own traces on the database) is made available through a Web interface allowing users to post-process and retrieve collected statistics.

In closing the paper we present some examples of the 88 different measurements available in `Tstat` as of today.

In the remaining of the paper we assume that the reader is familiar with the Internet terminology, that can be found in [21, 22, 23] for example.

2 Tstat

The lack of automatic tools able to produce statistical data from collected network traces was a major motivation to develop a new tool, called `Tstat` [1], which is able to offer network managers and researchers important information about classic and novel performance indices and statistical data about Internet traffic. `Tstat` has been developed starting from standard software libraries, and runs on "off the shelf" hardware.

`Tstat` is build up by a trace analyzer, which takes as input either previously recorded traces stored in various dump formats, or measurements collected in real time and pipelined into `Tstat`. It internally collects several statistical histograms that are dumped on files at regular interval, so that a measure database is made available to the user. A set of post-processing routines make it possible to retrieve statistics on a particular measurement, and to aggregate data (e.g., superimpose all the traffic collected during business hours). The post-processing routines can easily produce both time plots and distribution plots. A simple web interface can be used to retrieve information from the database and directly produce plots.

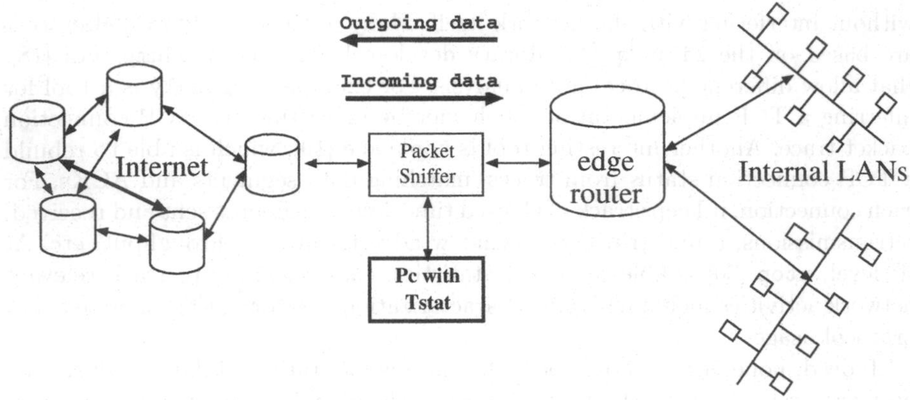

Fig. 1. Measuring setup

2.1 Trace Analyzer

The core features of `Tstat` are implemented in the trace analyzer, which is based on the source code of `TCPtrace` [19]. The basic assumption of the trace analyzer is that the traffic was measured on an edge node, so that the incoming and outgoing flows can be correlated searching for client-server bidirectional connections. Figure 1 show the measure environment used at the Politecnico di Torino to capture and analyze traces, which guarantees that the analyzer is fed by all the data coming in and going out the internal LANs.

A trace is the uninterrupted bidirectional flow of dumped packet headers, whatever it come out to be. For instance, in our measuring campaigns, since we could not pipeline the measurements in real-time, we used traces of 6 hours, to avoid the risk of excessive file sizes, but a trace can be anything from a short measurement on a slow link to weeks of seamless data collection on a fast interface pipelining the collected headers in the trace analyzer.

Whenever a new IP packet is observed, it is first analyzed at the IP level, which takes care of all the statistics at packet layer. Then, if it is a UDP or TCP segment, all the per-segment statistics are performed. In case it is a TCP segment, it goes through the TCP flow analyzer module. This is the most complex part of `Tstat`. First, the module tries to find if the segment belongs to an already identified flow, using the classic n-tuple. If not, then a new flow is identified only if the `SYN` flag is set, as `Tstat` processes only *complete* flows. Once the segment is successfully identified, all the statistics related to the flow analysis are performed. Finally, in case the segment correctly closes a flow, all the flow-level statistics are updated.

A very efficient memory management core has been implemented, based on reuse techniques that minimize memory usage. Hashing functions and dynamically allocated lists are used to identify if the packet currently under analysis belongs to any of the tracked flows. When a flow closing sequence is observed, the flow is marked as *complete*, and all the TCP flow-level measures are computed.

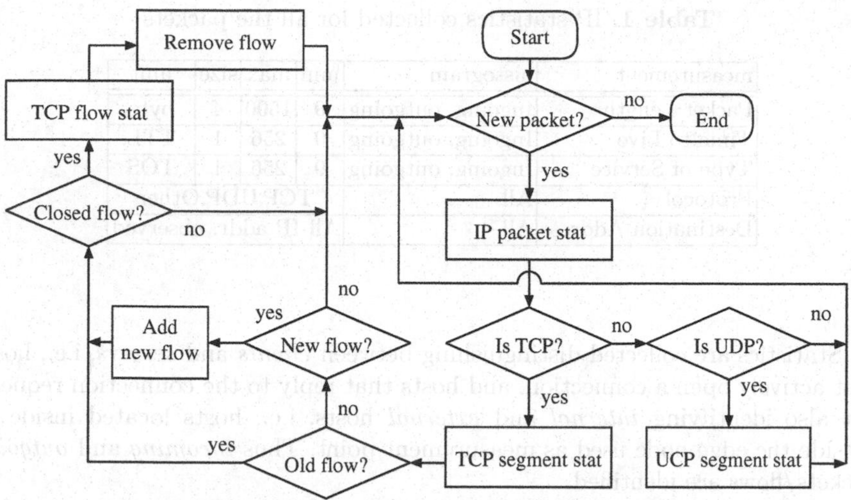

Fig. 2. Block diagram of the trace analyzer.

Given the possibility that a closing sequence of a flow under analysis is never observed (for example because of host or network misbehavior), a timer is used to trigger a garbage collection procedure to free memory: if no segments of a given flow are observed in between two garbage collections, the flow status memory used is freed, and the flow is considered abnormally terminated. Similarly, TCP segments that belong to flows whose opening sequence is not recorded (because either were started before running the trace analyzer, or early declared closed by the garbage procedure) are discarded and not analyzed. In this paper, we set the garbage collector timer to 15 minutes, which was shown to guarantee very good flow identification [24].

Figure 2 shows the schematic block diagram of the trace analyzer, and in particular the logical block through which the program moves for each analyzed packet. To give an intuition of the speed `Tstat` can guarantee, using a off-the-shelf 1GHz computer with 1Gb of RAM, a 6 hours long peak trace from a 24 Mbit/s link is precessed in about 15 minutes, with a total memory footprint of only 50MB. To process a 3 hours trace recorded on a OC48 backbone link (link utilization was about 100Mbps), Tstat used up to 900MB of RAM in about 1 hour of CPU time. In this second scenario, only 13% of observed flows were bidirectional, so that the majority of tracked flows are stored in memory until the garbage collector procedure destroys them, as a correct TCP closing procedure cannot be observed.

`Tstat` has been designed in a very modular way, so that it is easy to integrate modules for new measures. For example, we are currently considering the extension of `Tstat` to analyze also the RTP protocol with a “RTP segment stat,” so as to be able to analyze real-time, multimedia traffic.

Table 1. IP statistics collected for all the packets

measurement	histogram	min	max	size	unit
Packet Length	Ingoing, outgoing	0	1500	4	byte
Time to Live	Ingoing, outgoing	0	256	1	TTL
Type of Service	Ingoing, outgoing	0	256	1	TOS
Protocol	All	TCP,UDP,Other			
Destination Address	All	All IP addr. observed			

Statistics are collected distinguishing between *clients* and *servers*, i.e., hosts that actively open a connection, and hosts that reply to the connection request, but also identifying *internal* and *external* hosts, i.e., hosts located inside or outside the edge node used as measurement point. Thus *incoming* and *outgoing* packets/flows are identified.

Instead of dumping single measured data, `Tstat` builds a histogram for each measured quantity, collecting the distribution of that given quantity. Every 15 minutes, it produces a dump of all the histograms it collected. A total of 88 different histogram types are available, comprising both IP and TCP statistics. They range from classic measurements directly available from the packet headers, to advanced measurements, related to TCP. A complete log also keeps track of all the TCP flow analyzed, and is useful for post-processing purposes.

IP Level Statistics In more details, IP level measurements are collected looking at the relative field in each packet header. They refer to the packet length, Time to Live, Type of Service field, and Protocol type carried in the payload. `Tstat` also is able to keep track of all the destination IP addresses seen, and the hit number of each address, i.e., the number of packets with that address, is computed. Table 1 summarize the IP level histograms collected. It shows the collected measurement in the first column, the type of histogram generated in the second column, the histogram starting point, ending point, and bin size in the third, forth and fifth column, while the measuring unit is reported in the last column. For the protocol type histogram is slightly different, as it distinguish only between TCP, UDP or other upper level protocols. Also the Destination Address histogram is different from the previous one, as `Tstat` is able to collect the hit number for all the possible IP addresses.

TCP Level Statistics The TCP flow analysis allows the derivation of novel statistics, that ranges from the classic flow length analysis to the novel window evolution and out of sequence patterns observed. Being TCP a connection oriented protocol, interesting measurements are derived only by looking at the whole flow instead of each segment separately, and data segments must be correlated to backward acknowledgment segments. Having this in mind, `Tstat` is able to collect statistics on the following measurements:

Flow duration: These histograms collect the duration of the flows, i.e., the time elapsed from the first SYN segment up to the last segment (the FYN-ACK for correctly completed flows). The histogram starts from 0 seconds and ends at 72 seconds; the size of each bin corresponds to 5ms. Flows that are longer than 72 seconds are accumulated on the last bin. Both half duplex flows are separately measured.

Flow length in segments: These histograms collect the total length of the flow expressed in number of segments, from the initial SYN segment to the last segment. The histogram starts from 0 packets and end to 100 packets, while the size of each bin correspond to one packet. Note that the minimum number of a "well-formed" flow is 6 segments (three way handshake plus three segments for the bidirectional closing procedure). Thus flows that exhibits less than 6 packets are "malformed" flows. For example a refused connection counts 2 packets (SYN + RST segments), or a connection to a dead host counts only one segments (SYN), etc.

Flow length in bytes: These histograms collect the total length of the flow expressed in Bytes carried in the payload of the TCP packets. Given that length of flow can be relatively small (e.g., for Web pages) and very large (given the heavy tail distribution of flow length), two histograms are collected: the*small bin* one starts from 0 Bytes and ends at 50 kBytes; the size of each bin corresponds 50 Bytes; the *large bin* starts from 50 kBytes and ends at 50 MBytes; the size of each bin in the case corresponds to 50 kBytes.

Port number: These histograms collect Port used in each packets. It records the minimum number between the source port and the destination port, as usually this is the value associated to a particular service. The histogram starts from 0 and ends to 1024; each bin correspond to a particular port number. If the minimum port number exceeds 1024, it is accumulated in the last bin. A first histogram is updated every segment, while a second one is updated only on the first SYN segment. This allow to distinguish among number of flows, or number of segments per port.

Options: These histograms collect *Options* that are negotiated during the three way handshake. The options recognized by Tstat are SACK, Window Scale and Timestamp, as defined in [25, 26]. For each option, the number of flows that have successfully negotiated the option, flows in which only the clients offered the option, flows in which only the servers replied to the option without a client request and flows in which no option is offered/replied are distinguished.

Maximum Segment Size: These histograms collect statistics about the MSS negotiated in the three way handshake, as stated in [27].

Window estimate: These histograms collect statistics about both the *in-flight size* and the *advertised window* during the flow life. In particular, looking at the "in flight size", i.e., the difference between the sequence number and the corresponding reverse acknowledge number, it is possible to derive an estimation of the congestion window: they indeed correspond to the number of bytes already transmitted by the sender but not already acknowledged by the receiver. The advertised window on the contrary is directly derived from the TCP header analysis.

Duplicate bursts: These histograms collect statistics about number of Duplicated Burst size (in bytes) observed in each flow. A Duplicated Burst starts from the first byte already seen during the data transfer, and ends with the last duplicated, contiguous byte observed. Note that a Duplicated Burst can be observed if either the segment is retransmitted due to a loss that happened "after" the measurement point (i.e., on the external, wide area network side), or to useless retransmissions from the sender, caused either by ACK losses, or by a too aggressive recovery phase.

Out of Sequence burst: These histograms collect statistics about the number of Out of Sequence Burst size (in bytes) observed in each flow. An Out of Sequence Burst starts from the first out of sequence byte, and ends to the last contiguous byte observed before the missing data is received to fill the missing one. Note that a Out of Sequence can be observed if either the segment is reordered in the network, or as a consequence of a segment drop during the path from the source to the measurement point. In both cases, the Out of Sequence Burst can be identified only if at least one segment is correctly received after the missing one, that is if the congestion/receiver windows permit the transfer of at least two segments.

Round Trip Time: These histograms collect statistics derived by running the RTT estimation algorithm implemented in TCP on the observed data segments and their corresponding acknowledgment segments. Both the number of valid samples and the resulting average and standard deviation RTT estimation are collected. In particular, the number of valid samples according to the Karn's algorithm, the maximum, minimum and average Round Trip Time observed during flows' life, and the standard deviation are considered.

A detailed description of all the histograms is available in [1].

2.2 Post-processing Engine

The trace analyzer generates a database composed of all the histograms relative to an input trace. A number of post-processing tools that allow the user to extract the desired measurements from this database, are made available to the end user. They are made of a number of simple and flexible scripts that enable to aggregate in any desired way the recorded data, and to directly plot the relative performance figures. Moreover, a simple web page [1] is available as graphical user interface, to help the user use the post-processing engine.

In particular, the user can select a subset of the analyzed traces, specifying the starting date and the ending date, and the time slice for which the data is to be analyzed. For example it is possible to consider only the data collected from 10 AM to 2 PM, on days from Monday to Friday, or the traffic on weekend nights. Both time plots and cumulative/distribution plots can be produced.

After that a plot has been produced, a Java Applet allows the user to modify it, hiding lines, zooming in/out, changing the axis scales, etc. It is also possible to download the histogram data as a simple ASCII file, and thus easily usable by other programs, such as spreadsheets, or statistical tools.

3 Sample Measurements

We performed a trace elaboration using Tstat on data collected on the Internet access link of Politecnico di Torino, i.e., between the border router of Politecnico and the access router of GARR/B-TEN [28], the Italian and European Research network. The access link capacity is 28 Mbit/s. Within the Politecnico Campus LAN, there are approximately 7,000 access points; most of them are clients, but several servers are regularly accessed from outside institutions. Here we report two sample analysis that can be produced using Tstat, whose aim is only enable the reader to have a flavor of informations obtained with Tstat. We encourage all interested readers to connect to Tstat web page [1] and obtain any result they are interested in.

As first example, we report in Figure 3 the percentage of TCP, UDP or other protocols that were observed during a measurement time interval of one week. The traffic percentage is referred to the link capacity. As it can be expected, the plot exhibits a periodic behavior, which reflects the day/night usage of the network. Looking at the protocol distribution, the plot confirms that the majority of the traffic is composed of TCP connections, while UDP traffic is confined to a few percentage points; other protocols are practically negligible. Notice that the analysis is done at the IP level between two routers, hence all the ARP or non-IP traffic within the Politecnico LAN is not seen at all.

The results presented in Figure 3 are no news: the traffic level during business hours is nearly two orders of magnitude larger than during nights or week ends; however this allow the selection of business hours only (identified from 8 AM to 6 PM) for further traffic analysis.

As a second example of the available measurements, we selected a generally disregarded characteristic of the traffic, i.e., the Round Trip Time (RTT) between the measuring point and the TCP termination points, defined as the time elapsed from the instant Tstat observes a data segments and the instant Tstat observes the corresponding ACK on the reverse path. Given the measuring setup as described in Figure1, we can clearly identify two "half-connections[1]:" the first one is on the Politecnico LAN, while the second one is on the "big Internet," i.e., the WAN. The two can be merged to obtain the connection RTT, however, it is more interesting to analyze the different characteristics of the two components separately. We have defined four different figures for each flow: i) the minimum observed RTT; ii) the maximum observed RTT, iii) the average RTT (taking

[1] The term 'half-connection' here should not be confused with the half connection meaning of the TCP protocol, where an half connections is indeed a one-way connection. This term idetifies here the fact that the connection reconctruction done by Tstatis done somewhere between the source and the receiver: One half connction is between the source and Tstat and the other one is between Tstat and the receiver.

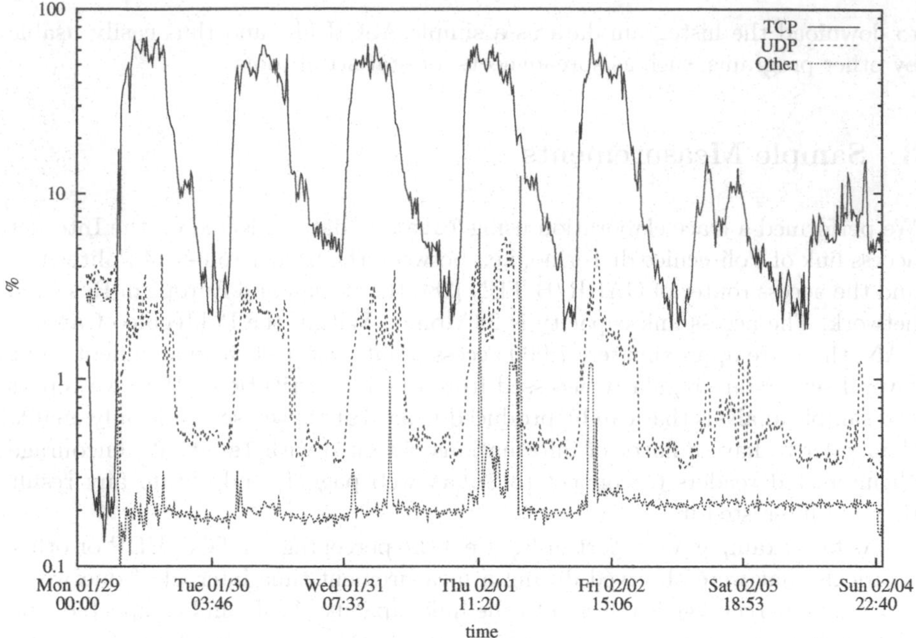

Fig. 3. Protocol type carried in IP packets measured during one week.

into accounts Karn's algorithm to discard ambiguous samples); iv) the standard deviation of the RTT computed as above.

Figure 4 reports three different plots, which are the measured histograms of the above defined figures; the histograms bins are 10 ms wide. The upper one is relative to the LAN RTT and the middle one to the WAN RTT (both plots reports the minimum, maximum and average RTT), while the lower one draws the standard deviation of both. The first interesting observation is that the LAN RTTs are not at all negligible. Only 60% of the flows exhibit an average RTT smaller than 10 ms, while the 90° percentile is around 60 ms and the 99° one is around 380 ms. The extended LAN of our institution is fairly complex, with several Ethernet switches and hubs that could increase the average delay. Another possible reason could be referred to the end host processing times. Moreover, looking at the distribution of the maximum RTT observed during flow life, there is a not negligible peak between 100 ms and 200 ms, as shown by the magnified inset. We have clues that this is due to the ARP request necessary to resolve the IP address into the Ethernet one. This might be exacerbated by the proxy-arp configuration of our LAN environment.

Considering the WAN RTT figures, we can distinguish roughly 3 classes of flows: flows that exhibit RTT below 50 ms, flows whose RTT is about 120 ms, and flows whose RTT is about 200 ms. These correspond for example to flows within Europe, flows from the US Est and West cost respectively.

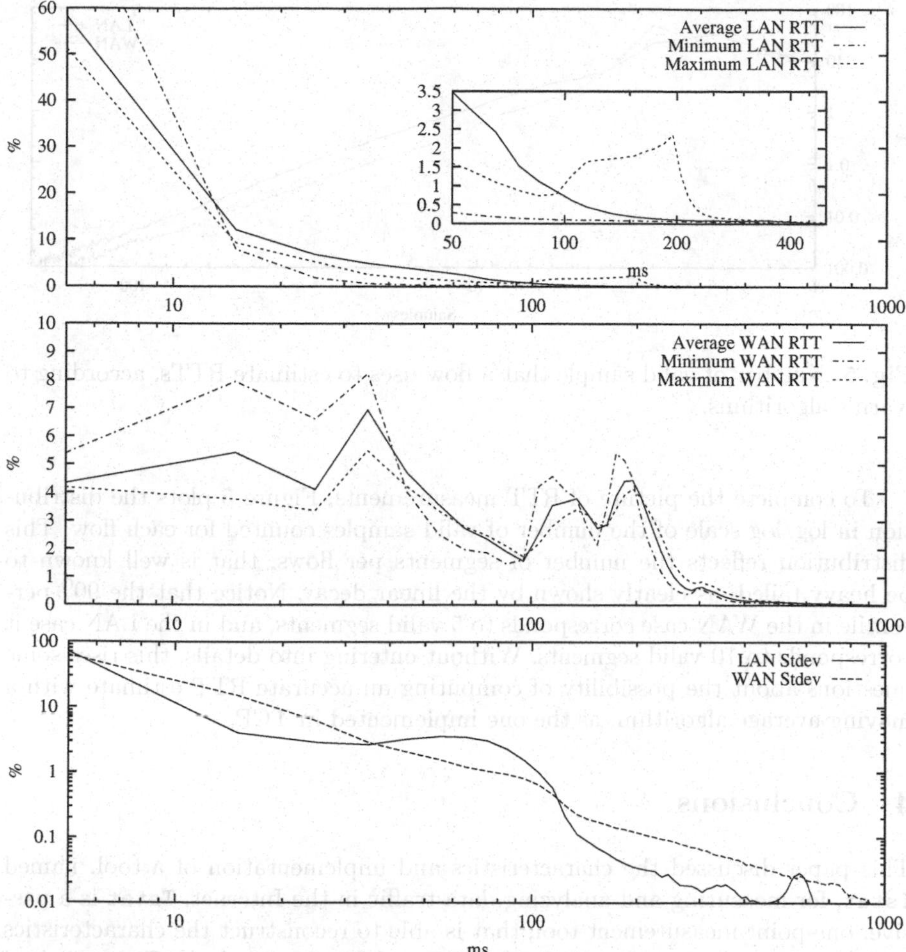

Fig. 4. Minimum, maximum, average and standard deviation of the Round Trip Time observed in the LAN and WAN.

Considering the standard deviation figure (reported in logarithmic scale), there are at least two considerations in order. First is quite clear that the standard deviation is generally small, as confirmed by the 90° percentile which is about 60 ms for both LAN and WAN. Indeed the WAN standard deviation distribution is linear in log/log scale, indicating a slow decaying distribution. In the LAN case, the shape is more complex, with peaks that are not easily explained but are probably routed in the same phenomena discussed for the LAN RTT distribution.

It must be noted that these are per-flow results, where all flows have the same weight in the distribution. Clearly this cannot be neither a measure of the Internet RTT distribution, since measures are taken from a given edge router, nor a measure of the RTT of each TCP segment.

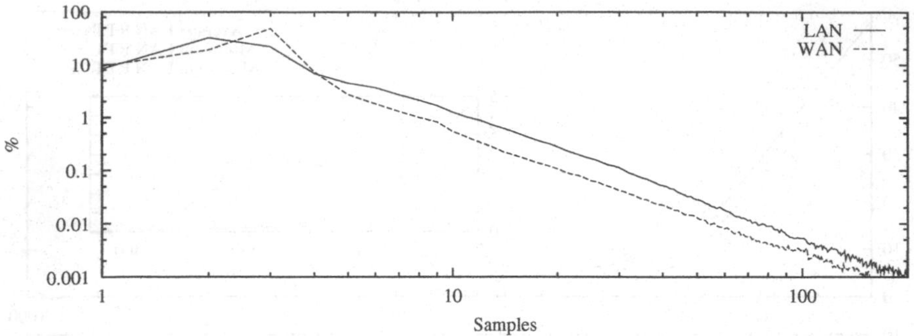

Fig. 5. Number of valid sample that a flow uses to estimate RTTs, according to Karn's algorithms.

To complete the picture of RTT measurements, Figure 5 plots the distribution in log/log scale of the number of valid samples counted for each flow. This distribution reflects the number of segments per flows, that is well known to be heavy tailed, as clearly shown by the linear decay. Notice that the 90° percentile in the WAN case corresponds to 5 valid segments, and in the LAN case it corresponds to 10 valid segments. Without entering into details, this rises some questions about the possibility of computing an accurate RTT estimate with a moving average algorithm, as the one implemented in TCP.

4 Conclusions

This paper discussed the characteristics and implementation of a tool, named `Tstat`, for measuring and analyzing data traffic in the Internet. `Tstat` is a passive, one-point measurement tool, that is able to reconstruct the characteristics of the Internet traffic at the IP, TCP/UDP and Application level. The main and novel characteristics of `Tstat` are its capability of perform on line measurements at the TCP level, its flexibility, and the availability of post-processing tools that allow the user to easily derive graphs using a simple Web interface. `Tstat` is part of a wider research project aiming at the analysis of Internet traffic with the purpose of gaining enough insight to allow the development of novel Internet planning tools. While data collection is going on both at our university premises and in other institutions, we plan to enhance the capabilities of `Tstat` to include other interesting figures.

References

[1] Mellia, M., Carpani, A., Lo Cigno, R.: `Tstat` web page. http://tstat.tlc.polito.it/ (2001)

[2] Paxson, V., Floyd, S.: Wide-Area Traffic: The Failure of Poisson Modeling. IEEE/ACM Transactions on Networking **3** (1995) 226–244

[3] Crovella, M.E., Bestavros, A.: Self Similarity in World Wide Web Traffic: Evidence and Possible Causes. IEEE/ACM Transactions on Networking **5** (1997) 835–846

[4] Paxson, V.: End-to-end routing behavior in the Internet. IEEE/ACM Transactions on Networking **5** (1997) 601–615

[5] Caceres, R., Danzig, P., Jamin, S., Mitzel, D.: Characteristics of Wide-Area TCP/IP Conversations. ACM SIGCOMM '91 (1991)

[6] Danzig, P., Jamin, S.: tcplib: A library of TCP Internetwork Traffic Characteristics. USC Technical report (1991)

[7] Danzig, P., Jamin, S., Caceres, R., Mitzel, D., Mestrin, D.: An Empirical Workload Model for Driving Wide-Area TCP/IP Network Simulations. Internetworking: Research and Experience **3** (1992) 1–26

[8] Paxons, V.: Empirically Derived Analytic Models of Wide-Area TCP Connections. IEEE/ACM Transactions on Networking **2** (1994) 316–336

[9] Duska, B.M., Marwood, D., Feeley, M.J.: The Measured Access Characteristics of World-Wide-Web Client Proxy Caches. USENIX Symposium on Internet Technologies and Systems (1997) 23–36

[10] Fan, L., Cao, P., Almeida, J., Broder, A.: Summary Cache: A Scalable Wide-Area Web Cache Sharing Protocol. ACM SIGCOMM '98 (1998) 254–265

[11] Feldmann, A., Caceres, R., Douglis, F., Glass, G., Rabinovich, M.: Performance of Web Proxy Caching in Heterogeneous Bandwidth Environments. IEEE INFOCOM '99 (1999) 107–116

[12] Mah, B.: An Empirical Model of HTTP Network Traffic. IEEE INFOCOM '97 (1997)

[13] Balakrishnan, H., Stemm, M., Seshan, S., Padmanabhan, V., Katz, R.H.: TCP Behavior of a Busy Internet Server: Analysis and Solutions. IEEE INFOCOM '98 (1998) 252–262

[14] Cleveland, W.S., LinD, D., Sun, X.: IP Packet Generation: Statistical Models for TCP Start Times Based on Connection-Rate Superposition. ACM SIGMETRICS 2000 (2000) 166–177

[15] Smith, F.D., Hernandez, F., Jeffay, K., Ott, D.: What tcp/ip protocol headers can tell us about the web. ACM SIGMETRICS '01 (2001) 245–256

[16] Padhye, J., Floyd, S.: On inferring tcp behavior. ACM SIGCOMM 2001, San Diego, California, USA (2001)

[17] McCanne, S., Leres, C., Jacobson, V.: libpcap. http://www.tcpdump.org (2001)

[18] McCanne, S., Leres, C., Jacobson, V: Tcpdump. http://www.tcpdump.org (2001)

[19] Ostermann, S.: tcptrace (2001) Version 5.2.

[20] Deri, L., Suin, S.: Effective traffic measurement using ntop. IEEE Communications Magazine **38** (2000) 138–143

[21] Postel, J.: RFC 791: Internet Protocol (1981)

[22] Postel, J.: RFC 793: Transmission control protocol (1981)

[23] Allman, M., Paxson, V., Stevens, W.: RFC 2581: TCP Congestion Control (1999)

[24] Iannaccone, G., Diot, C., Graham, I., McKeown, N.: Monitoring very high speed links. ACM Internet Measurement Workshop, San Francisco (2001)

[25] Mathis, M., Madhavi, J., Floyd, S., Romanow, A.: RFC 2018: TCP Selective Acknowledgment Options (1996)

[26] Jacobson, V., Braden, R., Borman, D.: RFC 1323: TCP extensions for high performance (1992)

[27] Postel, J.: RFC 879: The TCP Maximum Segment Size and Related Topics (1983)

[28] GARR: GARR - The Italian Academic and Research Network. http://www.garr.it/garr-b-home-engl.shtml (2001)

End-to-End QoS Supported on a Bandwidth Broker

José M. Calhariz[1] and Teresa Vazão[1,2]

[1] INOV, Rua Alves Redol, 9 5° Andar, 1000-029 Lisboa, Portugal
jose.calhariz@inesc.pt, teresa.vazao@inesc.pt
[2] Instituto Superior Tecnico, Av. Rovisco Pais, 1049-001 Lisboa, Portugal

Abstract. The support of Quality of Service (QoS) is an emerging problem in the Internet, as the new multimedia applications demand for QoS. Differentiated Service is foreseen as a solution to QoS provisioning in core networks due to its scalability characteristic. However, at the access IntServ is the solution most adequate, as it provides QoS to individual traffic flows. Should an end-to-end service model be defined, both service models must be considered and thus, interworking and service conversion aspects must be studied. This paper addresses the management aspects, not yet covered by the standards, that basically deals with the problem of end-to-end QoS support. It presents an architecture for a QoS Management Platform, which is able to deal with all the kind of Service Level Agreements (SLAs), ranging from static to dynamic, access to end-to-end. The proposed architecture is able to perform IntServ to Diffserv service conversion. The implementation on a LINUX environment and the results of the performance evaluation are also described.

1 Introduction

The current Internet network architecture comprises only one service class, the traditional Best Effort, where the network attempts to deliver all traffic as soon as possible within the limits of its abilities, but without any QoS guarantee. Thus, the responsibility to cope with network transport impairments is entirely left up to the end systems. However, this scenario is changing rapidly, as users and service providers are faced with a new reality: IP networks are becoming popular as multimedia backbone networks. The emerging applications, such as Voice over IP, Videoconference or Virtual Laboratories require significant bandwidth and/or strict timing that cannot be met with the current Internet paradigm. The present success of Internet happens because customers are attracted to the new type of services, but soon they will require higher performance and QoS [1].

In the past few years, different types of solutions aimed to solve this lack of QoS appeared, namely Integrated Service (IntServ) [2] and Differentiated Service (DiffServ) [3] models. IntServ provides QoS by mapping each incoming traffic flow into one of three service classes – Best Effort, Controlled Load and Guaranteed Service – where traffic characteristics are strictly defined and resources reserved through a signalling protocol, named Resource Reservation Protocol

M. Ajmone Marsan et al. (Eds.): QoS-IP 2003, LNCS 2601, pp. 158–170, 2003.

(RSVP) [4]. IntServ is particularly adequate for those applications with stringent QoS requirements. However, trials realised in different experimental test-beds shown its lack of scalability. In DiffServ QoS is supported by a set of policy-based mechanisms that may be applied either to individual or aggregated traffic flows, providing support to scalable network architectures.

Several QoS trials have been realized using DiffServ model, either by manufacturers or by the scientific community [5] [6]. The first results achieved have shown that service fulfilment must be supported on over provisioned network, as there is no signalling support to resource allocation defined in the model. Hence, the concept of a Bandwidth Broker (BB) – a dedicated network element responsible for resource allocation – has been proposed to support QoS in a more effective way. Internet Engineering Task Force (IETF) is standardizing a similar approach, under the scope of the Policy Decision Framework.

Different projects and test-beds are being realised to validate this new concept [7] [8]. However, those are mainly focused on core networks entirely based on DiffServ model. If end-to-end network architecture is considered, it is foreseen that IntServ model is also provided at the access network to support RSVP-aware applications.

To provide effectively end-to-end QoS, a heterogeneous model must be defined that comprises both access and core network solutions. This paper, whose work is being carried within MOICANE project, addresses the problem of end-to-end QoS provisioning. It proposes QoS Management Platform architecture, based on the concept of a Bandwidth Broker, which gives support to static or dynamic service requests. Using a trial test-bed, experimental results are also presented, which demonstrate the grade of service that the Platform is actually providing.

The remaining part of this paper is organised as follows: Sect. 2 describes the network architecture, Sect. 3 presents the QoS platform architecture, Sect. 4 identifies some implementation issues, Sect. 5 describes the performance results and Sect. 6 presents conclusions and future work.

2 QoS Network Architecture

The basic principals of MOICANE network architecture are depicted in Fig 1.

The network consists of a DS region, based on a set of contiguous DiffServ islands, each one of them located on a different place and comprising a set of access networks and a core network. The access networks are supported on different technologies, ranging from wired Ethernet LANs to wireless 802.11b LANs. End-to-end communication takes place over the MOICANE backbone constituted by the interconnection of its core networks.

In MOICANE, QoS is basically provided by DiffServ, which means that all the routers must be configured to handle traffic according to the classes of service supported by the network. A set of multimedia applications aimed to provide an e-learning and virtual laboratory environment has been developed in the project [9]. Some of them use RSVP to request QoS and thus, to properly support

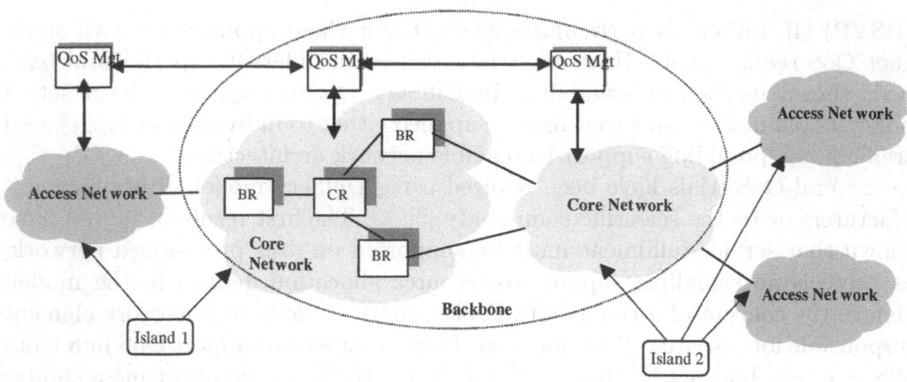

Fig. 1. The basic principals of MOICANE network architecture

them, IntServ is also used at some access islands. In such a case, RSVP messages will be tunnelled at the edge of DiffServ backbone and IntServ request will be mapped into DiffServ. Based on the current standardization work, the network supports one EF class [10] – for applications that require low loss, latency and jitter and assured bandwidth – and four AF classes [11], each one of them with three levels of drop precedence – for applications that require different grades of performance, which is expectable to be good, even if the network is overloaded.

Service Level Agreements (SLAs) are established at the edges of each core network, defining the committed performance objectives, regarding the negotiated DiffServ service classes [12]. Intra island agreements comprise the SLAs defined between each access and the associated core network and basically define the throughput reserved for EF and for AFx classes. Inter-island agreements involve the services of external service providers, such as GEANT that only support EF service class in some of the links. Thus, a tunnel is created between the two end-points of this connection, to carry MOICANE traffic and preserving its service class.

A core network comprises a set of Border Routers (BRs) that establish the interface with other networks and a set of Core Routers (CRs), used to interconnect BRs. All the BRs must be configured to classify, policy and mark the individual or aggregated flows, according to the set of rules defined in the SLAs. BRs and CRs perform queuing and scheduling.

Each core network has its own QoS Management Platform (QoS Mgt), which is responsible for service assurance and performance monitoring. The platform configures the routers DiffServ characteristics, monitors the QoS that is currently being provided and executes admission control procedures. If the SLA involves other domains, communication between peer QoS Management Platforms is also supported.

3 QoS Management Platform Functional Architecture

The architecture of the QoS Management Platform must support different types of SLAs and also core networks of different sizes and structures. Should all the types of customer expectations be considered, the platform must be able to provide the following SLAs:

- Static SLAs have to be supported to provide a basic set of services with different grades of quality, which are aimed to satisfy the vast majority of customer demands. The associated service classes must be specified to match a broad range of traffic that can be grouped due to its specific characteristics. This is the type of agreement used at the international connection.
- On demand SLAs must also be provided to satisfy those customers that temporarily want to request a service that is not available in the SLA they have defined with their provider.
- Finally, dynamic SLAs are needed to support to applications with stringent QoS requirements that request resources using a signalling protocol, such as RSVP.

To support core networks of different sizes, functional organisations and traffic volumes, the QoS Management Platform must be scalable, secure and, as much as possible independent of the network technology. Thus, an hierarchical structure with three levels have been defined, as it is depicted in Fig. 2:

- The top level of the hierarchy (service level) comprises the functional elements that directly deals with services, providing support to: (i) reception and response to service creation request's, modification or termination of existing ones, and (ii) evaluation the QoS that is actually being provided.
- The intermediate level (network level) comprises the functional elements that establish the interface between those at the top level that know the service, and those at the bottom that carry data traffic and implement the service.
- Finally, the bottom level (network element level) comprises functional elements that correspond to physical Network Elements (NEs) that are used to forward data, usually routers and switches.

Security mechanisms are supported in all the external interfaces and in the internal interfaces that establish communication between different physical elements. Within this scope, user authentication is provided, either to request services or to configure them, using public or private keys distributed in the fashion that best fits the structure of the network. Cipher mechanisms are also included in all the control information, namely: service and configuration.

4 Implementation Issues

One of the islands of MOICANE network has a core network entirely based on LINUX operating system, version 2.4, which includes DiffServ facilities in its kernel. Due to the small size of the network, the QoS Management Platform

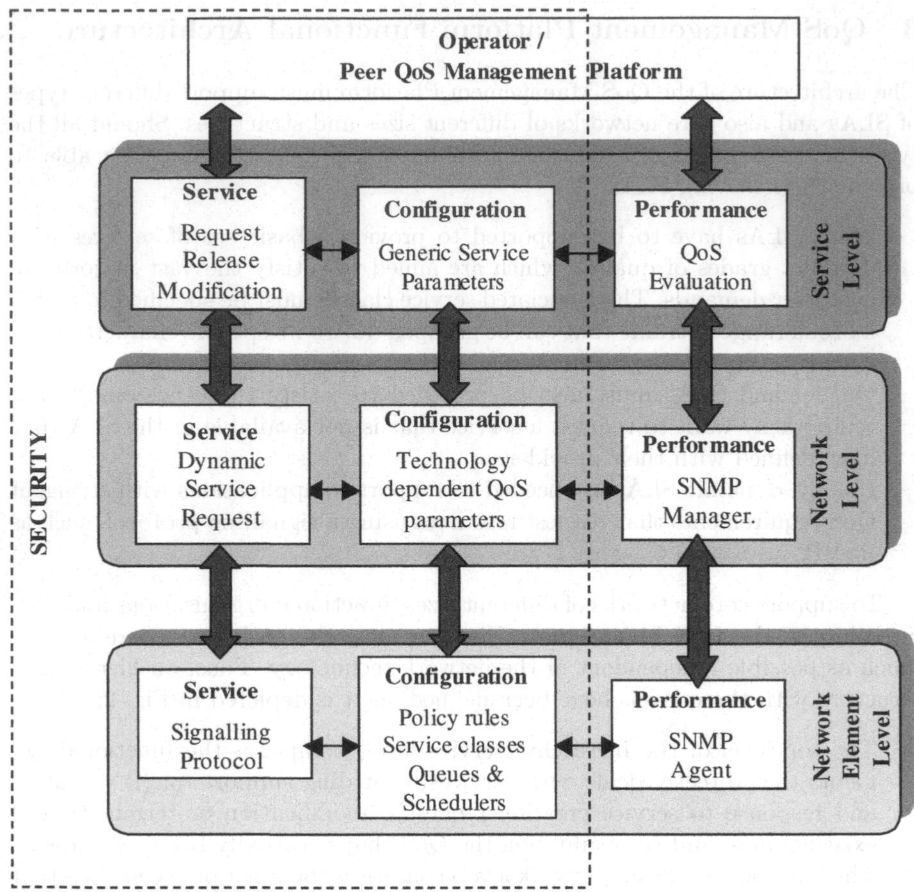

Fig. 2. QoS Platform Functional Architecture

was collapsed into three types of network elements: BB, BR and CR. The BB comprises all the service and network level functional elements; the BR the complete network element level and the CR a small part of the network element level.

4.1 The BB Network Element

According to Fig. 3, the BB is composed of modules (grey boxes), databases (cylinders) and interfaces (lines). All the SW was written in C++ language and is implemented using UNIX processes/threads that communicates using the UNIX TCP socket API, or the SSL interface, if configuration or service information are concerned.

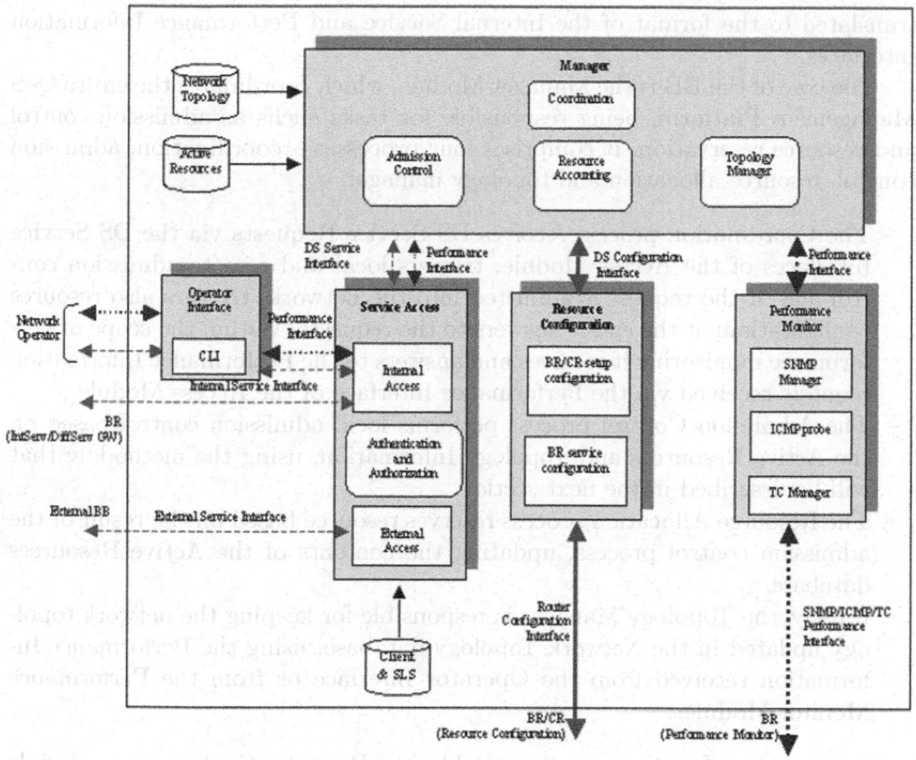

Fig. 3. BB architecture

All the Service and Performance requests are received at the Access Module that acts as the QoS Management Platform entry point. This module comprises two processes:

- Internal Access process that is responsible for internal communication, such as the one that happens with the Network Operator and the Dynamic Service Agent that implements the IntServ/DiffServ Gateway;
- External Access process where the communication with peer QoS Management Platforms existing in other domains, takes place.

Internal or External Service Requests are converted into DS services requests using the Client&SLS database information. Authentication and Authorization is also performed by both processes. Every time a new service request arrives at a process of this module, a new thread is created to deal with it.

As far as a local Network Operator emits requests, the Operator Interface Module is used. Basically it comprises a Web and Command Line Interfaces, which provide access to service and performance information. Using them, it is possible to request, modify or terminate services and to evaluate the QoS that is actually being provided. The information received at these Interfaces is internally translated to the format of the Internal Service and Performance Information Interfaces.

The core of the BB is the Manager Module, which coordinates the entire QoS Management Platform, being responsible for tasks suchs as admission control and resource reservation. It comprises four processes – coordination, admission control, resource allocation and topology manager.

- The Coordination process receives DS Service Requests via the DS Service Interfaces of the Access Module, triggers local and remote admission control and, if the request is admitted into the network, triggers also resource configuration; at the end it answers to the requests. Within the scope of performance monitoring it receives and answers to the Performance Information requests received via the Performance Interface of the Access Module.
- The Admission Control process performs local admission control based on the Active Resources and Topology Information, using the methodoly that will be described in the next section.
- The Resource Allocation process reserves resource based on the result of the admission control process, updating the contents of the Active Resources database.
- Finally, the Topology Manager is responsible for keeping the network topology updated in the Network Topology database, using the Performance Information received from the Operator Interface or from the Performance Monitor Modules.

Configuration functions are triggered by the Resource Configuration Module that translates DS service request into configuration commands and sends them to the routers. These functions are executed in two different instants of time, by two distinct processes:

- When the network starts operating, the basic DiffServ configuration is initialized in every router by BB/BR Setup Configuration process, which creates service queues, scheduler, filtering and policy rules for control traffic, such as RSVP, SNMP or ICMP.
- During the operational phase, when a service creation, modification or termination occur, resources are configured at the BRs associated to the request, by the BR Service Configuration process that creates/modifies or removes the corresponding filtering and policy rules.

The Performance Monitor Module interfaces the corresponding modules of the routers and is responsible for receiving performance information and analyzing them in terms of performance objectives defined in the SLA. It comprises the SNMP Manager process, the ICMP Probe process and the TC Manager process. The SNMP process computes generic performance information using SNMP protocol and MIB-II information. The ICMP Probe is a modified version of ping that supports service differentiation, which is used to measure end-to-end delay. Finally, the TC Manager computes performance information per service class, using the available capabilities of LINUX TC command.

The Admission Control Procedure. When a new service request arrives at the BB it has to perform admission control procedures, before admitting the request into the network. To improve the chances of accommodating all incoming service request, all the priority classes are over-provisioned. This means that either EF or AFx classes initially have 80% of available bandwidth. The remaining bandwidth is used to carry BE traffic, which is not submitted to any admission control procedure. As EF or AFx service requests are satisfied the available bandwidth of the associated class and of the link decreases.

The admission control is executed in two phases, the local and the remote, as represented in Fig. 4. During the local phase, the service parameters are identified and, as long as there are enough available resources in the class and in the link, the request is locally admitted into the network. If an intra-domain SLA is specified the request is send to a peer BB, using the Resource Admission Request (RAR) message of the X-COPS protocol, in order to perform remote admission control. Should the request be remotely admitted, a Request Admission Acknowledged (RAA) is sent back to the local BB, which triggers service configuration.

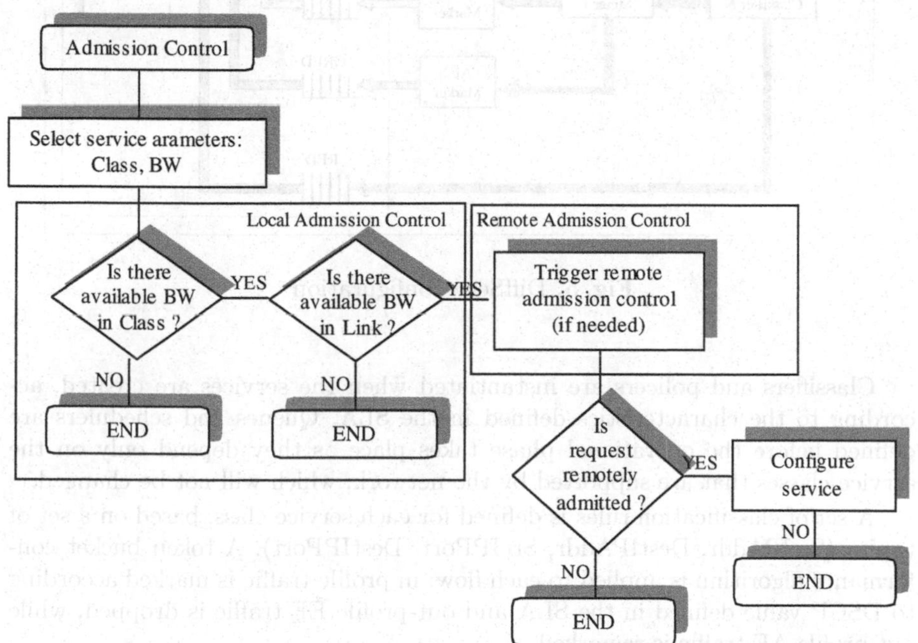

Fig. 4. Admission Control Procedure

4.2 The Routers NE

In LINUX, the DiffServ characteristics may be defined by running a set of scripts that configure its TC, defining its properties at the ingress and egress interfaces.

The DiffServ configuration of this core network, which is depicted in Fig. 5, comprises one EF class, four AF classes and one BE class. Classification, policing and marking are provided at the ingress interfaces of the BRs, while queuing and scheduling are supported at the egress interfaces of BRs and CRs.

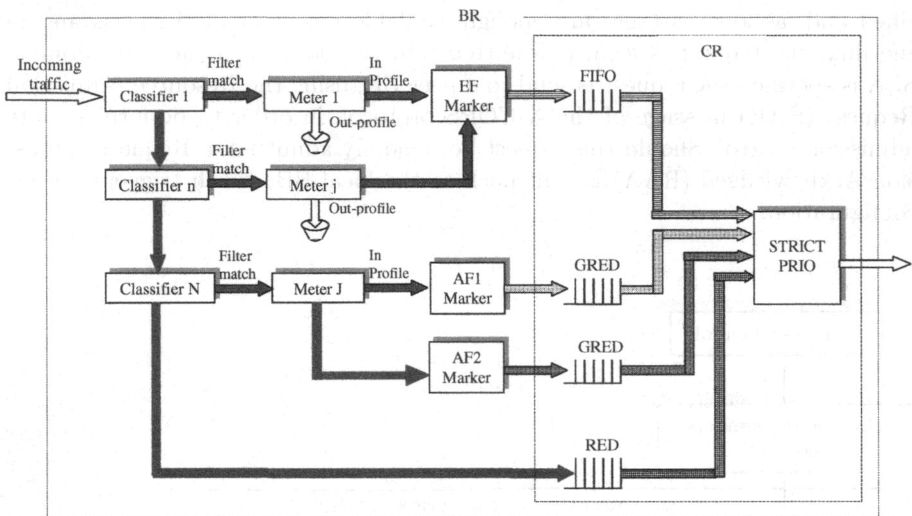

Fig. 5. DiffServ configuration

Classifiers and policers are instantiated when the services are created, according to the characteristics defined in the SLA. Queues and schedulers are defined before the operational phase takes place as they depend only on the service classes that are supported by the network, which will not be changed.

A set of classification rules is defined for each service class, based on a set of tuples (SrcIPAddr, DestIPAddr, SrcIPPort, DestIPPort). A token bucket conformance algorithm is applied to each flow: in profile traffic is marked according to DSCP value defined in the SLA and out-profile EF traffic is dropped, while out-profile AF traffic is remarked.

A Strict Priority Scheduler was installed as the main scheduler in all interfaces, being the bandwidth sharing defined by the BB admission control. Concerning the queues, EF uses a FIFO, AFx a Generic Random Early Drop (GRED) and BE a Random Early Drop (RED).

4.3 Dynamic End-to-End Service Request

QoS aware applications may request resources via a signalling protocol, such as RSVP. At the edge of the backbone network, the RSVP messages are put into a tunnel that ends at the other edge of the backbone. To translate the IntServ service requests into DiffServ, RSVP RESV messages are captured and converted to a DiffServ service request by a Dynamic Service Agent that lies at the BR. These requests are then sent to the local BB that triggers admission control in the entire DS region, using X-COPS protocol. Controlled Load IntServ class is mapped into AF1x class, while Guaranteed Service IntServ class is translated to EF class.

A detailed scheme of the dynamic end-to-end service request procedure is described in Fig. 6. According to it, the admission control procedure is split into local and remote admission control; resource allocation is done after the request is admitted into the network. Also RSVP messages are tunnelled after resource allocation is completed, in order to avoid allocating resources at the access networks, if the corresponding resources are not available at the backbone.

5 Performance Evaluation

The performance of the QoS Management Platform was evaluated concerning two main issues of the BB admission control procedure:

- The efficiency of network resource utilization.
- The overhead introduced by the BB in the network.

To perform these experiments a test application was developed that generates a set of flows whose properties are randomly setup: DiffServ class and bandwidth.

The efficiency of resource allocation is measured by the amount of available network resources e.g. bandwidth after performing a test. The results are presented for the BB algorithm and for an admission control algorithm that statically splits the resource per DiffServ QoS class. Both admission control algorithms are configured according to Table 1 and the results are presented in Fig. 7.

Table 1. Test configuration

Reserved BW	QoS	EF	AF1	AF2	AF3	AF4	BE
BB AC	90%	50%	80%	80%	80%	80%	Unlimited
Static AC	90%	10%	20%	20%	20%	20%	Unlimited

As depicted in the figure, BB admission control procedure leads to a more efficient use of network resources and a delayed rejection of QoS requests. To confirm these results, the tests have been randomly repeated several times and the results are shown in Table 2.

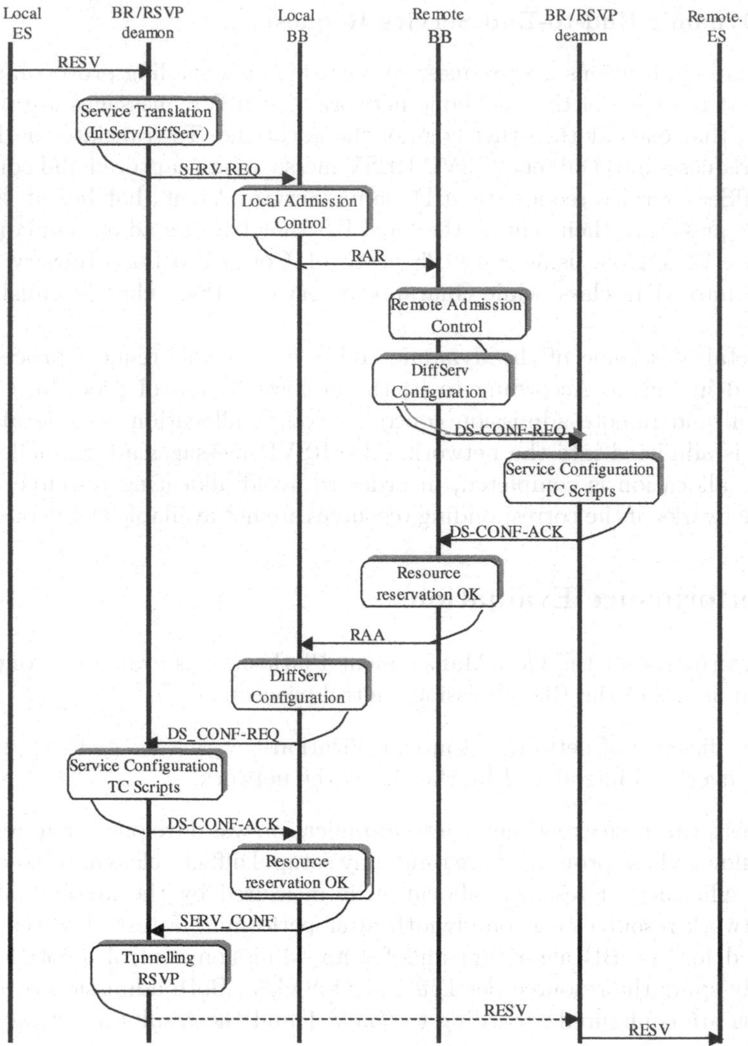

Fig. 6. Dynamic end-to-end service request procedures

Regarding the overhead introduced by the BB in the network, direct router configuration is performed and the time needed is compared to the one achieved by router configuration via BB. As shown in Fig. 8, BB introduces a very small overhead in the order of magnitude of 2.67 msec, which is approximately 3 % of the total time needed to perform the configuration task.

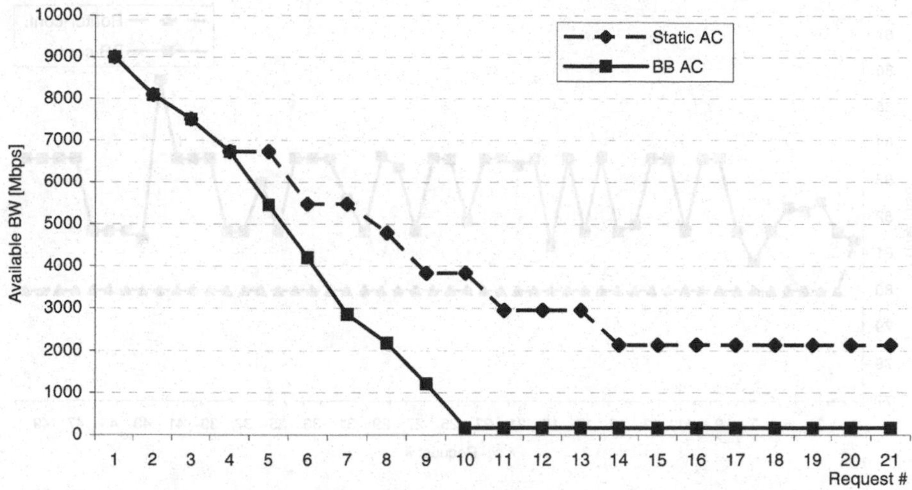

Fig. 7. Efficiency of network resource utilization – available bandwidth

Table 2. Efficiency of network resource utilization – available bandwidth

BW[Mb/s]	#1	#2	#3	#4	#5	#6	#7	#8	#9	#10
BB AC	129	124	304	240	261	256	336	96	274	52
Static AC	1482	3367	2276	3003	2653	2971	2869	1425	2706	2291

6 Conclusions

The paper presents a QoS management platform aimed to provide service assurance and performance monitoring in a DiffServ core network. The architecture described supports different types of SLAs, ranging from static to dynamic. The platform provides a scalable solution, which may be used in networks of different sizes and complexities. The platform relies on the concept of a Bandwidth Broker, which concentrates all the management tasks of a DiffServ domain, being also responsible for the communication with peer entities.

The platform was implemented in LINUX operating system and the performance results achieved show that the QoS platform uses efficiently the network resources and introduces a very small overhead. Future work will address two main issues: the platform scalability and the automatic re-configuration of the network to match SLA, in case of violations.

References

[1] Xipeng and Lionel M. Ni. Internet QoS: A Big Picture. *IEEE Network*, pages 8–18, March/April 1999.

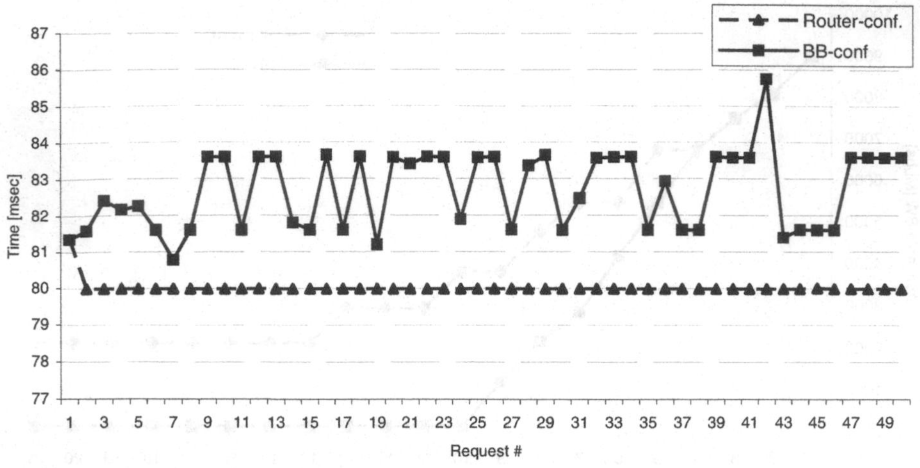

Fig. 8. Overhead introduced by the BB

[2] Y. Bernet et al. A Framework for Integrated Services Operation over Diffserv Networks. RFC 2998, IETF, Nov. 1998.

[3] S. Blake et al. An Architecture for Differentiated Services. RFC 2475, IETF, Dec. 1998.

[4] J. Wroclawski. The use of RSVP with the IETF Integrated Services. RFC 2210, IETF, Dec. 1997.

[5] http://www.ee.surrey.ac.uk/CCSR/IST/Tequila.

[6] http://www.alcatel.be/atrium.

[7] B. Teitelbaum. Qbone architecture (v1.0). http:// www.internet2.edu / qos / wg / papers / qbArch, 1999.

[8] Panos Trimintzios et al. A management and control architecture for providing ip differentiated services in mpls-based networks, May 2001.

[9] P. Polese, A. Martucci, F. Fondelli, N. Ciulli, A. Casaca, T. Vazão. Preparing the network infrastructure for the information economy in europe. In *SSGRR 2002 Conference*, 2002.

[10] V. Jacobson et al. An expedited forwarding phb. RFC 2598, IETF, Jun 1999.

[11] J. Heinanen et al. Assured forwarding phb group. RFC 2597, IETF, Jun 1999.

[12] Tele Management Forum. SLA Management Handbook, June 2001.

Multidomain End to End IP QoS and SLA

The Sequin Approach and First Experimental Results on a Pan-European Scale

Mauro Campanella[1], Michal Przybylski[2], Rudolf Roth[3], Afrodite Sevasti[4], and Nicolas Simar[5]

[1] INFN–GARR Via Celoria 16, 20133 Milano Italy
Mauro.Campanella@garr.it
[2] PSNC, Ul. Noskowskiego 12/14, 61-704 Poznan, Poland
Michalp@man.poznan.pl
[3] FhG FOKUS, Kaiserin-Augusta-Allee 31 10589 Berlin
Roth@fokus.gmd.de
[4] RA-CTI, 61 Riga Feraiou Str., 262 21 Patras, Greece
Sevastia@cti.gr
[5] DANTE, Francis House, 112 Hills Rd Cambridge CB2 1PQ UK
Nicolas.Simar@dante.org.uk

Abstract. This paper present the SEQUIN approach to the problem of providing end to end QoS services in the European framework of the GÉANT, the European 10Gb/s network backbone, which connects more than 30 National Research and Education Networks (NREN). A multidomain Premium IP service, based on the EF PHB, has been specified and implemented for that environment. Its architecture is summarised here and early experimental results are reported.

1 Introduction

The need to support Quality of Service in terms of delay, capacity, delay variation and packet loss has been recognised since the creation of the first European backbone network TEN-34 in 1997 that was delivered using ATM technology through the MBS (Managed Bandwidth Service). The advent of Gigabit networks at both national (NREN) and international level (GÉANT [1]) opens new approaches to network transmission, but some innovative application require QoS guarantees and ATM alone is no longer suitable and available.
An effort to develop and implement a multidomain QoS architecture at the IP layer in the European research environment was started through a close cooperation between DANTE, the operator of GÉANT, and a group of European NRENs. The activity has been mostly carried in the context of the IST project SEQUIN (Service Quality across Independently managed Networks)[2] in coordination with TF-NGN (Task Force for Next Generation Networking)[3].
TF-NGN is a working group composed of representatives of the NRENs and of research institutions with the aim to study and develop technologies, which are viewed as strategically important for the NRENs and GÉANT.

M. Ajmone Marsan et al. (Eds.): QoS-IP 2003, LNCS 2601, pp. 171–184, 2003.

The result was the specification of a completely scalable multidomain Premium IP service based on the Differentiated Services Expedited Forwarding Per Hop Behaviour model [4][5] [6][7]. A 'proof of concept'-testbed trials and the provision of this service in a early stage to international research groups were carried and provided first results and insight on the provisioning and monitoring processes. Next steps consist of more in depth performance and tuning analysis and the provision of the service to researchers in Europe on a wider scale.

2 Users' QoS Requirements

A set of research groups of users in Europe was interviewed to assess their perception and requirements for Quality of Service in the network. The groups were chosen in such a way to provide a non-homogenous significative sample of the users' base and ranged from large Universities to IST Projects on network research.
A questionnaire, articulated in 26 questions in 4 sections (geography, qualitative perception of QoS, quantitative perception of QoS and network options and expectation), was sent to 20 groups, out of which 11 provided complete answers. Figure 1 shows the application used and illustrates the users' requirements in terms of QoS parameters and capacity ranges requested. The result of the interviews is summarised in a list of desirable QoS services and their characteristics. The results of the interviews showed that three main QoS services were requested: the traditional Best Effort, a service focused on Guaranteed Capacity (IP+) and a leased line equivalent service called Premium IP service.

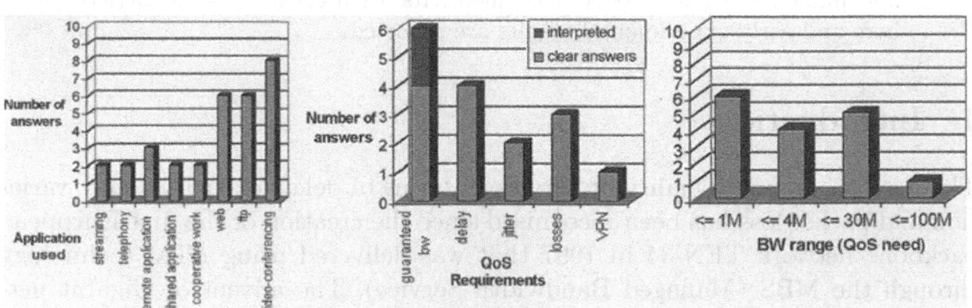

Fig. 1. Application used by users and User's QoS requirements

The interviewees also provided indicative values ranges for the QoS parameters, for each service, which are listed in the following table.

(1)	One-way delay	IPDV	Packet loss	capacity
best effort	unspecified	unspecified	< 5%	unspecified
IP+	distance delay +100 ms	< 50ms	< 2%	according to SLA
Premium IP	distance delay + 50 ms	< 20ms	0%	according to SLA

(1) The values are indicative and may vary for each application.

The table shows that the implementation of a Premium IP service can satisfy in practice all the users' QoS requirements. The implementation of a separate IP+ service was postponed, being considered a more technically complex service. A full description of interview results can be found in [8].

3 Premium IP Service Architecture and Implementation

The Premium IP architecture [9] set as objectives:

- Provisioning of an IP QoS service offering a profile comparable to the ATM-based Managed Bandwidth Service;
- End-to-end, modular and scalable service, operating across multiple management domains;
- Applicable to the router platforms currently installed in GÉANT and in the NRENs networks with minimal request to hardware;
- Fast service deployment and availability within the timeline of GÉANT
- Independence of underlying networking technology;
- The service should not interfere or inhibit the activation of other services, except in case of congestion. In any case it should not starve other services;
- Compliant to IETF standard tracks.

Additional criteria were adopted, to streamline the implementation:

- Premium IP should require only minimal additional processing per network node, to have the widest implementability on current production hardware;
- At least in the early introduction phase, it makes use of static manual configuration only, and it employs no additional dynamic signalling protocols;
- Introduction of Premium IP within a NREN can be accomplished in a stepwise manner, starting from a small Diffserv domain.

These requirements brought to the choice of the Diffserv architecture and the EF-PHB [4][5][6][7]. Diffserv achieves high scalability through aggregation of traffic flows and allows different service classes identified by the Diffserv Code Point (DSCP) value stored in the IP packet header.
The architecture implementation [10] specifies additions and extensions to the single domain Diffserv standard, to provide an end to end service, that crosses multiple administrative domains. Figure 2, shows the case of NRENs networks.

- The interface specification between different domains mandates that the inter domain hop behaves according to the EF PHB;
- Only a small fraction of the total core link capacity is assigned to Premium IP, between 5 and 10 and that deterministic bounds can be computed [11];
- The service will be destination aware, mandating the knowledge of the source and destination addresses in the provisioning and admission control phases.

A destination un-aware service, on the contrary, where the user specifies just the send and receive capacity, is not supported. Such a service would allow the user

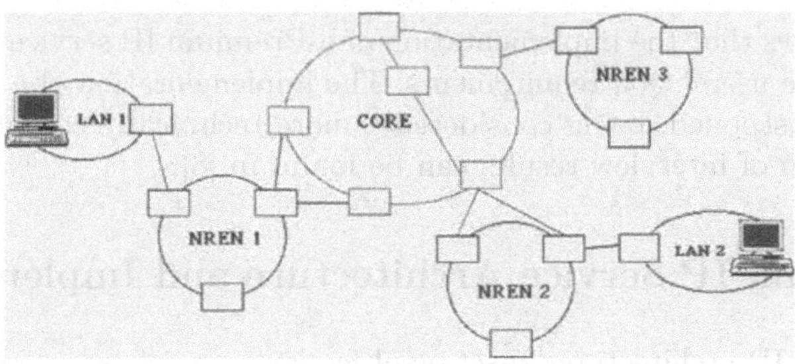

Fig. 2. Network Model for Premium IP

to freely send and receive up to an agreed capacity and would require a very high level of overprovisioning, without avoiding the risk of contract violation due to unpredictable congestion.
It should be noted here that destination-awareness is not an unrealistic assumption, since this has been the provision scenario on which the MBS service has been operation successfully so far.

3.1 Main Task Analysis

Shaping The Premium IP service does not provide shaping. It is the task of the service user to ensure that the generated flow does not violate the maximum contracted sending rate. It is recommendable to control the sending rate inside the end-system to prevent sending burst of packets that exceed the maximum service rate, which leads to packet loss. If such functionality is not available in the application or end-system it is recommendable to at least configure traffic shaping at the nearest network node to smooth out shorter burst and thus to comply to the service contract.
Merging of premium traffic aggregates may increase the burstiness within the wide-area networks. The high speed of the core links and the limit in the maximum capacity devoted to the Premium IP service aids in minimising the probability. Looser policy rules at domain boundaries will avoid any packet loss due to any unwanted burstiness.

Packet Marking In general it is assumed that the end-system is capable to generate the appropriate marking for Premium IP packets. Optionally a router within the local campus network could also perform that marking. In any case packet marks will be checked at the entrance in the first Diffserv domain. For Premium IP service, the DSCP decimal code 46 is chosen. Measurement on the current traffic in GÉANT has demonstrated that the current use of such value in the ToS byte of the IP Header is negligible, thus its use restricted to Premium IP pose no problem to normal traffic.

Admission Control In this initial approach, admission control with respect to the capability of the network to accept a new Premium IP connection is performed off-line by the service administrator. Still, after availability of network resources is verified and a contract for the connection is established, ingress routers to a network domain have to check incoming Premium traffic for admissibility. At the ingress to the first Diffserv domain the source and destination address prefixes pairs are checked (destination aware service). If there is no matching contract, the packet is not eligible for receiving Premium IP treatment, the DSCP field is reset to the best effort value and the packet will receive best-effort treatment. At the following boundaries between domains, just the DSCP value is checked.

Policing Packets for which there is a valid matching are submitted to policing at the ingress point of a networking domain to ensure that the incoming Premium IP traffic will not exceed the permitted maximum rate. Excess packets that violate this rate are discarded.
The Premium IP model assumes a strict policing at the ingress point closest to the customer network, whereas latter ingress nodes to subsequent domains will perform only a more lenient policing on the aggregate flows.
Policing is applied to the aggregated Premium IP traffic. Therefore policing at the ingress node between customer network and NREN will distinguish a granularity of source/destination IP prefixes, while at the ingress point to GÉANT only aggregated flows are regarded and policing is performed according to source and destination Autonomous System (AS) pairs.
The policing function uses a token bucket that is specified by rate and bucket depth. At the ingress point from the customer network the contracted maximum rate is configured and a small depth of 2 or 3 MTUs is set. At the hop between domains, a 20aggregated flows and deeper bucket depth is selected. The recommended depth at these nodes is 5 MTUs, which can be further increased, in longer paths. Experimental studies, described in [23] provide further insight on these configuration parameters.
Premium IP does not require policing at egress from a domain, but rather suggest avoiding it. An additional simplification might occur at domains that receive Premium traffic from the core backbone. As the traffic has already been checked at least twice and the provisioning model ensures a safe rate limitation, policing can be avoided also in ingress.

Router Scheduling Routers are required to forward Premium IP packets with minimal delay and delay variation. For the contracted Premium IP traffic volumes the implementation is such to avoid any packet loss due to queue overflow or too strict policing. Preferential treatment of Premium IP packets is realised by allocation of a separate hardware-processing queue, which is becoming a commodity in today's route. Routers support various scheduling mechanisms that may be used for the implementation of Premium IP [9], assigning in any case the highest priority to Premium traffic. For example:

- Strict priority queuing
- Weighted Fair Queuing.
- Modified Deficit Round Robin in strict or alternate priority mode

The architecture does not mandate a particular, mechanism, leaving the freedom of implementing the EF PHB using the hardware at hand.
Local lab experiments had been performed to validate those QoS mechanisms under heavy load conditions for high-end router types that have been deployed in GÉANT and the core network to the larger European NRENs. Experiments for rate limiting, packet marking, scheduling mechanisms and congestion control have been performed and a behaviour confirming the theoretical properties of these mechanisms could be observed. A summary and evaluation of those results is included in [12].
Figure 3 contains a pictorial view of the main Premium IP features for a one directional flow.

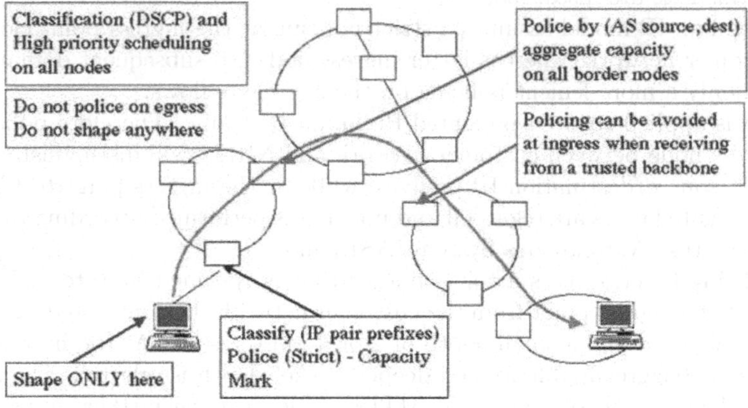

Fig. 3. Pictorial summary of main task for each node

3.2 QoS Monitoring and Measurement

An essential component for the deployment and operation of QoS services is QoS monitoring.
QoS monitoring requires the co-ordination between different domains and a correlation of results. Measurements have to be performed between the end-points, within particular subnetworks and between adjacent networks [13]. For Premium IP relevant metrics include capacity, packet loss, one way delay and IPDV (IP packet delay variation). Methodology is a mixture of active and passive measurement, i.e. measurement of artificial test traffic generated by hosts in the PoPs, direct measurements on user traffic and polling of counter variables at router interfaces. Figure 4 shows the measurement infrastructure between end

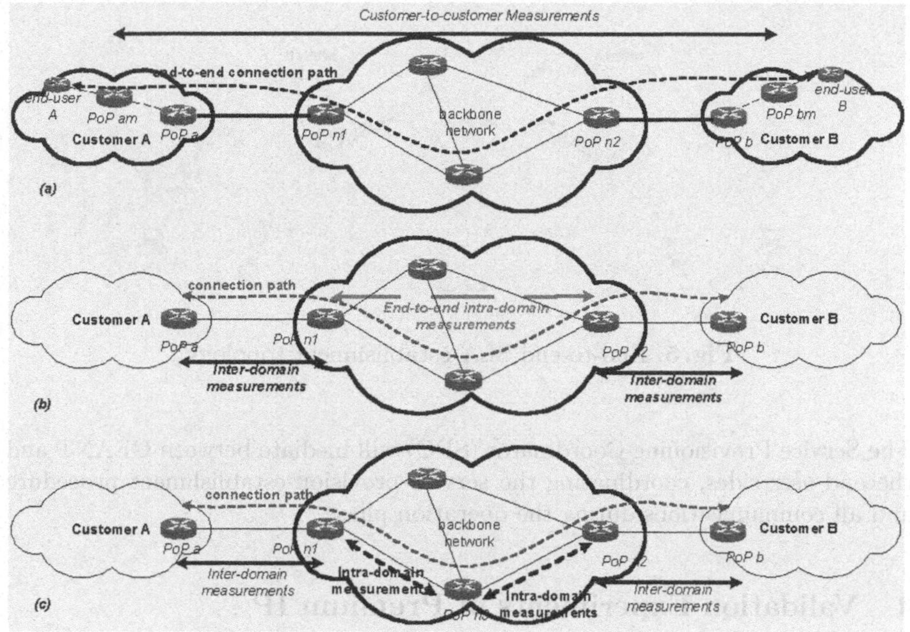

Fig. 4. Coordination of QoS Measurements

points (a), between domains (b) and for each link (c). Clock synchronisation using GPS receivers will be needed for accurate delay measurements, whilst NTP may provide a less accurate precision.

3.3 Provisioning Model and SLA

Service provisioning for QoS enabled networks comprises a process where intensive testing and probing of the available infrastructure has to take place, before being able to quantify the QoS offering in the service contract.

Dealing with provisioning across independently managed domains reinforces the need of a tight mechanism for the establishment of an end to end Service Level Agreements (SLA) based on bilateral SLAs and provisioning methodology.

SLA definition between two peers is the structural unit to create a chain that connects the two ends, which is mandatory complemented by an end to end SLA. Figure 5 provides a view of SLA topology. The provisioning process of IP Premium is detailed in [14]. Provisioning will be based on a static infrastructure in the first phase of deployment in the GÉANT environment. For the co-ordination of the processes and the orderly flow of information, a set of roles has been identified and responsibilities have been determined [17]. As part of this process, a consortium requesting Premium IP service appoints a common representative that acts as a single contact point.

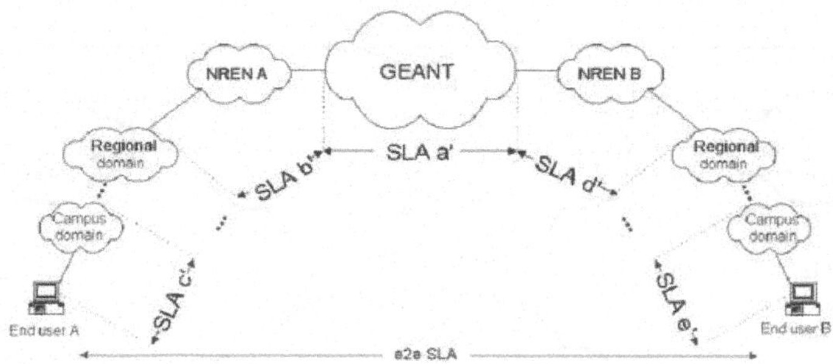

Fig. 5. End-to-end SLA establishment topology

The Service Provisioning Coordinator (SPC) will mediate between GEANT and the end-user sides, coordinating the service provision establishment procedure and all communications during the operation phase.

4 Validation Experiments of Premium IP

The proposed implementation model for Premium IP has been investigated in extensive test series. In a first testing phase Premium IP was examined in test scenario based on H.323 video conferencing. The objective of the experiments was the demonstration of a Premium IP implementation in production networks. The test environment included GÉANT and four NRENs [12][15][16].

Testbed Topology The testbed consisted of five sites in four countries that were connected to GÉANT via their NRENs. The testbed included research labs of SWITCH in Switzerland and GRNET in Greece. CINECA in Italy was connected via GARR-B, and the German test network QUASAR [18] with its sites at University of Stuttgart and Fraunhofer FOKUS in Berlin was attached across the DFN backbone G-WiN.
The SWITCH node served as a central hub and offered the functions of MCU (multipoint conferencing unit) and gatekeeper for connection set-up, call routing, addressing etc.
The testbed is depicted in Figure 6. It consists of networks operating at various speeds and applying different networking technologies. The GÉANT core network consists of Juniper M160 routers that are interconnected by 10 Gbit/s and 2.5 Gbit/s links. The NRENs are attached to GÉANT via a 2.5 Gbit/s POS interface with the exception of GRNET, which has access via 2 x 155Mbit/s ATM links. The QUASAR subnetwork has a dedicated STM-1 SDH between the two sites. The SWITCH testbed included two sites in Geneva and Zurich that had been connected by Gigabit Ethernet. Other test sites were connected via ATM PVCs. Routers in the NRENs and at the test sites covered a wide performance

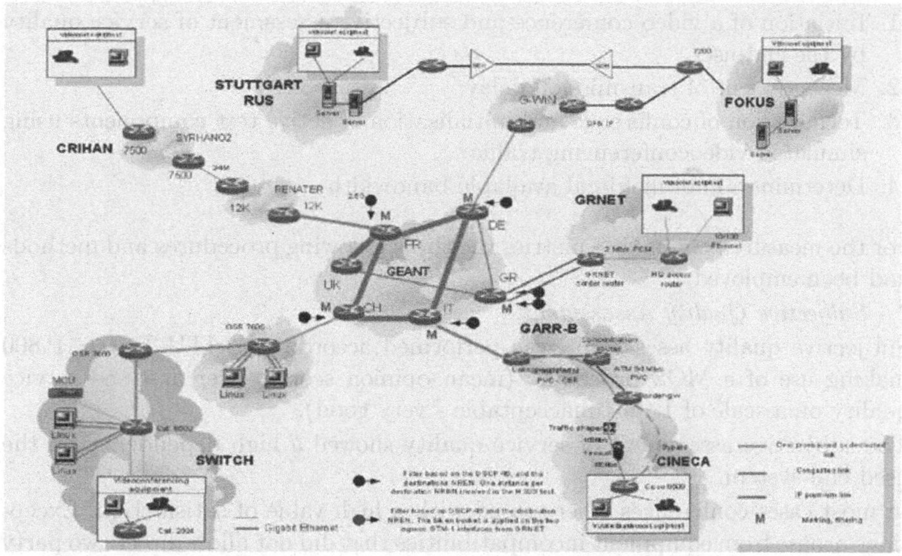

Fig. 6. The testbed topology

spectrum including devices of Cisco 12K, 7600, 7500 and 7200 series. Conferencing equipment from various vendors was used including VCON, Polyspan and SunVideo.
Premium IP was implemented on the Juniper M160 in GÉANT by activating weighted round robin (WRR) scheduling. Three queues were defined distinguishing between best effort, Premium IP and packets from signalling and control protocols, where the Premium IP queue received a weight of 90 and the two other classes of 5 each. At the access interfaces policing was configured that allowed 2 Mbit/s Premium IP capacity for each NREN pair. Excess Premium traffic would be discarded. Packet counters on these interfaces were installed to allow for monitoring on Premium packets. The token buckets were configured with a depth of 5 MTUs and a rate of 1.2 times the maximum allowed Premium capacity to allow for some burstiness in the Premium traffic.
Support for Premium IP in the NRENs was realised through dedicated ATM CBR PVCs, which would provide a policing function limiting the maximum rate of Premium traffic. The connection of the QUASAR testbed to GÉANT over G-WiN relied on over-provisioning. The access links carried only low load such that the Premium IP characteristics could be satisfied at those transitions points.

Measurement Tools and Test Scenarios The experiments included subjective quality assessment by conference participants and objective measurements of QoS parameters. Point-to-point conferences have been established between all 20 combinations of pairs. Active and passive measurements have been performed. The test scenario comprised the following steps:

1. Initiation of a video conference and subjective assessment of service quality by the end-user
2. Measurement of transmission delay
3. Termination of conference and initialisation of active test components using simulated videoconferencing traffic
4. Determination of maximal available bandwidth

For the measurement of QoS metrics the three following procedures and methods had been employed.
1 - Subjective Quality Assessment
Subjective quality assessment was performed according to ITU-T Rec. P.800 making use of a MOS procedure (mean opinion score). User assessed service quality on a scale of 1 - 6 (unacceptable - very good).
The subjective assessment of service quality showed a high dependency on the used end-system.
In most cases conferences where rated with a high value of satisfaction. Exceptions arouse from equipment incompatibilities that did not allow direct two party conferences but required the relay via a MCU, which introduced additional delays that lead to a significant down rating in the perceived quality. In general it appeared that application specific factors dominated the subjectively perceived quality to a much larger degree than network specific issues.
2 - IP Packet delay variation.
IPDV was derived using active measurements. The RUDE/CRUDE free software toolkit [20] was used, and measurements in both directions were performed. The measurement system on the sending side was parameterised to generate test traffic that replayed a recorded trace of a H.323 videoconference. In addition, test using traffic with CBR characteristics were performed to serve as cross check and reference values.
Figure 7 shws the result of the measurements in the case of a single congested link. It compares the distribution of IPDV values of VBR traffic when sent as Premium and as best effort traffic from the site in Stuttgart to Athens. This path had the highest number of hop and also included a critical high load link from GÉANT towards GRNET in Greece. The measurements show here an effective reduction of IPDV for Premium IP traffic against ordinary (congested) best effort traffic.
This is confirmed to be one of the beneficial results of referential treatment for IP Premium traffic.
3 - Packet Loss
Packet loss was also derived through active measurements by RUDE/CRUDE. Besides, there were performed control measurements by running a ping process parallel to the conferencing session sending a packet per second. For those parts of the network where the passive measurement system had been available, packet losses in the conferencing traffic could be determined. Monitoring counters at the core router interfaces of GÉANT provided the number of packets discarded by the policers for Premium traffic.

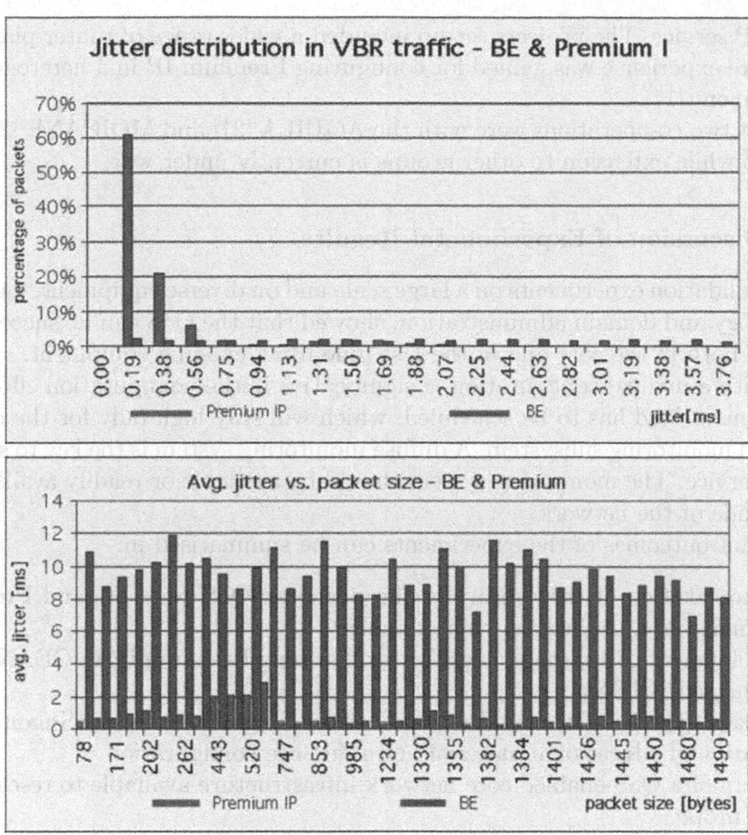

Fig. 7. IPDV Distribution Stuttgart to Athens as a function of packet size

Those packets could be correlated to the experiments with netperf [19], which generated short packet bursts that would exceed the maximum admissible bandwidth.
In nearly all cases, there occurred no or negligible packet losses (less than 0.02%) with the exception to the connection to CINECA, Italy, which had losses of up to 3%, most probably caused by an incorrectly configured shaping function at the connecting ATM CBR PVC. However the perceived quality of the videoconference had been rated high, even with these high loss rates. Apparently the application could well adapt to these losses.

4.1 Experiments with User Groups from IST Projects

IST projects from the European research programme with an interest in QoS enabled connections and European geographical coverage have been contacted to act as beta-tester for the Premium IP service. These tests allowed to better understand the architecture and the procedures involved in providing the Pre-

mium IP service. The projects set-up included a wider range of router platforms and more experience was gained for configuring Premium IP in a heterogeneous environment [17].
The first two co-operations were with the AQUILA [21] and MOICANE [22] IST projects while extension to other groups is currently under way.

4.2 Discussion of Experimental Results

These validation experiments on a large scale and on diverse equipment, data link technology and domain administration, showed that the QoS can be successfully enabled both in the core and in the last mile using existing equipment.
Even with a manual configuration, a significative initial configuration effort and management load has to be scheduled, which will stay high only for the debugging and monitoring subsystem. A diffuse monitoring system is the key to success of the service. The monitoring system should be available or readily available at every node of the network.
Important outcomes of the experiments can be summarised in:

- demonstration of the validity of the proposed QoS concepts and Premium IP model in a production environment
- development of a methodology for provisioning Premium IP in GÉANT and connected NRENs
- packaging of a set of modules that support diagnostics and QoS monitoring
- creation of educational material and reference configuration
- set-up of a QoS-enabled core network infrastructure available to researchers in Europe

The experiment confirmed that enabling QoS on a large network according to the Premium IP model does not interfere with normal BE traffic. It was shown that overprovisioning could be applied as a transition technique to fully QoS-enabled network.
It was not in the goals of the tests, due to time constraint, to make a detailed analysis of a QoS Premium IP performance. Due to the recent upgrade nearly in all involved research networks there was only a light traffic load and with exception to the transition point towards GRNET, there were no bottlenecks with risk of congestion. However QoS mechanisms are less intended to improve network service, but rather they should provide assurance that a certain minimal level of QoS will be preserved even during certain exceptional situation of heavy load, as e.g. under a DoS attack.

5 Conclusions and Future Work

In the trials described in this paper, for the first time, it was possible to test a Diffserv based approach to IP QoS in production networks, on a wide area scale. The viability of the proposed Premium IP model could be demonstrated and steps for wider deployment could be outlined.

The work performed gave stimulating impulses for new activities within the European research networks community. The deployment of a QoS monitoring infrastructure is an essential component for the European research network environment independent of the deployed QoS approach.
Testing of the QoS infrastructure, especially under congestion conditions, will further verify the capability of the Premium IP service to protect high-priority traffic's QoS. It will also continue to fine tune the network parameters and to evolve the provisioning system. The viability of such a service with respect to the provisioning costs and the price users are willing to pay for it should also be investigated in the future [24].
The provisioning of Premium IP across GÉANT and the European NRENs will create a valuable infrastructure enabling demanding experiments and innovative applications. GÉANT and some NRENs have confirmed their commitments for further support of Premium IP.

6 Acknowledgements

Most of the work presented in this paper has been performed within the context of the European research project IST SEQUIN and the project QUASAR with very valuable contributions from TF-NGN and the help of DANTE and NREN operations NOCs.
The authors would like to thank all participants of these projects for their valuable contributions.

References

1. GÉANT Home Page at http://www.dante.net/geant/
2. SEQUIN - Project IST-1999-20481, Home Page at http://www.dante.net/sequin/
3. TF-NGN Home Page at http://www.dante.net/tf-ngn/
4. Blake S. Black D. Carlson M. Davies E. Wang Z. Weiss W. "An Architecture for Differentiated Services", RFC2475, Informational, December 1998
5. Jacobson, V, Nichols, K., Poduri, K, "An Expedited Forwarding PHB", RFC2598 Standards Track, June 1999
6. B. Davie, A., Charny, J.C.R., Bennet, K., Benson, J.Y. Le, Boudec, W., Courtney, S, Davari, V., Firoiu, D., Stiliadis, "An Expedited Forwarding PHB (Per-Hop Behavior) ", RFC3246 March 2002, Obsoletes RFC2598
7. Charny, A., Bennett, J.C.R., Benson, K., Le Boudec, J.Y., Chiu, A., Courtney, W., Davari, S., Firoiu, V., Kalmanek, C., Ramakrishnan, K.K., "Supplemental Information for the New Definition of the EF PHB (Expedited Forwarding Per-Hop Behavior) ", RFC3247, March 2002
8. M. Campanella, P. Chivalier, A. Sevasti and N. Simar, SEQUIN D2.1 "Quality of Service Definition" April 2001 - http://www.dante.net/sequin/QoS-def-Apr01.pdf
9. GÉANT D9.1: Specification and implementation plan for a Premium IP service M. Campanella, T. Ferrari, S. Leinen, R. Sabatino, V. Reijs, April 2001. http://www.dante.net/tf-ngn/GEA-01-032.pdf

10. SEQ-D2.1 addendum 1: Implementation architecture specification for an Premium IP service, M. Campanella, October 2001, http://www.switch.ch/lan/sequin/GEA-D9.1-Addendum-1-v05.pdf/
11. A. Charny and J.Y. Le Boudec, "Delay bounds in a network with aggregate scheduling," in Proc. First International Workshop of Quality of future Internet Services (QofIS'2000), Sept. 25–26, 2000, Berlin, Germany
12. SEQ-D3.1 Definition of QoS Testbed M. Campanella, M. Carboni, P. Chivalier, S. Leinen, J. Rauschenbach, R. Sabatino, N. Simar, April 2001.
13. SEQ-D2.1 Addendum 3: Monitoring and Verifying SLAs in GEANT Athanassios Liakopoulos, April 2002
14. C. Bouras, M. Campanella and A. Sevasti, SLA definition for the provision of an EF-based service, in: Proc. 16th International Workshop on Communications Quality & Reliability (CQR 2002), Okinawa, Japan, 2002, pp. 17-21.
15. SEQ-D5.1 Proof of Concept Testbed M. Przybylski, R. Sabatino, N. Simar, J. Tiemann, S. Trocha, December 2001
16. SEQ-D5.1 Proof of Concept Testbed - Addendum: H.323 Testing Results, Jerzy Brzezinski, Tomas Bialy, Artur Binczewski, Michal Przybylski, Szymon Trocha PSNC, April 2002
17. SEQ-D6.1 Report on Users of QoS Testing Rudolf Roth, Afrodite Sevasti, Mauro Campanella, Nicolas Simar, April 2002
18. QUASAR Home Page http://www.ind.uni-stuttgart.de/quasar
19. Netperf Home Page: http://www.netperf.org/netperf/NetperfPage.html
20. RUDE/CRUDE at http://www.atm.tut.fi/rude/
21. AQUILA Home Page http://www-st.inf.tu-dresden.de/aquila/
22. MOICANE Home Page http://www.moicane.com/
23. Quantum EP 29212, D6.2 "Report on the Results of Quantum Test Programme", Chapter 4, "Differentiated Services and QoS Measurement" T.Ferrari June 23, 2000. http://www.dante.net/quantum/qtp/final-report.pdf
24. C. Bouras, A. Sevasti, "Pricing priority services over DiffServ-enabled transport networks", in Proceedings of IFIP INTERWORKING 2002 Conference, 6th International Symposium on Communications Interworking, Fremantle, Perth, Australia, 13-16 October 2002

Dimensioning Models of Shared Resources in Optical Packet Switching Architectures

Vincenzo Eramo, Marco Listanti

University of Rome "La Sapienza", INFOCOM Department,
00184 Rome, Italy
{eramo, marco}@infocom.uniroma1.it

Abstract. In the paper the comparison of some Optical Packet Switching architectures is carried out. The proposed architectures use the wavelength conversion technique to solve the packet contention problem and all of them share the wavelength converters needed in order to wavelength shift the arriving packets. In particular two architectures are considered: the Shared Per Output Line (*SPOL*) and the Shared Per Input Line (*SPIL*) architectures in which the wavelength converters are shared per output and output fiber respectively. The packet loss probability of the proposed architectures is evaluated as a function of the used number of wavelength converters by means of analytical models validated by simulations. The obtained results show that the *SPIL* architecture allow to obtain a larger saving of wavelength converters with respect to the *SPOL* architecture. In unbalanced offered traffic condition the *SPIL* architecture allows to save a number of wavelength converters in the order of 50% more than the *SPOL* architecture.

1 Introduction

The transmission capacity of optical fibers has been increasing at a tremendous rate as a result of DWDM technology. Although terabit capacity IP routers are now starting to appear, the mismatch between the transmission capacity of DWDM and the switching capacity of electronic routers remains remarkable. Emerging all-optical switching technologies will enable packets to traverse nodes transparently without conversion and processing at each node. The limited optical processing power and the lack of efficient buffers are the main obstacles in the realization of all-optical networks in which both the payload ad the header of the packets remain in the optical domain from the source node to the destination node. Although some primitive header processing forms begin to appear [1], the current approach to optical packet switching is to keep the payload in the optical domain and convert the header to the electronic domain for processing [2] [3] [4].

A critical issue in photonic packet switching is the contention resolution. Contention occurs when two or more packets contend for the same output port at the same time. In traditional electronic switches, packet contentions are handled through buffering; unfortunately, this is not feasible in the optical domain, since there is no optical

M. Ajmone Marsan et al. (Eds.): QoS-IP 2003, LNCS 2601, pp. 185-203, 2003.

equivalent of electronic random-access memories. Today, optical buffering can be only achieved through the use of fiber delay lines [5] [6] [7] [8] [9], however, the buffer size is severely limited, not only by signal quality concerns, but also by physical space limitations; as a matter of example, to delay a single packet for 5 μs requires over one kilometer of fiber.

In order to avoid optical buffering, Wavelength Conversion approach has been widely investigated to handle packet contentions in the optical domain [10],[11],[12]. The basic principle is very simple: packets addressed towards the same output are converted to different wavelengths by means of Tunable Optical Wavelength Converters (TOWC). So, the reference packet switch architecture foresees a TOWC for each output wavelength channel. Unfortunately, TOWCs introduce performance penalties and make switches complex and costly. For this reason in order to achieve a saving of TOWCs, switch architectures have been proposed in which the TOWCs are shared [13],[14]. In the *Share-Per-Node* (*SPN*) architecture, all of the TOWCs are collected in a converter bank. This bank can be accessed by any arriving packet by appropriately configuring an Optical Switch. In the *SPN* architecture, only the packets which require conversion are directed to the converter bank. The converted packets are then switched to the appropriate output fiber (OF) by a second Optical Switch. The *SPN* architecture allows the greatest saving of TOWCs to be obtained, in some cases even near to 90% [13] with respect to an architecture using a TOWC for each output wavelength channel. The disadvantage of the *SPN* architecture is to increase the complexity of the Switching Matrix in terms of needed number of Semiconductor Optical Amplifier (SOA) and this increase can be in the order of 40% [13]. For this reason, we propose architectures in which partial sharing technique are used, allowing to maintain low the complexity of the switching matrix. In particular two switching architectures are proposed: the *Shared-Per-Output-Link* (*SPOL*) and the *Shared-Per-Input-Line* (*SPIL*). In the *Share-Per-Output-Link* (*SPOL*) architecture, each OF is provided with a dedicated converter bank which can be accessed only by those packets directed to that particular OF. The *SPOL* architecture allows a TOWC saving smaller than *SPN*, but the complexity of its Switching Matrix is not increased with respect to the architecture having one TOWC for each output wavelength channel. In the *Shared-Per-Input-Link* (*SPIL*) a dedicated wavelength converter bank is located for each input fiber (IF). The performance of the *SPOL* and *SPIL* architectures is evaluated when the following traffic model is considered: i) no assumptions are done about the statistic of the traffic offered by each input wavelength channel; ii) the traffic offered by each input wavelength channel is the same and it is independent each other; iii) the traffic offered to each OF can be different according to a distribution characterizing the traffic unbalancing. The performance of the *SPOL* and *SPIL* architectures is evaluated by means of analytical models and the needed number of TOWCs to assure a given level of packet loss probability is compared in the two architectures.

The performance of the *SPIL* architecture heavily depends on the adopted contention resolution control algorithm, deciding which packets have to be wavelength shifted. In this paper two objectives are pursued. Initially our aim is to compare the *SPIL* architecture performance with those offered by *SPOL* architectures when the same control algorithm, denoted as "random" algorithm, is adopted; successively, the problem to define an "optimum" control algorithm so as to evaluate the best perform-

ance of the *SPIL* architecture is discussed. As the random algorithm is optimum for the *SPOL* architecture, but this is not true for the *SPIL* architecture, we will propose an ILP formulation of the packet scheduling problem allowing us to obtain, for a given number of TOWCs, the lowest value of Packet loss Probability (P_l) the *SPIL* architecture. This approach, though practically unfeasible, provides a useful bound of performance of the *SPIL* architecture, demonstrating the remarkable gain potentially obtainable with this kind of architecture.
The organisation of the paper is as follows. In Section 2 we describe the considered switching architectures; the control algorithms, under which the *SPIL*, *SPOL* architectures are compared, are discussed in Section 3; the analytical and simulation models, needed to evaluate the performance of the *SPOL* and *SPIL* architectures, are introduced in Section 4. Finally the Section 5 illustrates the obtained results and in particular, both the number of TOWCs and the optical gates, needed in the various proposed switching architectures, are evaluated and compared. The main conclusions and the further research topics are treated in Section 6.

2 Architectures of the WDM Packet Optical Switches

The two Optical Packet Switches, that we want to analyse and compare, are illustrated in this section. They allow to save in terms of TOWCs with respect to an architecture in which one TOWC is used for each output wavelength channel [13]. This is accomplished by sharing the TOWCs and allowing the access to them only to packets needing wavelength conversion. The proposed architectures adopt different sharing techniques, in particular, in the *SPOL* and *SPIL* architectures the TOWCs are shared per output/input fiber respectively. In the two following sections, the two architectures will be described. They has N input and output fibers, each fiber supports a WDM signal with M wavelengths, so an input (or output) channel is characterized by the couple (i,λ_j) wherein i $\{i=1,......,N\}$ identifies the input/output fiber and λ_j $\{j=1,......,M\}$ identifies the wavelength.

We assume that the operation mode of the architectures is synchronous, that is the arriving packets have a fixed size and their arrivals on each wavelength are synchronized on a time-slot basis [15] where a time slot is the time needed to transmit a single packet. We assume also to have r_l converters per input/output fiber in the *SPOL*/*SPIL* architectures.

2.1 *SPOL* Architecture

In the *SPOL* architecture, illustrated in Figure 1, the TOWCs are shared per OF. There are r_l TOWCs shared among the packets directed to a specific output fiber. The packets, needing wavelength conversion are sent to the output branches where there are TOWCs. On each output branch one optical filter is used in order to select the wavelength on which the packet is arriving. The selection of an output branch of the OF #i is accomplished by turning on a SOA gate of the switching matrix $S_{1,i}$. For example the circles in Figure 1 denote the SOA gates to be turned on in order to wavelength shift a packet directed to the OF #N and arriving at the IF #1 on the wavelength λ_M.

As mentioned in [13], the TOWCs partial sharing in the *SPOL* architecture, allows to obtain a TOWCs saving lower than an architecture in which all of the arriving packets share a TOWCs bank only [15]. However, the *SPOL* architecture has the advantage to need a lower number of SOA gates. Finally, notice as the complexity of the switching matrix in the *SPOL* architecture, expressed in terms of used number of SOA, is $C_{SPOL}=M^2N+N^2M$.

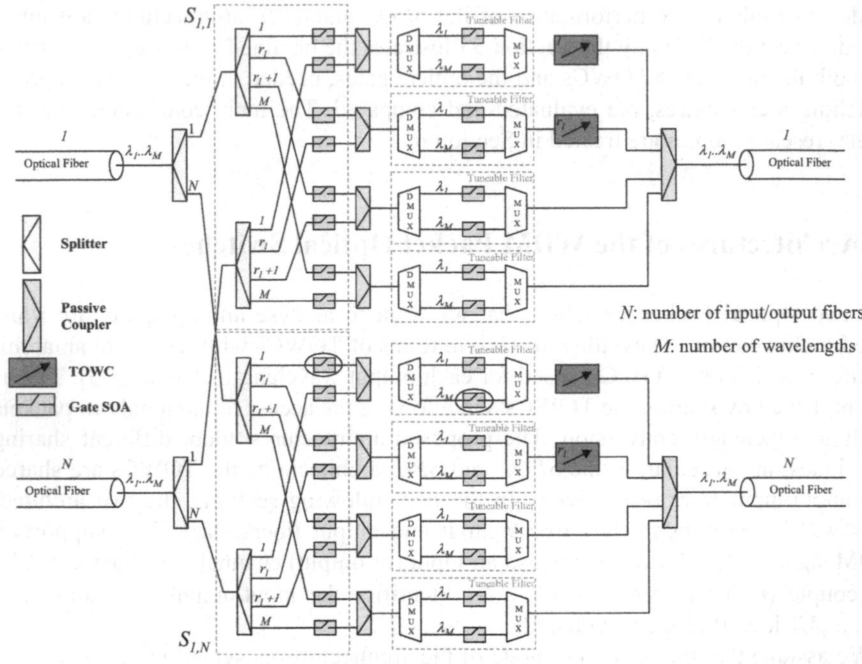

Fig. 1. Single-Per-Output-Line (*SPOL*) architecture

2.2 *SPIL* Architecture

The *SPIL* architecture is illustrated in Figure 2. In this architecture a dedicated TOWCs pool is located for each IF. The *SPIL* architecture allows to save a greater number of TOWCs with respect to the *SPOL* architecture, preserving the same level of complexity of the switching matrix, expressed by $C_{SPIL}=M^2N+N^2M$. This is due to a more efficient sharing of TOWCs, in fact, in line of principle, packets directed to a given output fiber can use all of the TOWCs of the switch, sharing them with packets destined to other output fibers; on the contrary in the *SPOL* architecture packets can use only the TOWCs placed on the output fiber which they are directed to.

The operation mode of the *SPIL* architecture is as follows. The arriving packets are first wavelength filtered, converted if it is needed and finally routed to the OF which they are directed to. The selection of the OF #*i* is accomplished by turning on a SOA gate of the switching matrix $S_{1,i}$. For example the circles in Figure 2 denote the SOA gates to be turned on in order to wavelength shift a packet directed to the OF #1 and arriving at the IF #*N* on the wavelength λ_1.

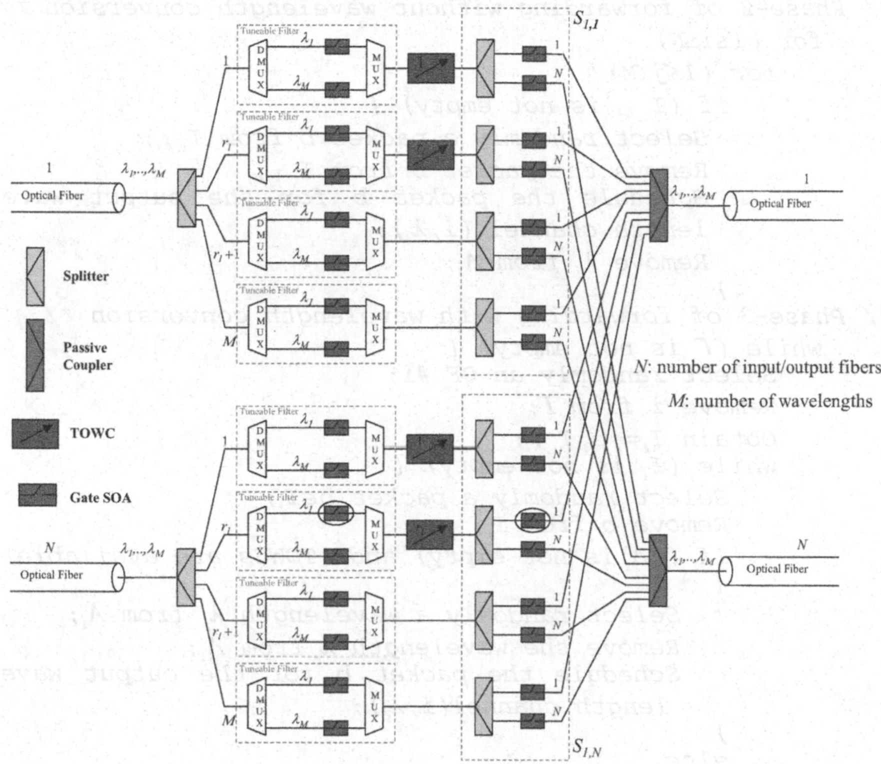

Fig. 2. Single-Per-Input-Line (*SPIL*) architecture

3 Control Algorithm

The *SPOL* and *SPIL* architectures will be evaluated and compared under the hypothesis that the same control algorithm, proposed in [13] is used. The switch control unit adopts a simple and fair technique in assigning in each time-slot the TOWCs to the various arriving packets. Let I_{i,λ_j} be the set of the packets contending for the output wavelength channel (i,λ_j), that is the packets arriving at the wavelength λ_j (j=1,…,M) and directed to the OF #*i* (i=1,…,N). Further we denote as Λ and Ω the sets of the wavelengths and the OFs respectively, that is $\Lambda \equiv \{\lambda_1,\ldots,\lambda_M\}$ and $\Omega \equiv \{1,\ldots N\}$.

In the following we report the considered control algorithm and referred to as "random" control algorithm:

```
*/ Initialization Phase-1 */
   for (1≤i≤N) {
      Determine the set I_{i,λj};
      Set Λ_i=Λ; Γ=Ω
   }
*/ Phase-2 of forwarding without wavelength conversion */
   for (1≤i≤N)
      for (1≤j≤M)
         if (I_{i,λj} is not empty) {
          Select randomly a packet b from I_{i,λj};
          Remove the packet b from I_{i,λj};
          Schedule the packet b for the output wave-
          length channel (i,λ_j);
          Remove λ_j from Λ_i;
         }
*/ Phase-3 of forwarding with wavelength conversion */
   while (Γ is not empty) {
      Select randomly an OF #i;
      Remove i from Γ;
      Obtain I_i= ∪_j I_{i,λj};
      while (I_i is not empty) {
         Select randomly a packet b∈I_i;
         Remove b from I_i;
         If ((Λ_i is not empty) and (TOWCs are available))
         {
            Select randomly a wavelength λ_p from Λ_i;
            Remove the wavelength λ_p from Λ_i;
            Schedule the packet b for the output wave-
             length channel(i,λ_p);
         }
         else
            Discard all the packets b∈I_i;
      }
   }
```

The algorithm is composed by three phases. In the *phase-1* the set I_{i,λ_j} (i=1,…,N; j=1,…,M) is evaluated; in the *phase-2* the control unit, for each output wavelength channel (i,λ_j),determine a packet, if there is one, to be transmitted without wavelength conversion. If more than one packet is addressed to a specific output channel, the control unit randomly selects one of them; moreover for each OF #i, the set Λ_i (i=1,…,M) of free wavelengths, i.e. the set of wavelengths on which no packet arrives, is also evaluated. In the *phase-3* the scheduling of the packets to be wavelength shifted is accomplished. The various OFs are randomly selected and, for each of them, the control unit tries, by means of wavelength conversions, to forward the packets belonging to the set I_i (i=1,…,N), that is the packets addressed to the OF considered and which have not been selected in the *phase-2* due to contentions. Obviously, these packets are forwarded only whether both free wavelengths and TOWCs are available. The availability of the TOWCs in the case of the *SPOL*/*SPIL* architectures refers to the availability of one TOWCs in the pool shared per output/input line respectively.

The "random" algorithm minimizes the needed number of wavelengths conversions, since it is assured that one of the packets contending for each output wavelength channel is forwarded without wavelength conversion. However, differently from the *SPOL* architecture case, in the *SPIL* architecture, in order to minimize P_l, a suitable selection of the packets needing wavelength conversion should be accomplished. That is illustrated in Figure 3, where we have reported an optical switch with 2 IF/OF, each one carrying two wavelengths λ_1,λ_2. The switch shares the TOWCs per IF and it is equipped with one TOWC per IF. We have denoted as $a_{i,j}$ $(i,j=1,2)$ the packet arriving at the IF #*i* and on the wavelength λ_j. If the packets are directed a s shown in Figure 3, the "random" control algorithm may choose $a_{1,2}$ and $a_{1,1}$ as packets to be wavelength shifted and $a_{2,2}$ and $a_{2,1}$ as packets to be forwarded without wavelength conversion. Now, notice as this choice would need two TOWCs from the IF#1 and, since only one TOWC is available for each IF, one packet would have to be lost. On the contrary a packet scheduling without loss may be accomplished by wavelength shifting the packets $a_{1,1}$ and $a_{2,2}$, so that one TOWC for both the IFs is needed and no packet loss occurs.

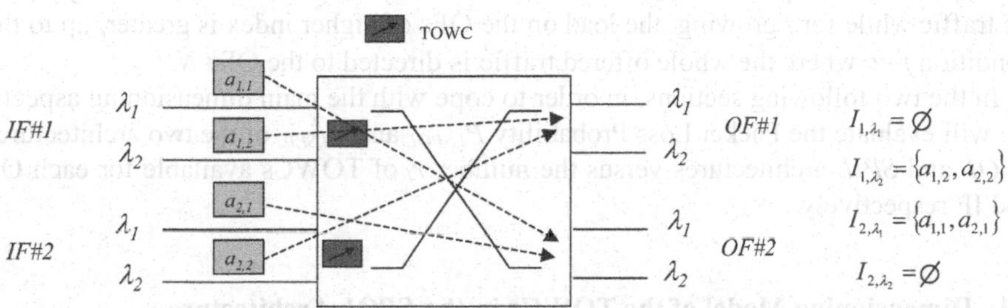

Fig. 3. Optimization of the used number of TOWC in a 2×2 *SPIL* switch with 2 wavelengths per IF/OF and 1 TOWC shared per IF. ∅ denotes the empty set.

As the "random" algorithm is optimum for the *SPOL* architecture, but this is not true for the *SPIL* architecture, we will propose a simulation/analytical hybrid model based on an ILP (Integer Linear Program) formulation of the packet scheduling problem and allowing us to obtain, for a given number of TOWCs, the lowest value of P_l of the *SPIL* architecture. This approach, though practically unfeasible, provides a useful bound of performance of the *SPIL* architecture, demonstrating the remarkable gain potentially obtainable with this kind of architecture.

4 Performance Evaluation of the *SPOL* and *SPIL* Architectures

In this section we present the models we have used to obtain the performances of the *SPOL* and *SPIL* architectures with unicast traffic. In our evaluation we assume that the operation of the switching node is synchronous [15] on time-slot basis. Denoting by M and N the number of wavelengths and the number of IF/OF respectively, we assume that the packet arrivals on the $N \cdot M$ input wavelength channels at each time-slot are independent and occurring with probability p. No assumption is done on the arrival dependence of the packets at different time-slot, in fact, due to bufferless nature of the switch, the performance and the TOWCs dimensioning procedure depends on p only [15]. In order to evaluate the performance of the *SPIL* architecture in both balanced and unbalanced traffic scenarios, we assume that the probabilities p_i (i=1,2,…N) of a packet to be directed to the various OFs can be different. We also assume the ratio $f=p_i/p_{i-1}$ (i=2,…N) to be constant; f is a factor characterizing the unbalancing of traffic. Furthermore we suppose that f can assume values in the range $[1,\infty)$ that is we consider scenarios in which the traffic is unbalanced on the OFs of higher index. It is easy to show that the probabilities p_i can be expressed as follows:

$$p_i = \frac{1-f}{1-f^N} f^{i-1} \qquad i = 1,2,\ldots\ldots N \tag{4.1}$$

From the last expression we can observe that for f=1 we have the case of symmetric traffic while for f growing, the load on the OFs of higher index is greater, up to the condition $f=\infty$ where the whole offered traffic is directed to the OF# N.

In the two following sections, in order to cope with the main dimensioning aspects, we will evaluate the Packet Loss Probability $P_{l,SPOL}$ and $P_{l,SPIL}$ of the two architectures *SPOL* and *SPIL* architectures versus the number r_l of TOWCs available for each OF and IF respectively.

4.1 Dimensioning Model of the TOWCs in the *SPOL* Architecture

Our aim is to evaluate $P_{l,SPOL}$ as a function of the number of TOWCs. We can calculate $P_{l,SPOL}$ by considering the packets offered and lost by the switch and hence we can write:

$$P_{l,SPOL} = \frac{\sum_{k=1}^{N} E[N_{l,k}]}{E[N_0]} \tag{4.2}$$

wherein:

-$E[N_o]=N \cdot M \cdot p$ is the average number of packets offered to the switch

-$E[N_{l,k}]$ is the average number of packets lost by the OF#k of the switch

In the evaluation of $E[N_{l,k}]$ we define the following variables:
- β_k: the random variable denoting the number of wavelengths on which no packets arrive at the OF#k at a given time-slot (for example in Figure 4 (a) we have β_k equal to 2 because there are two wavelengths, λ_3 and λ_4, on which no packets arrive). After some algebra, it is easy to show that β_k is distributed according to a binomial $(M,(1-p\cdot p_k)^N)$ distribution; hence:

$$\Pr\{\beta_k = j\} = \binom{M}{j}\left((1-p\cdot p_k)^N\right)^j\left(1-(1-p\cdot p_k)^N\right)^{M-j} \qquad j=0,1....M \tag{4.3}$$

- $\alpha_{k,j}$: the random variable denoting the number of packets arriving on the $(M-j)$ wavelengths on which at least on packet arrives. After some algebra, the probabilities of $\alpha_{k,j}$ can be expressed as follows:

$$\Pr\{\alpha_{k,j} = i\} = \overset{M-j}{\otimes}\, p_b^k(i) \qquad i = M-j+1,....,N\cdot(M-j) \tag{4.4}$$

wherein:

$$p_b^k(i) = \frac{\binom{N}{i}(p\cdot p_k)^i(1-p\cdot p_k)^{N-i}}{1-(1-p\cdot p_k)^N} \qquad i=1,....,N \tag{4.5}$$

and $\overset{h}{\otimes}$ denotes the convolution operator applied h-1 times.

According to the definition of the variables β_k and $\alpha_{k,j}$, we can evaluate the term $E[N_{l,k}]$ as follows:

$$E[N_{l,k}] = \sum_{j=0}^{M}\Pr\{\beta_k = j\}\sum_{p=M-j+1}^{N(M-j)}\Pr\{\alpha_{k,j} = p\}E[N_{l,k}/\beta_k = j, \alpha_{k,j} = p] \tag{4.6}$$

The conditioned expected value, appearing in (4.6) can be evaluated taking into account that: i) if β_k is equal to j, there are j wavelengths on which the arriving packets can be wavelength shifted; ii) if r_l is the number of TOWCs available for each OF, only $\min(r_l,j)$ of the j wavelengths can be utilized because for each converted packet we need one TOWC; iii) if p packets arrive, $(M-j)$ of them are forwarded without wavelength conversion, $\min(r_l,j)$ are wavelength shifted on the j available wavelengths and the remaining packets are lost; in particular in Figures 4,5 we show the lost packets when $r_l>j$ $(r_l=3, j=2)$ and $r_l<j$ $(r_l=1, j=2)$ respectively.

According to the earlier remarks we can write:

$$E[N_{l,k}/\beta_k = j, \alpha_{k,j} = p] = \max(p-(M-j)-\min(r_l,j),0) \tag{4.7}$$

and hence by (4.2), (4.6) and (4.7) we obtain:

$$P_{l,SPOL} = \frac{1}{M\cdot N\cdot p}\sum_{j=0}^{M}\Pr\{\beta_k = j\}\sum_{p=M-j+1}^{N(M-j)}\Pr\{\alpha_{k,j} = p\}\max(p-(M-j)-\min(r_l,j),0) \tag{4.8}$$

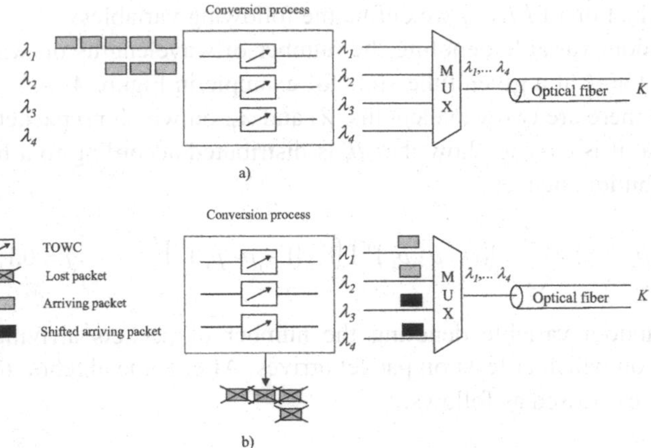

Fig. 4. Wavelengths used by the arriving packets before (a) and after (b) the conversion process is accomplished when the number of converters per output fiber is greater than the number of free wavelengths (M=4, β_k=2, r_T=3 and $\alpha_{k,j}$=8)

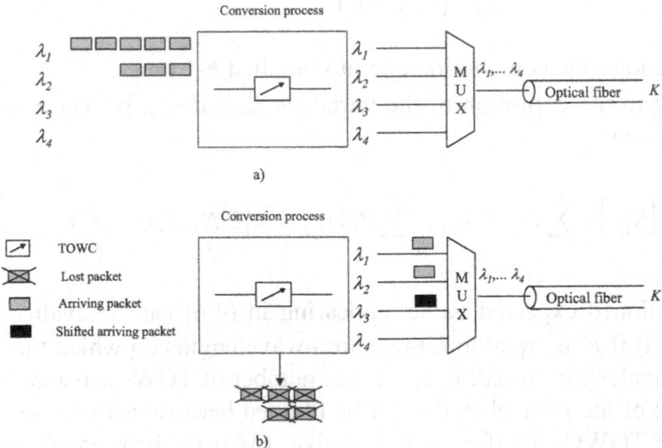

Fig. 5. Wavelengths used by the arriving packets before (a) and after (b) the conversion process is accomplished when the number of converters per output fiber is smaller than the number of free wavelengths (M=4, β_k=2, r_T=1 and $\alpha_{k,j}$=8)

4.2 Dimensioning Model of the TOWCs in the *SPIL* Architecture

In the following two sections we illustrate the models allowing us to evaluate the performance of the *SPIL* architecture when the "random" and "optimized" control algorithms are adopted respectively.

4.2.1 Performance Evaluation in the Case of "Random" Control Algorithm

Our aim is to evaluate P_l as a function of the used number of TOWCs when the control unit accomplishes the packet scheduling according to the "random" control algorithm mentioned in Section 3.

Because the IFs offer traffic to the OFs according to the same distribution, $P_{l,SPIL}$ can be expressed according to the following expression:

$$P_{l,SPIL} = \frac{E[N_{in}]}{E[N_{o,in}]} \tag{4.9}$$

-$E[N_{o,in}]=M \cdot p$ is the average number of packets offered by any IF
-$E[N_{in}]$ is the lost average number of packets arriving at any IF

Taking into account that a packet may be lost because of either the output wavelength unavailability or the TOWC lack, we can express $E[N_{in}]$ as follows:

$$E[N_{in}] = E[N_{in,wl}] + E[N_{in,cl}] \tag{4.10}$$

where $E[N_{in,wl}]$ and $E[N_{in,cl}]$ are the lost average number of packets at an IF because of the output wavelength unavailability and the TOWC lack respectively. In the following the two loss terms will be evaluated.

From the observation that the IFs offer traffic to the OFs according to the same distribution, the average number of packets lost because of the output wavelength unavailability is the same for all of the IFs. Hence we can calculate the term $E[N_{in,wl}]$ by evaluating the total number of packets lost due to the output wavelength unavailability and write:

$$E[E_{in,wl}] = \frac{\sum_{i=1}^{N} E[N_{out,wl,i}]}{N} \tag{4.11}$$

being $E[N_{out,wl,i}]$ the average number of packets directed to the OF#i and lost because of the output wavelength unavailability. According to the observation that if j packets are offered to an OF, $max(0,j\text{-}M)$ of them are lost and taking into account that the packets are directed to the OF #i with binomial $(N \cdot M, p \cdot p_i)$ distribution, we can write:

$$E[N_{out,wl,i}] = \sum_{j=M+1}^{N \cdot M} (j-M) \binom{N \cdot M}{j} (p \cdot p_i)^j (1-p \cdot p_i)^{N \cdot M - j} \tag{4.12}$$

The latter loss term in (4.10) takes into account the fact that even when there are free wavelengths some packets can be lost due to the lack of available TOWCs. This loss term depends on the adopted control algorithm and in particular can be minimized by assigning in an optimum way the TOWCs to the packets needing wavelength conversion. We evaluate the latter term in (4.10) when the "random" control algorithm, illustrated in Figure 3, is adopted.

The term $E[N_{in,cl}]$ is evaluated using the total probability law after defining the random variable C, representing the number of conversions required by the packets arriving at any IF. With this definition we obtain the following expression:

$$E[N_{in,cl}]=\sum_{j=0}^{M}E[N_{in,cl}\mid C=j]\mathrm{Prob}\{C=j\}=\sum_{j=0}^{M}(j-r_i)\mathrm{Prob}\{C=j\} \tag{4.13}$$

Denoting b as the probability that a wavelength conversion is needed on an input wavelength channel of an IF, we can express the probabilities of the random variable C as follows:

$$\Pr\{C=j\}=\binom{M}{j}b^j(1-b)^{M-j} \tag{4.14}$$

To evaluate b, we observe that according to the rules of the "random" control algorithm, a wavelength conversion occurs on an input wavelength channel when: i) a packet arrives at that input wavelength channel; ii) the arriving packet is not lost because of the lack of at least one free wavelength on the OF#i which the packet is directed to; iii) the packet is not selected by the control unit to be forwarded to the OF#i without wavelength conversion. The last event Σ occurs with probability:

$$\Pr\{\Sigma\}=\sum_{q=0}^{N-1}\frac{q}{q+1}p_i(q) \tag{4.15}$$

$$p_i(q)=\binom{N-1}{q}(p\cdot p_i)^q(1-p\cdot p_i)^{N-1-q} \tag{4.16}$$

being $p_i(q)$ the probability that q packets are directed to the same OF#i and arrive on the same wavelength of the considered packet.

According to the earlier remarks and the expressions (4.15) and (4.16), we can express b as follows:

$$b=\sum_{i=1}^{N}p\cdot p_i\left(1-\frac{E[N_{out,wl,i}]}{M\cdot p\cdot p_i}\right)\sum_{q=0}^{N-1}\frac{q}{q+1}\binom{N-1}{q}(p\cdot p_i)^q(1-p\cdot p_i)^{N-1-q} \tag{4.17}$$

Finally the terms from (4.10) to (4.17), once inserted in the expression (4.9) allows us to evaluate $P_{l,SPIL}$ as a function of the needed number of TOWCs and hence to perform the required dimensioning.

4.2.2 Performance Evaluation in the Case of "Optimum" Control Algorithm

In this section the methodology is presented to evaluate P_l when the "optimum" control algorithm is used in each time-slot. We have, in each time-slot, as input data a particular pattern of arrivals and, by means of an ILP formulation of the scheduling problem, the best way of wavelength shifting the arriving packets is determined so that the lost number of packets is minimized. The final results are obtained by simulation and they have a certain degree of uncertainty due to the pseudo-random procedure used to generate traffic in each time-slot. The solution of the ILP problem has been obtained using the ILOG CPLEX [16] software package; this tool allows us to obtain the optimal solution in each time slot to be found and, after a suitable simulation time, we can estimate the P_l of the switch. In the following we illustrate the ILP formulation of the packet scheduling problem.

Let $\{\alpha_{i,j,h}\}$ be binary data variables that take the value 1 if a packet is arriving at the input wavelength channel (i,λ_j) $(i=1,\ldots,N; j=1\ldots,M)$ and it is directed to the OF #h $(h=1,\ldots,N)$.

Let $\{b_{i,j,k}\}$ be binary optimization variables that take the value 1 if one packet arriving on the input wavelength channel (i,λ_j) $(i=1,\ldots,N; j=1,\ldots,M)$ is scheduled to be transferred on the output wavelength λ_k $(k=1,\ldots M)$. Obviously if $\lambda_j=\lambda_k$ the packet is forwarded without wavelength conversion, otherwise the use of a TOWC is needed.

With this definition, the MILP formulation of the problem can be expressed as follows:

Subject to:

$$\sum_{k=1}^{M} b_{i,j,k} \le 1 \qquad (i=1,\ldots,N;\ j=1,\ldots,M) \tag{4.18}$$

$$\sum_{i}\sum_{j} b_{i,j,k}\alpha_{i,j,h} \le 1 \qquad (h=1,\ldots,N;\ k=1,\ldots,M) \tag{4.19}$$

$$\sum_{j}\sum_{k\neq j} b_{i,j,k} \le r_l \qquad (i=1,\ldots,N) \tag{4.20}$$

maximise:

$$\sum_{i}\sum_{j}\sum_{k}\sum_{h} b_{i,j,k}\alpha_{i,j,h} \tag{4.21}$$

The constraint (4.18) states that a packet, arriving at an input wavelength channel, can be forwarded on one output wavelength λ_k only; that follows from the unicast traffic assumption.

The constraint (4.19) states that on each output wavelength channel (h,λ_k) cannot be forwarded more than one packet.

The constraint (4.20) states that in any IF #i, the number of packets that can be wavelength shifted have to be fewer than the available number r_l of TOWCs per IF.

Finally the expression (4.21) is the forwarded number of packets and it is the term that we want to maximize.

5 Numerical Results

This section reports results concerning the comparison among the various proposed switching architectures. In particular: i) we evaluate the goodness of the analytical models, introduced in Section 4, by means of simulation results; ii) the *SPOL* and *SPIL* architectures will be compared in terms of TOWCs needed in order to guarantee a specific value of P_l; iii) the two proposed architectures will be compared to the *SPN* (*Shared-Per-Node*) architecture, proposed in [15] and sharing the TOWCs among all of the arriving packets.

The Packet Loss Probability $P_{l,SPOL}$ and $P_{l,SPIL}$ of the *SPOL* and *SPIL* architectures as a function of the used total number $r=N{\cdot}r_l$ of TOWCs are reported in Figures 6,7 for a traffic offered to each IF $q=M{\cdot}p=0.8$, $N=16$, $M=6$ and $f=1,1.2,1.4,1.6$. We report both the simulation values and the results obtained by the analytical models proposed in Section 4. From Figures 6,7, we can notice as the analytical model provides results in good agreement with the simulation results and for this reason it will be used in the following to carry out the comparison analysis of the two architectures, *SPIL* and *SPOL*.

The needed number of TOWCs for the *SPIL* and *SPOL* architectures is evaluated according to the dimensioning curves sketched in Figure 8, where $P_{l,SPIL}$ and $P_{l,SPOL}$ are reported as a function of the used total number of TOWCs; the traffic and system parameters are $q=0.8$, $N=16$, $M=10$ and $f=1,1.2,1.4,1.6$. The main comments about the Figure 8 are the following: i) all of the sketched curves show the same trend, that is decreasing as r increases up to a threshold value r_{th} denoting the dimensioning value: ii) the dimensioning is more severe for the *SPOL* architecture with respect to the *SPIL* architecture, for example for $f=1.2$, *SPOL* needs $r_{l,SPOL}=80$ TOWCs, while *SPIL* architecture requires $r_{l,SPIL}=48$ TOWCs only; iii) more the traffic unbalanced, more the obtained saving of the *SPIL* architecture with respect to the *SPOL* architecture is great, in fact the percentage saving g, expressed as $g=100{\cdot}(r_{th,SPOL}-r_{th,SPIL})/r_{th,SPOL}$ is 33%,40%,50%,60% for $f=1,1.2,1.4,1.6$ respectively. The worse performance of the *SPOL* architecture, for f increasing, is due to an ineffectiveness use of the TOWCs collocated in those OFs in which fewer packets are directed.

The comparison of the two proposed architectures, *SPOL* and *SPIL* with the *SPN* architecture in terms of used number of TOWCs is reported in Figure 9 in the case of balanced offered traffic. In particular, we show P_l as a function of the total used number of TOWCs when $q=0.8$, $N=16$, $f=1$ and $M=7,11,15$.

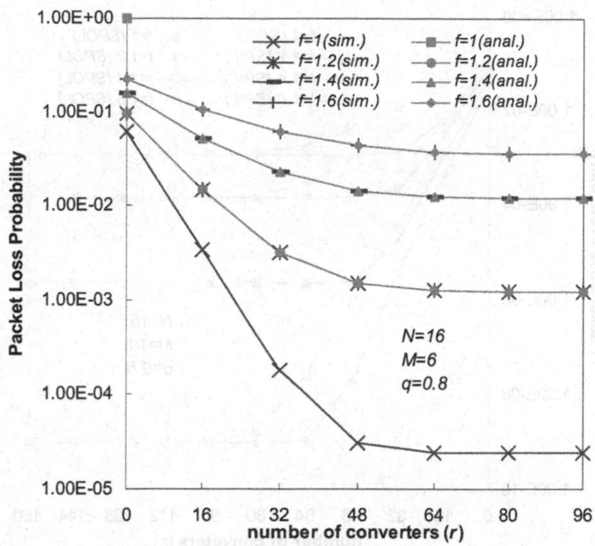

Fig. 6. Comparison between the analytical and simulation results in the *SPOL* architecture. The traffic and system parameters are q=0.8, N=16, M=6 and f=1,1.2,1.4,1.6.

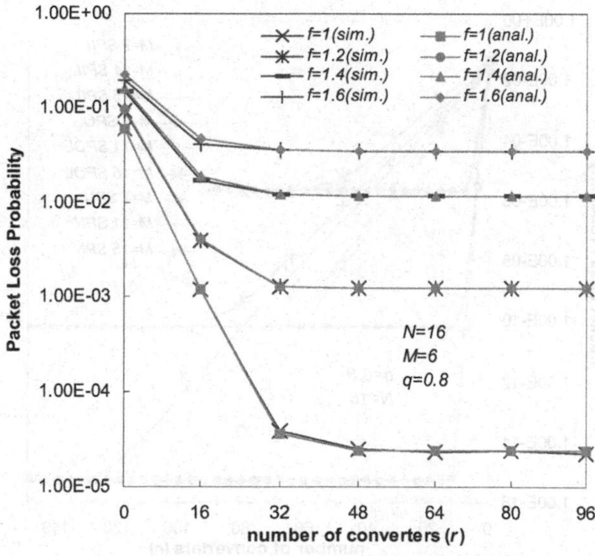

Fig. 7. Comparison between the analytical and simulation results in the *SPIL* architecture. The traffic and system parameters are q=0.8, N=16, M=6 and f=1,1.2,1.4,1.6.

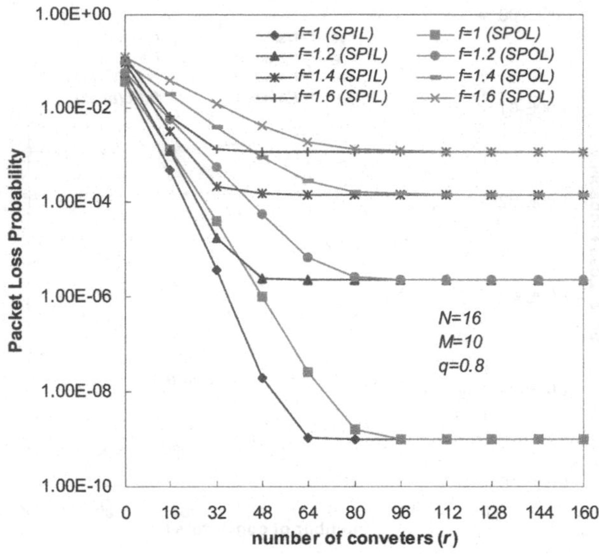

Fig. 8. Comparison between the *SPOL* and *SPIL* architectures. The traffic and system parameters are q=0.8, N=16, M=10 and f=1,1.2,1.4,1.6.

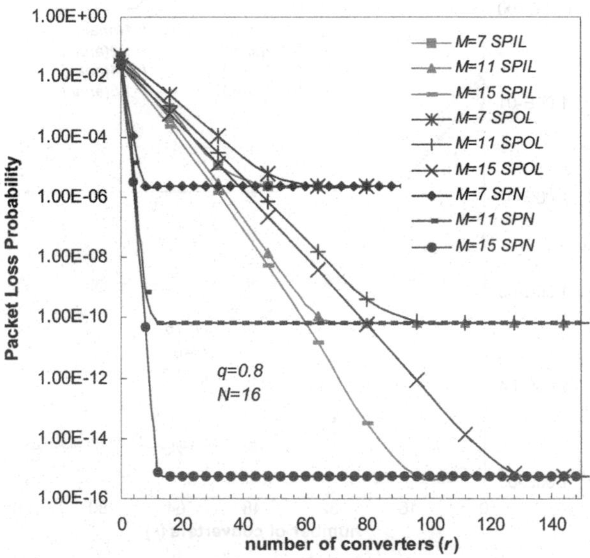

Fig. 9. Comparison among the *SPIL,SPOL* and *SPN* architectures in terms of the used number of TOWCs when the packets are scheduled according to the “random” control algorithm. The traffic and system parameters are q=0.8, N=16 and M=7,11,15

The switching architectures are compared when the packets are scheduled according to the "random" control algorithm.. The reported results of the *SPN* architecture are obtained by means of the analytical model proposed in [15]. Observing the curves we can see that the number of TOWCs required by the *SPOL* architecture to reach the saturation loss probability are 64, 96, 128 for *M*=7,11,15 respectively. Instead for the *SPIL* architecture the threshold values are 48, 64, 96 and this leads to percentage gains equal to 25%, 33% and 25% with respect to the *SPOL* architecture.

From Figure 9 we can also evaluate the threshold values for the number of TOWCs required by the *SPN* architecture to reach the saturation loss probability . The values, using 7, 11, 15 wavelengths per fiber, are 9, 12, 15 respectively. The correspondent percentage savings with respect to *SPOL* architecture are: 86%, 89%, 90%. This confirm the great advantage of a total sharing technique with respect to techniques that share the TOWCs per IF or OF.

This greater advantage of the *SPN* architecture is counterbalanced by the increase in number of SOA gates required in the space switching matrix with respect to the *SPIL* and *SPOL* architectures. In fact the *SPN* architecture requires additional switching and filtering structures in the shared converters pool. The other two architectures need both the same number of SOA gates and this is smaller than the required number in the *SPN* architecture. As a matter of example for *M*=11 the *SPN* architecture needs 5268 SOA gates differently from the *SPIL/SPOL* architectures that need 4752 SOAs only.

Now we provide a comparison among the *SPIL* and *SPOL* switching architectures when the packets are scheduled according to the "optimum" control algorithm mentioned in Section 3. To accomplish this comparison, the reported results for the *SPIL* architecture are obtained according to the simulation/analytical hybrid model illustrated in Section 4.2.2. On the contrary the reported results of the *SPOL* architecture are the ones obtained when the "random" control algorithm is adopted because in this case the algorithm is optimum too. The obtained results are reported in Figure 10 for *M* even and for the same traffic and system parameters of the earlier case study.

For the *SPIL* architecture we report also the values of P_l in the case in which the packet scheduling is accomplished according the "random" control algorithm. As it can be seen from the Figure 10 the adoption of the "optimum" control algorithm allows to obtain dimensioning values of the TOWCs lower and hence a TOWCs saving greater with respect to the *SPOL* architecture. For example for *M*=4,6 we need 1 TOWC per IF only against 3,5 in the *SPOL* architectures; this values lead to percentage gains in the order of 66% and 80% respectively. Hence, with respect to the "random" control algorithm, an optimized scheduling of the packets allows a further saving of TOWCs.

Finally we can conclude noticing as the control algorithm plays a very important role in the case of the *SPIL* architecture. In future, further efforts will be done in order to determine the procedure according to which an "optimized" scheduling of the packets can be accomplished.

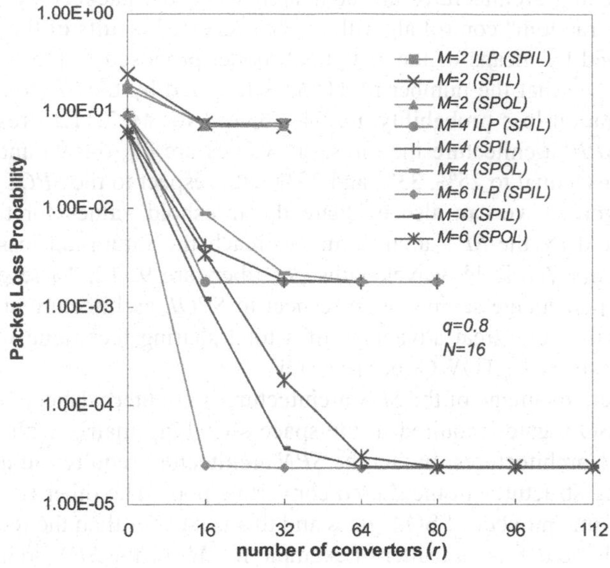

Fig. 10. Performance evaluation of the *SPIL* and *SPOL* architectures when the packets are scheduled according to both the "random" and "optimum" control algorithm. The traffic and system parameters are q=0.8, N=16 and M=2,4,6

6 Conclusions

In this paper we have discussed two WDM optical packet switching architectures denoted as *SPOL* and *SPIL* and equipped with TOWCs, shared per OF and IF respectively. In order to compare the proposed architecture with other ones already proposed in literature, some analytical and simulation models have been developed that allowed us to evaluate the needed number of TOWCs as a function of the main system and traffic parameters (offered traffic, number of IF/OF, number of wavelength,...). The performance evaluation of the *SPIL* architecture has been carried out when two control algorithms are adopted in order to schedule the arriving packets; the former one, simple computationally but not able to reach the best dimensioning values of TOWCs, the latter one allowing to schedule the packets in an optimized way but needing a computation effort greater and according to a procedure still to be determined. With respect to an architecture sharing the TOWC per OF and having the same complexity of the switching matrix of the *SPIL* architecture expressed in terms of used number of SOAs, the proposed models allowed us to evaluate for example for M=7, a saving of TOWCs equal to 25% and 50% for the two above control algorithms respectively. The best performance involved when an optimized scheduling of the arriving packets is accomplished, suggest further investigations about it, in fact is not only sufficient to evaluate its performance but also to determine the way in which the packets have to be scheduled in order to reach this performance. Further research items are the complexity evaluation of the "optimized" control algorithm and the determination of heuristics allowing to obtain performance both near the one of the "optimized" control algorithm and having low computation cost.

References

1. M. Murata and K. Kitayama: Ultrafast Photonic Label Switch for Asynchronous Packets of Varaiable Length, IEEE Infocom 2002, June 23-27, New York
2. C. Qiao and M. Yoo: Optical Burst Switching (OBS)- a new paradigm for an optical Internet, Vol.8, pp.69-84,1999
3. M. Renaud, C. Janz, P. Gambini and C. Guillemot: Transparent optical packet switching: The European ACTS KEOPS project approach, in Proc. LEOS '99, Vol.2,pp. 401-402, 1999
4. C. Guillemot et al.: Transparent Optical Packet Switching: The European ACTS KEOPS Project Approach, IEEE Journal of Lightwave Technology, Vol.16, No. 12, December 1998
5. M. Listanti, R. Sabella, V. Eramo: Architectural and Technological Issues for Future Optical Internet Networks, IEEE Communications Magazine, pp.82-92, September 2000
6. B. Li, Y. Qin, X. Cao, K. Sivalingam: Photonic Packet Switching: Architectures and Performance, Optical Network Magazine, January/February 2001, pag 27-39
7. L. Xu, H.G. Perros and G. Rouskas: Techniques for Optical Packet Switching and Optical Burst Switching, IEEE Communication Magazine- Jannuary 2001
8. S. Yao, B. Mukherjee and S. Dixit: "Advances in Photonic Packet Switching: an Overview, IEEE Communication Magazine- February 2000
9. M. Mahony, D. Simeoniddu, D. Hunter and A. Tzanamaki: The Application of Optical packet switching in Future Communications Networks, IEEE Communication Magazine- March 2001
10. S. Yao, S.J.B. Yoo, B.Mukherjee and S. Dixit: All-Optical Packet Switching for Metropolitan Area Networks: Opportunities and Challenges, IEEE Communication Magazine- March 2001
11. S. L. Danielsen, C. Joergensen, B. Mikkelsen and K. E. Stubkyaer: Optical Packet Switched Network Layer Without Optical Buffers, IEEE Photonic Technology Letters, Vol.10, No. 6, June 1998
12. S. L. Danielsen, C Jorgensen, B. Mikkelsen and K. E. Stubbkyaer: Analysis of a WDM Packet Switch with Improved Performance Under Bursty Traffic Conditions Due to Tuneable Wavelength Converters, IEEE Journal of Lightwave Technology, Vol.16, No. 5, May 1998
13. V.Eramo, M.Listanti: Comparison of Unicast/Multicast Optical Packet Switching Architectures using Wavelength Conversion, Optical Network Magazine, vol. 3, pp. 63-75, March-April 2002
14. V. Eramo, M. Listanti, P. Pacifici: A Comparison Study on the Wavelength Converters Number Needed in Synchronous and Asynchronous All-Optical Switching Architectures, accepted for publication to IEEE Journal of Lightwave Technology, February 2003
15. V. Eramo, M. Listanti: Packet Loss in a Bufferless WDM Switch Employing Shared Tuneable Wavelength converters, IEEE Journal Lightwave Technology, December 2000
16. http://www.ilog.com/products/ampl

Efficient Usage of Capacity Resources in Survivable MPλS Networks*

Lan Tran[1], Kris Steenhaut[1], Ann Nowé[1], Mario Pickavet[2], and Piet Demeester[2]

[1] Vrije Universiteit Brussel-Erasmus Hogeschool Brussel, COMO-INFO-TW, Pleinlaan 2,1050 Brussels, BELGIUM
{latranth,ksteenha,asnowe}@info.vub.ac.be
http://como.vub.ac.be/

[2] Ghent University-INTEC, Department of Information Technology, Sint-Pietersnieuwstraat 41, 9000 Ghent, BELGIUM
{mario.pickavet,demeester}@intec.rug.ac.be

Abstract. The Multi-Protocol Lambda Switching (MPλS) has been recently applied in the optical network control plane to provide fast lightpath provisioning. With the growth of traffic in optical network, the survivability of the network becomes extremely important. Therefore, an efficient protection mechanism is needed. Shared backup tree lightpath protection is a promising paradigm in MPλS networks because of its fast recovery from single failure and its efficiency in terms of consumed capacity. In this paradigm, a shared backup tree is used to protect a group of working lightpaths. In this paper, we propose a heuristic that yields an efficient placement of working paths and backup trees in polynomial time.

1 Introduction

Multi-Protocol Lambda Switching (MPλS) is an extension of Multi-Protocol Label Switching (MPLS) in which each wavelength on a WDM link represents an MPLS label. MPλS was proposed to provide a framework for real time, automatic provisioning of lightpaths [1]. The MPλS approach combines the recent improvements of control plane techniques for MPLS traffic engineering with optical cross-connect technology.

In MPλS network, the failure of a network element such as a fiber link or a cross-connect can cause the loss of a large amount of traffic. Therefore, lightpath restoration in optical networks becomes a key consideration [2][3][4].

Shared path protection is known as an efficient fast recovery mechanism. But current MPλS networks do not support that two backup optical paths, which protect two mutually disjoint working paths, share a wavelength. Therefore, they only allow dedicated protection; wasting a lot of capacity resources. Recently, Colle et al. proposed an improvement in the optical network elements for merging

* Research supported by the Fund for Scientific Research FWO, Belgium

M. Ajmone Marsan et al. (Eds.): QoS-IP 2003, LNCS 2601, pp. 204–217, 2003.

two backup optical paths to the same destination[5], enabling MPλS protection mechanisms based on Single Ended Backup Trees (SEBTs). A SEBT forms a directed tree (at most one wavelength per link) towards the root node. A SEBT can be used to protect a group of working lightpaths from the same or from different sources to the same destination.

Compared to the shared path protection approach [3] in the optical network, the shared backup tree protection approach has faster recovery from failure because it does not require the signaling to set up the backup connection, but it consumes more capacity resources. As mentioned earlier, the shared path protection cannot be applied in MPλS networks. In order to efficiently organize shared backup tree protection, the total cost of capacity resources used by working paths and backup trees has to be minimized. This is a computationally hard problem.

In this paper we propose the Successive Greedy Source Placement (SGSP) heuristic to solve the above optimization problem. This heuristic can find more efficient protection schemes compared to dedicated path protection, with small computational effort. We demonstrate that the heuristic is scalable with respect to topology and demand vector growth.

The paper is organized as follows. Section 2 defines the problem and discusses related work. Section 3 describes the proposed heuristic for organizing efficient shared backup protection. Section 4 compares the proposed heuristic with other existing approaches. Section 5 contains the conclusion.

2 Problem and Related Work

The problem of optimizing shared backup tree protection can be described as follows: given the working network demand vector, the objective is to find the working paths and the backup trees such that the total cost of these paths and trees is kept as low as possible.

In the dedicated path protection approach, a working path can be protected against single node failure by a node disjoint backup path. In the SEBT protection approach, SEVERAL disjoint working paths to the same destination (from the same source or from different sources) can be protected against single node failures by a single ended backup tree. Indeed, the SEBT consists of several backup paths, which share part of the route towards the destination node. The working paths must be mutually disjoint and each working path has to be node disjoint with its backup path.

The global protection problem can be split up per destination node. Several backup trees can be used for protecting working paths towards the same destination node. To do so, the demand vector to a destination has to be decomposed in subdemands. Working paths for a subdemand are protected by a single backup tree; and are called a group of working paths.

The optimization problem can be formulated as an Integer Linear Programming (ILP) problem as demonstrated in [6]. The numerical solution of the ILP becomes infeasible for networks of realistic size. For example, for a network with

14 nodes and 25 links and for the demand of 20 connections in total to a given destination node (from various sources), the ILP formulation consist of 1710 binary variables and 57430 constraints.

Groebbens et al. propose a heuristic approach [7] that divides the problem into two separate subproblems: the partitioning subproblem and the placing of working paths and backup trees for each subdemand. The partitioning subproblem refers to picking the cheapest combination of subdemands based on knowing the cost of placing the subdemands and their backup trees. Because the number of possible combinations of subdemands can be quite high, the method does not scale well for large networks and high working demands. Even when using a Subdemand Space Reducer, the number of considered combinations remains prohibitively high.

We propose the SGSP heuristic, which decreases remarkably the number of combinations to be taken into account. The method iterates towards a solution for the problem (per destination) by considering each source successively. For the placement of working paths and backup paths of a given source, groups of working paths and their respective backup trees obtained *from previously considered sources* are taken into account. To start the heuristic (first source), the set of groups is empty.

Per source, we generate a set of possible disjoint working paths. From this set, a group of working paths must be selected based on the possibility and the efficiency to join existing backup trees for the protection and the cost of the working paths themselves. The details of the steps to follow will be described in the next section.

3 Heuristic

3.1 The SGSP Heuristic

As mentioned previously, the global problem can be divided into a subproblem per destination (SEBT). The problem per destination is then divided into a subproblem per source following our approach. Therefore, the global problem can be solved as follows:

Repeat per *destination*
  Repeat per *source*
    Find working paths and backup paths (the Per_Source heuristic: details in next subsection).

3.2 The Per_Source Heuristic

Preliminaries. The construction of the heuristic is based on the possibility of adding new mutually disjoint working paths from the same source to an existing group of working paths. The working paths can become member of an existing group of working paths if the following conditions hold:

– The new mutually disjoint working paths must be disjoint with the working paths of the group they join.
– The new mutually disjoint working paths must be protectable by a disjoint backup path, which can merge with the backup tree of the considered group.

This concept is illustrated by an example shown in Fig. 1. It depicts a situation with one group of working paths WP_1 and its backup tree T_1. The considered destination is node 7. New working paths WP_1 and WP_2 from source 8 are node disjoint with the group of working paths W_1. Therefore, these working paths are added to the group W_1; the backup path BP joins the existing backup tree T_1 via merge node 4, since node 4 is closest to source node 8.

The *extra cost* of the backup path consists only of the cost of the links from the source node to the merge node (the cost of the part of the backup path BP that is shared with the existing tree T_1 is zero because it has already been counted). We will call this extra cost *the merge cost of the backup path.*

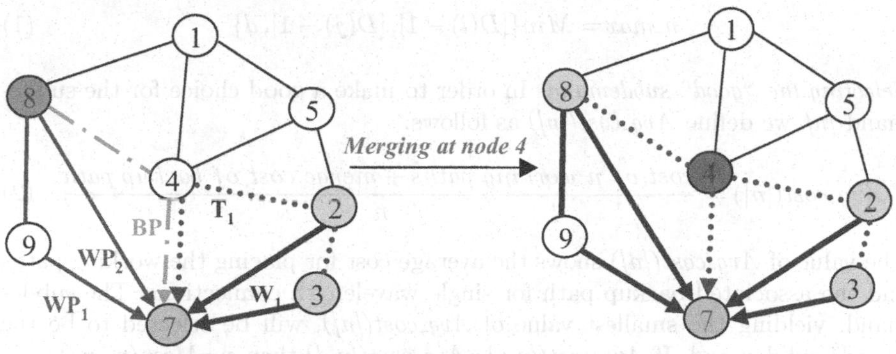

Fig. 1. Example of backup path merging with the backup tree (Working paths (thick) and backup tree (dotted)and backup path (dashed))

Problem Formulation. Suppose we already considered source *1, 2, ..,i-1* and that source *i* must now be taken into account. As mentioned previously, new working paths can be added to an existing group of working paths. Therefore, the problem of finding working paths and backup paths from source node *i* to destination node *j* can be described as follows: given a demand of WP_1 wavelength connections between node *i* and node *j* and given *N* groups of already existing working paths $(W_1, W_2, .., W_N)$ and their respective backup trees $(T_1, T_2, .., T_N)$ from precedent sources to destination *j*, the objective is to find new working paths and backup paths such that the total cost is as low as possible.

Basic Idea of the Heuristic. Each source has its connection requirement to the considered destination, denoted as demand *[d]*. Demand *[d]* can be decomposed into subdemands. The problem is how to place and select the “good” subdemand. Placing the subdemand means finding the working paths and organizing their protection. In order to organize their protection in a cost effective way, the protection path tries to join an existing backup tree. To evaluate the placement of a given subdemand, a parameter called *Avg_cost* will be introduced.

Placing the Subdemand. Let *[n]* be a subdemand of n connections for a given source and destination. Such n connections can join one of the existing groups or they can make a new group depending on the possibility of merging. Placing a subdemand *[n]* requires the existence of n mutually disjoint working paths from source to destination and one backup path for protecting them. This placement is only feasible if *(n+1)* disjoint paths are available. The maximum number of disjoint paths cannot be higher than the degree of source node *D(i)* and the degree of destination node *D(j)*. Since *[n]* is a subdemand of the demand *[d]*, n cannot be higher than d. Therefore, the upper bound for n, called n_max, can be calculated as follows:

$$n\_max = Min\{[D(i) - 1], [D(j) - 1], d\} \tag{1}$$

Selecting the "good" subdemand. In order to make a good choice for the subdemand *[n]*, we define *Avg_cost([n])* as follows:

$$Avg\_cost([n]) = \frac{cost\ of\ n\ working\ paths + merge\ cost\ of\ backup\ path}{n} \tag{2}$$

The value of *Avg_cost([n])* shows the average cost for placing the working paths and the associated backup path for single wavelength connections. The subdemand, yielding the smallest value of *Avg_cost([n])*, will be selected to be the "good" subdemand. If $Avg\_cost([n_1])=Avg\_cost([n_2])$ then $n=Max\ (n_1,n_2)$.

Main Heuristic. In this section, we will explain how the heuristic finds the "good" combination of subdemands and how it places the subdemands. We consider two types of the demand vectors between the considered source and destination: the demand of one connection and the demand of more than one connection.

For the demand of only one wavelength between node i and node j, there is only one possible combination. This combination consists of one subdemand. Therefore, the original problem reduces to the placement problem. The placement of the subdemand can be solved quite simply by finding a merge node as close as possible to the source node i. If the merge node is different from the destination node, a working path of the considered source node is added to the selected group of existing working paths and its partial backup path from the merge node to destination node is added to the backup tree, which protects this group. If not, a new group with one working path and an associated backup path is generated.

When the number of required connections between two nodes is two or more, there are several possible combinations. The evaluation of all the combinations in order to find the best one can be time-consuming. Therefore we need to find a good method to reduce the number of considered combinations. First, we evaluate all possible subdemands *[n]* *(n=1..n_max)* and greedily select the one which yields the best value of *Avg_cost([n])*. Based on the merge node for that subdemand, the working paths of that subdemand join the selected group or form a new group. After selecting the first subdemand, we take care of the remaining demand. The remaining demand is again split into a combination of subdemands (same problem as before). For the example shown in Fig. 2, the "good" subdemand is the subdemand *[d-3]*. The subdemands that need to be considered further are subdemands *[1]*, *[2]*, *[3]*. The idea is to repeat this solving method with a reduced number of considered subdemands as illustrated in Fig. 2.

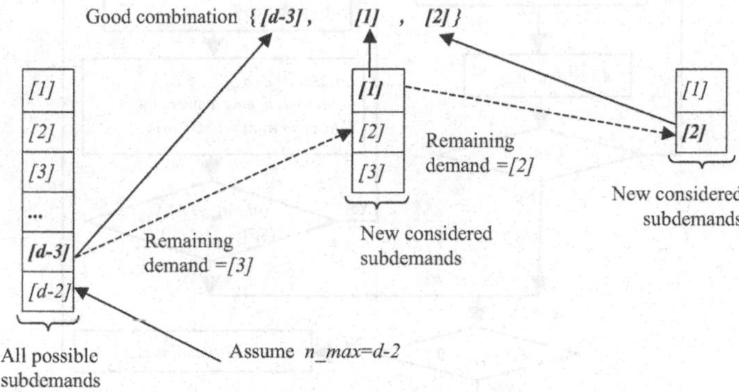

Fig. 2. The "good" combination for a demand *[d]*

Fig. 3 illustrates the main steps of the heuristic for finding working paths and backup trees per source. In order to evaluate each subdemand *[n]*, we need to find n disjoint working paths. Step 1 shows that n disjoint working paths can be drawn from the produced set of nw disjoint paths. The details can be found in Fig. 4. Our method of Selecting nw disjoint paths from K Random Paths is called SKRP. A comparison with the well-known Busacker-Gowen algorithm [10] for finding disjoint paths is presented in section 4.1.

Step 2 is done for each possible subdemand *[n]*. It finds the node to merge with the backup path. If the merge node is different from the destination node, the backup path can merge with a shareable group. The set of shareable groups is determined and one group will be selected. The value of *Avg_cost([n])* is calculated in order to evaluate the subdemand *[n]*. More details of step 2 are described in Fig. 4. Step 3 selects the subdemand which gives the smallest value of *Avg_cost[n]*. After knowing the best subdemand *[n_{best}]*, step 4 considers the merge node for this subdemand.

If the merge node is the destination node, a new group is created and the remaining demand is calculated (step 4.1). When the remaining demand is higher than *[n_{best}]*, the best subdemand is the same as the resulting best subdemand in

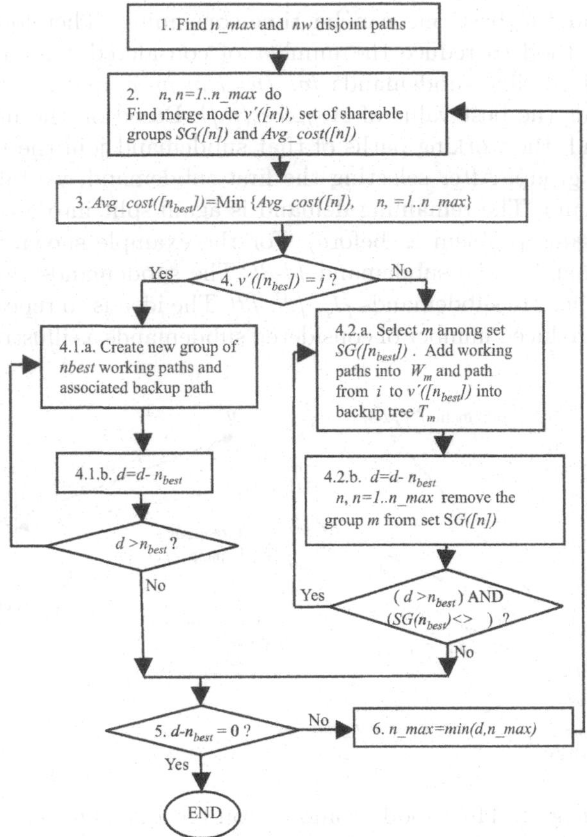

Fig. 3. Flowchart of the Per_Source heuristic for a demand *[d]*

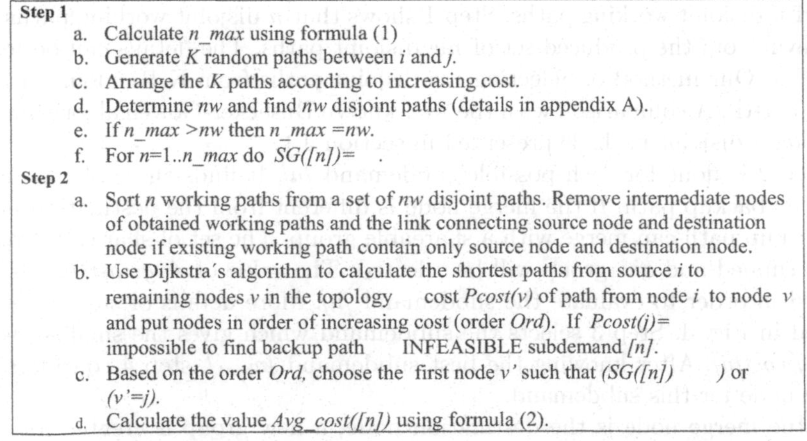

Step 1

a. Calculate *n_max* using formula (1)
b. Generate *K* random paths between *i* and *j*.
c. Arrange the *K* paths according to increasing cost.
d. Determine *nw* and find *nw* disjoint paths (details in appendix A).
e. If *n_max* >*nw* then *n_max* =*nw*.
f. For *n*=1..*n_max* do *SG([n])*= .

Step 2

a. Sort *n* working paths from a set of *nw* disjoint paths. Remove intermediate nodes of obtained working paths and the link connecting source node and destination node if existing working path contain only source node and destination node.
b. Use Dijkstra's algorithm to calculate the shortest paths from source *i* to remaining nodes *v* in the topology cost *Pcost(v)* of path from node *i* to node *v* and put nodes in order of increasing cost (order *Ord*). If *Pcost(j)*= impossible to find backup path INFEASIBLE subdemand *[n]*.
c. Based on the order *Ord*, choose the first node *v'* such that *(SG([n]))* or *(v'=j)*.
d. Calculate the value *Avg_cost([n])* using formula (2).

Fig. 4. The details of step 1 and step 2 in the flowchart in Fig. 3

step 4.1. Therefore step 4.1 is repeatedly applied until the remaining demand is smaller than *[n_{best}]*. If the merge node is different from the destination node, step 4.2 is applied. First, one group is selected from a set of shareable groups and the subdemand *[n_{best}]* is added to this group. Secondly, the merged group is removed from the set of shareable groups *SG([n])* for every possible subdemand *[n]* in order not to reuse this group for merging other subdemands from this source. After that, the remaining demand is calculated. For the same reason as for the repetition of step 4.1, step 4.2 is repeatedly applied until the remaining demand is smaller than n_{best} or the set of shareable groups is empty. If the remaining demand is higher than zero then step 6 recalculates the value of *n_max* due to the change of the demand value and the heuristic method returns to step 2 to continue to place another subdemand of the "good" combination.

The heuristic terminates when the remaining demand between the considered source and destination is zero.

Fig. 5 shows one example for finding working paths and backup paths from source 8 to destination 7 for a given demand d=3 starting from already existing working path W_1 and backup tree T_1 (see Fig. 5(a)).

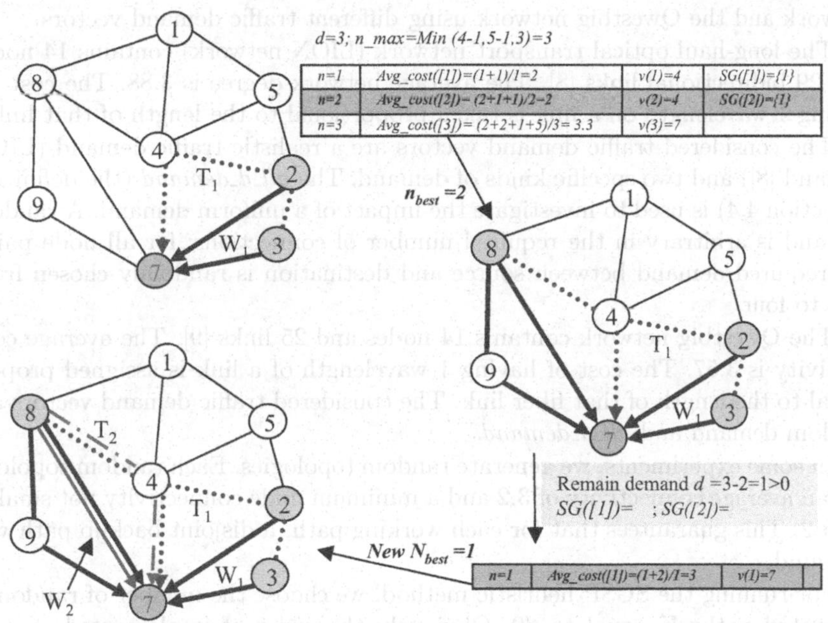

Fig. 5. Working and backup paths from source 8 to destination 7 for demand *[3]*

Based on the *Avg_cost([n])* values of all possible subdemands, we select n_{best}=*2* because *Avg_cost([2])* is the smallest *Avg_cost* value and is covering the highest number of working paths. The merge node v is node 4 and the shareable group *SG([1])=1*. We add two paths *(8,7)* and *(8,9,7)* into the group of working paths W_1 and add the partial backup path *(8,4)* into the backup tree T_1. Now we have found the working paths and the backup path for two working connections but we still have to find working paths and backup path for *(3-2)= 1* remaining working connection. New $n_{best}=1$, the merge node *v'* equals *7* being the destination node, we have to create a new group of working paths W_2 and a backup tree T_1.

4 Experimental Results

In this section, we first compare two different methods for finding disjoint paths. Secondly, we illustrate the gain of using Shared Backup Tree protection compared to the dedicated path protection. Thirdly, we introduce a new metric to investigate the efficiency of the Shared Backup Tree protection with respect to topology and demand. Afterwards, the performance of the SGSP heuristic is compared to the Groebbens's heuristic [7] in terms of cost and execution time. Lastly, the scalability of the SGSP heuristic is evaluated.

We applied the SGSP heuristic on two existing wide area networks: the LION network and the Qwestbig network using different traffic demand vectors.

The long-haul optical transport network (LION network) contains 14 nodes and 29 bidirectional links [8]. The average network degree is 3.88. The cost for having 1 wavelength on a link is taken proportional to the length of that link.

The considered traffic demand vectors are a realistic traffic demand (LION demand [8]) and two specific kinds of demand. The *all_d_demand* (the definition in section 4.4) is used to investigate the impact of a uniform demand. A random demand is arbitrary in the required number of connections: for all node pairs, the required demand between source and destination is randomly chosen from zero to four.

The Qwestbig network contains 14 nodes and 25 links [9]. The average connectivity is 3.57. The cost of having 1 wavelength of a link is assigned proportional to the length of that fiber link. The considered traffic demand vectors are random demand and *all_d_demand*.

In some experiments, we generate random topologies. Each random topology has an average connectivity of 3.2 and a minimum node connectivity not smaller than 2. This guarantees that for each working path, a disjoint backup path will be found.

For running the SGSP heuristic method, we choose the number of randomly generated paths K equal to 30. Obviously the order of implemented sources affects the result of the SGSP heuristic. The distance order is chosen because we expect this order to yield a high potential of merging. In distance order, the source sequence is chosen according to increased path cost from source to destination. The execution times are measured on a Processor PIII - 733 MHz with 128 MB RAM.

4.1 Comparison of Methods for Finding Disjoint Paths

Table 1 shows the results of comparison between SKRP method and Busacker-Gowen algorithm for five random topologies with the same random demand vector. For each network scenario, SKRP is run ten times and the minimum, maximum and average result in terms of cost and execution time are given. For small network size, Busacker-Gowen is best in both terms of cost and execution time. For large network, SKRP is much better in terms of execution time but about 1.5% less cost effective. SKRP yields a good trade-off between cost of the paths and execution time.

Table 1. Comparison between the SKRP method and Busacker-Gowen algorithm

Network size (# nodes)	Cost disavantage of using SKRP vs. Busacker-Gowen (%)			Execution time(ms)			
				SKRP method			Busacker -Gowen
	Min	Max	Avg	Min	Max	Avg	
10	0,04%	0,04%	0,04%	270	330	296	220
20	0,67%	2,13%	1,33%	320	440	378	330
30	1,72%	4,00%	2,86%	440	500	446	610
40	0,57%	1,55%	1,13%	440	550	518	770
50	0,14%	2,88%	1,41%	560	660	611	1050

4.2 Comparison with the Dedicated Path Protection

Table 2 shows the advantage of using shared backup tree protection for the LION demand in the LION network. Note that the working paths are longer when using shared backup tree protection. In dedicated protection, the working paths are indeed the shortest paths. But in shared backup tree protection, the working paths are placed as disjoint as possible because this leads to cost reduction of the shared backup tree; which will make the global result of placing both working paths and backup trees better. For the dedicated protection result, the working path and the backup path are calculated based on the efficient algorithm for finding a pair of node disjoint paths [11]. From the experiments, the cost advantage and wavelength advantage of the SGSP heuristic is about 15% compared to dedicated path protection.

4.3 Comparison with the Groebbens's Heuristic

For the LION network with the LION demand, the execution time of the SGSP heuristic is only 0.236s ; while the execution time of Groebbens's heuristic is 41.890 s. Groebbens's results are somewhat more optimal (see table 3) but at the expense of execution time. The fast execution time is especially important when considering dynamic demand placement.

Table 2. Advantage of Using Backup Trees vs. Dedicated Path Protection

	WPs wavelengths	Protection wavelengths	WPs + protection wavelengths	Cost WPs	Cost protection	cost WPs + protection
Relative advantage with SGSP	-12.57%	32.39%	15.97%	-5.64%	29.89%	14.94%

Table 3. Comparison between Groebens' s and the SGSP Heuristic

	Backup Trees (Groebbens)	SGSP's Backup trees	Difference
total cost	110084	111844	+ 1.60%
cost of WPs	57595	58447	+ 1.48%
cost of protection	52489	53397	+ 1.73%
total number of wavelengths	418	421	+ 0.72%
number of WPs wavelengths	200	206	+ 3.00%
number of wavelengths for protection	218	215	-1.38%

Fig. 6 gives results per destination node for the LION demand in the LION network. The results obtained with the SGSP heuristic are the same as Groebbens's in seven destinations, better in Rome and worse than Groebbens's in the six remaining destinations. SGSP yields better results than Groebbens's in Rome due to the selection process of the Subdemand Space Reducer in Groebbens's heuristic. For the random demand in the Qwestbig network, the cost advantage of using Groebbens's backup tree is 10.67% and of backup tree generated by the SGSP is 8.48% (see in Fig. 7).

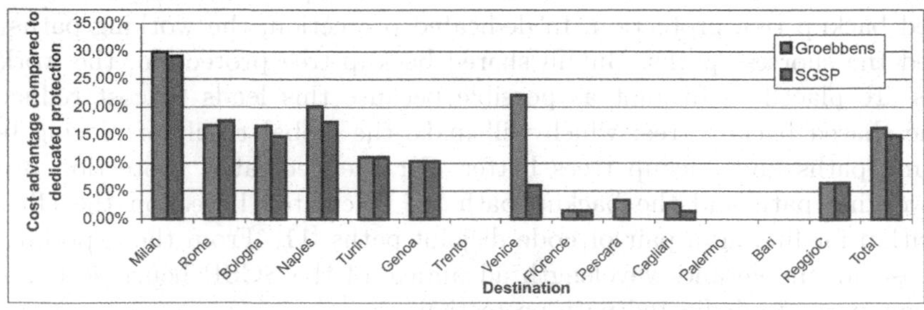

Fig. 6. Cost advantage per destination node for a long haul network

4.4 The Scalability of the SGSP Heuristic

In this section, we evaluate the scalability of the SGSP heuristic with respect to network topology and demand vector.

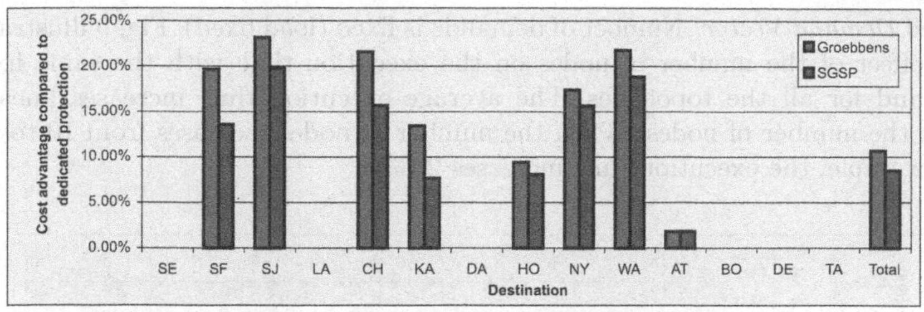

Fig. 7. Results per destination node for Qwestbig net

The Effect of the Number of Connections on the Execution Time. We consider the special demand vector in which demands of working connections between all node pairs are equal to *d*. We call this demand vector the *all_d_demand.* Fig. 8 illustrates the effect of the number of connections *d* of all node pairs on the execution time for the LION network and for the Qwestbig network. The execution time for the LION network is always higher than for the Qwestbig network because the average network degree and number of nodes of the LION network is higher than the Qwestbig network. Both execution times of the LION network and the Qwestbig network increase gradually with an increasing value of *d*. In the LION network, the execution time for the *all_1_demand* is 337 ms and for the *all_19_demand* is 683 ms. In the Qwestbig network, the execution time for the *all_1_demand* is 256 ms and for the *all_19_demand* is 607 ms. These numbers indicate that the execution time of SGSP heuristic is still acceptable when the demand per node pair is high.

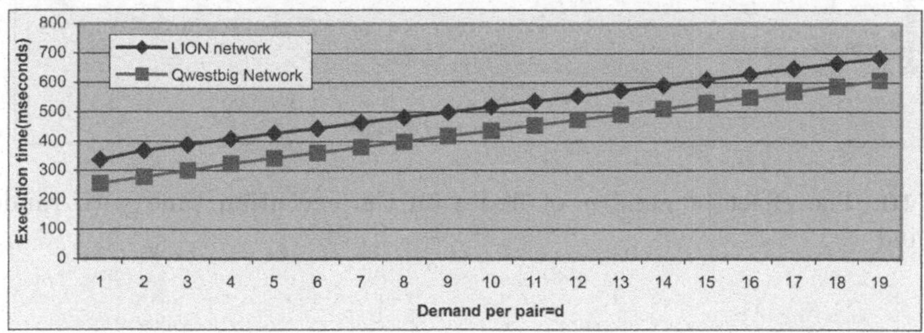

Fig. 8. The effect of number of connections on the execution time

The Effect of the Number of Nodes on the Execution Time. 20 random topologies are generated with 10, 20, 30, 40 and 50 nodes. The average execution time is taken over the 20 topologies with the same number of nodes; first for a fixed demandand afterwards for a random demand.

Fixed Demand Vector. Number of demands is fixed (load fixed). Fig. 9 illustrates the effect of the number of nodes on the execution time with the same fixed demand for all the topologies. The average execution time increases linearly with the number of nodes. When the number of nodes increases from 10 to 20, for example, the execution time increases 20%.

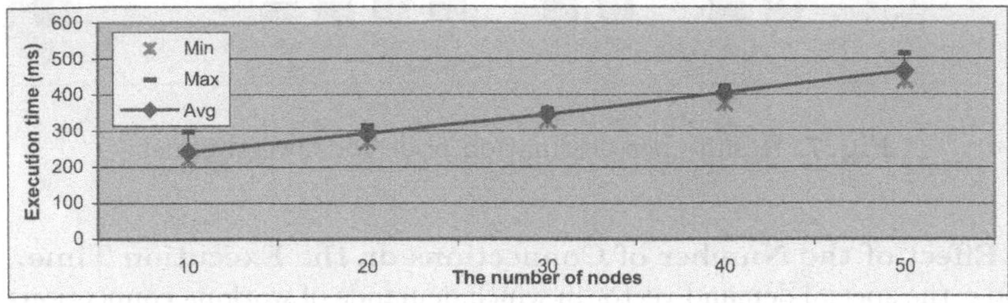

Fig. 9. The effect of number of nodes on the execution time with fixed demand

Random Demand Vector. For the random demand, the number of demands is proportional to the square of the number of nodes. Fig. 10 shows that the average execution time also increases quadratically with the number of nodes. For a large network with 50 nodes, for example, the execution time is 18s which is acceptable.

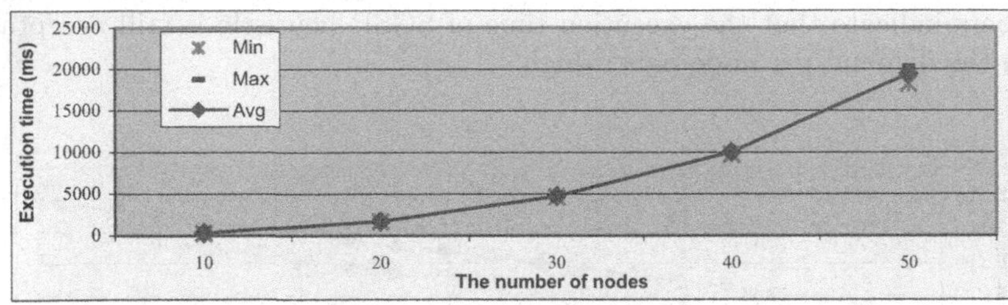

Fig. 10. The effect of number of nodes on the execution time with random demand

5 Conclusion

This paper reports on the application and the performance analysis of a new heuristic for organizing shared backup tree protection in MPλS networks. Shared backup tree protection is simulated on existing optical topologies for different working network demand vectors. The requirements in terms of cost for the placements of the working path and backup trees are compared with the cost consumed for dedicated protection. Two heuristics for finding optimized backup tree protection are compared in terms of cost gain and computational effort needed. The SGSP heuristic can find nearly the same results as the other heuristic, but a hundred times faster. The experiments show that the SGSP heuristic scales up to large topologies and high network demand vectors. The SGSP heuristic is designed to allow dynamic placement fo new connections; as such the execution time is very important.

6 Acknowledgments

The authors would like to thank the FWO for their support of this research project. We are particular indebted to Ir. Adelbert Groebbens for the interesting discussions and fruitful suggestions.

References

1. Awduche, D.et al.: Multi-protocol lambda switching: combining MPLS traffic engineering control with optical crossconnects. IETF Internet draft (1999)
2. Rubin, I., Ling, J.: Failure Protection Methods for Optical Meshed-Ring Communications Networks. IEEE JSAC,vol. 18,no. 10 (2000)
3. Zhou, D., Subramaniam, S.: Survivability in optical networks. IEEE Network, vol. 14,no. 6 (2000),16-23.
4. Mohan, G., Siva Ram Murthy, C.: Lightpath Restoration in WDM Optical Networks. IEEE Network, vol. 14, no. 6 (2000)
5. Colle, D. et. al.: Porting MPLS-recovery techniques to paradigm. Optical Network Magazine, vol .2, no.4 (2001)
6. Groebbens, A. et al.: Use of backup trees to improve resource efficiency of MPλS protection mechanisms. Proc. 3rd Design Reliable Communication Network (2001)
7. Groebbens, A.: Share capacity cuts in MPλS networks using backup trees. Proc. 6th INFORMS Telecommunication (2002)
8. LION Deliverable D10.: Multilayer resilient network planning and evaluation: preliminary results. (2001)
9. http://www.qwest.com/about/inside/network/nationip.html
10. Busacker, R.G., Gowen, P.J.: A procedure for determining minimal-cost network flow patterns. ORO Technical Report 15, Johns Hopkins University (1961)
11. Suurballe, J.W., Tarjan,R.E.: A quick method for finding shortest pairs of disjoint paths. Networks Magazine, Vol 14 (1984)

Appendix A

Pseudocode of Procedure for finding nw and selecting nw node disjoint paths

Input: K arranged paths P, source node s, destination node $d$and n_max

Output: nw and Set of nw node Disjoint Paths SDP

```
1.   S← ∅ ; SDP← ∅; i←0; nw←0
2.   While (nw≠n_max) or (i≠K)
3.   do i←i+1
4.       For each node v ∈ Path P_i and v ≠ s, v≠ d
5.           do If v∈ S
6.               then goto step 3.
7.       nw←nw + 1
8.       For each node v ∈ Path  P_i
9.           do S←S∪v
10.      SDP← Path P_i
11.  Return nw and SDP
```

Performance Evaluation of a Distributed Scheme for Protection against Single and Double Faults for MPLS

Fabio Ricciato, Marco Listanti, Angelo Belmonte, Daniele Perla

INFOCOM dept. – University of Rome "La Sapienza"
{fabio, listanti}@infocom.uniroma1.it

Abstract. MPLS can be used to provide network robustness to faults through path protection techniques. In this paper we present a dynamic model supporting different classes of end-to-end protection, including protection against Single Fault and Dual Fault, with and without sharing of backup bandwidth. Beyond link and node failures we also consider protection against Shared Risk Link Group (SLRG) failure. An interesting feature of the proposed scheme is the ability to offer service differentiation with respect to the recovery probability, by coupling the differentiation on the number of backup paths with bandwidth assignment policy. In this paper we describe the underlying algorithms for route selection and backup bandwidth sharing. The route selection is based on explicit load-dependent routing of service and backup paths. We show by simulation that the proposed route selection algorithm is effective in improving the network utilization. We discuss two alternative implementations of our model: distributed and partially centralized. The primary concern with the distributed approach is the message overhead implied by link-load dissemination, e.g. by flooding. However we show by simulation that message overhead can be taken under control by adopting a well-tuned adaptive overhead reduction algorithm. Our conclusion is that both distributed and partially-centralized implementation are feasible.

1 Introduction

Telecommunication networks are facing today an evolutionary trend towards a dynamic paradigm: more and more network functionalities are being shifted from the management plane - that implicitly foresees a sensible human intervention - towards the control plane – with a minimal human intervention. At the same time, the migration of new critical application on IP networks is changing the service paradigm associated to such networks. From delivering simple best-effort service, IP networks are challenged today to offer bandwidth and reliability guarantees, along with service differentiation. Given the dimensions of current networks in size and bandwidth, each network operation model aimed at answering this challenge should be scalable first, then effective in resource usage. In this framework, MPLS and the related dynamic protocols constitute a useful add-on to traditional IP platforms. MPLS is currently being implemented in several IP networks, and it can be exploited to implement connection-oriented traffic engineering techniques.

M. Ajmone Marsan et al. (Eds.): QoS-IP 2003, LNCS 2601, pp. 218-232, 2003.

In the framework of this evolutionary trends, this paper proposes a scheme to achieve service robustness to faults by means of end-to-end path-based protection schemes. Beyond traditional single link / single node failures, we also consider protection against failure of Shared Risk Link Group (SRLG, [1]). Additionally, we consider the possibility that some network links are not prone to failure at all at the MPLS layer. We denote them as No-Risk Links (NRL): they can model for example a link protected at lower layer (e.g. a Sonet/SDH ring). The service and the backup LSPs associated to a generic demand can share a same NRL. The adopted route selection algorithm is based on the joint computation of service and backup paths that are *opportunely* disjoint, i.e. they can not share a same link or SRLG, but can share a same NRL. We further extend our model to deliver service survivability guarantees under Dual-Fault. The basic idea is to associate to a service LSP *two* disjoint backup LSPs, rather than just one. Moreover, our model supports sharing of backup bandwidth.

The detailed description of the underlying algorithms (e.g. for disjoint-routes selection, for bandwidth sharing) can be found in [4]. Here we present the overall model and briefly revise the underlying algorithms, and focus on performances and applicability aspects. We discuss several issues regarding the applicability of this scheme to packet network. In particular we present two alternative implementations of our model: distributed and partially-centralized. The primary concern with the distributed approach is the message overhead implied by link-load dissemination, e.g. by flooding. However we show by simulations that the message overhead can be taken under control by adopting a well-tuned overhead reduction algorithm based on adaptive thresholds. Our conclusion is that both distributed and partially-centralized implementation are feasible.

The rest of the paper is organized as follows. First in section 2 we provide a global description of the proposed protection scheme. In section 3 we discuss the opportunity of load-aware routing also for protected demands, supported by results from simulations. In section 4 we present the distributed and the partially-centralized alternatives for implementation. In section 5 we present simulation results about the impact of a overhead reduction mechanism that can be exploited in the distributed approach. Finally, in section 6 we conclude.

2 An Architecture of Differentiated End-to-End Protection

In the proposed scheme each connection request can be associated to three different *protection types*: Unprotected (UP), Single-Fault Protected (SFP), Dual-Faults Protected (DFP). For UP demands only a service circuit P_s is established, and no service continuity is guaranteed after the occurrence of a fault along P_s. For SFP demands a service circuit P_s plus a single backup path P_r are allocated: upon failure of P_s the traffic can be readily switched on P_r by the ingress edge node. In case of DFP demands, a service circuit P_s plus *two* backup paths are allocated: a *primary* backup P_{r1} and a *secondary* one P_{r2}. Upon failure of P_s the traffic is switched on P_{r1}, and in case of contemporary interruption of P_s and P_{r1} it is switched to P_{r2}. Note that for DFP the order of preference between the two backup paths is fixed *a priori*.

Both SFP and DFP schemes can be implemented as Dedicated or Shared Protection (following the terminology in [2]) resulting in a total of five ***protection classes***. In our model both Dedicated and Shared alternatives are supported for SFP and DFP, resulting in four different protection classes in addition to the basic unprotected one, as summarized in Table 1.

The amount of reserved backup resources on each link and the reaction procedures to the faults will be designed in order to provide the following resilience guarantees with respect to the number φ of contemporary faults in place in the network:

- In case of one single fault (φ=1) all the affected SFP and DFP demands are guaranteed service continuity over the corresponding *primary* backup circuit.
- In case of two faults (φ=2) all the DFP demands experiencing interruption of the service circuit P_s are guaranteed service continuity over the primary P_{r1} or secondary P_{r2} backup circuit, depending on the integrity of P_{r1}. There are no guarantees of successful recovery for *all* the SFP traffic, but the network procedures are designed to attempt to recover SFP traffic to the possible extent, in a sort of "best-effort recovery" fashion.
- In case of three or more contemporary faults (φ>2) no demand is *guaranteed* service continuity, but all are recovered in a "best-effort" fashion. Nevertheless, DFP demands are more likely to be successfully recovered than SFP ones because they have more alternative paths (3 vs. 2) and prioritized access to backup bandwidth in case of contention (see section 2.3). This preserves a certain degree of service differentiation under multiple contemporary faults.

Table 1. The five considered Protection Classes

H	Acronym	Protection Class	number of End-to-end LSPs	backup bandwidth
0	**UP**	Unprotected	1 (service path)	n.a.
1	**Sh-SFP**	Shared Single-Fault P.	2	shared
2	**De-SFP**	Dedicated Single-Fault P.	(1 service + 1 backup)	dedicated
3	**Sh-DFP**	Shared Dual-Fault P.	3	shared
4	**De-DFP**	Dedicated Dual-Fault P.	(1 service + 2 backup)	dedicated

2.1 Functional Description

Hereafter we present a scheme for dynamic on-demand allocation of protected connections. Such scheme can be implemented either in a *distributed* or *partially-centralized* fashion. In both cases we have a decisional entity, the Route Selection Engine (RSE), which is in charge of computing the routes for the circuits (service and backup) associated to each new demand. The RSE has an associated Network-State Database (NSD), collecting information about the current network state. For each network link *m*, the NSD includes information about its capacity u_m and the totally reserved bandwidth b_m^{tot}.. In the *partially-centralized* model there is only a single RSE entity processing all the connection requests (hereafter called "demands") arriving to the network, while in the *distributed* approach each edge nodes has an associated RSE processing the demands arriving to it.

In both cases the edge nodes are in charge of the actual handling of such circuits (included their setup and release), as well as of the reaction procedures to the faults (i.e., traffic switching onto backup paths). Therefore, in the partially-centralized case some signaling is needed for the central RSE to communicate the selected routes for the new demands to the edge nodes. Also, in both cases the management of backup bandwidth sharing, i.e. the determination of backup reserved bandwidth and handling of bandwidth contention policies, is located at the intermediate nodes. In section 4 we discuss several aspects related to the two approaches, in particular regarding the updating of the NSD(s) associated to the RSE(s).

Each request i has the following associated attributes: ingress node S_i, egress node O_i, requested bandwidth B_i and protection class H_i. It can be initiated directly by the customer or by the network operator. In the following we assume that each demand i arrives to a RSE (centralized or local to S_i). Five different values of H are used to discriminate between the five different protection classes reported in Table 1. For each incoming request, the RSE computes a number of paths towards between S_i and O_i based on the current representation of the network state enclosed in the NSD. The number of computed paths depends on the requested protection class: one, two or three respectively for UP, SFP and DFP demands. For protected demands the service and backup paths must be *fault-disjoint*, i.e. they can not share a same link nor a same SRLG. The algorithm used for finding a minimum-cost set of fault-disjoint path can be found in [4]. Basically, the algorithm applies elementary graph transformations on each SRLG and NRL, then run classical disjoint-paths algorithms (e.g. Suurballe [5]) on the transformed graph. The same approach was independently proposed by other authors [6] limited to the case of single-fault protection. Additionally, in our model a particular graph transformation is used to handle the case of "No-Risk Links" (NRL). A NRL is a link which is not prone to any failure at all at MPLS level. This is for example the case of a transmission link that is protected at lower layer than IP/MPLS, typically a Sonet/SDH ring with Automatic Protection Switching. The fault-disjoint condition does not avoid that the service and backup paths of a single demand share the same link if it is not prone to be interrupted. Therefore the presence of NRLs in the network gives more flexibility in the route selection process.

Here follows a high-level description of the algorithm used by the RSE module to select the set of paths for the new request, further details can be found in [4]:

- **Step 1**. From the Network State Database produce a weighted graph G_1 by associating to each link m a weight w_m which is inversely proportional to the link load (see below). Links with no enough spare bandwidth are eliminated from the graph.
- **Step 2**. From the graph G_1 apply some basic transformations to each SRLG and RL, and produce the auxiliary virtual graph G_2. Such transformations are designed so that link-disjoint paths in G_2 correspond to fault-disjoint path in G_1.
- **Step 3**. Find the best set of link-disjoint routes (two for SFP, three for DFP) on the auxiliary virtual G_2, using the Suurballe algorithm - for a pair of disjoint paths - or its extension - for a triple.
- **Step 4**. Find the correspondent routes on the real graph G_1. Such routes are fault-disjoint. Order the paths with respect to ascending hop-length and assign the first, second and third respectively to P_s, P_{r1} and P_{r2}.

The path selection algorithm is designed in order to *minimize* and at the same time *balance* the overall resources usage. This is achieved by associating to each link a weight inversely proportional to the current spare bandwidth. As the route selection algorithms pick minimum-cost set of paths, the less loaded links are preferred in the selection of the new paths. This helps in avoiding saturation points (i.e. links exhausting their bandwidth) which could be critical for the allocation of future demands. The exact form of the link-weight function used by the RSE is:

$$w_m = \frac{1}{u_m - b_m^{tot} - B_i} + \varepsilon \tag{1}$$

wherein b_m^{tot} is the current value of reserved bandwidth on link m stored in the local TD, u_m the link capacity and B_i is the requested bandwidth by the new demand i. The term $\varepsilon \ll 1/u_m$ is a small value extracted randomly for each link at each run of the RSE. Its role is to scramble the path selection between the existing equal cost shortest paths for successive demands between the same node pair when the values of b_m^{tot} in the databases are null. This occurs at the very beginning of network operation, when all the links are unloaded, and during the whole network operation in lack of link-load dissemination. Such scrambling improves the level of backup bandwidth sharing. In fact, the potential bandwidth sharing is null when the active paths between a given node pair follow the same route, and increases with the spatial diversity of active paths between the same endpoints. Also, in case of inhibited flooding, such small randomization in the path selection is helpful in distributing the network load over the existing number of equal cost shortest paths, therefore achieving a minimum level of load-balancing ("blind balancing").

If the RSE fails in finding a suitable set of paths the new request is immediately rejected. This can be due to either a poor topological connectivity (i.e. no disjoint paths exist between the given nodes) or to current heavy resource usage. In case of successful path computation the ingress node initiates the setup signaling phase. The signaling setup (e.g. by RSVP-TE) for service and backup circuits can be run in parallel to speed up the overall procedure. During the setup phase each intermediate node along the involved paths checks for actual resources availability on the local link, and eventually rejects the setup attempt in case of lack of bandwidth. In case that the setup of one or more of the involved paths fails, the request is rejected.

In case of Shared Protection certain information must be conveyed in the setup messages along the *backup* path(s) in order to allow the intermediate nodes to compute the exact amount of additional backup bandwidth that needs to be reserved to support the new demand. Such information substantially identifies the *service* path, which was computed jointly with the backup one(s). Details about the sharing mechanism will be given below in section 2.3. For instance it can be noted that the RSE does not consider the potential bandwidth sharing in selecting the backup path, but lets the intermediate nodes to evaluate and implement bandwidth sharing independently. This mechanism is precise and robust with respect to routing information uncertainty: in facts the evaluation of the potential bandwidth sharing is done locally by intermediate nodes based on locally collected information. Furthermore, this mechanism fits well in a migration scenario where not all intermediate nodes implement backup bandwidth sharing. In fact, the handling of shared reservations requires additional capabilities to be installed at intermediate nodes (e.g., bandwidth evaluation, see section 2.3), which could be deployed incrementally node-by-node in an operational network.

2.2 Algorithm for Backup Bandwidth Sharing

Each network node n maintains the following state information for each outbound link m attached to it: the amount of resources (bandwidth) currently allocated to service circuits b_m^s, to dedicated backup circuits b_m^d, to shared backup circuits b_m^r, and the total link capacity u_m. The total reserved bandwidth is defined at any time as $b_m^{tot}= b_m^s+b_m^d+b_m^r$. The components b_m^s and b_m^d are updated during the signaling procedures in a very simple way: they are incremented / decremented by B_i at every circuit setup / release for a demand with associated B_i bandwidth. Instead, the determination of the component b_m^r relevant to the shared backup bandwidth requires the maintenance of additional data structures. In this subsection we provide a detailed description of such structures and the related update algorithms run during the signaling procedures. In the following E will denote the number of possible fault events (link, node and SRLG failures) that are considered for the specific network, and M the total number of network links. It is reasonable to expect that E is not much greater than M, or at least in the same order of magnitude.

In order to implement Shared-Single Fault Protection (Sh-SFP), a single data structure must be maintained by node n for each outbound link m: the vector FS_m of size E whose generic component $FS_m[k]$ represents the bandwidth needed on link m to carry all the Sh-SFP traffic rerouted on m after the *single* fault event k. Obviously, $FS_m[k]=0$ if k affects m itself. The vector FS_m is maintained by letting the ingress node include in the signaling messages along the *backup path* the list V_s of the links constituting the associated *service path*. From V_s and from the knowledge of the SRLGs associated to each network link the intermediate node can easily derive the list W_s of the *faults* affecting the service path. During the backup setup phase, the intermediate node will increment by B the components $FS_m[k]$ for each $k \in W_s$. Remarkably, the same information V_s should be advertised also during the circuit release phase, in order to decrement the relevant components of FS_m.

In order to implement Shared Dual-Faults Protection (Sh-DFP) two different data structure must be maintained by node n for each outbound link m: the vector $FD1_m$ and the $E \times E$ matrix $FD2_m$. The vector $FD1_m$ for Sh-DFP has exactly the same meaning as FS_m for Sh-SFP. Additionally, the generic component $FD2_m[k_1,k_2]$ represents the bandwidth needed on m to support the demands whose *service* and *primary backup* paths are interrupted by the fault events k_1 and k_2 *respectively*. The vector $FD1_m$ is updated in the same way as FS_m, therefore the ingress node has to advertise the list V_s of the *links* constituting the *service path* in the signaling messages along the *primary backup* path. On the other hand, along the *secondary backup* path it will advertise both the lists V_s and V_{r1}, the latter referring to the links constituting the *primary backup* path. Similarly to above, the intermediate node n will derive the fault lists W_s and W_{r1} from the link lists V_s and V_{r1}, and during the setup [release] phase will increment [decrement] by B the component $FD2_m[k_1,k_2]$ for each $k_1 \in W_s$, $k_2 \in W_{r1}$.

At every update of these data structures, the new value of the bandwidth reserved to shared backup circuits b_m^r is computed according to the following **allocation rule**:

$$b_m^r = \max_{\substack{k_1,k_2 \\ k_1 \neq k_2}} \{FS_m[k_1] + FD1_m[k_1] + FD1_m[k_2] + FD2_m[k_1,k_2] + FD2_m[k_2,k_1]\} \quad \textbf{(2)}$$

It can be seen that with the above allocation rule the amount of shared backup resources is slightly overestimated with respect to the minimum amount required to meet the service requirements expressed above. This is because the information included in the data structures introduced above is *partial* and *aggregated*. In fact, FS_m, $FD1_m$ and $FD2_m$ enclose global information about the *per-link* reserved resources, not about the full set of *per-demand* paths. While this is enough for evaluating *exactly* the bandwidth needed to protect all the SFP demands against a single network fault, the lack of complete information fatally leads to an imprecise evaluation of the minimum amount of bandwidth required to protect all the DFP demands against a dual network fault. Consider for example two network links a_1 and a_2, impacted by faults k_1 and k_2 respectively. Denote by D_i (i=1,2) the set of demands whose service path includes link a_i, and by $D_{1,2} = D_1 \cap D_2$ the set of demands whose service path include *both* a_i and a_2. Denote by *Size(D)* the sum of bandwidths associated to the demands in the set *D*. Consider a third link *m*: the allocation rule given above will reserve for b_m^r an amount of bandwidth equal to $Size(D_1)+Size(D_2)$, while just $Size(D_1 \cup D_2)$ would suffice. Therefore there is a potential over-reservation of $Size(D_{1,2})$ resources, as the demands in the set $D_{1,2}$ are counted twice. In order to eliminate such an inefficiency, the intermediate node *n* should maintain per-demand information, i.e. the vectors W_s and eventually W_{r1} for all the demands having a shared backup path routed through it. This would increase the amount of state information required at each network node, and add complexity (and time) to the computation of b_m^r. On the other hand we found by simulation that the advantage that can be expected from this refinement is modest. We run simulations in the same scenario described below for Fig. 3, with Sh-DFP demands. After *K* allocated demands, and for different values of *K* from 200 to 600 (i.e. just before the first blocked request) we compared the amount of backup bandwidth reserved by our scheme with the minimum amount needed to protect all the Sh-DFP demands from any possible pair of faults, as evaluated by simulating *all* the possible fault pairs and considering the worst-case for each link. The results show that our scheme reserves about 4÷6% more backup bandwidth than the minimum required, corresponding to about +2÷4% more total bandwidth usage. Such modest values mitigate the interest towards further refinements of the sharing mechanism based on complete state information.

It is a fact that under dual contemporary faults the allocation rule given above reserves enough bandwidth to protect *all the DFP demands, but not all the DFP and SFP demands*. Therefore, in case of dual faults there is a potential contention on backup reserved bandwidth, as each edge nodes will switch traffic onto the backup LSP without coordination with the other edge nodes. Such contention should be resolved by the intermediate nodes in favor of DFP demands. In a IP/MPLS network a possible way to achieve contention resolution is by means of packet level priority-scheduling: each router interface should support a number of queues served with priority scheduling, and Sh-SFP LSPs should be assigned to a lower priority queue than Sh-DFP ones. Further investigations on that topic are out of the scope of this paper.

The important point here is to show how different protection classes and packet-level bandwidth assignment prioritization can be combined to achieve a sensible degree of service level differentiation.

3 Effectiveness of Load-aware Routing for Protected Demands

The distributed model presented above is based on explicit load-aware routing. In our model a generic demand can be blocked during the route computation (we call it a "immediate blocking") or during the setup phase ("setup blocking"). The knowledge of network link loads by the RSE is helpful in a twofold sense. First, it diminishes the probability that a route without enough spare bandwidth is selected by the RSE, thus diminishing the incidence of setup blocking. In this sense the gain in resources usage is immediate: the "better" is the link load information available at the RSE – i.e., more accurate and more updated - the lower the total blocking rate. Secondly, the availability of link-load information allows to pursue load distribution policies that are helpful to prevent – or at least to delay - the emergence of saturated link that could block future demands. In this sense the achievable gain in resource usage depends not only on the quality of the link-load information but also on the effectiveness of the decisional allocation algorithms. In [7] it was noted that some algorithms could perform worse than the simple shortest-path criterion, particularly those underscoring the path length in favor of path width (e.g. shortest-widest-path). In our model the route selection relies on a minimum-cost paths with load-dependent link weights (eq. 1). With regards to this aspect, we are interested here in assessing the relative gain in resource usage that can be achieved by our model with respect to a simple load-unaware shortest-path routing.

The assessment of such gain is also important from the perspective of a cost-benefit analysis of link-load dissemination. In fact, the distribution of updated link-load information at the route decision point has an associated cost in terms of message overhead, especially for the distributed implementation. The issues related to link-load dissemination and associated overhead costs will be discussed in section 4. In this section we focus on the assessment of the achievable gain. Several previous works (see e.g. [7]) found that a well designed load-aware routing can outperform the load-independent shortest-path algorithm. However, those works apply to unprotected demands, and the applicability of such result is not straightforward when considering DFP (and SFP) demands. In facts, three fault-disjoint routes have to be selected for each DFP demand by the RSE. In generally meshed networks the disjointedness constraint could be so restrictive that very few (or even just one) alternative solutions exist between the requested node pair, so that the possibility to exploit link-load information to "optimize" the choice is very poor or null at all.

In order to assess the convenience of our load-aware routing algorithm we carried a comparative investigation by means of simulations. We considered the extreme cases of "with" and "without" link-load dissemination (LLD). In the first case an exact and update view of the network link loads (i.e. b_m^{tot}) is available to the RSE. In the second case instead the RSE always consider unloaded links (i.e. b_m^{tot}=0), then the link weight is fixed for all links as they all have the same capacity.

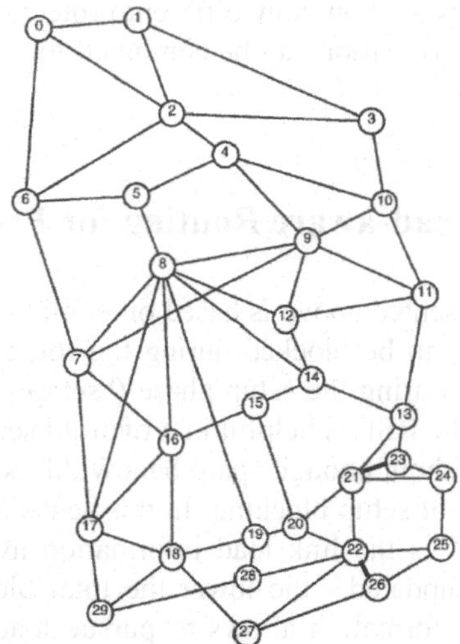

Fig. 1. Network topology used in simulations. It is the same found in [8] with arbitrary addition of two links to make it 3-connected as required for DFP demands.

The simulations were run on the network depicted in Fig. 1. We assumed a simple failure scenario where each bi-directional link is associated to a single elementary fault event affecting both directions at the same time. This choice was driven by the consideration that currently available commercial equipment maintains circuit symmetry in both directions, however this is not a requirement for our model. We fed each network with homogeneous traffic: for each node pair (i,j) requests arrive randomly with mean frequency $\lambda_{ij}=\lambda = 1/N(N-1)$ where N denotes the number of network nodes. In each experiment we considered the homogeneous case where all demands request the same protection class. The link capacity is fixed to 10 Gbps. Each demand is bi-directional and symmetric, and requests a random bandwidth extracted uniformly in [50,150] Mbps. We considered demands with infinite holding time, therefore after a transitory period the network gets to saturation. After the k-th accepted demand, we computed the total bandwidth reserved in the network (i.e. $b(k)=\Sigma_m b_m^{tot}$ after k allocated demands), and the number $r(k)$ of blocked request until that moment. In Fig. 2-above we plot $b(k)$ versus k for the cases De-DFP and Sh-DFP (Fig. 2-bottom shows the same for the remaining protection classes). Each curve represents the average values over ten different experiments. Let's focus for the moment on Sh-DFP demands. The upper and lower curves refer respectively to the cases with and without link-load dissemination. For each curve, we marked from left to right the points at the 1st, 10th, 50th, 100th, 500th and 1000th blocked demand.

It can be seen that the occurrence of the first blocking occurs after ~400 requests without LLD, and ~600 with LLD. We also notice that after the first blocking, the slopes of the curves diminishes, indicating that less additional bandwidth-per-demand is reserved for the subsequent arrivals. This is due to the selective filtering of "longer" requests, i.e. those between farther endpoints. In fact, when the network is close to saturation longer requests experience a higher blocking probability, so that residual

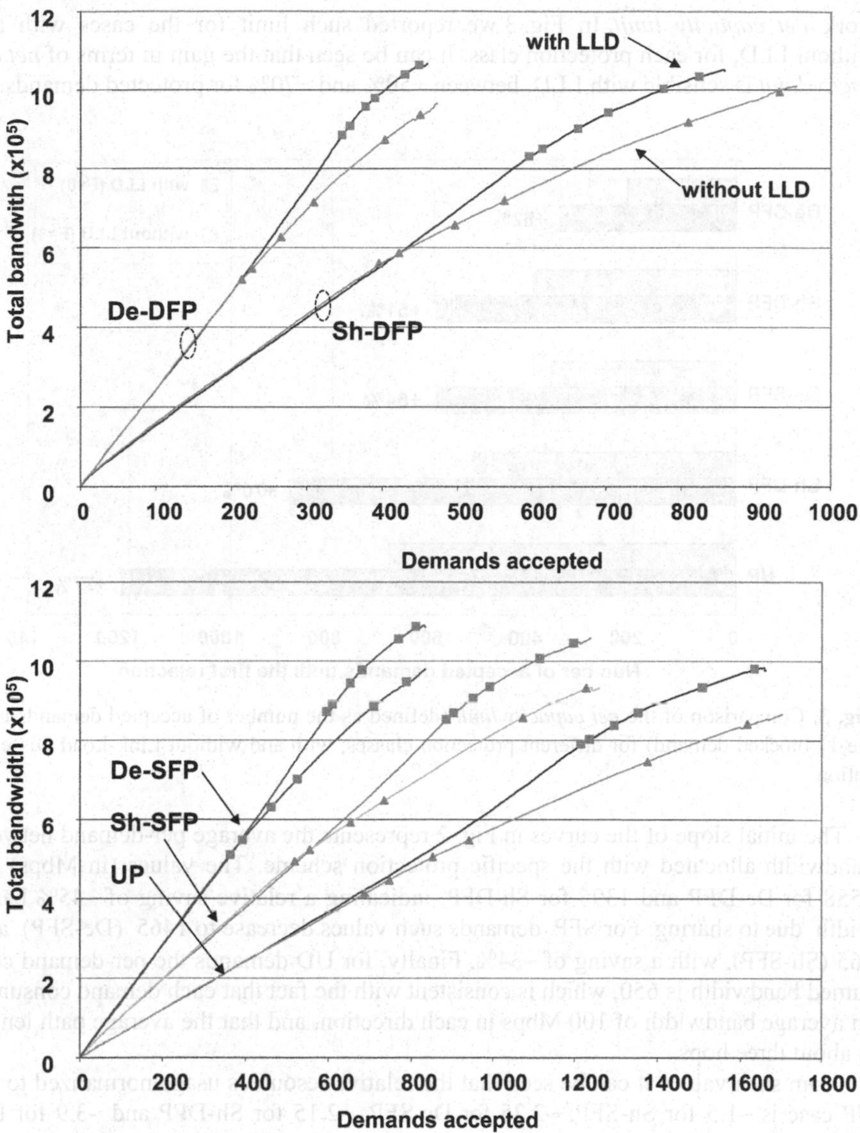

Fig. 2. Total bandwidth usage vs. number of accepted demands for each protection class, with (square spots) and without (triangle spot) Link-Load Dissemination. The spots indicate the 1st – 10th –50th –100th –500th –1000th blocked demand

shorter request can be admitted with lower bandwidth consumption. Clearly the selective blocking effect is highly undesirable, and the network operation point – in terms of allocated demands - should not exceed the region identified by the first few blocking. Therefore we could take the value of k at the 1st-10th blocking as an estimate of the number of demands that the network can support before the emergence of saturation effects. In other words, such value can be considered as an estimation of the network *net capacity limit*. In Fig. 3 we reported such limit for the cases with and without LLD, for each protection class. It can be seen that the gain in terms of *net capacity limit* is sensible with LLD, between +50% and +70% for protected demands.

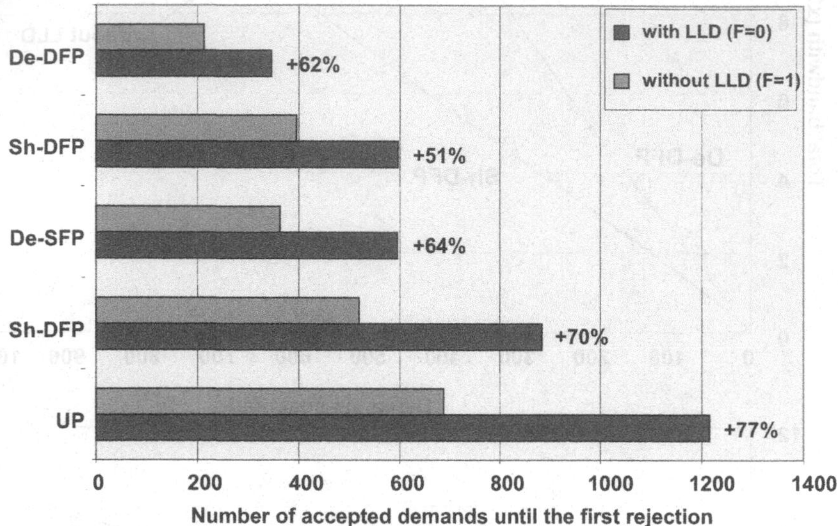

Fig. 3. Comparison of the *net capacity limit* (defined as the number of accepted demands until the 1st blocked demand) for different protection classes, with and without Link-Load Dissemination.

The initial slope of the curves in Fig. 2 represents the average per-demand network bandwidth allocated with the specific protection scheme. The values (in Mbps) are 2558 for De-DFP and 1395 for Sh-DFP, indicating a relative saving of ~45% bandwidth due to sharing. For SFP demands such values decrease to 1465 (De-SFP) and 965 (Sh-SFP), with a saving of ~34%. Finally, for UD demands the per-demand consumed bandwidth is 650, which is consistent with the fact that each demand consumes an average bandwidth of 100 Mbps in each direction, and that the average path length is about three hops.

From such values it can be seen that the relative resources usage normalized to the UP case is ~1.5 for Sh-SFP, ~2.25 for De-SFP, ~2.15 for Sh-DFP and ~3.9 for De-DFP. Such values should be carefully considered in the definition of the billing model for each protection class.

4 Distributed or Partially-Centralized ?

In the previous section we assessed that for each protection class the gain achievable with the proposed load-aware routing model is substantial. In a partially-centralized implementation of such model, an updated view of the network state can be maintained by the centralized route server, by keeping memory of each request arrival /

departure. This is possible because a single entity manages all the network requests. Note that in this case the central RSE has a complete knowledge of network state, and can also compute the amount of backup bandwidth sharing on each link in case of Shared protection. In this approach, that reminds the concept of *Bandwidth Broker* in the Diffserv context, some signaling is needed between the central server and the edge nodes that are in charge of the LSP setup.

On the other hand, in the distributed implementation link-load dissemination is needed to all edge-nodes. Each internal network node must advertise link load changes to the edge nodes hosting the RSEs, and this can be achieved in two ways: *flooding* or *multicasting*. Both approaches have pros and cons. Link-load flooding can be implemented with slight modification to the existing routing protocols, and this feature is already being considered for the traffic-engineering extensions to such protocols (e.g. OSPF-TE [3]). On the other hand, link-load flooding requires implies more processing overhead as *all* network nodes, not just edge ones, have to process flooded messages. On the other hand, multicasting link-load advertisements to edge nodes only requires support for multicast forwarding.

Independently on the dissemination technique, flooding or multicasting, an important overhead metric is the frequency of *link-load advertisements* (LLA), i.e. how many LLA are generated per unit of time. By assuming a mean arrival rate of R requests per second to the network, the mean rate of LLA generation is $G = L{\cdot}R$, with L denoting the average number of links involved in supporting each demand. For UP demands L roughly approaches the average shortest-path D, as most of the selected routes will not depart too much from the min-hop shortest paths (this is a convenient feature, as observed in [7]). On the other hand, for SFP and DFP demands that involve two and three paths respectively for each demand, the value of L should be close to $2{\cdot}D$ and $3{\cdot}D$. In case that requests with finite holding time are considered, the value of G must be doubled, as also the release of a circuit triggers a (negative) link-load change. Therefore we end up with a mean LLA frequency of $G \approx 2\ R{\cdot}n{\cdot}D$ (n=*1,2,3*). The volume of such overhead depends on the network size (factor D) and on the time-scale of demands arrivals (factor R), which in turn is likely to depend on the bandwidth granularity of the request. It is reasonable to expect that a high request rate R is associated to small requested bandwidth B, and *vice versa*. Therefore it is possible that the distributed model presented so far can not scale up to large networks delivering dynamic on-demand service at small bandwidth granularity, unless some mechanism is implemented to control such overhead.

5 Impact of Overhead Reduction Mechanism

In case the network operator is willing to maintain a distributed approach, some *advertisement reduction* mechanism must be implemented to control the LLA overhead. Several such mechanisms were proposed in [9] as *flooding reduction* algorithms. Basically, with such algorithms LLAs are not generated upon each link-load modification (i.e., for each LSP setup / release), but only when certain thresholds are crossed. Typically, hysteresis cycles (based on double thresholds) and/or hold-down timers are used to avoid that fluctuations around a single threshold trigger a large number of LSAs. Eventually, such algorithm can be *adaptive*, i.e. they include some parameter subject to dynamic adjustment.

With any flooding reduction mechanism the diminution in the overall flooding rate comes at the cost of a degradation in the quality of the link-load information disseminated in the network. In other words, the view of the network state maintained by the edge nodes get coarser, i.e. less precise and less updated. On the other hand such degradation does not necessarily results in a diminution of bandwidth allocation efficiency, as in most cases a coarse link-load information is enough to take "good" routing decisions [7]. Intuitively, in order to take a good routing decision it is often enough to know which links are lightly loaded and which ones are heavily loaded, rather than know the exact load value of each link. Also, the knowledge of the exact value of spare bandwidth is useful for heavily loaded links, much less for low loaded ones.

Based on such considerations, and comforted by some previous results for non-protected paths, we chose a particular algorithm based on double adaptive thresholds, and we analyzed its impact on our distributed SFP / DFP protection schemes. The algorithm was first presented in [9], to which the reader is referred for algorithmic details. Basically, the algorithm is based on two thresholds, t_{high} and t_{low}. For each new reservation, the new value of total bandwidth b^{tot} is computed (for Shared protection this involves the sharing algorithm presented in section 2.2) and compared with the thresholds. A new LLA advertising the new value of b^{tot} is generated only if $b^{tot} < t_{low}$ or $b^{tot} > t_{high}$. As the distance between the thresholds is non-null ($t_{high} > t_{low}$) an hysteresis cycle is introduced. The threshold values are dynamically updated after each reservation as (u denotes the link capacity):

$$
\begin{aligned}
t_{high} &= b^{tot} + \left(u - b^{tot}\right) \cdot F \\
t_{low} &= \max\left(0, b^{tot} - \left(u - b^{tot}\right) \cdot F\right)
\end{aligned}
\tag{3}
$$

where F is fixed parameter with $0 \le F \le 1$. The setting of parameter F governs the trade-off between link-load dissemination overhead and precision. At one extreme F=1, and the lower and upper thresholds are initialized to 0 and u (link capacity) respectively, so that no LLA will never be triggered. At the opposite extreme F=0, and each new reservation will trigger a new LSA. Therefore F=0 and F=1 denote respectively the "exact" and "null" link-load dissemination cases introduced above. Simulations were run to evaluate the impact of such overhead reduction algorithm on the performances of our distributed SFP / DFP scheme. We run simulations in the same scenario considered for Figg. 2-3, but with a random holding time exponentially distributed with mean T. We will denote the first set of simulations with infinite holding time as "static simulations", and this second set as "dynamic". In the dynamic simulations it is meaningful to determine the mean input traffic intensity as $Q_B = \sum_{ij}\lambda_{ij}\ T \cdot B$ (in Mbps), with B the mean requested bandwidth (B = 100 Mbps in our case), T the mean holding time and λ_{ij} the mean arrival rate between nodes i,j. In the simulations we normalized $\sum_{ij}\lambda_{ij} = 1$, and tuned T so that the mean input traffic was equal to the net capacity limit found in the first set of static simulation for the "without LLD" reported in Fig. 3 (corresponding to setting F=1 in the considered flooding reduction scheme). We considered this choice point as a satisfactory network operation point.

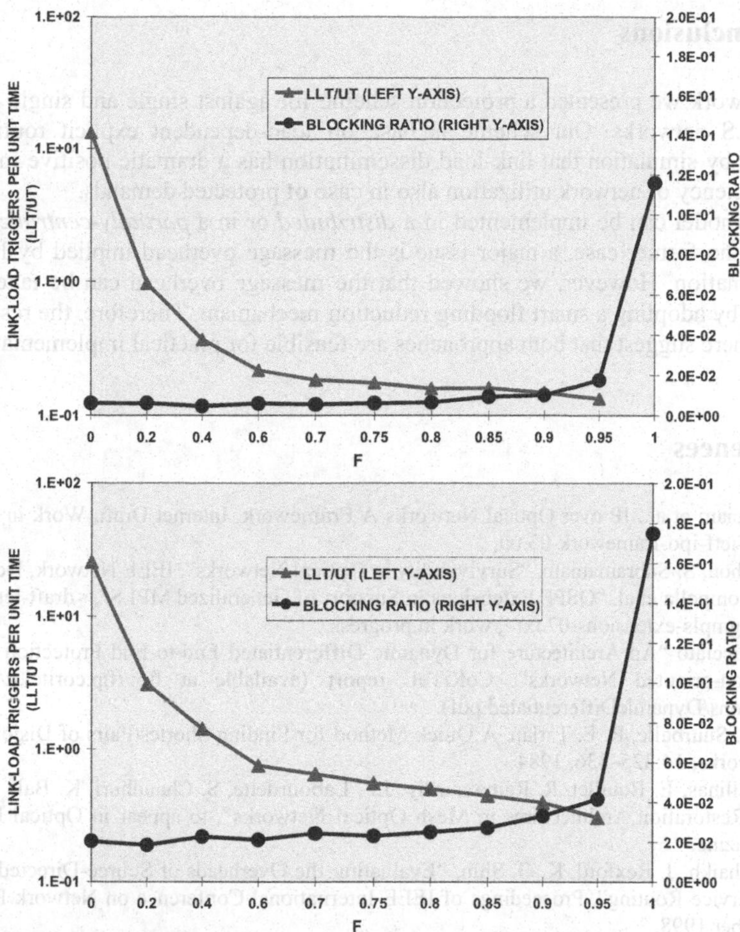

Fig. 4. Call blocking and message overhead vs. parameter F (above: Sh-DFP with T=600 time-units; bottom: De-DFP with T=350 time-units)

We repeated the simulation for different values of parameter F, and measured i) the experimental blocking ratio and ii) the mean frequency of LLAs. The values of both metrics, averaged over 10 different experiments, are reported in Fig. 4 for a network loaded with DFP requests. Very similar results were found for the remaining protection classes. The graphs show that with the appropriate setting of parameter F = 0.8 it is possible to achieve a reduction in LLA frequency of two order of magnitude with a negligible impact on the blocking probability. Notice that with the adopted overhead reduction mechanism the LLA rate has dropped below the request arrival rate. On the other hand, by completely eliminating LLAs (i.e., for F=1) the blocking probability increases by a factor of 10. This is a further confirmation about the usefulness of load-aware routing also in case of protection routing schemes.

6 Conclusions

In this work we presented a protection scheme for against single and single fault for IP/MPLS networks. Our scheme is base on load-dependent explicit routing. We showed by simulation that link-load dissemination has a dramatic positive impact on the efficiency of network utilization also in case of protected demands.

Our model can be implemented in a *distributed* or in a *partially-centralized* fashion. In the former case, a major issue is the message overhead implied by link-load dissemination. However, we showed that the message overhead can be taken under control by adopting a smart flooding reduction mechanism. Therefore, the results presented here suggest that both approaches are feasible for practical implementation.

References

[1] J.Luciani et al., IP over Optical Networks A Framework, Internet Draft, Work in progress, draft-ietf-ipo-framework-03.txt.

[2] D. Zhou, S. Subramaniam, "Survivability in Optical Networks", IEEE Network, Dec. 2000.

[3] K. Kompella et al. "OSPF Extensions in Support of Generalized MPLS", <draft-ietf-ccamp-ospf-gmpls-extensions-07.txt>, work in progress.

[4] F. Ricciato "An Architecture for Dynamic Differentiated End-to-End Protection for Connection-oriented Networks", CoRiTeL report (available at ftp://ftp.coritel.it/pub/Publications/DynamicDifferentiated.pdf).

[5] J. W. Suurballe, R. E. Tarjan. A Quick Method for Finding ShortestPairs of Disjoint Paths. Networks, 14:325–336, 1984

[6] G. Ellinas, E. Bouillet, R. Ramamurthy, J.F. Labourdette, S. Chaudhuri, K. Bala "Routing and Restoration Architectures in Mesh Optical Networks", to appear in Optical Networks Magazine.

[7] A. Shaikh, J. Rexford, K. G. Shin, "Evaluating the Overheads of Source-Directed Quality-of-Service Routing", Proceedings of IEEE International Conference on Network Protocols, October 1998.

[8] R. R. Irashko, W. D. Grover, M. H. MacGregor, "Optimal Capacity Placement for Path Restoration in STM or ATM Mesh-Survivable Networks", IEEE/ACM Transactions on Networking, June 1998.

[9] A.Botta, M. Intermite, P.Iovanna, S.Salsano, "Traffic Engineering with OSPF-TE and RSVP-TE: flooding reduction techniques and evaluation of processing cost", CoRiTeL report, available at ftp://ftp.coritel.it/pub/Publications/thresholds.pdf.

Edge Distributed Admission Control for Performance Improvement in Traffic Engineered Networks

Alessandro Bosco[1], Roberto Mameli[1], Eleonora Manconi[1], Fabio Ubaldi[2]

[1] Ericsson Lab Italy S.p.A., via Anagnina 203, 00040 – Roma (Italy)
{alessandro.bosco, roberto.mameli, eleonora.manconi}@eri.ericsson.se
[2] CoRiTeL, via Anagnina 203, 00040 – Roma (Italy)
ubaldi@coritel.it

Abstract. Admission Control and routing are key aspects in next generation networks supporting advanced Quality of Service and Traffic Engineering functionalities. However, traditional Traffic Engineering solutions usually rely exclusively on routing algorithms to achieve optimization in network resource usage, while Admission Control is simply limited to a local check of resource availability, without any optimization purpose. This paper proposes a novel approach for Admission Control in IP/MPLS networks, which applies at network edges by means of dynamic thresholds evaluated on the basis of network status. The proposed solution allows the achievement of more efficient usage of network resources, especially at medium/high load, with an increased robustness of the network and an overall performance improvement.

1 Introduction

In the recent past we have been assisting to the impressive growth of Internet traffic and correspondingly to the gradual migration of real time and multimedia services towards IP. However, traditional IP networks typically support only best effort service. As a consequence, advanced network scenarios require the introduction of technologies able to provide users with Quality of Service (QoS) and network operators with more dynamic and flexible resource utilization, i.e. with the capability to perform Traffic Engineering (TE) [1].

Multi Protocol Label Switching (MPLS) [2]-[4] certainly represents one of the most suitable technologies developed for the purpose. MPLS networks supporting advanced QoS and TE require proper Admission Control (AC) and routing functionalities. Admission Control and routing cooperates strictly whenever a new traffic flow with some specific QoS constraints is to be transmitted across the network. As a consequence, they represent a crucial point in such networks, since they affect both the user's perception of the offered service and the network's resource utilization, and therefore the operator's revenues.

Traditional QoS and TE solutions often focus exclusively on routing as a mean to provide advanced TE functionalities, while AC is simply limited to a local check of resource availability, without any optimization purpose. Specifically, TE optimization is obtained by means of proper routing algorithms, able to find routes according to some optimization criteria. Such routing algorithms, often termed Constraint Based Routing (CBR) perform path calculation by taking into account various constraints,

M. Ajmone Marsan et al. (Eds.): QoS-IP 2003, LNCS 2601, pp. 233-246, 2003.

typically related to resource requirements of the flow, network status and administrative permissions [4]. The choice of the CBR algorithm, which is out of the scope of this paper, depends on the optimization criteria preferred by the network administrator and will not be further discussed. However, an "optimal" algorithm does not exist, since different algorithms pursue optimization according to different criteria, and the choice is not completely independent of the network load [5]. In fact, most of the CBR strategies aim at even traffic distribution across the network and work fine at low/medium load, i.e. remarkably better than the traditional shortest path routing. Under high load conditions, instead, shortest path often behaves quite better, since it tries to route flows while minimizing the amount of resource consumption (typically the number of links).

This paper proposes a novel approach in which the CBR algorithm is integrated with a proper AC functionality, performed at network edges in a completely distributed fashion, at the purpose of limiting congestion and increasing both the performance and the network utilization.

The paper is structured as follows: section 2 describes the algorithm proposed by the authors, while section 3 describes the simulation activity performed for the purpose of evaluating its effectiveness and the corresponding results. Finally, section 4 reports conclusions and possible directions for future work.

2 Edge Distributed Admission Control in Traffic Engineered Networks

The novel approach described in this paper applies to an MPLS Traffic Engineered (MPLS-TE) network. MPLS is an innovative technology, able to enhance IP networks with advanced TE and QoS functionalities. In the MPLS architecture data plane and control plane are separate and independent. Data plane is based on a label-switching mechanism, therefore getting rid of the traditional hop-by-hop routing of IP networks and introducing a "connection-oriented" paradigm. Control plane, instead, comes directly from existing signaling and routing protocols of the TCP/IP suite, with proper extensions [2]-[4].

Fig. 1 provides an example of such a network. The main network elements are:

- *Label Switching Routers (LSR)*, internal nodes of the network, which forward packets on a label switching basis;
- *Label Edge Routers (LER)*, edge nodes of the network, which classify incoming IP packets attaching them a label at the ingress, and removing the label at the egress;
- *Label Switched Paths (LSP)*, virtual connections from ingress LER to egress LER represented by the concatenation of labels along the path.

There are some key requirements that must be fulfilled in order to let the network of Fig.1 support TE. First of all, we need the possibility to route LSP along explicitly specified paths, which is usually obtained by proper TE-extended Label Distribution Protocols (LDP), such as RSVP-TE [6] and CR-LDP [7]. Second, there must be an intelligent entity able to calculate "optimal" paths according to a predefined TE strategy; this is usually achieved by CBR algorithms, implemented in LER, which act upon each incoming request, looking for the "best" path for the incoming LSP.

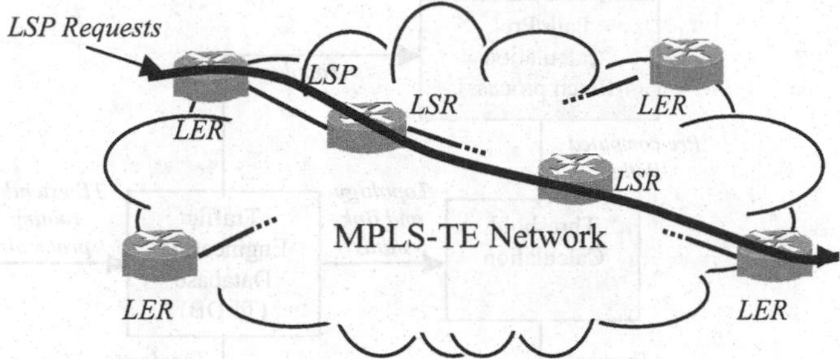

Fig. 1. Network scenario representing a typical MPLS network. Upon each incoming LSP request the ingress LER run a CBR algorithm in order to compute an explicit path, according to information dynamically gathered by means of TE-extended routing protocols. If computation succeeds, the LSP is established along the new path through TE-extended Label Distribution Protocols. During LSP establishment, each node checks availability of resources for the incoming LSP and eventually rejects the request notifying it to the originating node. This limited functionality is the only form of Admission Control currently performed in MPLS-TE networks.

Finally, these algorithms must be provided with information about the network status, in order to let them work properly; such information is distributed by means of TE-extended routing protocols, such as OSPF-TE [8] and ISIS-TE [9]. In this scenario, Traffic Engineering optimization is completely up to the routing algorithm, while the role played by Admission Control is very limited, since it consists only in the link-by-link check of resource availability.

The solution proposed in this paper gives more value to the AC functionality from the point of view of TE optimization. Specifically, we introduce edge distributed Admission Control, where AC is performed at the ingress node of the network before route computation, i.e. upon each incoming request, in a completely distributed fashion.

The AC strategy proposed hereafter consists of two distinct phases, respectively off-line and on-line. The off-line component of the algorithm is conceptually very simple. Specifically, denoting the number of edge nodes by N, the generic ingress LER pre-calculates and stores K shortest paths towards each of the N-1 possible destinations, by means of algorithms available in literature [10] (the value of K is chosen by the network operator during network configuration). The only input needed for the purpose is the network topology, which is known through routing protocol flooding. The $K{\cdot}(N\text{-}1)$ paths are stored in the ingress node and are used by the on-line component of the algorithm to evaluate proper thresholds for dynamic AC (Fig. 2). The computational complexity in this case does not represent a problem, since these paths are recomputed only when a failure or a topology change occurs (the implicit assumption is that this event happens only occasionally). Moreover, path re-computation can be performed by a batch process, which employs processing resources in the nodes only when available.

The on-line component of the method is applied in the ingress LER upon each incoming request, according to Fig.2. It is particularly simple, since it is based on the comparison of the bandwidth requirements of the incoming request with a threshold value suitably evaluated.

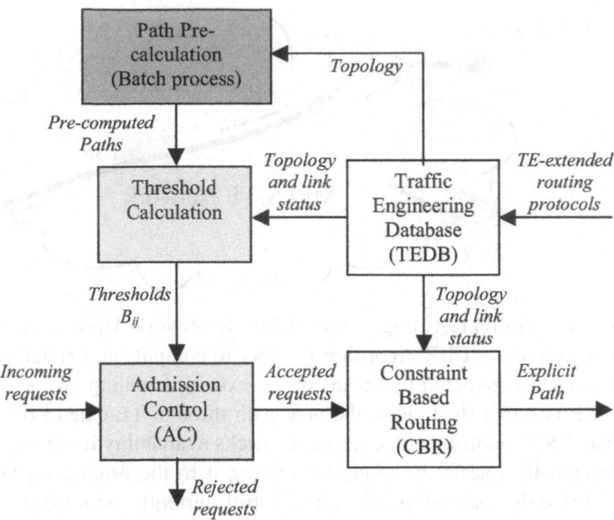

Fig. 2. Logical structure of the ingress LER. The dark gray block represents the batch process that pre-calculates paths off-line. Light gray blocks take part to the on-line behavior of the proposed algorithm. Each request presented to the ingress LER is filtered by the AC block and, if accepted, is processed by CBR in order to calculate the optimal path. Acceptance is decided according to a threshold, which in turn is evaluated on the basis of information received by TE-extended routing protocols and stored in the TEDB. Please note that this is the same database accessed by the pre-existing CBR algorithm and contains information about the topology and the links status.

Specifically, an incoming request for a new flow requesting a bandwidth r, which enters the network at node i towards node j, is accepted iff:

$$r \leq B_{ij} \, . \tag{1}$$

The right side term of Eq. (1) belongs to a set of thresholds dynamically evaluated and updated by the ingress node (Fig.2). Referring e.g. to the ith edge node we can denote such thresholds by:

$$\left[B_{i1} \quad B_{i2} \quad \ldots \quad B_{iN} \right] . \tag{2}$$

The generic threshold B_{ij} limits the total amount of traffic from the ith source node to the jth destination node. Thresholds evaluation is performed on-line using both the K paths pre-computed in the off-line phase and information obtained dynamically by TE-extended routing protocols (Fig. 2). Please note that, besides TE extensions already available in the MPLS-TE scenario of Fig. 1, neither architectural nor protocol extensions are further required. As a consequence, the proposed scheme fits perfectly in the current MPLS-TE architecture.

The algorithm proposed for threshold determination is run separately and independently by each edge node and relies on the pre-computed paths. The pre-computation approach has been recently proposed as a suitable solution for the routing problem [11]. Basically, in this approach paths are calculated off-line by a background process, which prepares a database that is subsequently accessed upon each connection request, through a simple and fast procedure. Usually this technique is used to find a path for the incoming request; however, in the proposed approach, the pre-computed paths are rather used with the aim of estimating the network load in order to evaluate thresholds B_{ij}.

The algorithm works as follows. Given the destination, for each of the K pre-computed paths we store the corresponding available bandwidth (i.e. the minimum available bandwidth over all the links that belong to that path). This value, denoted by $b_{ij}^{(k)}$, varies in time depending on the network resource usage, and is updated by means of information about links' available bandwidths flooded by TE-extended routing protocols (Fig.2). The threshold B_{ij} is simply given by:

$$B_{ij} = \max_{1 \le k \le K} \left\{ b_{ij}^{(k)} \right\}. \tag{3}$$

Whenever a new request is presented to the ingress node, condition (1) is checked; if that condition holds, then CBR is run in order to find a suitable path. Therefore route computation does not exploit the pre-computed paths and is completely independent from AC, allowing the choice of the preferred CBR strategy. If (1) holds, this just guarantees that at least a feasible path exists (among the K pre-computed ones) and, as a consequence, that CBR can be run with success. In contrast, if (1) does not hold, a path might exist (outside the set of the K shortest ones), but this cannot be guaranteed for sure. However, we decide anyway to reject the incoming request. In this way we do not try to route anyway a flow along an excessive number of hops; instead, we discard it for the purpose of limiting resource consumption, with an overall benefit for the whole network, especially at medium high load, as described in the following section.

It is important to note that scalability does not represent an issue in the proposed approach. In fact, the $(N\text{-}1)\cdot K$ paths stored by each edge node are not recomputed upon each request; they are pre-calculated and updated only when a failure or a topology change occurs. Moreover, the proposed method does not even require an increase in flooding, since no new information is distributed by routing protocols (evaluation of (3) requires the same information needed by the CBR algorithm, which is already available in the LER).

3 Performance Evaluation

The proposed method achieves better performance in comparison with traditional CBR based optimization. Intuitively, this effect can be explained as follows. In the traditional approach, each incoming LSP request triggers a CBR instance that, in turn, tries to find a feasible path taking into account the current network status. If the network is far from congestion, this mechanism works fine. However, as the network load increases, this approach leads to longer and longer paths, with an overall decrease of the network performance. The adoption of the proposed Admission Control, instead, limits this behavior. In fact, it acts so as to accept only requests for which a path not much longer than the shortest one is available. Implicitly, this limits the average length of accepted paths, and consequently the amount of resources needed to support the incoming LSP, with an overall benefit for the network.

From an analytical point of view this can be explained as follows. We define the *blocking probability* p_B as the ratio between the average number of rejected requests and the average number of offered requests. Therefore we have:

$$n_S = (1 - p_B) \cdot n_O \ . \tag{4}$$

In Eq. (4) n_O represents the average number of offered requests, while n_S is the average number of accepted requests. Moreover, defining:

- b, as the average bandwidth of an incoming LSP request(expressed in *Mb/s*) ;
- C, as the *total network capacity*, i.e. the sum of all links' capacities (expressed in *Mb/s*);
- Λ_S, as the *normalized throughput*, i.e. the cumulative bandwidth of all the accepted flows normalized against the *total network capacity*;

the following expression yields:

$$\Lambda_S = \frac{n_S \cdot b}{C} \ . \tag{5}$$

Multiplying Eq. (5) by l (average path length, expressed in number of hops) and combining it with Eq. (4) we find the normalized average amount of resources employed in the network:

$$\Lambda_S \cdot l = \frac{(1 - p_B) \cdot n_O \cdot b \cdot l}{C} \ . \tag{6}$$

As explained above, the ultimate effect of the proposed edge distributed AC is to limit the average path length l, and consequently to decrease the *blocking probability* p_B (given the same amount of resources employed in the network). In other words, applying edge distributed AC, the network tends to accept more incoming requests (on the average), routing them on shorter paths. This explains the performance improvement obtained.

Unfortunately, it is not easy to evaluate performance from an analytical point of view, since it depends on many factors, such as the network topology, the CBR algorithm employed, the offered load and, last but not least, the number of paths K. For this reason, the performance of the proposed method has been evaluated by means of simulations. The following results refer to the random generated network topology shown in Fig. 3, with $N=8$ nodes and $L=12$ bi-directional links (degree $2 \times L/N=3$).

Other simulations have been performed on different topologies, such as the NSFNET ($N=14$, $L=21$, degree $2 \times L/N=3$), with similar results. In such networks there are $N \cdot (N-1)$ possible Traffic Relations (TR), associated to all the possible ordered *(ingress, egress)* pairs. We have assumed a flat spatial distribution, with the same average total load $f_{ij} = f$ (expressed in *Mb/s*) for each TR, divided into an average number of n flows of b *Mb/s* each ($f = n \cdot b$). We have also assumed Poisson arrivals with average inter-arrival rate $\lambda_{ij} = \lambda = n/T = f/(b \cdot T)$ and flow duration exponentially distributed with mean duration $T=3600s=1h$. This choice is somehow arbitrary.

However, considering that we are modeling network user requests, it has been considered a reasonable approximation [12].

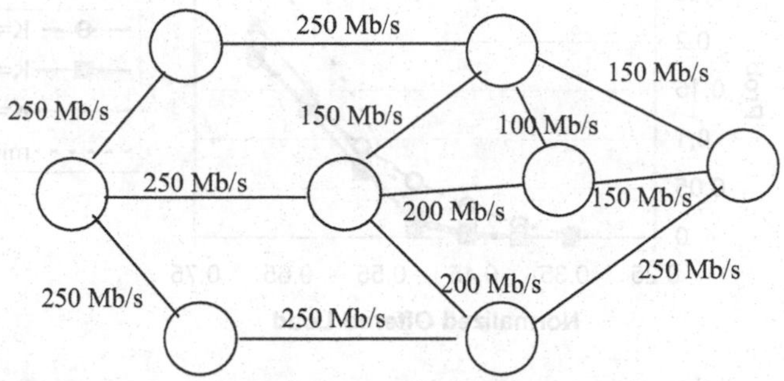

Fig. 3. Random generated network topology used in simulations. Links' capacities range from *100 Mb/s* to *250 Mb/s*, with a *total network capacity C=4900 Mb/s* and an average value $C/(2{\cdot}L)\approx 204$ *Mb/s*. Link's capacities have been evaluated assuming a flat spatial distribution of the offered traffic, i.e. assuming the same amount of offered traffic (on the average) between each pair of nodes.

In order to evaluate performance we define the *normalized offered load* as the sum of the offered loads over all the Traffic Relations, normalized against the *total network capacity*. The hypothesis of flat spatial distribution leads to the following expression for the *normalized offered load*:

$$\Lambda_O = \frac{N\cdot(N-1)\cdot f}{C} = \frac{N\cdot(N-1)\cdot n\cdot b}{C}\ . \tag{7}$$

The effectiveness of the proposed solution is evaluated from two different perspectives, both by the user's point of view (i.e. by means of the *blocking probability* against the *normalized offered load*) and by the network's point of view (i.e. by means of the *normalized throughput* against the *normalized offered load*). Results show that the adoption of the proposed edge distributed Admission Control provide benefits both for the user and for the network. Simulations are described in details below.

Case #1

In this case we assume that the single LSP request has a bandwidth uniformly distributed between 0.8Mb/s and 1.2Mb/s, with an average value $b = 1$ *Mb/s*. To have an idea, observe that a *normalized offered load* $\Lambda_O = 0.2$ means that each TR contributes with an offered load of $f = 0.2\times 4900/56 = 17.5$ *Mb/s*, i.e with $n=17.5$ flows of $b=1Mb/s$ each (on the average, see Eq. (7)). This case represents a dynamic situation in which a relatively high number of small flows are contemporarily active (on average) between each *(ingress, egress)* pair. The routing algorithm chosen is Min Hop routing, that consists in a shortest path applied to the network pruned from unfeasible links.

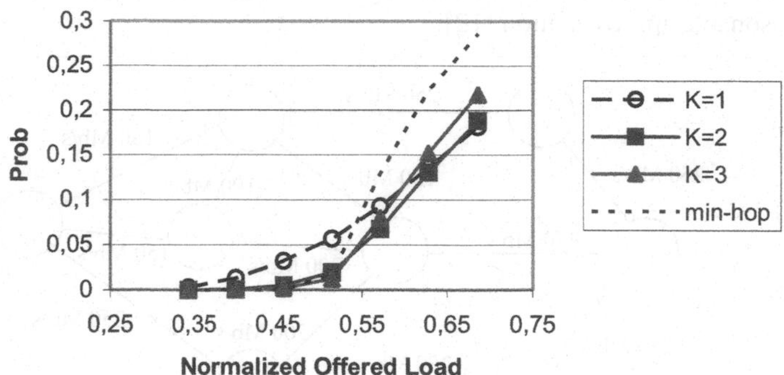

Fig. 4. *Blocking probability* against *normalized offered load* for the first case, plotted for different values of *K*.

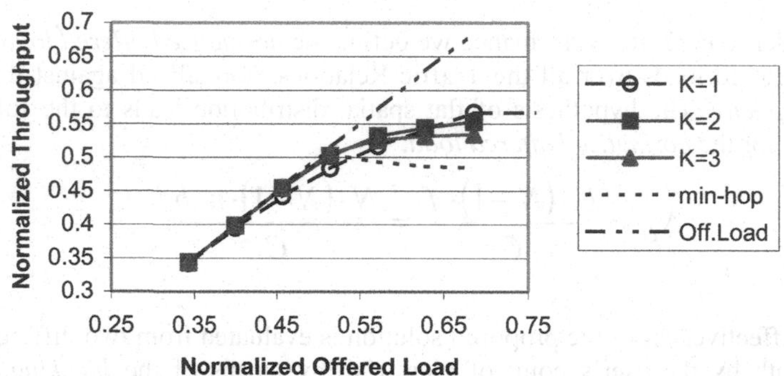

Fig. 5. *Normalized throughput* against *normalized offered load* for the first case, plotted for different values of *K*.

Fig. 4 and Fig. 5 show respectively the *blocking probability* and the *normalized throughput* against the *normalized offered load*, for *K=1,2,3* and for Min Hop without the application of the proposed AC. As expected, at low/medium load CBR performance is not affected, while at higher load AC filters the offered traffic, with an overall benefit for the network. In fact, for *K=2,3* we observe an improvement both in *blocking probability* and in *throughput* with respect to pure Min Hop (*K=1* does not exhibit the same behavior, but this is not surprising, since it turns back to the traditional shortest path routing). The choice of *K* in this simulation has been done considering that it should be neither too large, in order to have a real benefit in terms of filtering the incoming traffic, nor too small in order to avoid performance degradation at low load. However, the choice of *K* depends on many factors and is an optimization problem that is left for further study.

Case #2

This case is very similar to the previous one, with the only exception that now LSP requests are characterized by a bandwidth uniformly distributed between 8Mb/s and 12Mb/s, with an average value $b = 10\ Mb/s$. Therefore, with respect to case #1, the total offered load is the same, but now flows are a magnitude order bigger and are characterized by an inter-arrival rate ten times smaller.

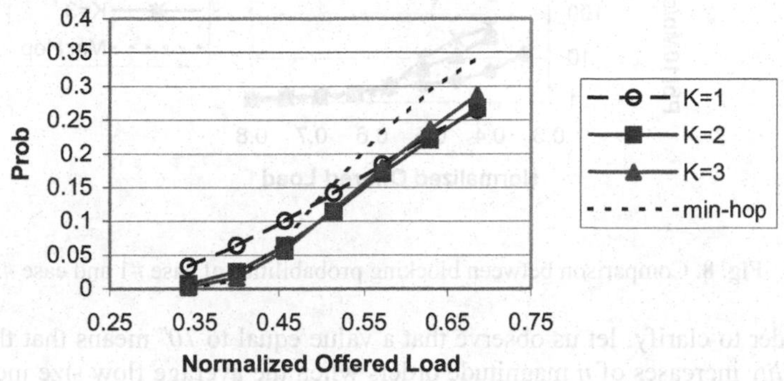

Fig. 6. *Blocking probability* against *normalized offered load* for the second case, plotted for different valucs of *K*.

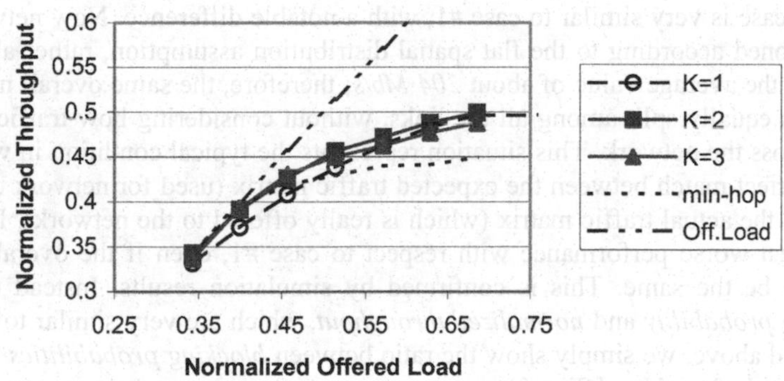

Fig. 7. *Normalized throughput* against *normalized offered load* for the second case, plotted for different values of *K*.

Blocking probability and *normalized throughput* against the *normalized offered load* are represented in Fig. 6 and Fig. 7. As expected, performance decreases with respect to case #1; in fact, given the same offered load, the routing algorithm obviously works better at the higher granularity (i.e. with smaller flows). However, even if the overall behavior is very similar to case #1, it is very interesting to compare case #1 and case #2, in order to quantify to what extent performance are affected by the different average size of the single flow. Fig. 8 plots the ratio between *blocking probabilities* of case #2 and case #1 against *offered load.*

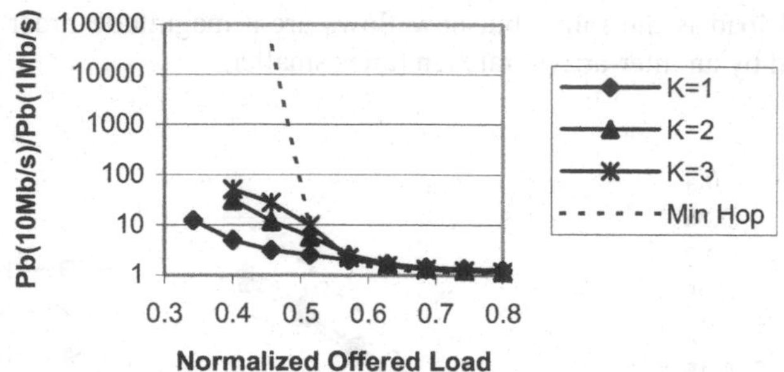

Fig. 8. Comparison between blocking probabilities of case #1 and case #2.

In order to clarify, let us observe that a value equal to 10^n means that the *blocking probability* increases of *n* magnitude orders when the average flow size increases tenfold. By looking at Fig. 8, we can easily see that at medium load pure Min Hop degrades up to 4 magnitude orders, while the adoption of the proposed edge distributed AC tends to limit this degradation effect and to preserve network resources, thus increasing network robustness with respect to uncertainty in the traffic model.

Case #3

This case is very similar to case #1, with a notable difference. Now network is not dimensioned according to the flat spatial distribution assumption, rather all links are equal to the average value of about *204 Mb/s*; therefore, the same overall network capacity is equally split among all the links, without considering how traffic is distributed across the network. This situation represents the typical condition in which there is no perfect match between the expected traffic matrix (used for network dimensioning) and the actual traffic matrix (which is really offered to the network). In this case we expect worse performance with respect to case #1, even if the overall behavior tends to be the same. This is confirmed by simulation results. Instead of plotting *blocking probability* and *normalized throughput*, which are very similar to the graphs presented above, we simply show the ratio between *blocking probabilities* against the *normalized offered load* (Fig. 9).

As in the previous case, a value equal to 10^n means that the mismatch between the expected and the offered traffic matrix may increase the *blocking probability* of *n* magnitude orders. From Fig. 9 we observe that at medium load pure Min Hop degrades up to 4 magnitude orders, while the proposed solution limits this degradation to a maximum of 1 magnitude order. This result confirms the conclusion expressed in the previous case, showing that one of the main benefits of the proposed AC method consists in the increase in network robustness.

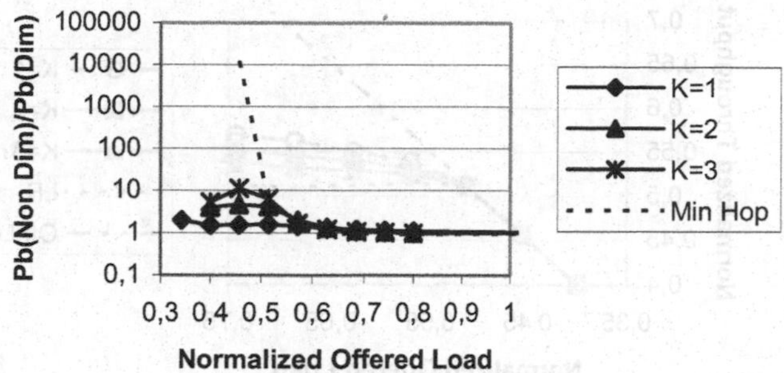

Fig. 9. Blocking probability ratios for traffic matrix mismatch.

Case #4

In order to show how the proposed method behaves with a different routing strategy we have run a simulation with the same parameters of case #1, but using the *Least Resistance* algorithm proposed in [13]. Such algorithm, which is known to perform quite better than Min Hop, is applied as follows. Each link is assigned a weight inversely proportional to the available bandwidth, according to the following expression:

$$w_i = \frac{C_{MAX}}{C_i - R_i}. \quad (8)$$

Eq. (8) represents the link's resistance; C_{MAX} is the maximum capacity over all the links and the difference $(C_i - R_i)$ represents the capacity available on the ith link. The Dijkstra algorithm is applied with weights given by Eq. (8), and the "shortest" path is found (shortest according to the previous metric). Finally the path so computed is checked to verify that all component links are feasible, i.e. have enough bandwidth to support the flow.

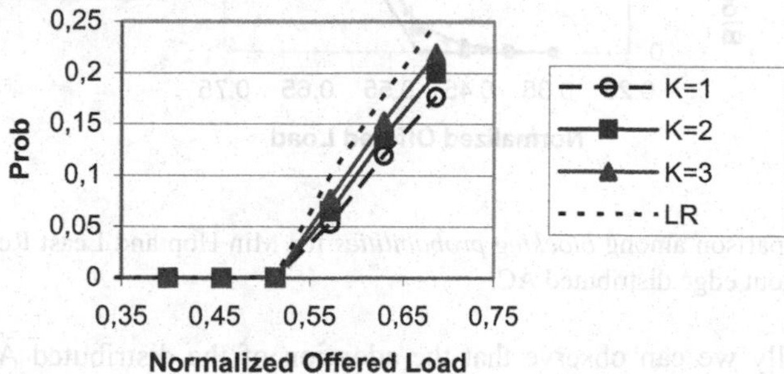

Fig. 10. *Blocking probability* against *normalized offered load* with the adoption of Least Resistance routing, plotted for different values of *K*.

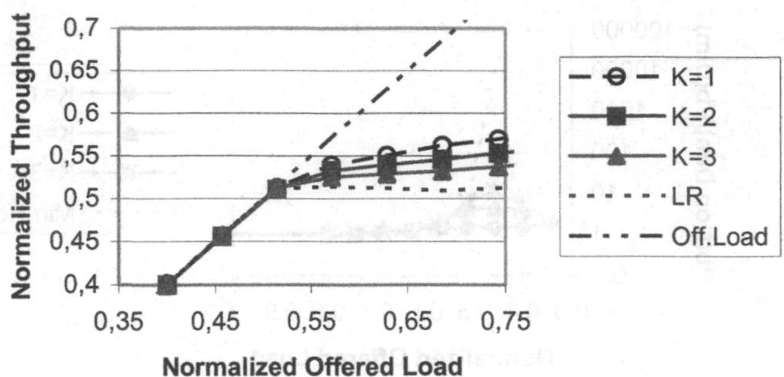

Fig. 11. *Normalized throughput* against *normalized offered load* with the adoption of Least Resistance routing, plotted for different values of *K*.

Simulation results are shown in Fig. 10 and Fig. 11; differently from case #1 there are no cross-points among curves. Instead, the adoption of the proposed invention increases performance for the whole range of offered traffic, both in terms of *blocking probability* and *normalized throughput*. Specifically, the best performance seems to be obtained in the case *K=1*. This is a very good result, which shows that the joint use of distributed AC and Least Resistance achieves a global optimization objective, outperforming the pure Least Resistance, i.e. one of the best routing algorithms available. Moreover an interesting result can be observed by comparing some curves for Min Hop and Least Resistance (Fig. 12).

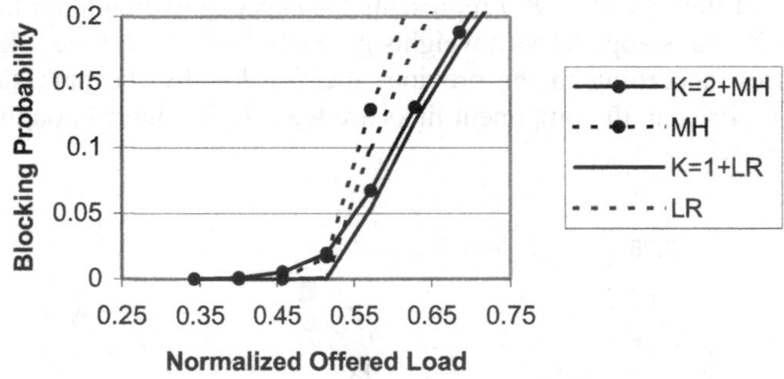

Fig. 12. Comparison among *blocking probabilities* for Min Hop and Least Resistance routing, with and without edge distributed AC.

Specifically we can observe that the adoption of the distributed AC always provides benefits (just compare dashed curves, associated to pure routing and the corresponding solid lines, associated to the adoption of AC). This is true especially at medium/high loads, where the filtering effect of the proposed method comes into play; in fact, in such conditions the *blocking probabilities* are quite independent from the routing algorithm employed (the curves "K=2+Min Hop" and "K=1+Least Resistance" tends asymptotically to converge). Therefore, at medium/high load, the usage of networks resources tends to be regulated and optimized by the distributed AC, while at medium/low load the optimization is left to the routing algorithm chosen by the operator, with advantages in terms of flexibility in the choice of the Traffic Engineering strategy.

4 Conclusions and Future Work

This paper has introduced a novel approach for Admission Control in IP/MPLS networks, which applies at network edges in a distributed and efficient way. The effect of the proposed algorithm consists mainly in the reduction of network congestion and, correspondingly, in the performance increase, especially at medium/high load. Moreover, it also provides increased robustness of the network, with respect to temporary situations of exceptional traffic and uncertainty in the traffic model. This is confirmed by the simulation results presented above. More efforts are needed to exploit all the benefits of the proposed method, and the future work will provide more simulation results under different network and traffic conditions. Specific attention will be paid to the performance evaluation with different network topologies, for the purpose of understanding to what extent some topological properties (e.g. number of nodes, degree) influence the method applicability and effectiveness. It is worth to notice that the strategy proposed herein is easily extensible to a variety of scenarios, e.g. ATM or even GMPLS networks. Moreover, even if the approach described above is distributed to network edges, basically the same algorithm could be applied in a centralized way by a single network entity, e.g. a centralized Resource Manager or a Bandwidth Broker.

References

1. Awduche, D., Chiu, A., Elwalid, A., Widjaja, I., Xiao, X.: Overview and Principles of Internet Traffic Engineering, IETF RFC 3272, May 2002
2. Rosen, E., Viswanathan, A., Callon, R.: Multiprotocol Label Switching Architecture, IETF RFC 3031, January 2001
3. Awduche, D., Malcolm, J., Agogbua J., O'Dell, M., Mcmanus, J.: Requirements for Traffic Engineering Over MPLS, IETF RFC 2702, September 1999
4. Boyle, J., Gill, V., Hannan, A., Cooper, D., Awduche, D., Christian, B., Lai, W.S.: Applicability Statement for Traffic Engineering with MPLS, IETF RFC 3346, August 2002
5. Casetti, C., Favalessa, G., Mellia, M., Munafo, M.: An Adaptive Routing Algorithm for Best-effort Traffic in Integrated-Services Networks, 16-th International Teletraffic Congress (ITC-16), Edinburgh, UK, June 1999
6. Awduche, D., Berger, L., Gan, D., Li, T., Srinivasan, V., Swallow, G.: RSVP-TE: Extensions to RSVP for LSP Tunnels, IETF RFC 3209, December 2001
7. Jamoussi, B., Andersson, L., Callon, R., Dantu, R., Wu, L., Doolan, P., Worster, T., Feldman, N., Fredette, A., Girish, M., Gray, E., Heinanen, J., Kilty, T., Malis, A.: Constraint-Based LSP Setup using LDP, IETF RFC 3212, January 2002

8. Kats, D., Yeung, D., Kompella, K.: Traffic Engineering Extensions to OSPF Version 2, IETF draft, work in progress
9. Li, T., Smit, H.: IS-IS extensions for Traffic Engineering, IETF draft, work in progress
10. Eppstein, D.: Finding the K Shortest Paths, 35-th IEEE Symp. Foundations of Computer Science, Santa Fe, 1994, pp. 154-165
11. Orda, A., Guerin, R., Sprintson, A.: QoS Routing: the Precomputation Perspective, Proc. of IEEE Infocom 2000, pp. 128-136
12. Floyd, S., Paxson V.: Difficulties in Simulating the Internet, IEEE/ACM Transactions on Networking, Vol.9, No.4, pp. 392-403, August 2001
13. Wen, B., Sivalingam K. M.: Routing, Wavelength and Time-Slot Assignment in Time Division Multiplexed Wavelength-Routed Optical WDM Networks, Proc. of IEEE Infocom 2002

Bandwidth Estimation for TCP Sources and Its Application

Rossano G. Garroppo, Stefano Giordano, Michele Pagano, Gregorio Procissi, Raffaello Secchi

Department of Information Engineering
University of Pisa Via Diotisalvi 2,
56126 Pisa (Italy)
Tel. +39-050-568661, Fax. +39-050-568522
{r.garroppo, s.giordano, m.pagano, g.procissi, r.secchi}@iet.unipi.it

Abstract. In this paper, we propose a novel bandwidth estimation algorithm for TCP connections and its possible application to congestion control mechanism. The estimation algorithm relies upon an analytic relation which expresses the connections' available bandwidth as a function of the inter-departure time of packets and of the inter-arrival time of ACKs. It is worth noticing that this approach can be extended to protocols other than TCP, as long as they support an acknowledgment mechanism. The bandwidth estimation performance is assessed through discrete event simulations under various network topologies, traffic scenarios and link error conditions. Bandwidth estimation is then applied to TCP congestion control to select the value of the congestion window after a packet loss episode. Performance of this modified version of TCP is validated by means of simulations and compared to the one achieved by TCP NewReno. Finally, the possible coexistence of the modified version of TCP and TCP NewReno is proved through a detailed analysis of fairness and friendliness of the new protocol.

1. Introduction

Transmission Control Protocol (TCP) provides a reliable and connection oriented transport service to the most part of today's Internet applications. It has been shown [8] that 90% of the total Internet data and 95% of packets are transmitted over TCP, which, in turn, plays a central role in the overall traffic control over the global network.

The massif diffusion of the protocol has determined a huge research effort towards the proposal of modified versions of TCP to cope with the continuously evolving contexts in which it is called to operate in. In particular, the new high speed wired/wireless environment expands the domain in which TCP was initially developed and validated.

To improve TCP performance, in recent years, TCP Westwood (or TCPW, in short) [5] introduced a novel approach to achieve a faster recovery after a loss episode. TCP Westwood takes advantage of two basic concepts: the end to end estima-

M. Ajmone Marsan et al. (Eds.): QoS-IP 2003, LNCS 2601, pp. 247-260, 2003.

tion of the available bandwidth and its use in the selection of both the slow start threshold and the congestion window. A TCPW source continuously estimates the available bandwidth by processing the received ACK time series. The estimation is then used to set the slow-start threshold and the congestion window to the value of the estimated pipe-size, when the Fast Retransmit algorithm is activated. TCPW estimates the connection available bandwidth by low pass filtering the sequence of rate estimates defined as the ratio between the amount of acknowledged data carried by the any ACK and the corresponding ACK inter-arrival times.

In this paper, we propose a novel bandwidth estimation algorithm together with its possible application to TCP congestion control mechanism. The proposed estimation technique is based on a single bottleneck network approximation and on the derivation of an analytic relation that expresses the available bandwidth as a function of inter-departure times of packets and inter-arrival-time of ACKs.

The paper is organized as follows. In section 2, we describe the proposed estimation algorithm along with possible drawbacks of the methodology. In section 3, the performance of the estimator is assessed through simulations under various network scenarios and traffic conditions. In the same section we also discuss the accuracy of the estimation algorithm. Section 4 is devoted to the application of the bandwidth estimation algorithm to TCP congestion control mechanism. The modifications to TCP (throughout the paper referred to as *Modified NewReno*) are validated through simulations, as reported in section 4.1. In particular, to assess the level of performance improvement that can be achieved by using the bandwidth estimation algorithm, a comparison of goodput and completion time of NewReno and Modified NewReno connections is given. Finally, in section 4.2, the analysis of fairness and friendliness of the new TCP version is presented in order to explore its possible coexistence with TCP NewReno. Finally, we summarize the main result of the paper in the Conclusions.

2. Bandwidth Estimation Technique

In this section we present the proposed bandwidth estimation technique for traffic sources that make use of the acknowledge mechanism. Throughout this paper, this method will be applied to the most obvious and natural case of TCP connections; nevertheless, it is worth noticing that it can be extended to more generic closed-loop traffic sources.

In the literature, many papers have dealt with bandwidth estimation, according to different approaches (see, for example, [1][2][3][4][6][7][9]).

Similarly to the above summarized works, in this section we present a bandwidth estimator that essentially relies upon the computation of the inter-arrival times of ACKs at the sender side and assumes a simple network model based on the hypothesis of a single bottleneck.

2.1 Bandwidth Estimator Description

To describe the bandwidth estimation technique, let us consider first Figure **1**, which depicts the transmission of a pair of packets from source to destination and the path of the corresponding ACKs back to the source.

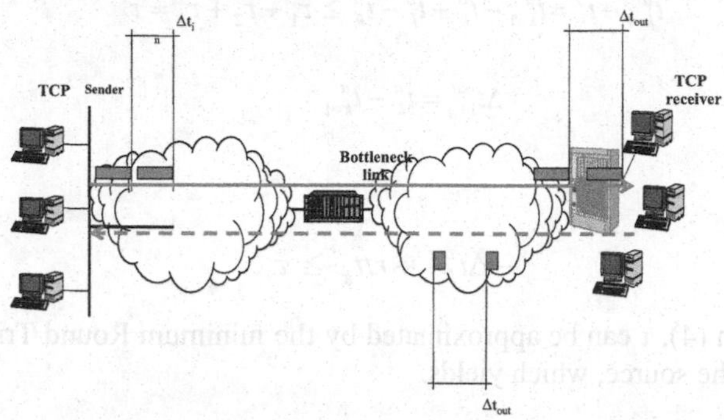

Fig. 1. Bandwidth Estimation scheme

Let us define Δt_{in} as the inter-departure time of the two packets transmitted back to back at the source side and Δt_{out} as the inter-arrival time of the two packets at the destination side. Clearly, while the packet pair transits across the network, the spacing varies depending on the cross-traffic sharing the same network nodes. In our approach, we adopt the approximation of single bottleneck, that is, we assume that packets can suffer from queuing delay only at a router; the remaining links of the path are supposed to be fast enough to exclude any packet buffering. Queuing delays experienced by ACKs are assumed negligible and will not be considered in the analysis.

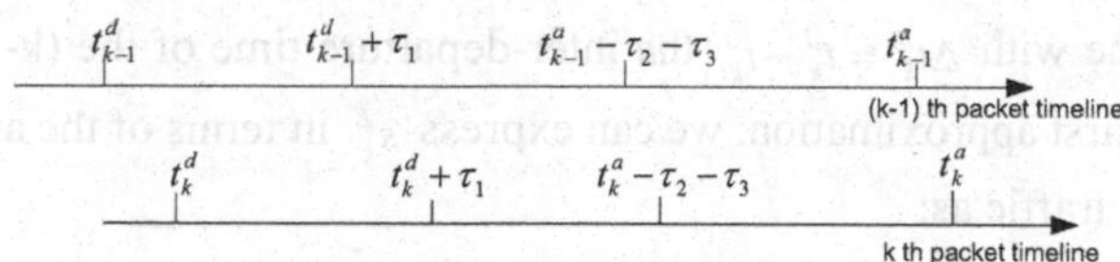

Fig. 2. Queuing events timeline

Figure **2** shows the timeline of a packet pair transmission event. Let us define with t^d_k and t^a_k the departure time from the source of the kth packet and the arrival time of the corresponding ACK back to the source, respectively. The (k-1)th packet leaves the source at time t^d_{k-1}. Then, τ_1 seconds later it reaches the bottleneck queue where it gets buffered and eventually completely transmitted at time $t^a_{k-1}-\tau_2-\tau_3$. The values τ_2 and τ_3 are the delays experienced by packets in the path bottleneck-destination and by ACKs in the path destination-source, respectively. According to the single bottleneck assumption, τ_1, τ_2 and τ_3 are assumed constant. The kth packet will find the (k-1)th packet in the buffer if:

$$t_{k-1}^{a} - \tau_2 - \tau_3 \geq t_k^d + \tau_1 \tag{1}$$

and thus:

$$t_{k-1}^{a} - t_k^d = t_{k-1}^{a} - t_k^a + t_k^a - t_k^d \geq \tau_1 + \tau_2 + \tau_3 = \tau \tag{2}$$

$$\Delta t_k^{out} = t_k^a - t_{k-1}^a \tag{3}$$

we obtain:

$$-\Delta t_k^{out} + rtt_k \geq \tau \tag{4}$$

In equation (4), τ can be approximated by the minimum Round Trip Time (RTT_{min}) measured at the source, which yields:

$$\Delta t_k^{out} \leq rtt_k - rtt_{\min} \tag{5}$$

If condition (5) holds, the gap between the (k-1)th and kth packets introduced by the bottleneck is given by the sum of the transmission time of one packet of the connection and the transmission time of packets belonging to cross connections which arrived at the bottleneck during the time interval $[t_{k-1}^d + \tau_1, t_k^d + \tau_1]$. If we denote with L the packet size (bits), with μ the bottleneck capacity (bps), and with s_k^E the amount of cross-traffic (bits) standing in the buffer between the (k-1)th and the kth packets, the following relation holds:

$$\Delta t_k^{out} = \frac{L}{\mu} + \frac{s_k^E}{\mu} \tag{6}$$

Let us define with $\Delta t_k^{in} = t_k^d - t_{k-1}^d$ the inter-departure time of the (k-1)th and the kth packets. As a first approximation, we can express s_k^E in terms of the average bit rate λ of the external traffic as:

$$s_k^E = \Delta t_k^{in} \cdot \lambda \tag{7}$$

Hence, Δt_k^{in} can be determined by:

$$\Delta t_k^{in} = \left(t_k^d - t_k^a\right) + \left(t_k^a - t_{k-1}^a\right) + \left(t_{k-1}^a - t_{k-1}^d\right) = -rtt_k + \Delta t_k^{out} + rtt_{k-1} \tag{8}$$

which allows us to express Δt_k^{out} as a linear function of Δt_k^{in}:

$$\Delta t_k^{out} = \frac{L}{\mu} + \frac{\lambda}{\mu} \cdot \Delta t_k^{in} \tag{9}$$

The algorithm collects as many samples Δt_k^{in} and Δt_k^{out} as possible, as long as they satisfy the constraint (5) and performs a linear regression to estimate the slope $a=\lambda/\mu$ and the constant term $b=L/\mu$. The bandwidth estimation is then given by:

$$B = \mu - \lambda = \frac{1-a}{b} \cdot L \tag{10}$$

Shortly, the bandwidth estimation algorithm acts as it follows:

1. Collects sample of RTTs and ACK inter-arrival times and discards samples that do not satisfy:

$$\Delta t_k^{out} \leq rtt_k - rtt_{\min} \tag{11}$$

2. Performs a linear regression on the sample pairs:

$$\begin{cases} \Delta t_k^{in} = rtt_{k-1} - rtt_k + \Delta t_k^{out} \\ \Delta t_k^{out} \end{cases} \tag{12}$$

3. Computes the bandwidth estimation according to (10)

2.2 Practical Issues

In this section we discuss some of the main issues that can affect the performance of the above described estimator. Basically, three main critical drawbacks can be outlined:

- single bottleneck approximation;
- TCP packet loss;
- Reverse path congestion.

Single Bottleneck Approximation

The first practical issue is the goodness of the single bottleneck assumption; in other words, in the real Internet, TCP packets may compete against cross traffic in more than one link. In this case, the spacing between packet pairs can be significantly larger than expected and that would lead to underestimate the actual available bandwidth of the connection. In the context of resource allocation, this type of error is in general not critical as it does not affect network stability.

TCP Packet Loss

Occasionally, TCP packets are dropped within the networks (for example, as a result of buffer overflow or link errors). In fact, this does not significantly affect the bandwidth estimation as packet losses are typically recovered by the Fast Retransmit algorithm of TCP. This will take more than one RTT and thus, constraint (5) will not be met. The estimation dynamic will only be slightly impaired, but the performance will not be affected.

Reverse Path Congestion

The most critical issue related to packet-pair like bandwidth estimators is represented by the possible congestion on the reverse (ACKs) path. This is the well known problem of ACK compression and occurs when ACKs get buffered at a network resource. In our approach, this drawback is somewhat alleviated by the low pass filtering performed by the linear regression. Typically, ACK compression causes reduced values of Δt^{out}, which will likely lead to over-estimate the available bandwidth. This kind of problem is essentially unavoidable when dealing with pure sender-side estimator and would only be sorted out by requesting the receiver side for cooperation (e.g. RBPP – Receiver Based Packet Pair [3]). However, extensive experimental studies [10] have recently proved that ACK compression only occur in about 10-15% of the packet pairs transmitted: the analysis of this phenomenon and its impact on the estimator performance will be elaborated upon in section 3.1 by using CBR UDP disturb traffic.

3. Numerical Validation

In this section we validate the performance of the bandwidth estimation technique by using *ns2* (Network Simulator ver. 2). The bandwidth estimator has been implemented as a module of TCP NewReno.

Figure **3** shows the network topology of simulations.

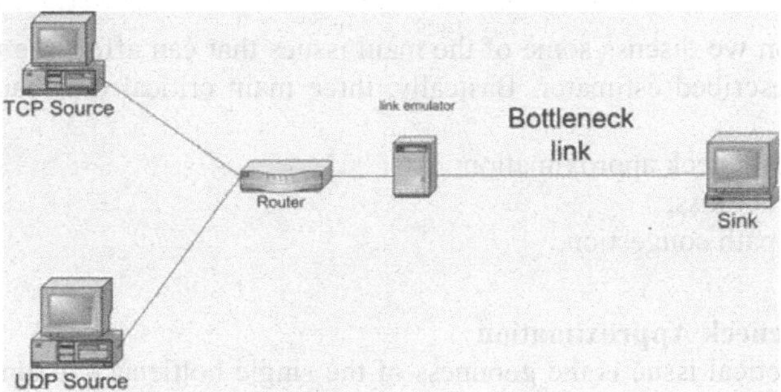

Fig. 3. Network topology in *ns* simulations

A TCP and a UDP sources share a common 2Mbps bottleneck link equipped with a FIFO DropTail buffer. The link emulator can emulate stochastic link errors. The TCP source is equipped with the bandwidth estimation module and delivers data generated by a greedy FTP application; at this stage of the work, the UDP source generates CBR disturb traffic. TCP and VBR UDP disturb traffic will be included in future papers. Table 1 summarizes the topology and connections' parameters adopted in the first set of simulations.

Table 1. Network topology and TCP connections' parameters

Parameter	Value
Packet Length	1000Bytes
Bottleneck Bandwidth μ	2Mbps
Access Link Capacity	10 Mbps
Round Trip Delay	100ms
Buffer Size	60pkt
Bottleneck link latency	40ms

The amount of available bandwidth is determined by varying the UDP source's bit-rate. The linear regression is performed over a number of samples $\lfloor \Delta t^{in}, \Delta t^{out} \rfloor$ set to 60. However, the selection of the window length (number of most recent pairs in the estimation) is quite a critical issue, and will be discussed in section 3.2.

Figure **4** shows the result of bandwidth estimation obtained in both cases of a loss-free and lossy channel. In the latter case, errors are assumed Bernoulli distributed with packet drop probability *p=1%*.

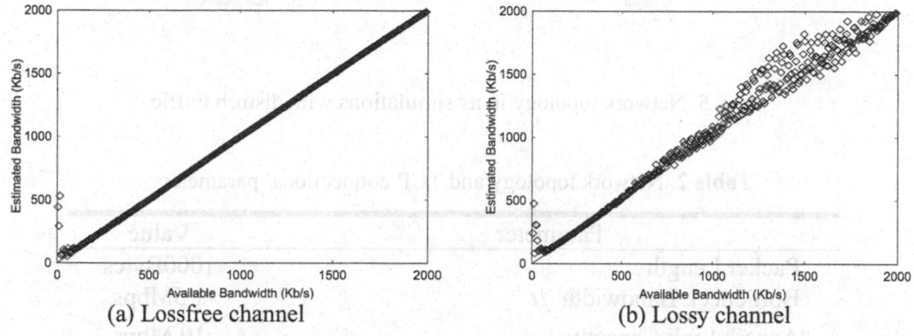

(a) Lossfree channel (b) Lossy channel

Fig. 4. Bandwidth Estimation for lossfree and lossy channel

In both cases, a qualitative inspection of the plots reveals that the bandwidth estimates match the actual value of the available bandwidth. More in detail, figure **4**.a shows an excellent match throughout the whole range of available bandwidth values, except for a negligible border effect at very low bandwidth. Figure **4**.b depicts two distinct regions instead. For low values of the available bandwidth, the behavior is analogous to the previous graph, while a slight dispersion of the estimates is present at higher ranges of actual bandwidth. The reasons of this phenomenon can be explained by noticing that link errors likely induce a smaller congestion window size which, in turn, reduces the interaction between TCP and UDP traffic. Indeed, what it may happen with low disturb traffic, is that small bursts of TCP packets never get interleaved with UDP packets. Thus, the samples of $\lfloor \Delta t^{in}, \Delta t^{out} \rfloor$ tend to accumulate in the neighborhood of the link capacity, producing an overestimation of the available bandwidth.

3.1 Reverse Path Congestion

In this section, we address the issue of the impact of traffic congestion on the reverse path on the performance of the bandwidth estimation algorithm. The network scenario depicted in Figure **5** shows the disturb traffic on the reverse path. Packets can get buffered in the forward link (forward bottleneck) and ACKs may be buffered on the reverse path (reverse path bottleneck).

Topology and connections' parameters are reported in Table **2**.

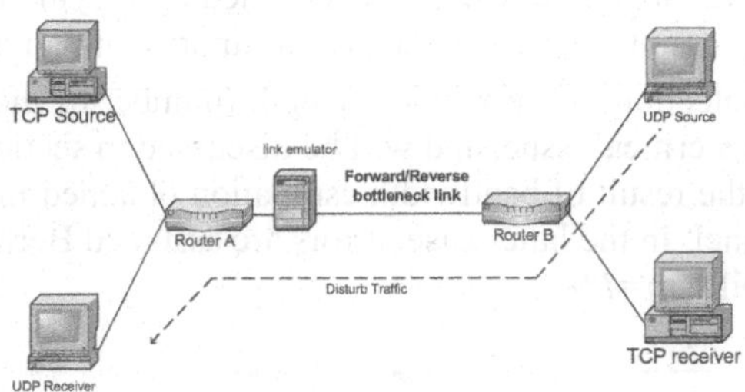

Fig. 5. Network topology in *ns* simulations with disturb traffic

Table 2. Network topology and TCP connections' parameters

Parameter	Value
Packet Length	1000Bytes
Bottleneck Bandwidth μ	1.5Mbps
Access Link Capacity	10 Mbps
Round Trip Delay	280ms
Buffer Size	60pkt
Bottleneck link latency	100ms

At this stage, we use CBR UDP disturb traffic flowing on the opposite direction. The case of more bursty disturb traffic (e.g. TCP or VBR UDP) on the reverse path is currently under investigation and it will be presented in future works. Figure **6**.a shows that the estimator returns correct estimates in the case of lossfree channel. Figure **6**.b reports the bandwidth estimates when the link suffers from Bernoulli random losses, with packet drop probability of $p=2\%$. The plot exhibits a dispersion around the actual value of available bandwidth: this phenomenon is likely due to the limited number of packets within the window that determine a noise effect on the estimates. Notice that this effect does not depend on the amount of disturb traffic on the reverse path and that the estimates look overall unbiased.

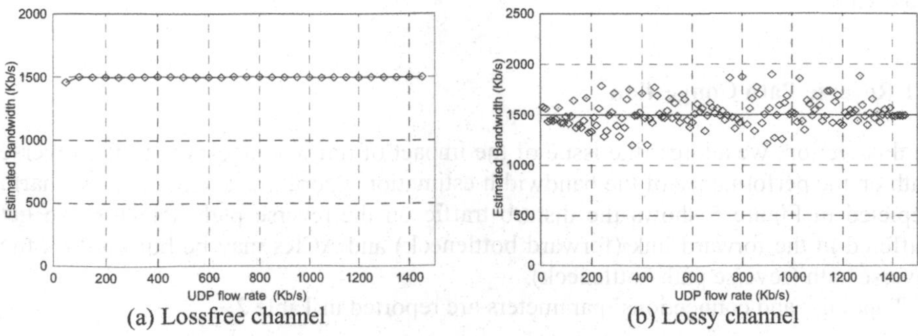

Fig. 6. Bandwidth Estimation for lossfree and lossy channel

3.2 Estimation Accuracy

In this section we focus on the analysis of the number of samples needed by the algorithm to obtain a reasonably accurate estimation of the available bandwidth. Indeed, it is worth recalling that in the current Internet, the number of packets sent over the net by common application exceeds 30 packets only in the 5% of the cases [11][12]. In other words, a reasonable bandwidth estimation algorithm has to produce accurate estimates by observing a few tens of ACKs. The above presented simulations rely upon the assumption of steady network conditions and long-lived connections. Here, we want to investigate the robustness of the proposed method in the case of short-lived connections and, possibly, to determine the minimum connection length that allows us to achieve proper estimates.

We begin the discussion by noticing that, in TCP context, bandwidth estimation is generally used to set the congestion window to the value of the pipe-size W_p in response to a packet drop event. Thus, minor errors on the bandwidth estimation that do not affect the correct value of the pipe-size are not relevant. For example, let us consider a TCP connection with round trip delay of 200ms, available bandwidth of 1 Mbps and packet size of 1000Bytes. Hence, the pipe-size is 25 pkts. The minimum error of 1 packet over 25 corresponds then to a relative bandwidth estimation error of $\Delta B/B = 4\%$. In other words, errors less than 4% in the bandwidth estimation do not produce any effect on the pipe-size estimation.

For our objectives, we arbitrarily assume that a sufficient degree of accuracy is achieved when the bandwidth estimation error is less than $1/W_p$, which leads to an estimated pipe-size value which differs from the actual one for *at maximum* 1 packet.

Figure 7 reports the percentage of bandwidth estimates that satisfy the previous requirement versus the number of ACKs used (i.e. versus the estimation window length) in the estimation algorithm. The results are averaged over 200 simulation runs, in the cases of both a lossfree and a lossy link.

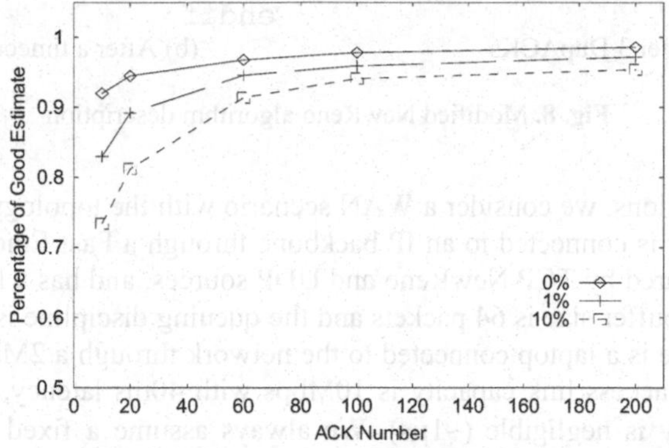

Fig. 7. Bandwidth estimation accuracy vs. number of ACKs

We can notice that, in the case of lossfree channel, even 10 ACKs are sufficient to achieve an accurate estimation of the available bandwidth. Obviously, as the link error rate increases, the number of needed samples grows. However, even for reasonable value of *PER* - packet drop rate - (of the order of 1-2%), the 90% of bandwidth estimates computed over 20 ACKs, are sufficiently accurate. Higher (and quite unrealistic) packet error rates definitely require a larger number of ACKs. For example, in the figure, about 60 ACKs are needed in the case of a packet drop rate of 10%, for the 90% of the estimates to be accurate with the above specified criterion.

Tests concerning the dynamic responsiveness of the estimation algorithm are under study and will be presented in future papers.

4. Application to TCP Congestion Control

In this section, we apply the previously described bandwidth estimation algorithm to the congestion control mechanism of TCP. Following the philosophy proposed by TCP Westwood [5], we use the information of the available bandwidth to set the congestion window (*cwnd*) and the slow start threshold to the value $cwnd = BWE^* \cdot RTT^*$ after a packet drop event which triggers the fast retransmit/fast recovery mechanism (figure **8**). The resulting TCP version will be referred to as *Modified Newreno* as it is obtained by modifying the Newreno class of *ns*.

The values BWE* and RTT* are the estimated available bandwidth and the Round Trip Time respectively.

```
if (3 DupACKs received )
    ssthresh = BWE * RTT_min / seg_size
    if (cwnd > ssthresh)
         cwnd = ssthresh
    endif
endif
```

(a) After 3 DupACKs

```
if ( timeout expires )
    cwnd = 1;
    ssthresh = BWE * RTT_min / seg_size
    if (ssthresh < 2)
         ssthresh = 2;
    endif
endif
```

(b) After a timeout expires

Fig. 8. Modified NewReno algorithm description

In the simulations, we consider a WAN scenario with the topology shown in figure **9**. A TCP server is connected to an IP backbone through a Fast Ethernet. The 2Mbps bottleneck is shared by TCP NewReno and UDP sources, and has a latency of 100ms. The bottleneck buffer size is 64 packets and the queuing discipline is a FIFO droptail. The receiver side is a laptop connected to the network through a 2Mbps wireless link. The server-side access link capacity is 10Mbps with 40ms latency, while the client-side link latency is negligible (~1μs). We always assume a fixed packet length of 1500 Bytes.

In the following sub-sections we present the results obtained by running ns2 simulations, according to the analyzed performance metrics.

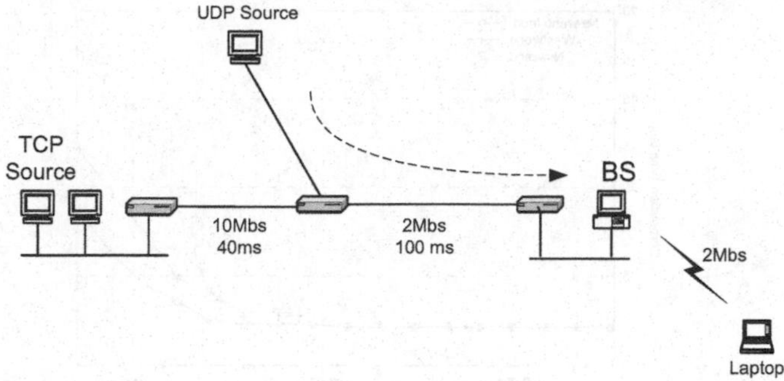

Fig. 9. Network Scenario

4.1 Goodput and Completion Time

In this section, we compare TCP NewReno, Modified NewReno and Westwood as far as their achieved goodput and completion time concern. The goodput is defined as the ratio between the amount of data successfully transmitted in a given time interval and the time interval itself. The completion time is defined as the time needed to transmit all the packets of a connection, that is the time elapsed between the transmission of the first packet of a connection and the reception of the last ACK.

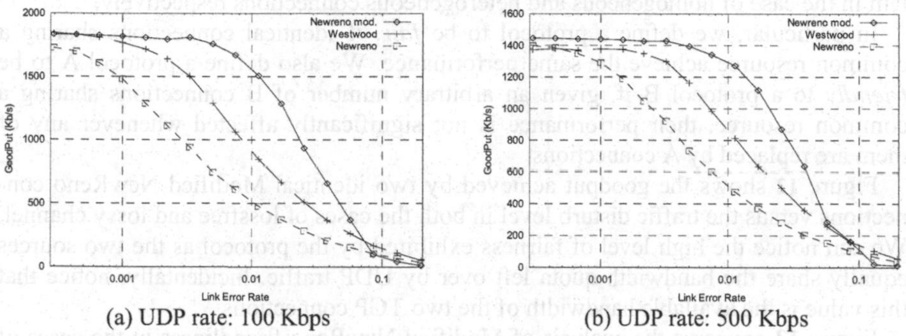

(a) UDP rate: 100 Kbps (b) UDP rate: 500 Kbps

Fig. 10. TCP goodput with disturb traffic

Figure **10** shows the goodput achieved by long-lived TCP NewReno, Westwood and Modified NewReno connections with respect to the packet drop rate, when the TCP source competes with a UDP source characterized by 100 and 500 Kbps rate respectively. No disturb traffic is present on the reverse path at this stage. Notice that Modified NewReno over-performs NewReno and Westwood over a wide range of Packet Error Rates (PER), and its goodput equals the available bandwidth up to values of PER of 0.2-0.5%. For increasing values of PERs, the higher number of timeouts affects the TCP performance that monotonically decreases; for very high value of PER (about 10%), the three TCP versions exhibit the same low performance.

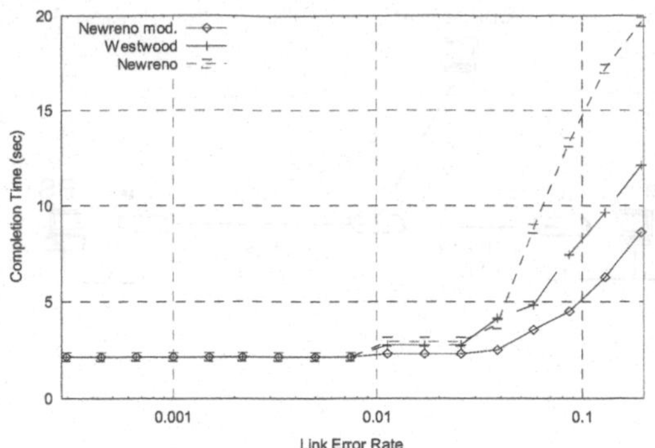

Fig. 11. TCP Average completion time

Similar results are obtained when we measure the completion time (Figure **11**) of 60 packets long NewReno, Westwood and Modified NewReno connections.

4.2 Fairness and Friendliness

Any proposal for new versions of TCP must guarantee the fair sharing of network resources. Commonly, we use the terms *fairness* and *friendliness* to address this problem in the case of homogeneous and heterogeneous connections respectively.

In particular, we define a protocol to be *fair*, if identical connections sharing a common resource achieve the same performance. We also define a protocol A to be *friendly* to a protocol B if, given an arbitrary number of B connections sharing a common resource, their performance is not significantly affected whenever any of them are replaced by A connections.

Figure **12** shows the goodput achieved by two identical Modified NewReno connections versus the traffic disturb level in both the cases of lossfree and lossy channel. We can notice the high level of fairness exhibited by the protocol as the two sources equally share the bandwidth quota left over by UDP traffic. Incidentally, notice that this value is the available bandwidth of the two TCP connections.

Figure **13** concerns the analysis of Modified NewReno friendliness in the cases of lossfree and lossy channel. Figure **13**.a reports the goodput versus the disturb traffic rate achieved by a TCP NewReno connection when it shares the lossfree bottleneck with a homogeneous NewReno connection (dashed line) and with a Modified NewReno connection (solid line) respectively. The results show that TCP NewReno performance is not significantly affected by the presence of Modified NewReno which, in turn, exhibits a good level of friendliness with respect to TCP NewReno.

Figure **13**.b shows the results obtained for a lossy bottleneck, with 1% PER. Notice first that the goodput of a TCP NewReno connection is solely determined by the random losses and that TCP NewReno cannot fill the whole available bandwidth left over by the UDP source. Moreover, figure **13**.b proves that TCP NewReno goodput is not relevantly decreased by the presence of the Modified NewReno connection.

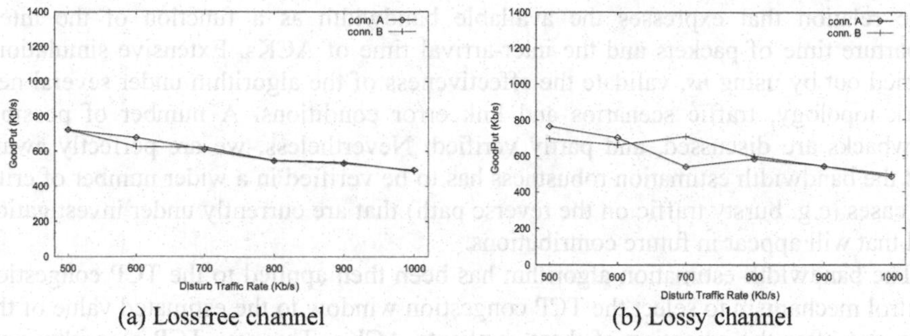

(a) Lossfree channel (b) Lossy channel

Fig. 12. Modified NewReno fairness

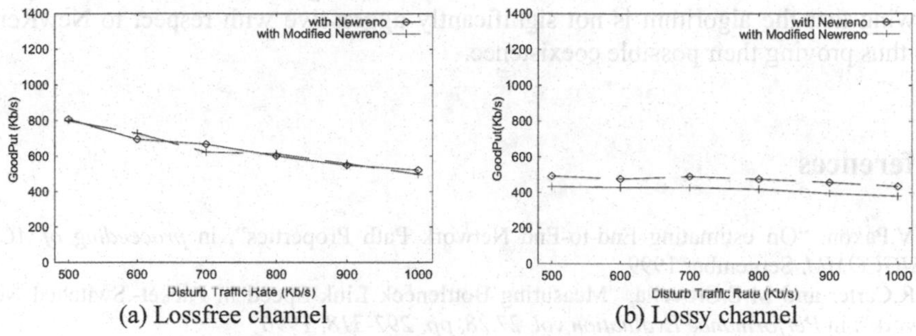

(a) Lossfree channel (b) Lossy channel

Fig. 13. Modified NewReno friendliness

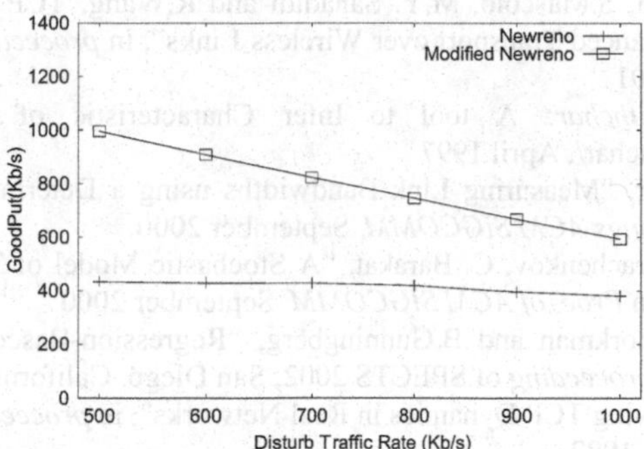

Fig. 14. NewReno vs. Mod. NewReno in lossy channel

5. Conclusions

In this paper, we propose an end-to-end bandwidth estimation technique for TCP connections that makes use of the ACKs time series. The algorithm relies upon an analytic relation that expresses the available bandwidth as a function of the inter-departure time of packets and the inter-arrival time of ACKs. Extensive simulations carried out by using *ns*, validate the effectiveness of the algorithm under several network topology, traffic scenarios and link error conditions. A number of possible drawbacks are discussed, and partly verified. Nevertheless, we are perfectly aware that the bandwidth estimation robustness has to be verified in a wider number of critical cases (e.g. bursty traffic on the reverse path) that are currently under investigation and that will appear in future contributions.

The bandwidth estimation algorithm has been then applied to the TCP congestion control mechanism to select the TCP congestion window to the estimated value of the pipe-size after the reception of three duplicate ACKs. The new TCP algorithm performance is assessed through simulations, showing a clear improvement with respect to TCP NewReno as far as goodput and completion time of connections concern. Fairness and friendliness to TCP NewReno of the algorithm are also investigated showing that the algorithm is not significantly aggressive with respect to NewReno and thus proving their possible coexistence.

References

[1] V.Paxon, “On estimating End-to-End Network Path Properties”, in *proceeding of ACM SIGCOMM,* September 1999

[2] R.Carter and M.E.Crovella, “Measuring Bottleneck Link Speed in Packet-.Switched Network”, in *Performance Evaluation,vol. 27,28, pp. 297-318, 1996*

[3] K. Lai and M.Baker,”Measuring Bandwidth”, in *proceeding IEEE INFOCOM,* April 1999.

[4] J.Hoe, “Improving the Start-up Behaviour of a Congestion Control Scheme for TCP”, in *Proceedings ACM SIGCOMM,*September 1996

[5] C.Casetti, M.Gerla, S.Mascolo, M.Y. Sanadidi and R.Wang,”TCP Westwood: Bandwidth Estimation for Enhanced Transport over Wireless Links”, in *proceedings of Mobicom 2001,* Rome, Ital, July 2001

[6] V.Jacobson, “*pathchar:* A tool to Infer Characteristic of Internet Paths.”, ftp: //ftp.ee.lbl.gov/pathchar/, April 1997

[7]K.Lai and M.Baker, “Measuring Link Bandwidths using a Deterministic Mode of Packet Delay”, in *proceedings ACM SIGCOMM,* September 2000.

[8] E. Altman, K. Avrachenkov, C. Barakat, “A Stochastic Model of TCP/IP with Stationary Random Losses“, in *Proc. of ACM SIGCOMM,* September 2000

[9] B.Melander, M.Bjorkman and B.Gunningberg, “Regression-Based Available Bandwidth Measurement”, in *proceeding* of SPECTS 2002, San Diego, California, July 2002.

[10] J. Mogul, "Observing TCP Dynamics in Real Networks", in *proceedings SIGCOMM'92,* in pag. 305-317, Aug. 1992.

[11] Hari Balakrishnan, Mark Stemm, Srinivasan Seshan, and Randy H. Katz, “Analyzing stability in wide-area network performance”, in *SIGMETRICS'97,* June 1997

[12] K.Claffy, Greg Miller, and Kevin Thompson, “The nature of the beast: Recent traffic measurement from an Internet backbone”, in *proceedings of INET'98, July 1998*

Best-Effort and Guaranteed Performance Services in Telecommunications Networks: Pricing and Call Admission Control Techniques

Marco Baglietto*, Raffaele Bolla*, Franco Davoli*, Mario Marchese°, Maurizio Mongelli°

* DIST - Department of Communications, Computer and Systems Science
mbaglietto, lelus, franco@dist.unige.it
° CNIT - Italian National Consortium for Telecommunications
University of Genoa
Via Opera Pia 13, 16145 Genova, Italy
mario.marchese, maurizio.mongelli@cnit.it

Abstract. Pricing for the use of telecommunication services has received in the last years a growing attention. The main objective is to establish various fairness criteria in the bandwidth allocation among different traffic classes. In the context of Quality of Service (QoS) networks (e.g. ATM, DiffServ, IntServ) the pricing scheme influences the Call Admission Control (CAC) rules. While, for Best Effort (BE) services, users accept a variable bandwidth allocation and are not subject to CAC and the pricing policies, according to the Proportional Fairnes Pricing, are integrated within the flow control. In this paper we consider both BE traffic and traffic explicitly requiring QoS (Guaranteed Performance, GP) and we propose three Call Admission Control rules for the GP traffic. The aim is to maximize the Internet Service Provider's overall revenue and to establish a bound over the GP traffic prices. Numerical results are presented to show the good performance of the proposed techniques.

1. Introduction

The exponential growth of the Internet, the pervasive diffusion of the TCP/IP paradigm for the transport of Best Effort (BE) services, and the emergence of Guaranteed Performance (GP) services (as provided by ATM and IP QoS mechanisms), have fostered the development of Internet pricing schemes. Besides differentiating prices according to QoS levels, such schemes should also be capable of achieving "globally optimal" bandwidth allocations and fairness for the users. A number of pricing models have been considered and analyzed in the context of Quality of Service (QoS) guaranteed networks, mainly with respect to the Asynchronous Transfer Mode (ATM) world (see, e.g. [1, 2, 3, 4, 5, 6]). Kelly in [4] bases the pricing scheme in a QoS context on the concept of effective bandwidth (i.e., the bandwidth that is necessary to satisfy QoS requirements). On the other hand, in a Best Effort (BE) context, where the user does not declare QoS parameters and is not subject to a Call Admission Control (i.e., flows are "elastic", as determined by TCP

M. Ajmone Marsan et al. (Eds.): QoS-IP 2003, LNCS 2601, pp. 261-275, 2003.

congestion control, or by TCP-friendly mechanisms at the application level), pricing policies should be different from those adopted in the above mentioned QoS environments. In the literature, models for CAC strategies aimed at maximizing the Internet Service Provider's (ISP) revenue ([19, 20, 21, 25, 26]) and models for the integration between the congestion control and the pricing ([7, 8, 9, 10, 13, 14, 29]) are usually analyzed separately. Our aim is to establish a unified model, able to manage both the GP and the BE traffic. We propose novel price-based CAC rules, whose main goal is the maximization of the ISP's revenue. However, at the same time, we obtain a fair bandwidth allocation for BE users (consistently with the PFP scheme). The papers that deal with the CAC in telecommunications networks propose models for the minimization of a weighted sum of the blocking probabilities of different traffic classes, each of which is associated to a particular reward for the ISP's revenue (see, for example, [19, 20]). The issue of modeling and analyzing the integration of BE and GP traffic pricing is addressed in rather few papers (see e.g. [11, 12]). In [11] an analytical model for the evaluation of the decay of the BE traffic performance as a function of the GP traffic blocking probability is proposed. The aim is to set the prices so that those users who are free to choose between GP or BE service (the so-called mixed users), would be inclined toward the traffic class that results more convenient, so as to reach the best global social welfare. In this paper, we do not decide upon the choice of the traffic class; rather, each incoming flow will be declared to belong either to GP or BE beforehand and we take on-line decisions by using different CAC techniques that are explicitly based on the presence of both traffic categories. The paper is organized as follows. In the next section, we introduce the main optimization problem for the BE environment (Proportional Fairness Pricing) following Low and Lapsley [9] and Kelly [7]. The third section is devoted to the description of our optimization model. Numerical results are presented in the fourth section and conclusions and future work are discussed in the fifth one.

2. Proportional Fairness Pricing

The concept of the Proportional Fairness Pricing was motivated by the desire to incorporate the notion of fairness into the allocation of network resources [3]. The PFP is dedicated to the BE service class, because the bandwidth allocation of a PFP user depends on the current network congestion conditions. In the PFP model the willingness to pay for such bandwidth allocation is also taken into account (see also [6], for an introduction to the congestion price models, or [29]). The PFP is aimed at maximizing the network social welfare and at guaranteeing a fair bandwidth allocation, either in terms of the Max-min fairness [22], adopted by the ATM forum for ABR services [27], or in terms of the "proportional fairness" proposed in [7, 28]. A BE user, in the context of the PFP, can represent a single domestic user, but also a "big one" such as an aggregation of domestic users or a group of LANs (for example a company, one of its branches, or a university campus [23]). Each of these aggregates includes one or more groups of single users that have the same routing path. Their willingness to pay is modeled by a single "big" utility function that can vary during the day ([23, 24]). With a notation that slightly differs from that in [7, 9], we consider a telecommunication network composed by a set J of unidirectional links j, each with capacity c_j. We call "BE user" r a connection established on a specific path consisting of a non-empty subset $J_{BE}(r)$ of J; R_{BE} is the set of active BE users, whose cardinality is denoted by $|R_{BE}|$. We indicate with $R_{BE}(j)$ the subset of BE users that use link j and with $A=\{A_{jr}, j\in J, r\in R_{BE}\}$ the matrix assigning resources to BE users ($A_{jr}=1$ if link j is used by user r, $A_{jr}=0$ otherwise). Moreover, let x_r be the rate of user r and $U_r(x_r):[m_r,M_r]\to\Re$ the utility function of such user, supposed to be strictly concave, increasing and continuously differentiable over $I_r=[m_r,M_r]$; m_r and M_r are the minimum and maximum transmission rates, respectively, required by user r. Such utility function describes the sensitiveness of user r to changes in x_r. In the context of pricing, one can think of $U_r(x_r)$ as the amount of money user r is willing to pay for a certain x_r. Finally, let $\mathbf{c}=[c_j, j\in J]$, $\mathbf{x}=[x_r, r\in R_{BE}]$, $\mathbf{U(x)}=[U_r(x_r), r\in R_{BE}]$ be the aggregate vectorial quantities. The main goal of the ISP can now be stated, consisting of the maximization of the sum of all users' utilities, under the link capacity constraints over the given paths [7, 9, 10, 13, 14]:

The SYSTEM Problem:

$$\mathbf{x}^o = \arg \max_{x_r\in I_r, \forall r\in R_{BE}} \sum_{s\in R_{BE}} U_s(x_s) \quad \text{with } \mathbf{A}\cdot\mathbf{x}\le\mathbf{c} \text{ and } \mathbf{x}\ge\mathbf{0} \tag{1}$$

It is shown in [7, 9, 10, 13, 14] that to formulate a distributed and decentralized solution of the **SYSTEM** problem it is convenient to look at its dual. Following [9], the capacity constraints can be incorporated in the maximization by defining the Lagrangian:

$$L(\mathbf{x},\mathbf{p}) = \sum_{r\in R_{BE}(j)} U_r(x_r) - \sum_{j\in J} p_j\Big(\sum_{r\in R_{BE}(j)} x_r - c_j\Big) = \sum_{r\in R_{BE}(j)} \Big[U_r(x_r) - x_r \sum_{j\in J_{BE}(r)} p_j\Big] + \sum_{j\in J} p_j c_j$$

(2)

where $\mathbf{p}=[p_j, j\in J]$. The objective function of the dual problem is:

$$D(\mathbf{p}) = \max_{x_r\in I_r,\forall r} L(x,p) = \Big\{\sum_{s\in R_{BE}(j)} B_s(p_s) + \sum_{j\in J} p_j c_j\Big\}$$

where:

$$B_r(p^r) = \max_{x_r\in[m_r,M_r]} \big(U_r(x_r) - x_r p^r\big) \tag{3}$$

$$p^r = \sum_{j\in J_{BE}(r)} p_j \tag{4}$$

Thus, the dual problem for (1) is:

$$\mathbf{p}^o = \arg\min_{\mathbf{p}\ge 0} D(\mathbf{p}) \tag{5}$$

The first term of the dual objective function $D(\mathbf{p})$ is decomposed into $|R_{BE}|$ separable subproblems (3) and (4). Define by $x_r(p), r\in R_{BE}$ the solution to (3) for a given p. Such $x_r(p)$ may not be the solution of (1), but if each subproblems (3) takes p^r from the solution of (5), then the primal optimal source rate $\mathbf{x}^o$ of (1) can be computed by each individual user r. The dual problem can be solved using a gradient projection method, where link prices are adjusted in opposite direction to the gradient $\nabla D(\mathbf{p})$:

$$p_j(t+1) = [p_j(t) - \eta \frac{\partial D(p(t))}{\partial p_j}] \tag{6}$$

Noting that $\frac{\partial D(p(t))}{\partial p_j} = c_j - \sum_{r \in R_{BE}(j)} x_r(p(t))$, also the solution of the dual problem (5) can be achieved in a decentralized way. Namely, (6) can be implemented by individual links using only the local information $\sum_{r \in R_{BE}(j)} x_r(p(t))$ that is the aggregate source rate at link j. At each iteration, user r individually solves (3) and communicates the results $x_r(p)$ to each link $j \in J_{BE}(r)$ on its path. Each link j then updates its price p_j according to (6), then communicates the new prices to user r, and the cycle repeats.

An alternative decomposition is proposed in [7]: the **SYSTEM** problem (1) can be decomposed, by separately considering an ISP part and a user part. Let $w_r = p^r x_r$, $r \in R_{BE}$, be the “shadow price” ([28, 29]) for user r, i.e., the price per time unit that user r is willing to pay. Let be $\mathbf{w}=[w_r, r \in R_{BE}]$; each BE user solves the following optimization problem:

The USER$_r$ Problem:

$$w_r^o = \arg \max_{w_r \in p^r I_r} \left[U_r\left(\frac{w_r}{p^r}\right) - w_r \right] \tag{5}$$

subject to: $w_r \geq 0$. In practice, a software agent periodically contracts with the network the bandwidth allocation x_r of each user r, it computes w_r in function of its utility, and sends it to the network ([24]). The ISP, instead, has to solve the following optimization problem:

The NETWORK Problem:

$$\mathbf{x}^o = \arg \max_{x_r \in I_r, \forall r \in R_{BE}} \sum_{s \in R_{BE}} w_s \log x_s \tag{8}$$

Given the vector $\mathbf{w}$ the network computes $\mathbf{x}$, and sends it as a feedback to the flow controller of each user r. Asynchronous distributed approaches to the **NETWORK** problem have been developed in several works (see e.g. [7, 10]). [10] suggests the use of feedback from the real system while in [7] it is shown that modeling flow control dynamics through suitable differential equations [8], can yield to an arbitrarily close approximation to the solution of the problem. More specifically, cost functions are defined for each link j, of the type:

$$\mu_j(t) = \gamma_j\left(\sum_{r \in R_{BE}(j)} x_r(t)\right) \tag{9}$$

where the arguments of the functions $\gamma_j(\cdot)$, $j \in J$, represents the total rates on the link j of the network. Such functions should set a penalty on an excessive use of the resource. The following dynamic system, including pricing and flow control is considered in [7, 8, 10]:

$$\frac{d}{dt}x_r(t) = k_r\left(w_r - x_r(t)\sum_{j \in J_{BE}(r)} \mu_j(t)\right) \tag{10}$$

It can be shown that, under not too restrictive hypotheses on the form of functions $\gamma_j(\cdot)$, the system of differential equations is globally stable and by adapting the prices w_r according to the solutions of the **USER$_r$** problems, the PFP optimum of (1) can be reached ([7]).

3. Guaranteed Performance Service: Call Admission Control and Pricing

In this section we shall consider both BE traffic and of traffic explicitly requiring guaranteed QoS (GP traffic). In this context, our goal is to influence the BE traffic flow control and to apply a CAC to the GP traffic in order to maximize the ISP's overall revenue. A GP user does not request a variable bandwidth allocation as in the PFP scheme, as it is interested in receiving a fixed bandwidth pipe in spite of the network traffic conditions. In this section we investigate the fact that, if BE PFP users are multiplexed together with GP users, a trade-off between GP and BE revenues should be taken into account. A GP user requests a service with strict QoS requirements, such as constraints over the end-to-end mean delay or over the delay jitter or in terms of loss probability of the packets. At the time of a GP call, these requirements can be translated in terms of the equivalent bandwidth necessary to satisfy the GP user's performance requests ([30, 31]). There are sereral methods to calculate the equivalent bandwidth based on analytical models or by means of simulation analysis, possibly also on the basis of on-line measurements (see e.g. [6 or 31]). In the last decade, the telecomunication network traffic has shown a "fractal" statistic behaviour at the packet level ([32]) and this has a dramatic impact over the resources that must be reserved to guarantee a QoS constraints (see e.g. [30, 33]). Also in presence of such statistic behaviour of the sources, it is possible to use analytic models for the calculation of the equivalent bandwidth ([30, 33]). For this reason, in several works, the equivalent bandwidth is the unique parameter used to formulate novel models of CAC or routing in telecommunication networks (see e.g. [6, 21, 19]). We follow this approach and we shall use the bandwidth as the unique QoS metric to manage the GP calls. In the following we shall propose three strategies that influence the amount of resources to allocate to the GP traffic, taking into account the bandwidth allocation of the BE one. In our first strategy, a price is given for all the

GP traffic and we decide whether it is suitable or not to accept the requests of incoming connections on the basis of a revenue derivative comparison. In the second one, all the requests that pass the CAC bandwidth availability check are accepted, but every GP user is assigned a new price. In the third one, we decide whether it is suitable to accept an incoming GP connection if it increases the estimated overall revenue at the end of its duration. In our model of the network, the total ISP's revenue per unit time G (for example expressed in €/s) is formed by the sum of two terms, concerning the GP and the BE traffic, respectively:

$$G = G_{GP} + G_{BE} \tag{11}$$

The revenue concerning the BE users is given by the $\mathbf{w}^o$ vector as the optimal solution of the corresponding proportional fairness pricing problem (1) and (7) (given the vector $\mathbf{x}^o$ from (1), the w_r^o of each user is obtained by (7) as $w_r^o = x_r^o U'(x_r^o)$), while that of the GP traffic is obtained by multiplying the Effective Bandwidth [4] by the assigned charge. The ISP assigns a reserved bandwidth y_r to each user $r \in R_{GP}$, where we denote by R_{GP} the set of active GP users (i.e., all of the GP connections accepted in the network and in progress). Every user r, $r \in R_{GP}$, pays an amount b_r per unit of sent GP traffic data per unit time (e.g., b_r could be €/Mbps per minute). Each time a new GP call asks to enter the network (i.e., a new GP user $\tilde{r}$ wants to start up a connection), the network is asked for a new amount of bandwidth $y_{\tilde{r}}$. We suppose the BE traffic to be regulated by a flow control mechanism such as in Section 2, so, before the new GP user $\tilde{r}$ enters the network, the rates x_r and the prices per unit time w_r of the current BE users $r \in R_{BE}$, have reached the stationary optimal values x_r^o and w_r^o. If the new bandwidth $y_{\tilde{r}}$ will be reserved for the new GP user $\tilde{r}$, the BE traffic rates x_r^o and price w_r^o, $r \in R_{BE}$, will move to the new optimal values $\tilde{x}_r^o$ and $\tilde{w}_r^o$ according to (1) and (7), where the capacity constraints in (1) becomes:

$$\mathbf{A} \cdot \mathbf{x} \le \tilde{\mathbf{c}} \tag{12}$$

where $\tilde{\mathbf{c}} = [\tilde{c}_j, j \in J]$ is the residual capacity matrix, with $\tilde{c}_j = c_j - \sum_{r \in R_{GP}, j \in r} y_r$ the residual capacity (capacity not reserved to GP traffic) of link j.

3.1 First CAC Control Rule: CACPricing1

It is clear that the revenue's rates with the traffic change. If a new GP user $\tilde{r}$ enters the network, the GP revenue rate, G_{GP}, increases, while the BE traffic rates, decrease (less bandwidth available for BE), so the BE revenue rate, G_{BE}, decreases, too. In this respect, a possible Acceptance Control Rule for the requests of increasing the GP traffic reserved bandwidth is to accept the new GP bandwidth reservation only if the total revenue rate increases with respect to the current situation. So, in our first proposal, we use the revenue rate to decide whether to accept a new GP request. In particular, let y_r, $r \in R_{GP}$, be the current GP bandwidth reservations, and $y_{\tilde{r}}$ a new bandwidth request for the new GP user $\tilde{r}$ with associated tariff $b_{\tilde{r}}$. The ISP accepts the new bandwidth request if:

$$\sum_{r \in R_{GP}} b_r y_r + b_{\tilde{r}} y_{\tilde{r}} + \sum_{r \in R_{BE}} \tilde{w}_r^o \geq \sum_{r \in R_{GP}} b_r y_r + \sum_{r \in R_{BE}} w_r^o \tag{13}$$

where $\tilde{w}_r^o$ represents the optimal values of the BE price in the presence of the new GP allocation $y_{\tilde{r}}$. This results in the following CAC rule:

$$b_{\tilde{r}} y_{\tilde{r}} \geq \sum_{r \in R_{BE}} \left(w_r^o - \tilde{w}_r^o \right) \tag{14}$$

3.2 Second CAC Control Rule: VariableGPPrice

If the GP prices can be freely assigned by the ISP every time a connection is accepted, it is possible to assign them in order to leave the total revenue derivative unchanged:

$$\sum_{r \in R_{GP}} b_r y_r + b_{\tilde{r}} y_{\tilde{r}} + \sum_{r \in R_{BE}} \tilde{w}_r^o = \sum_{r \in R_{GP}} b_r y_r + \sum_{r \in R_{BE}} w_r^o \tag{15}$$

Imposing condition (15) leads to the following pricing scheme for GP users:

$$\bar{b}_{\tilde{r}} = \sum_{r \in R_{BE}} \left(w_r^o - \tilde{w}_r^o \right) \Big/ y_{\tilde{r}} \tag{16}$$

In a sense, the "VaribleGPPrice" strategy can be interpreted as a way to fix a lower bound to the GP prices. Any $b_{\tilde{r}} > \bar{b}_{\tilde{r}}$ would augment the ISP's global revenue.

For both strategies presented so far, every time the CAC block acts, it is necessary to foresee the revenue $\tilde{G}_{BE} = \sum_{r \in R_{BE}} \tilde{w}_r^o$ which will be obtained in the future on the BE traffic after the bandwidth reallocation. The basic idea is to use the **SYSTEM** problem (1) to calculate the new value of the **x** vector after the bandwidth reallocation; then, using the **USERr** problem (7), it is possible to calculate the new BE prices and the new BE revenue and evaluate the total revenue after the possible bandwidth reallocation. (1) and (7) are mathematically fairly tractable ([7]), so it seems they can be actually applied on-line by the CAC block during the network evolution. "CACPricing1" and "VariableGPPrice" are only based on the current PFP bandwidth and prices allocations. The PFP model does not consider the dynamic nature of the network ([25]), namely, call arrivals and departures are not taken into account. So, in (13) and (15) only the revenues per unit time and not the total revenues are compared. The model proposed is clearly based on a centralized approach. Even if it is raccomandable to propose decentralized CAC techniques (see for example [20, 34]), centralized CAC models are not novel in the CAC literature, see for example [35, 36]. In our model, it is necessary that each node of the network is able to know tha "state" of the BE traffic, i.e. the current BE users' routing paths and the utility functions. Only in this way it is possible to apply correctly the previously proposed algorithms. To maintain in each node a perfect knowledge of the network, it is necessary to establish something similar to the Link State (LS) information exchange of the QoS routing in a MPLS enviroment. In LS routing, network nodes should be aware of the state of the links, possibly located several hops away. This calls for a flooding periodic exchange of LS information, which contributes extra traffic to the network. Among the cost factors of QoS routing, the cost of LS exchange is the dominant contributor and can severely limit the scalability of the QoS routing. Our model needs a periodic exchange of Node State (NS) information concerning the state of the BE users actually present in the network. This because, at the time of the CAC of an incoming GP call, it is necessary to evaluate the impact of the new GP bandwidth reservation that can influence, according to the Proportional Fairness Pricing scheme, the bandwidth reservation of all BE users. Due to such NS information exchange, it is necessary to further investigate the possibility of decentalize the proposed CAC strategies in order to guarantee a higher scalability, for example according to a model based on a team theory framework ([34]). Futhermore, in order to avoid periodic exchange of NS information it could be possible to foresse the variability of the BE traffic demands during a certain period and to use on-line this knowledge to apply the proposed CAC and possibly to update it by an on-line estimate of the current network traffic conditions ([23]).

3.3 Third CAC Control Rule: Accept the Incoming GP Requests if, at the End of Its Duration, It Increases the Total Revenue

If the lengths of the connections were explicitly considered, the ISP's decisions would be different from those induced by "CACPricing1" and "VariableGPPrice". Taking into account the length of the connections would lead to maximize the total revenue, rather than the revenue rate. In order to accomplish this task, it would be necessary to identify all the possible events that happen during each new incoming GP connection lifetime and to solve the **SYSTEM** problem (1) for each time interval between them. In the following, we propose a heuristic technique that considers this additional information and takes into account the possible future opening of new BE and GP connections. Again, let $\tilde{r}$ be the new incoming user asking, at time $\tilde{t}$, for a GP connection, expected to terminate at $\tilde{T}$. Suppose $\tilde{r}$ to be subject to the CAC based on the total revenue comparison and let $G_{TOT}(\tilde{t},\tilde{T})$ and $\tilde{G}_{TOT}(\tilde{t},\tilde{T})$ be the total revenues the ISP expects to obtain in $[\tilde{t},\tilde{T}]$ in the case $\tilde{r}$ is refused or accepted, respectively . To calculate the two terms $G_{TOT}(\tilde{t},\tilde{T})$ and $\tilde{G}_{TOT}(\tilde{t},\tilde{T})$ it is necessary to break $[\tilde{t},\tilde{T}]$ in all of the sub-intervals where there are no changes in the **w** and **x** vectors. To take into account also the arrival of new requests from GP and BE traffic, we have used a heuristic approach based on Montecarlo simulation: when a new request of GP connection occurs, we generate n different simulation runs of length $[\tilde{t},\tilde{T}]$. At the end of each simulation we calculate the overall revenue, considering the terminations of the BE and GP connections within $[\tilde{t},\tilde{T}]$ and the arrivals of new BE and GP connections within the same time interval. In this way, all of the bandwidth reallocations occurring during the new connection are considered. In the n different simulations, as to the new GP connections starting after $\tilde{t}$, a CAC strategy based only on the bandwidth availability is applied and an estimate of the expectation of the revenue is computed. Two situations are considered: the case in which the new GP connection is refused (G_{TOT}), and the case in which new GP connection is accepted ($\tilde{G}_{TOT}$). The final choice is to accept the incoming GP user $\tilde{r}$ only if it increases the estimate of the mean value of the total revenue obtained at the end of the recursive procedure ($\tilde{G}_{TOT} \geq G_{TOT}$). This strategy (called "CACPricing2" in the following), as the "CACPricing1", is a static pricing policy and it is related to the family of the so-called "Receding Horizon" techniques. A performance index (the revenue) that is referred to a finite temporal window (the duration of the new GP connection) is maximized. The perfect information on the termination instants of all the connections that will end during the new GP connection is exploited and averaging is performed on the arrivals of the new GP and BE calls. It is clear that this approach is very time consuming and cannot be applied in a real scenario where the CAC block acts on-line, but it could be very useful to test the performance of the previous techniques, where only an optimization on the revenue rate is applied and the events that can occur during the new GP connection life time are ignored.

4. Numerical Results

We have developed a simulation tool that describes the behaviour of the network at call level to verify the performance of the proposed price-based CAC mechanisms. The simulator does not model the packet level. The COST 239 experimental network [15] (depicted in Fig. 1) is utilized for the tests; it is composed by 20 links and by 11 nodes. We consider a subset of 10 active routes, where each active route can generate both BE and GP traffic connections:

Route 1: { 0, 1, 5, 7, 10 }, Route 2: { 4, 8, 12, 19, 15 }, Route 3: { 14, 17, 16 }, Route 4: { 2, 5, 6, 12 }, Route 5: { 3, 8, 9, 10 }, Route 6: { 5, 7, 9, 11 }, Route 7: { 3, 4, 13 }, Route 8: { 1, 19, 12 }, Route 9: { 17, 18, 11 }, Route 10: { 3, 4, 5, 7, 10, 16 }

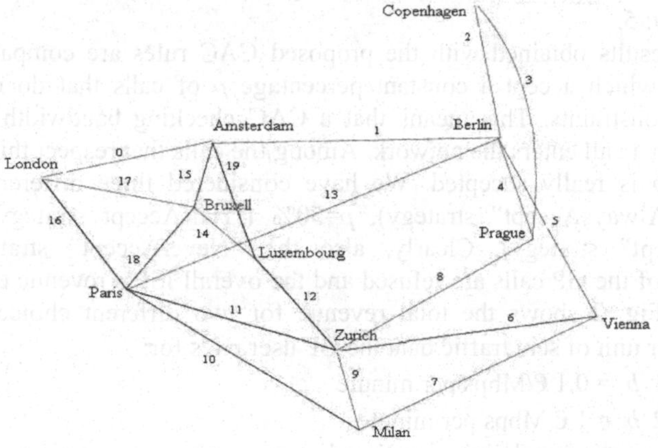

Fig. 1 Topology of the test network.

Our test scenarios are very close to the simulations tests of some among the most relevant papers on CAC and Pricing in telecommunication networks (see for example [19, 20] for the pure CAC and [23, 25, 26] for the Pricing). We have imposed a probability distribution over all of the significant variables of the problem: interarrival times of the BE and GP users, required bandwidth and utility functions, to produce variable traffic conditions. The simulations performed fall in the category of the so-called "finite time horizon" or "terminating" simulations [16].

4.1 The Static Scenario

Connections are generated following a Poisson distribution with mean rate $\lambda_r^{(BE)}$ and $\lambda_r^{(GP)}$ for each route r, for BE and GP traffic, respectively. The call durations follow an exponential distribution with mean value $1/\mu_r^{(BE)}$ and $1/\mu_r^{(GP)}$. The bit rate of the BE traffic is controlled according to the Proportional Fairness Pricing scheme using eq. (1) and eq. (7). Every GP call requires an amount of bandwidth generated with exponential distribution with mean value γ. We use the following utility function for the BE traffic:

$$U(x) = \alpha\sqrt{x} \tag{17}$$

where the parameter α is generated with an exponential distribution with mean $\overline{\alpha} = 1$.

The simulation data are summarized in the following:

- $\lambda_r^{(BE)} = \lambda_r^{(GP)} = \lambda = 10$ calls per minute $\forall\ r \in \{1, \ldots, 10\}$
- $1/\mu_r^{(BE)} = 1/\mu_r^{(GP)} = 1/\mu = 1$ minute $\forall\ r \in \{1, \ldots, 10\}$
- $c_j = c = 5$ Mbps (link capacity), $\forall\ j \in \{0,\ldots,19\}$
- $\gamma = 1$ Mbps (average bandwidth required by a GP call)
- Time of simulation: 100 minutes
- n: number of simulations of the procedure used by the "CACPricing2" strategy: 5.

The results obtained with the proposed CAC rules are compared with fixed CAC rules, which accept a constant percentage *p* of calls that do not violate the bandwidth constraints. This means that a CAC checking bandwidth availability is applied when a call enters the network. Among the calls that respect this rule, only the percentage *p* is really accepted. We have considered three different percentages: *p*=100% ("AlwaysAccept" strategy), *p*=50% ("HalfAccept strategy") and *p*=0% ("NeverAccept" strategy). Clearly, also the "NeverAccept" strategy generates revenue: all of the GP calls are refused and the overall ISP's revenue comes from the BE traffic. Fig. 2 shows the total revenue for two different choices of the price *b* (money per unit of sent traffic data the GP user pays for:

- Case 1 $b = 0.1$ €/Mbps per minute
- Case 2 $b = 1$ €/Mbps per minute.

These two price choices are aimed at evaluating, in a static scenario, the two typical revenue situations of the system. As we can see from Fig. 1, the best fixed strategy in terms of achieved revenue are: the "AlwaysAccept" strategy for b=1.0 and the "NeverAccept" strategy for b=0.1. In fact, in the case b=0.1 the BE traffic contributes for the largest part of the total revenue; on the other hand, GP traffic represents the largest part of the total revenue in the case b=1.0. When the b is very low (e.g., b=0.1), the maximum revenue is obtained when the GP traffic is not accepted ("NeverAccept" technique); if the value of b is higher (e.g., b=1.0), the best performance is obtained using the "AlwaysAccept" technique. This behaviour is due to the fact that, if the price paid by the GP users is low, it is convenient to refuse all of the GP calls, leaving all the bandwidth to the BE traffic. On the other hand, in the opposite situation, giving all the bandwidth to the GP traffic is the most convenient choice in terms of the ISP's revenue. The "CACPricing1" technique offers a good level of performance in both cases, showing a good adaptation to the GP price changes. The behaviour of the "VariableGPPrice" technique does not depend on b; the tariff of every connection is dynamically decided on the basis of formula (16). The obtained performance turns out to be the best one in absolute in the b=0.1 situation, while it appears quite poor if b=1.0. This shows that the b=0.1 value is too low (GP users pay less than they would have to), while b=1.0 is slightly too high. If the revenue performance of CACPricing1 is compared with that obtained by "CACPricing2", the latter seems not to be able to produce significant improvements, despite the greater computational complexity.

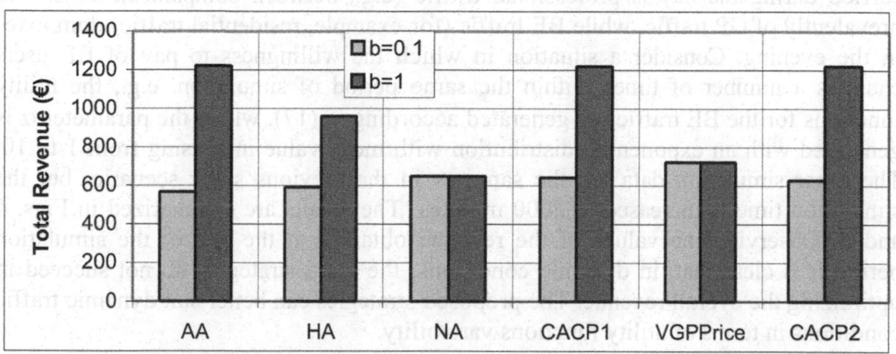

Fig. 2 Static simulation scenario #1. Total revenue [€] with two values of GP prices.

Concerning the Blocking Probability (depicted in Fig. 3), "CACPricing1" offers good performance if b=1.0. In the b=0.1 case, instead, more than 67% of the GP connections are refused. However, it is important to remind that from the point of view of the revenue, in this situation, the best solution was "NeverAccept": "CACPricing1" obtains similar revenue, but with a much lower blocking probability. In terms of the GP blocking probability "CACPricing2" offers lower values than the "CACPricing1" ones. So, we can see that the greater computational complexity of the "CACPricing2" has more impact on the blocking probability than directly on the revenue. As obvious, the technique "VariableGPPrice" has the same blocking probability as the "AlwaysAccept" strategy, because it accepts all the connections that pass the first CAC level based on the bandwidth availability.

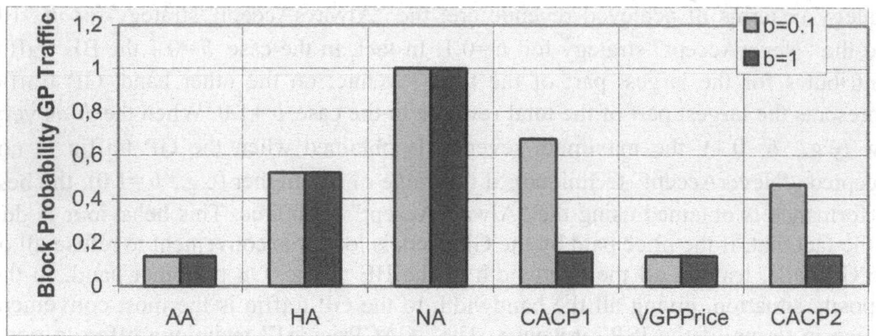

Fig. 3 Static simulation scenario #1. GP traffic Blocking Probability, with two values of GP prices.

4.2 The Dynamic Scenario

Now we consider a situation in which the volume of the traffic and the users' behaviour can change within the same simulation. In fact, it is well known that the traffic profile depends on the period of the day (see e.g. [17, 18]), most of the traffic carried during the day is professional traffic (e.g., between companies), consisting prevalently of GP traffic, while BE traffic (for example, residential traffic) dominates in the evening. Consider a situation in which the willingness to pay of BE users changes a number of times within the same period of simulation, e.g., the utility functions for the BE traffic are generated according to (17), where the parameter α is generated with an exponential distribution with mean value increasing from 1 to 10. The other simulation data are the same as in the previous static scenario, but the simulation time is increased to 2000 minutes. The results are summarized in Figs. 4 and 5. Observing the values of the revenue obtained at the end of the simulation period it is clear that, in dynamic conditions, the fixed strategies do not succeed in optimizing the overall revenue. The proposed strategies can better suit dynamic traffic conditions in terms of utility functions variability.

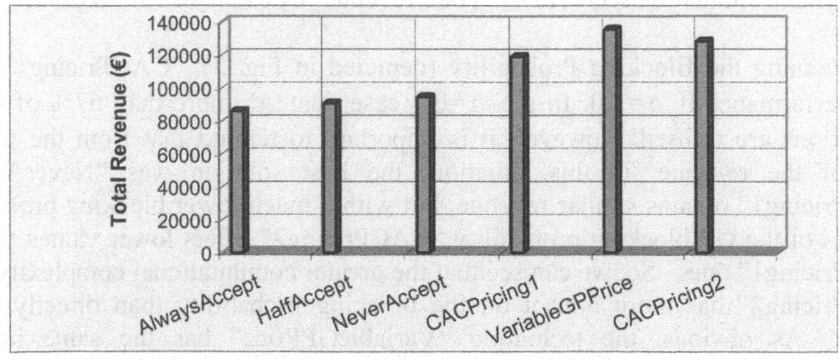

Fig. 4 Dynamic simulation scenario #1. Total Revenue (€).

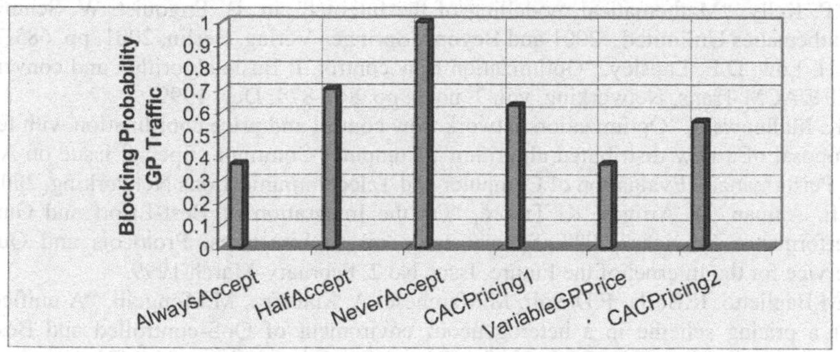

Fig. 5 Dynamic simulation scenario #1. GP Traffic Blocking Probability.

5. Conclusions and Future Work

In this paper we have proposed three price optimization mechanisms that operate in networks with both Best Effort and Guaranteed Performance traffic types. We use a decentralized flow control method (Proportional Fairness Pricing) for the Best Effort traffic and three Call Admission Control rules for Guaranteed Performance calls. The simulation results presented show that all of the proposed mechanisms adapt well to traffic changes in order to maintain the best global revenue for the Internet Service Provider. Future work could include the analysis of the dynamics of the Proportional Fairness Pricing optimum in the presence of fluctuations in the bandwidth allocation of the BE traffic and how this can influence the revenue forecast applied in our Call Admission Control rules. A model for the maximization of a unified social welfare for both the GP and the BE users is under investigation, too.

References

[1] C. Courcoubetis, V.A. Siris, G. Stamoulis, "Integration of pricing and flow control for Available Bit Rate services in ATM networks", Proc. IEEE Globecom, London, UK, 1996, pp. 644-648.

[2] L.A. DaSilva, "Pricing for QoS-enabled networks: a survey", IEEE Commun. Surveys, 2nd Quart. 2000, pp. 2-8.

[3] M. Falkner, M. Devetsikiotis, I. Lambadaris, "An overview of pricing concepts for broadband IP networks", IEEE Commun. Surveys, 2nd Quart. 2000, pp. 2-8.

[4] F.P. Kelly, "Charging and accounting for bursty connections", in: L.W. McKnight, J.P. Bailey, Eds., Internet Economics, MIT Press, Cambridge, MA, 1996.

[5] J. Murphy, L. Murphy, E. Posner, "Distributed pricing for embedded ATM networks", Proc. IFIP Conf. Broadband Commun. (BB-94), Paris, France, March 1994.

[6] J.Walrand, P. Varaiya, High-Performance Communication Networks, 2nd Ed., Morgan-Kaufmann, San Francisco, CA, 2000.

[7] F.P. Kelly, A.K. Maulloo, D.K.H. Tan, "Rate control for communication networks: shadow prices, proportional fairness and stability", J. Operat. Res. Soc., vol. 49, no. 3, pp. 237-252, 1998.

[8] F.P. Kelly, "Mathematical modelling of the Internet", in: B. Engquist, W. Schmid, Eds., Mathematics Unlimited - 2001 and Beyond, Springer-Verlag, Berlin, 2001. pp. 685-702.
[9] S.H. Low, D.E. Lapsley, "Optimization flow control, I: Basic algorithm and convergence", IEEE/ACM Trans. Networking, vol. 7, no. 6, pp. 861-874, Dec. 1999.
[10] K. Malinowski, "Optimization network flow control and price coordination with feedback; proposal of a new distributed algorithm", Computer Commun., Special Issue on Advances in Performance Evaluation of Computer and Telecommunications Networking, 2002.
[11] E. Altman, D. Artiges, K. Traore, "On the Integration of Best-Effort and Guaranteed Performance Services", ETI, Special Issue on Architectures, Protocols and Quality of Service for the Internet of the Future, Issue No.2, February-March 1999.
[12] M.Baglietto, R.Bolla, F.Davoli, M.Marchese, A. Mainero, M.Mongelli, "A unified model for a pricing scheme in à heterogeneous enviroment of QoS-controlled and Best Effort connections", Proc. SPECTS 2002, San Diego, CA, July 2002, pp. 414-421.
[13] S. Low "Optimization Flow Control with on-line measurement or multiple paths", 16th International teletraffic congress, Edinburgh UK, June 1999
[14] S.Low "Optimization flow control with on-line measurement" Proceedings of the ITC, volume 16 June, 1999.
[15] Cost 239 Ultra high capacity optical transmission networks, final report of Cost project 239, ISBN 953-184-013-X, 1999.
[16] K. Pawlikowski. "Steady state simulation of queueing processes: A survey of basic problems and solutions". ACM Computing Surveys, June 1990, 22(2): 123-170.
[17] Ben-Ameur W, Kerivin H., "New economical virtual private networks", Communications of the ACM, 2001.
[18] Ben-Ameur W, Gourdin E, Liau B, Michel N. "Dimensioning of Internet networks", Proc. of DRCN2000, pages 56-61, Munich, April 2000.
[19] P. Marbach, O. Milhatsch, J. Tsitsiklis "Call Admission Control and Routing in Integrated Services Networks using Neuro-Dynamic Programming" Submitted to IEEE Journal on Selected Areas in Communications 1999.
[20] K. Gokbayrak, C. Cassandras "Adaptive Call Admission Control in Circuit-Switched Networks", subitted to IEEE trans. On Automatic Control September 2000.
[21] K. Ross "Multiservice Loss models for Broadband Telecommunication Networks", Berlin: Springer, 1995.
[22] D. Bertsekas, R. Gallager Data Networks 1987, Prentice-Hall.
[23] M. Malowidzki, K. Malinowski, "Optimization Flow Control in IP Networks: Distributed Pricing Algorithms and Reality Oriented Simulation", SPECTS 2002, San Diego, CA, July 2002, pp. 403-413.
[24] C. Courcoubetis, G. D. Stamoulis, C.Manolakis, and F.P.Kelly "An Intelligent Agent for Optimizing QoS-for-Money in Priced ABR Connections", Telecommunication Systems, Special Issue on Network Economics, 1998.
[25] X. Lin, B. Shroff "Pricing-Based Control of Large Networks", Evolutionary Trends of the Internet 2001 Tyrrhenian International Workshop on Digital Communications, IWDC 2001, Taormina, Italy.
[26] I.C. Paschlidis, J.N. Tsitsiklis "Congestion-Dependent Pricing of Network Services", IEEE/ACM Transactions on Networking, vol. 8, pp. 171-184, April 2000.
[27] Traffic management specification, version 4.0 Technical Report AF-TM-0056.000, ATM Forum Traffic Management Working Group, April 1996.
[28] F.P. Kelly, "Charging and rate control for elastic traffic", European Transactions on Telecommunications, volume 8 (1997) pages 33-37
[29] H.Yaiche, R. R. Mazumdar, C. Rosenberg "A Game Theoretic Framework for Bandwidth Allocation and Pricing in Broadband Networks", IEEE/ACM Transactions On Networking, vol. 8, NO.5, October 2000.
[30] J. M. Pitts and J. A. Schormans Introduction to IP and ATM Design and Performance, Second Edition, Wiley Ed. 2000.

[31] H. J. Chao, X. Guo, "Quality of Service Control in High-Speed Networks," John Wiley & Sons, New York, 2002.
[32] W.E. Leland, M.S. Taqqu, W. Willinger, D.V. Wilson, "On the Self-Similar nature of Ethernet traffic," IEE/ACM Transaction on Networking, 1994, 2(1): 1,15.
[33] B. Tsybakov, N.D. Georganas, "Overflow Probability in ATM Queue with self-similar Input Traffic," Proceedings IEE International Conference Communication (ICC'97), Montreal, Canada, 1997.
[34] M. Baglietto, T. Parisini, R. Zoppoli, "Distributed-Information Neural Control: The case of Dynamic Routing in Traffic Networks," IEEE Transactions on Neural Networks, Vol. 12, NO. 3, May 2001.
[35] C. M. Barnhart, J. E. Wieselthier, A. Ephremides, "Admission control policies for multihop wireless networks," Wireless Networks, 1(4):373-387, 1995.
[36] N. Celandroni, F. Davoli, E. Ferro, "Static and Dynamic Resource Allocation in a Multiservice Satellite Network with Fading," International Journal of Satellite Communications, Special Issue of Quality of Service (QoS) over Satellite Network, to appear.

TCP Smart Framing: A Segmentation Algorithm to Improve TCP Performance

Marco Mellia, Michela Meo, and Claudio Casetti*

Dipartimento di Elettronica, Politecnico di Torino, 10129 Torino, Italy
{mellia, michela, casetti}@mail.tlc.polito.it

Abstract. In this paper we propose an enhancement to the TCP protocol, called *TCP Smart Framing*, or TCP-SF for short, that enables the Fast Retransmit/Recovery algorithm even when the congestion window is small. TCP-SF is particularly effective for short-lived flows, as most of the current Internet traffic is. Without modifying the TCP congestion control based on the additive-increase/multiplicative-decrease paradigm, TCP-SF adopts a novel segmentation algorithm: while Classic TCP starts sending one segment, a TCP-SF source is allowed to send an initial window of 4 smaller segments, whose aggregate payload is equal to the connection's MSS. This key idea can be implemented on top of any TCP flavor, from Tahoe to SACK, and requires modifications to the server behavior only.

Analytical results, simulation results, as well as testbed implementation measurements show that TCP-SF sources outperforms Classic TCP in terms of completion time.

1 Introduction and Work Motivation

Balancing greediness and gentleness has always been the distinctive feature of congestion control in the TCP protocol [1]. Mindful of the presence of other traffic sharing the same network resources, TCP tries to grab as much bandwidth as possible, eventually causing congestion and data loss. Data lost by TCP is used as congestion signal, and cause the source to slow down its transmission rate. Thus, lost data can actually be seen as bandwidth used to control and regulate the network, since every segment the network discards is an indication that a TCP source has been congesting the network and should temporarily back off.

This scheme has been successfully employed over the years, while the traffic pattern has shifted from long file transfers and short, persistent connection, typical of terminal-emulation traffic, to the "Click-and-Run" paradigm found in Web interactions [2].

In this paper, we investigate a new approach to data segmentation in the early stages of Slow Start that i addresses the nature of today's Internet traffic: short, spotty client-server interactions between a Web client and a Web server. We will refer to this variant of TCP as "TCP Smart Framing", or TCP-SF for short. TCP-SF was first proposed in [3]; here, we further investigate TCP-SF performance by proposing an analytical model

* This work was supported by the Italian Ministry for University and Scientific Research under the PlanetIP project and by the Center for Multimedia Radio Communications (CERCOM).

M. Ajmone Marsan et al. (Eds.): QoS-IP 2003, LNCS 2601, pp. 276–291, 2003.

of its behavior and by showing results collected from testbed measurements under real Web traffic.

As will be detailed below, TCP-SF increases the number of segments transmitted by the TCP source, without increasing the amount of application data actually sent in the congestion window. This will be done whenever the congestion window is "small", i.e., at the beginning of each Slow Start phase, and in particular at the flow startup.

The main observation is that TCP's congestion control is only marginally driven by the rate at which the *bytes* leave the source but, rather, by the rate at which *segments* (and their respective ACKs) are sent (or received) at the source.

TCP infers that a segment is lost whenever one of the following two events occurs: a Retransmission Time Out (RTO) expiration, or the arrival of three duplicate ACKs that triggers the Fast Retransmit (FR) algorithm. Of these two events, RTO is the most undesirable one as the RTO period is usually much larger than the Round Trip Time (RTT) Indeed, regardless of the actual amount of bytes transmitted, a coarse RTO expiration can be prevented only if enough segments are sent in the transmission window (i.e., at least three more following the lost segment). This situation can occur only if i) the congestion window is larger that 4 MSS (Maximum Segment Size) and ii) the flow is long enough to allow the transmission of at least 4 back-to-back segments (i.e., it is not a so-called *short-lived* flow).

Also, it should be pointed out that repeatedly forcing a short-lived connection into RTO often results in excessive penalty for the connection itself, that would otherwise be finished in few more segments, rather than in actual network decongestion. Since today's Internet traffic is heavily represented by short-lived connections [2], the need is felt to address their requirements in the design of TCP's congestion control.

While Classic TCP[1] starts sending one segment, in our scheme a TCP-SF source is allowed to send N_{sf} segments, whose aggregate payload is equal to the MSS associated to the connection. Thus, the resulting network load is, byte-wise, the same of a Classic TCP connection (except for the segmentation overhead). The ACK-driven window increase law employed by TCP-SF affects the amount of data per segment, rather than the number of segments, until a threshold is reached, after which TCP-SF resumes the classic behavior. The Classic TCP algorithms (Slow Start, Congestion Avoidance, Fast Retransmit, Fast Recovery) are not otherwise affected. However, the modification introduces a number of key advantages:

- the lengthy first-window RTO (set to 3 seconds) is no longer the only outcome if a loss occurs at the onset of a connection;
- when Delayed ACKs are employed and the congestion window is 1 segment large, the receiver has not to wait for 200 ms before generating an ACK; several current TCP implementations start a connection with a window of 2 segments, a widely-employed acknowledged workaround to the Delayed ACK initial slowdown;
- the RTT estimate, which is updated upon the reception of every ACK, and is used to set the retransmission timer, improves its accuracy early on, thanks to the increased number of returning ACKs in the first window already;

[1] unless otherwise specified, by "Classic" TCP we refer to any TCP version currently implemented in standard TCP stacks (i.e., TCP Tahoe [4], TCP Reno [5], TCP NewReno [6], TCP SACK [7])

- short-lived flows, for which the completion time is paramount, are less likely to experience a coarse RTO expiration, since the number of transmitted segments grants a bigger chance of triggering FR;
- shorter segments can exploit pipelining transmission, completing the transfer in a shorter time because of the store-and-forward mechanism at the routers; this is especially useful in slow links;
- not requiring any contribution from the receiver, the scheme can quite easily be deployed on a source-only basis; furthermore, it can equally benefit well-established Classic TCP flavors, such as TCP Reno, NewReno, SACK, and also works coupled with ECN (Early Congestion Notification).

2 TCP Smart Framing

As is well known, when the TCP congestion window size (*cwnd*) is smaller than four segments, TCP has no other means to recover segment losses than by RTO expiration. Indeed, since ACK transmission is triggered by the reception of a segment, the receiver has no chance to send three duplicated ACKs when the congestion window size is smaller than four segments. Being the time to recover a loss by RTO expiration much longer than the time needed by FR, this behavior deteriorates TCP performance, especially when connections are short-lived. In particular, when the flow length is shorter than 7 segments (i.e., about 10 Kbytes using a 1460-bytes MSS), there are no chances for the transmitter to trigger a FR. Note that if the Delayed-ACK option is implemented, the flow must be longer than 10 segments.

In the scheme we propose, *TCP-SF*, we enhance TCP behavior in the operating region where RTO is the only way to recover losses (i.e., when $cwnd < 4MSS$) making FR possible, as for example at the beginning of each Slow Start phase. The region in which we enhance TCP behavior is shadowed in Figure 1 which shows the dynamics of *cwnd*. This region is commonly known as the *small window regime*.

TCP-SF is based the following idea: increasing the upstream flow of ACKs by sending downstream a larger number of segments whose size is smaller than the MSS. While maintaining unchanged the amount of data injected into the pipe, a larger number of segments received at the other end triggers a larger number of ACKs in the backward channel and thus a larger probability that the transmitter can recover losses without waiting for the RTO to expire. In other words, this procedure gives the transmitter the chance to obtain more feedback about what is happening at the receiver. Increasing the number of ACKs will therefore enable FR when the congestion window is smaller than four segments (in particular, any flow larger than $3(MSS/N_{sf})$ will benefit from this) and it will also help the RTT estimation algorithm to converge quickly to the correct values of the RTO, thus alleviating the first RTO penalty of 3 seconds.

We now illustrate our approach by means of an example, using $N_{sf} = 4$, as will be done in the rest of the paper. Upon the onset of a connection, the congestion window size is equal to one segment, i.e., MSS bytes. If this segment is lost, the transmitter gets no information back, and waits for the RTO to expire before sending the segment again; this behavior can be observed in the right part of Figure 1 in the diagram labeled by "Classic TCP". Now, if instead of sending one MSS bytes long segment, the transmitter sends four segments whose size is $MSS/4$, the loss of the first segment can be

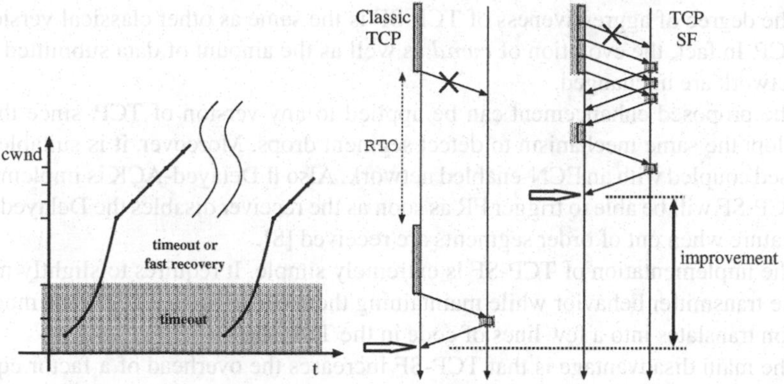

Fig. 1. *cwnd* growth and small window region (on the left). Error recovery for Classic TCP and TCP-SF when congestion window size is equal to MSS bytes (on the right).

recovered by FR after the reception of three duplicated ACKs, as shown in the right part of Figure 1 in the diagram labeled by "TCP SF".

Since the enhancement introduced by TCP-SF is needed when only RTO can be used to detect segment losses, we only activate the smart framing option when the window size is smaller than a given threshold, denoted by $cwnd_{\min}$, and we switch back to Classic TCP behavior (i.e., segment size equal to MSS) as soon as the congestion window is large enough to enable FR. Of course, this behavior applies both at connection start and whenever the congestion window is shrunk to a size smaller than $cwnd_{\min}$.

Let us now elaborate a bit on the small-segment option. We define $cwnd_0$ as the initial congestion window size in bytes [2]. We consider two possible behaviors:

- **Fixed-Size (FS-) TCP-SF**. When $cwnd < cwnd_{\min}$, the segments size is equal to $MSS\frac{cwnd_0}{cwnd_{\min}}$; otherwise, the segment size is equal to MSS.
- **Variable-Size (VS-) TCP-SF**. The initial segment size is equal to $MSS\frac{cwnd_0}{cwnd_{\min}}$; then, while $cwnd < cwnd_{\min}$, the segment size is increased by a factor α, until the segment size is equal to MSS. The value of α can be determined by imposing that the amount of data sent by the TCP-SF is equal to the one sent by a Classic TCP. After some calculation, we obtain $\alpha^5 = 4$; $\alpha = 1.32$.

Given that the FR algorithm is triggered by the reception of 3 DUP-ACKs, we suggest that $cwnd_{\min}$ be set to $4MSS$.

One advantage of using TCP-SF with fixed-size segments relies in its simplicity: two segment sizes only are possible, either MSS or $MSS\frac{cwnd_0}{cwnd_{\min}}$. However, the overhead introduced increases with *cwnd*. On the contrary, when using variable-size segments, TCP-SF keeps the overhead constant but a more careful implementation is required to deal with variable-size segments.

Let us point out and summarize some critical issues related to the implementation of TCP-SF.

[2] We consider a TCP implementation where initial window (IW) and loss window (LW), as defined in [1], take the same value.

- The degree of aggressiveness of TCP-SF is the same as other classical versions of TCP. In fact, the evolution of *cwnd* as well as the amount of data submitted to the network are unchanged.
- The proposed enhancement can be applied to any version of TCP, since they all adopt the same mechanism to detect segment drops. Moreover, it is suitable to be used coupled with an ECN-enabled network. Also if Delayed-ACK is implemented, TCP-SF will be able to trigger FR as soon as the receiver disables the Delayed-ACK feature when out of order segments are received [8].
- The implementation of TCP-SF is extremely simple. It requires to slightly modify the transmitter behavior while maintaining the receiver unchanged. This modification translates into a few lines of code in the TCP stack.
- The main disadvantage is that TCP-SF increases the overhead of a factor equal to the segment size reduction factor; i.e., using four segments per MSS, the TCP-SF overhead is four times the Classic TCP overhead. In particular, when no losses occur, FS-TCP-SF will send 28 small-size segments before switching back to large-size segments. VS-TCP-SF, on the contrary, sends 12 segments. Instead, Classic TCP always sends 7 segments. It should also be pointed out that a larger number of segments can nominally slow down router operations.

3 Throughput Gain

In this Section we analytically evaluate the gain obtained by employing TCP-SF instead of other classic versions of the protocol. The gain is expressed in terms of throughput and is derived given the network conditions, i.e., given a value of the average round-trip time and of the segment loss probability.

We restrict our analysis to the working region of TCP where TCP-SF acts on segment size in order to improve the protocol performance; i.e., the small window regime. The region is defined by the window size ranging between 1 and $cwnd_{\min}$. Out of this region, all the TCP versions behave the same, and no further gain can be achieved by TCP-SF. For simplicity, we assume that the congestion window grows always according to the Slow Start algorithm, and that the segment transmission time due to the store and forward mechanism implemented in the routers is negligible with respect to the RTT.

In order to describe the behavior of Classic TCP and of TCP-SF for small values of the window size, we develop two models based on the TCP state machine which are presented in the next two Subsections. The idea is to evaluate the time spent (and the data sent) by the different versions of TCP in the small window region. For this purpose, a continuously backlogged source is assumed.

For simplicity, we consider the TCP-SF option with fixed segment size equal to $MSS/4$ and $cwnd_{\min}$ =4. Similar conclusions can be drawn for the variable segment size case.

3.1 Model of Classic TCP

In this Subsection we present the State Machine (SM) which describes the behavior of classic versions of TCP when the window size is comprised between 1 and the threshold

$cwnd_{\min}$ =4. When developing the SM, we focus on the state changes neglecting the time spent in each state, which will be introduced later.

The state diagram of the classic TCP SM is shown in the left part of Figure 2. At the connection set up or when Slow Start is entered, the window size is equal to 1, in the SM this is represented by state 1. One segment is transmitted and, after a round-trip time, if the segment was not lost, TCP increases the window size to 2. Correspondingly, the SM moves from state 1 to state 2 with probability $(1-P)$, where P is the segment loss probability. Similarly, the transition from state 2 to state 4 occurs with probability $(1-P)^2$.

State TO represents the timeout. If the segment retransmitted after timeout expiration is lost again, the SM remains in TO and the backoff mechanism applies. When the retransmitted segment is successfully delivered, the SM leaves state TO and enters state 2, because the new ACK makes the window slide and widen.

We now want to evaluate the average time D_c needed by the SM to move from state 1 to state 4, since from that moment on the behavior of TCP-SF is the same as the Classic TCP. From the SM, we observe all the possible paths between 1 and 4 and we evaluate their probability. We then weight each path with its duration which is the sum of the times needed to visit all the states along the path.

The duration of a visit to states 1 and 2 is equal to the round-trip time RTT; a visit to state TO, instead, takes a time equal to the average timeout duration, T. Taking into account the backoff mechanism, T results to be,

$$T = RTO \sum_{i=1}^{\infty} 2^{i-1} P^{i-1} (1-P) = \frac{RTO(1-P)}{1-2P}. \tag{1}$$

We obtain,

$$\begin{aligned} D_c &= (1-P)r + P(T+r) + \sum_{i=0}^{\infty} (1-P)^2 \left[1-(1-P)^2\right]^i \left[i(T+r)+r\right] = \\ &= (1-P)r + P(T+r) + \frac{(T+r)\left[1-(1-P)^2\right]}{(1-P)^2} + r. \end{aligned}$$

where T is the average timeout as in (1) and r is the round trip time. The infinite sum accounts for losses occurring in state 2.

In a similar way, it is possible to compute N_c, the average number of segments transmitted in the small window region. We again consider all possible paths from state 1 to state 4 and for each visited state along a path we compute the number of transmitted segments. N_c results,

$$\begin{aligned} N_c &= P + (1-P) + (1-P)^2 \sum_{i=0}^{\infty} \left[1-(1-P)^2\right]^i \left[i(\overline{N}+1)+2\right] \\ &= 3 + \frac{(\overline{N}+1)\left[1-(1-P)^2\right]}{(1-P)^2} \end{aligned}$$

where $\overline{N} = 2P(1-P)/(1-(1-P)^2)$ is the average number of segments successfully transmitted in state 2 given that some losses occur.

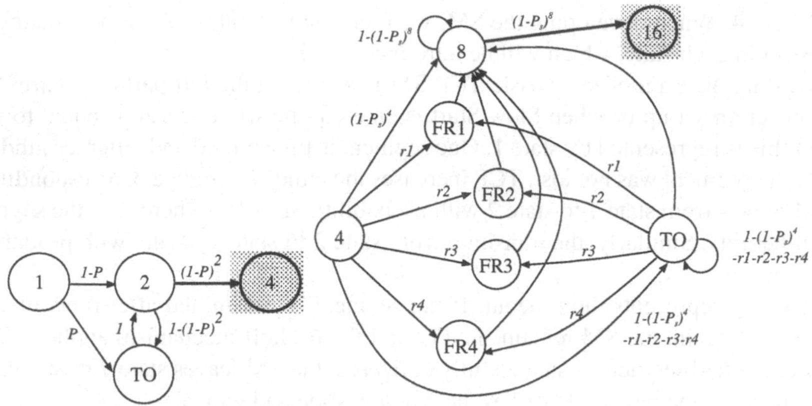

Fig. 2. State machine which describes the behavior of Classic TCP (on the left) and FS-TCP-SF (on the right) when the window size ranges between 1 and $cwnd_{\min}$.

We can now compute the throughput in the small window region as,

$$X_c = MSS \cdot N_c / D_c \,. \tag{2}$$

3.2 Model of TCP Smart Framing

The state machine diagram which describes FS-TCP-SF behavior is shown in the right plot of Figure 2.

When entering Slow Start the congestion window size is equal to $cwnd_0$ and TCP-SF transmits 4 small segments. In the SM, we represent this situation by state 4.

When some losses occur, TCP-SF can either enter FR or RTO; this choice depends on the loss pattern, and in particular it changes on the basis of which segment within the window is the first one to be lost. In order to distinguish different cases, we introduce in the SM four states labeled as FR_i with $i = 1, \cdots, 4$, where FR_i represents the FR algorithm entered by TCP when the first lost segment is the $i-$th one.

State FR_1 represents the case in which the first segment of the window is lost. In this case, FR can be entered only if the following three segments are successfully delivered. Therefore, the transition probability from state 4 to FR_1 is equal to $r_1 = P_s(1-P_s)^3$, where P_s is the small-size segment loss probability.

In case the second segment of the window is lost, the ACK of the first segment makes the window slide and widen so that a total number of four segments are transmitted after the lost one. Therefore, FR is entered if three out of the four segments transmitted after the lost one are successfully delivered so that a total number of 3 duplicated ACKs are generated. In the SM, the state FR_2 represents this case. Similar situations are represented by states FR_3 and FR_4. We denote by r_i the transition probability from state 4 to FR_i, and we have,

$$r_i = (1-P_s)^{i-1} P_s \sum_{j=0}^{i-1} \binom{i+2}{j} (1-P_s)^{i+2-j} P_s^j \qquad i > 1 \,.$$

Since in state FR_i the number of segments which follow the lost one is $i+2$, the sum accounts for the probability that at least 3 segments following the lost one are correctly received, so that FR can be triggered.

Once the FR is completed, the SM will always move to the state labeled 8, where 8 small segments are sent. This is due to the clipping of the Slow Start threshold to $2MSS$.

The transition from state 4 to state 8 represents the case in which all the four segments transmitted when window size is equal to 4 are successfully delivered so that, after roughly a round-trip time, the window size is equal to 8 small segments. The transition probability is equal to $(1-P_s)^4$.

When some losses occur and FR can not be entered, the SM moves from state 4 to state TO; the probability of this transition is, $1-(1-P_s)^4 - r_1 - r_2 - r_3 - r_4$.

The SM can then leave the state TO and move to state 8 if no loss occurs, or, similarly to what happens from state 4, the SM can move to a state FR_i, if three duplicate ACKs arrive, or remain in the state TO otherwise.

When the window size is 8, the probability that RTO is used instead of FR is extremely small, therefore, we assume that in case of losses FR only is possible. Loss occurrence and recovery when the window size equal to 8 is represented by the transition from state 8 to itself; the transition probability is equal to $1-(1-P_s)^8$. Notice that the window size does not change due to the minimum Slow Start threshold which is equal to $2MSS$.

Similarly to (2) in previous model, the throughput is evaluated as the ratio of the average number of transmitted segments, N_s, over the time spent by TCP-SF in the considered region, D_s.

In order to evaluate D_s, all the paths from state 4 to 16 are considered; along a path, each visit to a state is weighted by the time spent in that state. The time spent in state 4 is equal to RTT. The time to recover a loss in FR_1 is RTT, while in states FR_i with $i > 1$, two round trip times are needed. Indeed, one RTT is needed to receive the ACKs which enlarge the window and allow the transmission of new segments; one more RTT is needed to receive the ACKs of these new segments. The time spent in TO is equal to the average timeout T, which is computed by substituting P_s to P in (1).

The average number of transmitted segments, X_s, can be derived again considering all the possible paths; along a path each visit to a state is weighted by the number of segments transmitted in that state. We associate four segment transmissions to states 4 and TO, so that states FR_i account for transmissions due to window enlargement only. Thus, 0 segments are transmitted in FR_1, 2, 4, and 6 are respectively transmitted in states FR_2, FR_3 FR_4. Eight segment transmissions are associated to state 8.

For the sake of brevity we do not report here the formula for the D_s and N_s. The throughput is given by,

$$X_s = MSS/N_{sf} \cdot N_s/D_s \tag{3}$$

where N_{sf} is equal to 4 in the considered case.

3.3 Comparisons

Loss probability has a remarkable impact on the achievable performance gain. In order to estimate the performance gain we need to evaluate the ratio $R = P/P_s$ between the

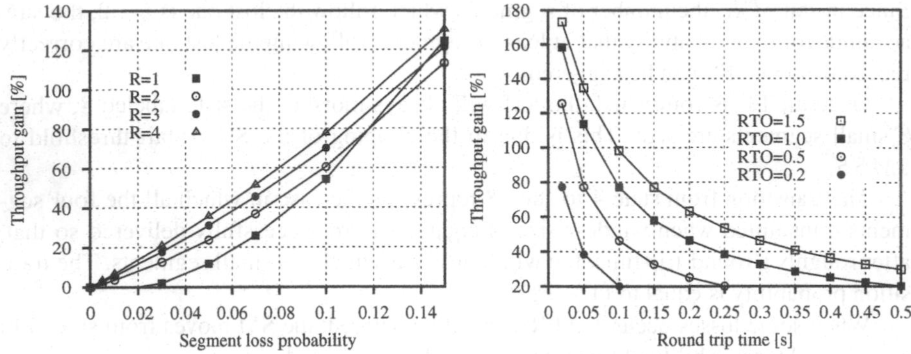

Fig. 3. Throughput gain in the small window region for TCP-SF over Classic TCP versus the full-size segment loss probability (on the left) and versus round trip time (on the right).

full-size segment loss probability, P, and the small-size segment one, P_s. We expect that the gain achieved by TCP-SF grows with R.

Since the segment loss probability depends on network congestion, on buffer management policies at routers and on the segment size, it is in general difficult to estimate R. Therefore, instead of setting a scenario and deriving R accordingly, we consider different values of R which are representative of a wide set of cases. As limiting cases we use $R = 1$ and $R = 4$. The rationale behind this choice is illustrated below.

When RED buffer management schemes are adopted, the queue length is managed in bytes, thus it is reasonable to assume that the segment loss probability is proportional to the segment size, as suggested in [9]. Then, the probability to lose a small segment, P_s, is about 4 times smaller than the probability of losing a full-size segment P; R is equal to 4. On the contrary, as a worst case, we consider a buffer management scheme which handles packets as units, independently of their size. In this case, P_s is equal to P, and $R = 1$.

Left plot of Figure 3 shows the throughput gain of TCP-SF over Classic TCP in the small window regime versus the full-size segment loss probability and for different values of R. The average round trip time is equal to 200 ms, RTO is set to 1.5 s. The chosen values of segment loss probability were taken from simulation results for different values of offered load, see Section 4. Points in the plot are computed in the following way. Given a value of the segment loss probability P, we first compute the probability P_s as P/R. Then, we obtain the throughput of TCP-SF and of Classic TCP from the models as in (2) and in (3). We compute the throughput gain G from, $G = \frac{100(X_s - X_c)}{X_c}$. As expected, the throughput gain increases with the segment loss probability and with the loss probability ratio R. In the small window region, TCP-SF achieves up to more than double the throughput of Classic TCP.

Since the advantage of TCP-SF is mainly due to the smaller time spent in FR rather than waiting for RTO expiration, we now focus on the impact of the values of RTT and RTO on the throughput gain. In the right plot of Figure 3 the throughput gain is plotted versus RTT for different values of RTO. The full-size segment loss probability P is set to 0.1; small segments are lost with probability $P_s = P/2$. Given a value of RTO, as the round trip time increases the difference between RTO and RTT decreases; consistently, the throughput gain decreases. Notice however, that for the considered scenario, the throughput gain is remarkable also for the minimum admissible value of RTO, i.e., RTO= 2RTT.

Finally, notice that the considerations drawn above hold for the small window region only. Short-lived flows benefit of further advantages in using TCP-SF which are not considered in this analysis, such as, for example, the higher chance not to use the initial RTO. For longer flows, on the contrary, the benefits of adopting the smart framing option are observed only at the connection set up and when Slow Start is entered due to some losses.

4 Performance Evaluation

We have chosen to investigate the performance of TCP-SF using both simulation and actual testbed measurements. Simulation gave us full control over specific scenarios. On the other hand, the testbed implementation allowed a more realistic evaluation featuring actual traffic patterns. Specifically, our approach splits the measurements into two stress tests: *real network, synthetic traffic*: these tests were performed using a client and a server on remote systems, respectively requesting and sending pools of data over the Internet; such tests allowed us to control the amount of traffic injected into the network, and, in general, to have greater control on the type of results we wish to obtain; *real traffic, synthetic network*: tests using real traffic involved a proxy server on a departmental subnet fed through a link emulator; these tests gave us control on such key network parameters as packet loss and latency, while the traffic reflected actual usage patterns, as will be detailed below.

We implemented both flavors of TCP-SF in the ns-2 simulator [10]. For our testbed measurements, we implemented the Fixed-Size version in the Linux kernel 2.2.17.

4.1 Simulation Results

We report results for a network scenario in which both long-lived FTP-file transfer and Web-like connections share a common bottleneck link. In particular we derived the flow length distribution reported in the second column of Table 1 from the AT&T Labs' traffic estimate [2]. The distribution was obtained by dividing the flows in 10 groups with the same number of flows per group, and computing the average flow length for each group. As can be seen, more than 70% of the flows are shorter than 11 Kbytes, and thus FR can not be triggered if the MSS is 1.5 Kbytes, which is the most common value used today for MSS.

To model Web-like traffic, TCP-SACK traffic generators are connected to a bottleneck link of 10 Mbit/s capacity and 50 ms delay. A Poisson process drives the setup of new connections, whose length (in bytes) is randomly set according to the previous traffic distribution. The MSS is set to 1,460 bytes. Long-lived connections are

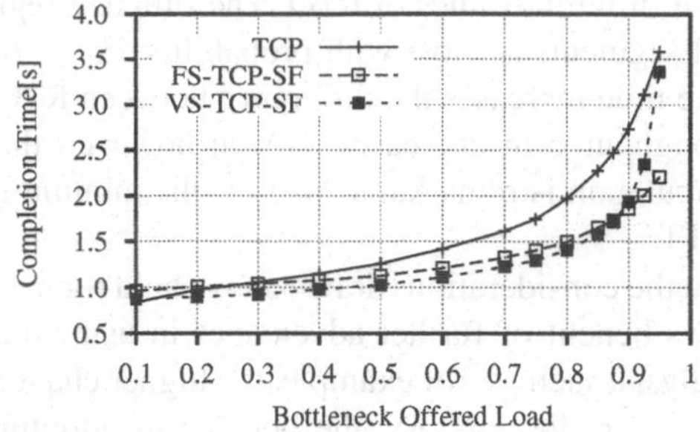

Fig. 4. Average Completion Time: Comparison between TCP-SF and normal TCP (SACK enabled).

accounted for using 10 greedy TCP sources. The bottleneck link is managed by a byte-wise Droptail buffer, whose capacity is set to 150 Kbytes. To get rid of transient effects, the simulation time is 4,000 seconds.

Figure 4 plots the average Completion Time, i.e., the time that is required by the source to successfully transfer all the data, for the classic TCP-SACK implementation, and for the proposed VS- and FS-TCP-SF, versus the average offered load normalized with respect to the bottleneck capacity. As expected, the average completion time required to end the transfer is larger when the offered load grows. This is due to two main reasons: first, the average window size shrinks as the capacity of the link is shared among a larger number of flows. Second, the probability of segment loss is larger, causing either RTO or FR to occur.

On the contrary, both VS- and FS-TCP-SF outperform Classic TCP, as they are quicker to react to segment losses. In particular, when the offered load is smaller than 0.9, VS-TCP-SF performs better than FS-TCP-SF, as the overhead introduced is smaller. When, instead, the offered load is larger than 0.9, the probability of multiple drops per RTT increases, and thus VS-TCP-SF is no longer able to avoid RTO. On the contrary FS-TCP-SF shows the best performance. To give further insight into the performance that both flavors of TCP-SF are able to obtain, Table 1 reports, for different flow groups, the average completion time, the percentage of FR and RTO instances used to detect segment loss for Classic TCP, VS- and FS-TCP-SF. The offered load is fixed and equal to 0.8. It should be noted that, for some flow lengths, FR is not observed since there are not enough segments to trigger it.

As can be seen, both flavors of TCP-SF outperform the Classic TCP performance in terms of completion time for all flow lengths. Specifically, different completion times are observed, even if the RTO and FR percentage are the same, because each scenario exhibits a slightly different loss probability. For example, although all group-0 flows do not benefit from the TCP-SF enhancement, their completion times differ because dropping probabilities for TCP-SF were slightly smaller.

In particular, Table 1 underlines the benefits obtained by TCP-SF in increasing the number of FR, while at the same time reducing the RTO occurrences. The benefits are clearly visible starting from flows that have to send 1,349 bytes (group 3): using TCP-

Table 1. Average Completion Time (CT) and Percentage of FR and RTO occurrences per group.

Group	bytes	Classic			VS-TCP-SF			FS-TCP-SF		
		CT	% FR	% RTO	CT	% FR	% RTO	CT	% FR	% RTO
0	61	0.25	–	100.00	0.22	–	100.00	0.24	–	100.00
1	239	0.37	–	100.00	0.28	–	100.00	0.31	–	100.00
2	539	0.46	–	100.00	0.34	–	100.00	0.43	–	100.00
3	1349	0.81	–	100.00	0.42	8.02	91.98	0.49	10.84	89.16
4	2739	1.14	–	100.00	0.53	33.52	66.48	0.64	35.57	64.43
5	4149	1.33	–	100.00	0.54	43.26	56.74	0.86	32.00	68.00
6	6358	1.71	–	100.00	0.77	52.12	47.88	0.93	30.02	69.98
7	10910	1.96	5.54	94.46	1.00	64.78	35.22	1.16	37.26	62.74
8	18978	2.17	21.91	78.09	1.23	62.98	37.02	1.57	49.01	50.99
9	90439	4.43	46.34	53.66	3.01	69.68	30.32	3.87	64.64	35.36

SF, they manage to trigger the FR algorithm. On the contrary, the first class of flows that can use FR for the Classic TCP is the one that has to send 10,910 bytes (group 7), where only less than 6% of dropped segments trigger FR. This is reflected by a smaller completion time required to successfully end the transfer (i.e., reduced by a half).

A counterintuitive result is observed for connections belonging to classes 5 and higher: the VS flavor appears to run into FR more frequently than FS-TCP-SF, even though FS sends more small-size segments (thus increasing the chance to trigger FR). This is due to the burstier traffic generated by FS-TCP-SF, which causes multiple drops that cannot be recovered but through RTO.

4.2 Testbed Measurements: Real Network, Synthetic Traffic

To gauge the impact of our new scheme in a real Internet environment, we have implemented FS-TCP-SF on a Linux workstation acting as file server. We have then tested the implementation using several clients (adopting different OSs) which requested 10-Kbytes file uploads to the server, the clients being located both on the local network (in Italy) and in the Internet at large (in the U.S.A.). Each client repeated the requests some 20,000 times over a time span of 4 hours during peak business hours in Italy.

Results in Table 2 are reported in terms of FR and RTO occurrences, similarly to those in Table 1. In our testbed configuration, the dropping probability is small, since the Italy-to-U.S.A. path is usually lightly loaded, most of the traffic flowing in the opposite direction. Thus, a small fraction of the flows suffered from segment losses. This is reflected in a average completion time measure that is almost constant, and we opted not to report it. On the contrary, the frequency of FR and RTO occurrences are a more significant metric because they are a relative measure.

On the whole, the testbed results confirm what we observed by simulation: TCP-SF connections resort to FR four times more frequently than Classic TCP does. This behavior, in presence of a larger percentage of drops, would have lead to a significantly smaller completion time, as shown by simulations.

Table 2. Percentage of FR and RTO occurrences measured per host.

Host	Classic		FS-TCP-SF	
	% FR	% RTO	% FR	% RTO
sigcomm.cmcl.cs.cmu.edu	24.2	75.8	53.2	46.8
manchester-1.cmcl.cs.cmu.edu	11.1	88.9	72.5	27.5
ssh.workspot.net	19.0	81.0	72.3	27.7
wavelan24.cs.ucla.edu	34.3	65.7	69.3	30.7

4.3 Testbed Measurements: Real Traffic, Synthetic Network

The second set of testbed experiments involved a realistic traffic pattern (i.e., Web client-server transactions) routed over a link emulator. The topology of the testbed is described in Figure 5. Its functional elements are the following:

- a switched 100 Mb/s LAN carrying Department traffic;
- a Web Proxy Server used by clients on the Department Subnet;
- a Router R connecting to the Internet and carrying outbound and inbound traffic over a 32 Mb/s link;
- a Router R_L implemented on a Linux machine which also runs a Link Emulator, affecting only traffic *from* the Proxy Server;
- all links except for the outgoing ones are Full Duplex Fast-Ethernet.

In order to generate realistic traffic, every Web browser in our Department Subnet was configured so as to use the Proxy Server. TCP-SF was only implemented over the machine running the Proxy Server, and a link emulator was added on the return path between the Proxy and the Department Subnet. The link emulator was configured so as to enforce a specific latency and byte-wise drop probability (i.e., longer packets have a higher probability of being dropped than smaller ones). The above configuration allows a substantial amount of traffic (namely, the Web objects fetched by the Proxy and returned to the requesting clients on the Department Subnet) to be sent over the link emulator using TCP-SF as transport layer.

Performance metrics were collected for different values of emulated latency and drop probability, over a period of one month in June 2001; in order to collect a meaningful set of data, each latency/drop pair was set for a whole day, and the Proxy Server had its transport layer switched between Classic TCP and TCP-SF every five minutes. Only connections between the Proxy server and a local client were considered. Statistics were later collected for each version of TCP, and for each day. Unlike the results in the previous Subsection, we had little control over the actual client sessions: the amount of data transferred during each transaction depended on the browser used and the operating system installed on each user's machine. Also, each browser has its own session-level behavior, including idle time between back-to-back requests, or connection shut-down after a timeout. Unfortunately, this prevented us from obtaining reliable estimates of completion times. Indeed, completion times are computed at the Proxy Server as the time elapsed between the reception of a SYN segment from a client, until the last FIN (or RST) segment is received by the server, or sent by the server, whichever occurred last. Flows that did not transport any payload are not taken into account. We have thus

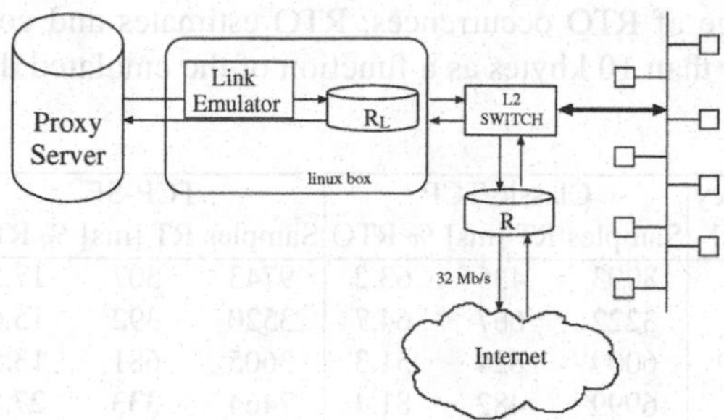

Fig. 5. Topological description of the Real Traffic-Synthetic Network experiment.

chosen to report just the *difference* between the completion times of TCP-SF and Classic TCP, assuming that, on the whole, whatever discrepancies between the behaviors of different clients' OSs and browsers would equally affect the two versions of TCP being switched on the Proxy Server. We are aware that these statistics are hardly reliable, but we chose to report them anyway, since they might give an indication as to which version ensures a faster completion.

Also, the percentages of Timeout expiration are reported, as in Subsection 4.2. Unlike completion times, statistics on the occurrence of RTOs and triple duplicate ACKs are less dependent on browser behavior, and are therefore more reliable metrics. These results are shown in Tables 3, 4 and 5, each Table referring to a result set collected by file size (smaller than 10 kbytes, between 10 and 100 kbytes, and larger than 100 kbytes). Columns in Table show: emulated drop probability; emulated latency; samples number (number of flows); estimate of the average Retransmission Timer per connection; percentage of times a loss resulted in an RTO expiration instead of triggering Fast Recovery; completion time difference between the two examined TCP versions, computed only for flows that have experienced at least one loss event. Observe that positive values for the completion time difference indicate that TCP-SF exhibited a faster completion time, while negative times indicate the opposite.

First of all, it should be observed that the number of samples is significantly larger for flows smaller than 10 kbytes, confirming that a majority of the Internet traffic today is made up of short-lived flows. Also, we remark that the number of samples allows some confidence in the statistical examination of the output only for file sizes smaller than 100 kbytes. Statistics regarding larger file sizes are reported here for completeness.

Overall, results confirm the findings shown in previous Sections via simulation and, specifically, identify TCP-SF as less prone to RTO expirations than Classic TCP. Estimating the proper value of the Retransmission Timer also benefits from the features of TCP-SF: the larger number of segments sent, compared to Classic TCP, accounts for a larger number of samples used in the estimation of the round trip time, thus refining the estimate and providing a smaller, more accurate value for the timer. The combined effects of fewer RTOs and smaller values of the retransmission timer shortens the completion time in presence of a loss ΔCT_{loss}; although we remarked that completion times are affected by the unpredictable behavior of different browser types, the trend nonetheless confirms a sizable reduction of completion times for TCP-SF.

Table 3. Percentage of RTO occurrences, RTO estimates and completion times for web objects smaller than 10 kbytes as a function of the emulated drop probability and latency.

drop probability	latency [ms]	Classic TCP Samples	RT [ms]	% RTO	TCP-SF Samples	RT [ms]	% RTO	ΔCT_{loss} [s]
0.01	20	8093	435	63.2	9743	307	17.5	8.32
	50	3222	667	64.7	3520	392	15.0	14.24
	100	6009	824	51.3	3605	681	13.5	-0.25
0.05	20	6999	482	81.4	7464	333	27.8	5.63
	50	4994	669	63.9	3406	429	20.3	5.11
	100	7058	1206	59.7	8227	652	16.0	3.48
0.10	20	5022	477	80.5	3302	397	24.0	4.23
	50	2859	725	79.7	4601	428	22.4	4.63
	100	1535	1008	69.5	3011	661	22.3	3.80

Table 4. Percentage of RTO occurrences, RTO estimates and completion times for web objects sized between 10 kbytes and 100 kbytes, as a function of the emulated drop probability and latency.

drop probability	latency [ms]	Classic TCP Samples	RT [ms]	% RTO	TCP-SF Samples	RT [ms]	% RTO	ΔCT_{loss} [s]
0.01	20	2575	387	12.2	2970	263	8.4	4.45
	50	1228	458	19.0	1349	345	12.2	0.35
	100	1885	485	21.1	1827	319	17.1	-1.02
0.05	20	2083	392	26.8	2289	263	13.7	1.93
	50	1558	466	28.3	1339	345	14.0	-0.08
	100	2120	722	27.2	2912	319	19.8	2.60
0.10	20	1876	476	34.5	1604	274	14.7	1.97
	50	1350	555	37.2	1987	298	18.5	2.14
	100	895	741	39.7	1424	375	22.8	2.73

5 Conclusions

We proposed an enhancement to the TCP protocol that is based on the key idea of transmitting smaller size segments when the congestion window is smaller than 4 MSS, without changing the degree of aggressiveness of the source. This allows the sender to receive more feedback from the destination, and thus use the Fast Recovery algorithm to recover from segment losses, without waiting for a Retransmission Timeout expiration to occur. TCP-SF is particularly effective for short-lived flows, but improves the responsiveness of long file transfers also. Coupled with the current Internet traffic, TCP-SF outperforms Classic TCP in terms of both completion time, and probability to trigger Fast Recovery to detect segment losses.

The proposed modification is extremely simple and can be implemented on top of any TCP flavor; moreover, it requires changes on the server side only.

Table 5. Percentage of RTO occurrences, RTO estimates and completion times for web objects larger than 100 kbytes, as a function of the emulated drop probability and latency.

drop probability	latency [ms]	Classic TCP Samples	RT [ms]	% RTO	TCP-SF Samples	RT [ms]	% RTO	ΔCT_{loss} [s]
0.01	20	139	253	10.4	150	254	6.1	3.19
	50	47	276	15.9	46	261	6.9	9.62
	100	76	288	22.7	86	253	17.3	0.76
0.05	20	88	285	19.4	88	219	7.3	12.34
	50	47	298	14.4	64	291	15.5	-66.17
	100	84	370	29.1	87	280	22.2	22.72
0.10	20	79	290	28.0	67	270	17.0	-3.69
	50	104	334	22.0	152	273	15.4	-6.04
	100	105	657	48.8	97	330	29.8	-85.63

6 Acknowledgments

Hands-on measurements of Internet traffic could not have been possible without the cooperation of several people from our Department who willingly (or unaware), left their browsers at our mercy for a month or so. In acknowledging their help, we also wish to thank the Politecnico di Torino Network Facilities (CESIT) that allowed us to take the network measurements, and Mario Gerla at UCLA and the Computer Science Facility at CMU for letting us run remote experiments on their workstations.

References

1. M. Allman, V. Paxson, W. Stevens. TCP Congestion Control. *RFC-2581*, April 1999.
2. A. Feldmann, J. Rexford, R, Caceres. Efficient policies for carrying Web traffic over flow-switched networks. *IEEE/ACM Transactions on Networking*, Vol: 6, NO: 6, Dec. 1998.
3. Marco Mellia, Michela Meo, Claudio Casetti, TCP Smart Framing: using smart segments to enhance the performance of TCP. *Globecom 2001,* San Antonio, TX, 25-29 November 2001.
4. V. Jacobson, "Congestion Avoidance and Control", *Sigcomm 88*, Standford, CA, pp. 314-329, Aug. 1988.
5. W.R. Stevens. *TCP/IP Illustrated, vol. 1*. Addison Wesley, Reading, MA, USA, 1994.
6. S. Floyd, T. Henderson, "The NewReno Modification to TCP's Fast Recovery Algorithm", *RFC 2582*, Apr. 1999.
7. M. Mathis, J. Mahdavi, S. Floyd, S., A. Romanow, "TCP Selective Acknowledgment Options", *RFC 2018*, Apr. 1996.
8. D. Clark. Window and Acknowledgment Strategy in TCP. *RFC-813*, July 1982.
9. S. De Cnodder, O Elloumi and K Pauwels. RED Behavior with Different Packet Sizes. *Proceedings of the Fifth IEEE Symposium on Computers and Communications (ISCC 2000)*.
10. ns-2, network simulator (ver.2). http://www.isi.edu/nsnam/ns/.

Live Admission Control for Video Streaming

Pietro Camarda, Domenico Striccoli

Politecnico di Bari – Dip. di Elettrotecnica ed Elettronica
Via E. Orabona, 4 – 70125 Bari (Italy)
Tel. +39 080 5963642, Fax +39 080 5963410,
camarda@poliba.it

Abstract. The core aspect of most multimedia applications is the transmission of Variable Bit Rate (VBR) video streams requiring a sustained relatively high bandwidth with stringent Quality of Service (QoS) guarantee. In such systems, a statistical bandwidth estimation is especially relevant as it is indispensable for implementing an efficient admission control. In this paper a novel measurement-based admission control algorithm for video distribution systems has been developed. Such an algorithm estimates the bandwidth occupied by the actual video streams, observing their temporal evolution in a chosen measurement window. In particular, for each video stream, the estimation procedure needs an online measurement of the bandwidth values assumed successively and their persistence time. An exponential weighted moving average filter is used for smoothing the estimated values. Finally, some measurements of video traffic are shown comparing the results with simulation and other analytical results, discussing the effectiveness of the considered solution.

1 Introduction

In the world of telecommunications several live multimedia applications like video conferencing, online distance learning, live video events, etc, are assuming a growing importance [1, 2]. The core aspect of such applications is the transmission of video streams requiring a sustained relatively high bandwidth with stringent Quality of Service (QoS) guarantee, in terms of loss probability, delay, etc.

The scenario presented in this paper refers to a video distribution system where a large number of heterogeneous video streams are transmitted in real-time across a network. All video streams are supposed codified with various standards (MPEG, H323, etc.) which usually produce a Variable Bit Rate (VBR) traffic. As an example, in Fig. 1 the bit rate of the first 32.000 frames of a MPEG-1 coded trace of the film "Asterix" is reported. The adopted frame rate is 25 frames/s, thus each video frame is transmitted in 40 milliseconds.

Given the nature of VBR traffic, a static conservative assignment of bandwidth in the case of statistically multiplexed video streams produces a non negligible waste of network resources. Supposing, in fact, that all starting points of video streams are randomly shifted in time, the sum of bit rates of all streams will usually be lower than the sum of peak rates. Thus, a statistical approach is needed for evaluating the

M. Ajmone Marsan et al. (Eds.): QoS-IP 2003, LNCS 2601, pp. 292-305, 2003.

bandwidth resources needed by a given number of video streams, respecting the QoS requirements.

Smoothing techniques can be exploited to reduce the total amount of bandwidth assigned to video streams [3-9]. These techniques are based on the reduction of peak rate and rate variability of every stream present in the network; they consist in transmitting, as long as possible, pieces of the same film with a constant bit rate that varies from piece to piece according to a scheduling algorithm that smoothes the bursty behavior of video streams. On the transmission side a buffer regularizes the transmission, while on the receiving side the frames are temporarily stored in a client buffer and extracted during the decoding process. Obviously, the bit rate must be chosen appropriately in order to avoid buffer overflow and underflow. The client smoothing buffer size determines the number and duration of the CBR pieces that characterize the smoothed video stream. An increase of the smoothing buffer size produces a smaller number of CBR segments and a reduction of peak rate and rate variability of smoothed video streams, ensuring, as will be shown later, a consistent gain in network resource utilization [10, 11]. Different types of smoothing algorithms have been studied: online smoothing [3, 4] and offline smoothing [5-9]. The study developed in this paper considers only online smoothing and does not rely on specific details of smoothing techniques; consequently, any online smoothing technique can be exploited. In addition the developed technique can be adapted also to unsmoothed video streams. As an example, in Figure 2 the "Asterix" film, smoothed exploiting the FAR online smoothing algorithm [3], is reported. The adopted client smoothing buffer is 1024 Kbytes and the time windows in which the smoothing algorithm is applied are 120 frames large.

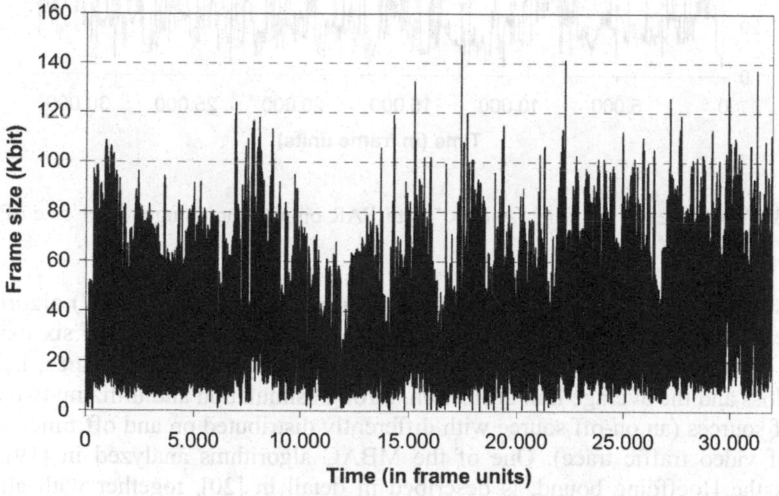

Fig. 1. 32.000 frames of the film "Asterix".

From Figure 2 it can be noted that the online smoothing algorithm reduces peak rate and rate variability of the unsmoothed film, although it is less efficient than the corresponding offline smoothing algorithm.

In the considered systems, a statistical bandwidth estimation is especially relevant as it is indispensable for implementing efficient admission control algorithms, sharing, without waste, bandwidth resources while respecting QoS specifications [10, 12, 13], under the hypothesis of bufferless multiplexer. This last hypothesis has been taken into account in [10, 14, 15] and can be supported by the following considerations. First of all, it has been shown [13, 16] that large network buffers are of substantial benefit in reducing losses in the high frequency domain and the only effective way to reduce losses in the low frequency domain is to allocate sufficient bandwidth for each network link. Since the smoothed traffic has mainly a slow time scale variability, the adoption of large network buffers is not useful in this scenario. A second consideration is that the employment of large buffers introduces delays and jitters along the network nodes that can be intolerable for the correct delivery of video streams guaranteeing the QoS specifications.

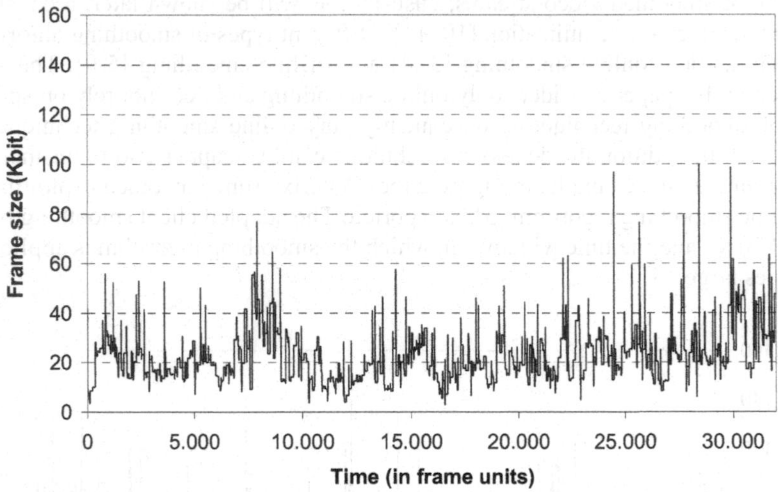

Fig. 2. 32.000 frames of the film "Asterix" with FAR online smoothing (buffer size of 1024 bytes).

Various approaches of Measurement-Based Admission Control (MBAC) algorithms have been proposed in the literature [19-21]. In [19] a broad sample of six existing MBAC algorithms is analyzed and discussed, comparing their performance, i.e., the packet loss and the average link utilization, through simulation and utilizing two basic types of sources (an on-off source with differently distributed on and off times, and a piece of video traffic trace). One of the MBAC algorithms analyzed in [19], that exploits the Hoeffding bound, is described in detail in [20], together with another approach based on the normal distribution. In [21], a framework for efficient MBAC schemes is introduced and discussed, taking into account the impact of some parameters, i.e., errors in estimating parameters that could bring to erroneous admission control decisions, flow dynamics to correctly calculate parameters that vary with time, memory, to take into account past information of the flows present in the system, etc.

The algorithm proposed in this paper estimates, for each measurement window, the aggregate bandwidth by a novel statistical approach, modeling the entire system of video streams as a multi chain, multi class network of queues [17, 18], whose solution is suitable for real time evaluation, as shown in the next sections. The average aggregate bandwidth is calculated using an exponential-weighted moving average filter. The obtained solutions are compared with the results provided by other analytical approaches [19-21] and by simulation.
The paper is structured as follows. In section 2 the analytical model for evaluating the aggregate bandwidth considering heterogeneous streams is presented. Numerical results are provided in Section 3. In Section 4 some conclusions will be given.

2 Admission Control Procedure

In this section a novel approach based on a statistical aggregate bandwidth estimation is exploited for implementing a measurement-based admission control procedure. The aggregate bandwidth is evaluated such that a stationary arrival rate for the considered video traffic exceeds the evaluated value with probability at most p_l. The admission control scheme admits a new connection if the aggregate bandwidth evaluated adding the new connection does not exceed the available bandwidth for the video service.
The considered video traffic is composed by F heterogeneous smoothed streams. For each stream, the measurement-based admission control procedure relies on the measurement of the needed bandwidth levels, their temporal duration and evolution. These parameters can be measured in real time rather easily. In fact, what is needed are just two vectors: the first one contains the bandwidth levels assumed progressively by the smoothed video, while the second vector contains the corresponding number of frames in which such a bandwidth has been assumed.

2.1 Estimating the Aggregate Bandwidth

State Probability Evaluation. Let us suppose to have a generic number of F heterogeneous smoothed streams. A smoothed stream is generically characterized by a given number of bandwidth levels, as can be noted in Figure 2. The novel approach introduced in this paper consists on modeling the aggregation of video streams as a multi chain network of queues with different classes of customers [17, 18, 22], in which each of the bandwidth levels represents a service center. We suppose that the system of F streams is characterized by M service centers, corresponding to the total number of different bandwidth levels λ_m (with $1 \le m \le M$). Since a given bandwidth level can be occupied in different time intervals of the same smoothed stream, we suppose that the corresponding service center has a number of classes equal to the number of times in which that bandwidth level is occupied.
Given these assumptions, the system state can be represented by the following vector:

$$(n_{111}, n_{112}, \ldots, n_{11R_{11}}, \ldots, n_{1F1}, \ldots, n_{1F1}, \ldots, n_{1FR_{1F}}, \ldots, n_{MF1}, \ldots, n_{MF1}, \ldots, n_{MFR_{MF}}) = \{ n_{ijr} \}$$

where there are generically n_{ijr} users in the ith bandwidth level λ_i of the jth stream, of class r, for $1 \leq i \leq M$, $1 \leq j \leq F$ and $1 \leq r \leq R_{ij}$. Obviously, we have

$$\sum_{i=1}^{M} \sum_{r=1}^{R_{ij}} n_{ijr} = n_j = 1, \qquad j = 1,2,......., F \tag{1}$$

The model is general enough to consider also $n_j > 1$ in (1), as developed in [23], but in this paper we focus on the rather intuitive case of a single live video transmission $(n_j = 1)$ multicasted to all interested users. The relative arrival rate to the ith bandwidth level of the jth stream of class r, called e_{ijr}, can be found solving the homogeneous system of flow balance equations [17, 18]:

$$e_{ijr} = \sum_{k=1}^{M} \sum_{l=1}^{F} \sum_{s=1}^{R_{ij}} e_{kls} p_{k,l,s;i,j,r} \tag{2}$$

where $p_{k,l,s;i,j,r}$ represents the transition probability from the kth bandwidth level of the lth stream, of class *s*, to the ith bandwidth level of the jth stream, of class *r*. These probabilities can assume only the values 0 or 1 and can be easily derived observing the sequence of the bandwidth levels in each stream. In particular, from the analysis of the temporal evolution of the single video trace, it can be noted that the probability $p_{k,l,s;i,j,r}$ assumes the value 1 only if $l=j$ and the two bandwidth levels k of class s and i, of class r are temporally consecutive. The same probability assumes the value 0 otherwise. Since it is supposed that the number of video streams does not change, it is clear that when a stream finishes, another one begins. Moreover, it is supposed that if the bandwidth level k is the last one of the measurement window, the probability $p_{k,l,s;i,j,r}$ assumes the value 1 only if the bandwidth level i is the first one of the same measurement window, and assumes 0 otherwise.

Since only the temporal evolution of the single source is analyzed in detail, disregarding all temporal dependencies among video streams, it can be argued that the proposed method is suitable only for independent streams. As a matter of facts, in this paper all issues of dependent streams [24, 25] have not been taken into account.

It can be noted that there is no limitation to the number of streams that can assume the same bandwidth level, thus the system can be modeled as a network of queues with infinite servers. From queueing networks theory, the state probability can be expressed in product form as follows [18]:

$$P(\{n_{ijr}\}) = \frac{1}{C} \prod_{i=1}^{M} \prod_{j=1}^{F} \prod_{r=1}^{R_{ij}} \left(\frac{e_{ijr}}{\mu_{ijr}} \right)^{n_{ijr}} \frac{1}{n_{ijr}!} \tag{3}$$

where the generic term $1/\mu_{ijr}$ represents the time of permanence in the level λ_i, of the jth stream, of class *r*. Grouping together all classes belonging to the same service center and to the same stream, a more concise representation of the state probability can be obtained:

$$(n_{11}, n_{12}, \ldots, n_{1F}, n_{21}, \ldots, n_{2F}, \ldots, n_{M1}, \ldots, n_{MF}) = \{ n_{ij} \}$$

where $n_{ij} = \sum_{r=1}^{R_{ij}} n_{ijr}$. The marginal state probability can be evaluated exploiting the multinomial formula:

$$P(\{n_{ij}\}) = \frac{1}{C} \prod_{i=1}^{M} \prod_{j=1}^{F} \left(\sum_{r=1}^{R_{ij}} \frac{e_{ijr}}{\mu_{ijr}} \right)^{n_{ij}} \frac{1}{n_{ij}!} \tag{4}$$

The value of the constant C can be calculated imposing that:

$$\sum_{n_{111}} \sum_{n_{112}} \cdots \sum_{n_{MFR_{MF}}} P(\{n_{ijr}\}) = 1$$

from which derives the value of C:

$$C = \prod_{j=1}^{F} \left(\sum_{i=1}^{M} \sum_{r=1}^{R_{ij}} \frac{e_{ijr}}{\mu_{ijr}} \right)^{n_j} \frac{1}{n_j!} \tag{5}$$

The calculation of the state probability given by (4) can be greatly simplified considering that the relative arrival rates e_{ijr} must be all equal, since the network of queues, for each stream, is a simple ring. So, we can choose $e_{ijr} = 1/T_j$ for each $1 \le j \le F$. Let $\sum_{r=1}^{R_{ij}} (e_{ijr} / \mu_{ijr}) = \rho_{ij}$. We have:

$$\rho_{ij} = \frac{1}{T_j} \sum_{r=1}^{R_{ij}} \frac{1}{\mu_{ijr}} = \frac{T_{ij}}{T_j}$$

where $T_{ij} = \sum_{r=1}^{R_{ij}} \frac{1}{\mu_{ijr}}$ represents the permanence time in the ith bandwidth level of the jth video stream, expressed in number of frames, while T_j is the global duration of the jth video stream. It derives that the evaluation of the coefficients ρ_{ij} is very simple and can be obtained without solving the system (2).

Aggregate Bandwidth Estimation. The state probability evaluated by (4) can be exploited for evaluating the aggregate bandwidth, that, in a given state $\{n_{ij}\}$, is $\Lambda = \sum_{i=1}^{M} \sum_{j=1}^{F} n_{ij} \lambda_i$. The evaluation of the aggregate bandwidth Λ_S in correspondence of a loss probability p_l can be evaluated by the sum of all state probabilities that satisfy the constraint on loss probability:

$$\begin{cases} P(\Lambda \le \Lambda_S) = (1-p_l) = \sum_{n_{11}} \sum_{n_{12}} \cdots \sum_{n_{MF}} P(\{n_{ij}\}) \\ n_{1j} + n_{2j} + \ldots + n_{Mj} = n_j \quad \forall 1 \le \mathrm{j} \le \mathrm{F} \\ \sum_{i=1}^{M} \lambda_i \sum_{j=1}^{F} n_{ij} \le \Lambda_S \end{cases} \tag{6}$$

The solution of (6) is discussed in detail in [23] and [25]. In particular, the solution of the equation:

$$P(\Lambda \le \Lambda_S) = (1-p_l) = \frac{\frac{1}{\Lambda_S!} \frac{d^{\Lambda_S}}{dz^{\Lambda_S}} \left[\frac{z^{\Lambda_S+1}-1}{z-1} \prod_{j=1}^{F} \left(\sum_{i=1}^{M} \rho_{ij} z^{\lambda_i} \right)^{n_j} \right]_{z=0}}{\prod_{j=1}^{F} \left(\sum_{i=1}^{M} \sum_{r=1}^{R_{ij}} \frac{e_{ijr}}{\mu_{ijr}} \right)^{n_j} \frac{1}{n_j!}}$$

has to be found, where the numerator is the coefficient of z^{Λ_s} of the polynomial $\frac{z^{\Lambda_S+1}-1}{z-1} \prod_{j=1}^{F} \left(\sum_{i=1}^{M} \rho_{ij} z^{\lambda_i} \right)^{n_j}$, while the denominator represents the constant C evaluated by (5). Let us remember that Λ_S is the unknown, so the problem can not be easily solved a priori. To overcome it, let us suppose that the M bandwidth levels are ordered in such a way that $\lambda_1 < \lambda_2 < \ldots < \lambda_M$. Thus it can be generically written that:

$$\prod_{j=1}^{F} \left(\sum_{i=1}^{M} \rho_{ij} z^{\lambda_i} \right)^{n_j} = c_{N\lambda_1} z^{N\lambda_1} + c_{N\lambda_1+1} z^{N\lambda_1+1} + \ldots + c_{N\lambda_M} z^{N\lambda_M} \tag{7}$$

The term (7) has to be multiplied by the factor:

$$\frac{z^{\Lambda_S+1}-1}{z-1} = \left(1 + z + z^2 + .. + z^{\Lambda_s}\right) \tag{8}$$

Let us suppose, for the moment, that $\Lambda_S = N\lambda_1$. After the product between (7) and (8) we obtain that the coefficient of z^{Λ_s} is given by the product between the first term of (7), $c_{N\lambda_1}$, and the first term of (8), that is 1. For this reason we obtain $P(\Lambda \le \Lambda_S) = c_{N\lambda_1} / C$. Now let us suppose that $\Lambda_S = N\lambda_1 + 1$. The coefficient of $z^{N\lambda_1+1}$ is given by the sum of two terms. The first of them is the product of $c_{N\lambda_1} z^{N\lambda_1}$ of (7) by the term z of (8), while the second is the product of $c_{N\lambda_1+1} z^{N\lambda_1+1}$ of (7) by the term 1 of (8). For this reason, the coefficient of z^{Λ_s} is $c_{N\lambda_1} + c_{N\lambda_1+1}$ and $P(\Lambda \le \Lambda_S) = \left(c_{N\lambda_1} + c_{N\lambda_1+1}\right)/C$. If we repeat the same procedure for increasing values of Λ_S, we obtain the general formula:

$$P(\Lambda \le \Lambda_S) = (1-p_l) = \frac{1}{C} \sum_{B=N\lambda_1}^{\Lambda_S} c_B \tag{9}$$

Since $P(\Lambda \leq \Lambda_S)$ is a known value, to find Λ_S we simply have to add the coefficients of (7) until their sum reaches or is superior to the value $C \cdot (1-p_l)$, that is a priori known.

Since all the exponents n_i of (7) are equal to 1, the coefficients of (7) can be obtained by making F convolutions of the vectors that contain the coefficients of the polynomial $\sum_{i=1}^{M} \rho_{ij} z^{\lambda_i}$, for $1 \leq j \leq F$. The convolutions can be evaluated in an efficient way if we transform each of the F vectors with the Fast Fourier Transform (FFT). To complete the derivation of (7), we have to multiply term by term the F transformed vectors and then transform the obtained vector back to the original domain. It is very important to note that the polynomial (7) is formed by $N\lambda_M + 1$ elements, so each of the vectors obtained from the polynomials $\sum_{i=1}^{M} \rho_{ij} z^{\lambda_i}$ for $1 \leq j \leq F$ is padded with $(N-1)\lambda_M$ zeros after the $(\lambda_M + 1)^{th}$ term, before transforming them with the FFT.

2.2 Description of the Admission Control Procedure

The bandwidth estimation obtained with the proposed algorithm can be used for implementing Admission Control procedures, establishing if a further request of a new stream can be accepted or not, based on the estimated available bandwidth. The Admission Control procedure can be implemented as follows. Let us suppose F smoothed video streams in the network node. We choose a sampling period T, in which the aggregate bandwidth occupied by the F flows is calculated. In particular, let $B_n(p_l)$ be the aggregate bandwidth calculated in the nth sampling window, by the proposed algorithm. The aggregate bandwidth estimated in each sampling window T is averaged by considering also past bandwidth values, to keep track of the past history of the system. This operation is performed by using an exponential-weighted moving average with weight w [20]:

$$\overline{B}_n(p_l) = (1-w) \cdot \overline{B}_{n-1}(p_l) + w \cdot B_n(p_l)$$

in which $\overline{B}_n(p_l)$ and $\overline{B}_{n-1}(p_l)$ are respectively the average bandwidth values obtained in the nth and $(n-1)^{th}$ sampling windows. The weight w is calculated by $w = 1 - e^{-T/\tau}$ [20].

The choice of the time constant τ is very important, since a time constant too short does not take into account adequately past bandwidth values, that have to be taken into account for a correct admission control policy. If the time constant is too large, on the other side, strong bandwidth variations due to another accepted flow are taken into account long after they effectively occur.

If there is a further bandwidth request of the $(F+1)^{th}$ flow, it is supposed that this flow is admitted in the network if and only if the sum of the peak rate of the new flow

$r_{F+1}^{\max}$ and the estimated bandwidth $\overline{B}_n(p_l)$, in the specified sampling window T, is less than channel capacity C, that is:

$$\overline{B}_n(p_l) + r_{F+1}^{\max} \leq C$$

If the new flow is admitted, the algorithm estimates the bandwidth occupied by the aggregation of F+1 flows in the subsequent sampling windows, considering the same value of p_l.

3 Numerical Results

In this section some numerical results are presented, in order to test the effectiveness of the admission control policy based on the proposed algorithm. The obtained results are compared with simulation and other alternative algorithms that estimate the aggregate bandwidth utilizing the Hoeffding bounds [20], the Normal distribution [20], and the Chernoff bound [10], for each time window.

Eleven different MPEG1 encoded video streams have been considered: the "Asterix", "James Bond: Goldfinger", and "Jurassic Park" movies, "The Simpsons" cartoon, an ATP tennis final, two different soccer matches, a Formula 1 race, a super bowl final and two different talk shows. Each video stream has a length of 40.000 frames (approximately half an hour) with a transmission rate of 25 frames/s. The Group of Pictures (GOP) pattern is IBBPBBPBBPBB, i.e., each GOP is composed by 12 video frames. All video streams have been smoothed exploiting the FAR online smoothing algorithm [3], with a smoothing window of approximately 5 seconds and a client smoothing buffer size of 1024 Kbytes.

In the proposed scenario it is supposed that 10 different types of video streams are aggregated in a shared network link. Their starting points have been randomly chosen along the entire duration of the video streams. To guarantee the total superimposition of all streams in each frame, it is supposed that when a stream finishes, it starts again from the beginning. Then, a further bandwidth request made by the eleventh stream, chosen as "The Simpsons" cartoon, is supposed to happen in a sampling window randomly chosen among all. Each of the four mentioned admission control algorithms establishes if the eleventh flow can be admitted in the network, based on the specified QoS parameter (i.e., the chosen loss probability), in each of the sampling windows in which the algorithm is applied. In particular, the new stream is admitted only if the sum of its peak rate and the estimated bandwidth of the aggregation of streams is less than channel capacity. In our experiments a sampling window of 50 GOP has been used, so that the total number of aggregate bandwidth samples is 67. Then the exponential-weighted moving average filter, as reported in Section II.2, has been applied for bandwidth estimation, with a chosen time constant τ of 1 minute. In Figure 3 a comparison among the different algorithms for the aggregate bandwidth estimation is illustrated, supposing to have ten different types of video streams in the network link (all the types mentioned above in this section, except "The Simpsons" video stream) and assuming a loss probability of 10^{-4}.

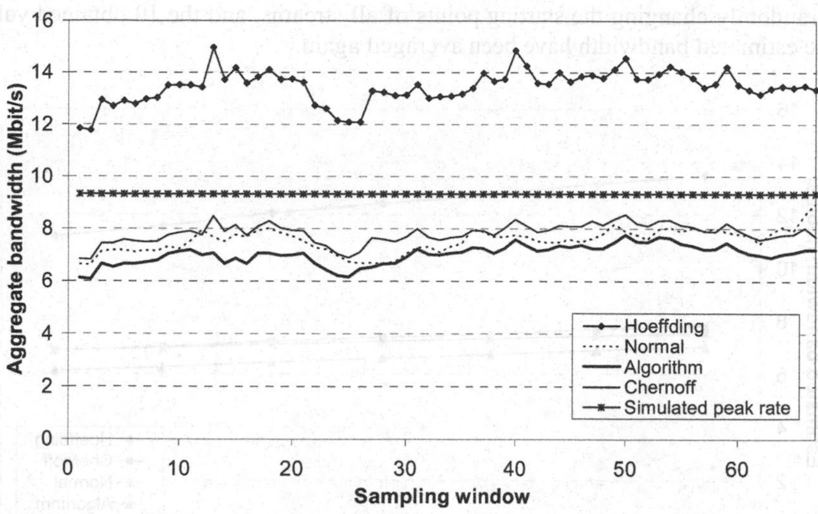

Fig. 3. Comparison among different bandwidth estimation methods, with loss probability 10^{-4} and sampling window of 50 GOP.

The "Simulated peak rate" curve represents the peak rate reached by the aggregation of streams along the entire duration of the simulation; for this reason, the peak rate value is constant for each temporal window.

It can be noted that in the considered scenario the bandwidth estimation obtained with the Hoeffding bound always overestimates the simulated peak rate curve, resulting in this way in an excessively conservative bandwidth estimation. While estimation performed with the three other algorithms, instead, remain under the peak rate curve providing in this way a better estimation. Nevertheless, the bandwidth values estimated with the proposed algorithm are, on average, lower than bandwidth values obtained with the three other algorithms. This analysis is confirmed by the results illustrated in Figure 4, where a different comparison among the four admission control algorithms is reported.

In Figure 4 is represented a comparison among aggregate bandwidth estimations performed with the four mentioned admission control algorithms, for different loss probability values. Referring to each of the four algorithms, each of the bandwidth values has been derived as follows. Ten different types of video streams, all of length 40.000 frames, have been aggregated randomly choosing their starting point uniformly along the entire duration of the streams and cyclically repeating each stream to guarantee the total superimposition of all the streams in each frame. Then, the aggregation of streams has been divided into 67 temporal windows, 66 of length 50 GOPs (600 frames) and the last of length 400 frames. In each of these windows, the aggregate bandwidth has been estimated in correspondence of a fixed value of p_l. All the bandwidth values have then been averaged on the total number of temporal windows. This kind of simulation has then been repeated 10 times, each time randomly changing the starting points of all streams, and the 10 obtained values of the estimated bandwidth have been averaged again.

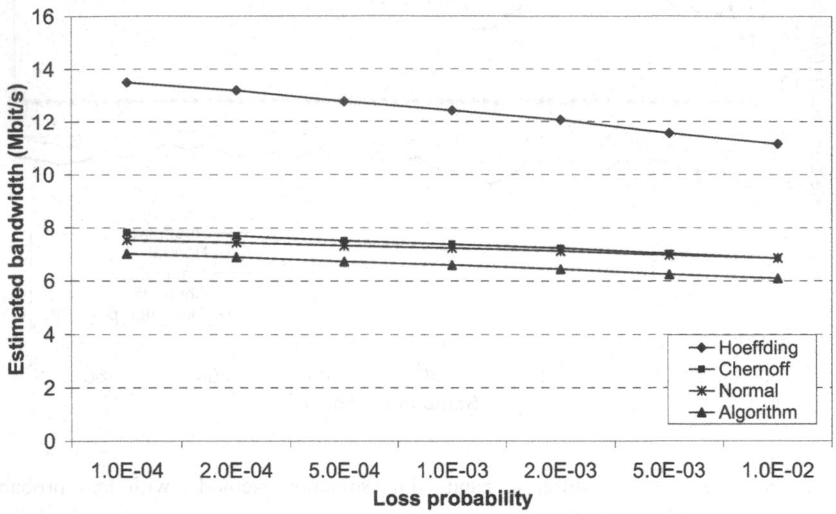

Fig. 4. Average bandwidth estimation vs loss probability, with different algorithms.

From Figure 4, it can be noted that, as expected, bandwidth estimation performed with each of the algorithms grows if loss probability decreases. This is due to the fact that more bandwidth resources have to be allocated to obtain lower losses. Again, it can be noted that the average bandwidth estimation performed with the Hoeffding bound is considerably higher than the three other algorithms, while the proposed algorithm presents, on average, the lowest values of estimated bandwidth.

Nevertheless, a lower bandwidth estimation can result in an underestimation of available bandwidth resources, that would bring to accept new streams with more losses than the specified loss probability parameter. To test the effectiveness of the proposed algorithm, an admission control test has been performed, comparing the proposed algorithm with the "Chernoff bound" and the "Normal distribution" algorithms, and with simulation results, as can be seen in Figure 5. The comparison with the Hoeffding bound has been omitted because it overestimates aggregate bandwidth, as can be clearly seen in Figure 4. As regards algorithm results, in each of the temporal windows the aggregate bandwidth has been estimated, then the peak rate of the new flow ("Simpsons") has been added and the channel capacity necessary to accept the new flow has been derived, for each loss probability value. The so found channel capacity values have then been averaged on the total number of the temporal windows of interest. This procedure has been repeated 10 times and the 10 corresponding channel capacity values have been averaged, for each loss probability.

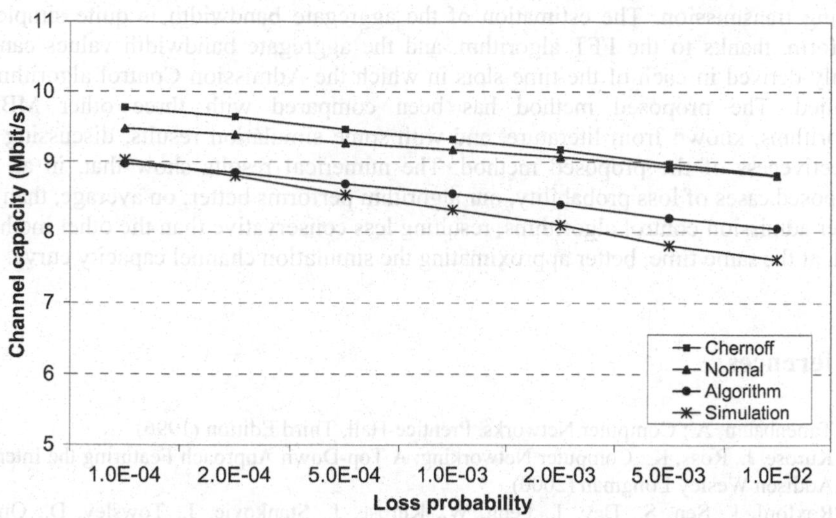

Fig. 5. Average channel capacity vs loss probability with different algorithms and comparison with simulation.

Simulation results have been derived as follows. 10 different experiments have been carried out. In each of them, in correspondence of a fixed value of loss probability, it has been supposed to accept a request of the new stream which randomly starts in each of the 67 temporal windows in which the aggregation of streams is subdivided. The new flow is then aggregated with the existing flow in a circular way, to ensure the total superimposition of all streams, as reported at the beginning of this section. A bandwidth value is then appropriately chosen in such a way that the number of frames in which the aggregate bandwidth is higher than the chosen bandwidth, divided by the total number of frames considered in the simulation, gives us the loss probability fixed previously. All the 67 bandwidth values have been averaged, and a further average has been made on the 10 performed simulations, like the comparison in Figure 4.

From Figure 5, it can be noted that, on average, the proposed algorithm performs better than the Chernoff bound and Normal distribution algorithms, since the estimated channel capacity curve is closer to the simulation curve, especially for lower loss probability values.

4 Conclusions

In this paper, a novel online algorithm to perform an Admission Control for smoothed video streams has been developed. The proposed method estimates the aggregate bandwidth into temporal windows of specified size, requiring only as input data, in the specified time slot, all bandwidth levels of the different types of video streams and the loss probability specification. These data can be directly derived from video traces during transmission. The estimation of the aggregate bandwidth is quite simple to perform, thanks to the FFT algorithm, and the aggregate bandwidth values can be easily derived in each of the time slots in which the Admission Control algorithm is applied. The proposed method has been compared with three other MBAC algorithms, known from literature, and with some simulation results, discussing the effectiveness of the proposed method. The numerical results show that, in all the proposed cases of loss probability, our algorithm performs better, on average, than the other admission control algorithms, resulting less conservative than the other methods and, at the same time, better approximating the simulation channel capacity curve.

References

1. Tanenbaum, A.: Computer Networks. Prentice-Hall, Third Edition (1996)
2. Kurose, J., Ross, K.: Computer Networking: A Top-Down Approach Featuring the Internet. Addison Wesley Longman (2000)
3. Rexford, J., Sen, S., Dey, J., Feng, W., Kurose, J., Stankovic, J., Towsley, D.: Online Smoothing of Live, Variable-Bit-Rate Video. Workshop on Network and Operating Systems Support for Digital Audio and Video. St. Louis, Missouri (1997) 249-257
4. Rexford J., Sen S., Basso A.: A Smoothing Proxy Service for Variable-Bit-Rate Streaming Video. IEEE Global Internet Symposium, Rio de Janeiro (1999)
5. Feng, W., Rexford, J.: Performance Evaluation of Smoothing Algorithms for Transmitting Prerecorded Variable-Bit-Rate Video. IEEE Transactions on Multimedia (1999) 302-313
6. Feng, W., Sechrest, S.: Critical Bandwidth Allocation for Delivery of Compressed Video. Computer Communications, Vol. 18 (1995) 709-717
7. Feng, W., Jahanian, F., Sechrest, S.: Optimal Buffering for Delivery of Compressed Prerecorded Video. IASTED/ISMM International Conference on Networks (1995)
8. Feng, W., Rexford, J.: A Comparison of Bandwidth Smoothing Techniques for the Transmission of Prerecorded Compressed Video. IEEE INFOCOM. Kobe, Japan (1997) 58-66
9. McManus, J.M., Ross, K.W., Video on Demand Over ATM: Constant-Rate Transmission and Transport. IEEE INFOCOM (1996) 1357-1362
10. Zhang, Z.L., Kurose, J., Salehi, J.D., Towsley, D.: Smoothing, Statistical Multiplexing, and Call Admission Control for Stored Video. IEEE Journal on Selected Areas in Communications Vol. 15 N° 6 (1997) 1148-1166
11. Salehi, J.D., Zhang, Z.L., Kurose, J., Towsley, D.: Supporting Stored Video: Reducing Rate Variability and End-to-End Resource Requirements Through Optimal Smoothing. IEEE/ACM Transactions On Networking Vol. 6 N° 4 (1998) 397-410
12. Knightly, E.W., Shroff, N.B.: Admission Control for Statistical QoS: Theory and Practice. IEEE Network Vol. 13 N° 2 (1999) 20-29
13. Li, S., Chong, S., Hwang, C.L.: Link Capacity Allocation and Network Control by Filtered Input Rate in High Speed Networks. IEEE/ACM Transactions on Networking Vol. 3 N° 1 (1995) 10-25
14. Reisslein, M., Ross, K.W.: Call Admission for Prerecorded Sources with Packet Loss. IEEE Journal on Selected Area in Communications (1997) 1167-1180
15. Heszberger, Z., Zátonyi, J., Bíró, J., Henk, T.: Efficient Bounds for Bufferless Statistical Multiplexing. IEEE GLOBECOM 2000 (2000) San Francisco, CA
16. Hwang, C.L., Li, S.: On Input State Space Reduction and Buffer Noneffective Region. IEEE INFOCOM (1994) 1018-1028
17. Kleinrock, L.: Queueing Systems, Volume 1: Theory. John Wiley & Sons New Jork (1975)

18. Baskett, F., Mani Chandy, K., Muntz, R.R., Palacios, F.G.: Open, Closed, and Mixed Networks of Queues with Different Classes of Customers. Journal of the Association for Computing Machinery Vol.22 N° 2 (1975) 248-260
19. Breslau, L., Jamin, S., Shenker, S.: Comments on the Performance of Measurement-based Admission Control Algorithms. IEEE Infocom 2000 (2000) Tel Aviv, Israel
20. Floyd, S.: Comments on Measurement-based Admission Control for Controlled-Load Services. Technical report, Laurence Berkeley Laboratory (1996)
21. Grossglauser, M., Tse, D.: A Framework for Robust Measurement-Based Admission Control. IEEE / ACM Transactions on Networking Vol. 7 N° 3 (1999) 293-309
22. Gross, D., Harris, C.M.: Fundamentals of Queueing Theory. John Wiley & Sons, Third Edition (1998)
23. Camarda, P., Striccoli, D., Trotta, L.: An Admission Control Algorithm based on Queueing Networks for Multimedia Streaming. International Conference on Telecommunication Systems – Modeling and Analysis. Monterey, CA (2002)
24. Zhang, Z.L., Kurose, J., Salehi, J.D., Towsley, D.: Smoothing, Statistical Multiplexing, and Call Admission Control for Stored Video. IEEE Journal on Selected Areas in Communications Vol. 15 N° 6 (1997) 1148-1166
25. Camarda, P., Striccoli, D.: Admission Control in Video Distribution Systems Exploiting Correlation Among Streams. SoftCOM 2001. Split, Dubrovnik (Croatia), Ancona, Bari (Italy) (2001)
26. Manjunath, D., Sikdar, B.: Integral Expressions for Numerical Evaluation of Product Form Expressions Over Irregular Multidimensional Integer State Spaces. Symposium on Performance Evaluation of Computer and Telecommunication Systems. Chicago, Illinois (1999) 326-333

DQM: An Overlay Scheme for Quality of Service Differentiation in Source Specific Multicast

Ning Wang, George Pavlou

Centre for Communication Systems Research, University of Surrey
Guildford, United Kingdom
{N.Wang, G.Pavlou}@eim.surrey.ac.uk

Abstract. In this paper we propose a new scheme named *DQM* (Differentiated *QoS* Multicast) based on the Source Specific Multicast (*SSM*) [7] model in order to provide limited and qualitative *QoS* channels to support heterogeneous end users. Similar to the DiffServ paradigm, in *DQM* the network is configured to provide finite *QoS* service levels to both content provider and receivers. Based on the Service Level Agreement, both the content provider and group members should select a specific *QoS* channel available from the network for data transmission, and in this case arbitrarily quantitative *QoS* states are eliminated. Moreover, we use the group address *G* contained in the (*S*, *G*) tuple in *SSM* service model to encode *QoS* channels, and data packets that belong to the same *QoS* channel identified by a common class *D* address can be treated aggregately, and this can be regarded as an overlay solution to Differentiated Services, specifically for source specific multicast applications.

1 Introduction

In contrast to the current Internet, applications with Quality-of-Service (*QoS*) requirements will be an important aspect in the next generation of the Internet. In addition, applications based on group communication will also become widespread. Given this expected evolution, the situation in which the Internet is uniquely dominated by point-to-point applications based on the Best Effort (*BE*) service level will change in the future.

Multicasting is an efficient approach for group communications, and the recently proposed Source Specific Multicast (*SSM*, [7]) model has been considered to be a promising solution for the development of one-to-many applications in a large scale. In *SSM* each group is identified by an address tuple (*S*, *G*) where *S* is the unique *IP* address of the information source and *G* is the destination channel address. Direct join requests from individual subscribers create a unique multicast tree rooted at the well-known information source, i.e., *SSM* defines (*S*, *G*) channels on per-source basis. In this model, the scalability problems of *IP* multicast such as address allocation and inter-domain source discovery are not deployment obstacles any more. Due to its simplicity and scalability, *SSM* is expected to see significant deployment on the Internet in the near future, especially for single source applications.

On the other hand, the provisioning of *QoS* requirements in a scalable manner is another major research dimension towards the next generation of the Internet. An

M. Ajmone Marsan et al. (Eds.): QoS-IP 2003, LNCS 2601, pp. 306-319, 2003.

efficient solution is to classify traffic into finite service levels and treat packets that belong to the same *QoS* service level in an aggregate manner. Differentiated Services (*DiffServ*) [1] is a typical example of this approach and is considered a scalable scheme for deploying *QoS* widely. Research efforts have also addressed the problem of applications with heterogeneous *QoS* requirements given the potentially different capacity of individual receivers. D. Yang *et al* proposed Multicast with *QoS* (*MQ*) [12] as an integrated mechanism with the consideration of *QoS* routing, resource reservation and user heterogeneity. This genuine receiver-initiated approach inherits some basic characteristics of *RSVP* [3], such as quantitative *QoS* guarantee and resource reservation merging from heterogeneous end users. It should be noted that *MQ* also requires that on-tree routers maintain state on a per-flow basis for end-to-end *QoS* guarantees, and this aspect still leaves the scalability issue problematic.

In this paper, we propose a new framework called Differentiated *QoS* Multicast (*DQM*) to support *qualitative* service levels (e.g., Olympic Services) based on the Source Specific Multicast model. The basic characteristic of *DQM* is as follows: First, qualitative *QoS* states are directly encoded in the class D address and is centrally managed by the *ISP*, so that core routers inside the network remain stateless regarding *QoS* service classes. Second, differentiated level of *QoS* demands for the specific information source is merged in the distribution tree, and data packets from different sources but belonged to the same *QoS* service level identified by a common multicast group address can be treated aggregately. Moreover, a pre-defined number of classes of services by the *ISP* make it easier to provision network resources for each *QoS* aggregate, and this is in the same flavor of the classical Differentiated Services. From this point of view, the proposed *DQM* model can be regarded as an overlay solution of DiffServ, specifically for source specific multicast applications.

2 Supporting Applications with *QoS* Heterogeneity

2.1 The *MQ* Approach

Being an integrated solution, *MQ* sets up a multicast distribution tree with quantitative *QoS* requirements, and makes explicit bandwidth reservation for each group member during the phase of tree construction. When there exist heterogeneous receivers, resources are reserved up to the point where the paths to different receivers diverge. When a join request propagates upstream towards the source, it stops at the point where there is already an existing *QoS* reservation that is equal to or greater than that being requested. Fig. 1 basically illustrates how different resource reservations are merged along the multicast join procedure. Suppose the requests from receiver *A*, *B* and *C* demands 10Mbps, 512kbps and 56kbps bandwidth respectively, their reservations are merged to the highest request at each hop as shown in the figure. *MQ* can also adapt to resource consumption with dynamic group membership. For example, if an on-tree router detects that the departing receiver originally requested the highest *QoS*, it will automatically shrink its reservation or even reshape the distribution tree to exactly satisfy the remaining participants. In Fig. 1(b), we can find that when receiver *A* with the bandwidth requirement of 10Mbps wants to leave the multicast session, the remaining receiver *B* with 512kbps requirement will switch from the original "shared" path (S→R1→R2→R4) with the capacity of 10Mbps to a shorter one (S→R3→R4) which still satisfies its *QoS* demand for bandwidth optimization purpose.

On the other hand, the mechanism for network resource allocation is in an accumulative fashion, i.e., bandwidth is reserved in sequence for various incoming *QoS* requests until the link becomes saturated. This approach is simple but might not be efficient in bandwidth allocation especially in case of highly dynamic group membership. From the deployment point of view, each on-tree router needs to maintain not only group states but also the quantitative *QoS* demands for its downstream receivers, and this imposes heavy overhead in a similar fashion to *RSVP*.

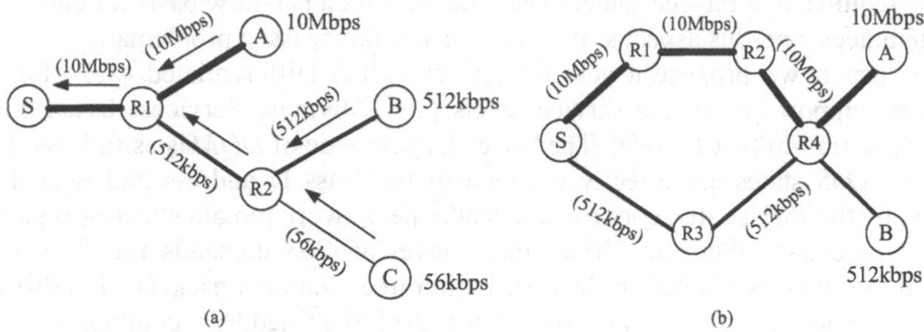

Fig. 1 *MQ* group join and tree reshaping

2.2 Layered Transmission

This approach is particularly useful for Internet *TV* applications since it relies on the ability of many video compression technologies to divide their output stream into layers: a *base layer* as well as one or more *enhancement layers*. The base layer is independently decoded and it provides a basic level of quality. The enhancement layers can only be decoded together with the base layer to improve the video quality. The source can send individual layers to different multicast groups and a receiver can join the group associated with the base layer and as many layers for enhancement as its capability allows. Receiver-Driven Layered Multicast (*RLM*) [8] is a typical example for layered video transmission. Fig. 2 illustrates the working scenario of *RLM*, and Fig. 2(b) describes how receiver *R2* "probes" to subscribe to additional enhanced layers.

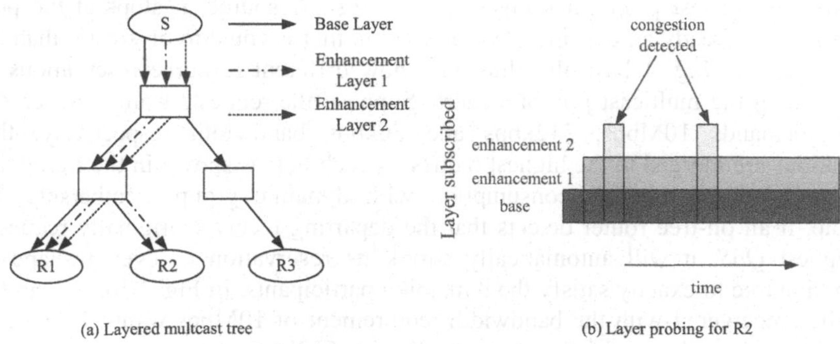

Fig. 2 Layered transmission (*RLM*)

2.3 Replicated Transmission

It should be noted that not all types of multimedia streams can be encoded into layers as described above. An alternative approach is replicated transmission that is applicable to generalized type of multimedia applications. In this approach, the information source keeps a finite number of streams carrying the same content but each targeted at receivers with different capabilities. In a similar fashion to layered transmission, the data source assigns different multicast groups to each of the maintained streams, and receivers may move among them by subscribing to the corresponding group address. A typical example of replicated transmission is Destination Set Grouping (*DSG*) [4]. Fig. 3 describes how *DSG* works in a heterogeneous environment.

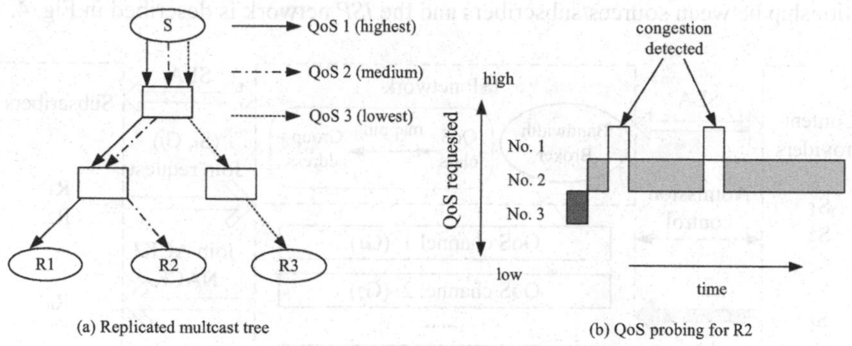

Fig. 3 Replicated transmission (*DSG*)

3 Basic *DQM* Framework

From the previous section, we can find that *MQ* provides a type of arbitrary bandwidth guaranteed service, while *DSG* and *RLM* adopts differentiated services to heterogeneous end users. From the viewpoint of scalability, the latter two approaches incur lighter group state overhead inside the network. On the other hand, in both *RLM* and *DSG* it is external sources that decide individual *QoS* classes, and thus it is difficult to perform aggregated data treatment for multiple sources inside the network since the *QoS* definition and configuration of each source/group session is different.

In this paper we propose a new transmission scheme, called Differentiated *QoS* Multicast (*DQM*), which can be regarded as the integration of Source Specific Multicast (*SSM*) and the Olympic Service model in DiffServ. From a viewpoint of the *ISP*, it provides external source/receivers with a finite class of services (e.g., gold service, silver service and bronze service), each of which is uniquely encoded into a class *D* address (within subset of 232/8) in the *SSM* model. In such a situation, the interpretation of *SSM* address tuple (*S, G*) becomes straightforward: *S* identifies the address of the information source and *G* stands for the *QoS* service level (we name it *QoS* channel) that is available from *S*. Once receivers have decided the source address *S* and the class address *G*, they will directly send (*S, G*) join requests to the source.

Since the provided service levels are centrally managed by the *ISP* instead of individual sources, this type of mapping between *QoS* class and group address does not introduce any scalability problem when the number of external sources increases (i.e., total number of *QoS* channels is independent of the number of external sources). The Bandwidth Broker (*BB*) in the network is responsible for computing available bandwidth resource and deciding whether or not to accept new data injections from sources as well as join requests from group members. When a new subscriber wants to receive data from the source *S*, it will first negotiate with *BB* on *QoS* channel selection. If the *BB* is able to allocate sufficient bandwidth to graft the subscriber to the existing source specific tree in the requested channel, the (*S, G*) join request is accepted. If the *BB* cannot allocate the required bandwidth for including the new subscriber into the *QoS* channel, the join request will be rejected, and the subscribers may subscribe to other *QoS* channels by sending different (*S, G'*) join requests. The relationship between sources/subscribers and the *ISP* network is described in Fig. 4.

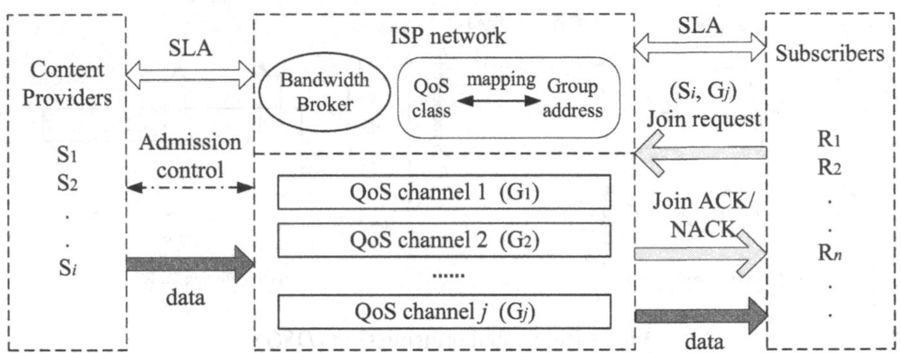

Fig. 4 *DQM* Framework

When the join request has been approved, the new subscriber will combine the *IP* address of *S* and group address *G* identifying the desired *QoS* channel into a (*S, G*) tuple, and send this join request towards the source *S*. If this (*S, G*) join reaches an on-tree router with group state (*S, G'*) where *G'* identifies a higher *QoS* channel, then this join request is terminated at this router. In this case, a new flow is branched from the existing source specific tree and leads to the new subscriber. A sender simply needs to inject a single data stream into the network with the highest *QoS* channel requested. Fig. 5 presents the basic description of the proposed *QoS* merging in *SSM* service model. It is worth mentioning that this type of *QoS* merging is significantly different from both *MQ* and *RSVP* explicit reservation. *MQ* and *RSVP* target to achieve end-to-end *QoS* guarantee with quantitative reservation style and hence the relevant state needs to be maintained for each flow with various *QoS* demands. In the scenario of differentiated *QoS* classes, not only group members have to select one of the available channels provided by the *ISP*, but also information sources should conform to the *ISP's* channel configuration profile. In this case group states scalability inside the network is achieved since external sources are not allowed to provide their own *QoS* channel configuration as they do in *DSG* and *RLM*. Also note that data treatment inside the network is decided by the class *D* address *G*, i.e., within each router data packets from different sources *S* but with common class *D* address *G* can be treated aggregately. In this sense, the group address *G* takes additional role similar to that of the Differentiated Services Code Point (*DSCP*) in DiffServ networks.

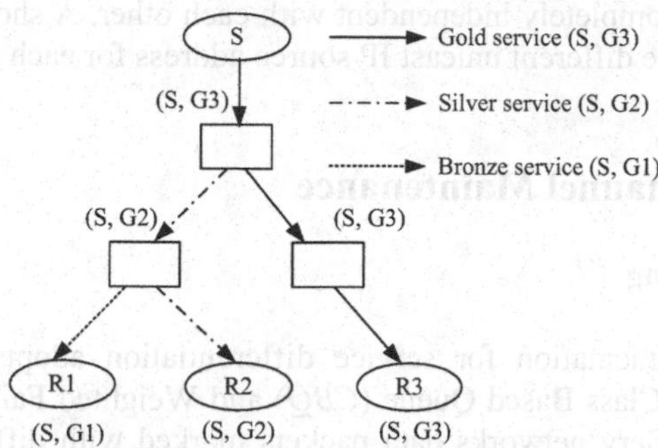

Fig. 5 *DQM* group state maintenance

The advantages of the proposed *DQM* scheme are as follows. Most importantly, it solves the fundamental confliction between stateless DiffServ service model and state based IP multicast semantics. In DiffServ model, core routers do not maintain any *QoS* states for individual applications/flows, and data treatment is indicated inside each packet header. On the other hand, the basic mechanism of IP multicast is to keep group states inside the network so as to enroute data to active group members. In *DQM*, *QoS* states are directly encoded into multicast address and maintained inside the network, as originally needed in IP multicast, and hence no other *QoS* information need to be kept at core routers. Otherwise, if we don't use group address to identify *QoS* channels in the join request, this information should be contained elsewhere, for instance, in the option field of the join request packet. When this join request is heading for the information source, each of the routers it has passed through should record the *QoS* requirements information and associate it with the downstream interface from which the join request has been received. This is necessary because otherwise *DSCP* contained in the data packet cannot be modified when the packet reaches the branching point of the source specific tree where heterogeneous *QoS* classes meet each other. By recording the *QoS* class information at core routers, when the group data comes from the upstream interface, the on-tree router exactly knows through which *QoS* class it should forward the packets on its different downstream interfaces. No doubt, this approach requires that core routers maintain *QoS* information at its downstream interfaces in addition to the plain (*S, G*) state, but this does not conform to the basic *QoS* stateless requirement of DiffServ service model. Second, service differentiation is centrally defined and managed by the *ISP* instead of individual sources, as it is done in *DSG* and *RLM*, traffic from different sources but with identical *QoS* class can be treated in an aggregated fashion inside the network. Finally, in contrast with the "come and use" strategy of bandwidth allocation in *MQ*, *DQM* allows an *ISP* to allocate network resources specifically to individual *QoS* channels according to the forecasted traffic demands from each of them, so that the traffic loading can be improved by the more flexible bandwidth configuration.

However, there is a restriction regarding this approach. Since the *QoS* channel is source specific, it is impossible for a single source with a unique *IP* address *S* to send multiple data streams with different contents. In the classic *SSM* model, an information source can be simultaneously involved in multiple groups because (*S, G1*) and (*S, G2*) are completely independent with each other. A short-term solution to this restriction is to use different unicast IP source address for each group session.

4 *DQM QoS* Channel Maintenance

4.1 Data Forwarding

Current router implementation for service differentiation adopts priority queuing technologies such as Class Based Queue (*CBQ*) and Weighted Faired Queue (*WFQ*). For example, in DiffServ networks data packets marked with different *DSCP* value are treated in queues with different priority for scheduling. Similarly, in *DQM* core network bandwidth is divided specifically for each *QoS* channel and data packets from different channels (distinguished by class D address) are scheduled in the corresponding priority queues. In this section we basically describe the working mechanism of routers that can support the relevant functionality of *QoS* channel differentiation.

Once a router receives (*S, G*) join requests with different values of *G* that are associated with various *QoS* channels from subscribers, it will merge all of them and only send a single (S, G_m) join request towards *S*, where G_m is the class *D* address associated with the highest *QoS* channel being requested. Here we define the interface from which a join request is received as the *downstream interface* and the one used to deliver unicast data to the source as the *upstream interface*. When the router receives group data from its upstream interface it will take the following actions to forward the packets (also shown in Fig. 6(a), suppose *QoS*(*G*)>*QoS*(*G*')):

(1) Look up the group state(s) associated with the source *S* on each downstream interface and duplicate the packet where necessary.
(2) Copy the value of *G* contained in the (*S, G*) state at each downstream interface to the *IP* destination field in the duplicated packet (if the two are not consistent).
(3) Assign the data packet to the priority queue associated with relevant *QoS* channel at the downstream interface based on the (*S, G*) channel state.

Step (2) is necessary since the value of *G* contained in the packet indicates how this packet will be treated in the next hop of on-tree router. Remember that the group states are created by (*S, G*) join requests for different *QoS* channels, and the way data packets are treated at each router is uniquely identified by the value of *G* contained in the (*S, G*) state, and in this way data can be forwarded according to the *QoS* requests from individual users. On the other hand, data packets from different sources *S* but with the same class *D* address in their (*S, G*) address tuples are treated aggregately in the corresponding queues. To achieve this, the *ISP* should also make the configuration such that each priority queue is associated with a group address on downstream interfaces (shown in Fig. 6(b)). This figure also illustrates how data from different sources but with a common group address is treated aggregatively in a specific queue of a core router.

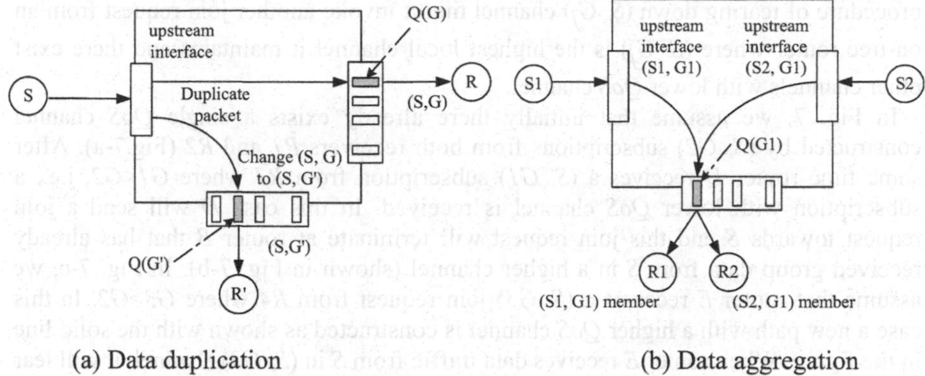

(a) Data duplication (b) Data aggregation

Fig. 6 *DQM* Forwarding behaviour at core routers

4.2 Dynamic Group Membership

In most multicast applications, subscribers join and leave frequently throughout a session. In this section we will discuss on how *QoS* channel merging for a common source is performed with dynamic group membership. On the other hand, individual priority queue should be implemented with proper bandwidth allocation for each *QoS* channel, thus the joining path of same source-destination pair might not be exactly the same for all (*S, G*) channels. Path computation should also consider the specific bandwidth availability of the subscribed *QoS* channel, and we name this Per Channel *QoS* routing in *DQM*.

(1) *QoS* channel subscription

When a (S, G_i) join request reaches a router that (i) has already received traffic from the source *S* with the same or a higher *QoS* channel, i.e., with group state of (S, G_j) where $G_i \leq G_j$[1] and (ii) the corresponding priority queue at the interface from which the join request is received has sufficient bandwidth, then the join procedure terminates and this interface obtains group state (S, G_i). Thereafter, data packets from *S* are duplicated and forwarded to this interface with the class *D* address of the new packets modified from G_j to G_i. In this way, a new tree branch is grafted to the current *QoS* channel that has equal or higher service level.

If the (S, G_i) join request reaches a router with the highest available *QoS* channel (S, G_j) where $G_i > G_j$ (i.e., a router with lower *QoS* channel for *S*), the join will continue to explore a new path that satisfies the requirement of the (S, G_i) channel subscription. Once a path with desired *QoS* channel has been set up and this particular router has received the traffic from (S, G_i) channel, it will tear down the (S, G_j) channel on the original path with lower *QoS* level. It should also be noted that the

---

[1] In this paper we assume that higher class *D* address is associated with higher *QoS* channel, i.e., $G_i > G_j \leftrightarrow QoS(G_i) > QoS(G_j)$

procedure of tearing down (S, G_j) channel might invoke another join request from an on-tree router where (S, G_j) is the highest local channel it maintains and there exist other channels with lower *QoS* channel.

In Fig. 7, we assume that initially there already exists a single *QoS* channel constructed by *(S, G2)* subscriptions from both receivers *R1* and *R2* (Fig.7-a). After some time router *D* receives a *(S, G1)* subscription from *R3* where *G1*<*G2*, i.e., a subscription with lower *QoS* channel is received. In this case *D* will send a join request towards *S* and this join request will terminate at router *B* that has already received group data from *S* in a higher channel (shown in Fig. 7-b). In Fig. 7-c, we assume that router *E* receives a *(S, G3)* join request from *R4* where *G3*>*G2*. In this case a new path with a higher *QoS* channel is constructed as shown with the solid line in the figure. When router *E* receives data traffic from *S* in *(S, G3)* channel, it will tear down the original *(S, G2)* channel back to *S*. When router *B* has detected the pruning, it finds that it has also maintained a lower *QoS* channel for *R3*, namely *(S, G1)*. Therefore, router *B* will first send a *(S, G1)* join request back to *S*. When detecting that group data from *S* comes in the new channel *(S, G1)*, router *B* will tear down the original *(S, G2)* channel on link *AB* as shown in Fig. (7-d). More detail on how *QoS* channels are torn down is presented below.

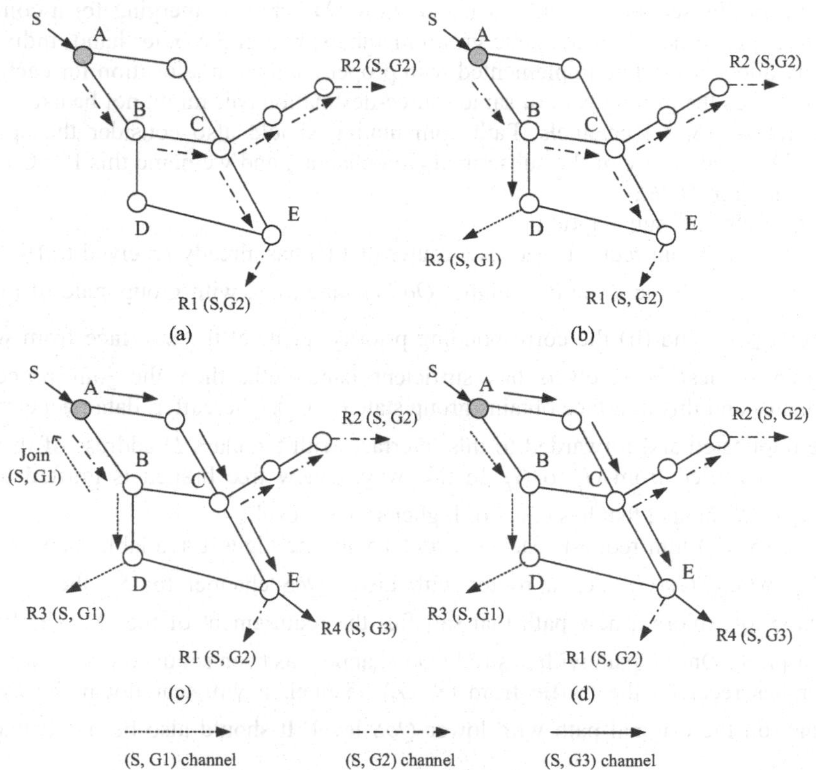

Fig. 7 Dynamic *QoS* channel subscription

(2) *QoS* channel unsubscription

Suppose that a particular router is currently receiving traffic from source *S* with *QoS* channel (S, G_i). When it detects no (S, G_i) subscribers attached and wants to leave the channel, it will stop sending (S, G_i) join requests towards the source *S*. When the (S, G_i) state times out, the on-tree router will check all its downstream interfaces with *QoS* channels associated with *S*. There exist three possible cases as follows (illustrated in Fig. 8):

a. There exists at least one (S, G_j) state where $G_j \geq G_i$, the router simply stops forwarding traffic on (S, G_i) channel at the corresponding downstream interface, and it needs not to take any further pruning actions;

b. There does not exist any (S, G_j) state where $G_j > G_i$, the router will check the status of all the remained *QoS* channels associated with *S* and select class *D* address G_m that is mapped to the highest *QoS* channel being currently requested, and send (S, G_m) join request towards the source *S*. Once this router has received data traffic from the (S, G_m) channel, it will stop sending (S, G_i) join requests on its upstream interface;

c. If this is the last subscriber attached on the router, the router simply stop sending any (S, G) join requests towards the source and hence it breaks from the tree.

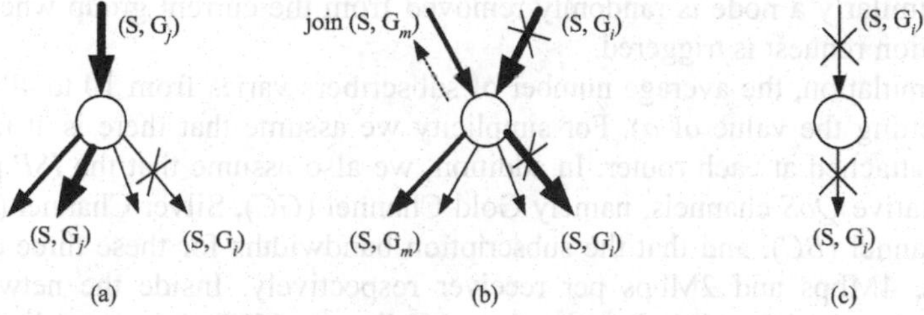

Fig. 8 Dynamic *QoS* channel unsubscription

Finally, it should be noted that, a boundary router issues join/leaving requests only when the first receiver for a new (S, G) session is joined or the last member leaves the group. This strategy of pushing group management to the edge of the network significantly reduces the frequency of reshaping delivery trees inside of the domain.

5 Simulation Results

5.1 Simulation Model

In this section, we evaluate the proposed scheme through simulation. We adopt the commonly used Waxman's random graph generation algorithm [11] that has been implemented in *GT-ITM*, for constructing our network models. This approach distributes the nodes randomly on the rectangular grid and nodes are connected with the probability function:

$$P(u,v) = \lambda \exp(\frac{-d(u,v)}{\rho L})$$

where $d(u,v)$ is the distance between node u and v and L is the maximum possible distance between any pair of nodes in the network. The parameters λ and ρ ranging (0, 1] can be modified to create the desired network model. A larger value of λ gives node with a high average degree, and a small value of ρ increase the density of shorter links relative to longer ones. In our simulation we set the values of λ and ρ to be 0.3 and 0.2 respectively, and generate a random network with 100 nodes with the source node being randomly selected.

In order to generate group dynamics, a sequence of events for *QoS* subscription/unsubscription are also created. A probability model is used to determine whether a request is for *QoS* subscription or unsubscription. The function

$$P_c = \frac{\alpha(N-m)}{\alpha(N-m)+(1-\alpha)m}$$

is defined for this purpose [11]. The function P_c is the probability that a *QoS* subscription is issued. In the function, m indicates the current number of subscribers while N identifies the network size. α ranging (0, 1) is the parameter that controls the density of the group (i.e., average number of subscribers). When a *QoS* subscription is issued, a node that is not in the multicast group is randomly selected for joining the session. Similarly a node is randomly removed from the current group when a *QoS* unsubscription request is triggered.

In our simulation, the average number of subscribers varies from 10 to 40 in steps of 5 (by setting the value of α). For simplicity we assume that there is at most one subscriber attached at each router. In addition, we also assume that the *ISP* provides three qualitative *QoS* channels, namely Gold Channel (*GC*), Silver Channel (*SC*) and Bronze Channel (*BC*), and that the subscription bandwidths for these three channels are 8Mbps, 4Mbps and 2Mbps per receiver respectively. Inside the network, the bandwidth capacity of each link varies from 10Mbps to 45Mbps in even distribution. The bandwidth capacity of each link is configured according to the following proportion: 50% for *GC*, 30% for *SC* and 20% for *BC* respectively. Among all the receivers, we suppose that 20% of them subscribe to *GC*, 30% to *SC* and 50% to *BC*. In our simulation we adopt *QOSPF* for receiver-initiated multicast routing that is introduced in [12] as the underlying routing protocol for each *QoS* channel. According to [12], *QOSPF* based multicast routing does not support user heterogeneity within a particular group, but in *DQM* such type of *QoS* heterogeneity is reflected by different (*S, G*) group identification. In this sense, *QOSPF* can still apply to per *QoS* channel routing in *DQM*, and different tree branches can be merged if the same source address *S* is discovered.

5.2 Performance Analysis

First of all, we investigate bandwidth conservation performance, and comparisons are made between *DQM* and that of building independent trees for each service level with disjoined *QoS* channels (e.g., *DSG* [4] in which the source maintains independent data streams for heterogeneous users simultaneously). Following that we focus on the capability of traffic engineering in terms of load balancing between *DQM* and *MQ* approaches. The simulation results regarding network utilization are also compared between *DQM* and *MQ* but overhead for group states maintenance is incomparable since the latter involves quantitative states for user heterogeneity.

In order to evaluate the network utilization, we define the bandwidth saving overhead for a particular channel *C* (*C* could be gold, silver and bronze service etc.) as follows:

$$O_C = 1 - \frac{U^C_{DQoM}}{U^C_{DSG}}$$

where U^c_{DQoM} is the bandwidth utilization of channel *C* by *DQM*, and U^c_{DSG} is that by using non-hybrid tree schemes such as *DSG*. Similarly, we define the overhead for all channels as:

$$O_T = 1 - \frac{U_{DQoM}}{U_{DSG}}$$

where U_{DQoM} is the overall link utilization by *DQM* and U_{DSG} is that by *DSG*.

Fig. 9 illustrates the overhead performance for both individual *QoS* channels and overall bandwidth conservation. From the figure we can find that in *DQM* bandwidth for non-gold channels can always be saved and the corresponding overhead varies from 0.33 to 0.46. However, bandwidth for gold channel is not conserved at any time since it cannot be merged into any other *QoS* channel. Regarding the overall bandwidth conservation, from the figure we notice that the aggregated overhead varies from 0.19 to 0.23, i.e., by using *QoS* channel merging in *DQM*, the average bandwidth consumption is 81.3% to 84% that of non-*QoS* merging approaches.

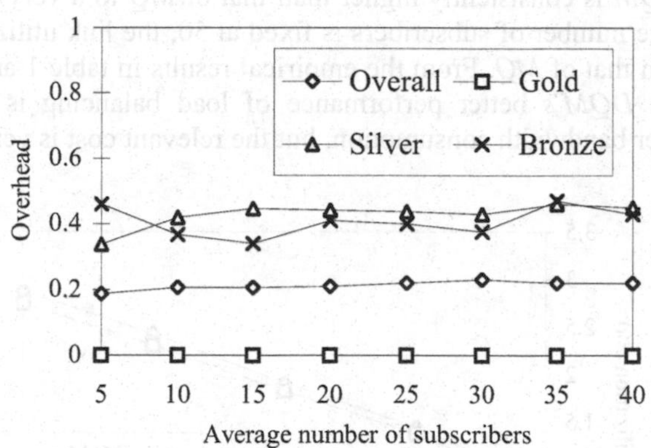

Fig. 9 Overhead of bandwidth conservation

Another interesting empirical study is the traffic engineering capability of *DQM* and *MQ*. In *DQM* network bandwidth is pre-allocated to specific traffic aggregates of individual *QoS* channels, and this is very similar to the strategy of DiffServ. In contrast, *MQ* and *RSVP* allows the overall bandwidth to be accumulatively reserved by various quantities of *QoS* demands until the link has become saturated. In the following simulation, we will examine the performance of load balancing semantics of *DQM* and *MQ/RSVP*. Basing on bandwidth utilization, we classify network links into the following three categories: (1) High load link with overall utilization above 50%; (2) Medium load link with overall utilization between 20% and 50%; and (3) Low load link with overall utilization below 20%. Table 1 presents the proportion of these three types of links inside the network with the average number of subscribers varying from 10 to 50. From the table we can find that *DQM* performs better capability of traffic engineering in that data traffic is more evenly distributed. For example, when the average number of subscribers is below 30, none of the network links become highly loaded by using *DQM*. In contrast, *MQ* always results in hotspots with utilization above 50% even when the average number of subscribers is 10. From the table we can also see that the proportion of low load link in *DQM* is consistently higher than that in *MQ*.

Table 1. Traffic distribution comparison with *MQ*

		10	20	30	40	50
DQM	High load link	0.00%	0.00%	0.00%	0.07%	0.12%
	Medium load link	1.2%	2.6%	4.1%	4.9%	5.4%
	Low load link	98.8%	97.4%	95.9%	95.0%	94.5%
MQ	High load link	0.23%	0.41%	0.86%	1.33%	1.58%
	Medium load link	1.7%	3.1%	4.1%	4.7%	6.7%
	Low load link	98.1%	96.5%	95.0%	94.0%	91.7%

We also investigate the overall link utilization of *DQM* and *MQ*, and the simulation results are presented in Fig. 10. From the figure we notice that the average link utilization of *DQM* is consistently higher than that of *MQ* to a very small scale, e.g., when the average number of subscribers is fixed at 50, the link utilization of *DQM* is 4.7% higher than that of *MQ*. From the empirical results in table 1 and Fig. 10, it can be inferred that *DQM's* better performance of load balancing is in effect at the expense of higher bandwidth consumption, but the relevant cost is very small.

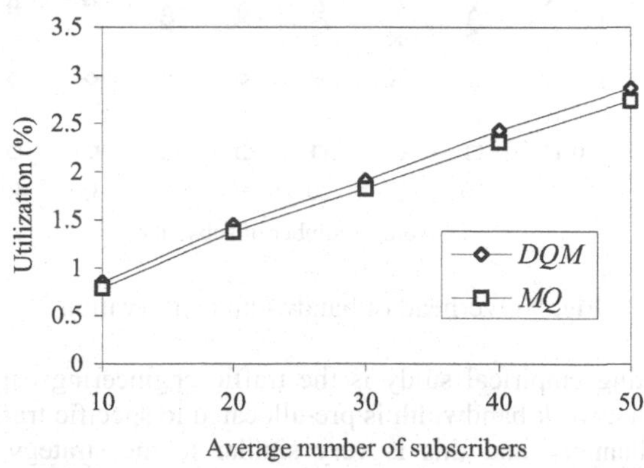

Fig. 10 Overall link utilization comparison with *MQ*

6 Summary

In this paper we proposed a novel scheme called *DQM* that provides differentiated *QoS* channels based on the Source Specific Multicast (*SSM*) service model. This approach efficiently supports heterogeneous *QoS* requirements applications on a qualitative basis. By means of Per channel *QoS* routing and merging mechanism, not only router overhead for maintaining group states is alleviated, but also network bandwidth consumption is reduced compared with traditional solutions such as multicast layered transmission and replicated transmission. Moreover, per *QoS* channel based bandwidth management also contributes to improvements in terms of traffic load distribution compared with the *MQ/RSVP* approaches.

Our future work will address dynamic configuration and management of network resources (e.g., bandwidth pre-emption between *QoS* channels) based on forecasted traffic condition of *QoS* aggregates, and also algorithms for Diffserv-aware multicast traffic engineering as well.

References

1. Blake, S. *et al*: An Architecture for Differentiated Services, *RFC* 2475
2. Bless, R., Wehrle, K.: Group Communication in Differentiated Services Networks, Proc. *IQ2001*, pp. 618-625
3. Braden, R., *et al*: *Resource* ReSerVation Protocol (RSVP) -- Version 1 Functional Specification, RFC 2205, Sept. 1997
4. Cheung, S. *et al*: On the Use of Destination Set Grouping to Improve Fairness in Multicast Video Distribution, Proc. *IEEE INFOCOM'96*, pp553-560
5. Diot, C. *et al*: Deployment Issues for the IP Multicast Service and Architecture, *IEEE* Network, Jan./Feb. 2000, pp 78-88
6. Holbrook, H. W., Cheriton, D. R.: IP Multicast Channels: EXPRESS Support for Large-scale Single-source Applications, Proc. *ACM SIGCOMM'99*
7. Holbrook, H. W., Cain, B.: Source-Specific Multicast for IP, Internet Draft, draft-holbrook-ssm-arch-*.txt, Mar. 2001, work in progress
8. McCanne, S. *et al*: Receiver-Driven Layered Multicast, Proc. of *ACM SIGCOMM'* 96
9. Striegel, A., Manimaran, G.: A Scalable Approach for DiffServ Multicasting, Proc. *ICC'2001*
10. Wang, B., Hou, J.: QoS-Based Multicast Routing for Distributing Layered Video to Heterogeneous Receivers in Rate-based Networks, Proc. *INFOCOM* 2001
11. Waxman, B. M.: Routing of multipoint connections, *IEEE JSAC* 6(9) 1988, pp1617-1622
12. Yang, D. *et al*: MQ: An Integrated Mechanism for Multimedia Multicasting, *IEEE* Trans. on Multimedia, Vol. 3, No. 1., 2001, pp 82-97

Directed Trees in Multicast Routing

Maria João Nicolau[1], António Costa[2], Alexandre Santos[2], and Vasco Freitas[2]

[1] Departamento de Sistemas de Informação,
Universidade do Minho, Campus de Azurém,
4800 Guimarães, Portugal
joao@uminho.pt
[2] Departamento de Informática,
Universidade do Minho, Campus de Gualtar,
4710 Braga, Portugal
{costa,alex,vf}@uminho.pt

Abstract. Traditional multicast routing protocols use RPF (Reverse Path Forwarding) concept to build multicast trees. This concept is based upon the idea that an actual delivery path to a node is the reverse of the path from this node to the source. This concept fits well in symmetric environments, but in a routing environment where Quality of Service is considered the guarantee that a symmetrical path will exist between two network addresses is broken. Available network resources impose specific Quality of Service asymmetries, therefore reverse path routing may not be used.

In this paper a new multicast routing strategy is proposed, enabling directed trees establishment, instead of reverse path ones. This new strategy, DTMP- Directed Trees Multicast Routing, is then implemented and simulated using Network Simulator. Simulation results, driven from several scenarios are presented, analyzed and compared with PIM-SM.

1 Introduction

Low communication costs, rapid deployment, and the ability to deal with almost every media type (namely audio, video and even video conferencing) make multicast applications very useful tools and so, large ISPs, are now supporting native multicast access inside their backbones.

Many applications in the Internet, such as video-conference, distance learning and other Computer Supported Cooperative Work (CSCW) applications require multicast support from the underlying network. These applications interconnect multiple users (several sources and receivers), exchanging large streams of data and thus an efficient use of network resources is needed. Multicast communication is the best way to send the same data simultaneously, and in an efficient way, to a non-empty set of receivers without incurring into network overloads. Hence, at each multicast-interested router, only a single copy (per group) of an incoming multicast packet is sent per active link, rather than sending multiple copies (per number of receivers accessed) via that link.

M. Ajmone Marsan et al. (Eds.): QoS-IP 2003, LNCS 2601, pp. 320–333, 2003.

Routing multicast traffic requires building a distribution tree (or set of trees). Data packets are delivered using that tree, thus the major goal of the routing protocol is to build a tree with minimum cost (in what respects to some set of parameters). The problem of finding such a tree is NP-complete and is known as *Steiner Tree Problem*[1] and plenty of heuristics have been proposed to efficiently find multicast trees. The most commonly used heuristic consists of building a spanning tree by adding each participant at a time, by means of finding the shortest path from the new participant into the nearest node of the spanning tree. Such a tree is called *Reverse Path Tree*. This heuristic assumes that links connecting any two nodes are symmetric, in other words, assuming that link costs, in either direction, are equal.

However, when routing constraints are introduced there is no guarantee that this would be the case. Links may be asymmetric in terms of the quality of service they may offer, thus link costs are likely to be different in each direction. Therefore reverse path routing is not adequate to address Quality of Service Routing.

The Protocol Independent Multicast-Sparse Mode (PIM-SM)[2] is a widely deployed multicast routing protocol, designed for groups where members are sparsely distributed over the routing domain. It is based upon the concept of Rendez-Vous Points (RP), pre-defined points within the network known by all routers. A router with attached hosts interested in joining a multicast group will start a multicast tree by sending a join message on the shortest path to the RP. This join message is processed by all the routers in between the new receiver and the first in-tree node and a new branch for the new member is setup within the multicast tree.

PIM-SM has important advantages when compared to other multicast routing protocols: it does not depend on any particular unicast routing protocol and source rooted trees may be used, instead of the shared tree, if the data rate of a source exceeds a certain threshold. However, PIM-SM assumes symmetric routing paths as it uses reverse-path routing and thus it is not suited for use in conjunction with Quality of Service Routing.

In this paper, a new multicast routing protocol is proposed called DTMP (Directed Trees Multicast Protocol), inspired in PIM-SM that takes into account link asymmetry. Here, *directed-tree* based routing strategy as opposite to a *reverse-path-tree* based one is defined and tested.

2 Related Work

Most of previous works in this area assume that links connecting any two nodes are symmetric. The underlying networks are usually modeled by undirected graphs and the heuristics used address the Steiner Tree Problem in symmetric networks.

Finding a minimal multicast tree in asymmetric networks, called the *Direct Steiner Tree Problem*, is also NP-complete. There are some theoretical studies [3], [4] focusing on directed graphs, aiming to present approaches to this problem.

However, most of the deployed multicast routing protocols, like DVRMP[5], CBT[6] and PIM-SM are based upon reverse path routing. Only MOSFP[7] handles asymmetric networks topologies, since the topological database in MOSFP is stored as a directed graph. In PIM-SM, the packet deliver path is set-up as PIM-join messages propagates towards the RP or the source. Due to asymmetric links, the path taken by the join message may not be the shortest path that actual traffic toward the receiver should follow. Thus the resulting shared or source based trees may not be optimal. Tree construction for Core Based Trees (CBT) in asymmetric networks, also suffers from a similar problem.

In [8] a new directed tree construction mechanism is proposed based upon CBT. The join process is similar to the proposed CBT approach A new participant (sender or receiver) joins the group by propagating a join request to the core node. When the join request reaches the core node a join-ack is sent back to the participant along the shortest path from core node to the new participant (likely to be different from the path taken by the join request). As well as CBT, this approach may concentrate traffic in fewer links, thus increasing the network load, than protocols that use source-based tree schemes.

REUNITE[9] implements multicast distribution based on the unicast routing infrastructure. Although the focus of the REUNITE approach is to implement multicast distribution using recursive unicast trees, it potentially implements source shortest path trees. Besides the *Join Message*, REUNITE proposes the use of a *Tree message* that travels from source to destination nodes, thus installing forwarding state. Nevertheless REUNITE may fail to construct Shortest Path Trees in certain situations and may lead to unneeded packet duplications on certain links. In [10] some modifications to the REUNITE propose are presented in order to solve these problems, and a new protocol is proposed: the Hop-by-Hop Multicast Routing Protocol (HBHP). But only Source Based Trees are considered both in REUNITE and HBHP.

3 DTMP Overview

There are two basic approaches to implement multicast tree construction: the first one is to build a shared tree to be used by all participants, and the other is to construct multiple sources based trees, one for each sender. The shared tree is rooted at some pre-defined center and because it is shared by all senders, fewer resources are used. However for large groups it may concentrate too much traffic and certain links may become bottlenecks. With the source based trees approach, each sender builds a separate tree rooted at itself.

In PIM-SM the use of both, shared and source based trees, is proposed. It allows nodes to initially join a shared tree and then commute to source based trees if necessary. The same idea is used in the Directed Trees Multicast Protocol (DTMP), herein presented.

3.1 DTMP Tree Construction

First, a shared tree is proposed in order to give receivers the ability to joining the group without knowing where are the sources located. Explicit join requests must be sent by the receivers towards the Rendezvous Point (RP) router. When RP router receives a join request it must send back to the new receiver an acknowledgment packet. This acknowledgment packet is sent back to the receiver along the shortest path between RP router and the new receiver which may be different from the path followed by the join request.

Routers, along this path, receiving such an acknowledgment packet may then update their routing tables in order to build new multicast tree branches. Updating is done basically by registering with the multicast routing entry for that tree, the acknowledge packet's incoming and outgoing router interfaces.

The join to shared tree mechanism proposed by DTMP, is illustrated at Figure 1.

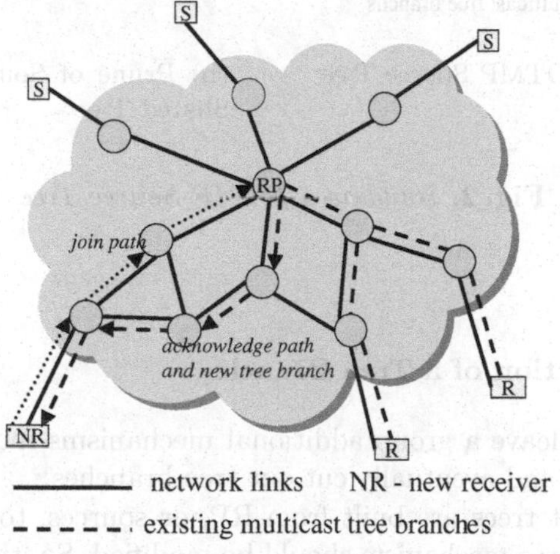

Fig. 1. *Building a DTMP Shared Tree*

After receiving a given set of data packets from a source, a receiver may decide to join a source-based tree. This procedure is similar to the one described above. An explicit join request must be sent from the receiver to the source. When accepting a join, the source must generate an acknowledgment packet, addressed to the corresponding receiver. This acknowledgment packet will signal the necessary routing table updates that will lead to the construction of a new source based tree branch. All the routers along this source tree branch should stop receiving data from that source through the shared tree in order to prevent duplicate of data packets. To accomplish this, a mechanism similar to "prune of source S in shared tree", proposed in PIM-SM specification, must be implemented, as described in section 4.

Figure 2 illustrates the mechanisms used to switch from a shared to a source base tree, as well as pruning that source from the shared tree.

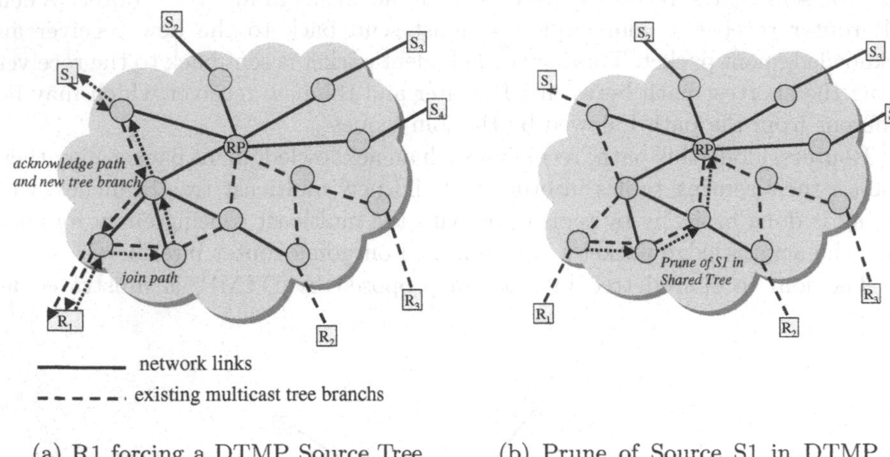

(a) R1 forcing a DTMP Source Tree rooted at S1

(b) Prune of Source S1 in DTMP Shared Tree

Fig. 2. *Building a DTMP Source Tree*

3.2 De-construction of a Tree Branch

When participants leave a group additional mechanisms must be implemented to tear down state, and eventually cut out tree branches.

As the multicast trees are built from RP, or sources, toward the receivers, the PIM-SM leave tree mechanism should be modified. So, the leave group functionality is implemented by explicit *triggered prune messages* toward the shared tree and source based trees, as detailed in section 4.

Another possible approach would be to implement some kind of tree refresh mechanism instead of an explicit action to tear down state. In this case, when a receiver wants to leave the group it simply stops sending *periodic join request messages*. If tree routers do not receive *join ack messages* within a time-out period, the corresponding entry is deleted. This alternative mechanism is not yet included in DTMP implementation.

4 DTMP Implementation

Network Simulator (NS)[11] has been used to simulate the DTMP proposal and to analyze its characteristics and control overhead.

NS includes a multicast routing protocol able to construct shared trees and source trees with the same structure as the trees constructed by the PIM-SM protocol. However, as the NS's implementation is centralized, a distributed version of PIM-SM has been defined and implemented, in addition to DTMP implementation.

With PIM-SM, tree construction is based on explicit join requests issued by receivers. When a receiver wishes to join a multicast group it should send a *join-request* towards the RP of the respective group. Each upstream router creates or updates its multicast routing table when receiving a *join-request.* The interface where the join-request arrives is added to the list of outgoing interfaces of the corresponding entry. The routers in the shortest path between the new receiver and the RP only forward the *join-request* if they do no yet belong to the shared tree.

In DTMP this behavior has been modified. The *join-request* sent by the new receiver is just forwarded towards the RP by all the routers along the way. When the RP receives the *join request*, it sends back a *join-ack message* towards the new receiver. This *join-ack* will cause the construction of the new tree branch. All the routers in shortest path between the RP and the new receiver will process and forward the *join-ack*, updating their routing tables. The interface added in the corresponding outgoing interface list is the one that has been used to forward the *join-ack message* to the new receiver. Notice that no resource reservation is performed in any on-tree router, so there is no guarantees that there will be any dynamic route adaption besides the adaptation granted by underlying unicast routing protocols.

The process of joining the shared tree in DTMP is detailed in Figure 3, where variables and flags have the same meaning as defined in PIM-SM[2].

The routing table entries have the same fields as the PIM-SM ones, and an extra one: the upstream neighbor in the tree. This field has been introduced in order to be able to implement the prune mechanism.

The process of commuting to a source based tree is similar to the above described one. However, after the construction of the new branch, when a router between the source and the receiver starts to receive data from that source, it must issue a prune of that source on the shared tree. This prune indicates that packets from this source must not be forwarded down this branch of the shared tree, because they are being received through the source based tree. This mechanism is implemented by sending a special prune to the upstream neighbor in the shared tree. When a router at the shared tree receives this type of prunes, it creates a special type of entry (an (S,G)RPT-bit entry) exactly like a PIM-SM router. In DTMP the outgoing interface list of the new (S,G)RPT-bit entry is copied from the (*,G) entry and the interface deleted is the one being used to reach the node that had originated the prune, which may not be the arriving interface of the prune packet[1]. This is because in DTMP there are directed trees not reverse path ones.

[1] In PIM-SM the outgoing interface list of the new (S,G)RPT-bit entry is copied from the (*,G) entry and the arriving interface of the prune packet is deleted.

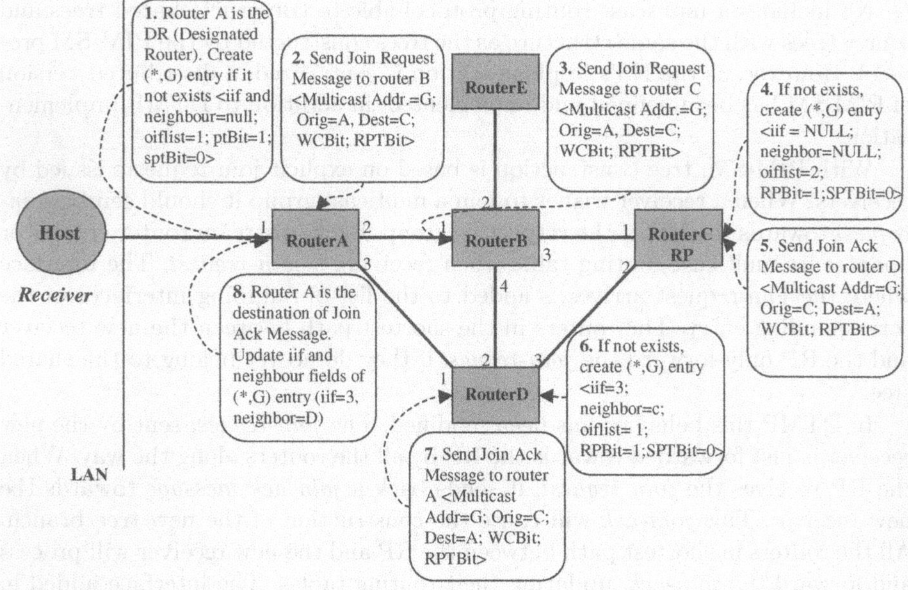

Fig. 3. *Set Up Shared Tree Implementation Actions are numbered in the order they occur*

These (S,G)RPT-bit entries must be updated too when a join-ack arrives in order to allow the join of a new receiver on a shared tree with source-specific prune state established.

The process of switching from the shared tree to a source based tree in DTMP is detailed in Figure 4.

As the actual NS implementation does not include the periodic Join/Prune process proposed in PIM-SM specification to capture member-ship changes, explicit prunes requests to tear down state when receivers wish to leave the group, had to be implemented. For that purpose the implementation of the DTMP uses the additional field that has been added to the routing table entries. As stated before, this field contains the identification of the upstream neighbor in the tree. So, the prunes must be sent toward that router. When the upstream neighbor receives this type of prunes (that are different from the "prune of a source on the shared tree") it must delete the interface used to reach the node that had originated the prune from the outgoing interface list of the corresponding (*,G) or (S,G) entry. Again, this interface may be not the arriving interface of the prune packet. If the outgoing interfaces list become empty the entry may be deleted and the prune should be forward to the upstream neighbor in the tree.

This way, although the construction process of the multicast tree was inverted, from RP or source toward the new receiver, the de-construction process follows the same way that in PIM-SM: from the bottom to the top of the multicast tree. This fact saves a lot of unnecessary control messages in the de-

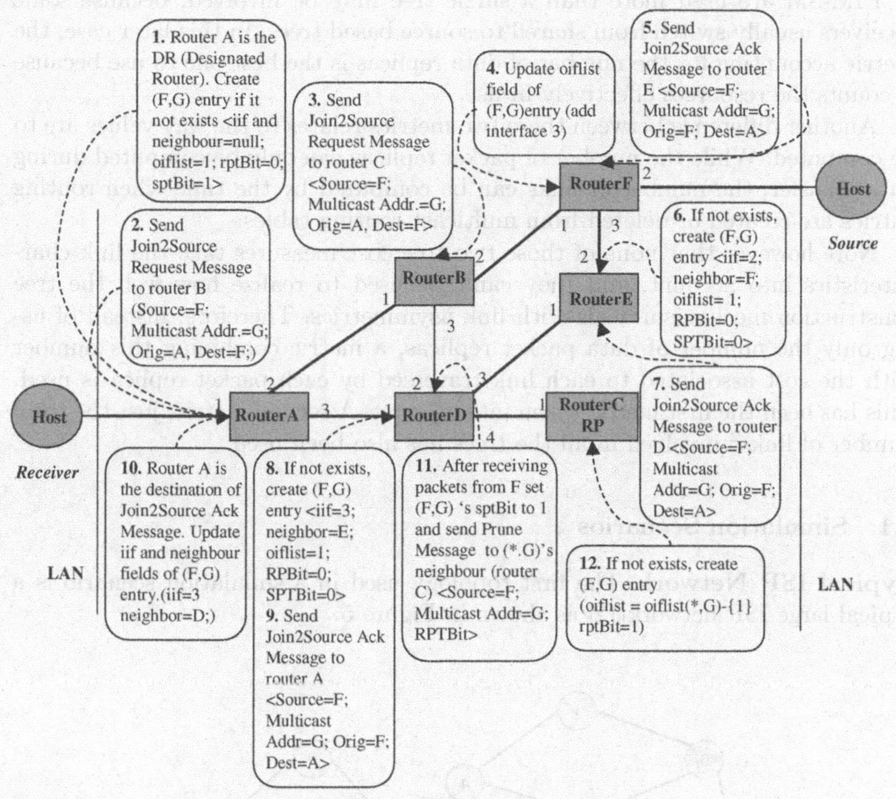

Fig. 4. *Switching from Shared Tree to Source Based Tree Implementation*

construction tree process. We believe that this field may be used to implement the periodic tree refresh, too.

5 Simulation Analysis

NS has been used in order to simulate DTMP and to compare DTMP with PIM-SM.

There are several different ways to measure the quality of multicast trees. One possible way is to count the number of links in the topology belonging to the tree; another way is to count the number of data packet replicas that are originated by nodes while forwarding those packets across the distribution tree. Since each node should receive no packet duplicates from any source, the number of packet replicas it sends, is in fact the number of different outgoing interfaces which can be used to reach receivers. Therefore, both values are the same (except for transients) if a single tree is involved. Although, when DTMP or PIM-SM are used more than a single tree may be involved, because some receivers usually switch from shared to source based trees. In this later case, the metric accounting for the number of data replicas is the best one to use because it counts the resources effectively in use.

Another difference between these two metrics relates to the way values are to be computed. While the number of packet replicas can only be computed during data transfer, the number of links can be computed by the time when routing entries are created or deleted from multicast routing tables.

Note however that none of those two tree cost measures take the link characteristics into account, and they can't be used to realize how well the tree construction mechanism deals with link asymmetries. Therefore, instead of using only the number of data packet replicas, a metric combining this number with the cost associated to each link traversed by each packet replica is used. This has been the first metric taken into account. A second metric, just the total number of links[2] involved in all the trees has also been used.

5.1 Simulation Scenarios

Typical ISP Network The first topology used in a simulation scenario is a typical large ISP network[12] as shown in Figure 5.

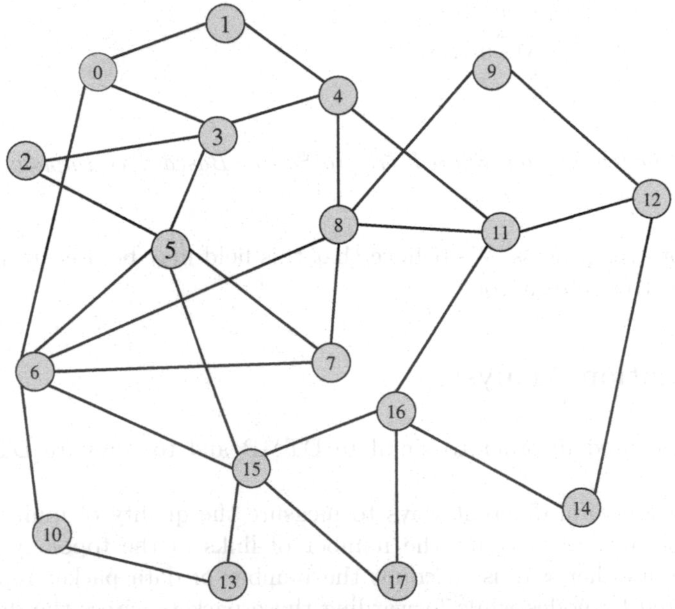

Fig. 5. *Network Topology: Typical large ISP*

[2] This value is not multiplied by the associated link cost because it is only used in order to assure that DTMP does not build larger trees than PIM-SM.

This topology includes 18 nodes and 30 links. Associated to each link there are two link utilization costs, one for each direction. Each cost is an integer randomly chosen from different intervals, as specified later. In this scenario simulations consider only one group with two fixed sources, at node 3 and node 9. It is assumed that a single receiver is connected to each node in the topology and that all nodes have one potential receiver attached. For each simulation run, the RP node is randomly chosen within the set of all the nodes. At the beginning there are no receivers joined to the group. After an initial period, receivers start to join the group building a shared tree rooted at RP.

After all the receivers have joined, one receiver, randomly chosen from all the receivers, issues a join to the source attached to the node 3 and later another receiver (randomly chosen as well) issues a join to the source attached to the node 9. This scenario (one shared tree and two source based tree) is then kept till the end of simulation. Before the simulation ends, all the receivers abandon the group.

Several experiments have been made with this topology. In the first one each link cost is an integer randomly chosen from the interval [1, 5], from the interval [1, 10] in the second one and finally the cost is randomly chosen from the interval [1, 20]. For each experiment a set of 100 independent simulations have been used and the results shown are the average from those 100 simulations.

Hundred Nodes Network Another experiment scenario, with a second topology randomly generated using GT-ITM[13][3], with 100 nodes and 354 links has been analyzed. With this topology a set of 10 independent simulations have been used. In each, the RP is randomly chosen within the set of all the 100 nodes, and the link costs are also randomly chosen from the interval [1, 10]. Like in the other experiments there is a potential receiver connected to each node (100 receivers) and two fixed sources in node 3 and node 9. All the receivers start by joining the shared tree and, some time after that, two receivers randomly chosen join source 3 and source 9 respectively. The results shown were taken from the average of those 10 simulations.

One final note about tree cost, just to to mention that it includes the aggregate cost of all the trees created during simulation: one shared tree rooted at RP (node 0) and two source trees rooted at nodes 3 and 9. Also note that since all receivers join and leave the group during simulation, there are always two different measures for each number of active receivers. For example, there are 0 receivers when starting and also 0 receivers when finishing. Presented values are averages of all observations, grouped by the number of active receivers.

---

[3] Using Pure Random edge generation method, 100 nodes, scale 100 and edge probability of 0.033

5.2 Simulation Results and Analysis

Simulations results are presented in Figures 6 and 7. Figures 6(a) to 6(f) show the average cost of the trees constructed by the two protocols (PIM-SM and DTMP) for the first topology. Tree cost reflects the quality of the constructed tree and thus provides a good way for comparison among different tree construction mechanisms, but as stated before there are several ways to measure those costs. The curves presented in Figure 6(a), 6(c) and 6(e) show results when the first metric, number of replicas `times` the link cost, is used. The curves presented in Figure 6(b), 6(d) and 6(f) show results using the second metric: the total number of links in the topology that are involved in the multicast trees.

These results demonstrate that DTMP constructs trees with costs smaller than those created by PIM-SM without enlarging the size of the trees. These results are more evident as link asymmetries became more significant. The average gain of DTMP over PIM-SM is 13,7% when the links costs are randomly chosen from the interval [1, 5], 20,8% when the links costs are randomly chosen from the interval [1, 10], and 27,5% when the links costs are randomly chosen from the interval [1, 20].

Figures 7(a) and 7(b) show the average cost of the trees constructed by the two protocols (PIM-SM and DTMP) for the second topology (the 100 nodes randomly generated topology).

With this second topology the advantage of DTMP over PIM-SM is also clear and the number of links shows also a similar value both in DTMP and in PIM-SM. This result fact indicates that DTMP builds up trees with similar number of links than those created by PIM-SM but with the advantage of being directed trees. In this experiment the gain of DTMP over PIM-SM is 21%, which led us to conclude that the advantage of DTMP does not depend on the topology, neither on the number of nodes or receivers involved. But of course it is straight related with the link asymmetries.

6 Conclusions and Future Work

A new proposal is presented in this paper, DTMP - a multicast routing protocol that implements directed trees construction in opposite of reverse path ones. The original idea is based on PIM-SM protocol, a widely deployed multicast routing protocol in the Internet. The PIM-SM, as the majority of multicast routing protocols, builds reverse path trees. This fact may lead to poor routes in the presence of asymmetric networks and problems may arise when trying to implement QoS Routing, as the links usually have different characteristics in each direction.

DTMP uses both shared trees and source based trees. Receivers begin joining a shared tree, rooted in a pre-defined point called Rendez-Vous Point. After having received a certain amount of data packets from a source, a receiver may switch to a source based tree. The protocol allows for an easy way of constructing a source based tree, pruning unnecessary tree branches within the shared tree.

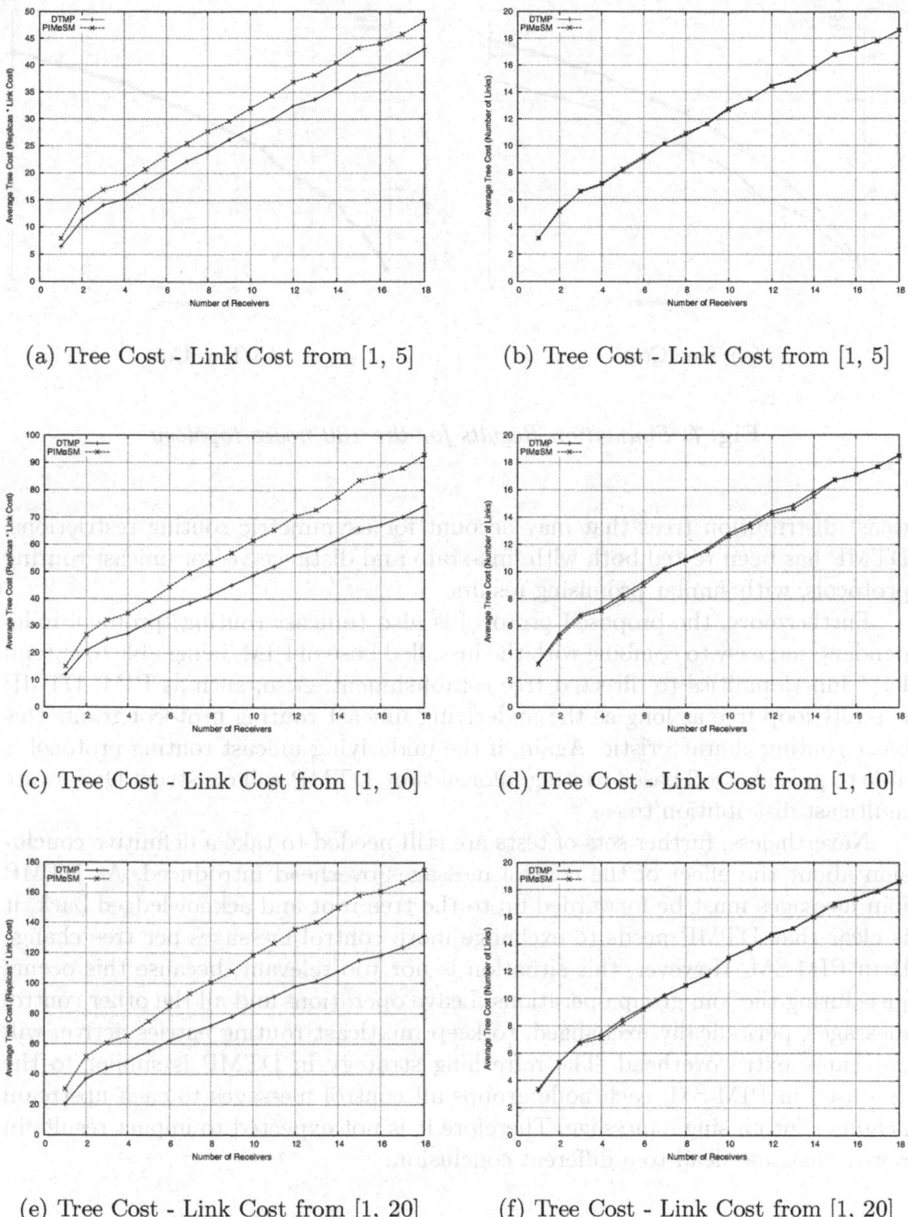

(a) Tree Cost - Link Cost from [1, 5]

(b) Tree Cost - Link Cost from [1, 5]

(c) Tree Cost - Link Cost from [1, 10]

(d) Tree Cost - Link Cost from [1, 10]

(e) Tree Cost - Link Cost from [1, 20]

(f) Tree Cost - Link Cost from [1, 20]

Fig. 6. *Simulation Results for the 18 nodes ISP topology*

DTMP has been implemented and tested with Network Simulator. The simulations results show that in presence of asymmetries within the network the DTMP is a promising approach, enabling the establishment of directed mul-

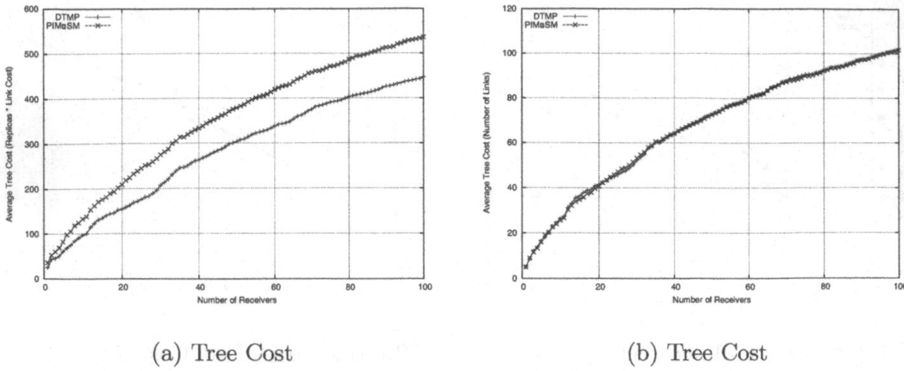

(a) Tree Cost

(b) Tree Cost

Fig. 7. *Simulation Results for the 100 nodes topology*

ticast distribution trees that may account for asymmetric routing restrictions. DTMP has been tested both with link-state and distance-vector unicast routing protocols, with similar promising results.

Furthermore, the proposed protocol is also (unicast routing) protocol independent and easy to combine with the installed base of PIM, being able to extend PIM functionalities to directed tree establishment. Also, such as PIM, DTMP is really loop free as long as the underlying unicast routing protocol grants this basic routing characteristic. Again, if the underlying unicast routing protocol is able to provide QoS based routing information, DTMP will construct QoS-aware multicast distribution trees.

Nevertheless, further sets of tests are still needed to take a definitive conclusion about the effect of the control messages overhead introduced. As DTMP join messages must be forwarded up to the tree root and acknowledged back, it is clear that DTMP needs to exchange more control messages per tree change than PIM-SM. However, this situation is not too relevant, because this occurs just during the join group operations. Leave operations and all the other control messages, periodically exchanged, to keep multicast routing entries active, will not cause extra overhead. The refreshing strategy in DTMP is similar to the one used in PIM-SM: each node groups all control messages to each upstream neighbor into a single message. Therefore it is not expected to impact results in a way that may lead to a different conclusion.

7 Acknowledgments

This work has been partially funded by FCT under the Project QoS II, POSI EEI/10168/98.

References

[1] P. Winter. Steiner problem in networks: A survey. *Networks*, 17:129–167, 1987.
[2] D. Estrin, D. Farinacci, A. Helmy, D. Thaler, S. Deering, M. Handley, V. Jacobson, C. Liu, P. Sharma, and L. Wei. Protocol independent multicast-sparse mode (PIM-SM): protocol specification. Request for Comments 2362, Internet Engineering Task Force, June 1998.
[3] Moses Charikar, Chandra Chekuri, To yat Cheung, Zuo Dai, Ashish Goel, Sudipto, and Ming Li. Approximation Algorithms for Directed Steiner Problems. *Journal of Algorithms*, 33(1):73–91, October 1999.
[4] S.Ramanathan. Multicast Tree Generation in Networks with Asymmetric Links. *IEEE/ACM Transations on Networking*, 4(4):558–568, 1996.
[5] D. Waitzman, C. Partridge, and S. E. Deering. Distance vector multicast routing protocol. Request for Comments 1075, Internet Engineering Task Force, November 1988.
[6] A. Ballardie. Core based trees (CBT version 2) multicast routing. Request for Comments 2189, Internet Engineering Task Force, September 1997.
[7] J. Moy. MOSPF: analysis and experience. Request for Comments 1585, Internet Engineering Task Force, March 1994.
[8] J.Eric Klinker. Multicast Tree Construction in Direct Networks. In *IEEE MILCOM*, Whashington DC,USA, October 1996.
[9] Ion Stoica, T. S. Eugene Ng, and Hui Zhang. REUNITE: A recursive unicast approach to multicast. In *INFOCOM (3)*, pages 1644–1653, 2000.
[10] Luís Henrique M.K. Costa and Serge Fdida and Otto Carlos M.B. Duarte. Hop-by-hop multicast routing protocol. In *ACM SIGCOMM'2001*, pages 249–259, August 2001.
[11] K. Fall and K. Varadhan. *The NS Manual*, Jan 2001. URL=http://www.isi.edu/nsnam/ns/ns-documentation.html.
[12] George Apostolopoulos, Roch Guerin, Sanjay Kamat, and Satish K. Tripathi. Quality of service based routing: A performance perspective. In *SIGCOMM*, pages 17–28, 1998.
[13] K. Calvert and E.W. Zegura. *GT-ITM: Georgia Tech internetwork topology models* (software), 1996. URL=http://www.cc.gatech.edu/fac/Ellen.Zegura/gt-itm/gt-itm.tar.gz.

A Proposal for a Multicast Protocol for Live Media

Yuthapong Somchit, Aki Kobayashi, Katsunori Yamaoka, and Yoshinori Sakai

Sakai Laboratory, Department of Communications and Integrated Systems,
Graduated School of Science and Engineering, Tokyo Institute of Technology,
2–21–1 Oookayama, Meguro-ku, Tokyo, 152–8852, Japan
{yutha,koba,yamaoka,ys}@net.ss.titech.ac.jp

Abstract. Live streaming media is delay sensitive with small allowable delay. Current conventional multicast protocols do not provide a loss retransmission mechanism. Even there are some researches on a reliable multicast providing a loss retransmission mechanism but a long delay and a high packet loss rate make them inefficient for live streaming. This paper proposes a new multicast protocol based on a protocol relay concept. The proposed protocol focusing on allowable delay provides Quality of Service (QoS) for live streaming. Relay nodes are placed along the multicast tree. Data recovery is done between relay nodes. We propose the methods that enable protocol relays to request retransmission immediately and reduce the number of duplicate packets. Finally, we make a mathematical analysis of the proposed protocol and compare it with other multicast protocols. The results show that the proposed protocol is more efficient for live streaming than the conventional protocols.

1 Introduction

Live streaming media, such as streaming video or audio, is delay sensitive and its allowable delay is an important factor. If data are not delivered before a certain point in time, they become useless and are discarded by receivers [1], [2]. Live streaming media can tolerate some amount of loss. However, it still requires loss as small as possible.

Multicast is used in one-to-many communication without a lost packet retransmission capability. It enables a better use of the available bandwidth [3] and the best effort in data delivery without a reliability guarantee. On the other hand, the reliable multicast has been widely studied [1], [4]. It provides a lost packet retransmission mechanism. Nevertheless, it is not efficient for streaming media due to a long delay and a high packet loss rate.

In this paper, we propose a new protocol enhancing the Quality of Service (QoS) for the multicast of live streaming media based on a protocol relay concept which focuses on the allowable delay of the packet. It reduces loss and the amount of unnecessary packets sent to receivers. The next section explains the problems of conventional multicast. The rest of paper explains the concepts of the protocol relay and the proposed protocol. Finally, we show the results of

M. Ajmone Marsan et al. (Eds.): QoS-IP 2003, LNCS 2601, pp. 334–346, 2003.

the mathematical evaluation of the proposed protocol and compare it to other conventional protocols.

2 Current Problems of Multicast Protocols with Live Streaming Media

Multicast provides the best-effort data delivery. However, it does not guarantee that the receivers always receive the data. Some packets are lost along the way sending to the receivers. In this research, we call this multicast "unreliable multicast"and call this kind of loss a "network layer loss". Several developments have enhanced the reliability of multicast. We call this kind of multicast "reliable multicast". Receiver-initiated reliable multicast has been proved to be efficient [5]. *Negative-Acknowledgement* (NACK) is used at end receivers to request the retransmission of lost packets from the sender or from proxy servers. A proxy server has shown to improve the efficiency of multicast [6]. However, a long delays and high packet-error rates cause long total packet delays. Packets with delay exceeding the allowable delay cannot be used even through they arrive at the end receivers. We call this loss an "application layer loss". The traffic volume also increases due to these non-usable data.

In conventional multicast, problems with NACK implosion and duplicated packets may occur. Both problems arise when more than one receiver detect that the same packets are lost. Each receiver with the same lost packet sends a NACK to its data recovery sources. If those receivers have the same data recovery source, that data recovery source receives many NACKs requesting the same packet. In addition, many retransmitted packets are repeatedly sent to all receivers when the retransmission is done by multicast. The *NACK suppression* technique is applied to solve these problems. When any end receiver detects a lost packet, it sets off a timer at a random time. If this receiver receives a NACK for the same packet before its timer is up, it suppress its own NACK. By contrast, if the timer is up without a NACK for the same packet from other receivers, this receiver sends a NACK to the data recovery source. It also sends NACKs to other receivers to suppress their NACKs for this packet. With this technique, only one NACK is sent. Moreover, retransmission is done in one time. A receiver can request retransmission again in case the previous retransmission has failed. Nevertheless, a data recovery becomes longer because a receiver has to wait before it can request retransmission. This results a delay of packet becoming longer, hence the application layer loss increases.

In this research, we propose a new protocol that reduces the total loss in multicast. We also propose new techniques to replace NACK suppression that should be applicable to live streaming.

3 Protocol Relay

The Protocol relay concept is introduced in [2] for the UDP/IP unicast communication. Protocol relays are placed along the path. Each relay node stores the forwarded packets in its memory for a finite time. An extra sequence number is assigned to each packet for protocol relay. The protocol relay analyzes this extra sequence number on every packet that propagates through it. When a loss is detected by a gap in the sequence, the relay node immediately requests retransmission of the missing packets from the previous relay node. It uses a NACK to request the retransmission. In this research, we call a relay node that request retransmission a "lower relay node" and a relay node that provides retransmission an "upper relay node".

However, in multicast, a relay node may receive more than one NACK requesting retransmissions from its lower nodes. The same problems in reliable multicast is likely to occur.

4 Proposed Protocol

4.1 Protocol Overview

In this proposed protocol, protocol relays are placed along the multicast tree. Each relay node has only one upper relay to request a retransmission. However, one relay node may have several lower relay nodes. When a relay node detects a loss, it sends a NACK back to its upper relay node. The upper relay node retransmits the packet by multicast. Because retransmitting by unicast to only the requesting node requires fine-grained router [7]. Moreover, the router has to manage the information of all its child nodes, which may change dynamically. In this proposed protocol, the relay node only has to manage the address of its upper relay node. The number of lower relay nodes per one upper relay node is usually small compared with the number of all receivers. Therefore, we allow any lower relay node to send a NACK immediately when a loss is detected. We reduce the workload of the lower relay node checking whether a NACK requesting the same packet has already been sent from other nodes or not when it requests retransmission. Consequently, the workload of the upper relay node increases. The node has to check whether it has already retransmitted the packet when it receives the NACK. However, the waiting time of the lower relay node before it can send the NACK is eliminated.

Figure 1 shows an example. Node B and node C are lower relay nodes of relay node A. When packet 5 is lost at both node B and node C, they send NACKs requesting retransmission of packet 5 to node A. In this case, the NACK from node B arrives at node A first. The packet is then retransmitted to all nodes and thus the NACK from node C, which arrives later, is ignored.

4.2 A Lower Relay Node Process

When a relay node detects a loss, it sends a NACK back to its upper relay node. Then, it renumbers the current packets a continuous sequence and forwards down. The next hop relay nodes do not detect this loss. Therefore, the data recovery is done in only one hop of protocol relays. In case that the retransmission fails, after waiting for a certain of time, it sends a NACK requesting the

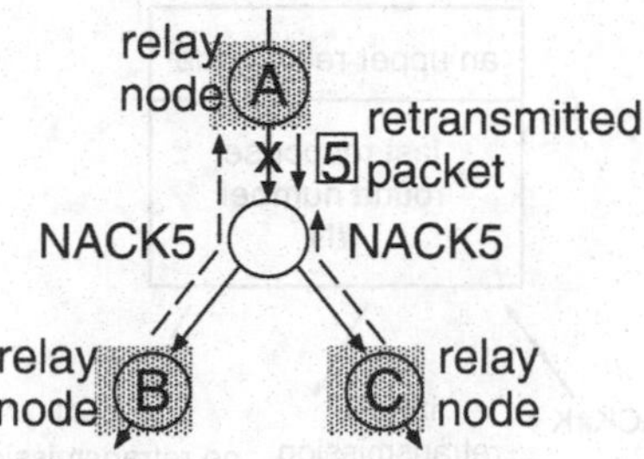

Fig. 1. The example of a relay node process when packet 5 is lost

retransmission again. We limit the number of NACKs that one relay node can send for each packet in order to reduce the packets exceeding the allowable delay being sent.

When a retransmitted packet is forwarded from the upper node, its sequence number remains unchanged. Therefore, the lower node can recognize whether an incoming packet is a new packet or a retransmitted packet. To save network bandwidth, any relay node that does not request this retransmitted packet drops this retransmitted packet. The node that has requested this packet stores the packet in its buffer when it receivers the packet. Then, it renumbers the packet into the current continuous sequence and forwards it downwards.

4.3 An Upper Relay Node Process

Because retransmission is done by multicast, the retransmitted packet is delivered to all lower nodes. This includes any node that has issued a NACK which has not yet arrived at the upper relay node. Therefore, the upper relay node only has to respond to the first arriving NACK. Since we allow a relay node to request retransmission again if the previous retransmission has failed, the subsequent NACK may arrive later. The upper relay node also has to respond to only the first arriving NACK of this subsequent request. Therefore, the upper relay node has to decide which NACK it should respond to and which NACK it should ignore. In this paper, we propose two methods for relay nodes to make this decision. One method is a timer technique and the other is a round number technique. In both methods, a retransmission number is limited.

A Timer Model

When a relay node receives the first arriving NACK, it retransmits the packet immediately and it starts a timer. Until the timer is up, NACKs for the same packet that arrive later are ignored. This is based on the assumption that those NACKs come from other relay nodes that retransmitted packets have been already delivered to. This method is implement into relay nodes. However, determining a suitable waiting time when the multicast tree can change dynamically

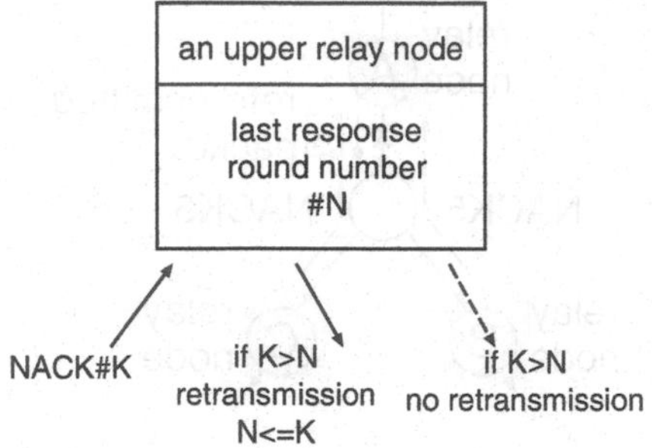

Fig. 2. A round number model

is difficult. If the time is too short, unnecessary duplicate packets are sent. Conversely, if the time is too long, the subsequent NACK for a later request, in case the previous retransmission has failed, is ignored. The latter case results in a network layer loss.

A Round Number Model

With this method, a relay node that requests retransmission implicitly tells the upper node how many times it has requested the retransmission of the same packet. We assign a round number attached to the NACK. The upper relay node can simply decide if it should respond to or ignore the incoming NACK by considering the round number. If that round of request has already been responded, it is not necessary to respond again.

When the first NACK arrives at the relay node, the packet is retransmitted immediately. The upper relay node memorizes the round number of the NACK it responds to. When other NACKs arrive later, only the one with a round number greater than one in the upper node memory is responded to. The upper relay node updates the NACK round number in its memory every time it responds to the incoming NACK. This is shown in fig. 2.

5 Protocol Evaluation

In this paper, we evaluated the proposed protocol by mathematical analysis.

5.1 Protocols to Be Compared

We compared the proposed protocol to a non-reliable multicast protocol and a reliable multicast protocol. For the proposed protocol, all intermediate nodes are protocol relays. The number of the transmission for any packet in each relay node is limited to L. In the non-reliable multicast, a sender transmits each packet only one time . Neither a sender nor any intermediate node retransmits the lost packet. In the reliable multicast, all intermediate nodes are proxies. A proxy can restransmits a packet but only end receivers can request retransmission. A retransmission request is sent to a proxy. If the proxy does not have the requested packet, it forwards request upstream until it finds the proxy that has the requested packet. The retransmission is done until the packet arrives at the end receivers.

5.2 Issues to Be Compared

We evaluated the proposed prtocol by comparing the following issues -

1. *Expected Traffic*: It shows the total of the packets that pass through the links in multicast tree. In case of no loss at all, expected traffic is equal to the number of links in multicast tree.
2. *Loss Probability*: A loss probability shows probability that the packet is lost. In non-reliable protocol, a loss probability comes from only a network layer loss. In reliable protocol, a loss probability comes from only an application layer loss. In proposed protocol, a loss probability combines with a network layer loss and an application layer loss.

5.3 Mathematical Analysis

We analyzed the performance on a symmetrical multicast tree, where the depth of the tree is N and every node except the leaf nodes has m links. Each link has a delay T and a packet loss probability p. All receivers are at the leaves of the tree in N^{th} level. The tree is shown in fig. 3. We calculated the expected traffic of all protocols. Then, we calculated the network layer loss (P_{net}) of the proposed protocol and the non-reliable protocol. Next, we calculated the application layer loss (P_{app}) of the proposed protocol and the reliable protocol. Finally, we calculated the total loss (P_{loss}) of all protocols.

Non-reliable Multicast

The probability that a packet is send from the node at $i-1^{th}$ depth to the node at i^{th} is $(1-p)^{i-1}$. Therefore, the expected traffic can be calculated from

$$\text{expected traffic} = m + (1-p)m^2 + \ldots + (1-p)^{N-1}m^N$$

$$\text{expected traffic} = \sum_{i=1}^{N}(1-p)^{i-1}m^i. \quad (1)$$

Total loss only results from network layer loss. The tree is symmetrical so we considered one receiver. Any receiver at depth N receives the packet if and only if all its intermediate nodes receive the packet. The success probability (Q) is determined from

$$Q = (1-p)^N,$$

so the loss probability (P_{loss}) is calculated from

$$P_{\text{loss}} = 1 - (1-p)^N. \quad (2)$$

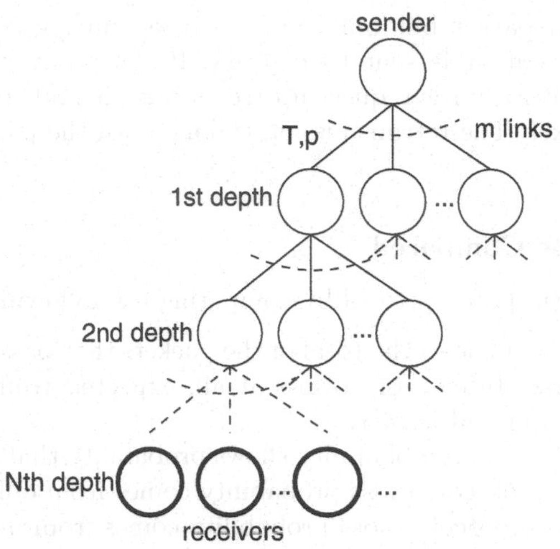

Fig. 3. The multicast tree used for analysis

Reliable Multicast

The *expected transmission*($E(X)$) is used in the reliable multicast analysis ($E(X)$) [8], [9]. The $E(X)$ tells how many times the transmitting node has to transmit the packet until it succeeds. Since every intermediate node is a proxy with a retransmission capability, the calculation becomes simple. Each node transmits a packet until all its child nodes in a next hop away receive the packets. Any intermediate node has m children nodes and transmits a packet j times. The probability that all children receive the packet at least 1 time if the proxy transmits that packet j times is $(1-p^j)^m$. The $E(X)$ can be calculated from

$$E(X) = \sum_{i=1}^{\infty} i \cdot \Pr(i), \tag{3}$$

and

$$\Pr(i) = (1-p^i)^m - (1-p^{i-1})^m. \tag{4}$$

Each intermediate node sends a packet $m \cdot E(X)$ copies into the network, so the total expected traffic is calculated from

$$\text{expected traffic} = E(X) \cdot \sum_{i=1}^{N} m^i. \tag{5}$$

Then, we defined D as the total delay of a packet propagating from a sender to a receiver. The D_i is the delay of a packet propagating from node $i-1$ to node i with n_i transmissions ($n_i \geq 1$). Note that only the receiver can request a retransmission. We make a calculation in case that the interval time between each packet is ignored as follows.

$$D = D_1 + D_2 + D_3 + \ldots + D_N$$
$$D_i = 2(n_i - 1)(N + i - 1)T + T,$$

The probability that a packet propagate from $i-1^{th}$ node to i^{th} node with delay D_i, $(Pr(D_i))$, is calculated from

$$Pr(D_i) = p^{n_i - 1}(1-p) \cdot \delta(D_i - (2(n_i - 1)(N + i - 1)T + T)). \quad (6)$$

Therefore,

$$Pr(D) = Pr(D_1) \otimes Pr(D_2) \otimes Pr(D_3) \otimes \ldots \otimes Pr(D_N). \quad (7)$$

note: $\otimes$ is convolute.

Finally, only the application layer loss occurs. Therefore, the total loss is

$$P_{loss} = 1 - \int_{D_{min}}^{\text{allowable delay}} Pr(D)dD. \quad (8)$$

Proposed Protocol

In the proposed protocol, the number of transmission per one packet in each relay node is limited to L. Using $\Pr(i)$ from eq.(4), the $E(X)$ is

$$E(X) = \sum_{i=1}^{L} i \cdot \Pr(i) + \sum_{i=L+1}^{\infty} L \cdot \Pr(i). \quad (9)$$

Since the number that one relay node can transmit a packet is limited to L, each packet is sent from the $i-1^{th}$ depth node to the ith node with a probability $(1-p^L)^{i-1}$. Therefore, the expected traffic with network layer loss is

$$\text{expected traffic} = E(X) \cdot \sum_{i=1}^{N} m^i (1-p^L)^{(i-1)}. \quad (10)$$

The total loss is a combination of a network layer loss (P_{net}) and an application layer loss (P_{app}). We first calculated the network layer loss. Any receiver, as in the non-reliable protocol, receives the packet if and only if all its intermediate nodes receive this packet. Since all nodes are relay nodes, each node receives a packet from its upper node with probability $(1-p^L)$. Therefore, P_{net} can determined from

$$P_{net} = 1 - (1-p^L)^N. \quad (11)$$

To calculate the application layer loss, we first calculated the probability that the packet arrives at the end receiver within the allowable delay (Q).

$$D = D_1 + D_2 + D_3 + \ldots + D_N$$
$$D_i = (2n_i - 1)T$$

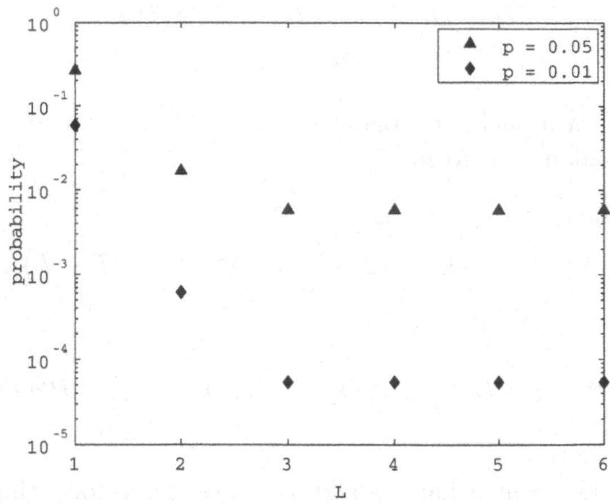

Fig. 4. The total loss probability of the proposed protocol when L varies

Note that $1 \leq n_i \leq L$.

$$Pr(D_i) = p^{n_i-1}(1-p) \cdot \delta(D_i - (2n_i - 1)T) \quad (12)$$

$$Pr(D) = Pr(D_1) \otimes Pr(D_2) \otimes Pr(D_3) \otimes \ldots \otimes Pr(D_N) \quad (13)$$

$$Q = \int_{D_{min}}^{\text{allowable delay}} Pr(D)dD \quad (14)$$

Finally, we have

$$P_{app} = 1 - Q - P_{net} \quad (15)$$

$$P_{loss} = P_{net} + P_{loss} = 1 - Q. \quad (16)$$

5.4 Results

Figure 4 shows the total loss probability of the proposed protocol when L varies. The multicast tree has $N = 6$, $m = 2$, allowable delay $= 11T$, and $p = 0.01$ and 0.05. The result shows that the total loss of the proposed protocol is minimized when L is equal to or more than 3. Thus, we used $L = 3$ in the proposed protocol to compare it with other protocols.

The multicast tree has $N = 6$, $m = 2$, and allowable delay $= 11T$. Figure 5 shows the expected traffic of all protocols when p varies. Figure 6 shows the network layer loss probability of the proposed protocol and the non-reliable protocol when p varies. Figure 7 shows the application layer loss probability of the proposed protocol and the reliable protocol when p varies. Figure 8 shows the total loss of all protocols when p varies. We made a calculation of total loss probabilities of the proposed protocols when the number of relay node varies and the reliable protocol when the number of proxy varies. Each node in the

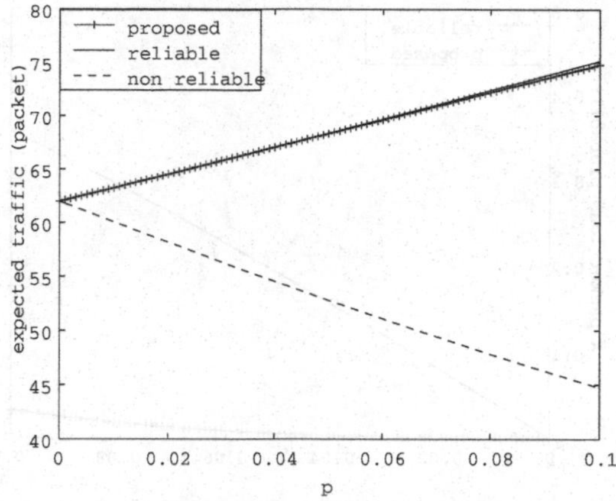

Fig. 5. The expected traffic volumes of all protocols

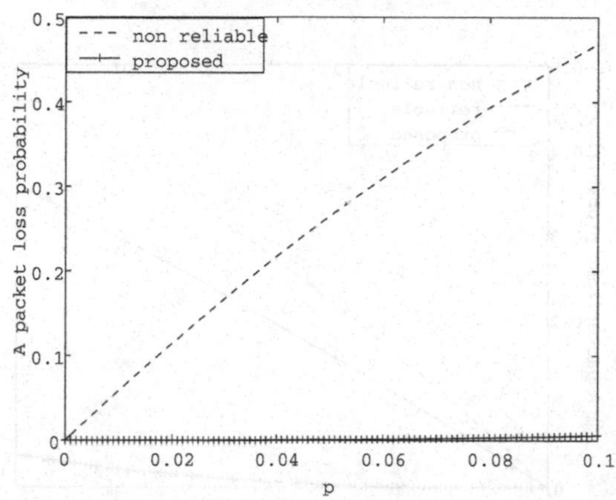

Fig. 6. The network layer loss probabilities of the proposed protocol and a non-reliable multicast protocol

proposed protocol is still limited to transfer any packet up to 3 times. This is shown in fig. 8.

From fig. 5 to fig. 9, we obtained the results for these protocols as follows.

1. The non-reliable protocol has the smallest expected traffic volume. However, it drops below the total number of links in the multicast tree when p increases. The expected traffic of the proposed protocol and the reliable protocol increase when p increases. In addtion, the proposed protocol has an expected traffic slightly lower than the reliable multicast. However, the real traffic in the reliable protocol should be higher than the value from calcuation because it cannot terminate the duplicated packet sending to the end receivers.

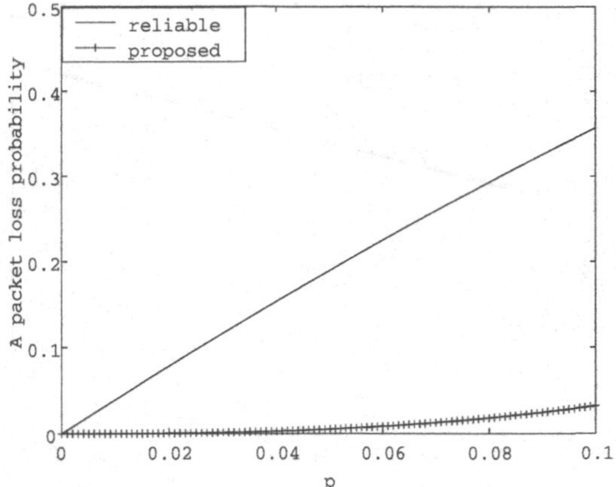

Fig. 7. The application layer loss probabilities of the proposed protocol and a reliable multicast protocol

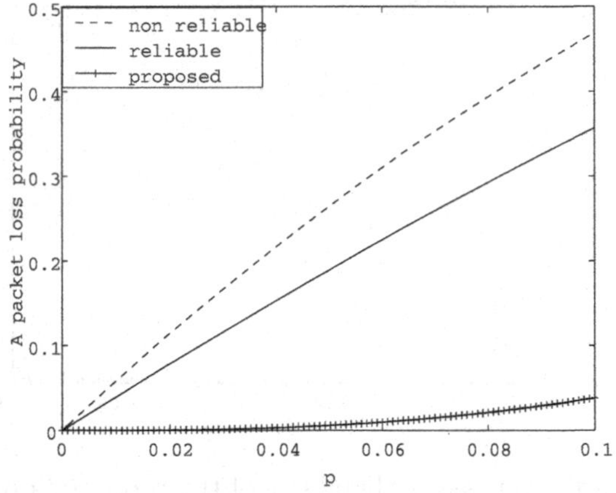

Fig. 8. The total loss probabilities of all protocols

2. The proposed protocol has a significantly lower network layer loss probability than the non-reliable protocol.
3. The proposed protocol also has a considerable lower application layer loss than the reliable protocol. This means that with a slightly smaller traffic, there are more useful packets arriving at receivers in the proposed protocol.
4. The total loss of the proposed protocol is a conbination of the network layer loss and the application layer loss. However, the proposed protocol has the smallest loss out of all protocols. The non-reliable protocol has the highest loss probability.

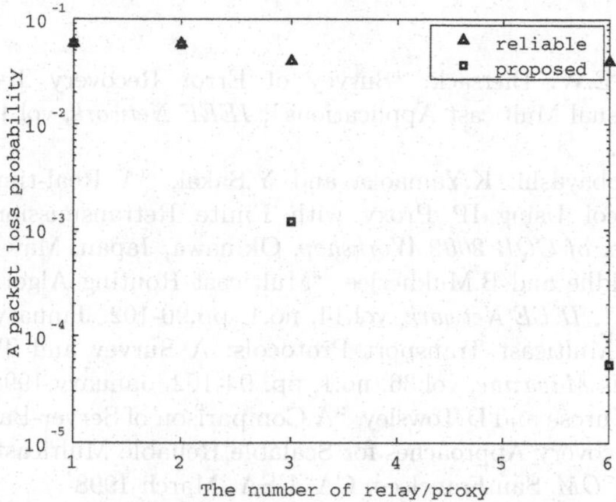

Fig. 9. The total loss probabilities of the proposed protocol and a reliable protocol when number for relay node/proxy varies

5. The total loss probabilies in both reliable protocol and proposed protocol increase when the number of relay node or proxy decreases. However, the proposed protocol has loss equal or less than the reliable protocol.

Finally, we concluded that the proposed protocol provides the smallest loss probability without a large increase in traffic compared to the non-reliable protocol and the reliable protocol.

6 Conclusion

Live streaming media is delay sensitive and it has a small allowable delay. The data must be delivered before the time that they are played. The conventional multicast protocols do not meet the QoS requirement due to a long delay and a high loss rate. We have proposed a new protocol for multicast of live streaming media, which is based on a protocol relay concept and focuses on the delay. A relay node detects loss and recovers the loss. We analyzed the performance by mathematically analysis. The results show that the proposed protocol has loss smaller than the conventional multicast protocols and its traffic volume is slightly lower than the the traffic volume of the reliable protocol that we compared with. This concludes that the proposed protocol can provide better QoS for streaming media with acceptable traffic increment.

Future research should simplify NACK management by using buffer management and apply the proposed protocol to the actual Internet. The dynamical multicast issue that the receivers join or leaves the group at any time will also be studied.

References

1. G. Carle and E.W. Biersack, "Survey of Error Recovery Techniques for IP-based Audio-Visual Multicast Applications", *IEEE Network*, vol.11, no.6, pp.24-36, Nov./Dec. 1977.
2. H.Otsuki, A.Kobayashi, K.Yamaoka and Y.Sakai, "A Real-time Media Stream Transfer Protocol Using IP Proxy with Finite Retransmission between Relay Nodes", In *Proc. of CQR 2002 Workshop*, Okinawa, Japan, May 2002.
3. L.H.Sahasrabuddhe and B.Mukherjee, "Multicast Routing Algorithms and Protocols: A Tutorial", *IEEE Network*, vol.14, no.1, pp.90-102, January/February 2000.
4. K. Obraczka, "Multicast Transport Protocols: A Survey and Taxonomy", *IEEE Communications Magazine*, vol.36, no.1, pp. 94-102, January 1998.
5. S.K.Kasera, J.Kurose and D.Towsley, "A Comparison of Server-Based and Receiver-Based Local Recovery Approaches for Scalable Reliable Multicast", In *Proceedings of IEEE INFOCOM*, San Francisco, CA, USA, March 1998.
6. J.Nonnenmacher, M.Jung, E.W.Biersack and G.Carle, "How Bad is Reliable Multicast without Local Recovery?", In *Proc. of INFOCOM '98*, San Francisco, April 1998.
7. P.Radoslavov, C.Papadopoulos, R.Govindan, D.Estrin, "A Comparison of Application-Level and Router-Assisted Hierarchical Schemes for Reliable Multicast", In *Proc. INFOCOM '01*, Anchorage, Alaska, USA, 2001.
8. M.Schuba and P.Reichl, "An Analysis of Retransmission Strategies for Reliable Multicast Protocols", In *Proc. International Conference on Performance of Information and Communication Systems PICS'98*, Lund, Sweden, May 1998. http://www.tik.ee.ethz.ch/~reichl/
9. A.P.Markopoulou and F.A.Tobagi, "Hierarchical Reliable Multcast: Perfomance Analysis and Placement of Proxies", In *Proc. of International Workshop on Networked Group Communication NGC 2000*, Standford University, Palo Alto, California, USA, November 2000. http://www.cs.ucsb.edu/ngc2000/

On the Use of Sender Adaptation to Improve Stability and Fairness for Layered Video Multicast[1]

Ping Wu, Ji Xu, Feng Chen, Jiangchuan Liu[2], Jihui Zhang[3], and Bo Li

Department of Computer Science, The Hong Kong University of Science and Technology
{csjenny, xuji, chenf, csljc, jhzhang, bli}@cs.ust.hk

Abstract. Layered multicast has been shown as a promising technique for distributing a video program to a potentially large number of heterogeneous receivers. While several layered multicast approaches have been proposed, prior work has almost exclusively focused on improving the perceived video quality in terms of PSNR value or/and the fairness among different receivers. In this paper, we first argue that stability is another major index for the video playback quality; in an extreme case even with the best possible achievable PSNR value, the frequent switching between different layers still can lead to significant quality degradation. We then introduce an end-to-end adaptation framework that improves the stability and fairness for receivers by employing a dynamic layer rate allocation scheme on the sender's side. Specifically, we propose a new metric, called Stability-aware Fairness Index (SFI), which can capture both the stability as well as fairness. We then formulate the optimal layer rate allocation problem with the objective of minimizing the expected SFI for all the receivers in a multicast session, and derive an efficient algorithm to solve the problem. Simulation results demonstrate that our scheme can significantly improve the degree of stability and fairness, thus leading to better video playback quality.

1 Introduction

The simultaneous multicast of video to many receivers is complicated by variations in the amount of bandwidth available throughout the network. For example, a video session distributed over the Internet may be watched by users who are connected either via a high-speed Ethernet or a low-speed modem. The use of layered video is commonly recommended to address this heterogeneity problem [1,3,4]. In this approach, the raw video is first encoded into a base layer representing a basic quality at a low rate; it will then be compressed into several enhancement layers that are combined with the base layer to improve the overall perceived quality. Heterogeneous receivers thus can obtain a quality of video commensurate with their available bandwidths.

However, multi-layered encoding of video itself is not sufficient to provide ideal video quality and effective bandwidth utilization. As bandwidth changes over time, an

---

[1] This research was supported in part by grants from Research Grant Council (RGC) under contracts AoE/E-01/99, HKUST6196/02E and N_HKUST605/02.

[2] J. Liu 's work is partially supported by a Microsoft fellowship.

[3] J. Zhang 's work is partially supported by a Microsoft fellowship.

M. Ajmone Marsan et al. (Eds.): QoS-IP 2003, LNCS 2601, pp. 347-357, 2003.

adaptation algorithm is necessary to adapt to the dynamic network conditions, in particular, the bandwidth fluctuations. Receiver-driven Layered Multicast (RLM) is often recommended as a promising solution [1]. In RLM, layers are mapped to multicast groups and receivers perform adaptation by joining and leaving groups (layers). As decisions are made by receivers, it can be scaled to large multicast sessions.

Note that joining and leaving multicast groups are time-consuming operations. Adding or dropping layers to react to changes in a network's load can result in sudden changes in video quality [2]. Moreover, in RLM, the layer rates are fixed. As the number of layers is limited in practical coders, the adaptation granularity for receivers is very coarse. There can be significant mismatches between the fixed transmission rates and the dynamic and heterogeneous bandwidth requirements from the receivers.

Existing studies show that the use of priority dropping can improve the stability of layered multicast [5]. It is, however, not suitable for the current best-effort Internet where FIFO drop-tail routers are widely used. In this paper, we propose an adaptation framework that improves stability and fairness of receivers by employing dynamic layer rate allocation on the sender's side. Our framework, called Sender-Assisted Receiver-driven Layer Multicast (SARLM), only requires best-effort service. We first propose a novel metric, called Stability-aware Fairness Index (SFI), which can represent the system's stability as well as fairness. We then formulate the optimal layer rate allocation problem with the objective of minimizing the expected SFI for all the receivers in a multicast session, and derive an efficient algorithm to solve the problem. Simulation results demonstrate that our scheme can significantly improve the degree of stability and fairness compared to static allocation based schemes.

The remainder of this paper is organized as follows. Section II describes the system model of SARLM. Section III formulates the optimal allocation problem and presents efficient allocation algorithms. Simulation results are shown in Section IV. Finally, Section V concludes our study and discusses some possible future directions.

2 System Model

SARLM works on top of the existing IP multicast model. It uses pure end-to-end control where all functionalities are implemented at end systems (sender or receiver). Therefore it is fully compatible with the current best-effort Internet infrastructure.

SARLM performs adaptations on the sender's side as well as the receiver's side. As illustrated in Figure 1, a sender encodes the raw video into l cumulative layers using a layered coder, where layer 1 is the base layer and layer l is the least important enhancement layer. The layers are then distributed to the receivers using separate multicast channels.

Each receiver continuously monitors its available bandwidth and performs an adaptation by joining and leaving layers (multicast groups). Assume the session size is N and the receivers' expected bandwidth at time t are $\{r_1(t), r_2(t),\ldots, r_N(t)\}$. The receivers also periodically report these bandwidths to the sender. Assume at time t, the layer rates are given by $b_i(t)$, i=1,2,…, l. Let $c_j(t)$ denote the cumulative layer rate up to layer j at time t, that is, $c_j(t)= \sum_{i=1}^{j} b_i(t), j = 1,2,\ldots, l$, and $\rho_l\ (t)$ denote the rate vector at time t, $\rho_l\ (t)= (c_1(t), c_2(t),\ldots, c_l\ (t))$. With the cumulative subscription policy, this discrete set offers all possible video rates that a receiver in a session could receive. In particular, the maximum rate that a receiver with an expected bandwidth r can receive is given by $\Gamma(r,\rho_l)=\max\{c : c \le r, c \in \rho_l\}$. To improve stability and minimize bandwidth mismatch, the sender will allocate these layer rates based on the receivers' bandwidth reports.

There has been much work on receiver-based bandwidth estimation [3,5,8] and scalable report (feedback) [4,5,8]in a multicast network. In this paper, we assume the bandwidth reports are available to the sender using some existing feedback algorithm and thus focus on the optimal layer rate allocation mechanism.

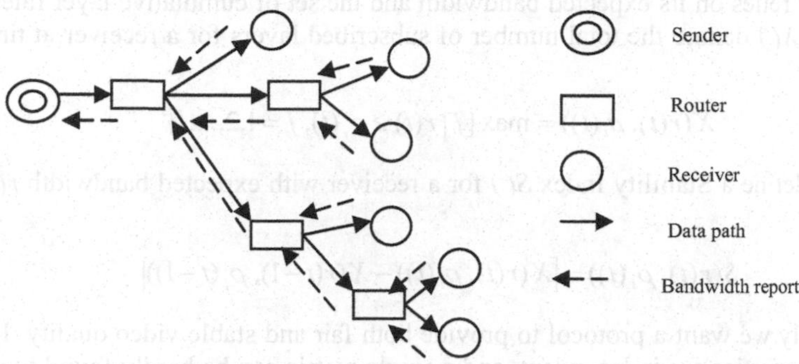

Fig. 1. System diagram of SARLM

3 Problem Formulation and Solution

In this section, we consider the layer rate allocation strategy on the sender's side. Two fundamental issues are addressed. First, what is an optimal allocation? Second, how is the optimal allocation achieved? We present a Stability-aware Fairness Index (SFI) that is suitable for evaluating the stability and fairness of receivers in a heterogeneous environment. We then formulate the problem of minimizing the expected SFI and provide efficient algorithms to solve this problem.

3.1 Optimization Criteria

Existing studies suggest a normalized fairness index can be used to evaluate the bandwidth mismatch for a receiver. In [4], a fairness index for a receiver with expected bandwidth r is defined as

$$F(r,\rho_l)=\frac{\Gamma(r,\rho_l)}{r}. \tag{1}$$

$1-F(r,\rho_l)$ is essentially the bandwidth mismatch of the receiver normalized by its expected bandwidth.

Some algorithms that maximize the expected fairness index have been presented [4,5,8]. However, these algorithms simply perform a first-order optimization, that is, determining the layer rates according to the instant bandwidth distribution of receivers. They do not consider the impact of such dynamic source rate allocations to the frequency of joining/leaving actions. Although the specific change in perceived quality depends on the encoding scheme being used, adding/dropping a layer typically results in a noticeable fluctuation in video quality [2]. Therefore, it is necessary to consider the stability of a system in optimization. As in previous studies, we use the average number of layers being added/dropped per unit time for all N receivers as the stability measure.

With a cumulative subscription policy, the total number of subscribed layers of a receiver relies on its expected bandwidth and the set of cumulative layer rates. Therefore let *X(.)* denote the total number of subscribed layers for a receiver at time *t*, that is,

$$X(r(t), \rho_l(t)) = \max\{j \mid r(t) \geq c_j(t), j = 1,2,...,l\} \tag{2}$$

We define a Stability Index *S(.)* for a receiver with expected bandwidth *r(t)* as follows:

$$S(r(t), \rho_l(t)) = \left| X(r(t), \rho_l(t)) - X(r(t-1), \rho_l(t-1)) \right| \tag{3}$$

Ideally we want a protocol to provide both fair and stable video quality. However, these properties are independent, and a single metric can be hardly found to subscribe both these two properties. In our paper, we use a linear combination of the two indices for optimal allocation. We call this metric the Stability-aware Fairness Index (SFI), which is defined as

$$SFI(r(t), \rho_l(t)) = \alpha \cdot S(r(t), \rho_l(t)) + \beta \cdot (1 - F(r(t), \rho_l(t)) \tag{4}$$

where α and β are the weights for the Stability Index and Fairness Index independently.

3.2 Optimal Allocation Problem

For a multicast session, a natural optimization objective is to minimize the expected compound metric $\overline{SFI}(r(t), \rho_l(t))$, for all the receivers in the session by choosing an optimal rate vector. We state the optimization problem as follows:

$$\begin{aligned} &\textit{Minimize} \quad \overline{SFI}(r(t), \rho_l(t)) = \frac{1}{N}\sum_{i=1}^{N} SFI(r_i(t), \rho_l(t)) \\ &\textit{Subject to} \quad l \leq L, \\ &\qquad 0 < c_{i-1}(t) < c_i(t), i = 2,3,...,l. \end{aligned} \tag{5}$$

The complexity of this optimization problem can be further reduced by considering some characteristics of a practical layered coder. First, it is a fact that every lossy data compression scheme has only a finite set of admissible quantizers. Therefore, there are only a finite number of possible rates for any given source. These rates, called operational rates [6], depend only on the compression algorithm and source characteristics. Second, to avoid the undesirable situation where a receiver cannot join any layer, the base layer should adapt to the minimum expected bandwidth. However, the dynamic range of a layered coder is limited, and usually there is a lower bound of the base layer rate [7]. Taking these two characteristics into account, we assume there are M operational points. The set of operational rates is given by $\pi = \{R_1, R_2, ..., R_M : R_i < R_{i+1}\}$, and R_1 is the lower bound of the base layer rate. We can then re-formulate the optimization problem as follows:

$$\begin{aligned} &Minimize \quad \overline{SFI}(r(t), \rho_l(t)) = \frac{1}{N}\sum_{i=1}^{N} SFI(r_i(t), \rho_l(t)), \\ &Subject \ \ to \quad l \le L, \\ &\qquad c_1(t) = \max_j \{R_j : R_j \le \min_i \{r_i(t) \ge R_1\}\}, \\ &\qquad c_i(t) \in \pi, c_{i-1}(t) < c_i(t), i = 2,3,...,l \end{aligned} \tag{6}$$

3.3 Optimal Allocation Algorithms

Note that the receivers can be divided into l sets according to their subscription levels; in each set the receivers have the same subscription level or cumulative layer rate. Assume $c_{l+1} \to \infty$, then the expected compound metric can be calculated as follows:

$$\begin{aligned} &\overline{SFI}(r(t), \rho_l(t)) \\ &= \frac{1}{N}\sum_{j=1}^{l} \sum_{c_j(t) \le r_i(t) < c_{j+1}(t)} SFI(r_i(t), \rho_l(t)) \\ &= \frac{1}{N}\sum_{j=1}^{l} \sum_{c_j(t) \le r_i(t) < c_{j+1}(t)} [\alpha \cdot S(r_i(t), \rho_l(t)) + \beta \cdot (1 - F(r_i(t), \rho_l(t)))] \\ &= \frac{1}{N}\sum_{j=1}^{l} \sum_{c_j(t) \le r_i(t) < c_{j+1}(t)} \left[\begin{array}{l} \alpha \cdot \left| X(r_i(t), \rho_l(t)) - X(r_i(t-1), \rho_l(t-1)) \right| \\ + \beta \cdot (1 - \dfrac{\Gamma(r_i(t), \rho_l(t))}{r_i(t)}) \end{array} \right] \\ &= \frac{1}{N}\sum_{j=1}^{l} \sum_{c_j(t) \le r_i(t) < c_{j+1}(t)} [\alpha \cdot \left| j - X(r_i(t-1), \rho_l(t-1)) \right| + \beta \cdot (1 - \frac{c_j(t)}{r_i(t)})]. \end{aligned} \tag{7}$$

Let $\varphi(m,l) = \min_{c_1(t)=R_m} \overline{SFI}(r(t), \rho_l(t))$, i.e., the minimum expected SFI when $c_l(t)$ is set to the m-th operational point, R_m. If l=1, this is trivial, since the base layer rate $c_1(t) = \max_j \{R_j : R_j \le \min_i \{r_i(t) \ge R_1\}\}$, and the fairness part is thus $\frac{1}{N}\sum_{r_i(t)\geq R_m}(1-\frac{R_m}{r_i(t)})$, where $R_m = c_1(t)$, but in order to calculate the stability part we need to clarify three situations as follows.

- For each receiver whose subscription level is above two at time *t*-1, if its rate is $r_i(t) \geq R_m(R_m = c_1(t))$, it will drop to the base layer, as now there is only one layer. The level change is $|1 - X(r_i(t-1), \rho_L(t-1)|$.
- For each receiver that has been in the session at time *t*-1, if its rate is $r_i(t) < R_m(R_m = c_1(t))$, it will leave the session at time *t*. The level change is $|0 - X(r_i(t-1), \rho_L(t-1)|$.
- For each receiver that has not been in the session as $r_i(t-1) < c_1(t-1)$, if its rate is $r_i(t) \geq R_m(R_m = c_1(t))$, it will join the session at time *t*. The level change is $|1-0|$.

If *l*>1, this can be viewed as adding a new layer to the case where only *l*-1 layers are used.

Considering all the situations, we have the following recurrence relation:

$$\varphi(m,l) = \tag{8}$$

$$\begin{cases} \frac{1}{N}\{\alpha\cdot[\sum_{j=2}^{l}\sum_{r_i(t)\geq R_m}|1-X(r_i(t-1),\rho_L(t-1))| \\ +\sum_{j=1}^{l}\sum_{r_i(t)<R_m}|0-X(r_i(t-1),\rho_L(t-1))| + \sum_{r_i(t-1)<c_1(t-1)}\sum_{r_i(t)\geq R_m}|1-0|] \\ +\beta\cdot\sum_{r_i(t)\geq R_m}(1-\frac{R_m}{r_i(t)})\}, & \text{if } l=1, R_m = c_1(t), \\ \min_{1\leq j<m}\{\varphi(j,l-1)+\alpha\cdot[-\frac{1}{N}\sum_{r_i(t)\geq R_m}\sum_{X(r_i(t-1),\rho_L(t-1))\geq l} 1 \\ +\frac{1}{N}\sum_{r_i(t)\geq R_m}\sum_{X(r_i(t-1),\rho_L(t-1))<l} 1] \\ +\beta\cdot\frac{1}{N}\sum_{r_i(t)\geq R_m}[(1-\frac{R_m}{r_i(t)})-(1-\frac{R_j}{r_i(t)})]\}, & \text{if } l>1, m>1, \\ +\infty & \text{Otherwise.} \end{cases}$$

where L is the maximum number of layers that the sender has before we do this adaptation, i.e., at time *t*-1. Also we do the initialization of the layer rates vector at time 0, i.e., we give the value of $\rho_L(0)$. Below we show a pseudo code of the algorithm.

```
Algorithm for Optimal Allocation

Input: the number of the subscribed layers for receiver
i at time t-1: X(ri(t-1),ρL(t-1)), the set of operational
```

rates $\Pi=\{R_1,R_2,...,R_M:R_i<R_{i+1}\}$ and the receivers' expected bandwidths at time t $\{r_1(t),r_2(t),...,r_N(t)\}$.

Output: the optimal layer rate vector at time t: $\rho_L(c_1(t),c_2(t),...,c_L(t))$

Begin

1. Let $c_1(t)=\max_j\{R_j:R_j\leq \min_i\{r_i(t)\geq R_1\}\}$, $l=1$ and $R_{m_0}=c_1(t)$.

2. Let $\varphi(m_0,1)$ be the value according to equation (1). (This is the initialization.)

3. Let $l=2$

4. Repeat as long as $l\leq L$:

   4.1 For $m_0\leq m\leq M$, according to equation (1), compute $\varphi(m,l)$ and record the corresponding value of j.

   4.2 Set $i=i+1$

5. Identify a value of m with the least $\varphi(m,l)$.

6. Let $c_L(t)=R_m$

7. Let $l=L-1$

8. Repeat as long as $l>1$

   8.1 According to the value of m, find the corresponding value of j recorded in step 4.1.

   8.2 Let $c_l(t)=R_j$

   8.3 Set $m=j$

   8.4 Set $l=l-1$

End

4 Simulation Results

In this section, we conduct simulations to examine the performance of SARLM. We also compare it with existing static allocation based RLM.

4.1 Simulation Setting

SARLM targets video multicast for large groups. It is hard to use existing network simulators, such as ns-2, to simulate its behavior. In our study, we use network topology generators to generate a large-scale network and apply network traffic models to produce cross traffic over different links. Specifically, we use a popular network topology generator GT-ITM to generate a 1000-node network with default settings as shown in [9]. We then place one ON-OFF source at each link to represent the dynamic cross traffic. The average ON and OFF periods are both 1 second. In our study, the receivers' available bandwidths are from 100Kbps to 2Mbps. This range covers the bandwidth of many available network accesses and video compression standards. It is also a typical dynamic range of existing layered coders, such as MPEG-4 PFGS coder [7].

For SARLM, we assume there are 512 uniformly spaced operational points. The maximum number of layers L is set to 4, and the cumulative layer rates of a sender are initialized to {128,512,1024,1792 Kbps}, which is also the cumulative layer rates of the static allocation based RLM.

All of our simulations were run for 200 seconds, which is long enough for observing transient and steady-state behaviors. The rate of state transition in a Markov chain is set to 1 time per second. We run the adaptation every 10 seconds.

4.2 Results and Explanations

In Figure 2, we show the effects of the value of weights α and β in three different settings: (a) $\alpha = 0.5, \beta = 0.5$; (b) $\alpha = 1.0, \beta = 0.0$; (c) $\alpha = 0.0, \beta = 1.0$. In all three settings, the optimal allocation in SARLM exhibits much better performance and often outperforms the static scheme by 40%-50% in terms of the expected SFI. This is because the optimal allocation algorithm allocates the layer rates according to the receivers' expectations. In contrast, the static scheme may set the layer rates naively, e.g., set to a point with few receivers.

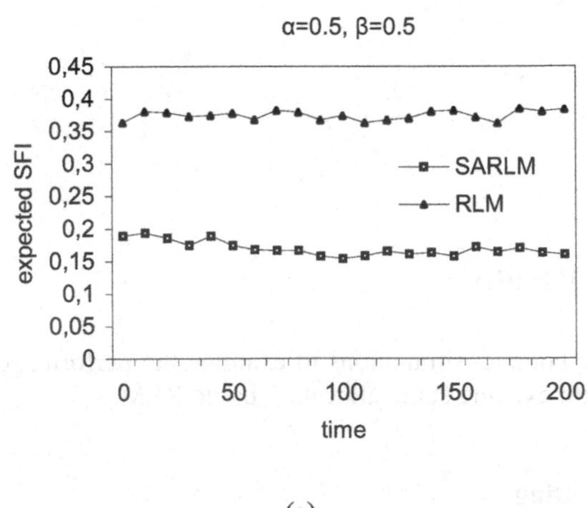

(a)

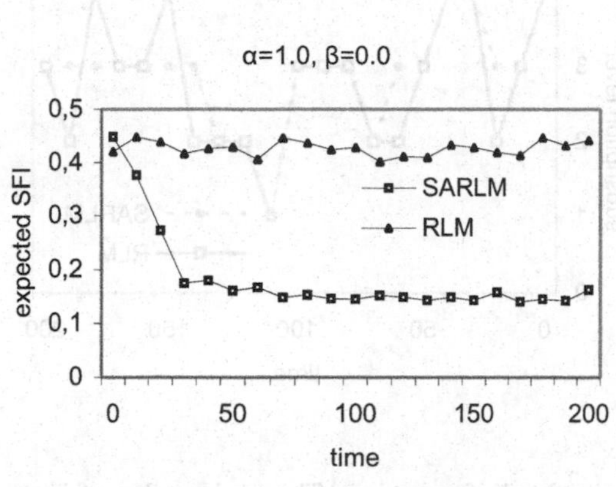

(b)

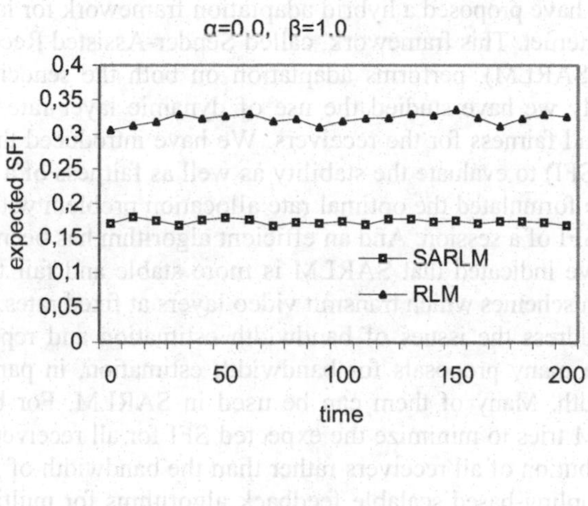

(c)

Fig. 2. Average SFI of different weight settings. (a) α=0.5,β=0.5; (b) α=1.0,β=0.0; (c) α=0.0,β=1.0.

To gain a better understanding of our effort to make the scheme more stable and fair, we also examine one special receiver. Figure 3 shows how the subscription level changes over time for one receiver. In the static allocation scheme, the receiver experiences many layer changes throughout the simulation. For SARLM, the variances in the subscription level are reduced because our optimal algorithm always tries to make the subscription level changes close to 0, the minimum value.

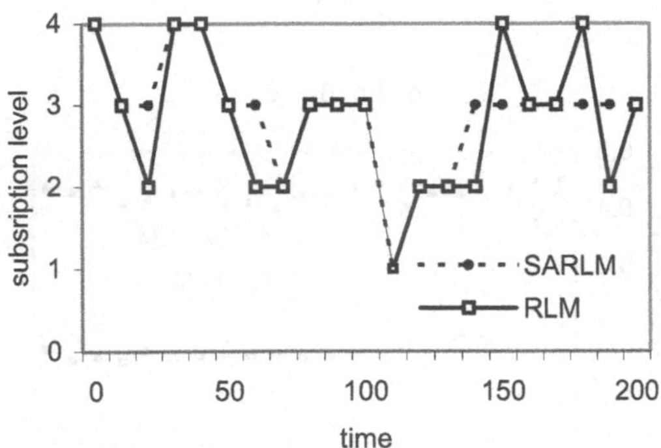

Fig. 3. The subscription level of a receiver. The total number of layers L = 4, and weights α=0.5, β=0.5.

5 Conclusions and Future Work

In this paper, we have proposed a hybrid adaptation framework for layered video multicast over the Internet. This framework, called Sender-Assisted Receiver-driven Layered Multicast (SARLM), performs adaptation on both the sender's and receivers' sides. Specifically we have studied the use of dynamic layer rate allocation to improve stability and fairness for the receivers. We have introduced the Stability-aware Fairness Index (SFI) to evaluate the stability as well as fairness of a layered multicast system. We have formulated the optimal rate allocation problem with the objective of minimizing the SFI of a session. And an efficient algorithm has been proposed. Simulation results have indicated that SARLM is more stable and fair than the receiver-driven adaptation schemes which transmit video layers at fixed rates.

We do not address the issues of bandwidth estimation and report in this paper. There have been many proposals for bandwidth estimation, in particular, the TCP-friendly bandwidth. Many of them can be used in SARLM. For bandwidth report, note that SARLM tries to minimize the expected SFI for all receivers; it relies on the bandwidth distribution of all receivers rather than the bandwidth of individual receivers. A lot of sampling-based scalable feedback algorithms for multicast sessions can thus be applied. We are currently conducting experiments over the Internet to investigate the performance of SARLM, in particular, the interactions with TCP traffic. To observe the improvement from the viewpoint of subjective video quality, we will also embed advanced layered video coders that support real-time rate allocation, e.g., the MPGE-4 PFGS video coder [7].

References

1. S. McCanne, V. Jacobson, and M. Vetterli, "Receiver-driven Layered Multicast," in Proceeding of ACM SIGCOMM 96, pp.117-130, August 1996.
2. R. Gopalakrishnan, J. Griffioen, G. Hjalmtysson, and C.Sreenan, "Stability and Fairness Issues in Layered Multicast," in Proceeding of NOSSDAV 99, June 1999.
3. L. Vicisano, L. Rizzo, and J. Crowcroft, "TCP-like congestion control for layered multicast data transfer," in Proceeding of the Conference on Computer Communications (IEEE Infocom), San Rancisco, USA, Mar. 1998.
4. J. Liu, B. Li, and Y.-Q. Zhang, A Hybrid Adaptation Protocol for TCP-friendly Layered Multicast and Its Optimal Rate Allocation, in Proceedings of IEEE INFOCOM'02, June 2002.
5. B.Vickers, C. Albuquerque, and T. Suda, 'Source-Adaptive Multi-Layered Multicast Algorithms for Real-Time Video Distribution," IEEE/ACM Transaction on Networking, Vol. 8, No. 6, 770-733, December 2000.
6. G. Schuster and A. Katsaggelos, Rate-Distortion Based Video Compression, Kluwer Publishers, 1997.
7. S. Li, F. Wu, and Y.-Q. Zhang, "Experimental Results with Progressive Fine Granularity Scalable (PFGS) Coding," ISO/IEC JTCI/SC29/WG11, MPEG98/M3988, October 1998.
8. D. Sisalem and A. Wolisz, "MLDA: A TCP-friendly Congestion Control Framework for Heterogeneous Multicast Environments," in Proceeding of the 8th International Workshop on Quality of Service (IWQoS), June 2000.
9. E. W. Zegura, K. Calvert, and S. Bhattacharjee, "How to Model an Internetwork," *IEEE Infocom 96*, April 1996.

A New Fluid-Based Methodology to Model AQM Techniques with Markov Arrivals

Mario Barbera[1], Antonio Laudani[2], Alfio Lombardo[1], and Giovanni Schembra[1]

[1] DIIT - University of Catania
(mbarbera, lombardo, schembra)@diit.unict.it
[2] DEES - University of Catania
alaudani@dees.unict.it

Abstract. In the present Internet, due to the popularity of AQM techniques, great effort is being devoting to analyzing their performance and optimizing their parameters. In this perspective the target of the paper is to provide an analytical framework to evaluate the performance of an AQM router loaded by any kind of traffic which can be modeled with a Markov arrival process. In order for complexity of the model to be independent of the queue size, the model uses a fluid-flow approach. In the paper, just to provide an example of application, we will consider RED routers. After the model definition, with the aim of demonstrating the correctness of the proposed methodology, the case of a RED buffer loaded by a constant bit-rate (CBR) source has been studied. In this particular case, where the classical theory of probability can be applied, the solution of the problem is derived, verifying that obtained analytical equations coincide with those obtained in the general case.

1 Introduction

In the last few years the Internet has registered an explosive increase in the number of both users and applications. Traffic has consequently become heavier and more heterogeneous: today not only data traffic loads the Internet, but also real-time multimedia traffic. This has led the Internet Engineering Task Force (IETF) to recommend the use of active queue management (AQM) in Internet routers [1,10], and recently AQM has taken on an increasingly important role in network telecommunications research.

AQM in Internet nodes has been demonstrated to be useful for all kinds of traffic. For traffic generated by data applications based on TCP, in fact, AQM is able to avoid global synchronization of the TCP connections sharing the same congested router, and to reduce the bias against bursty connections [5]. AQM therefore increases the overall throughput for this kind of traffic. For real-time multimedia applications, on the other hand, AQM schemes are aimed to avoid node buffer saturation and bursty losses, and therefore decreases average delay, delay jitter, and loss concentration.

One of the most widely implemented AQM techniques is Random Early Detection (RED), proposed by Floyd and Jacobson [4] as an effective mechanism to control congestion in network routers/gateways.

M. Ajmone Marsan et al. (Eds.): QoS-IP 2003, LNCS 2601, pp. 358–371, 2003.

Due to the popularity of AQM techniques, great effort is being devoting to analyzing their performance and optimizing their parameters, mainly following simulation or measurement-based approaches. In more recent literature a simple analytical model of this mechanism has been developed in [9] and [2], but the input traffic is modeled in too simplistic a way. A more complex model is introduced in [11,6,7], where a fluid-flow approach is used to model a RED node loaded by TCP traffic. The strength of this approach, being fluid-based, is the high degree of scalability which allows the model of one node to be replicated for each node in a network, thus modeling the whole of the Internet. However, although TCP traffic may at a first glance appear more complex to model, due to the dependence of the source behavior on the buffer behavior, it requires a deterministic model because the Additive Increase Multiplicative Decrease (AIMD) mechanism is absolutely deterministic. In this way results can easily be achieved. Most of the traffic in the current Internet, however, is not deterministic and it is more appropriate to model them using stochastic processes.

The challenge now is to be able to model AQM routers accurately with the aim of evaluating their performance and optimizing their parameters. Models should also be flexible, in order easily to incorporate any kind of AQM technique and traffic source, including traffic sources which have to be modeled with Markov-based arrival processes. In this perspective the target of the paper is to provide an analytical framework to evaluate the performance of an AQM router loaded by any kind of traffic which can be modeled with a Markov arrival process. In order for complexity of the model to be independent of the queue size, the model uses a fluid-flow approach [8,3]. The proposed model has been derived to work with any AQM technique, provided that it works by monitoring any set of buffer queue parameters. In the paper, to provide an example of application, we will focus on RED routers which consider average queue length only to make their discard decision. After the model definition, with the aim of demonstrating the correctness of the proposed methodology, we will study the case of a RED buffer loaded by a constant bit-rate (CBR) source. In this particular case, where we can apply the classical theory of probability, we derive the solution of the problem, verifying that the analytical equations obtained coincide with those obtained in the general case.

The paper is structured as follows. Section 2 describes the system modeled in the paper, made up of a RED router loaded by integrated traffic modeled with a Markov chain, and introduces some notation. Section 3 proposes the model. Section 4 analyzes the case of a RED router loaded by CBR traffic. Section 5 considers a case study and applies the model to evaluate performance, with the aim of demonstrating the flexibility and accuracy of the model. Finally, Section 6 concludes the paper.

2 System Model

Let us now describe the system we will study in the rest of the paper. We consider a buffer of a RED router [4]. Let C be the output link capacity, expressed in

IP packets/sec. RED routers are used to prevent congestion, discarding packets at the input of buffers with a probability which increases as long as the estimated average buffer queue length increases. The average buffer queue length is estimated by means of a low-pass filter with an Exponential Weighted Moving Average (EWMA). The discard function applied by RED routers is the following:

$$\hat{p}(y) = \begin{cases} 0 & \text{if } y < t_{\min} \\ \frac{(y-t_{\min})\cdot p_{\max}}{t_{\max}-t_{\min}} & \text{if } t_{\min} \leq y \leq t_{\max} \\ 1 & \text{if } y > t_{\max} \end{cases} \quad (1)$$

where y is the average queue length, estimated with the EWMA using the parameter α: if we indicate the queue length immediately after the arrival of the $(n-1)$-th packet as q_{n-1}, and the estimated average queue length at the time of the $(n-1)$-th arrival as m_{n-1} , the average queue length at the time of the n-th arrival is estimated as follows:

$$m_n = (1-\alpha) \cdot m_{n-1} + \alpha \cdot q_{n-1} \quad (2)$$

Let us note that if the RED policy algorithm and the queue dimension are well designed, the queue length never reaches the maximum value, and therefore losses are only due to application of the RED policy. Assuming the queue dimension to be infinite does not influence system performance. So we will make this assumption.

Let $q(t)$, $m(t)$ and $p(t)$ be the three stochastic processes characterizing the buffer, that is, the instantaneous queue length, the estimated average buffer queue length, and the instantaneous probability of loss due to the RED discarding function, respectively.

Let us model the input traffic with a fluid Markov-based arrival process [8,3]. It is a fluid emission process whose rate, $\lambda(t)$, is modulated by an underlying Markov chain. Let M be the number of states of this chain. The process $\lambda(t)$ can be characterized by the state space of the underlying Markov chain, $\Im^{(S)} = \{1, \ldots, M\}$, and the parameter set $(Q^{(S)}, \Lambda^{(S)})$, where $Q^{(S)}$ is the transition rate matrix of the underlying Markov chain, and $\Lambda^{(S)}$ is the row array containing the emission rate for each state of the Markov chain: when the state of the underlying Markov chain is $S(t) = i$, the emission rate is $\lambda(t) = \lambda_i$, for each $i \in \{1, \ldots, M\}$. From $Q^{(S)}$ the steady-state probability array can be calculated, the generic element of which is $\pi^{(S)}_{[i]} \equiv \text{Prob}\{\lambda(t) = \lambda_i\}$, as the solution of the following system of linear equations:

$$\underline{\pi}^{(S)} Q^{(S)} = \underline{0} \qquad \text{with the condition:} \quad \sum_{i=1}^{M} \pi^{(S)}_{[i]} = 1 \quad (3)$$

where $\underline{0}$ is a row array whose elements are all equal to 0. The buffer processes are governed by three equations, one for each stochastic process characterizing it. The equation describing the queue length process can be obtained from the Lindley relationship, taking into account that the queue is filled at an instantaneous rate given by the source emission rate multiplied by the probability that the arriving information units will not be discarded by the AQM policy applied:

$$\frac{dq}{dt} = \begin{cases} 0 & \text{if } (1-p(t)) \cdot \lambda(t) < C \text{ and } q(t) = 0 \\ -C + (1-p(t)) \cdot \lambda(t) & \text{otherwise} \end{cases} \tag{4}$$

The second equation needed to describe the queue behavior regards the estimated average queue length process. It can be derived from (2), converting it into a differential equation [11]:

$$\frac{dm}{dt} = \ln(1-\alpha) \cdot \lambda(t) \cdot (m(t) - q(t)) \tag{5}$$

Finally, the third equation regards the loss process, and links it to the estimated average queue length process through the AQM discarding function:

$$p(t) = \hat{p}(m(t)) \tag{6}$$

Given the deterministic relationship linking $p(t)$ to $m(t)$, it will be sufficient to study the statistics of the processes $q(t)$ and $m(t)$ alone.

3 Statistical Analysis of the System

The target of this section is to describe the behavior of the system illustrated in the previous section through a set of differential equations, and to provide an analytical derivation of its performance, in terms of mean delay, delay jitter and loss probability due to the RED discarding policy.

We can describe the temporal behavior of the statistics of the processes $q(t)$ and $m(t)$ with a set of differential equations of the following probability function:

$$P_i(t,x,y) \equiv \text{Prob}\{q(t) \leq x, m(t) \leq y, \lambda(t) = \lambda_i\} \qquad \forall i \in \Im^{(S)} \tag{7}$$

The target is therefore to derive the function:

$$\frac{\partial}{\partial t} P_i(t,x,y) \equiv \lim_{\Delta t \to 0} \frac{P_i(t+\Delta t,x,y) - P_i(t,x,y)}{\Delta t} \tag{8}$$

In order to calculate $P_i(t+\Delta t,x,y)$ in (8), let us indicate the state of the underlying Markov chain of the source at the instant t as j. From (7), ignoring the second-order terms Δt^2, we have:

$$\begin{aligned} &P_i(t+\Delta t,x,y) = \\ &= \textstyle\sum_{j=1,j\neq i}^{M} \text{Prob}\{q(t+\Delta t) \leq x, m(t+\Delta t) \leq y, \lambda(t) = \lambda_j\} \cdot Q_{[j,i]}^{(S)} \cdot \Delta t + \\ &+ \text{Prob}\{q(t+\Delta t) \leq x, m(t+\Delta t) \leq y, \lambda(t) = \lambda_i\} \cdot \left(1 + \Delta t\, Q_{[i,i]}^{(S)}\right) \end{aligned} \tag{9}$$

In (9), we need to calculate the term:

$$\hat{P}_i^{(\Delta t)}(t,x,y) \equiv \text{Prob}\{q(t+\Delta t) \leq x, m(t+\Delta t) \leq y, \lambda(t) = \lambda_i\} \tag{10}$$

To this end, let us note that, from (4) and (5), we have:

$$q(t+\Delta t)-q(t)=\begin{cases}0 & \text{if}\,(1-p(t))\,\lambda(t)\leq C \text{ and } q(t)=0\\ [-C+(1-p(t))\,\lambda(t)]\,\Delta t & \text{otherwise}\end{cases}$$
$$m(t+\Delta t)-m(t)=\ln(1-\alpha)\cdot\lambda(t)\cdot(m(t)-q(t))\,\Delta t \tag{11}$$

Thus the conditions $q(t+\Delta t)\leq x$ and $m(t+\Delta t)\leq y$, for any $x\geq 0$ and $y\geq 0$, are obtained if:

$$q(t)\leq\begin{cases}x & \text{if}\,(1-p(t))\cdot\lambda(t)\leq C \text{ and } q(t)=0\\ x+[C-(1-p(t))\cdot\lambda(t)]\;\Delta t & \text{otherwise}\end{cases}$$
$$m(t)\leq\frac{y+\ln(1-\alpha)\cdot\lambda(t)\cdot q(t)\cdot\Delta t}{1+\ln(1-\alpha)\cdot\lambda(t)\cdot\Delta t} \tag{12}$$

It can be demonstrated that the first part in the $q(t)$ definition is only useful to establish the problem's boundary condition and the term $\hat{P}_i^{(\Delta t)}(t,x,y)$ can be written as follows:

$$\hat{P}_i^{(\Delta t)}(t,x,y)=\text{Prob}\left\{\begin{array}{l}q(t)\leq x+[C-(1-p(t))\cdot\lambda(t)]\cdot\Delta t,\\ m(t)\leq\frac{y+\ln(1-\alpha)\,\lambda_i\,q(t)\,\Delta t}{1+\ln(1-\alpha)\,\lambda_i\,\Delta t},\lambda(t)=\lambda_i\end{array}\right\} \tag{13}$$

and, applying the theorem of total probability, finally we have:

$$\hat{P}_i^{(\Delta t)}(t,x,y)=$$
$$=\int\int_{-\infty}^{+\infty}F_{qmi|qm}\left(\begin{array}{l}t,x+[C-(1-p(\tilde{y}))\,\lambda_i]\,\Delta t,\\ \frac{y+\ln(1-\alpha)\,\lambda_i\,\tilde{x}\,\Delta t}{1+\ln(1-\alpha)\,\lambda_i\,\Delta t}\end{array}\middle|\,\tilde{x},\tilde{y}\right)f_{qm}(t,\tilde{x},\tilde{y})\,d\tilde{x}\,d\tilde{y} \tag{14}$$

where we have set:

$$F_{qmi|qm}(t,x,y|\tilde{x},\tilde{y})=\text{Prob}\,\{q(t)\leq x,m(t)\leq y,\lambda(t)=\lambda_i\mid q(t)=\tilde{x},m(t)=\tilde{y}\} \tag{15}$$

Therefore, from (8), we have:

$$\frac{\partial}{\partial t}P_i(t,x,y)=\sum_{j=1}^{M}P_j(t,x,y)\cdot Q_{[j,i]}^{(S)}+\lim_{\Delta t\to 0}\int\int_{-\infty}^{+\infty}\frac{1}{\Delta t}\cdot f_{qm}(t,\tilde{x},\tilde{y})\cdot$$
$$\cdot\Big[F_{qmi|qm}\left(t,x+[C-(1-p(\tilde{y}))\cdot\lambda_i]\cdot\Delta t,\frac{y+\ln(1-\alpha)\,\lambda_i\,\tilde{x}\,\Delta t}{1+\ln(1-\alpha)\,\lambda_i\,\Delta t}\middle|\,\tilde{x},\tilde{y}\right)-$$
$$-F_{qmi|qm}\left(t,x,y|\,\tilde{x},\tilde{y}\right)\Big]\,d\tilde{x}\,d\tilde{y} \tag{16}$$

Calculating the limit for $\Delta t\to 0$ in the right-hand side of (16), we obtain:

$$\frac{\partial}{\partial t}P_i(t,x,y)=\sum_{j=1}^{M}P_j(t,x,y)\cdot Q_{[j,i]}^{(S)}+$$
$$+\int\int_{-\infty}^{+\infty}\Big[[C-(1-p(\tilde{y}))\cdot\lambda_i]\,\frac{\partial}{\partial x}F_{qmi|qm}\left(t,x,y|\,\tilde{x},\tilde{y}\right)+$$
$$+\ln(1-\alpha)\,\lambda_i\,(\tilde{x}-y)\frac{\partial}{\partial y}F_{qmi|qm}\left(t,x,y|\,\tilde{x},\tilde{y}\right)\Big]\cdot f_{qm}(t,\tilde{x},\tilde{y})\,d\tilde{x}\,d\tilde{y} \tag{17}$$

and finally we obtain the objective set of differential equations describing the transient behavior of the system:

$$\frac{\partial}{\partial t}P_i(t,x,y)=\sum_{j=1}^{M}P_j(t,x,y)Q_{[j,i]}^{(S)}+\int_{-\infty}^{y}[C-(1-p(\tilde{y}))\,\lambda_i]f_{qmi}\left(t,x,\tilde{y}\right)d\tilde{y}+$$
$$+\ln(1-\alpha)\,\lambda_i\int_{-\infty}^{x}(\tilde{x}-y)\,f_{qmi}\left(t,\tilde{x},y\right)\,d\tilde{x} \tag{18}$$

Now, in order to calculate the steady-state description of the system, we calculate the limit for $t \to +\infty$ of both sides of (18). To this end let us set:

$$F_{qmi}^{(\infty)}(x,y) \equiv \lim_{t\to+\infty} P_i(t,x,y) \qquad f_{qmi}^{(\infty)}(x,y) \equiv \lim_{t\to+\infty} f_{qmi}(t,x,y) \quad (19)$$

and observe that:

$$\begin{array}{c} f_{qmi}^{(\infty)}(x,y) = \frac{\partial^2}{\partial x \partial y} F_{qmi}^{(\infty)}(x,y) \\ \int_{-\infty}^{x} f_{qmi}^{(\infty)}(\tilde{x},y)d\tilde{x} = \frac{\partial}{\partial y} F_{qmi}^{(\infty)}(x,y) \text{ and } \int_{-\infty}^{y} f_{qmi}^{(\infty)}(x,\tilde{y})d\tilde{y} = \frac{\partial}{\partial x} F_{qmi}^{(\infty)}(x,y) \end{array} \quad (20)$$

Given that both the cumulative and the density probability functions $F_{qmi}^{(\infty)}(x,y)$ and $f_{qmi}^{(\infty)}(x,y)$ are non null for $x \geq 0$ and $y \geq 0$ only , and that in the steady state the partial derivatives with respect to t are null, from (18) after some calculation we obtain:

$$\begin{array}{l} (\lambda_i - C)\frac{\partial}{\partial x} F_{qmi}^{(\infty)}(x,y) - \lambda_i \int_0^y p(\tilde{y}) \frac{\partial^2}{\partial x \partial y} F_{qmi}^{(\infty)}(x,\tilde{y})\, d\tilde{y} + \\ + \ln(1-\alpha)\,\lambda_i\,(y-x)\,\frac{\partial}{\partial y} F_{qmi}^{(\infty)}(x,y) + \ln(1-\alpha)\,\lambda_i \int_0^x \frac{\partial}{\partial y} F_{qmi}^{(\infty)}(\tilde{x},y)\, d\tilde{x} = \\ = \sum_{j=1}^{M} Q_{[j,i]}^{(S)}\, F_{qmj}^{(\infty)}(x,y) \end{array} \quad (21)$$

This set of equations characterizes the steady-state behavior of the system, allowing us to calculate the joint cumulative distribution of the instantaneous queue, the estimated average queue and the source state. Let us note that, up to this moment, we have not explicitly used the RED discarding function, but only the fact that it depends on the estimated average queue length only. In other words, (21) can be used for any AQM policy with the only constraint that the discard function is based on the estimated average queue length only. Now, let us specify (21) for the RED AQM policy, using its discard function definition:

$$\left\{ \begin{array}{l} \text{- if } y < t_{min}\text{:} \\ \quad \ln(1-\alpha)\,\lambda_i \int_0^x \frac{\partial}{\partial y} F_{qmi}^{(\infty)}(\tilde{x},y)\, d\tilde{x} + \ln(1-\alpha)\,\lambda_i (y-x) \frac{\partial}{\partial y} F_{qmi}^{(\infty)}(x,y) + \\ \quad +(\lambda_i - C)\,\frac{\partial}{\partial x} F_{qmi}^{(\infty)}(x,y) = \sum_{j=1}^{M} Q_{[j,i]}^{(S)}\, F_{qmj}(x,y) \\ \text{- if } t_{min} \leq y \leq t_{max}\text{:} \\ \quad \frac{\lambda_i\, p_{max}}{t_{max}-t_{min}} \cdot \int_{t_{min}}^{y} \frac{\partial}{\partial x} F_{qmi}^{(\infty)}(x,\tilde{y})\, d\tilde{y} + \ln(1-\alpha)\,\lambda_i\,(y-x)\,\frac{\partial}{\partial y} F_{qmi}^{(\infty)}(x,y) + \\ \quad + \ln(1-\alpha)\,\lambda_i \int_0^x \frac{\partial}{\partial y} F_{qmi}^{(\infty)}(\tilde{x},y)\, d\tilde{x} + \\ \quad + \left[(\lambda_i - C) - \frac{\lambda_i\, p_{max}\,(y-t_{min})}{t_{max}-t_{min}} \right] \frac{\partial}{\partial x} F_{qmi}^{(\infty)}(x,y) = \sum_{j=1}^{M} Q_{[j,i]}^{(S)}\, F_{qmj}(x,y) \\ \text{- if } y > t_{max}\text{:} \\ \quad -C\,\frac{\partial}{\partial x} F_{qmi}^{(\infty)}(x,y) + \lambda_i\,(1-p_{max})\,\frac{\partial}{\partial x} F_{qmi}^{(\infty)}(x,t_{max}) + \\ \quad + \frac{\lambda_i\, p_{max}}{t_{max}-t_{min})} \int_{t_{min}}^{t_{max}} \frac{\partial}{\partial x} F_{qmi}^{(\infty)}(x,\tilde{y})\, d\tilde{y} + \ln(1-\alpha)\,\lambda_i\,(y-x)\,\frac{\partial}{\partial y} F_{qmi}^{(\infty)}(x,y) + \\ \quad + \ln(1-\alpha)\,\lambda_i \int_0^x \frac{\partial}{\partial y} F_{qmi}^{(\infty)}(\tilde{x},y)\, d\tilde{x} = \sum_{j=1}^{M} Q_{[j,i]}^{(S)}\, F_{qmj}^{(\infty)}(x,y) \end{array} \right. \quad (22)$$

This system of differential equations can be solved numerically, by using the following boundary conditions:

- by definition, the probability density function tends to zero for values tending to infinity of both the estimated average queue length and the instantaneous queue length, that is:

$$\lim_{x\to+\infty} f_{qmi}^{(\infty)}(x,y)=0 \qquad \text{and} \qquad \lim_{y\to+\infty} f_{qmi}^{(\infty)}(x,y)=0 \qquad \forall i \quad (23)$$

- when the arrival rate at the buffer is greater than the service rate, both the estimated average queue length and the instantaneous queue length cannot be zero, that is:

$$\begin{array}{lll} F_{qmi}^{(\infty)}(0,y)=0 & \forall y\geq 0 & \forall i \text{ such that } \lambda_i > C \\ F_{qmi}^{(\infty)}(x,0)=0 & \forall x\geq 0 & \forall i \text{ such that } \lambda_i > C \\ \lim_{x,y\to+\infty} F_{qmi}^{(\infty)}(x,y)=\pi_{[i]}^{(S)} \ \forall i\in\Im^{(S)} & & \end{array} \qquad (24)$$

Once the array of functions $\underline{F}_{qm}^{(\infty)}(x,y)=\left[F_{qm1}^{(\infty)}(x,y),F_{qm2}^{(\infty)}(x,y),\ldots,F_{qmM}^{(\infty)}(x,y)\right]$ has been calculated, it is easy to obtain the marginal probability density functions $f_q^{(\infty)}(x)$, $f_m^{(\infty)}(y)$ and $f_p^{(\infty)}(z)$ as follows:

$$f_q^{(\infty)}(x)=\frac{\partial}{\partial x}F_q^{(\infty)}(x) \qquad \text{where: } F_q^{(\infty)}(x)=\lim_{y\to\infty}\sum_{i\in\Im^{(S)}} F_{qmi}^{(\infty)}(x,y) \quad (25)$$

$$f_m^{(\infty)}(y)=\frac{\partial}{\partial y}F_m^{(\infty)}(y) \qquad \text{where: } F_m^{(\infty)}(y)=\lim_{x\to\infty}\sum_{i\in\Im^{(S)}} F_{qmi}^{(\infty)}(x,y) \quad (26)$$

$$\begin{array}{l} f_p^{(\infty)}(z)=F_m^{(\infty)}(t_{min})\cdot\delta(z)+\left(1-F_m^{(\infty)}(t_{max})\right)\cdot\delta(z-1)+ \\ +\frac{t_{\max}-t_{\min}}{p_{\max}}\cdot f_m^{(\infty)}\left(z\,\frac{t_{\max}-t_{\min}}{p_{\max}}+t_{min}\right)\cdot rect\left(\frac{z-p_{max}/2}{p_{max}}\right) \end{array} \qquad (27)$$

where $\delta(\cdot)$ is the Dirac impulse function, and $rect(t)$ is a rectangular impulse centered in 0 with amplitude and width 1.
Finally, from these marginal distribution functions, we can easily calculate the performance of a RED buffer system, in terms of mean queueing delay, delay jitter, defined as the variance of the queueing delay, and the loss probability due to the RED discarding policy.

4 A Simple Case: A Red Router Loaded by CBR Traffic

In order to demonstrate the validity of equations (22), we will now consider a CBR source with an emission rate λ greater than the service rate C, and we will demonstrate that if we derive the equations that regulate the dynamics of the system with a classical approach (the simplicity of the source allows us to do so), these equations are identical to those that we obtain from (22).
Indeed, a CBR source can be modeled as a Markov chain with only one state, and its transition rate matrix $Q^{(S)}$ is a constant equal to 0. Consequently, from (22) we immediately obtain:

$$\begin{cases} \text{- if } y < t_{min}: \\ \quad (\lambda - C)\,\frac{\partial}{\partial x} F_{qm}^{(\infty)}(x,y) + \ln(1-\alpha)\,\lambda\,(y-x)\,\frac{\partial}{\partial y} F_{qm}^{(\infty)}(x,y) + \\ \quad + \ln(1-\alpha)\,\lambda \int_0^x \frac{\partial}{\partial y} F_{qm}^{(\infty)}(\tilde{x},y)\,d\tilde{x} = 0 \\ \text{- if } t_{min} \le y \le t_{max}: \\ \quad \left[(\lambda - C) - \frac{\lambda p_{max}(y - t_{min})}{t_{max} - t_{min}}\right] \frac{\partial}{\partial x} F_{qm}^{(\infty)}(x,y) + \frac{\lambda p_{max}}{t_{max} - t_{min}} \int_{t_{min}}^{y} \frac{\partial}{\partial x} F_{qm}^{(\infty)}(x,\tilde{y})d\tilde{y} + \\ \quad + \ln(1-\alpha)\lambda(y-x)\frac{\partial}{\partial y} F_{qm}^{(\infty)}(x,y) + \ln(1-\alpha)\lambda \int_0^x \frac{\partial}{\partial y} F_{qm}^{(\infty)}(\tilde{x},y)d\tilde{x} = 0 \\ \text{- if } y > t_{max}: \\ \quad -C\,\frac{\partial}{\partial x} F_{qm}^{(\infty)}(x,y) + \lambda\,(1 - p_{max})\,\frac{\partial}{\partial x} F_{qm}^{(\infty)}(x,t_{max}) + \\ \quad + \frac{\lambda\, p_{max}}{t_{max} - t_{min})} \int_{t_{min}}^{t_{max}} \frac{\partial}{\partial x} F_{qm}^{(\infty)}(x,\tilde{y})\,d\tilde{y} + \ln(1-\alpha)\lambda(y-x)\frac{\partial}{\partial y} F_{qm}^{(\infty)}(x,y) + \\ \quad + \ln(1-\alpha)\lambda \int_0^x \frac{\partial}{\partial y} F_{qm}^{(\infty)}(\tilde{x},y)d\tilde{x} = 0 \end{cases} \tag{28}$$

where $$F_{qm}^{(\infty)}(x,y) \equiv \lim_{t \to +\infty} P(t,x,y) \equiv \text{Prob}\,\{q(t) \le x, m(t) \le y\} \tag{29}$$

As previously, the target is to derive the following function, but this time with a classical and well-known approach:

$$\frac{\partial}{\partial t} P(t,x,y) \equiv \lim_{\Delta t \to 0} \frac{P(t+\Delta t,x,y) - P(t,x,y)}{\Delta t} \tag{30}$$

In order to calculate $P(t+\Delta t, x, y)$ in (30), we can write it as the sum of 3 terms:

$$P(t+\Delta t,x,y) = P^{(1)}(t+\Delta t,x,y) + P^{(2)}(t+\Delta t,x,y) + P^{(3)}(t+\Delta t,x,y) \tag{31}$$

where

$$\begin{aligned} P^{(1)}(t+\Delta t,x,y) &\equiv \text{Prob}\{q(t+\Delta t) \le x, m(t+\Delta t) \le y, 0 \le m(t) \le t_{min}\} \\ P^{(2)}(t+\Delta t,x,y) &\equiv \text{Prob}\{q(t+\Delta t) \le x, m(t+\Delta t) \le y, t_{min} \le m(t) \le t_{max}\} \\ P^{(3)}(t+\Delta t,x,y) &\equiv \text{Prob}\{q(t+\Delta t) \le x, m(t+\Delta t) \le y, m(t) \ge t_{max}\} \end{aligned} \tag{32}$$

Now we will calculate each term separately. For the first term we can write:

$$\begin{aligned} &P^{(1)}(t+\Delta t,x,y) = \text{Prob}\{q(t+\Delta t) \le x, m(t+\Delta t) \le y, 0 \le m(t) \le t_{min}, q(t) = 0\} + \\ &+ \text{Prob}\{q(t+\Delta t) \le x, m(t+\Delta t) \le y, 0 \le m(t) \le t_{min}, q(t) > 0\} \end{aligned} \tag{33}$$

In our special case, because $\lambda > C$, we have $Pr\{q(t) \le 0\} = 0$ and $Pr\{m(t) \le 0\} = 0$. Consequently the first term in (33) is equal to 0, while the second term, using (12), can be written as:

$$\begin{aligned} &P^{(1)}(t+\Delta t,x,y) = \\ &= \text{Prob}\{q(t+\Delta t) \le x, m(t+\Delta t) \le y, 0 \le m(t) \le t_{min}, q(t) > 0\} = \\ &= \text{Prob}\{0 < q(t) \le x + (C-\lambda)\Delta t, m(t) \le \tfrac{y + \ln(1-\alpha)\lambda\Delta t \cdot q(t)}{1 + \ln(1-\alpha)\lambda\Delta t}, 0 \le m(t) \le t_{min}\} \end{aligned} \tag{34}$$

To calculate the probability in (34) we can consider Fig.1.
If we define the probability density joint function $f_{qm}(t,x,y)$ as:

$$f_{qm}(t,x,y) = \frac{\partial^2 P(t,x,y)}{\partial x \partial y} \tag{35}$$

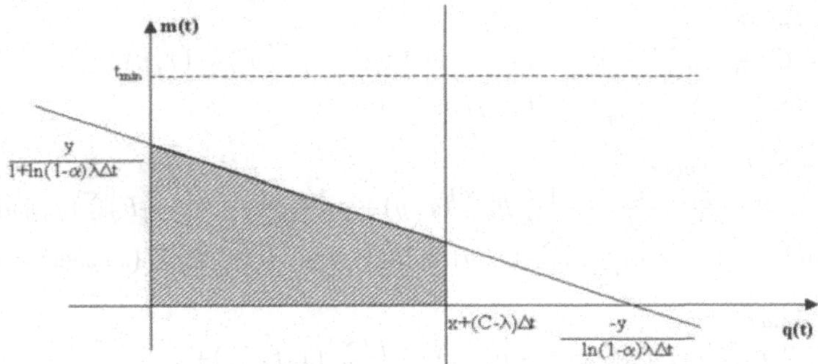

Fig. 1. Couples of points $(q(t), m(t))$ that verify the condition in (34)

from Fig.1 we obtain:

$$P^{(1)}(t+\Delta t, x, y) = \\ = \begin{cases} 0 & \text{if } x \leq 0 \text{ or } y \leq 0 \\ \int_0^{x+(C-\lambda)\Delta t} d\tilde{x} \int_0^{\frac{y+\ln(1-\alpha)\lambda\Delta t\cdot\tilde{x}}{1+\ln(1-\alpha)\lambda\Delta t}} f_{qm}(t,\tilde{x},\tilde{y})d\tilde{y} & \text{if } x > 0 \text{ and } 0 \leq y \leq t_{min} \\ \int_0^{x+(C-\lambda)\Delta t} d\tilde{x} \int_0^{t_{min}} f_{qm}(t,\tilde{x},\tilde{y})d\tilde{y} & \text{if } x > 0 \text{ and } y > t_{min} \end{cases} \quad (36)$$

For the second term in (31) we have:

$$P^{(2)}(t+\Delta t, x, y) = \\ = \text{Prob}\{q(t+\Delta t) \leq x,\ m(t+\Delta t) \leq y,\ t_{min} < m(t) \leq t_{max},\ q(t) > 0\} = \\ = \text{Prob}\begin{Bmatrix} q(t) \leq x + \Delta t\left[C - \lambda\left[1 - \frac{(m(t)-t_{min})p_{max}}{t_{max}-t_{min}}\right]\right], \\ m(t) \leq \frac{y+\ln(1-\alpha)\lambda\Delta t\cdot q(t)}{1+\ln(1-\alpha)\lambda\Delta t}, t_{min} < m(t) \leq t_{max}, q(t) > 0 \end{Bmatrix} \quad (37)$$

To calculate the probability in (37) we can consider Fig.2.

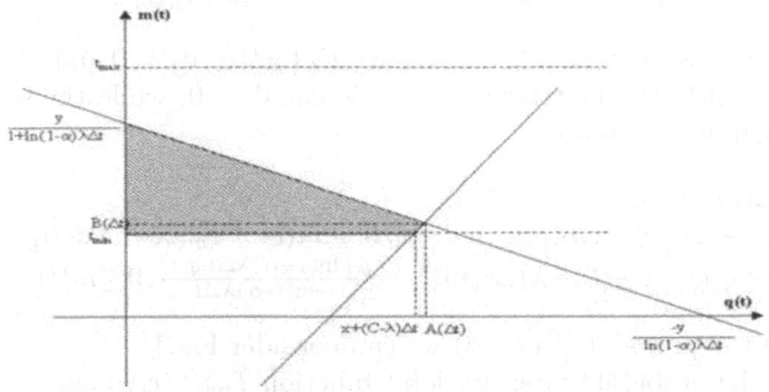

Fig. 2. Couples of points $(q(t), m(t))$ that verify the condition in (37)

If we call the coordinates of the intersection point between the two straight lines in Fig.(2) $A(\Delta t)$ and $B(\Delta t)$, we have:

$$A(\Delta t) = \frac{x + \left[C - \lambda - \frac{p_{max}\lambda}{t_{max}-t_{min}}\left(t_{min} - \frac{y}{1+\ln(1-\alpha)\lambda\Delta t}\right)\right]\Delta t}{1 - \frac{\ln(1-\alpha)\lambda^2 p_{max}\Delta t^2}{(t_{max}-t_{min})(1+\ln(1-\alpha)\lambda\Delta t)}} \tag{38}$$

$$B(\Delta t) = \frac{y + \ln(1-\alpha)\lambda\Delta t \cdot A(\Delta t)}{1 + \ln(1-\alpha)\lambda\Delta t} \tag{39}$$

So we obtain:

$$P^{(2)}(t+\Delta t, x, y) = \\ = \begin{cases} 0 & \text{if } x \le 0 \text{ or } y \le t_{min} \\ \int_0^{A(\Delta t)} d\tilde{x} \int_{t_{min}}^{\frac{y+\ln(1-\alpha)\lambda\Delta t\cdot\tilde{x}}{1+\ln(1-\alpha)\lambda\Delta t}} f_{qm}(t,\tilde{x},\tilde{y})d\tilde{y} + & \text{if } x > 0 \text{ and} \\ -\int_{t_{min}}^{B(\Delta t)} d\tilde{y} \int_{x+\Delta t\left[C-\lambda\left[1-\frac{(\tilde{y}-t_{min})p_{max}}{t_{max}-t_{min}}\right]\right]}^{A(\Delta t)} f_{qm}(t,\tilde{x},\tilde{y})d\tilde{x} & t_{min} < y \le t_{max} \\ \int_0^{x+(C-\lambda)\Delta t} d\tilde{x} \int_{t_{min}}^{t_{max}} f_{qm}(t,\tilde{x},\tilde{y})d\tilde{y} + & \text{if } x > 0 \text{ and} \\ +\int_{t_{min}}^{t_{max}} d\tilde{y} \int_{x+(C-\lambda)\Delta t}^{x+\Delta t\left[C-\lambda\left[1-\frac{(\tilde{y}-t_{min})p_{max}}{t_{max}-t_{min}}\right]\right]} f_{qm}(t,\tilde{x},\tilde{y})d\tilde{x} & y > t_{max} \end{cases} \tag{40}$$

Finally for the third term in (31) we have:

$$\begin{aligned} &P^{(3)}(t+\Delta t, x, y) = \\ &= \text{Prob}\{q(t+\Delta t) \le x,\ m(t+\Delta t) \le y,\ m(t) > t_{max},\ q(t) > 0\} = \\ &= \text{Prob}\{q(t) \le x + C\Delta t, m(t) \le \tfrac{y+\ln(1-\alpha)\lambda\Delta t\cdot q(t)}{1+\ln(1-\alpha)\lambda\Delta t}, m(t) > t_{max}, q(t) > 0\} \end{aligned} \tag{41}$$

To calculate the probability in (41) we can consider Fig.3, and we obtain:

$$P^{(3)}(t+\Delta t, x, y) = \\ = \begin{cases} 0 & \text{if } x \le 0 \text{ or } y \le t_{max} \\ \int_0^{x+C\Delta t} d\tilde{x} \int_{t_{max}}^{\frac{y+\ln(1-\alpha)\lambda\Delta t\cdot\tilde{x}}{1+\ln(1-\alpha)\lambda\Delta t}} f_{qm}(t,\tilde{x},\tilde{y})d\tilde{y} & \text{if } x > 0 \text{ and } y > t_{max} \end{cases} \tag{42}$$

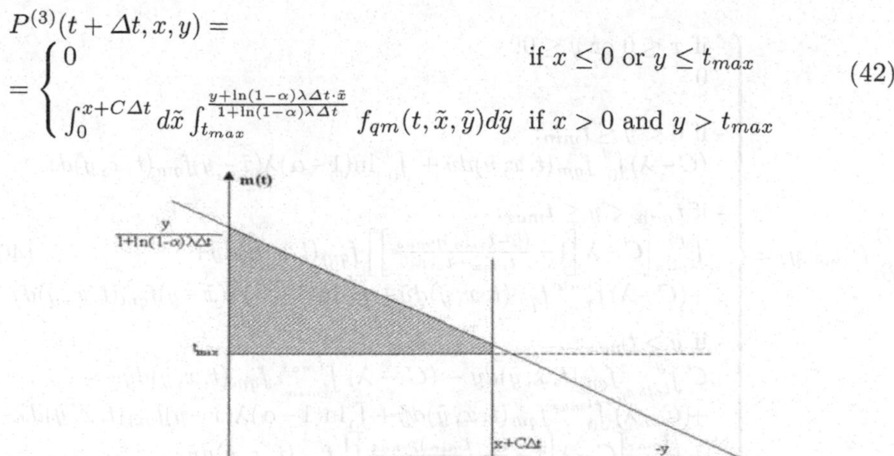

Fig. 3. Couples of points $(q(t), m(t))$ that verify the condition in (41)

Now, putting together all the parts of $P(t+\Delta t, x, y)$:

$$P(t+\Delta t,x,y)=\begin{cases} \text{- if } x\le 0 \text{ or } y\le 0: \\ \quad 0 \\ \text{- if } y\le t_{min}: \\ \quad \int_0^{x+(C-\lambda)\Delta t} d\tilde{x}\int_0^{\frac{y+\ln(1-\alpha)\lambda\Delta t\cdot\tilde{x}}{1+\ln(1-\alpha)\lambda\Delta t}} f_{qm}(t,\tilde{x},\tilde{y})d\tilde{y} \\ \text{- if } t_{min}<y\le t_{max}: \\ \quad \int_0^{x+(C-\lambda)\Delta t} d\tilde{x}\int_0^{t_{min}} f_{qm}(t,\tilde{x},\tilde{y})d\tilde{y}+ \\ \quad +\int_0^{A(\Delta t)} d\tilde{x}\int_{t_{min}}^{\frac{y+\ln(1-\alpha)\lambda\Delta t\cdot\tilde{x}}{1+\ln(1-\alpha)\lambda\Delta t}} f_{qm}(t,\tilde{x},\tilde{y})d\tilde{y}+ \\ \quad -\int_{t_{min}}^{B(\Delta t)} d\tilde{y}\int_{x+\Delta t\left[C-\lambda\left[1-\frac{(\tilde{y}-t_{min})p_{max}}{t_{max}-t_{min}}\right]\right]}^{A(\Delta t)} f_{qm}(t,\tilde{x},\tilde{y})d\tilde{x} \\ \text{- if } y>t_{max}: \\ \quad \int_{t_{min}}^{t_{max}} d\tilde{y}\int_{x+(C-\lambda)\Delta t}^{x+\Delta t\left[C-\lambda\left[1-\frac{(\tilde{y}-t_{min})p_{max}}{t_{max}-t_{min}}\right]\right]} f_{qm}(t,\tilde{x},\tilde{y})d\tilde{x}+ \\ \quad +\int_0^{x+(C-\lambda)\Delta t} d\tilde{x}\int_0^{t_{max}} f_{qm}(t,\tilde{x},\tilde{y})d\tilde{y}+ \\ \quad +\int_0^{x+C\Delta t} d\tilde{x}\int_{t_{max}}^{\frac{y+\ln(1-\alpha)\lambda\Delta t\cdot\tilde{x}}{1+\ln(1-\alpha)\lambda\Delta t}} f_{qm}(t,\tilde{x},\tilde{y})d\tilde{y} \end{cases} \tag{43}$$

In this way we can calculate (30) using the De l'Hopital theorem. In fact, if we consider the limit:

$$\lim_{\Delta t\to 0}\frac{d}{d\Delta t}P(t+\Delta t,x,y) \tag{44}$$

and this limit exists and is finite, we can conclude that:

$$\frac{\partial}{\partial t}P(t,x,y)=\lim_{\Delta t\to 0}\frac{d}{d\Delta t}P(t+\Delta t,x,y) \tag{45}$$

The calculus of (44) leads to:

$$\frac{\partial}{\partial t}P(t,x,y)=\begin{cases} \text{- if } x\le 0 \text{ or } y\le 0: \\ \quad 0 \\ \text{- if } 0<y\le t_{min}: \\ \quad (C-\lambda)\int_0^y f_{qm}(t,x,\tilde{y})d\tilde{y}+\int_0^x \ln(1-\alpha)\lambda(\tilde{x}-y)f_{qm}(t,\tilde{x},y)d\tilde{x} \\ \text{- if } t_{min}<y\le t_{max}: \\ \quad \int_{t_{min}}^{y}\left[C-\lambda\left[1-\frac{(\tilde{y}-t_{min})p_{max}}{t_{max}-t_{min}}\right]\right] f_{qm}(t,x,\tilde{y})d\tilde{y}+ \\ \quad +(C-\lambda)\int_0^{t_{min}} f_{qm}(t,x,\tilde{y})d\tilde{y}+\int_0^x \ln(1-\alpha)\lambda(\tilde{x}-y)f_{qm}(t,\tilde{x},y)d\tilde{x} \\ \text{- if } y>t_{max}: \\ \quad C\int_{t_{max}}^{y} f_{qm}(t,x,\tilde{y})d\tilde{y}-(C-\lambda)\int_{t_{min}}^{t_{max}} f_{qm}(t,x,\tilde{y})d\tilde{y}+ \\ \quad +(C-\lambda)\int_0^{t_{max}} f_{qm}(t,x,\tilde{y})d\tilde{y}+\int_0^x \ln(1-\alpha)\lambda(\tilde{x}-y)f_{qm}(t,\tilde{x},y)d\tilde{x}+ \\ \quad +\int_{t_{min}}^{t_{max}}\left[C-\lambda\left[1-\frac{(\tilde{y}-t_{min})p_{max}}{t_{max}-t_{min}}\right]\right] f_{qm}(t,x,\tilde{y})d\tilde{y} \end{cases} \tag{46}$$

and, if we simplify some terms and develop some integrals, calculating the limit for $t\to\infty$, we again obtain the equations in (28).

5 Numerical Results

Let us now apply the analytical model proposed in the previous sections to evaluate the performance of a RED router loaded by aggregated traffic. As far as input traffic is concerned, we measured the aggregated video traffic at the output of the subnetwork of the Multimedia Laboratory of the University of Catania, and we modeled it with an 8-state fluid Markov model described by the following parameter set:

$$Q^{(S)} = \begin{bmatrix} -14.4055 & 2.5473 & 2.5377 & 5.4271 & 1.5498 & 1.7570 & 0.5513 & 0.0352 \\ 2.2177 & -7.8412 & 1.6822 & 1.5680 & 0.5740 & 0.9188 & 0.8091 & 0.0714 \\ 2.7310 & 1.7831 & -8.8645 & 2.0980 & 0.6523 & 0.5672 & 0.9362 & 0.0967 \\ 6.5533 & 1.2668 & 1.3482 & -12.9124 & 1.1934 & 1.1723 & 1.2553 & 0.1230 \\ 1.5142 & 0.4999 & 1.3069 & 1.2964 & -6.9579 & 0.8952 & 1.2930 & 0.1523 \\ 0.9529 & 1.2404 & 1.0502 & 1.3149 & 1.0184 & -7.4561 & 1.6981 & 0.1811 \\ 0.4260 & 0.4541 & 0.7766 & 0.7771 & 1.4952 & 1.6897 & -6.7980 & 1.1794 \\ 0.0104 & 0.0496 & 0.1627 & 0.4309 & 0.4748 & 0.4559 & 0.2550 & -1.8391 \end{bmatrix}$$

$$\Lambda^{(S)} = \begin{bmatrix} 10\ 40\ 70\ 100\ 130\ 160\ 190\ 220 \end{bmatrix} \text{ packets/sec}$$

For the RED router parameters we used $t_{min} = 150$ packets, $t_{max} = 175$ packets, and $p_{max} = 0.1$. We used the model to carry out two kinds of analysis, evaluating mean delay, and loss probability by varying the output link capacity, C, and the RED parameter t_{max}, respectively.

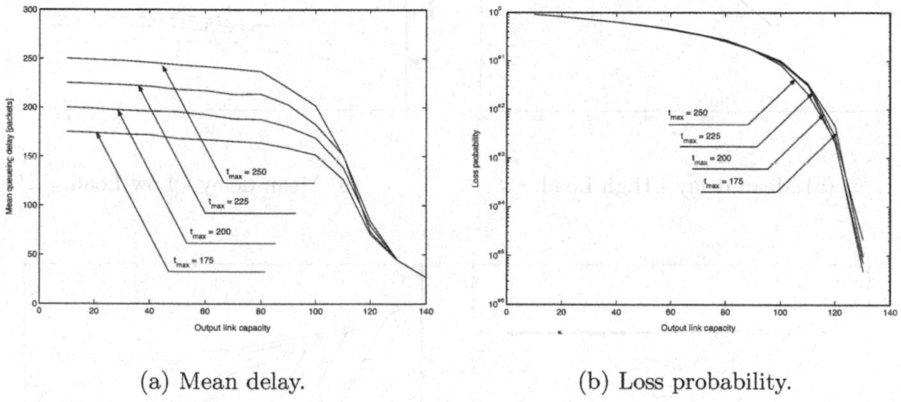

(a) Mean delay. (b) Loss probability.

Fig. 4. Performance versus output link capacity

Performance versus the output link capacity was calculated for 4 different values of t_{max}, and is shown in Figs. 4(a), and 4(b). In Fig. 4(a), we can observe that, as expected, the mean delay decreases when the output link capacity increases. Moreover, we can observe that differences given by different values of t_{max} are observable only for low capacities, that is, when the system is overloaded. On the contrary, for high output link capacity values, the value of t_{max}

is not significant. Regarding loss probability, we can see that it is strongly influenced by output link capacity; t_{max} instead is not relevant.

The second analysis was carried out for two different link capacities, $C = 70$ and $C = 120$ packets/sec, by varying the value of t_{max}, and for 4 different values of t_{min}. Figs. 5(a) and 5(c) show the results obtained for $C = 70$ packets/sec. In this case the system is characterized by utilization coefficient greater than 1 and the system is therefore maintained stable thanks to the RED policy only. This is the reason why the value of t_{min} is not relevant, and the mean delay increases linearly. For the same reason loss probability decreases very slowly, remaining very high. Finally, Figs. 5(b) and 5(d) show the results obtained for $C = 120$ packets/sec. In this case the system is stable even without RED, given that we have utilization coefficient of $\rho = 0.95$. Now we can observe that for a $t_{\text{max}} > 200$ packets/sec all the curves become flat, that is, performance is not influenced by t_{max} if it is greater than 200 packets/sec. Moreover, let us note that, as expected, the greater t_{min}, the worse the loss, since the system discards more packets. Of course, as a consequence, the system behaves better in terms of mean delay.

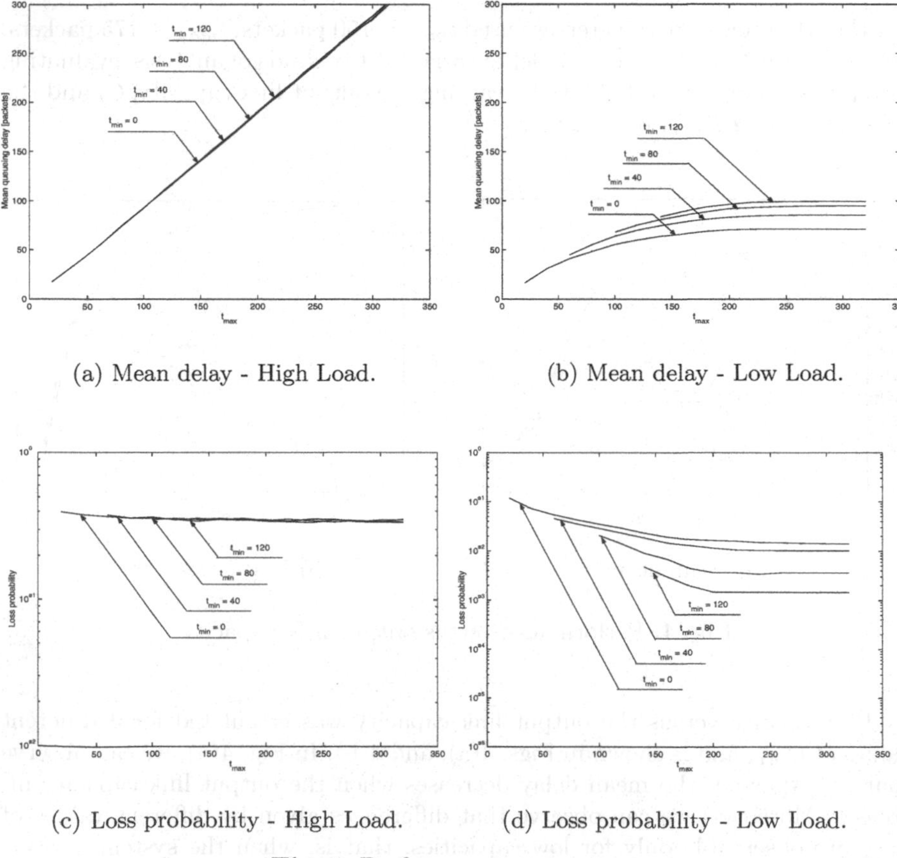

(a) Mean delay - High Load.

(b) Mean delay - Low Load.

(c) Loss probability - High Load.

(d) Loss probability - Low Load.

Fig. 5. Performance versus t_{max}

6 Conclusions

The paper introduces an analytical model of a RED router loaded by any kind of traffic which can be modeled with a Markov arrival process, but this model can be adapted easily for a generic AQM router that works monitoring any set of buffer queue parameters. In order for the complexity of the model to be independent of the queue size, the model uses a fluid-flow approach.

The model is applied to a case study in which a RED router is loaded by a fluid model representing the aggregated video traffic at the output of the subnetwork of the Multimedia Laboratory of the University of Catania. In this case study the model was applied to evaluate the influence of the output link capacity and the RED parameter t_{max} on performance.

References

1. R. Braden, D. Clark, J. Crowcroft, B. Davie, S. Deering, D. Es- trin, S. Floyd, V. Jacobson, G. Minshall, C. Partridge, L. Peter- son, K. Ramakrishnan, S. Shenker, J. Wroclawski, and L. Zhang. Recommendations on Queue Management and Congestion Avoidance in the Internet. Internet Draft draft-irtf-e2e-queue-mgt-00.txt, March 1997.
2. T. Bonald, M. May, "Drop Behavior of RED for Bursty and Smooth Traffic," *Proceedings of the* 7th *International Workshop on Quality of Service (IWQoS'99)*, London, UK, Jun 1-4, 1999.
3. A. Elwalid, D. Mitra, "Analysis, Approximations And Admission Control Of A Multi-Service Multiplexing System With Priorities," *Proceedings IEEE INFOCOM '95*, Boston, Massachusetts, USA, April 2-6, 1995.
4. S. Floyd, V. Jacobson, "Random Early Detection Gateways for Congestion Avoidance," *IEEE/ACM Transactions on Networking*, Vol. 17, No. 1, vol. 1, no. 4, August 1993.
5. S. Floyd, "RED: Discussion of Setting Parameters," available at *http://www.aciri.org/floyd/REDparameters.txt*.
6. C. V. Hollot, V. Misra, D. Towlsey, W. Gong, "A Control Theoretic Analysis of RED," *Proceedings of IEEE INFOCOM 2001*, Anchorage, Alaska, April 23 - 27, 2001.
7. C. V. Hollot, V. Misra, D. Towlsey, W. Gong, "On Designing Improved Controllers for AQM Routers Supporting TCP Flows," *Proceedings of IEEE INFOCOM 2001*, Anchorage, Alaska, April 23 - 27, 2001.
8. B. Maglaris, D. Anastassiou, P. Sen, G. Karlsson, J. Robbins, "Performance models of statistical multiplexing in packet video communications," *IEEE Transactions on Communications*, vol. 36, no. 7, July 1988.
9. M. May, J. Bolot, A. Jean-Marie, C. Diot, "Simple Performance Models of Differentiated Services Schemes for the Internet," *IEEE Proc. of INFOCOM'99*, 21-25 March 1999, New York, NY, USA.
10. A. Mankin, K. K. Ramakrishnan, "Gateway Congestion Control Survey," *Internet RFC-1254*.
11. V. Misra, W. Gong, D. Towsley, "A Fluid-based Analysis of a Network of AQM Routers Supporting TCP Flows with an Application to RED," *Proceedings of ACM SIGCOMM'00*, Stockholm, Sweden, August 28 - September 1, 2000.

Stochastic Petri Nets Models for the Performance Analysis of TCP Connections Supporting Finite Data Transfer*

Rossano Gaeta, Matteo Sereno, and Daniele Manini

Dipartimento di Informatica, Università di Torino,
Corso Svizzera, 185, 10149 Torino, Italy

Abstract. This paper proposes the Stochastic Petri Nets (SPN) formalism to model IP networks loaded with traffic resulting from a set of ON-OFF finite TCP-Reno connections. The approach is based on separate model descriptions of the TCP connection latency to transfer a finite number of packets and the IP network links; the two models are iteratively solved using a fixed point method. The overall model parameters are the primitive network characteristics; the model solution yields an estimation of the packet loss probability, the average completion time to transfer a finite number of packets over a TCP connection as well as the distribution of the completion time. The validation of the proposed approach and the future extensions to this work will also be discussed.

1 Introduction

TCP has been subject to numerous performance studies based on simulations, measurements, and analytical models. Modeling the TCP behaviour using analytical paradigms is the key to obtain more general and parametric results and to achieve a better understanding of the TCP behaviour under different operating conditions. At the same time, the development of accurate models of TCP is difficult, because of the intrinsic complexity of the protocol algorithms, and because of the complex interactions between TCP and the underlying IP network. Although numerous papers proposed models and/or methods to obtain the *expected* throughput rates or transfer time, the variability of QoS requirements for different types of services, hence different types of TCP connections, and the need to provide *Service Level Agreements (SLA)* to end users, require the computation of more sophisticated performance metrics, such as transfer time distributions, and quantiles of transfer time for TCP connections.

In this paper we address the problem of modeling and analyzing the average and the distribution of the completion time as well as the packet loss probability of a set of homogeneous TCP connections that behave as ON-OFF data sources; the activity begins when the connection starts to send a finite amount of data.

* This work was partially supported by the Italian Ministry for Education, University and Research (MIUR) in the framework of the PLANET-IP project.

M. Ajmone Marsan et al. (Eds.): QoS-IP 2003, LNCS 2601, pp. 372–391, 2003.

Each activity period begins after an exponentially distributed silence interval. We base this analysis on separate model descriptions for the latency of TCP connections and for the IP network that are iteratively solved using a fixed point method.

The balance of this paper is outlined as follows: Section 2 provides a short overview of previous works that most related to our proposal, Section 3 describes the modeling assumptions and the proposed iterative approach, Section 4 describes the TCP connection latency model, Section 5 presents the model we developed to describe the IP network link, Section 6 discusses validation of the proposed technique, and finally Section 7 draws some conclusions and discusses future development.

2 Related Works

The literature on analytical models of TPC is vast therefore it is difficult to provide here a comprehensive summary of previous contributions. In this section we only summarise some of the approaches that have been successfully used so far in this field and that are most closely related to our work.

One of the first approaches to the computation of the latency of short file transfers is presented in [3]. In this paper, an analytical model is developed to include connection establishment and slow start to characterise the data transfer latency as a function of transfer size, average round trip time and packet loss rate.

The work in [13] also deals with finite TCP data transfer; it analyses the cases when only one packet is lost and when multiple packets are lost providing a single model to predict both short and steady-state TCP transfers latency.

Works in [1, 4] too cope with finite TCP connections; their peculiarity is the analysis of connections that exhibit a on-off behaviour, following a Markov model. In these works, the description of the protocol behaviour is decoupled from the description of the network behaviour and the interaction between the two submodels is handled by iterating the solution of the submodels until the complete model solution converges according a fixed point algorithm.

The approach presented in [2, 6, 7] is based on the use of queueing network models for modeling the behaviour of the window protocol. These works are based on the description of the protocol through a queueing network model that allows the estimation of the load offered to the IP network (that is also represented by means of a queueing network). The merit of the queueing network approach lies in the product form solution that these models exhibit thus leading to more efficient computation of the interesting performance indexes.

On the other hand the approach presented in [9], allows the computation of the completion time distribution by using an analysis based on the possible paths followed by a TCP connections. The main problem of this approach is that the number of possible paths grows according to an exponential law with respect to the number of packets to be transferred over the connection.

3 The Modeling Approach and Assumptions

In this section we describe the network scenario we considered as well as the assumptions we made on the behavior of the TCP connections. Furthermore, we describe the decomposition technique and iterative solution method we used.

3.1 The Network Scenario

In this paper we consider a set of K homogeneous TCP Reno connections characterized by:

- an ON-OFF behavior where the OFF periods are exponentially distributed with average $T_{silence}$;
- when in the ON (active) state, a connection is characterized by a packet emission distribution $g = \{g_i\}$ where
  $g_i = Prob\{\#\text{ Packets to transfer} = i\}$, and $i = N_{\min}, N_{\min}+1, \ldots, N_{\max}+1$;
- same routing in the network;
- similar round trip times (RTT) and packet loss probabilities;
- same maximum value for the congestion window (W_{max});
- same fixed packet size (PK).

The topology we consider is a bottleneck link connecting K TCP senders to their receivers. The link has a capacity of C Mb/s, propagation delay of PD ms, and is associated with a drop-tail queue whose size is denoted as B.

3.2 The Modeling Approach

The proposed technique is based on separate descriptions of the TCP connection latency to transfer a finite amount of packets and the IP network links. The TCP latency submodel describes the behavior of one of the set of K connections sharing homogeneous characteristics.
The TCP connection latency submodel is presented in Section 4; it is a stochastic model yielding the average time required to transfer N packets over a route through the network experiencing a given round trip time RTT and a packet loss probability p.
The IP network link submodel is presented in Section 5; it is the description of a finite buffer queue with batch arrivals modulated by an ON-OFF process. A high-level description of the proposed modeling approach is depicted in Figure 1. The TCP connection latency submodel receives as inputs from the IP network link submodel the following parameters:

- estimate of the packet loss probability;
- estimate of the RTT summing up the average queuing delay for accepted packets, packet transmission time, and constant propagation delays.

The TCP connection latency model provides an estimate of the average completion time, the average number of batches (rounds) required to complete the packets transfer, the batch size distribution. The IP network link model receives as inputs:

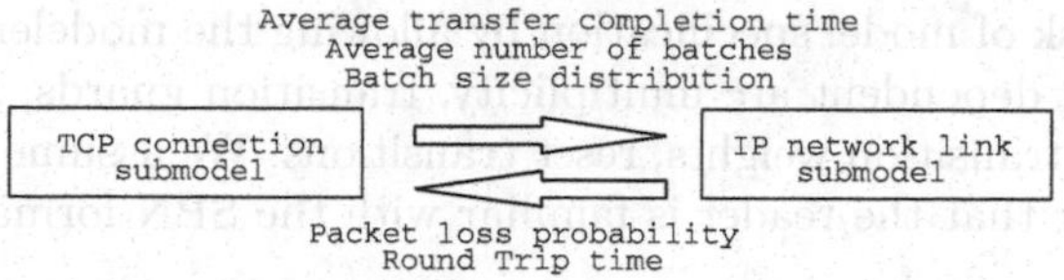

Fig. 1. High level description of the proposed approach

– estimate of the average completion time for the modulating batch arrival process;
– average number of batches required to complete the packets transfer to be used to set the arrival rate of batches at the link queue;
– batch size distribution.

The estimates produced by the IP network link model are fed back to the TCP connection latency model in an iterative process until convergence is reached.

3.3 Notation

The SPN models we developed use extensions of the classical Petri Nets formalism to allow an easier and more readable specification of the system models. In particular, the extensions we use are:

– *predicates*: we use basic predicates, i.e., logical expressions to compare two quantities using operators $=, \neq, <, \leq, >, \geq$. Quantities may be integer numbers, marking parameters, number of tokens of places in a given marking (that we denote as #place). General predicates may be obtained by combining basic predicates using the logical operators $\wedge$ and $\vee$. We denote predicate by enclosing them in square brackets, e.g., $[\#W < \#THRESHOLD \wedge \#W \geq 1]$.
– *marking dependent arc multiplicities*: in classical Petri Nets, arcs connecting places and transitions are assigned an integer, constant multiplicity. We allow an arc to have a marking dependent multiplicity obtained as an integer arithmetic expression involving integer constants, marking parameters, number of token in places. Arc expressions may also be guarded using predicates as defined above.
– *marking dependent immediate transition weights*: immediate transitions may have marking dependent weights; the general format is a list of case conditions and a default case. Each case is associated with a general predicate as defined above and the weight value is an arbitrary arithmetic expression involving parameters (rate and marking), rates, number of token in places, constants.

It must be pointed out that the extensions we used do not increase the modeling power of the original formalism: any system described using our extended SPNs can be described using a (more complex) SPN. The purpose of these extensions is to ease the task of model specification by allowing the modeler to use features such as marking dependent arc multiplicity, transition guards, marking dependent immediate transition weights, reset transitions. We assume throughout the rest of the paper that the reader is familiar with the SPN formalism.

4 The TCP Connection Latency Submodel

The model we developed to represent the TCP congestion control in case of short lived connections has been inspired by the work presented in [11, 12] where the model was based on the Reno flavor of TCP focusing on the congestion avoidance mechanism for greedy (infinite) sources neglecting the contribution of the connection establishment phase. We extended previous work by considering the case of finite TCP connections, and including the slow start phase of the congestion control mechanism. In the following, we briefly summarize the assumptions underlying our approach.

The TCP connections are assumed to use the Reno congestion control algorithm; delays arising at the end points from factors such as buffer limitations are not considered for the latency computation. The sender sends full-sized packets as fast as its congestion window allows and the receiver advertises a consistent flow control window. The congestion control mechanism (both slow start and congestion avoidance) of TCP is modeled in terms of *rounds*; a round begins with the transmission of a window of W packets and ends when the sender receives an acknowledgment for one or more of these packets. At the beginning of the next round, a set of $W^{'}$ new packets are sent; in the absence of loss, the window size at the beginning of the new round is $W^{'} = 2 * W$ if the connection is in the slow start phase, and is $W^{'} = W + 1$ if the connection is in congestion avoidance (both expressions are derived assuming delayed ACKs are not used). The initial value for the congestion window is assumed to be $W = 1$.

The time to transmit all packets in a round is assumed to be smaller than the duration of a round; the duration of a round is independent of the window size and is assumed to be equal to the round trip time. Losses in one round are assumed to be independent of losses in other rounds; on the other hand, losses in one round are assumed to be correlated among the back-to-back transmission within a round. If a packet is lost, all remaining packets transmitted until the end of that round are also lost. This loss model more adequately describes the behavior arising from drop-tail queueing policy.

As the work in [11, 12], we do not model all the aspects of TCP's behavior, e.g., fast recovery; nevertheless, the essential elements are captured by our model as witnesses by the excellent agreement between our model predictions and the outputs of detailed simulations under several different network settings.

To formally define our model we describe the set of places that define the state space of the modeled system:

- place W: models the congestion window size;
- place LOST: models the number of packets lost in the previous round;

- place TIMEOUT: models the timeout backoff value if timeout occurred in the previous round; if the number of tokens in place TIMEOUT is equal to zero then timeout did not occur as a result of packet loss in the previous round;
- place INIT_MODEL: models the start of the model dynamics;
- place CHOICE: this place is input to a (large) set of immediate transitions that probabilistically choose the future behavior of the packets in a round, e.g., no packet loss, the first packet loss, and so on;
- place DATA: models the number of packets yet to be transmitted;
- place BATCH_SIZE: models the number of packets actually sent in round; this place is necessary since the number of packets yet to be transmitted at the beginning of a round might be lesser than the congestion window size;
- place THRESHOLD: models the threshold value used to enter the congestion avoidance phase after starting in slow start.

The complete model is composed of two auxiliary transitions named start_model and restart_model. All places are empty in the initial marking except for place INIT_MODEL that contains one token and is input to transition start_model. The firing of transition start_model puts one token in place W, one token in place CHOICE, W_{max} tokens in place THRESHOLD, and N tokens in place DATA to represent the transfer of N packets. When place DATA becomes empty the firing of reset transition restart_model brings the model back to the initial marking.

Several other auxiliary places are part of the model definition and will be described as the model transitions will be presented. The rest of this section will describe the model transitions that allow the model to evolve in time. The notation we use in the remainder is the following: immediate transitions will always bear a name starting with the letter i lower case while timed transitions will always bear a name starting with the letter t lower case. Place CHOICE is input to all immediate transitions therefore this description will be omitted. All timed transitions have place CHOICE as output. A graphical description of the general structure of the SPN model is depicted in Figure 2; in the following subsections we describe each of the four submodels.

4.1 Case of No Packets Loss

- Immediate transition i_no_loss: models the probabilistic choice (transition weight is equal to $(1-p)^{\min(\#W,\#DATA)}$) of having no losses in a round. This transition is enabled if places LOST and TIMEOUT are empty; its firing sets the number of tokens in place BATCH_SIZE as the marking dependent expression $\min(\#W, \#DATA)$, adds one token in the control place NOL, and sets the number of token in place NOLOSS according to the following case expression ($f2$):

$$\begin{array}{c}[\#W >= W_{max}] : 1\\ [\#W < W_{max} \wedge \#W >= \#THRESHOLD] : 2\\ \text{default} : 3\end{array}$$

  This expression results in one token in place NOLOSS if the current congestion window size is greater than or equal to the maximum allowed value, two

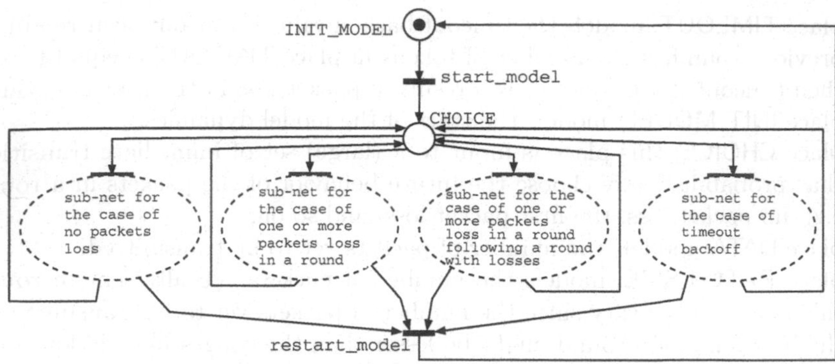

Fig. 2. Structure of the complete SPN model of TCP connection latency.

tokens if the current congestion window size is allowed to grow and the connection is in the congestion avoidance phase, three token is the connection is otherwise in the slow start phase. It is necessary to store this information to appropriately update the marking of place W.

– Timed transition t_no_loss: models the duration of a round hence the average firing delay is equal to RTT. Its firing decrease the number of tokens in place DATA by the marking dependent expression $\min(\#W, \#DATA)$, and empties places BATCH_SIZE, NOLOSS, and NOL. Depending on the number of tokens in place NOLOSS, the number of tokens in place W is incremented according to the following case expression $(f1)$:

$$[\#NOLOSS = 1 : 0 \#NOLOSS = 2] : 1$$
$$[\#NOLOSS = 2] : 1$$
$$[\#NOLOSS = 3 \wedge (\#THRESHOLD >= 2 * \#W)] : \#W$$
$$\text{default} : \#THRESHOLD - \#W$$

the above expression either leaves the marking of place W unchanged or increases it by one (congestion avoidance case) or by $\#W$ (slow start phase).

Figure 3 describes the submodel structure for the case of no packet loss.

4.2 Case of One or More Packets Loss in a Round

For the sake of clarity, the description of the submodel representing the case of packet loss in a round is further divided into separate descriptions of the case of the loss of the first packet in a round and the case of other packets loss.

Case of Loss of the First Packet in a Round

– Immediate transition i_lose_first: models the probabilistic choice (transition weight is equal to p) of losing the first packet in a round. This transition is

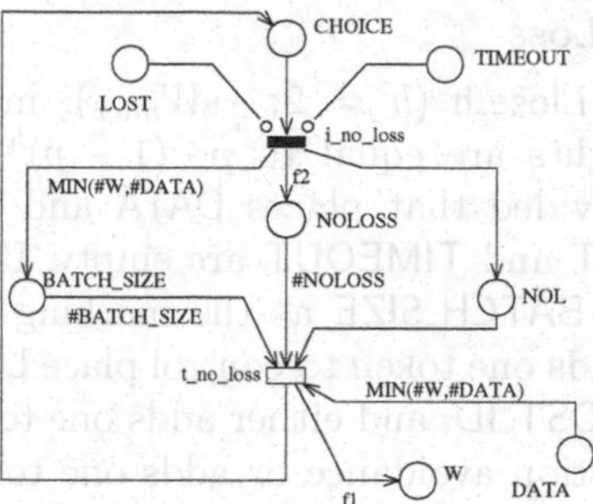

Fig. 3. Structure of the SPN submodel for the case of no packet loss.

enabled if places LOST and TIMEOUT are empty and place W contains at least one token. Its firing sets the number of tokens in place BATCH_SIZE as the marking dependent expression $\min(\#W, \#DATA)$ and adds one token to the control place LOSE_FIRST.

– Timed transition t_lose_first: following our loss model assumption, if the first packet in a round is lost all the following packets in that round are also lost. This means that no ACKs will ever return to the sender that must then wait for a timeout to expire before retransmitting. The firing of this transition will empty place W leaving one token; the marking of place THRESHOLD is updated by the following marking dependent expression $\max(2, \frac{\#W}{2})$, one token is put in place TIMEOUT, while places BATCH_SIZE and LOSE_FIRST are emptied. The average firing delay is equal to $T_0 = \frac{3}{\text{TCP tick}}$ if the lost packet is also the first packet of the connection , while in all other cases it is equal to $T_0 = 2 \cdot \text{TCP tick}$.

Figure 4 describes the submodel structure for the case of the loss of the first packet in a round.

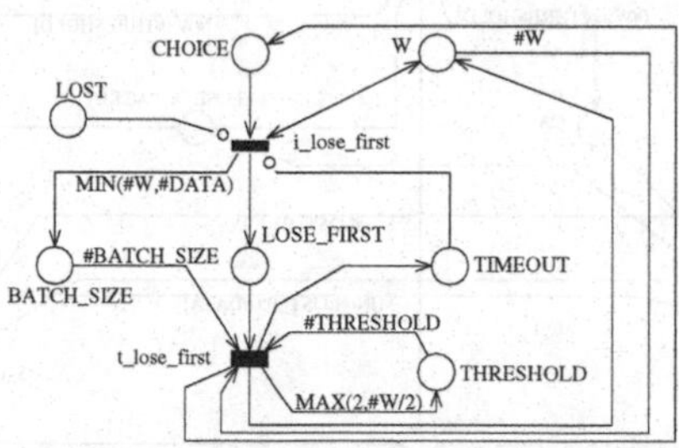

Fig. 4. Structure of the SPN submodel for the case of the loss of the first packet in a round.

Case of Other Packets Loss

- Immediate transitions i_lose_h ($h = 2, \cdots W_{max}$): model the probabilistic choice (transition weights are equal to $p \cdot (1-p)^{h-1}$) of losing the hth packet in a round provided that places DATA and W contain at least h tokens and places LOST and TIMEOUT are empty. Its firing sets the number of tokens in place BATCH_SIZE as the marking dependent expression $\min(\#W, \#DATA)$, adds one token to control place LOSE_A_PACKET, adds $h-1$ tokens to place LOST_ID, and either adds one token to place CA if the connection is in congestion avoidance or adds one token to place SS if the connection is in slow start.
- Timed transition SS_t_lose: following our loss model assumption, if the hth packet in a round is lost then $h-1$ packet will arrive at their destination and $h-1$ ACK will be received by the source. The firing empties places BATCH_SIZE, LOSE_A_PACKET, and SS, decrease the marking of place DATA by the marking dependent expression $\min(h-1, \#DATA)$ (marking of place LOST_ID is equal to $h-1$ upon firing of transition i_lose_h), increases the marking of place W by the marking dependent expression $\min(\#W + h - 1, \#THRESHOLD)$, and puts $\min(\#W, \#BATCH\_SIZE) - h + 1$ tokens in place LOST. The average firing delay is equal to RTT.
- Timed transition CA_t_lose: the same as timed transitions SS_t_lose but in this case place CA must be considered instead of place SS; furthermore, the marking of place W is left unchanged.

Figure 5 describes the submodel structure for the case of the loss of other packets in a round; as an example, only the case of the loss of the 8th packet in a round is depicted.

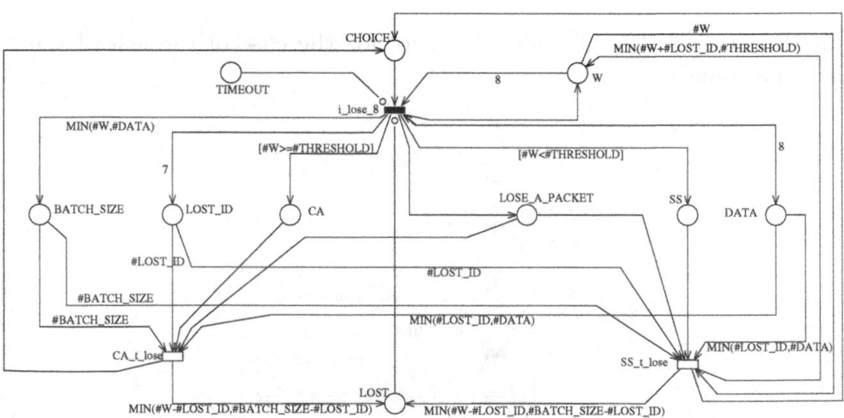

Fig. 5. Structure of the SPN submodel for the case of the loss of other packets in a round.

4.3 Case of Zero or More Packets Loss in a Round Following a Round with Losses

For the sake of clarity, the description of the submodel representing the case of zero or more packets loss in a round following a round with losses is further divided into separate descriptions of the case of no losses, losses leading to a timeout and the case of other packets loss.

Case of Zero Losses

- Transitions i_no_loss_next_round_W_gte_3 and t_no_loss_next_round_W_gte_3: model the choice of no packet loss in round following a round with losses and halving of the congestion window size, respectively, in case of a number of transmitted packets greater than or equal to three. In this case, congestion avoidance is entered by the source and the threshold is halved accordingly. In particular, transition i_no_loss_next_round_W_gte_3 has a weight equal to $(1-p)^{\min(\#W, \#DATA)}$ while t_no_loss_next_round_W_gte_3 has an average firing delay is equal to RTT. The firing of transition i_no_loss_next_round_W_gte_3 sets the number of tokens in place BATCH_SIZE as the marking dependent expression $\min(\#W, \#DATA)$ while the firing of t_no_loss_next_round_W_gte_3 empties place BATCH_SIZE and LOST, decreases the number of tokens of place DATA by $\min(\#W, \#DATA)$, changes the marking of place W to $\frac{W}{2}$, and the marking of place THRESHOLD to $\max(2, \frac{\#W}{2})$.
- Transitions i_no_loss_next_round_W_lt_3 and t_no_loss_next_round_W_lt_3: describe the above case but when the congestion window size is less than three. In this case i_no_loss_next_round_W_lt_3 has the same characteristics as i_no_loss_next_round_W_gte_3 while t_no_loss_next_round_W_lt_3 has average firing delay equal to $T_0 - RTT$. Furthermore, the firing of transition t_no_loss_next_round_W_lt_3 updates places DATA, LOST, and BATCH_SIZE the same way t_no_loss_next_round_W_gte_3 does, while it changes the marking of place W to 1, and puts 1 token in place TIMEOUT.

Figure 6 describes the submodel structure for the case of zero losses in a round following a round with losses.

Case of Losses Leading to Timeout

- Transitions i_lose_h_next_round ($h = 1, 2, 3$), these transitions model the case when the number of packets transmitted in a round following a round with a loss and that have arrived at destinations are less than three, i.e., not enough to get three duplicate ACKs and then enter the fast retransmit phase. In this case the timeout occurs. Transition weights are equal to $p \cdot (1-p)^{h-1}$; these transitions are enabled provided that places DATA and W contain at least h tokens, place LOST contains at least one token, and place TIMEOUT is empty. Its firings set the number of tokens in place BATCH_SIZE as the marking dependent expression $\min(\#W, \#DATA)$ and adds one token to control places LOSE_H_PACKET_NEXT_ROUND.

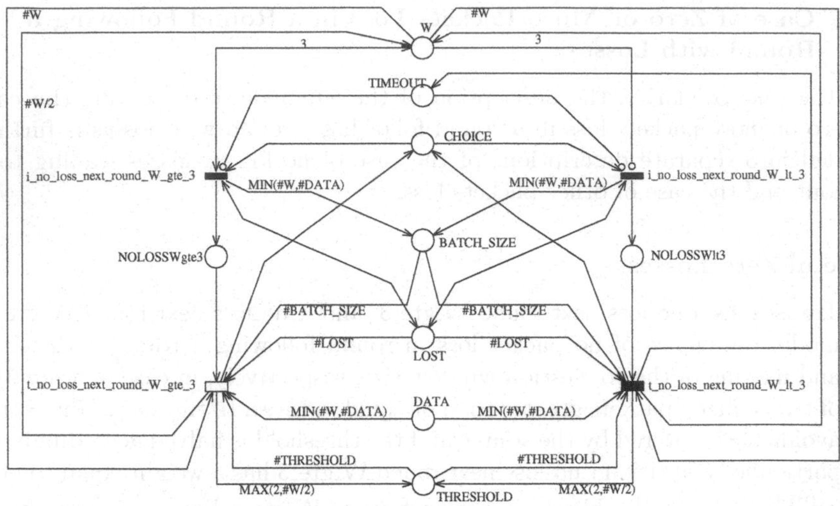

Fig. 6. Structure of the SPN submodel for the case of zero losses in a round following a round with losses.

- t_lose_h_next_round ($h = 1, 2, 3$): its firing empties places BATCH_SIZE, LOSE_H_PACKET_NEXT_ROUND and LOST; it decrease the marking of place DATA by the marking dependent expression $\min(h - 1, \#DATA)$ leaves one token in place W, and sets the marking of place THRESHOLD to $\max(2, \frac{\#W}{2})$ The average firing delay is equal to $T_0 - RTT$.

Figure 7 describes the submodel structure for the case of losses leading to timeout in a round following a round with losses. As an example, the case of the second packet loss is presented.

Case of Other Packets Loss

- Immediate transitions i_lose_h_next_round ($h = 4, \ldots W_{max}$): these transitions model the case when the number of packets transmitted in a round following a round with a loss and that have arrived at destinations are greater than or equal to three, i.e., enough to get three duplicate ACKs and the enter the fast retransmit phase. In this case the congestion avoidance is entered. Transition weights are equal to $p \cdot (1 - p)^{h-1}$; these transitions are enabled provided that places DATA and W contain at least h tokens, place LOST contains at least one token, and place TIMEOUT is empty. Its firings set the number of tokens in place BATCH_SIZE as the marking dependent expression $\min(\#W, \#DATA)$, adds one token to control places ISR and LOSE_A_PACKET, adds $h - 1$ tokens to place LOST_ID.
- Timed transition t_lose_next_round: its firing empties places BATCH_SIZE, LOSE_A_PACKET, ISR, LOST_ID, and LOST; it decrease the marking of place

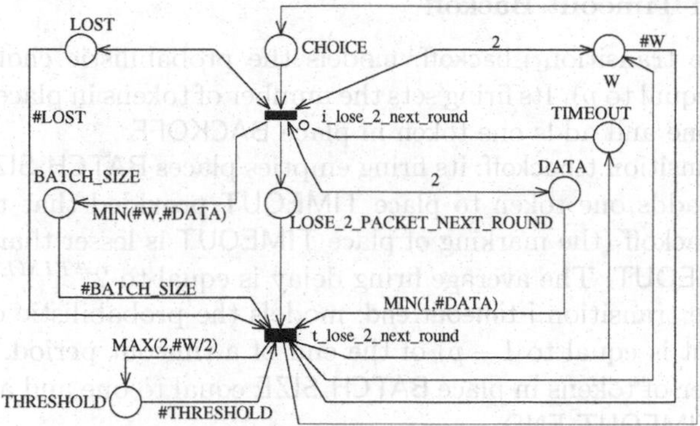

Fig. 7. Structure of the SPN submodel for the case losses leading to timeout in a round following a round with losses (case of the second packet loss).

DATA by the marking dependent expression $\min(h-1, \#DATA)$ (marking of place LOST_ID is equal to $h-1$ upon firing of transition _lose_h_next_round), halves the marking of place W, and sets the marking of place THRESHOLD to $\max(2, \frac{\#W}{2})$ The average firing delay is equal to RTT.

Figure 8 describes the submodel structure for the case of losses leading to timeout in a round following a round with losses; as an example, only the case of the loss of the 7th packet in a round is depicted.

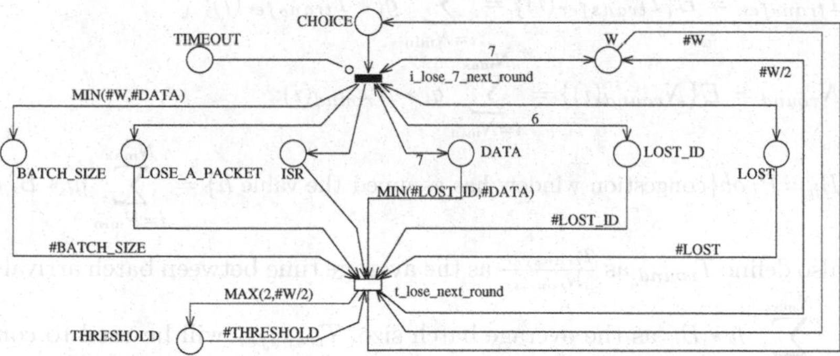

Fig. 8. Structure of the SPN submodel for the case other losses in a round following a round with losses.

4.4 Case of Timeout Backoff

- Immediate transition i_backoff: models the probabilistic choice (transition weight is equal to p). Its firing sets the number of tokens in place BATCH_SIZE equal to one and adds one token in place BACKOFF.
- Timed transition t_backoff: its firing empties places BATCH_SIZE and BACKOFF and adds one token to place TIMEOUT provided that the number of timeout backoff (the marking of place TIMEOUT is lesser than the constant MAX_TIMEOUT. The average firing delay is equal to $2^{\#TIMEOUT-1} \cdot T_0$.
- Immediate transition i_timeout_end: models the probabilistic choice (transition weight is equal to $1-p$) of the end of a timeout period. Its firing sets the number of tokens in place BATCH_SIZE equal to one and adds one token in place TIMEOUT_END.
- Timed transition t_timeout_end: its firing empties places BATCH_SIZE, TIMEOUT_END, and TIMEOUT, adds one token to place W, and withdraws one token from place DATA. The average firing delay is equal to RTT.

4.5 Model Exploitation

The average latency of the TCP connection to transfer N packets is computed as $T_{transfer}(N) = \frac{1}{X(\text{start\_model})}$ ($X(\text{t})$ denotes the steady-state throughput of transition t); the average number of rounds during the activity period is computed as $N_{round}(N) = \frac{\sum X(\text{t})}{X(\text{start\_model})}$ where the sum of throughput is carried over all immediate transitions of the model; the batch (round) size distribution is obtained as $B_{N,h} = P\{\#\text{BATCH\_SIZE} = h\}/(1 - P\{\#\text{BATCH\_SIZE} = 0\})$.

Starting from these definitions we compute the following results:

- $T_{transfer} = E\{T_{transfer}(i)\} = \sum_{i=N_{\min}}^{N_{\max}} g_i * T_{transfer}(i)$;
- $N_{round} = E\{N_{round}(i)\} = \sum_{i=N_{\min}}^{N_{\max}} g_i * N_{round}(i)$;
- $B_h = Prob\{\text{congestion window has assumed the value } h\} = \sum_{i=N_{\min}}^{N_{\max}} g_i * B_{i,h}$.

We also define T_{round} as $\frac{T_{transfer}}{N_{round}}$ as the average time between batch arrivals, and $\overline{B} = \sum_{h=N_{\min}}^{N_{\max}} h * B_h$ as the average batch size. $T_{transfer}$ will be used to compute the average activity duration of a TCP connection, T_{round} will be the used to set the batch arrival process rate, and B_h values will be the batch size distribution to be fed to the IP network link model.

At the end of the iterative solution when parameters p and RTT are computed, the TCP latency submodel is also used to compute the completion time distribution by performing a transient analysis of the probability that place INIT_MODEL is marked.

5 The IP Network Link Submodel

The model used to represent an IP network link traversed by the K ON-OFF TCP connections is a SPN representation of a finite capacity, single server queue with exponentially distributed service times whose average is equal to the packet transmission time. The queue arrival process is the superposition of K homogeneous, two state Markov Modulated Poisson process with batch arrivals; the mean time spent in the OFF state is one of the network primitives: the OFF periods are exponentially distributed with average $T_{silence}$ and transition T_{off_on} models the awakening of a TCP source. The mean time spent in the ON state is input from the TCP latency model and is given by $T_{transfer}$: transition T_{on_off} models the connection termination. The mean time between batch arrivals is given by T_{round}: transition batch_arrival models the batch generation. The batch size distribution is given by B_h: the set of immediate transitions batch$_h$ whose weights are equal to B_h models the probabilistic choice of a batch size.
Batch arrivals have been introduced in the IP network link model to consider the traffic burstiness but packets in the same window do not enter the link queue at the same time; instead, they are spaced in time by at least a transmission time, denoted as $T_{transmit}$. We devised an expedient to cope with this feature: we introduced a delay in the model using an exponential transition named deliver_packets whose firing rate is equal to $\frac{1}{T_{transmit}}$. Transition batch_arrival is enabled to fire only when all packets in a batch have entered the queue.

Since the enabling of transition batch_arrival is inhibited by the inhibitor arc from place GENPACKETS we defined a multiplicative coefficient for the rate of this transition based on the following observation: the average time between batch arrival and the completion of batch entering the queue is equal to $\overline{B} \cdot T_{transmit}$. The firing of transition batch_arrivals is prevented during this period of time; during the same period of time the average number of packets that might enter the queue is equal to $\frac{\overline{A} \cdot \overline{B}}{T_{round}}$ where $\overline{A}$ denotes the average number of active connections defined as $K \cdot \frac{T_{transfer}}{T_{transfer}+T_{off}}$. Therefore we multiply the rate of transition batch_arrival by $c \cdot (1 + \frac{\overline{A} \cdot \overline{B}}{T_{round}})$ where $c = \frac{\overline{B} \cdot \overline{A} \cdot T_{transmit}}{T_{round}}$ to smooth the multiplicative coefficient as the time to let a batch enter the queue becomes negligible with respect to the time between successive batch arrivals.

The queue is modeled by places QUEUE and CAPACITY that represent the queue occupancy and the queue capacity, respectively. Immediate transitions admit and loss model the admission and the loss of a packet, respectively. Finally, transition transmit_packet models the packet transmission along the link. An example of the resulting SPN model where only batches of size $\{1, 2, 3, 4\}$ are modeled, is depicted in Figure 9. In Section 6 we also present results obtained by using different batch sizes.

The SPN model is solved computing the steady state solution of the associated CTMC yielding two performance indexes: the packet loss probability defined as

$$p = \frac{X(loss)}{X(loss) + X(admit)},$$

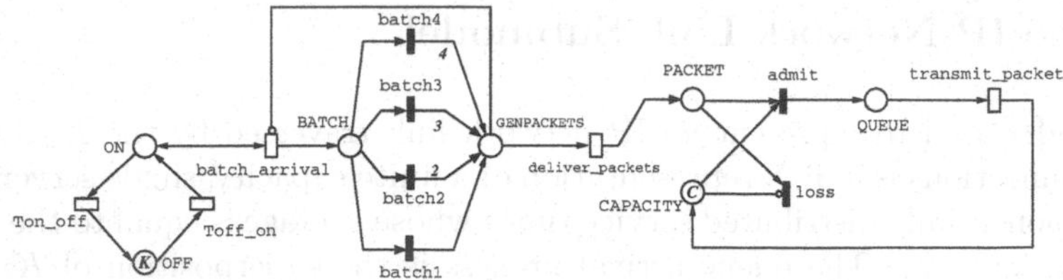

Fig. 9. SPN model of the IP network link

and the round trip time defined as

$$RTT = 2 \cdot PD \cdot \frac{\overline{QUEUE}}{C},$$

where $\overline{QUEUE}$ denotes the mean number of tokens in place QUEUE and $X(loss)$ and $X(admit)$ denote the throughput of transitions loss and admit, respectively. These two computed values are then fed back to the TCP connection latency model until convergence.

6 Results

This section provides a validation of the proposed modeling approach through extensive comparisons of the model's results with the output of the simulator ns−2 [10] that provides detailed descriptions of the dynamics of the Internet protocols.

Estimates of the average completion times (i.e., the average time for the TCP connection to transfer a finite amount of packets) and of the average packet loss probability output by the model have been compared with point estimates produced by the simulation of TCP Reno including all its features. Furthermore, the simulations provided estimation of the distribution of the completion times by discretizing simulated time in intervals whose width was equal to the two-way propagation delay.

All simulations were run using a batch method, with a confidence level equal to 97.5%, and accuracy equal to 10% for the interval estimation of the packet loss probability. An initial transient equal to 300 seconds of simulated time was discarded for each run. Batch width was uniformly distributed between 50 and 70 seconds of simulated time.

All scenarios we considered share the following characteristics: maximum congestion window size W_{max} equal to 21, drop tail buffer whose size B is equal to 32 packets, and one-way link propagation delay PD equal to $10ms$.

The TCP latency submodel has been solved using the TimeNET software tool [14] that includes all the modeling features we used to develop our models, while the IP network link submodel has been solved using the GreatSPN software tool [5] that allows a very efficient solution of GSPN models.

Scenario 1

The first scenario we consider is characterized by the following parameters: link capacity C equal to 45Mb/s, packet size PK equal to 512 bytes, packet emission probability distribution $g = \{g_1, g_3, g_5, g_8, g_{12}, g_{19}, g_{35}, g_{52}\}$ where $g_1 = 0.3$, $g_3 = g_5 = g_8 = g_{12} = g_{19} = g_{35} = g_{52} = 0.1$ for both 30 and 50 TCP connections. In Figure 10 the average completion time (left graph) and the average packet loss probability (right graph) are plotted versus increasing values of the average OFF period T_{off}. It can be seen that the model provides an extremely accurate representation of both performance metrics both for 30 and for 50 TCP connections.

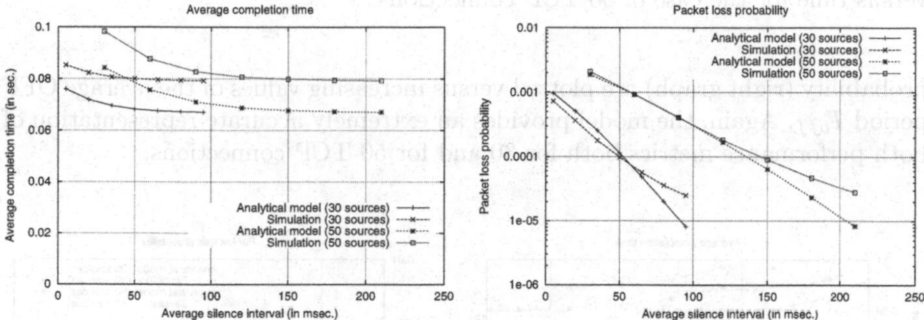

Fig. 10. Average completion times (left) and average packet loss probability (right) versus average OFF periods, link speed equal to 45Mb/s, packet size equal to 512 bytes, packet emission probability distribution $g_1 = 0.3$, $g_3 = g_5 = g_8 = g_{12} = g_{19} = g_{35} = g_{52} = 0.1$.

Furthermore, for the case of 50 TCP connections, Figure 11 the CDF of the completion time for transmitting 35 packets (left graph) and the CDF of the completion time for transmitting52 (right graph) are plotted versus increasing values of time. It can be seen that the model provides an extremely accurate representation of both CDFs. Due to space limitations other CDF graphs will not be presented; nevertheless, it must be pointed out that similar accurate results were obtained for the remaining number of packets as well as for all the following scenarios.

Scenario 2

In order to further assess the accuracy and robustness of the analytical performance estimates for different network settings, we considered the previous scenario with a different packet emission probability distribution: $g = \{g_1, g_8, g_{35}, g_{52}\}$ where $g_1 = 0.3$, $g_8 = g_{35} = g_{52} = 0.2$ for both 30 and 50 TCP connections. In Figure 12 the average completion time (left graph) and the average packet loss

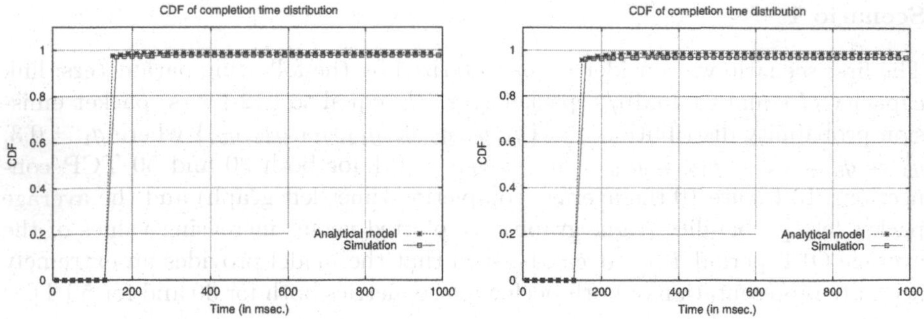

Fig. 11. Completion times CDF for 35 packets (left) and 52 packets (right) versus time for the case of 50 TCP connections.

probability (right graph) are plotted versus increasing values of the average OFF period T_{off}. Again, the model provides an extremely accurate representation of both performance metrics both for 30 and for 50 TCP connections.

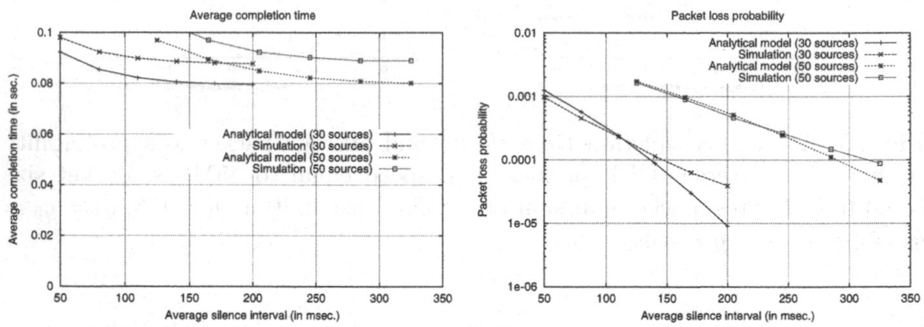

Fig. 12. Average completion times (left) and average packet loss probability (right) versus average OFF periods, link speed equal to 45Mb/s, packet size equal to 512 bytes, packet emission probability distribution $g_1 = g_8 = 0.3$, $g_{35} = g_{52} = 0.2$.

Scenario 3

Models of faster links, also yield very good analytical estimates. We considered the following scenario: link capacity C equal to 100Mb/s, packet size PK equal to 1024 bytes, packet emission probability distribution $g = \{g_1, g_3, g_5, g_8, g_{12}, g_{19}, g_{35}, g_{52}\}$ where $g_1 = 0.3$, $g_3 = g_5 = g_8 = g_{12} = g_{19} = g_{35} = g_{52} = 0.1$ for 30 and 50 TCP connections. In Figure 13 the average completion time (left graph) and the average packet loss probability (right graph) are plotted versus increasing values of the average OFF period T_{off}. Also in this case, the model accurately represents both performance metrics.

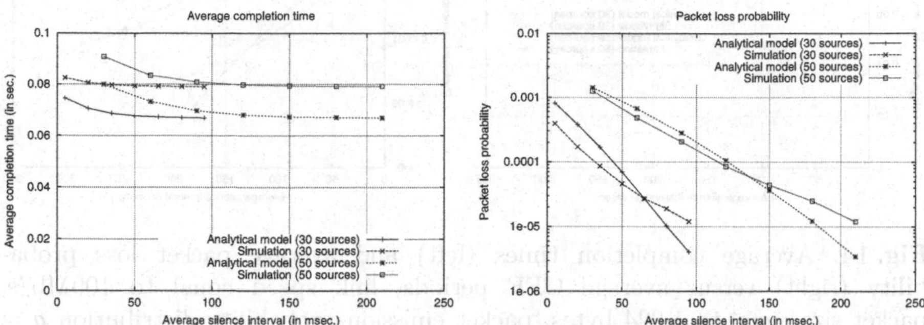

Fig. 13. Average completion times (left) and average packet loss probability (right) versus average OFF periods, link speed equal to 100Mb/s, packet size equal to 1,024 bytes, packet emission probability distribution $g_1 = 0.3$, $g_3 = g_5 = g_8 = g_{12} = g_{19} = g_{35} = g_{52} = 0.1$.

Scenario 4

The accuracy and robustness of the analytical performance estimates have been tested on the previous scenario but with a different packet emission probability distribution: $g = \{g_1, g_8, g_{35}, g_{52}\}$ where $g_1 = g_8 = 0.3$, $g_{35} = g_{52} = 0.2$ for 30 and 50 TCP connections. In Figure 14 the average completion time (left graph) and the average packet loss probability (right graph) are plotted versus increasing values of the average OFF period T_{off}. Again, the model provides an extremely accurate representation of both performance metrics.

Model Solution Complexity

As a final remark, it should be noted that the number of iterations before convergence depends on the accuracy imposed between successive computations of the average packet loss probability (that we set to 10^{-2}). The number off iterations in the settings we considered are usually quite small, ranging from 2 to 26 for one point of a curve.

The total CPU time for obtaining the model curves presented in the paper was 12,995 seconds (less than four hours) on a Pentium III notebook running Linux operating system that must be compared with ten days of CPU times on a bi-processor Pentium III computer running Solaris operating system for the *ns* simulator to produce point estimates.

The number of states of the TCP latency submodel grows for increasing number of packets to be transmitted; for the case of 52 packets, the number of

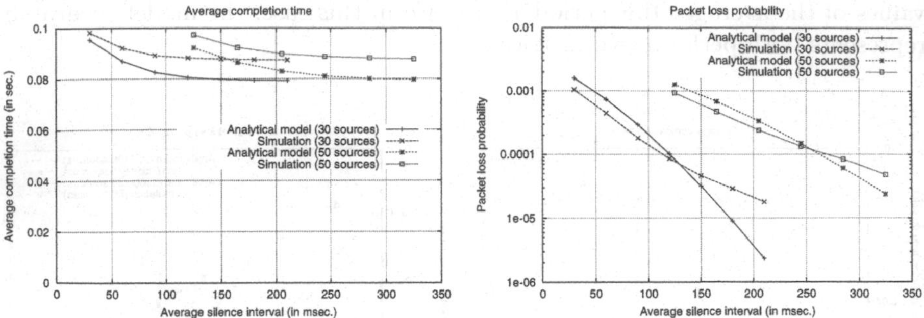

Fig. 14. Average completion times (left) and average packet loss probability (right) versus average OFF periods, link speed equal to 100Mb/s, packet size equal to 1,024 bytes, packet emission probability distribution $g = \{g_1, g_8, g_{35}, g_{52}\}$ where $g_1 = g_8 = 0.3$, $g_{35} = g_{52} = 0.2$.

tangible states was equal to 43,667 while for the case of 35 packets the number of tangible states was equal to 14,125. The number of states for the IP link submodel depends on the number of connections, the buffer size and the maximum window size: for the scenarios considering 50 connections, the number of tangible states was equal to 37,026.

7 Discussion and Future Directions

In this paper we have described and validated a new analytical modeling approach to study the behavior of a finite number of TCP Reno connections exhibiting an ON-OFF behavior; the TCP connections are short-lived and support the transmission of a finite number of packets following a given packet emission probability distribution.

The modeling approach is based on Markovian assumptions and on two SPN descriptions of the system behavior, the former considering the TCP connection latency to transfer a finite amount of packets, the latter considering the IP link queue loaded with packet arrivals from the TCP sources. From the two SPN descriptions, using a fixed point algorithm, performance metrics such as packet loss probability, average transfer completion time as well as completion time distribution are derived. As for the extensions to the present work, we are currently investigating several interesting research topics:

- Employ an alternative strategy for the IP network link model based on the Fluid Stochastic Petri Net (FSPN) [8]; this formalism allows the definition of places where the number of tokens is described by continuous values thus allowing to consider much larger systems with larger buffer size and/or larger number of TCP connections.
- Characterize the output process of the IP network link model to consider multibottleneck scenarios.

References

[1] M. Ajmone Marsan, C. Casetti, R. Gaeta, and M. Meo. Performance analysis of TCP connections sharing a congested internet link. *Performance Evaluation*, 42(2–3), September 2000.

[2] E. Alessio, M. Garetto, R. Lo Cigno, M. Meo, and M. Ajmone Marsan. Analytical Estimation of Completion Times of Mixed New-Reno and Tahoe TCP Connections over Single and Multiple Bottleneck Networks. Technical report, Politecnico di Torino, 2001. Submitted for publication.

[3] N. Cardwell, S. Savage, and T. Anderson. Modeling TCP latency. In *Proc. IEEE Infocom 2000*, pages 1742–1751, Tel Aviv, Israel, March 2000. IEEE Comp. Soc. Press.

[4] C. Casetti and M. Meo. A New approach to Model the Stationary Behavior of TCP Connections. In *Proc. IEEE Infocom 2000*, Tel Aviv, Israel, March 2000. IEEE Comp. Soc. Press.

[5] G. Chiola, G. Franceschinis, R. Gaeta, and M. Ribaudo. GreatSPN1.7: GRaphical Editor and Analyzer for Timed and Stochastic Petri Nets. *Performance Evaluation*, November 1995. Special issue on Performance Modeling Tools.

[6] M. Garetto, R. Lo Cigno, M. Meo, and M. Ajmone Marsan. A Detailed and Accurate Closed Queueing Network Model of Many Interacting TCP Flows. In *Proc. IEEE Infocom 2001*, Anchorage, Alaska, USA, 2001. IEEE Comp. Soc. Press.

[7] M. Garetto, R. Lo Cigno, M. Meo, and M. Ajmone Marsan. Queuing Network Models for the Performance Analysis Multibottleneck IP Networks Loaded by Short-Lived TCP Connections. Technical report, Politecnico di Torino, 2001. Submitted for publication.

[8] M. Gribaudo, M. Sereno, A. Bobbio, and A. Horvath. Fluid Stochastic Petri Nets augmented with Flush-out arcs: Modelling and Analysis. *Discrete Event Dynamic Systems*, 11(1 & 2), 2001. To appear.

[9] E. Király, M. Garetto, R. Lo Cigno, M. Meo, and M. Ajmone Marsan. Computation of the Completion Time Time Distribution of Short-Lived TCP Connections. Technical report, Politecnico di Torino, 2002.

[10] S. MCanne and S. Floyd. ns-2 network simulator (ver.2). Technical report, 1997. URL http://www.isi.edu/nsnam/ns/.

[11] J. Padhye, V. Firoiu, and D. Towsley. A Stochastic Model of TCP Reno Congestion Avoidance and Contro. Technical report, Department of Computer Science, University of Massachusetts, 1999. Technical Report 99-02.

[12] J. Padhye, V. Firoiu, D. Towsley, and J. Kurose. Modeling TCP Reno performance: a simple model and its empirical validation. *IEEE/ACM Transaction on Networking*, 8(2):133–145, 2000.

[13] B. Sikdar, S. Kalyanaraman, and K. S. Vastola. An integrated model for the latency and steady-state throughput of TCP connections. *Performance Evaluation*, 46:139–154, September 2001.

[14] A. Zimmermann, R. German, J. Freiheit, and G. Hommel. TimeNET 3.0 Tool Description. In 8th *Intern. Workshop on Petri Nets and Performance Models*, Zaragoza, Spain, Sep 1999. IEEE-CS Press.

A Queueing Network Model of Short-Lived TCP Flows with Mixed Wired and Wireless Access Links*

Roberta Fracchia, Michele Garetto, and Renato Lo Cigno**

Dipartimento di Elettronica – Politecnico di Torino
Corso Duca degli Abruzzi, 24 – I-10129 Torino, Italy
{fracchia,garetto,locigno}@mail.tlc.polito.it

Abstract. We present an analytical model, based on a Fixed Point Approximation (FPA) solution, that can be used to derive the performance of different sets of TCP connections that share, and compete for, a common resource, typically a link and its associated buffer. A set of TCP connections is a group of connections that can be considered homogeneous, e.g., they have similar RTTs and all have a wireless access. TCP connections are modeled through the OMQN (Open Multiclass Queueing Network) paradigm. The conditions that define the feasibility of the solution and allow the convergence of the model are discussed and an application example with a RED buffer where wired and wireless connections converge is presented.

1 Introduction

After the explosion of cellular telephony, wireless and mobile networks are becoming more and more popular also for data transmission. Specialized network architectures, like WAP (Wireless Application Protocol) [1,2,3,4] and i-mode [5,6], have been proposed and deployed trying to overcame the inevitable impairments of wireless channels. These new architectures may finally find a position in the telecommunication market, but they do not seem able to substitute the traditional TCP/IP architecture for data transmission. The fast diffusion of wireless LANs and the standardization of new ultra-fast radio interfaces [7,8,9], together with the introduction of 3G cellular systems, is a strong indication that the use of standard, PC-based terminals running TCP/IP and accessing the Internet through a wireless connection will be a common scenario in a very short time.

The interest in the performance of TCP over wireless channels is thus on the rise, and the issue of how well wireless TCP connections will be able to compete with wired ones to share networks resources is matter of debate. Several simulation-based studies on wireless TCP have appeared in recent years [10,11], and some effort was also placed in trying to model the TCP behavior on wireless channels, capturing the essential feature

* This work was supported by the Italian Ministry for Education and Scientific Research through the PLANET-IP Project and by CERCOM, the Center for Multimedia Radio Communications of Politecnico di Torino

** Starting Nov. 1 2002 R. Lo Cigno is with the Department of Computer Science and Telecommunications (DIT) of the University of Trento, Italy

M. Ajmone Marsan et al. (Eds.): QoS-IP 2003, LNCS 2601, pp. 392–404, 2003.

of random losses [12]. TCP in fact can't discriminate between packets lost due to lossy links, and those dropped due to congestion in the network.

The aim of this work is the extension of the modeling technique based on OMQN (Open Multiclass Queueing Network) models of TCP, combined with a Fixed Point Approximation (FPA) solution, to cover cases when different TCP connections experience different network conditions before competing for resources on one or more bottlenecks. One of the most interesting case is the wireless/mobile access, that can significantly modify the loss probability perceived by TCP connections, leading to unfair sharing of resources. Though this phenomenon is not new, to the best of our knowledge there are no other analytical approaches that allows to predict the performance and unfairness index in this situation.

2 Queuing Networks Models of TCP

Models of TCP in recent literature are numerous, with methods as different as heuristic analysis with empirical validation as in [13], or sophisticated mathematical abstractions as in [14]. A quick discussion of different approaches for TCP modeling can be found in [15] and its repetition here is superflous. We have successfully used queueing models of end-to-end protocols, TCP being a particular case, in several works [15,16,17,18,19], so that the basic modeling method we adopt in this work is already extensively described, and we only recall here its main features.

An OMQN model is a queueing network model in which all queues are $M/G/\infty$. The customers of the OMQN represent TCP connections. The state of a TCP connection (customer) is uniquely identified by a pair (q, c), where q represents a specific state of the protocol, described with a queue, and c identifies the number of remaining packets to be sent before the completion of the connection. An OMQN model is capable of describing any protocol whose dynamics can be described with a Finite State Machine (FSM). The use of classes to describe the backlog of the connection allows modeling short lived connections, which is indeed one of the main innovative characteristics of OMQNs. Given the average RTT of connections and the average packet loss probability, the OMQN model defines the load offered to the IP network by the aggregate of the connections represented by all the customers currently in the queueing network. New TCP connections enter the system with rate Λ_{ext} and exit when reach class 0, i.e., they have completed the transmission. The arrival rate, together with the initial class of customers, that define the amount of information to be transferred, represent the nominal load of the system.

An OMQN model must be complemented with a suitable description of the network that TCP connections exploit to transfer packets. The network model must be capable, given the load, to compute the basic parameters that drive TCP perfomance: the average loss probability and the average RTT. The overall solution is obtained iterating the solution of the two sub-models with a Fixed Point Approximation (FPA) technique (see Fig. 1). When convergence is reached, the OMQN allows the computation of the throughput and the duration of connections.

Previous works on OMQNs concentrated mainly on the TCP model, somewhat disregarding the underlying IP network. As discussed in Sect. 3, this works concentrates

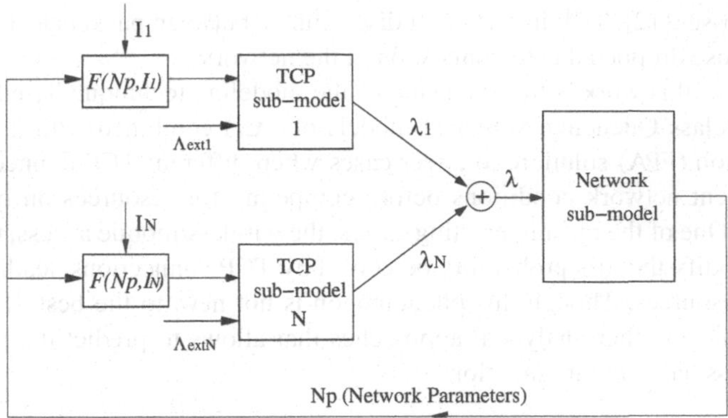

Fig. 1. Schematic representation of the Fixed Point Approximation method for the model decomposition, when N sets of TCP connections load the network

instead on the interaction of the OMQN models with the network, exploring the possibility of analysing the performance of TCP connections that experience different network conditions, but still have to compete for resources on a bottleneck. Sect. 4 presents a simple application of the method where two sets of TCP connections, one with a wired access and the other with a wireless access, compete for network resources on a RED-managed buffer. Results are validated against *ns-2* [20] based simulations.

We refer the reader to [15,16,17,18,19] for any further detail on the use of queueing networks in performance evaluation modeling of TCP connections.

3 Merging Different OMQN Models on a Bottleneck

One of the strengths of FPA techniques, is allowing the decomposition of an extremely complex modeling problem into simpler and partially decoupled sub-models, that interacts only through the iteration procedure. This is a classic application of the *divide at impera* (divide and conquer) principle. For instance the model decomposition allows using different modeling technique to tackle different tasks in the modeling process. Indeed, parts of the system could also be simulated or realized with a testbed instead of being analytically modeled.

Aiming at protocol performance evaluation, the natural problem decomposition is describing the protocol with one sub-model and the transport network with a different one. In the TCP/IP case, TCP becomes the protocol under analysis and the IP network, with everything that goes with it, the information transfer infrastructure. We stress the fact that the network sub-model is not a model of the IP protocol, but of the way the IP network reacts to the load offered by the TCP protocol.

An OMQN model allows the description of an *homogeneous* set of TCP connections, i.e., an ensamble of connections that share a similar propagation delay, the same maximum window size, the same version of the protocol (e.g., Tahoe, NewReno or SACK), etc. Here we use the OMQN TCP-NewReno model presented in [15].

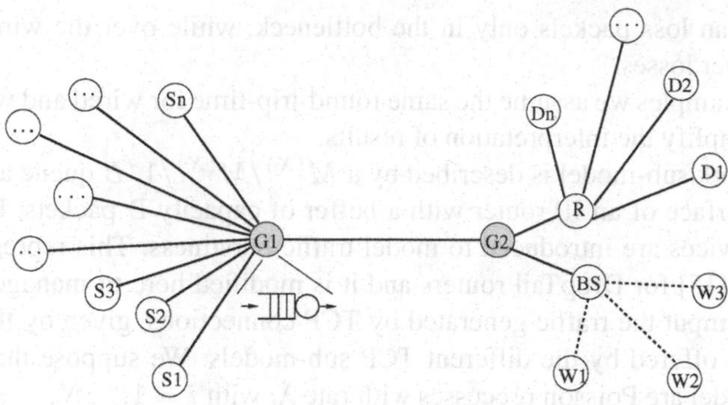

Fig. 2. Network topology with a RED ($G1$) router and wired (D_i) and wireless (W_i) destinations

Let $\mathcal{N}()$ be the model of the network and $\mathcal{T}()$ be a generic OMQN model of a set of TCP connections. Given a synthetic workload $\overline{\lambda}$, a network model returns packet dropping probabilities, and packet transfer delays, in general some network parameters $\overline{N_p}$, while a TCP model returns a synthetic workload $\overline{\lambda}$ given $\overline{N_p}$. If TCP connections loading the network are not homogeneous, then $\overline{\lambda}$ and $\overline{N_p}$ are, in the general case, all vectors.

The conditions for an FPA approach to converge is that there exist a single combination of $\overline{\lambda}$ and $\overline{N_p}$ solving the system

$$\begin{cases} \overline{N_p} = \mathcal{N}(\overline{\lambda}) \\ \lambda_i = \mathcal{T}_i({N_p}^i) \end{cases} \tag{1}$$

Fig. 1 represents a case where the workloads offered by TCP connection sets to the network simply sum one another and N_p^i are obtained from average values yielded by the network model $\mathcal{N}$ through a modifying function $F(N_p, I_i)$ that includes some additional interference I_i, that does not enter the FPA convergence process. In every TCP sub-model there are independent external arrivals Λ_{ext_i}; $i = 1, .., N$.

In the following we shall use this general framework to model TCP connections mixed on the same bottleneck but with etherogeneus access links. Every TCP model is used to describe the behavior of connections following routes with different characteristics, represented by the functions $F(N_p, I_i)$.

4 Sample Applications

4.1 Network Topology

Consider a single bottleneck with a generic AQM (Active Queue Management) system, say RED to focus on a specific case, with TCP connections converging from different subnetworks: for instance a wired network through router **R** and a wireless LAN through a base station **BS**. Fig. 2 represents this simple, yet interesting scenario. Wired connections can loss packets only in the bottleneck, while over the wireless channel there are further losses.

In these examples we assume the same round-trip-time for wired and wireless paths, in order to simplify the interpretation of results.

The network sub-model is described by a $M^{(X)}/M^{(X)}/1/B$ queue and represents an output interface of an IP router with a buffer of capacity B packets. Batch arrivals and batch services are introduced to model traffic burstiness. This representation was introduced in [15] for DropTail routers and it is modified here to manage RED AQM. It receives as input the traffic generated by TCP connections, given by the sum of the average loads offered by the different TCP sub-models. We suppose that the arrivals from each model are Poisson processes with rate λ_i with $i = 1, .., N$.

To model the clustering of transmissions due to TCP behavior we use batch arrivals. We assume that batches of packets arrive at the router with distribution $G = \{g_j\}$ where g_j is the probability that the size of a group is equal to j. The load generated by every model is:

$$\lambda_i = \sum_{j=1} j g_{j_i} \qquad i = 1, .., N \tag{2}$$

where g_{j_i} is the dimension of the groups for each i TCP model. The average load λ offered by TCP is the sum of the number of the groups with the same dimension, in order to keep the traffic burstiness. It results:

$$\lambda = \sum_{i=1}^{N} \lambda_i = \sum_{i=1}^{N} \sum_{j=1} j g_{j_i} \tag{3}$$

So we have batch arrivals at the network queue and the probability gn_j that the size of a group is equal to j is given by:

$$gn_j = \sum_{i} g_{j_i} \tag{4}$$

TCP burstiness should be represented with an arrival correlation function within an RTT, and not simply by bursts arrivals, since packets never arrives at the same instant at the router buffer. Batch arrivals are a convenient mathematical approximation, that, however, ends up in a very pessimistic assumption. In order to compensate this pessimistic assumption, we use batch services. In [15] and [19] there are further explanations and validation tests for this assumption.

The IP network sub-model estimates the packet loss probability and the average round-trip-time, comprising queuing delay at the router as well as propagation delays. These parameters are fed back to the TCP sub-model in an iterative procedure that is stopped only when convergence is reached. We study the convergence of both loss probability calculated by the IP sub-model and loads offered to the network. As can be seen in Fig. 1, the input parameters of TCP sub-models are functions of the results of the network model (the same for each model) and of the scenario parameters I_i with $i = 1, .., N$, that may be different for every model. For these reasons if in two following steps the network parameters are the same, so that convergence is reached, and if the input scenario parameters I_i don't change, also the loads λ_i are the same. So when the loss probability converges, also the different loads of the different TCP models reach the convergence since they are deterministic functions of the network parameters. It's important to stress that every analyzed parameter reach the convergence at the same time even if the whole model is asymmetric: at the entrance of the IP network model there is a sum of the results of different TCP models while the results of IP sub-model are the input parameters of each TCP sub-model.

Thanks to the use of different TCP sub-models, we can run the OMQN to calculate the average loss probability and the completion time of wired as well wireless connections, even if they are mixed.

4.2 Network and TCP Parameters

In this single-bottleneck topology we use a 45 Mb/s link whose length is equal to 5,000 km. We assume one-way TCP connections with uncongested backward path, so that ACKs are never lost or delayed at queues. The distance of servers from the congested router is assumed to be uniformly distributed between 200 and 3,800 km, while the distance of the wired clients on the other side of the link is assumed to be uniformly distributed between 200 and 2,800 km. The wireless clients distance from BS is negligible. Connections are established choosing at random a server-client pair, and are generated following a Poisson process. The packet size is constant, equal to 1,000 bytes; the maximum window size is 32 packets. The TCP tic is equal to 500 ms. The buffer of the $M^{(X)}/M^{(X)}/1/B$ queue is a RED buffer to avoid global synchronization and biases against bursty traffic. The size of the buffer is 128 packets, we use the 'gentle' algorithm with a minimum threshold of 10 packets, a maximum threshold of 50 packets and a queue weight $w_q = 10^{-5}$.

4.3 Bernoulli Loss Process

In this case, the loss probability P_{Lw} of the wireless link is considered Bernoulli. The average loss probability for wireless connections is:

$$P_W = 1 - (1 - P_L)(1 - P_{Lw}) = P_L + P_{Lw} - P_L P_{Lw} \tag{5}$$

where P_L is the loss probability estimated by the network model and due to the bottleneck. In other words the further interference I_i for wireless connections consists of an additional packet loss probability only, and the function $F()$ has the form expressed in (5). We show the behavior of the protocol for different rates of wireless links and for different number of wireless connections. The random loss probability on wireless channels is $P_R = 10^{-2}$.

Numerical Results Performance figures are plotted versus the total normalized nominal load (link utilization that would result without any retransmissions), for transfers of 100 segments objects. Results are validated against point estimates and 95% confidence intervals obtained with $ns-2$ simulations [20] with a confidence level of 90%. Figs. 3 and 4 show the loss probability and the average completion time of both wired

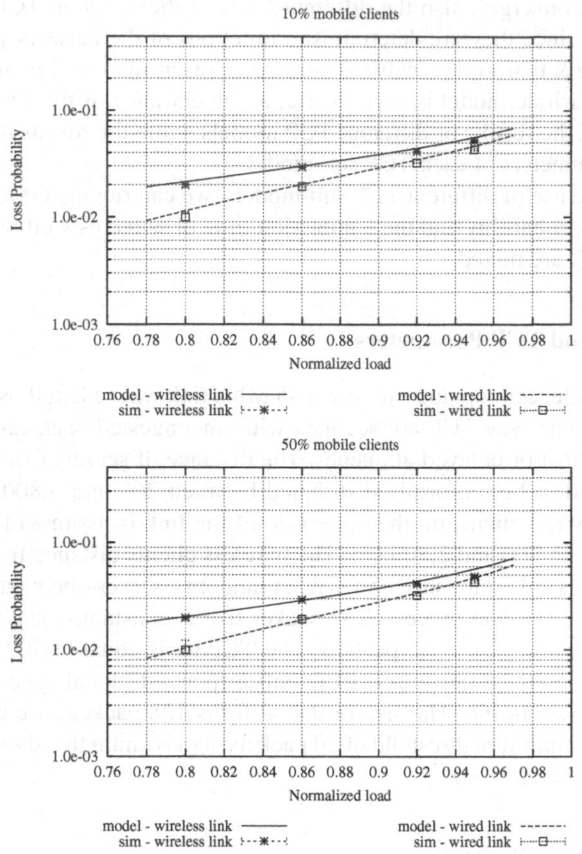

Fig. 3. Loss probability; Bernoulli wireless link: 45 Mbit/s, $P_R = 10^{-2}$

and wireless connections when the rate of wireless link is 45 Mbit/s, like in the recently standardized 802.1q VLAN (Virtual Local Area Network). Upper plots refer to a case when wireless clients are 10% of the total number while in bottom plots they are 50%. The model give accurate results differentiating wireless and wired connections, following closely values obtained with *ns-2* simulations. Fig. 5 and Fig. 6 presents the same figures when the rate of wireless links is 11 Mbit/s, like in standard 802.11b LANs.

We notice incidentally that RED AQM is not able to enforce any degree of fairness between wired and wireless connections. RED droppings should be proportional to the route of the TCP flows, thus balancing the performance among connections. With finite flows, however, the number of packets to be transferred remains constant and RED is not able to differentiate between wireless, low throughput users, and wired ones, that have a throughput 40-50% higher. This result may sound obvious; however it was never stressed in RED literature and the OMQN/FPA modeling approach offer a theoretical explanation of the phenomenon. In fact it is easy to demonstrate through it that a well-

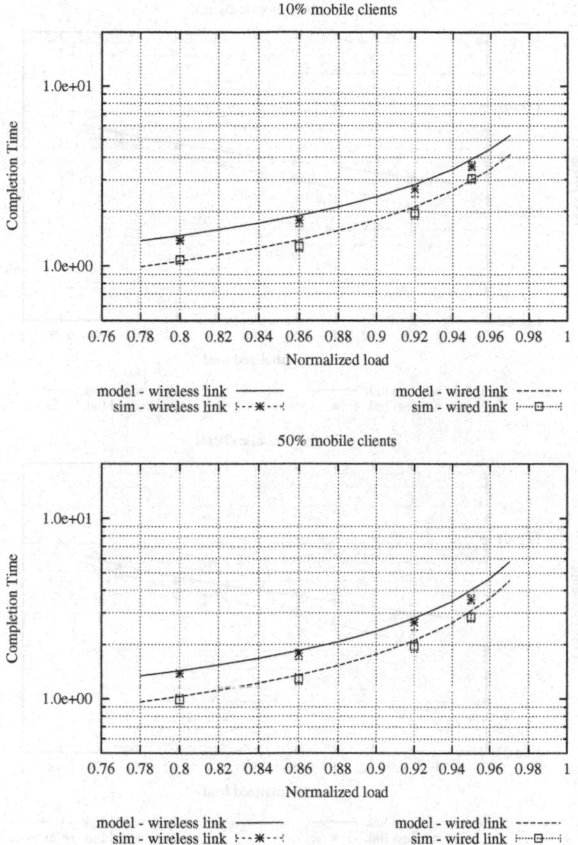

Fig. 4. Average Completion time; Bernoulli wireless link: 45 Mbit/s, $P_R = 10^{-2}$

tuned RED buffer (i.e., one where forced losses never occur) introduces losses that are uncorrelated, hence introducing a Bernoulli loss process. This imply dropping on average an equal number of packets from any connections with the same length, regardless of the actual throughput. Flow lengths of 20, 50 and 200 packets yield similar results, not reported to avoid cluttering the figures.

4.4 Two State Gilbert Wireless Channel

Over many real channels the errors occur in bursts, separated by fairly long error free gaps. To represent this situation, we use the Gilbert channel model: a binary channel described by a two state Markov chain. As shown in Fig.7, we model it with a Continuous Time Markov Chain (CTMC), in which in the Good State (G) errors occur with probability P_G, while in the Bad State (B) they occur with higher probability P_B. The transition matrix Q is:

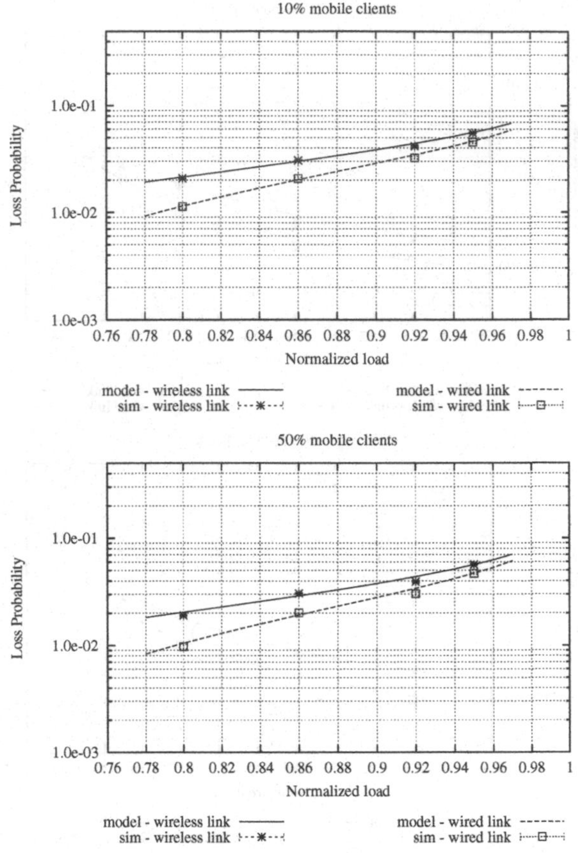

Fig. 5. Loss probability; Bernoulli wireless link: 11Mbit/s, $P_R = 10^{-2}$

$$Q = \begin{bmatrix} -\alpha & \alpha \\ \beta & -\beta \end{bmatrix}$$

where α is the rate of transition from G to B, while β from B to G. The probabilities of being in Good and Bad States are: $\pi_G = \dfrac{\alpha}{\alpha+\beta}$ $\quad \pi_B = \dfrac{\beta}{\alpha+\beta}$ The sojourn times for states B and G are exponentially distribuited with means: $E\{\tau_G\} = \dfrac{1}{\alpha}$ $\quad E\{\tau_B\} = \dfrac{1}{\beta}$ We consider $P_G = 0.01$ and $P_B = 0.1$ and different probabilities of being in Good or Bad.

Referring to Fig. 1, a Gilbert channel requires the use of three different TCP models: one for wired connections, one for wireless connections in the bad state and one for the good state. The average loss probability P_{W_G} for flows being in Good or Bad channel state are computeed using (5). Hence, the average loss probability for wireless connections is $P_W = P_{W_G}\pi_G - P_{W_B}\pi_B$

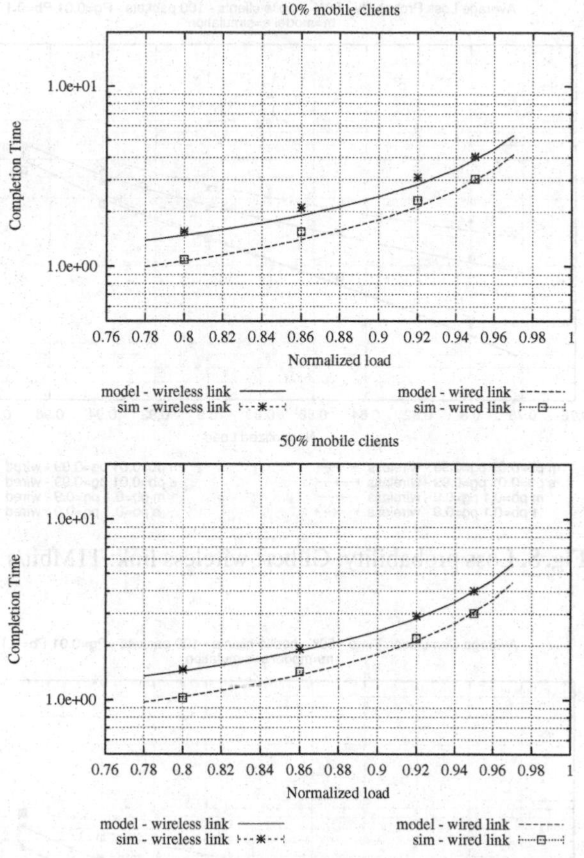

Fig. 6. Average Completion time; Bernoulli wireless link: 11Mbit/s, $P_R = 10^{-2}$

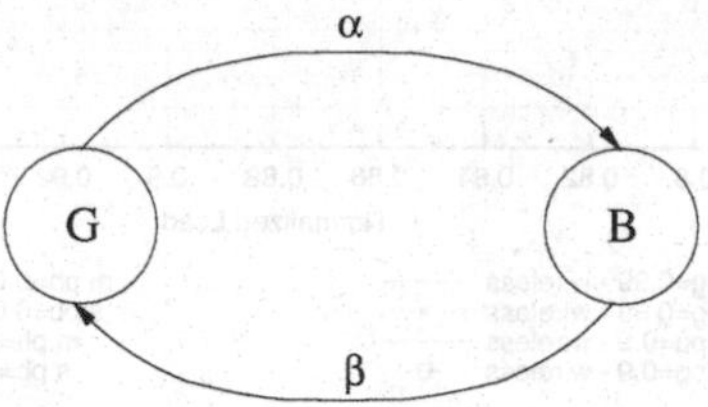

Fig. 7. Gilbert Model

Fig. 9 and Fig. 8 show the loss probability and the average completion time of both wired and wireless connections, 100 packets long, when the rate of wireless link is 11 Mbit/s and mobile clients are 50% of the total number. We consider two different cases: in the first, the probability of being in B is $\pi_B = 0.01$, in the second it is $\pi_B = 0.1$. Also

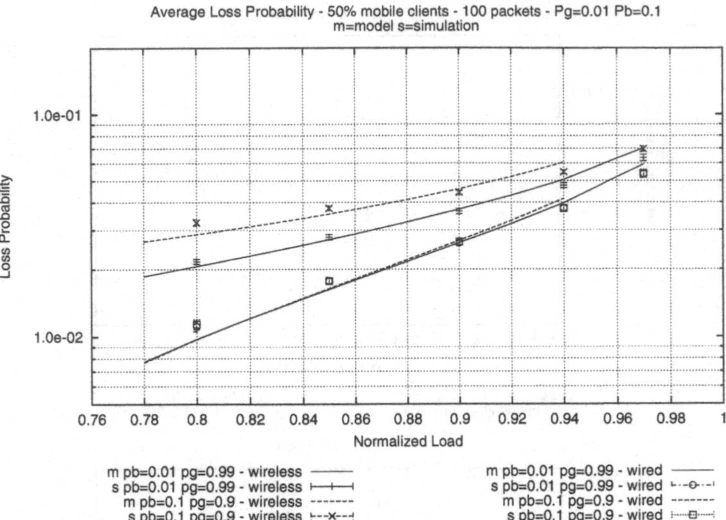

Fig. 8. Loss probability; Gilbert wireless link: 11Mbit/s

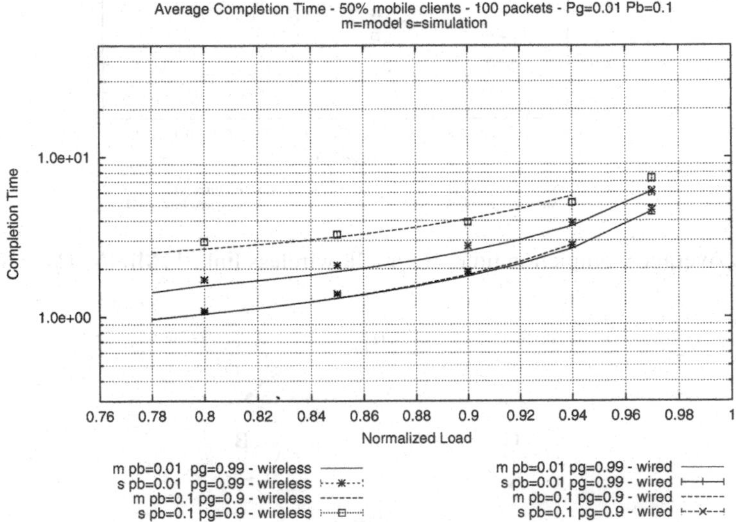

Fig. 9. Average Completion time; Gilbert wireless link: 11Mbit/s

in this more complex case, results provided by the model match closely the simulation points.

5 Conclusions and Future Work

This paper has presented a method to derive the performance of different sets of TCP connections when are competing for a common resource in the Internet. A *set* of TCP connections is defined as an ensamble of connections that share common characteristics, such as the average RTT, the maximum window size, etc. and each set of connections can be modeled through an independent OMQN model. Workloads offered to the network by different connections are summed together to obtain the performance of the IP network.

We have shown an application of the method with connections that compete for a RED buffer. Different OMQN models represent sets of connections that either access the network through a lossy, wireless link or through a traditional, reliable wired network. Results were validated against *ns-2* simulations, showing that the method is viable and yield accurate results. Results show that, as expected, wireless connections are heavily penalized at any load of the bottleneck. Simple AQM disciplines like RED, do not seem able to provide any protection against this penalization.

References

1. O. Kone, "Experiment with the Validation of WAP Systems," *in Proc. IEEE LCN'00*, Nov. 2000, Tampa. Florida
2. D. Maierhofer, "Internet and WAP Technology as Key for Easy Service Management and Convenient Call Control," *in Proc. Intelligent Network Workshop'00*, 2000
3. C.L.C. Wong, M.C. Lee, R.K.W. Chan, "GSM-Based Mobile Positioning Using WAP," *in Proc. Wireless Communications and Networking Conference'00*, Sept. 2000, Chicago
4. A. Andreadis, G. Benelli, G. Giambene, B. Marzucchi, "Performance Analysis of the WAP Protocol over GSM-SMS," *in Proc. IEEE ICC'01*, June 2001, Helsinki, Finland
5. G. Vincent, "Learning from i-mode [Packet-based Mobile Network]," IEEE Review, Vol. 47, Issue 6, Nov. 2001
6. K. Enoki, "i-mode: the Mobile Internet Service of 21st Century," *in Proc. Solid-State Circuits Conference'01*, Feb. 2001, San Francisco, CA
7. M. Natkaniec, "Wireless LAN Standards and Applications," IEEE Communications Magazine, Vol. 40, Issue 2, Feb. 2000
8. T. Ichikawa, "IEEE802: Compliant High-Speed Wireless LAN", *in Proc. High Performance Switching and Routing'02*, May 2002, Kobe, Japan
9. Haitao Wu, Yong Peng, Keping Long, Shiduan Cheng, Jian Ma, "Performance of Reliable Transport Protocol over IEEE 802.11 Wireless LAN: Analysis and Enhancement," *in Proc. IEEE INFOCOM'02*, June 2002, New York
10. A. Chockalingam, M. Zorzi, V. Tralli, "Wireless TCP Performance with Link Layer FEC/ARQ," *in Proc. IEEE ICC'99*, June 1999, Vancouver
11. G. Racherla, S. Radhakrishnan, C.N. Sekharan, "Performance Evaluation of Wireless TCP Schemes under Different Rerouting Schemes in Mobile Networks," *in Proc. IEEE TENCON'98*, Dec. 1998, New Delhi
12. H. Balakrishnan, N. Padmanabhan, S. Seshan, R.H. Kats, "A Comparison of Mechanisms for Improving TCP Performance over Wireless Links", *in Proc. SIGCOM'96*, Oct. 1996, Stanford, CA, USA
13. J. Padhye, V. Firoiu, D. Towsley, and J. Kurose, "Modeling TCP Throughput: A Simple Model and its Empirical Validation," *in Proc. ACM SIGCOMM'98*, Sept. 1998, Vancouver

14. M. Grosslauser, J. Bolot, "On the Relevance of Long-Range Dependencies in Network Traffic", IEEE/ACM Transaction on Networking, Vol.7, N.5, pp.629–640, Oct. 1999
15. M. Garetto, R. Lo Cigno, M. Meo, E. Alessio, M. Ajmone Marsan, "Modeling Short-Lived TCP Connections with Open Multiclass Queueing Networks," *7th International Workshop on Protocols For High-Speed Networks (PfHSN 2002)*, Berlin, Germany, April 22-24, 2002.
16. R. Lo Cigno, M. Gerla, "Modeling Window Based Congestion Control Protocols with Many Flows", *Performance Evaluation*, No. 36–37, pp. 289–306, Elsevier Science, Aug. 1999
17. M. Garetto, R. Lo Cigno, M. Meo, M. Ajmone Marsan, "A Detailed and Accurate Closed Queueing Network Model of Many Interacting TCP Flows," *in Proc. IEEE Infocom'01*, April 22–26, Anchorage, Alaska, USA
18. E. Alessio, M. Garetto, R. Lo Cigno, M. Meo, M. Ajmone Marsan, "Analytical Estimation of the Completion Time of Mixed NewReno and Tahoe TCP Traffic over Single and Multiple Bottleneck Networks," *in Proc. IEEE Globecom 2001*, San Antonio, TX, USA, Nov. 25–29, 2001
19. M. Garetto, R. Lo Cigno, M. Meo, M. Ajmone Marsan, "On the Use of Queueing Network Models to Predict the Performance of TCP Connections," *Proc. 2001 Tyrrhenian International Workshop on Digital Communications*, Taormina (CT), Italy, Sept. 17–20, 2001
20. ns-2, *Network Simulator (version 2)*, Lawrence Berkeley Laboratory, *http://www.isi.edu/nsnam/ns.*
21. T. Bu, D. Towsley, "Fixed Point Approximation fot TCP Behavior in an AQM Network," *in Proc. ACM SIGMETRICS 2001*, June 16-20, Cambridge, MA, USA
22. W. R. Stevens. *TCP/IP Illustrated, vol. 1.* Addison Wesley, Reading, MA, USA, 1994.

TCP-SACK Analysis and Improvement through OMQN Models*

Marco Bagnus and Renato Lo Cigno

Dipartimento di Elettronica – Politecnico di Torino
Corso Duca degli Abruzzi, 24 – I-10129 Torino, Italy
{bagnus,locigno}@mail.tlc.polito.it

Abstract. Protocol design and modification was traditionally based on heuristic processes and decisions. In particular several new versions of TCP have been proposed which not always stand up to their expectations when they are deployed or simply simulated in a real scenario. The closed-loop, nonlinear characteristics of TCP makes it very difficult to predict its performance in even simple scenarios. This papers analyses the process of modeling TCP-SACK with an Open Multiclass Queueing Network (OMQN) model, showing that sometimes simple implementation choices can have a non-marginal impact on the performance and discussing how an analytical model can be used to design and study TCP (or other protocols) modifications and improvements. As an example we present a modified, more robust implementation of TCP-SACK, named R-SACK, that significantly reduces the protocol timeout probability in presence of bursty losses.

1 Introduction

The quest for improved versions of TCP seems endless and new flavors continue to appear in the literature. Aside with well established (deployed or not) versions like SACK [1], Vegas [2], FACK [3] or ECN [4,5], more innovative solutions (e.g., Westwood and its variations [6,7,8], or the use of the receiver window field to convey feedback [9,10]) continue to appear, yet none of the proposals seem to gain widespread acceptance and use, though they often offer interesting potential gains over the traditional Tahoe and Reno/NewReno versions [11].

Reasons for high inertia to modifications are varied and difficult to analyze, but possibly they have a common denominator in the lack of an accepted theoretical framework for protocol performance analysis. This lack makes it difficult to compare different proposals, and makes it even more difficult to fully comprehend the scope and consequences of protocol changes and new algorithms introduction. New TCP versions are normally based on heuristic considerations, or, in cases like [10] on theoretical approaches that are based on models that are too simplistic to be accepted as accurate representations of the whole Internet complexity.

A basic component of a performance analysis framework is a modeling technique that allows *insight* in the dynamic protocol operation and *predictability* of protocols

* This work was supported by the Italian Ministry for Education and Scientific Research through the PLANET-IP Project.

M. Ajmone Marsan et al. (Eds.): QoS-IP 2003, LNCS 2601, pp. 405–418, 2003.

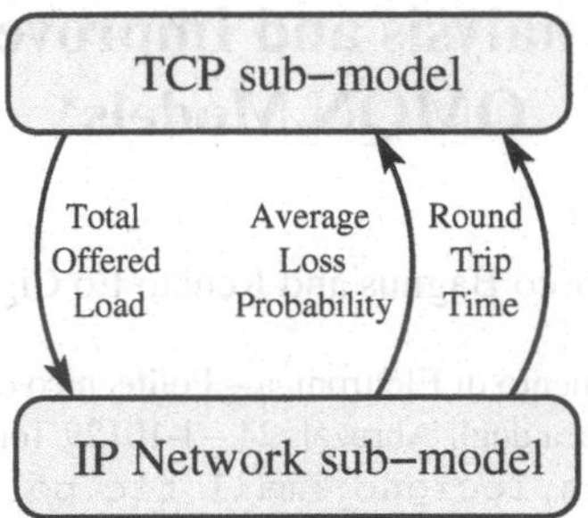

Fig. 1. Schematic representation of the Fixed Point Approximation method

performance in varied scenarios. Insight means that the model is helpful in identifying the protocol flaws, weaknesses and performance bottlenecks. Predictability means that the model can be used to design a new protocol (or a protocol modification) with high confidence of the design impact on performance.

Some recent modeling techniques that can be combined together have gone in the above direction. In [12] the authors show that the overall model representing the protocol under analysis and the underlying packet transport network can be decomposed in separate sub-models, and then jointly solved with a Fixed Point Approximation (FPA) technique. Fig. 1 schematically represent this approach. In [13,14,15,16,17] instead, the authors explored the use of queueing network models to represent the behavior of end-to-end protocols, and TCP in particular, to be coupled with generic network models and solved with FPA techniques. Open Multiclass Queueing Network (OMQN) models are by now the most refined modeling paradigm within this category, allowing the modeling of short lived TCP connections. This paper adopts OMQN models with aims similar to those presented in [18], where a first attempt to use OMQN and FPA techniques to study possible TCP modifications was presented with some success.

The main contributions of this paper are: i) the presentation of an OMQN model of TCP-SACK, highlighting how this process allowed the identification of possible flaws in BSD derived SACK implementations (see also the recent work in IETF on this subject [20]); and ii) the use of OMQNs to devise an opportune modification and predict its performance before implementing it.

1.1 OMQN Background

Following the approach first adopted in [13,14,15,16,17], in this paper we present an Open Multiclass Queueing Network (OMQN) model of TCP-SACK. An OMQN is a queueing network in which all queues are $M/G/\infty$. Each customer (namely a TCP connection) is uniquely identified by a pair (q, c), where q represents a specific state of the protocol and c identifies the number of remaining packets to be sent before the completion of the flow. An OMQN model is capable of describing any protocol whose dynamics can be described with a Finite State Machine (FSM). The use of classes to describe the backlog of the connection allows modeling short lived connections, that open in a specific protocol state, represented by a specific queue q^* and close when all the information is successfully transfered. Given the average RTT of connections and the average packet loss probability, the OMQN model defines the load offered by TCP connections to the IP network, as well as the throughput and the duration of connections. The model is complemented with any suitable description of the IP network loaded with the TCP connections, that, given the load, computes the average loss probability. The network model is not necessarily based on queueing networks, since the interaction of the two sub-models is based solely on the models results. The overall solution is obtained iterating with the FPA technique (see Fig. 1).

Any further detail about the OMQN modeling technique is superfluous in this context and we refer the interested reader to the available literature [13,14,15,16,17]. Additional information on the use of OMQNs to design protocol modifications can be found in [18], where this technique was used to analyze the impact on performance of several possible TCP modifications.

Given an existing model of a protocol, say TCP-NewReno, modifications are modeled in two possible ways. The first one implies modifying the protocol FSM, hence either adding or deleting queues in the OMQN. The second one does not affect the protocol states, but only its dynamics, and corresponds to the modification of the service rates of queues and the transition probabilities, rather than modifying the OMQN structure.

1.2 SACK Background

The Selective Acknowledgement is a TCP option introduced with the purpose of efficiently reacting to multiple losses within the same window of data.

When packets are dropped in bursts, holes appear in the receiver window. When a SACK receiver gets out-of-order packets, it sends duplicate ACKs containing additional information concerning the receiver buffer status in the Option Field. The information is structured based on the sequence numbers of the correctly received chunks of information. For each non-contiguous set of data correctly received the upper and lower sequence numbers (four bytes each) are sent in a *SACK block* of 8 bytes. Since the maximum size of the Option Field in the TCP header is 40 bytes, including 10 bytes for the commonly used *timestamp* option, the maximum number of SACK blocks allowed is three.

The SACK information is used only when the transmitter enters the fast retransmit phase. In this phase, a TCP-SACK tries to keep the amount of information transmitted but not acknowledged constant (we shall call this amount *in-flight data*) and equal to half the size of the congestion window when the fast retransmit is initiated. In order to achieve this result, the sender side maintains a data structure, usually called *scoreboard*, updated based on the receiver sent information, and the choice on whether to transmit new information or retransmit old one is based on the scoreboard and an estimation of the in-flight data.

Notice that this behavior is extremely similar to TCP-NewReno, that, however, is not able to sustain the retransmission rate in presence of multiple losses since the transmitter lacks the information about the lost packets. The result is the well known property of NewReno, that, after the recovery of the first lost packet, retransmits one segment per RTT without transmitting new information until the fast recovery procedure is concluded.

2 TCP-SACK Queueing Model

The SACK modeling approach presented in this paper is the direct extension of the OMQN used in [17] for TCP-NewReno. Fig. 2 shows the portion of the OMQN model that requires modifications to model the SACK option. Indeed, from the overview of SACK discussed in Sect. 1.2, it is clear that TCP-SACK and NewReno shares the same protocol states, and only the time spent in the states, as well as the transition from one state to another are modified. This implies that the model modifications affect only the queues service times and transition probabilities.

Let's first of all recollect the queue meaning. In Fig. 2, each rectangle represents one of the $M/G/\infty$ queues of the model. All queues in Fig. 2 are structured in a matrix pattern: queues in the same row correspond to similar TCP protocol states, while queues in the same column have the same congestion window size. New connections arriving from the outside (with arrival rate denoted by λ^e) enter an initial *slow start* phase described by queues FE_i. Queues L_i represent the *congestion avoidance* phase, while the states in which the protocol reacts to a loss event are described by queues FF_i, LF_i (related to fast recovery) and FT_i, LT_i (in case of timeout). For the moment, the reader should ignore the TS_1 queue, whose function will be clear in Sect. 2.1.

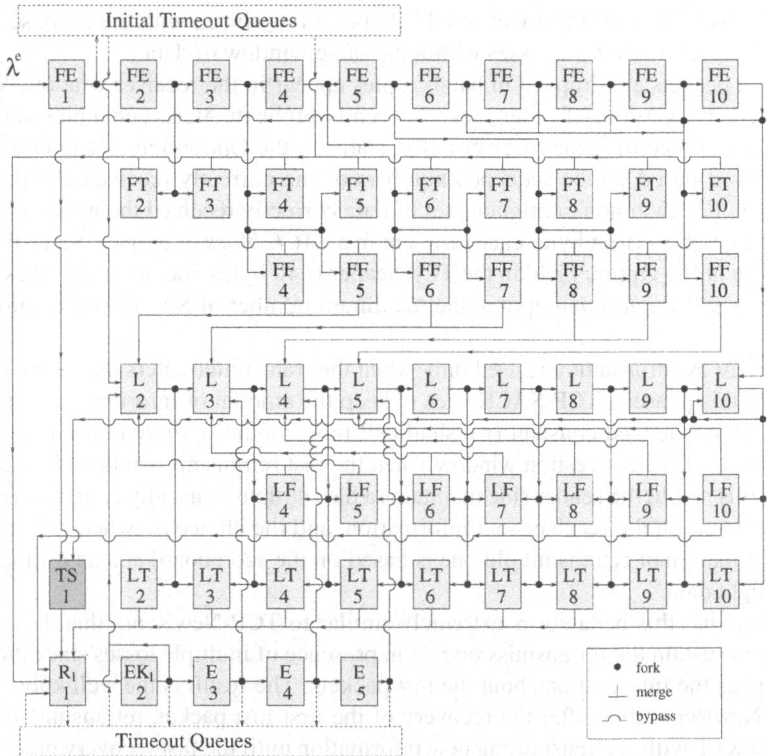

Fig. 2. OMQN model of TCP-SACK that is different from TCP-NewReno

NewReno recovers losses with a rate of one retransmission per RTT. This aspect was modeled assigning service times to queues FF_i and LF_i which are linearly proportional to the average number of losses given the window size. Moreover, the fast recovery ends successfully unless one of the retransmitted segments is lost again, in which case the receiver will not send a new partial ACK[1], and the transmitter is blocked until a timeout expires.

The main difference between SACK and the previous TCP versions lies in the behavior during the fast recovery phase. A selective repeat window protocol should recover all losses more quickly than a Go-back-N protocol, ideally within one RTT. Implementation and protocol constraints, however, prevent TCP-SACK from reacting so fast, thus we have to characterize SACK fast recovery completion time, which depends on the number of segments lost in a window (`nloss`), and on the size of the sender congestion window (`wnd`).

Since the fast recovery completion time is not defined in SACK standard, we decided to characterize three values of the ratio $\frac{\texttt{nloss}}{\texttt{wnd}}$ and then interpolate the points with a simple parabola. We use the following notation: $\texttt{cwnd} = \texttt{wnd}/2$ is the size of the new congestion window halved at the beginning of a fast recovery, `pipe` is the estimated number of packets outstanding in the path (the in-flight segments), and `nrtt` is the duration of the fast recovery expressed in number of RTTs.

1. $\texttt{nloss} = 1 \Rightarrow \texttt{nrtt} = 1$: when only one segment is lost in a window of data of any size, the protocol takes a single RTT to retransmit the segment and wait for the ACK.
2. $\texttt{nloss} = (\texttt{wnd} + 2)/4 \Rightarrow \texttt{nrtt} = 2$: when less then a quarter of the sender window is lost, the number of duplicate ACKs received is large enough to allow the retransmission of all lost packets during the first RTT and receive the total ACK (the one that ends the fast recovery) after two RTTs. In fact, let $\texttt{dup} = \texttt{wnd} - \texttt{nloss}$ be the number of duplicate ACKs received for the first packet lost in a window of `wnd` packets. Then we have that the sender can recover all losses in exactly two RTTs only when `dup`, decreased by the 3 ACKs needed to begin the fast recovery, is greater than the number of packets to retransmit, since the number of ACKs received are the only means the sender have to try to estimate the number of in-flight segments (which SACK tries to keep constant during a fast recovery). In formulae
$$(\texttt{dup} - 3) - (\texttt{pipe} - \texttt{cwnd}) \geq (\texttt{nloss} - 1).$$
Since `pipe` takes the initial value of $(\texttt{wnd} - 3)$ and `cwnd` is $(\texttt{wnd}/2)$:
$$\texttt{nloss} \leq \frac{\texttt{wnd}/2 + 1}{2} = \frac{\texttt{wnd} + 2}{4}.$$
3. $\texttt{nloss} = \lfloor \texttt{wnd}/2 \rfloor \Rightarrow \texttt{nrtt} = \log_2(\texttt{nloss}) + 1$: when exactly half of the packets in the sender window is lost, the protocol is able to recover the losses with a geometric progression (1 in the first RTT, 2 in the second RTT, 4 in the third RTT and so on). In this case the fast recovery completion time follows a logarithmic trend,

[1] In TCP jargon a *partial ACK* is an ACK that advances the lower bound of the transmission window, but does not acknowledge new data sent after entering the fast recovery procedure.

easy to compute. The reason lies again in the information available to the sender to compute `pipe`.

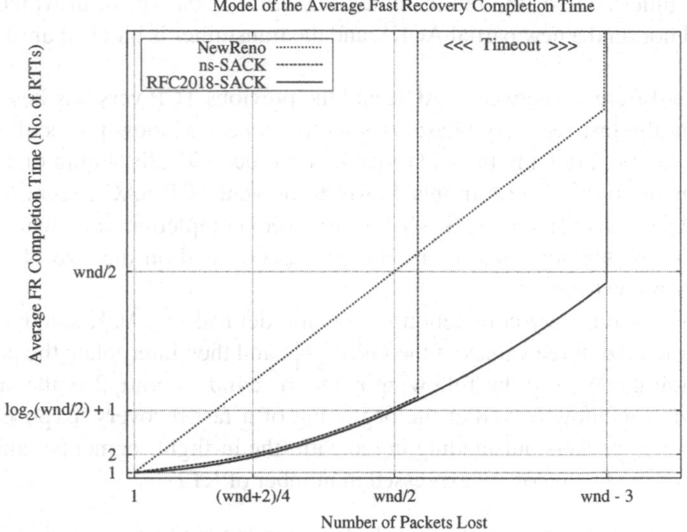

Fig. 3. Model of the average fast recovery completion time for NewReno, the OMQN model of SACK as derived from RFC 2018 [1] and the *ns-2* implementation

Fig. 3 reports (with a solid line) the parabolic interpolation of the three fixed points derived from the protocol specifications, with a "vertical jump" for $\texttt{nloss} = \texttt{wnd} - 3$, where the fast recovery cannot be triggered and TCP waits a retransmission timeout. The fast recovery queues FF_i and LF_i in the SACK model assume service times that follow this function, depending on the value of `wnd` and `nloss`. They are always smaller than respective service times in NewReno since in NewReno the function is the straight line shown with a dotted line in Fig. 3.

The model of TCP-SACK with this estimate of the average duration of Fast Recovery does not lead to satisfying results comparing them with those obtained from *ns-2* simulations. The reason of this mismatch can be searched in the particular way the implementation in *ns-2* computes `pipe`. This implementation is generally called the "*pipe algorithm*."

Analyzing the actual behavior of *ns*-SACK (which is, to the best of our knowledge, coherent with BSD implementations), it is easy to notice that when $\texttt{nloss} > \lfloor \texttt{wnd}/2 + 1 \rfloor$, a timeout occurs deterministically and the fast recovery is interrupted, as shown in Fig. 4 for several values of `wnd`. This behavior can have significant consequences on performance and overshadow the advantages of SACK information in presence of multiple losses. Indeed, the effect of this behavior was already observed in [22], though no detailed explanation on the causes were given, while recent work in the IETF [20] indicates that the problem is receiving attention by the scientific community. The *ns-2* average fast recovery completion time trend is reported with a dashed line in Fig. 3, highlighting the difference with the behavior of SACK as derived from RFC 2018 [1].

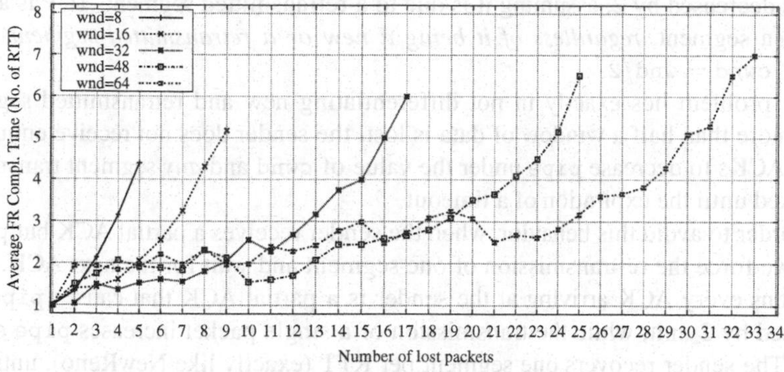

Fig. 4. Average fast recovery completion time measured with *ns-2* for several values of wnd

2.1 Model of *ns*-SACK

We are faced with two possibilities: i) derive an OMQN model including the behavior of *ns*-SACK; ii) devise a SACK implementation that have the same performance the model tells us SACK can have. In this section we explore the first alternative. Sect. 2.2 presents R-SACK: a different, more Robust implementation of SACK.

To model *ns*-SACK, we assign to queues FF_i and LF_i service times that follows the function plotted in Fig. 3, but introducing a new timeout event, represented by the queue TS_1 in Fig. 2. When more than half of a window worth of packets are lost ($\texttt{nloss} > \lfloor \texttt{wnd}/2 + 1 \rfloor$) queue TS_i is visited.

This modification affects transition probabilities from queues FE_i and L_i for $i > 8$. If $\texttt{wnd} \leq 8$ and $\texttt{nloss} > \lfloor \texttt{wnd}/2+1 \rfloor$ TCP does not enter fast recovery. From the above defined queues a customer has three possible transitions:

- towards FT_i (or LT_i) queues, if less than three packets are correctly received in a window of data and a timeout occurs;
- towards FF_i (or LF_i) queues, when the recovery of all lost packets is successful and fast recovery ends correctly;
- towards TS_1 queue, in presence of a timeout related to *ns-2* implementation as described above.

2.2 R-SACK Loss Recovery Algorithm

We are interested in finding a simple implementation as compliant as possible with RFC2018, whose performance is never worse than NewReno.

The problem in *ns*-SACK originates from the *pipe algorithm* implemented in *ns-2* and the retransmission policy that follows. During the fast recovery the sender updates the estimated number of packets outstanding in the path and stores it in a variable called pipe. pipe is initialized to ($\texttt{wnd} - 3$), it is incremented by 1 every time a segment is transmitted (i.e., it is injected into the pipe) and decremented every time a duplicated ACK is received (assuming a segment left the pipe). When a partial ACK is received, pipe is decreased by 2, assuming it is due to a retransmitted segment. TCP is allowed to send a segment, *regardless of it being a new or a retransmitted segment* only if $\texttt{pipe} < \texttt{cwnd} = \texttt{wnd}/2$.

The problem lies exactly in not differentiating new and retransmitted segments. When more than half a window of data is lost, the sender does not receive enough duplicate ACKs to decrease pipe under the value of cwnd and *any* segment transmission is blocked until the expiration of a timeout.

In order to avoid this behavior, when the sender receives a partial ACK but $\texttt{pipe} \geq \texttt{cwnd}$, we force the re-transmission of one segment and wait for the next ACK. In this conditions every ACK arriving at the sender is a partial ACK that causes pipe to be decreased by 2 unit, while the transmission of a single packet increases pipe only by 1 unit. The sender recovers one segment per RTT (exactly like NewReno), until pipe slowly drops below cwnd, then the fast recovery finishes to retransmit missing packets with SACK faster rate.

The R-SACK average fast recovery completion time shows a trend coinciding with SACK for $\texttt{nloss} \leq \lfloor \texttt{wnd}/2 + 1 \rfloor$ and a linear trend (parallel to NewReno) for greater values of nloss. This behavior can be easily visualized in Fig. 3 drawing a straight line parallel to NewReno recovery rate starting from the point $\left[\lfloor \frac{\texttt{wnd}}{2} + 1 \rfloor, \log_2(\frac{\texttt{wnd}}{2}) + 2\right]$. The fast recovery duration is just marginally longer that the ideal SACK behavior.

3 Numerical Results

In this section we present some numerical results, concerning the validation and accuracy of the *ns*-SACK model and the performance of R-SACK compared with *ns*-SACK and NewReno.

The simulation scenario is based on a single-bottleneck topology with a 45Mb/s link whose length is equal to 5000 km. We consider one-way homogeneous TCP connections with uncongested backward path. The distance of servers from the congested router is uniformly distributed between 200 and 3800 km, while the distance of the clients on the other side of the link is uniformly distributed between 200 and 2800 km. Connections are established choosing at random a server-client pair, and are generated following a Poisson process. This scenario represents quite well a portion of an Internet with continental span with a single congested point.

The Maximum Segment Size (MSS) is 1500 bytes, the maximum window size is 64 packets. We consider a DropTail buffer of 128 packets.

First, we verify the accuracy of OMQN model of *ns*-SACK considering several different file sizes as a function of the nominal load. The nominal load is obtained considering the connections arrival rate multiplied by the amount of information they have to transfer. Fig. 5 reports the average packet loss probability and average connection completion time for connections of 20 and 50 packets. Results are validated against point estimates reported with 95% confidence intervals obtained with *ns-2* simulations. The confidence level is 90%.

For what R-SACK is concerned, it is interesting to compare its performance with *ns*-SACK and NewReno versions in the case of longer connections, since for very short connections the performance is dominated by the initial slow start of TCP, which is

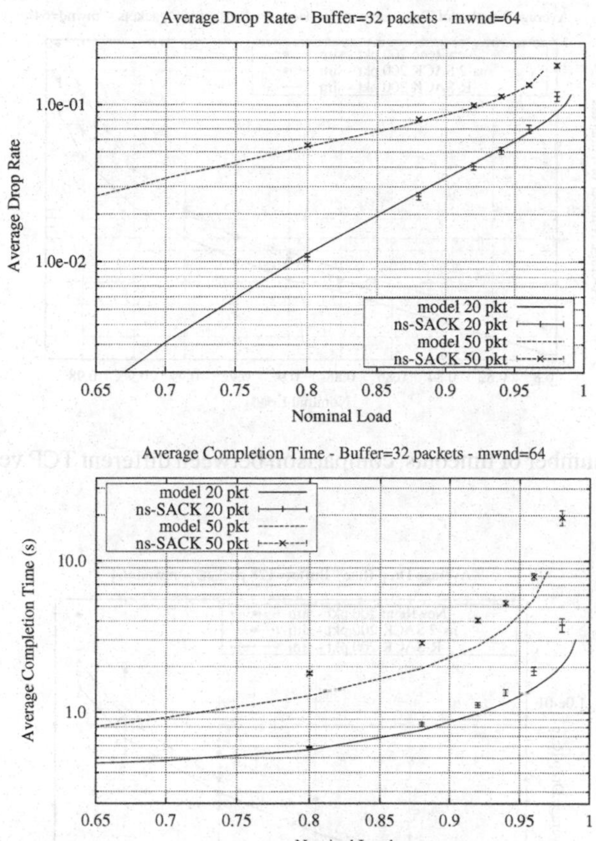

Fig. 5. Average drop rate and completion time

equal for all versions. Fig. 6 and Fig. 7 report results with connection length equal to 200 packets. We only report simulation results for the sake of brevity.

As we expected, R-SACK significantly reduces the average number of timeouts per connection especially at high load, as shown in Fig. 6. At medium load we observe a better behavior of NewReno, compared with *ns*-SACK, due to the more aggressive SACK attitude during fast recovery.

This feature heavily affects other performance indicators. the average drop rate and average completion time highlight the good results obtained with R-SACK, which achieves better results with any file transfer that is of medium-high length (100 kB or more) and in conditions of high congestion in the network.

4 Conclusions

This paper discussed modeling TCP-SACK with the OMQN approach, which ended up being a simple modification of the OMQN model already available for TCP-NewReno.

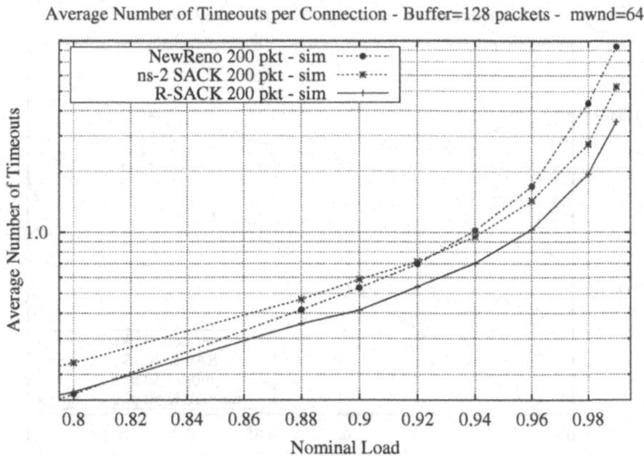

Fig. 6. Average number of timeouts, comparison between different TCP versions at high load

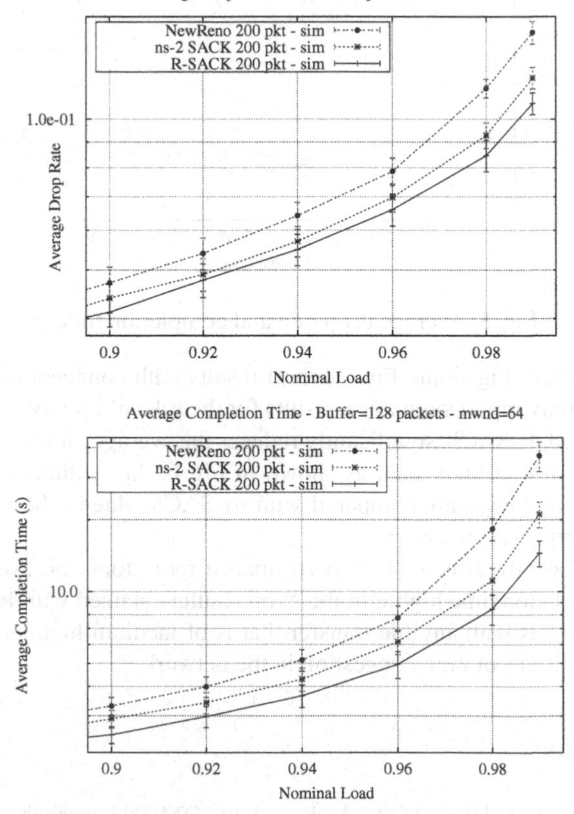

Fig. 7. Average drop rate and completion time, simulation results comparing different TCP versions

The modeling process highlighted the fact that slightly different implementations of the SACK principle may lead to fairly different performances, since they affect the probability of correctly terminate a fast recovery or being blocked until a timeout expires.

In particular it was possible to easily explain why BSD based SACK implementations, like the one found in *ns-2*, suffer a timeout rate higher than NewReno, or even Tahoe, when the network load is high.

The OMQN modeling process allowed to quickly devise and introduce in *ns-2* a novel implementation of SACK which is more robust than the traditional BSD based implementations.

Acknowledgments

The authors wish to thank Michele Garetto for providing the code of the TCP-NewReno model and a lot of useful discussions, and Marco Ajmone Marsan for his enduring support.

References

1. M. Mathis, J. Mahdavi, S. Floyd, A. Romanow, "TCP Selective Acknowledgement Options," RFC 2018, IETF, Oct. 1996.
2. L. Brakmo, L. Peterson, "TCP Vegas: End to End Congestion Avoidance on a Global Internet," *IEEE Journal on Selected Areas in Communications*, 13(8), Oct. 1995.
3. M. Mathis, J. Mahdavi, "Forward Acknowledgement: Refining TCP Congestion Control," Computer Communications Review, 26(4), 1996.
4. S. Floyd, "TCP and Explicit Congestion Notification," *ACM Computer Communication Review*, Vol. 24 No. 5, Pages 10–23, Oct. 1994.
5. K. K. Ramakrishnan, S. Floyd, "A Proposal to add Explicit Congestion Notification (ECN) to IP," RFC 2481, IETF, Jan. 1999.
6. M. Gerla, M.Y. Sanadidi, R. Wang, A. Zanella, C. Casetti, S. Mascolo, "TCP Westwood: Congestion Window Control Using Bandwidth Estimation," *In Proc. of IEEE Globecom 2001*, San Antonio, Texas, USA, Nov. 25–29, 2001.
7. C. Casetti, M. Gerla, S. Mascolo, M.Y. Sanadidi, R. Wang, "TCP Westwood: End-to-End Congestion Control for Wired/Wireless Network," *ACM Wireless Networks (WINET)* N. 8, pp. 467-479, 2002.
8. L.A. Grieco, S. Mascolo, R. Ferorelli, "Additive Increase Adaptive Decrease Congestion Control: a mathematical model and its experimental validation," *In Proc. IEEE Symposium on Computer and Communications*, Taormina, Italy, July 1–4, 2002.
9. L. Kalampoukas, A. Varma, K. K. Ramakrishnan, "Explicit Window Adaptation: A Method to Enhance TCP Performance," *In Proc. IEEE Infocom'98*, San Francesco, CA, USA, March 29 – April 2, 1998.
10. M. Gerla, R. Lo Cigno, S. Mascolo, W. Weng, "Generalized Window Advertising for TCP Congestion Control," *European Transaction on Telecommunication*, Accepted for Publication, preprint available, `http://www.tlc-networks.polito.it/locigno/papers/ett01.ps`.
11. W. Stevens, "TCP Slow Start, Congestion Avoidance, Fast Retransmit, and Fast Recovery Algorithms," RFC 2001, IETF, Jan. 1997.

12. T. Bu, D. Towsley, "Fixed Point Approximation for TCP Behavior in an AQM Network," *ACM SIGMETRICS 2001*, June 16-20, Cambridge, MA, USA.
13. R. Lo Cigno, M. Gerla, "Modeling Window Based Congestion Control Protocols with Many Flows", *Performance Evaluation*, No. 36–37, pp. 289–306, Elsevier Science, Aug. 1999.
14. M. Garetto, R. Lo Cigno, M. Meo, M. Ajmone Marsan, "A Detailed and Accurate Closed Queueing Network Model of Many Interacting TCP Flows," *in Proc. IEEE Infocom'01*, April 22–26, Anchorage, Alaska, USA.
15. E. Alessio, M. Garetto, R. Lo Cigno, M. Meo, M. Ajmone Marsan, "Analytical Estimation of the Completion Time of Mixed NewReno and Tahoe TCP Traffic over Single and Multiple Bottleneck Networks," *in Proc. of IEEE Globecom 2001*, San Antonio, Texas, November 25-29, 2001.
16. M. Garetto, R. Lo Cigno, M. Meo, M. Ajmone Marsan, "On the Use of Queueing Network Models to Predict the Performance of TCP Connections," *Proc. 2001 Tyrrhenian International Workshop on Digital Communications*, Taormina (CT), Italy, Sept. 17–20, 2001.
17. M. Garetto, R. Lo Cigno, M. Meo, E. Alessio, M. Ajmone Marsan, "Modeling Short-Lived TCP Connections with Open Multiclass Queueing Networks," *7th International Workshop on Protocols For High-Speed Networks (PfHSN 2002)*, Berlin, Germany, April 22-24, 2002.
18. R. Lo Cigno, G. Procissi, M. Gerla, "Sender-Side TCP Modifications: An Analytical Study," *IFIP Networking 2002*, Pisa (Italy), May 19-24, 2002.
19. K. Fall, S. Floyd, "Simulation-based Comparisons of Tahoe, Reno and SACK TCP," *ACM Computer Communication Review*, July 1996.
20. E. Blanton, M. Allman, K. Fall, "A Conservative SACK-based Loss Recovery Algorithm for TCP," *IETF Internet Draft (work in progress)*, July 2002, `http://www.ietf.org/internet-drafts/draft-allman-tcp-sack-12.txt`
21. ns-2, *Network Simulator (version 2.1b9)*, Lawrence Berkeley Laboratory, `http://www.isi.edu/nsnam/ns.`
22. B. Sikdar, S. Kalyanaraman, K. S. Vastola, "Analytic Models and Comparative Study of the Latency and Steady-State Throughput of TCP Tahoe, Reno and SACK," *in Proc. of IEEE Globecom 2001*, San Antonio, Texas, November 25-29, 2001.
23. W. R. Stevens. *TCP/IP Illustrated, vol. 1.* Addison Wesley, Reading, MA, USA, 1994.

A Details on R-SACK Loss Recovery Algorithm

The implementation of R-SACK is straightforward, since it only requires to always send a retransmitted packet, even when the "*pipe algorithm*" indicates that the in-flight data is larger than the currently estimated congestion window (notice that during fast recovery a TCP sender can only estimate the congestion window, since the presence of possibly unacknowledged data prevents an exact computation). In this appendix we present two examples that highlight the different behavior of R-SACK and *ns*-SACK. It is important to notice that the modified protocol never has worse performance than *ns*-SACK and concerns only the sender side, leaving unchanged the receiver's behavior.

Noteworthy differences in average flow completion time may be observed only when more than half a window of data is lost, but this condition is not as uncommon as one might think, at least until stable AQM techniques are not widely deployed. We consider a flow of 70 packets transferred from a sender host to a receiver host both supporting SACK. After about a second we force a burst of losses and compare the different behaviors.

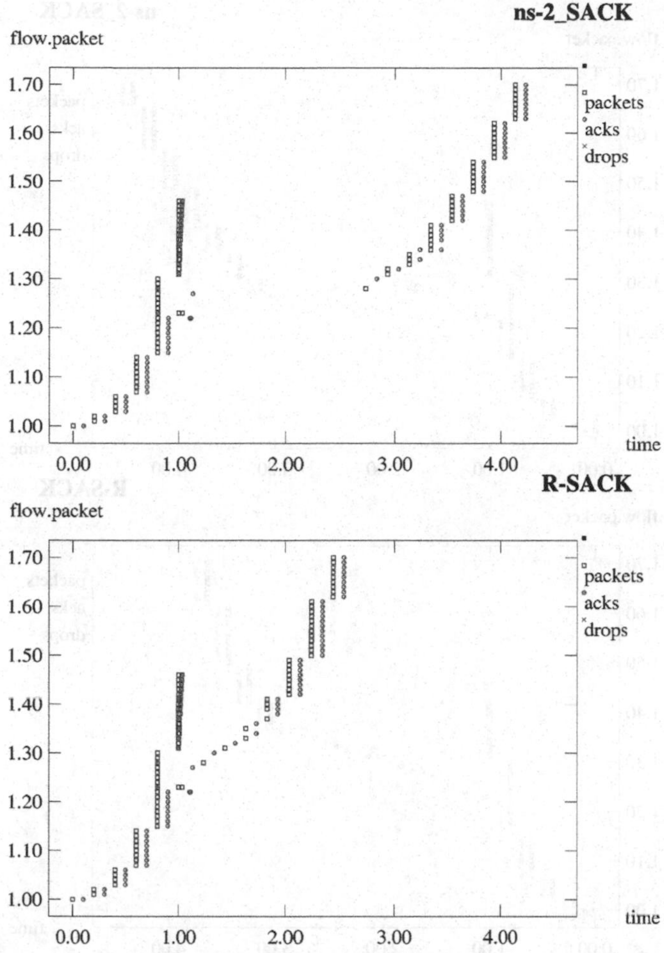

Fig. 8. Detailed trace of packets of two independent flows (ns-2 SACK and R-SACK): flow length 70 packets, 100Mbit/s 100ms link, 14 losses in a 24-packet window

Fig. 8 shows the *ns-2* trace of packets in presence of a loss event in which 14 packets from a window of 24 are lost. The sender window grows exponentially in the *slow start* phase until the first loss, then *ns*-SACK experiences a timeout just after the fast recovery is entered. On the other hand, the R-SACK algorithm allows the flow to continue without interruptions.

The "Reno-like" recovery method of R-SACK is more evident in the example traced in Fig. 9, with a greater number of losses. In this case 20 packets are lost from a window of the same size. While in *ns*-SACK a timeout occurs again, R-SACK allows the retransmission of a packet per RTT, waiting for the *pipe algorithm* to let a faster recovery of remaining packets.

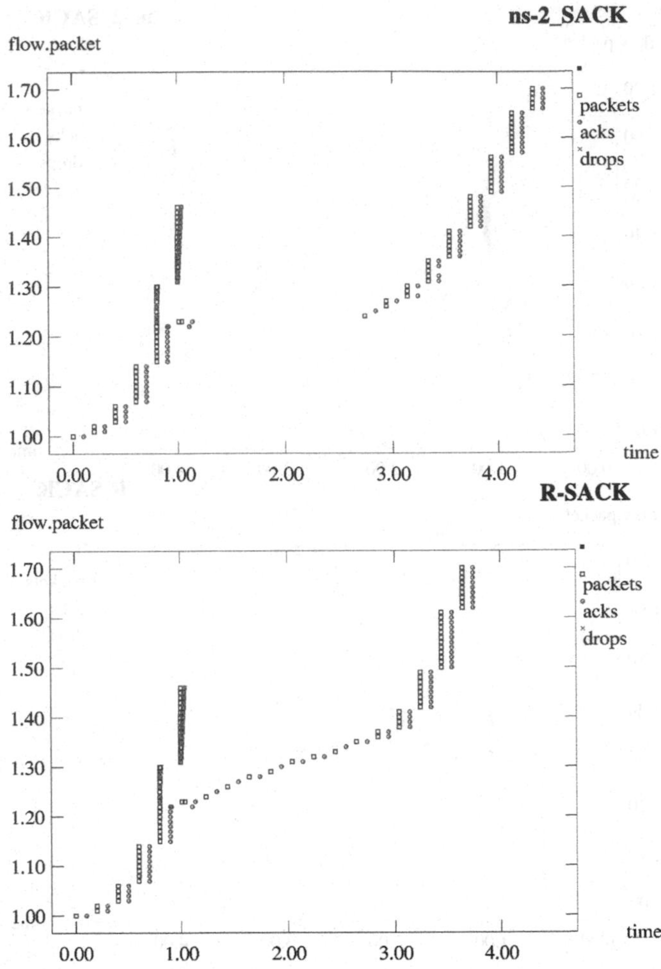

Fig. 9. Detailed trace of packets of two independent flows (ns-2 SACK and R-SACK): flow length 70 packets, 100Mbit/s 100ms link, 20 losses in a 24-packet window

The gain in performance can be estimated in 40% in the first example and 15% in the second example.

QoS Provision in Optical Networks by Shared Protection: An Exact Approach

Andrea Concaro[1], Guido Maier[1], Mario Martinelli[1], Achille Pattavina[2], and Massimo Tornatore[2]

[1] CoreCom
Via Colombo 81 - 20133 Milan, Italy
{t_andcon,maier,martinelli}@corecom.it
[2] Department of Electronics and Information, Politecnico di Milano
Piazza Leonardo da Vinci 32 - 20133 Milan
pattavina,tornatore@elet.polimi.it

Abstract. This paper considers planning and optimization of WDM networks by means of Integer Linear Programming (ILP), which is the most used exact method to perform this task. Since survivability is a very crucial issue in order to guarantee QoS in optical WDM networks, new ILP formulations are investigated here to design WDM networks under protection strategy with shared transmission resources. These formulations are applied to multifiber mesh networks with or without wavelength conversion. After presenting the formulations we discuss the results we obtained by exploiting them in the optimization of a case-study network.

1 Introduction

Optical networks based on Wavelength Division Multiplexing (WDM) are rapidly evolving towards higher capacity and reliability. The most advanced photonic technology available on the market is exploited in WDM to switch and route optical circuits in space and wavelength domains. The increase in WDM complexity brought the need for suitable network planning strategies into foreground. Problems such as optimal dimensioning routing and resource allocation for optical connections must be continuously solved by new and old telecom operators, to plan new installations or to update and expand the existing ones.

In this paper WDM network design is developed in order to guarantee network survivability against a link failure. The issue of survivability of optical connections has become of outstanding importance today: a loss of a high speed connection operating at such bit rates as Gbit/s or higher, even for few seconds, means huge waste of data. Moreover, when a customer wants a leased line, he usually needs a certain bandwidth, but also he looks for the "five nines" (i.e., an availability of 99.999 percent) [1]. The network operator can guarantee this availability only if the network is equipped with mechanism that are able to cope with network failures. The relationship between QoS, availability and protection technique is an important issue currently under study [2, 3, 4].

M. Ajmone Marsan et al. (Eds.): QoS-IP 2003, LNCS 2601, pp. 419–432, 2003.

Undoubtedly the adoption of a protection technique implies extra complexity of network design since spare capacity must also be taken into account in the optimization process.

The work we are presenting concerns exact methods to plan and optimize multifiber WDM networks. In particular we focus on Integer Linear Programming (ILP), a widespread exact solution technique.

The paper summary is as follows. In section 2 we introduce our network model and present a short review of the literature regarding ILP applications to WDM optimization. In section 3 the ILP formulations are presented and explained into details, in the two versions for networks with and without wavelength conversion. We investigate the shared-protection case applying two different approaches to the problem. Finally, in section 4 results obtained by applying the formulations to a case-study network are shown; using these results we are able to point out the advantages of the different approaches we are going to analyze.

2 Resilient WDM Network Optimization by Integer Linear Programming

Network design and planning is carried out with different techniques according to the type of traffic the network has to support. We investigate the static traffic case in which a known set of permanent connection requests is assigned *a priori* to the network. The connections requested by the nodes at a given time to a WDM network all together form the offered traffic matrix. In the simplified reference model we use in this work each request is for a point-to-point optical circuit (lightpath) able to carry a given capacity from the source optical termination to the destination termination. It is understood that each node pair may request more than one connection.

We assume that the channels composing the lightpath may have different wavelengths or may be all at the same wavelength, according to the availability of the wavelength conversion function in the transit OXCs. To simplify, we have considered two extreme cases:

- Virtual Wavelength Path (VWP) network: all the OXC's are able to perform full wavelength conversion, i.e. an incoming optical signal on any wavelength can be converted to an outgoing optical signal on any possible transmission wavelength;
- Wavelength Path (WP) network: no wavelength conversion is allowed in the whole network.

WDM networks today are often designed in order to be resilient to failures that may occur to switching or transmission equipment. This study examines shared path protection, an approach that guarantees survivability to a single-link failure. This approach is based on path protection survivability paradigm: a connection request between the source and the destination node is satisfied by setting up a working lightpath and reserving resources for a backup lightpath on an end-to-end basis. The two lightpaths must be link or node disjoint. In dedicated path protection each spare lightpath is used for one and only connection. In shared path protection two spare lightpaths of different connections can share some common WDM channels. As a result, backup channels are multiplexed among different failure scenarios (which are not expected to occur simultaneously) and therefore shared protection is more efficient when compared with dedicated protection [5]. From an availability point of view the effect of path protection is apparent: the protected lightpath unavailability can be obtained multiplying the unavailability of working path and the unavailability of spare path, so obtaining a significant improvement of total availability.

Linear Programing (LP) has been widely exploited in literature to optimize WDM networks. Its main advantage is a great flexibility in the description of various network environments: different cost functions (fiber or channel number, node complexity, link lengths, etc.), wavelength conversion setups (WP, VWP), protection schemes (dedicated or shared, path or link) and so on can be taken into account by slightly changing a basic mathematical model.

According to a well established terminology [6], Integer Linear Programming (ILP) is most commonly used in two versions, based on the Flow Formulation (FF) and the Route Formulation (RF), respectively. In the former, problem variables are associated to traffic flows, while in the latter the variables represent the distinct paths along which lightpaths can be routed.

Several papers in literature present and compare the two formulations. In Ref. [6] they are applied to unprotected networks, showing that they have roughly the same computational complcxity. However RF has the advantage that the number of variables can be drastically reduced by imposing a restriction on the admissible paths between connection end-points that can be used to route lightpaths (e.g. only the k shortest paths). Moreover, authors show that the size of the problem is independent of the number of wavelengths in the VWP case, while both numbers of variables and constraints linearly increase with the number of wavelengths in the WP case.

Formulation number of flow solution of some

The paper in Ref.[7] deals with networks having shared spare resources, specifically considering link and path restoration and path protection. Only the RF is presented; the model is oriented to the minimization of spare capacity, solely. Working lightpaths are not considered in the minimization procedure. This greatly simplifies the problem, especially in the WP case, in which the information regarding wavelengths assigned to the working paths becomes irrelevant. The model developed in that paper employes binary instead of integer variables, in order to exploit efficient ILP solution algorithms. This however comes at the cost of an increased number of variables and constraints and may become a disadvantage in networks with huge traffic load.

RF-based planning of networks adopting shared path protection is also described in Ref. [8]. The proposed model requires a large number of variables and constraints. The set of variables is in fact composed of two subsets: the first representing only working-ligthpath routes, while the second representing the combined routing of working/spare ligthpath pairs. In this paper we are going to show how a different model can be built leading to a RF based only on the second subset of variables, proving that the first subset is redundant. Moreover, compared to Ref. [8], we propose a different method to apply the RF to WP networks. Ref. [8] assumes a uniform traffic matrix with one single connection between each node pair.

An exhaustive analysis of MILP models in shared mesh protected network scenario can be found in Ref. [9]. The authors present an ILP formulation for combined routing and wavelength assignment and, after having verified the huge waste of time and computing resources in this combined approach, they propose some divide-and-conquer technique to make the problem more tractable. The paper also describes the Shared Risk Group constraints and more precisely applies the so-called duct-constraints in its mathematical models.

Undoubtedly the massive need of computational resources (i.e. time and memory occupation) represents the main obstacle to an efficient application of ILP in optical networks design. There are some simplification techniques able to overcome this limitation, but the solution they produce is only an approximation of the actual optimal network design. The great advantage of ILP over heuristic methods is the ability to guarantee that the obtained solution is the absolute optimum value. Any of the above techniques aimed at reducing the computational burden implies that the ILP approach loses its *added value*, even if approximate solutions may be close to the exact one. Our work is aimed at finding smart formulations of RFWA problem without introducing any approximation, thus preserving the *added value* of mathematical programming. Then after having verified computational limits of ILP models, some methods to soothe the problem complexity are proposed and analyzed. In the following we present route and flow formulation based ILP models for shared path protection.

3 Shared Path Protection

Let us consider a multifiber WDM network environment under static traffic, in which the number of wavelengths per fiber W is given *a priori*, while the fiber number on each physical link is a variable of the problem. We solve this problem in the path protection case using both flow and route formulation.

In shared path protection an optical channel can be shared between more spare lightpaths, only if their associated working lightpaths are link-disjoint. In other words, if some working lightpaths are routed on a common bidirectional link, their corresponding spare lightpaths can't share any channel. In this way, if a link fails, it will always be possible to reroute traffic on spare paths, because it will never happen that two connections require to be rerouted on the same channel. We don't set any constraint on number of spare lightpaths that can share a common link.

3.1 Flow Formulation in the Shared Case

In this section we try to apply ILP flow formulation to the shared path protection problem. The physical topology is modeled by the graph $\mathcal{G} = \mathcal{G}(\mathcal{N}, \mathcal{A})$. Physical links are represented by the undirected edges $l \in \mathcal{A}$ with $|\mathcal{A}| = L$, while the nodes $i \in \mathcal{N} = \{1, 2, ...N\}$, with $|\mathcal{N}| = N$, represent the OXCs. Each link is equipped with a certain amount of unidirectional fibers in each of the two directions; fiber direction is conventionally identified by the binary variable k ($k = 0$ for forward direction, $k = 1$ for backward direction). Finally, the static offered traffic matrix is represented by the array of known terms $C_{i,j}$, each one expressing the number of optical connections that must be established from the source node i to the destination node j. Only unidirectional point-to-point connections are considered (thus, in the general case, $C_{i,j} \neq C_{j,i}$). Each source destination node couple requiring connections is associated to an index c. In the case more than one connection is requested for a node couple c, they are distinguished by a second index t assuming values between 1 and $C_{i,j}$.

Let us define a first set of variables in this protected flow formulation:

- $x_{l,k,c,t,p}$ is a boolean variable indicating whether a WDM channel on link l on a fiber having direction k has been allocated to the connection t requested by node couple c. The index p is used to distinguish between working path ($p = w$) and spare path ($p = s$).
- $F_{l,k}$ is the number of fibers on link l in direction k.

The following additional symbols are also defined:

- (l, k) identifies the set of fibers of link l that are directed as indicated by k; in the following (l, k) is named "unidirectional link";
- I_i^+ is the set of unidirectional links having the node i as one extreme and leaving the node; analogously, I_i^- is the set of unidirectional links having the node i as a one extreme and pointing towards the node;
- (c, t) identifies a single connection request: c identifies the connection source-destination couple, while t identifies one particular connection request associated to a node pair;

Two different versions of this formulation have been defined for networks with or without wavelength conversion capability, respectively.

VWP Network The extension from dedicated to shared flow formulation model is based on the new binary variable $y_{l,k,c,t,\lambda,l'}$. This variable assumes value 1 if working path of connection (c, t) crosses unidirectional link (l, k) and its corresponding spare path crosses bidirectional link l'. Let's note that working and spare lightpaths have to satisfy different constraints: just a single working path can be routed on a channel, while different spare lightpaths can share a channel. Link (l, k) must be equipped with such a number of fibers to support working lightpaths spare lightpaths rerouted on (l, k) when link l' fails: the former are identified by variables $x_{l,k,c,t,w}$, the latter by variables $y_{l,k,c,t,l'}$.

$$\min \sum_{(l,k)} F_{l,k}$$

$$\sum_{(l,k)\in I_i^+} x_{l,k,c,t,p} - \sum_{(l,k)\in I_i^-} x_{l,k,c,t,p} = \\ = \begin{cases} 1 & \text{if } i = s_c \\ -1 & \text{if } i = d_c \\ 0 & \text{otherwise} \end{cases} \qquad \forall\, i, c, t, p; \qquad (1)$$

$$\sum_{k,p} x_{l,k,c,t,p} \leq 1 \qquad \forall\, l, c, t; \qquad (2)$$

$$y_{l,k,c,t,l'} \geq x_{l',f,c,t,w} + x_{l',b,c,t,w} + x_{l,k,c,t,s} - 1 \\ \forall\, (l,k), c, t, l';\ (l \neq l'); \qquad (3)$$

$$\sum_{c,t} (x_{l,k,c,t,w} + y_{l,k,c,t,l'}) \leq W_l F_{l,k} \\ \forall\, (l,k), l';\ (l \neq l'); \qquad (4)$$

$$F_{l,k} \text{ integer} \qquad \forall\, (l,k); \qquad (5)$$

$$x_{l,k,c,t,p} \text{ binary} \qquad \forall\, (l,k), c, t, p; \qquad (6)$$

$$y_{l,k,c,t,l'} \text{ binary} \qquad \forall\, (l,k), c, t, l';\ (l \neq l'); \qquad (7)$$

Solenoidality constraint routes each traffic request both on a working path and on a spare path (1). Constraint (2) imposes the link-disjointness condition: working and spare path associated to connection (c,t) cannot be routed on the same bidirectional link l. Disequation (3) fixes the value of variable y: constraint (3) returns $y_{l,k,c,t,l'} = 1$ if the working lightpath associated to connection (c,t) crosses the bidirectional link l' (i.e. $x_{l',f,c,t,w} + x_{l',b,c,t,w} = 1$) and its spare lightpath crosses unidirectional link (l,k) (i.e. $x_{l,k,c,t,s} = 1$). It is important noting that, if the previous two conditions (i.e. working on l' and spare on (l,k)) are not jointly satisfied, the value of y is not set to 1, because the constraint returns $y \geq 0$ or $y \geq -1$ (so y could be equal to 0 or 1 indifferently). Anyway the optimal solution search process will choose $y = 0$ instead of $y = 1$. Finally disequation (4) sets the minimum fiber number to be assigned on each link: by imposing that fibers on link (l,k) must support working lightpaths routed on this link ($\sum_{c,t}(x_{l,k,c,t,w})$ and spare lightpaths rerouted on this link when link l' fails ($\sum_{c,t} y_{l,k,c,t,l'}$). In other words constraint (4) expresses the capacity needed to support working traffic and spare traffic rerouted after a failure affecting any other link: the channel number in the worst case obtained varying l' will provide the value of $F_{l,k}$. The remaining (5,6,7) are integrality constraints.

WP Network The previous ILP model can be extended to WP networks by introducing an index λ in flow variables. Binary variable $x_{l,k,c,t,\lambda,p}$ assumes value 1, if the associated lightpath crosses link (l,k) on wavelength λ. Similarly $y_{l,k,c,t,\lambda,l'} = 1$ if working lightpath of connection (c,t) crosses bidirectional link l' (it doesn't matter on which wavelength) and its spare lightpath crosses unidirectional link (l,k) on wavelength λ.

$$\sum_{(l,k)\in I_i^+} x_{l,k,c,t,\lambda,p} - \sum_{(l,k)\in I_i^-} x_{l,k,c,t,\lambda,p} = \begin{cases} 1 \text{ if } i = d_c \\ -1 \text{ if } i = s_c \end{cases}$$

$$\forall\, i \in (s_c, d_c), c, t, p; \tag{8}$$

$$\sum_{(l,k)\in I_i^+} x_{l,k,c,t,\lambda,p} - \sum_{(l,k)\in I_i^-} x_{l,k,c,t,\lambda,p} = 0$$

$$\forall\, i, c, t, \lambda, p;\ (i \neq s_c, d_c); \tag{9}$$

$$\sum_{k,p,\lambda} x_{l,k,c,t,\lambda,p} \leq 1 \quad \forall\, l, c, t; \tag{10}$$

$$y_{l,k,c,t,\lambda,l'} \geq \left(\sum_{\underline{k},\underline{\lambda}} x_{l',\underline{k},c,t,\underline{\lambda},w} \right) + x_{l,k,c,t,\lambda,s} - 1$$

$$\forall\, (l,k), c, t, \lambda, l';\ (l \neq l'); \tag{11}$$

$$\sum_{c,t} (x_{l,k,c,t,\lambda,w} + y_{l,k,c,t,\lambda,l'}) \leq F_{l,k}$$

$$\forall\, (l,k), l';\ (l \neq l'); \tag{12}$$

$$\begin{array}{rll} F_{l,k} & \text{integer} & \forall\, (l,k); \\ x_{l,k,c,t,\lambda,p} & \text{binary} & \forall\, (l,k), c, t, \lambda, p; \\ y_{l,k,c,t,\lambda,l'} & \text{binary} & \forall\, (l,k), c, t, \lambda, l';\ (l \neq l'); \end{array}$$

Solenoidality constraints are split into (8) and (9): the former routes a couple of working-spare paths for each traffic request (c, t); the latter imposes the wavelength continuity constraint. The other equations (10,11, 12) follow as logical extension of VWP constraints to WP networks.

3.2 Route Formulation in the Shared Case

In order to apply route formulation, we have to carry out a preprocessing operation in order to prepare the set of route variables for the actual ILP processing. Preprocessing is done only for the node couples which actually request connections, but allows for all the cycles connecting the two nodes. We call cycle the working-spare couple of link-disjoint or node disjoint routes, say A and B, connecting the nodes through the network. Each cycle is identified assuming that the network is completely idle, i.e a WDM channel is always available between two nodes provided that a physical link exists. Each of the identified cycles gives origin to two distinct route options, which correspond to assigning route A to the working lightpath and route B to the protection and viceversa. We call these distinct options working-spare routes.

We have assigned a variable to each working-spare couple and not to each single path to make route formulation more competitive: if we would assign a variable to every single path, we should have used a variable analogous to $y_{l,k,c,t,l'}$ seen in flow formulation. In this way, the new approach to route formulation that we propose allows a substantial reduction of both variable and constraints number and allows route formulation to outperform flow formulation. Thanks to this new application of route formulation we obtain optimal solution (i.e. precomputing all possible working-spare couples) in networks with about ten nodes, with some hundred of connection requests and a medium connectivity index. If we want to obtain results in larger networks and if we accept sub-optimal solutions, we can however compute only a subset of couples, for example the first m shortest couples for each source-destination. Let's observe anyway that preprocessing time can not be totally neglected: the computation of all possible working-spare routes in the NSFNET network (we'll show it in the following) requires 10 hours run of a Matlab routine. Nevertheless also the classical approach to route formulation (one variable associated to a single path) would require a remarkable preprocessing time in order to precalculate all the variables and a more complex set of constraints.

Let's analyze VWP model and two possible extensions to WP networks.

VWP Network Let's consider a source-destination couple c and suppose we have precomputed n working-spare routes between these two nodes. Now we can identify a working-spare route using index (c,n). The variable $r_{c,n}$ indicates how many protected connections are routed on the nth working-spare route between the nodes of couple c. The subset $\mathcal{R}_{(l,k)}$ includes all the working-spare routes whose working is routed on link (l,k); the subset $\mathcal{R}_{l'}^{(l,k)}$ includes all the working-spare routes whose working path is routed on bidirectional link l' and whose spare path is routed on link (l,k).

$$\min \sum_{(l,k)} F_{l,k}$$

$$\sum_{n} r_{c,n} = v_c \qquad \forall\, c; \tag{13}$$

$$\sum_{(c,n)\in\mathcal{R}_{(l,k)}} r_{c,n} + \sum_{(c,n)\in\mathcal{R}_{l'}^{(l,k)}} r_{c,n} \leq W_l F_{(l,k)} \qquad \forall\,(l,k), l';\ (l \neq l'); \tag{14}$$

$$F_{l,k} \text{ integer} \qquad \forall\,(l,k);$$
$$r_{c,n} \text{ integer} \qquad \forall\, c, n;$$

Constraint (13) ensures that number of working-spare routes established between each source and destination of couple c satisfies offered load v_c. Constraint (14) ensures that number of fibers on link (l,k) can support working traffic routed on this link (i.e. the first sum) and spare traffic rerouted on this link when link l' fails (i.e. the second sum).

WP Network Two different approaches are possible to extend shared protection route formulation to WP networks: we can employ a single index λ and duplicate variables r (model WP1) or we can use two indices λ_1 and λ_2 (model WP2).

- Model WP1. $r^w_{c,i,\lambda}$ is the number of connections routed on working-spare couple i between the nodes of couple c, and whose working lightpath is routed on wavelength λ; similarly, $r^s_{c,i,\lambda}$ is the number of connections routed on working-spare couple i between node couple c, whose spare lightpath is routed on wavelength λ.
- Model WP2. We introduce a unique variable $r_{c,i,\lambda_w,\lambda_s}$ where λ_w indicates the wavelength of the working path and λ_s indicates the wavelength of the spare path.

The number of variables increases as λ^2, while in WP1 it increases linearly. However this model has less constraints than WP1. Let's show the two models in details.

In both models the subset $\mathcal{R}_{(l,k,\lambda)}$ includes all the working-spare couples whose working is routed on link (l,k) on wavelength λ; the subset $\mathcal{R}^{(l,k,\lambda)}_{l'}$ includes all the working-spare couples whose working is routed on bidirectional link l' and whose spare is routed on link (l,k) on wavelength λ.

Model WP1

$$\min \sum_{(l,k)} F_{l,k}$$

$$\sum_{n,\lambda} r^w_{c,n,\lambda} = v_c \qquad \forall\ c; \tag{15}$$

$$\sum_{\lambda} r^w_{c,n,\lambda} = \sum_{\lambda} r^s_{c,n,\lambda} \qquad \forall\ c, i; \tag{16}$$

$$\sum_{(c,n)\in\mathcal{R}_{(l,k,\lambda)}} r^w_{c,n,\lambda} + \sum_{(c,n)\in\mathcal{R}^{(l,k,\lambda)}_{l'}} r^s_{c,n,\lambda} \leq F_{(l,k)} \qquad \forall\ (l,k), l', \lambda;\ (l \neq l'); \tag{17}$$

$$F_{l,k} \text{ integer} \qquad \forall\ (l,k);$$
$$r^w_{c,n,\lambda} \text{ integer} \qquad \forall\ c, n, \lambda;$$
$$r^s_{c,n,\lambda} \text{ integer} \qquad \forall\ c, n, \lambda;$$

Equation (15) ensures that all the requests are satisfied by setting up protected connections. Equation (16) imposes r^w and r^s to be consistent. Equation (17) is the capacity constraint.

Model WP2

$$\min \sum_{(l,k)} F_{l,k}$$

$$\sum_{n,\lambda_1,\lambda_2} r_{c,n,\lambda_1,\lambda_2} = v_c \qquad \forall\ c; \tag{18}$$

$$\sum_{(c,n,\lambda_2)\in\mathcal{R}_{(l,k,\lambda_1)}} r_{c,n,\lambda_1,\lambda_2} + \sum_{(c,n,\lambda_2)\in\mathcal{R}_{l'}^{(l,k,\lambda_1)}} r_{c,n,\lambda_2,\lambda_1} \le F_{(l,k)}$$

$$\forall\ (l,k), l', \lambda_1;\ (l \neq l'); \tag{19}$$

$$F_{l,k} \text{ integer} \qquad \forall\ (l,k);$$

$$r_{c,n,\lambda_1,\lambda_2} \text{ integer} \qquad \forall\ c, n, \lambda_1, \lambda_2;$$

Equation (18) routes protected connections for each traffic request. Equation (19) is the capacity constraint. As we have pointed out before, WP1 model contains less constraints (in particular constraints 19), but more variables (for $N_\lambda > 2$) than WP2 model.

3.3 Complexity Comparison between Flow and Route Formulation

Before discussing results obtained by the previous formulations, let's focus our attention on the difference between route and flow formulation in shared case, in terms of ILP problem complexity. We will use the following notation:

- N and L are respectively the nodes and links number
- N_λ is the wavelength number per fiber
- T is the total offered traffic (number of connection requests)
- C is are the node pairs requiring connections
- R is the average number of working-spare paths of a node pair

In the VWP case (Table 1), the number of variables of flow formulation, determined by the number of y variables, grows with the product of the number of connection requests by the square of the number of links. In the route formulation model the number of variables grows with the product of the number of node pairs requiring connections by R, defined as R$=\sum_{j=1}^{C} r_j/C$, where r_j is the number of working-spare routes of the j-th source-destination couple. So if (T/C) is the average number of connections requested by a node pair, then we can notice that the route formulation becomes more convenient than flow formulation when $R \le 2L^2\,(T/C)$. Another important observation concerns the constraint number that in route formulation does not depend on T factor: this represents a significant advantage, especially in real networks, where offered traffic for ach node pair could be high. In Table 1 the values between brackets clearly show route formulation convenience also in a real network case-study, the NSFNET that is characterized by the following parameters $N = 14$, $L = 22$, $T = 360$, $C = 108$, $R = 415$.

In conclusion in shared protection case route formulation is more efficient for the following reasons:

- lower number of variables and constraints in low connected networks (in which R is small);
- possibility of pruning the variable number using only a working-spare route subset;
- independence of constraints and variables number from the offered traffic T (in route formulation it's possible to join in a single integer variable all the traffic associated to a node pair);
- time to precompute paths (i.e variables) negligible if compared to ILP solving time;
- good approximation obtained by relaxing integrity constraints (in flow formulation the set of constraints work so that relaxing integrity constraint the solution looses its significance).

Table 1. Complexity of VWP models.

Model	Variables	Constraints
Flow Formulation	$2L^2T + 4LT + 2L$ (312008)	$2L^2T + 2NT + LT + 2L^2$ (314730)
Route Formulation	$RC + 2L$ (44892)	$2L^2 + C$ (1032)

4 Case Study and Result Comparison

In this section we present and discuss the results we obtained by performing ILP optimization exploiting the previous models.

To solve the ILP problems we used the software tool CPLEX 6.5 based on the branch-and-bound method [10]. As hardware platform a workstation equipped with a 1 GHz processor was used. The available memory (physical RAM+ swap) amounted to 460 MB. This last parameter plays a fundamental role in performing our optimization. The branch-and-bound algorithm progressively occupies memory with its data structure while it is running. When the optimal solution is found, the algorithm stops and the computational time and the final memory occupation can be measured. In some cases, however, all the available memory is filled up before the optimal solution can be found. In this cases CPLEX returns the best but non-optimal solution branch-and-bound was able to find and forces the execution to quit. These cases are identified by the *out-of-memory* tag (O.O.M.) and the computational time measures how long it has taken to fill up memory. We have clarified this particular aspect of ILP to allow better understanding of the reported data.

4.1 Results in Path Shared Protected Networks

ILP models are tested on NSFNET, a well-known network with 14 nodes and 22 links, loaded by 360 unidirectional connection requests [11].

All the results are obtained by means of the route formulation models presented in section 3. We could not obtain any result by applying the flow formulation. We in fact experimented that it fails to find an optimal solution even when applied to much smaller examples; for instance a six-node VWP ring and a five node WP ring, even if just one connection is requested per node couple. Our experiments have confirmed that flow formulation is less convenient than route formulation in low connected networks (for example the NSFNET which is characterized by a low connectivity index). Table 2 shows the number of variables and constraints in VWP case: the first and second columns show flow and route complexity without approximations, while in the third and fourth rows some constraints are set on feasible paths. In particular we have restricted link-disjointness into node-disjointness hypothesis so implying a reduction of about 1/3 of variables multiplicity. Another constraint has been analyzed: the adoption of constrained routing, which consists in limiting the length of working and spare paths to predetermined number of links. Setting an upper bound of 5 links both on working and spare path lengths allows us to save an order of magnitude on the variables number: obviously with this approximation the optimality of the solution is no longer guaranteed.

Table 2. Complexity of VWP models - NSFNET

Model	Variables	Constraints
Flow VWP	312008	314730
Route VWP Link-Disjoint	44892	1032
Route VWP Node-Disjoint	30028	1032
Route VWP Link-Disjoint - Max 5 Link	3280	1032

Table 3 compares results obtained on NSFNET VWP (with 2 and 4 wavelengths per fiber) using the different approaches.

Table 3. Results on NSFNET VWP: final fiber number (with lower bound when optimal solution is not proved), memory occupation, execution time

N_λ	Link-disjoint Max 5 hop	Link-disjoint No approximation	Node-disjoint No approximation	Heuristic tool
2	557 21.07 MB 5 hour	558 (≥ 549) O.O.M. 5 days	553 (≥ 550) O.O.M. 7.5 days	575 5 MB 51 min
4	283 (≥ 281) 139.30 MB 4.6 days	306 (≥ 275) O.O.M. 3.5 days	285 (≥ 277) O.O.M. 7.5 days	295 14 MB 72 min

The best integer solution between the various hypotheses investigated is obtained under the node-disjoint assumption (553 fiber) for $W = 2$ and with constrained routing for $W = 4$ (283 fiber). This non intuitive result is due to the fact that our computational resources are limited (in particular the available memory). Generally if the branch-and-bound algorithm succeeds in finding optimal solution, the optimal fiber number obtained under link-disjointness hypothesis is less than or equal to the correspondent optimal solution obtained under node-disjointness or constrained routing hypothesis; in fact, the number of variables and the associated admissible solution field in the link-disjoint case is larger than in the other two approaches. Unfortunately also problem complexity and the resulting memory occupation is larger; consequently the optimization run fills up available memory before branch-and-bound algorithm has explored the region of the solutions tree containing an efficient solution. So as in our runs it could happen that best integer solution in node-disjoint case is more effective than solution in link-disjoint case.

For $W = 2$ the ILP resolution succeeds in finding optimal value when the number of feasible paths is constrained: in this case the run needs about 20 MB of memory space and two hours of computational time to obtain the optimal solution. The drawback is that the obtained solution is worst than in node-disjoint hypothesis, even if the difference in the fiber number is small: the percent difference between the two solutions (553 and 557) is about 0.7%, while the percent difference of the solution from the absolute lower bound returned by CPLEX (549) is about 1.4%. In $W = 4$ case, the best integer result is obtained under constrained routing hypothesis, but the percent distance from node-disjoint solution is again very small.

Table 4. Comparison results and performance between constrained ILP and heuristic approach.

N_λ	2	4	8	16	32	64
Link-disjoint Max 5 hop	557 21.07 MB 5 hours	283 (≥ 281) 139.30 MB 4.6 days	146 (≥ 143) 330 13 days	77 (≥ 76) 85.47 MB 27 hours	47 213.77 MB 6 days	35 17.59 MB 9 hours
Heuristic tool	575 5 MB 51 min	295 14 MB 72 min	155 14 MB 39 min	85 14 MB 33 min	48 14 MB 30 min	34 17 MB 38 min

Finally in Table 4 we compare constrained ILP solution with the results obtained with the heuristic tool [12]. Let's notice the trend of ILP execution time in function of N_λ. Processing is faster with $N_\lambda = 2$ and $N_\lambda = 64$, while in the other cases the algorithm often does not succeed. The branch-and-bound algorithm has been deliberately stopped when the integer solution was not changing for some days.

The results obtained by heuristic tool are very similar to exact results in terms of fiber number (the difference is always less than 6%). On the contrary, the difference in terms of computational time and memory occupation is apparent: while execution in the exact approach takes a time in the order of magnitude of the day, in the heuristic case execution time is in the order of minutes. Branch and bound algorithm often fills up the available memory, while the heuristic tool occupies just 17 MB in the worst case.

5 Conclusions

We have presented and discussed some novel formulations to solve static-traffic optimization by ILP in WDM resilient network. These formulations have been defined for multifiber networks supporting unidirectional protected optical connections. We have explored the shared path protection strategy: a detailed mathematical formulation has been investigated using both route and flow formulation as basilar paradigms. Tested on a low connected network (the NSFNET), the second alternative proved to be far better performing. Moreover, the possibility to set routing constraints allows us to gain a flexibility, which proved to be essential in problems that require a heavy computational burden.

References

[1] Nokia: Five-nines at the IP edge. Tutorial (2002) Available at http://www.iec.org.

[2] O.Gerstel, G.Sasaki: Quality of protection (qop): A quantitative unifying paradigm to protection service grades. In: Proceedings of Opticomm 2001, 12–23

[3] Chu, K.H., Mezhoudi, M., Hu, Y.: Comprehensive end-to-end reliability assesment of optical network transports. In: Proceedings of the Optical Fiber Comunications (OFC) 2002 Conference, 228–230

[4] Fumugalli, A., Tacca, M., Unghváry, F., Farago, A.: Shared path protection with differentiated reliability. In: Proceedings of IEEE ICC 2002. (Volume 4) 2157–2161

[5] Ramamurthy, S., Mukherjee, B.: Survivable WDM mesh networks, Part I - Protection. In: Proceedings of INFOCOM 1999. (Volume 2) 744–751

[6] Wauters, N., Deemester, P.M.: Design of the optical path layer in multiwavelength cross-connected networks. IEEE Journal on Selected Areas in Communications **14** (1996) 881–891

[7] Caenegem, B.V., Parys, W.V., Turck, F.D., Deemester, P.M.: Dimensioning of survivable WDM networks. IEEE Journal on Selected Areas in Communications **16** (1998) 1146–1157

[8] Baroni, S., Bayvel, P., Gibbens, R.J., Korotky, S.K.: Analysis and design of resilient multifiber wavelength-routed optical transport networks. Journal of Lightwave Technology **17** (1999) 743–758

[9] Zhang, H., Ou, C., Mukherjee, B.: Path-protection routing and wavelength-assignment in WDM mesh networks under shared-risk-group constraints. In: Proceedings of Asia-Pacific Optical and Wireless Communications (APOC) 2001, 49–60

[10] ILOG: ILOG CPLEX 6.5, User's Manual (1999)

[11] Miyao, Y., Saito, H.: Optimal design and evaluation of survivable WDM transport networks. IEEE Journal on Selected Areas in Communications **16** (1999) 1190–1198

[12] Dacomo, A., Patre, S.D., Maier, G., Pattavina, A., Martinelli, M.: Design of static resilient WDM mesh-networks with multiple heuristic criteria. In: Proceedings of INFOCOM 2002. (Volume 3) 1793–1802

Design of WDM Networks Exploiting OTDM and Light-Splitters*

Paolo Petracca, Marco Mellia, Emilio Leonardi, and Fabio Neri

Dipartimento di Elettronica - Politecnico di Torino, Italy
{mellia,leonardi,neri}@mail.tlc.polito.it

Abstract. Wavelength routed optical networks allow to design a logical topology, comprising lightpaths and routers, which is overlayed on the physical topology, comprising optical fibers and optical cross-connects, by solving a Routing and Wavelength Assignment (RWA) problem.

In this paper we extend the concept of lightpath to the one of Super-LightTree, which uses a simple bit level Time Division Multiplexing that can be directly implemented in the optical domain, to split the wavelength bandwidth over a tree among more than one traffic flow. This allows to design logical topologies with an increased number of logical links, reducing the average distance among nodes, i.e., the number of electro-optic and opto-electronic conversions, and the traffic congestion on logical links. This also reduces the number of wavelengths required to solve the RWA problem. Being the Super-LightTree RWA problem computationally intractable, we propose two heuristics for which we show that the number of wavelengths required to overlay a given logical topology on a given physical topology is reduced by more that 70% using Super-LightTrees.

1 Introduction

Internet is facing a constant increase in bandwidth demand, due to the growth of both the services available on-line, and the number of connected users. The fact that new users are increasingly attracted by new services, causes a positive feedback, whose consequence is the need for upgrades of the network infrastructure. Wavelength-Routed (WR) optical networks [1], based upon Wavelength Division Multiplexing (WDM), are considered the best candidate for the short-term implementation of a high-capacity IP infrastructure, since they permit the exploitation of the huge fiber bandwidth, and do not require complex processing functionalities in the optical domain.

In WR networks, high-capacity (electronic) routers are connected through semi-permanent optical pipes called *lightpaths*, that may extend over several physical links. Lightpaths can be seen as chains of physical channels through which packets are moved from a router to another toward their destinations. At intermediate nodes, incoming channels belonging to in-transit lightpaths are (transparently) switched to outgoing channels through an optical cross-connect that does not process in-transit information. Instead, incoming channels belonging to terminating lightpaths are converted to the

* This work was supported by the Italian Ministry for University and Scientific Research under the IPPO (IP Over Optics) project.

M. Ajmone Marsan et al. (Eds.): QoS-IP 2003, LNCS 2601, pp. 433–446, 2003.

electronic domain, so that packets can be extracted and processed, and possibly retransmitted on outgoing lightpaths after electronic IP routing.

In a WR network, a *logical topology*, whose vertexes are the IP routers and whose edges are the lightpaths, is therefore overlayed to the *physical topology*, made of optical fibers and optical cross-connects. In recent years, a large effort has been devoted by the research community to solve the problem of identifying the best logical topology that minimizes a given cost function, while accommodating the traffic pattern given as input to the problem. Traffic that cannot be directly transfered from the source node to the destination node (because no direct link is present in the logical topology) will be using a multi-hop routing approach. This problem is generally called in the literature Logical Topology Design (LTD).

Once the LTD problem has been solved, the resulting logical topology must be overlayed over the physical topology. This procedure must i) identify the set of physical optical fibers over which each lightpath will be routed from the source node to the destination node, (i.e., a suitable physical path must be identified), ii) choose the correct wavelength that will be used for each lightpath, such that two lightpaths that are routed over the same optical fiber use two different wavelengths. This problem is usually referred to in the literature as Routing and Wavelength Assignment (RWA) problem [2,3,4,5,6,7,8]. Wavelength converters can be instrumental to simplify the RWA problem.

Once the LTD and RWA problems are solved, and the logical network is set up, IP routers are able to transfer data among them, using the classic IP approach.

A possible bottleneck in a WDM system is the need of electronic conversion whenever a lightpath is terminated in a node. Data must be converted to the electronic domain, packets must be reassembled, processed, forwarded and then possibly converted again to the optical domain if the current node is not the final destination of the data (hence is implementing a multi-hop routing strategy for the considered traffic flow). While the current optical technology allows to transmit data at very high speed, the electronic conversion and processing is more speed limited and very costly at high speeds. Electronic processing is however still required at lightpaths endpoint, due to the lack of optical memory and processing power, but the minimization of the number of electronic conversions should be a major goal in the future optical networks. Moreover, a traffic flow that is accommodated on a multihop electronic path (because there is not a direct optical link between the source and the destination) uses more electronic resources in the network, possibly increasing the network bandwidth requirements.

There is a trade off between the number of links in the logical topology and the RWA problem: it is quite intuitive that the larger the number of lightpaths, the smaller is the number of required electronic conversions. However, this increases the number of wavelengths that are necessary to solve the RWA problem, and the number of transmitters/receivers that each node must have, resulting in a more costly solution both for optical cross-connects and for node interface.

We proposed in [9] the joint use of WDM and Optical Time Division Multiplexing (OTDM) technologies [10], which offers an interesting opportunity of splitting the bandwidth of a lightpath into a fixed number of sub-channels, using a Time Division Multiplexing (TDM) scheme directly in the optical domain.

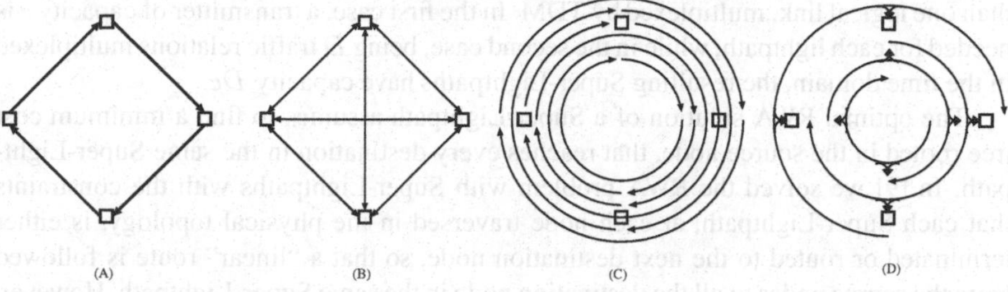

Fig. 1. Example: (A) - Physical Topology; (B) Logical Topology; (C) Classic RWA; (D) Novel Super-Lightpath RWA

Given that the bandwidth available on a wavelength channel is very large with respect to the bandwidth required by individual information flows, a bit level, fixed framing is determined, such that each bit in a given position in the frame, called bit slot, identifies a particular sub-channel. Using a bit interleaver, the transmitter multiplexes sub-channels into the frame, and transmits the resulting stream into one lightpath, using the same wavelength. We called this TDM lightpath a *"Super-Lightpath"*. Each receiver can then synchronize a tunable receiver to a particular bit slot, and receive only data on that particular sub-channel, directly in the optical domain, thanks to OTDM capabilities. Note that, to avoid synchronization problems, only the transceiver of the node where the Super-Lightpath is rooted transmits in the optical pipe.

A Super-Lightpath can travel through many nodes, and a node can receive one (or many) sub-channel(s) from it, instead of converting the whole bit stream to the electronic domain, while transparently routing the entire Super-Lightpath toward another node. It is thus possible to split a lightpath into many sub-channels, each having a bandwidth that is fraction of the lightpath bandwidth.

Figure 1 shows a simple scenario: in (A) the physical topology is plotted, comprising four nodes, connected in a unidirectional ring by means of optical fibers. The logical topology is reported in (B), where each node is connected with all other nodes, thus forming a fully connected logical topology. Figure (C) shows a possible solution of the RWA problem, where each node is source of three lightpaths that are terminated at each destination node; six wavelengths are used on each optical fiber. The solution using the Super-Lightpath approach is shown in (D), where each node is equipped with only one transmitter, and each Super-Lightpath is (partly) converted to the electronic domain in each destination node.

From a networking point of view, a logical link is completely identified by the wavelength of the Super-Lightpath and the bit slot of the TDM frame. From an optical point of view, instead, the union of the bit slots of the TDM frame forms a Super-Lightpath, which, instead of being a-point-to-point link between two nodes (as a classic lightpath), is a point-to-multipoint link, which originates from the source node, and sequentially reaches all the destination nodes in the TDM frame using the same wavelength.

Note that the use of the Super-Lightpath affects only the RWA problem, because the same logical topology can be overlayed by using point-to-point lightpaths carrying one logical link, or using point-to-multipoint Super-Lightpaths, each one carrying more than one logical link, multiplexed by TDM. In the first case, a transmitter of capacity c is needed for each lightpath, while in the second case, being D traffic relations multiplexed in the time domain, the resulting Super-Lightpaths have capacity Dc.

The optimal RWA solution of a Super-Lightpath assumes to find a minimum cost tree rooted in the source node, that reaches every destination in the same Super-Lightpath. In [9] we solved the RWA problem with Super-Lightpaths with the constraints that each Super-Lightpath, at each node traversed in the physical topology, is either terminated or routed to the next destination node, so that a "linear" route is followed from the source nodes to all the destination node in the same Super-Lightpath. However, the availability of *Light-Splitters* [1] allows us to remove this constraints. Several design of optical cross connects are based upon broadcast-and-select structures (see e.g. [11]), which naturally offer the light splitting function studied in this paper.

Light-Splitters have been already proposed as a mean to allow multicast transmission in the optical domain. The resulting optical tree is called usually *LightTree* [12]. In this paper, thus, we extend the concept of Super-Lightpath to the one of Super-LightTree.

2 Routing and Wavelength Assignment

In order to solve the RWA problem, we must identify a physical path and a color for each logical link. The source node then tunes a tunable transmitter to that wavelength, while each node along the path configures its optical cross-connect, and the destination node tunes a tunable receiver. Data can thus travel from the source node to the destination node using the resulting lightpath.

The classic RWA problem, without using wavelength converters, can be formulated as follows.

GIVEN

1. a physical topology, comprising nodes connected by optical fibers;
2. a logical topology, comprising nodes connected by logical links;

FIND

for each logical link:

1. a route from the source node to the destination node on the physical topology;
2. wavelength colors such that two logical links using the same optical fiber have two different colors;

such that the required total number of wavelengths is minimized.

The optimal routing solution of a Super-Lightpath assumes to find a minimum-cost tree rooted in the source node, that reaches every destination in the same Super-Lightpath. This implies the capability of splitting in the optical layer an incoming lightpath into many outgoing lightpaths.

In this paper we consider this opportunity, combining the OTDM capability in the source/destination nodes, with the ability of wavelength-routers to split an incoming wavelength into many outgoing fiber. The resulting object is called *Super-LightTree*.

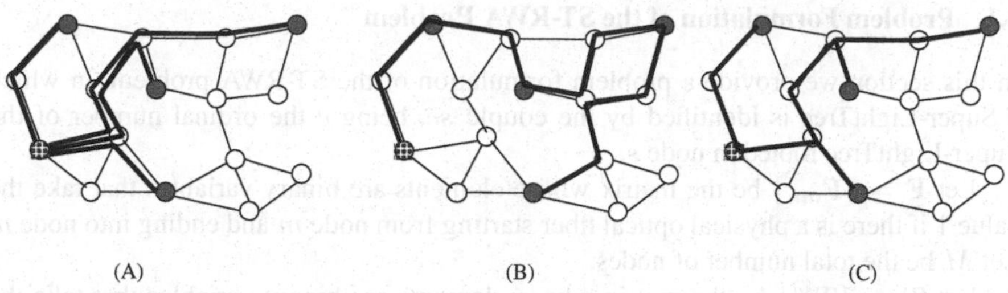

Fig. 2. Example: One source node (dotted), four destination nodes (black). Possible solution using: (A) - RWA; (B) S-RWA; (C) ST-RWA

Note that the use of Super-LightTrees only affects the RWA solution. We call the RWA problem with Super-LightTrees the SuperTree-Routing and Wavelength Assignment (ST-RWA). We will compare the results obtained using the classic RWA approach with the new ST-RWA solution. We will also explicitly compare the performance improvements obtained by using LightTrees with respect to the solution that only considers "linear" trees, i.e., supposing that a lightpath cannot be split in a node, but either terminated or routed to the next node. We call the RWA problem with linear Super-Lightpaths only the Super-Routing and Wavelength Assignment (S-RWA).

To give the intuition of the different approaches, Figure 2 shows a physical topology in which a node has 4 logical links to 4 different destinations. In (A), a possible classic RWA solution is depicted, which involves 3 wavelengths and 9 optical fibers. In (B), one possible S-RWA solution is presented, which involves only one wavelength and 11 optical fibers. In (C), one possible ST-RWA solution is presented, which requires one wavelength, and 9 optical fibers.

The problem can be generally formulated as follows.

GIVEN

1. a physical topology, comprising nodes connected by optical fibers;
2. a logical topology, comprising nodes connected by logical links;

FIND

for each source:

1. the set of destinations that, using TDM, are multiplexed in a Super-LightTree;
2. the order in which those destination must be reached by the Super-LightTree;

and for each Super-LightTree:

1. a tree from the source node to all the destination nodes in the Super-LightTree;
2. a wavelength color such that two Super-LightTrees using the same optical fiber have two different colors;

such that the total number of required wavelengths is minimized.

2.1 Problem Formulation of the ST-RWA Problem

In this section we provide a problem formulation of the ST-RWA problem, in which a Super-LightTree is identified by the couple sc, being c the ordinal number of the Super-LightTree rooted in node s.

Let $\mathbf{F} = \{\mathbf{F}_{mn}\}$ be the matrix whose elements are binary variables that take the value 1 if there is a physical optical fiber starting from node m and ending into node n. Let M be the total number of nodes.

Let $\mathbf{T} = \{\mathbf{T}^{sd}\}$ be the matrix whose elements are binary variables that take the value 1 if there is a logical connection starting from node s and ending in node d. Thus, $\mathbf{T}$ is the logical topology that must be superposed to the physical topology $\mathbf{F}$.

Let W_{max} be the maximum number of wavelengths that can be used in the optimization. Let β_w be a binary variable that takes the value 1 if the wavelength w is used in the solution.

Let a^{sdc} be binary variables that take the value 1 if the source s is using the Super-LightTree c to reach destination d.

Let z^{sdc} be binary variables that take the value 1 if the source s is using Super-LightTree c to reach destination d, and d is a leaf of the Super-LightTree, i.e., d is the last destination served by Super-LightTree c.

Let p^{sc}_{mnw} be binary variables used to describe the route of Super-LightTree c from s. p^{sc}_{mnw} is equal to 1 if the wavelength w is used by Super-LightTree c from s on the optical fiber from node m to node n.

Finally, let k^{sdc}_{jw} be a binary variable that assumes the value of 1 if, in node d, the wavelength w of the tree c rooted in source s is split and routed on j or more outgoing fibers. For example, consider the c tree rooted in s that has been assigned wavelength w. In node d, the incoming wavelength can be split to two outgoing fibers. Then k^{sdc}_{1w} will assume the value of 1, while $k^{sdc}_{jw} = 0 \ \forall j > 1$. If instead three branches are generated, will have $k^{sdc}_{1w} = 1,\ k^{sdc}_{2w} = 1, k^{sdc}_{jw} = 0 \, \forall j > 2$.

The following ILP formulation describes the ST-RWA problem

$$\min W_{max} \tag{1}$$

Subject to:

$$\sum_w \beta_w = W_{max} \tag{2}$$

$$\beta_w \leq \beta_{w-1} \qquad 0 \leq w \leq W_{max} \tag{3}$$

Equation (3) states that the wavelengths used must be turned on "in order". This is not strictly required, but it helps the solver to find a suitable solution.

$$\sum_{c=1}^{\gamma_O^{(s)}} a^{sdc} = \mathbf{T}^{sd} \quad \forall s, d = 1, \ldots, M \tag{4}$$

$$\sum_{d=1}^{M} a^{sdc} \leq D \quad \forall s = 1, \ldots, M, \;\; c = 1, \ldots, \gamma_O^{(s)} \tag{5}$$

Equation (4) states that a destination d can be reached from source s only if it is present in the logical topology and that d is reached at most by one Super-LightTree; $\gamma_O^{(s)}$ is the maximum number of Super-LightTrees that can depart from source s, i.e., the connectivity degree. Equation (5) limits the number of the destinations that can be reached by one Super-LightTree; D is the *multiplexing factor* that represents the bandwidth of the Super-LightTree in terms of destinations, i.e., the degree of multiplexing that can be used on a Super-LightTree using the TDM approach. Thus the connectivity degree for node s in the logical topology is $\delta_O^{(s)} = \gamma_O^{(s)} D$.

$$z^{sdc} \leq a^{sdc} \quad \forall s, c, d \tag{6}$$

The previous inequalities state that a node d can be the last node in a Super-LightTree only if it is also reached by the same Super-LightTree.

$$p_{mnw}^{sc} \leq \mathbf{F}_{mn} \quad \forall s, c, w \tag{7}$$

$$p_{mnw}^{sc} \leq \beta_w \quad \forall m, n, s, c, w \tag{8}$$

$$\sum_{s=1}^{M} \sum_{c=1}^{\gamma_O^{(s)}} p_{mnw}^{sc} \leq 1 \quad \forall m, n, w \tag{9}$$

$$\sum_{w=1}^{W} p_{mnw}^{sc} \leq 1 \quad \forall m, n, s, c \tag{10}$$

$$\sum_{w=1}^{W} \sum_{m=1}^{M} p_{mdw}^{sc} \leq a^{sdc} \quad \forall s, c, d \tag{11}$$

The inequalities in this set state that a Super-LightTree must be routed over an existing physical link (Eq. (7)), that a wavelength used on such a link must be turned on (Eq. (8)), that no more than a Super-LightTree can use the same wavelength on the same link (Eq. (9)), and that at most one wavelength can be used by each Super-Lightpath on each optical fiber (Eq. (10)). Finally, (Eq. (11)) states that a destination of a Super-LightTree must be reached only by one wavelength from all the incoming links (i.e., no merging of LightTree branches is allowed).

$$\sum_{d=1}^{M} \sum_{j=1}^{D-1} k_{jw}^{sdc} \leq D \quad \forall s, c, w \tag{12}$$

$$k_{(j+1)w}^{sdc} \leq k_{jw}^{sdc} \quad \forall s, d, c, w, j \tag{13}$$

Inequality (12) states that no more than D splitting points can be present in a tree . Eq. (13) simply forces that splitting points are chosen in increasing order.

$$1 + \sum_{j=1}^{D-1} \sum_{d=1}^{M} k_{jw}^{sdc} = \sum_{d=1}^{M} z^{sdc} \quad \forall s \neq s, c, w \tag{14}$$

$$\sum_{j=1}^{M} p_{jnw}^{sc} - \sum_{i=1}^{M} p_{niw}^{sc} + \sum_{j=1}^{D-1} k_{jw}^{snc} = z^{snc} \quad \forall s \neq n, c, w, n \tag{15}$$

These equations involves the construction of the tree, enforcing the number of splitting nodes (Eq. (14)), and the classic continuity equality that forces the Super-LightTree to be a continuous tree (Eq. (15)).

The above formulation considers physical links as unidirectional, forces each Super-LightTree to use the same wavelength (thus not assuming the presence of wavelength converters) and allows the logical topology to have at most one logical link between two nodes.

The ST-RWA problem is NP-hard, since it falls in the class of general MILP problems, and thus is numerically intractable, even for networks with a moderate number of nodes. Moreover, the ST-RWA problem represents a generalization of the well-known NP-hard RWA problem, in the sense that it includes RWA as a particular instance. More specifically, if $D = 1$, the ST-RWA formulation reduces to a classic RWA formulation, and the variables a_c^{sd} and z_c^{sd} assume the same meaning (being $c \leq 1$).

Indeed, the problem of finding a minimum-cost tree is also a well-known NP-hard problem (usually referred as "minimum Steiner Tree problem"), but several heuristic are known to find good solutions to this problem. Among all the available solutions, we selected the "*Selective Closest Terminal First*" ($SCTF$) algorithm proposed in [13], that, given the tree source s, the group definition d, and the graph describing the physical topology, finds a Steiner Tree with a good trade-off between performance and computational complexity.

A brief description of the algorithm follows.

$SCTF$ *algorithm*

Let s be the root of the tree, and $\mathcal{D}$ the set of destinations to be reached. Let $\mathcal{T}$ be a dynamic structure in which the Steiner Tree is gathered:

1. set $S = \{s\}$; $D = \mathcal{D}$; $\mathcal{T}$=NULL;
2. *while* $(D \neq \emptyset)$ *do*
3. select the node $d \in D$ closest to nodes in S;
4. select s', the node in S closest to d;
5. extract d from D;
6. select the shortest path route r from s' to d;
7. add the links of r to $\mathcal{T}$;
8. insert all the nodes crossed by r in S;
9. keep the latest z nodes in S;
10. *done*

The algorithm requires at each step the knowledge of all the shortest paths from nodes in S to nodes in $\mathcal{D}$, which can be easily solved by running for example the Dijkstra's shortest path algorithm. To reduce the algorithm complexity, the algorithm considers only the latest z nodes when selecting the closest destination d to S (step 9).

2.2 ST-RWA Heuristic

We present two greedy heuristics to solve the ST-RWA problem. The first one is called *SuperTree-Shortest Path First Fit* (ST-SPFF), and is a very simple algorithm that, starting from the source node, builds a Super-LightTree selecting from the current node the closest one among the destination nodes still not used, until D destinations are reached. The resulting Super-LightTree is then routed building a tree using the $SCTF$ algorithm. The wavelength assignment is done then looking for the first available wavelength on the resulting path.

The second algorithm presented, called *SuperTree-Maximum Fill* (ST-MF) is more complex, and aims to better pack the wavelengths, as it first selects a wavelength, and then tries to route all the possible Super-LightTrees on such a wavelength.

We are interested in studying the impact of the joint WDM and OTDM approach compared to the classic WDM approach, as well as in the impact of the splitting capabilities on allocating the Super-LightTree. We are not interested in evaluating and comparing the various known techniques to solve the RWA problem. Thus we will compare solutions obtained with a particular RWA algorithm and its natural extension to the ST-RWA problem.

Here we give a more detailed description of the two heuristics. Let

- $\mathcal{F}(\mathcal{V}, \mathcal{E})$ be a graph, where $\mathcal{V}$ is the set of vertexes, and $\mathcal{E}$ is the set of arcs
- $\mathcal{D}(s)$ be the set of all the destinations logically connected to s, i.e., $\mathcal{D}(s) = \{d \in \mathcal{V},\ \mathbf{T}_{sd} = 1\}$
- $\Pi(\mathcal{F}, s, d)$ be a function that returns the shortest path from s to d on graph $\mathcal{F}$.
- $SCTF(\mathcal{F}(\mathcal{V}, \mathcal{E}), s, \mathcal{D})$ be a function that returns a tree on graph $\mathcal{F}(\mathcal{V}, \mathcal{E})$, rooted in s, and that reaches all the destinations in $\mathcal{D}$
- $\Phi(l, w)$ be a function that takes the value of 1 if wavelength w is already used on physical link l, and 0 otherwise

ST-SPFF:
SuperTree - Shortest Path - First Fit

1. Let $\mathcal{F}(\mathcal{V}, \mathcal{E})$ be the physical topology
2. For all $s \in \mathcal{V}$
   Build the destination set
3. $c = 1,\ d_1 = s,\ \mathcal{S} = \emptyset$
4. find $d_{c+1} \in \mathcal{D}(s)$ such that $|\Pi(\mathcal{F}, d_c, d_{c+1})| \leq |\Pi(\mathcal{F}, d_c, d_i)|\ \forall d_i \in \mathcal{D}(s)$
5. $\mathcal{D}(s) = \mathcal{D}(s) \setminus \{d_{c+1}\}$; $\mathcal{S} = \mathcal{S} \cup d_{c+1}$;
6. if ($c < D$ and $\mathcal{D}(s) \neq \emptyset$) then $c = c + 1$; goto 4
   Build the LightTree
7. $\mu(s) = SCTF(\mathcal{F}(\mathcal{V}, \mathcal{E}), s, \mathcal{S})$
   Find a wavelength
8. $w = 0$
9. for all links $l \in \mu(s)$
10. if ($\Phi(l, w) \neq 0$) then
11. $w = w + 1$; goto 9
12. for all links $l \in \mu(s)$ let $\Phi(l, w) = 1$

13. route $\mathcal{S}$ on $\mu(s)$ using wavelength w
14. if $(\mathcal{D}(s) \neq \emptyset)$ then goto 3
15. end for all $s \in \mathcal{V}$

Items from 3 to 6 build the set $\mathcal{S}$ of destinations to be reached by the Super-LightTree. A Steiner Tree is then built (step 7), and a first-fit coloring algorithm is used (steps 8 to 12) to find a suitable wavelength.

ST-MF:
SuperTree - Maximum Fill

1. $w = 0$
2. Let $\mathcal{F}(\mathcal{V}, \mathcal{E})$ be the physical topology
3. For all $s \in \mathcal{V}$

Build the destination set

4. $c = 1,\ d_1 = s,\ \mathcal{S} = \emptyset$
5. find $d_{c+1} \in \mathcal{D}(s)$ such that $|\Pi(\mathcal{F}, d_c, d_{c+1})| \leq |\Pi(\mathcal{F}, d_c, d_i)|,\ \forall d_i \in \mathcal{D}(s)$
6. if d_{c+1} is not connected to $\mathcal{F}(\mathcal{V}, \mathcal{E})$ then $\mathcal{D}(s) = \mathcal{D}(s) \cup \mathcal{S}$; goto 14
7. $\mathcal{D}(s) = \mathcal{D}(s) \setminus \{d_{c+1}\}; \mathcal{S} = \mathcal{S} \cup d_{c+1};$
8. if $(c < D$ and $\mathcal{D}(s) \neq \emptyset)$ then $c = c + 1;$ goto 5

Build the tree

9. $\mu(s) = SCTF(\mathcal{F}(\mathcal{V}, \mathcal{E}), s, \mathcal{S})$

Color it and remove edges from $\mathcal{F}(V, E)$

10. for all links $l \in \mu(s)$
11. $\mathcal{E} = \mathcal{E} \setminus \{l\}, \Phi(l, w) = 1$
12. route $\mathcal{S}$ on $\mu(s)$ using wavelength w
13. if $(\mathcal{D}(s) \neq \emptyset)$ then goto 4
14. end for all $s \in \mathcal{V}$
15. if $\mathcal{D}(s) = \emptyset\ \forall s$ then exit
16. else
17. $w = w + 1$; goto 2

In this case the algorithm look for a suitable Super-LightTree on the reduced graph, which includes all the arcs in which wavelength w is still free. If such a tree does not exists for all the Super-LightTrees not already routed (step 6), then a new wavelength is tested (step 17).

Given this approach, it is not sufficient to run the Dijkstra algorithm once, because the graph $\mathcal{F}$ is modified every time a Super-LightTree is successfully routed (Line 11). Thus, for each Super-LightTree that is routed, we must run the shortest path algorithm again. The resulting asymptotic complexity is thus increased when compared to the $ST - SPFF$ algorithm, but the resulting wavelength utilization should be higher, as the Super-LightTrees are better packed.

Note that the above algorithms can be used to solve a classic RWA problem imposing $D = 1$.

Table 1. Average number of wavelengths required to solve the ST-RWA, S-RWA and RWA problem over the US network

	$D=1$	$D=4$		$\eta_w(4)[\%]$		$D=1$	$D=4$		$\eta_w(4)[\%]$	
Δ	SPFF	S-SPFF	ST-SPFF	S-SPFF	ST-SPFF	MF	S-MF	ST-MF	S-MF	ST-MF
8	24.1	17.0	15.2	29.5	37.0	18.2	11.2	10.4	38.5	42.9
12	34.5	21.0	18.1	39.1	47.6	26.8	14.1	13.5	47.4	43.3
16	44.9	24.6	21.3	45.2	52.6	35.4	16.9	15.9	52.3	55.1

3 Experimental Results

We selected two physical topologies that will be used in this section to compare the Super-LightTree approach to the classic lightpath solution. The first one is derived from the U.S. Long-Distance Network [14] comprising 28 nodes and 45 links. The second topology comprises 51 nodes and 166 links, and is derived from a possible evolution of the Japanese WDM network [15].

For each physical scenario, and for a selected connectivity degree $\delta_O^{(s)} = \Delta, \forall s$, we randomly generated 1000 logical topologies. Then, for each logical topology, we evaluated the number of wavelengths required to solve the ST-RWA problem, with either $D = 1$, i.e., classic approach, or $D = 4$, i.e., using Super-LightTrees that multiplex 4 logical links each. The solution of the Super-LightTree problem has been solved considering both the tree solution (ST-RWA), and considering the Super-Lightpath solution (S-RWA), which has been obtained by imposing $z = 1$ in the $SCTF$ algorithm. The reported results are averaged over the 1000 logical topologies with the same connectivity degree. We also report results in term of percentage gain, defined similarly to the congestion gain as $\eta_w(D) = 100\frac{w(1)-w(D)}{w(1)}$ where $w(D)$ is the average number of wavelengths required to solve the S-RWA problem using Super-LightTrees with a multiplexing factor of D.

Table 1 reports results for the US physical topology. As it can be seen, increasing the connectivity degree also increases the number of wavelengths required to solve the RWA problem. Moreover, it shows that the SPFF algorithm is outperformed by the MF. Comparing the results obtained using the classic approach with the Super-LightTree one, we see that using a multiplexing factor of 4 greatly reduces the number of wavelengths required by both algorithms. In particular, the higher the connectivity degree, the larger is the gain observed multiplexing more logical channels in the same Super-LightTree. Looking at the impact of the splitting capability in the optical domain instead, we observe a little improvement when comparing the ST-RWA solution (which exploits the splitting capabilities) to the S-RWA solution (which considers only "linear" trees). Moreover, the impact of the tree generation capability is larger when considering the SPFF algorithms, while can be almost neglected when the MF family of algorithms are considered.

Table 2 reports result for the Japanese physical topology with 51 nodes. The same considerations as above apply. In particular, when a connectivity degree of 16 is considered, we can see that the average number of wavelengths required to solve the classic RWA problem when using the SPFF algorithm can be too large, as it requires to have op-

Table 2. Average number of wavelengths required to solve the ST-RWA, S-RWA and RWA problem over the Japanese network

	$D=1$	$D=4$		$\eta_w(4)[\%]$		$D=1$	$D=4$		$\eta_w(4)[\%]$	
Δ	SPFF	S-SPFF	ST-SPFF	S-SPFF	ST-SPFF	MF	S-MF	ST-MF	S-MF	ST-MF
8	68.8	30.3	27.2	55.7	60.5	46.2	20.7	19.3	55.2	58.3
12	102.4	38.2	35.6	62.7	65.3	69.3	26.9	25.6	61.2	63.1
16	135.7	45.6	40.2	66.3	70.4	91.8	33.3	30.8	63.7	66.5

tical transmission equipments that can handle more than 128 wavelengths on the same physical link. Using instead the Super-LightTree approach, the number of wavelengths required is reduced to about 40, corresponding to a gain of more than 70%. A reduction of approximately the same order is also observed using the more efficient S-MF algorithm: the same logical topology can be overlayed on the same physical topology with only 31 wavelengths when using $D = 4$ instead of 92 wavelengths required with the classic LightTree approach. Also in this case, the impact of the splitting capabilities is little, as a very marginal gain (in the order of 3%) can be obtained when compared to the simpler S-RWA problem.

To give more insight to the reader, Figure 3 plots the number of wavelengths used to solve the S-RWA problem on the Japanese physical network, for all the 1000 logical topologies generated with $\Delta = 16$, sorted in decreasing order of w. As it can be seen, the advantages of using the Super-LightTree solution is homogeneous, and moreover, the variance of the number of wavelengths required is less dependent on the logical topology instance.

Looking at the number of wavelengths required on each optical fibers of a given topology, Figure 4 plots, for each fiber f of the physical topology, the number of wavelengths used on that fiber, usually referred as "load", $W(f)$. A single RWA problem is considered, and the ST-MF algorithm has been used to solve the problem. To better appreciate the plots, the results are sorted in decreasing order, and compared to the average number of wavelengths used on all links, and the maximum number required to solve the complete RWA problem W_{max}, i.e., after the solution of the coloring problem.

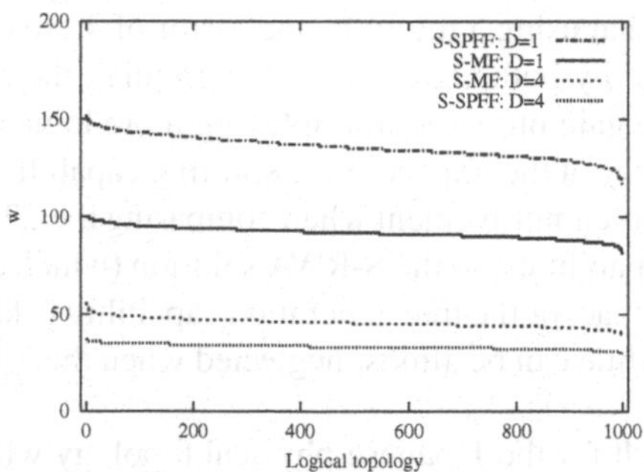

Fig. 3. Distribution of the number of wavelengths required to solve a ST-RWA problem on the Japan network

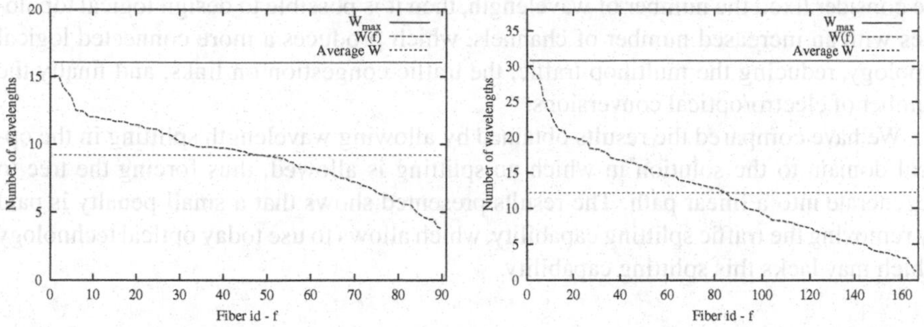

Fig. 4. ST-MF: wavelengths required on the optical link for the USA topology (left) and Japan topology (right)

As it can be observed the ST-MF heuristic finds solutions in which the distance among the most loaded optical fiber (the one with index $f = 0$) and W_{max} is quite small. This shows that the coloring constraints imposed by the RWA have little impact on the solution. Indeed, W_{max} can be larger than or equal to the number of wavelengths on all the optical fibers. On the contrary, the distribution of $W(f)$ shows that there are few fibers that are crossed by many Super-LightTrees, while a large number of fibers are lightly used. This suggests that there are few "bottleneck" fibers on the physical topology, which must be used to connect most source/destination traffic relations. Therefore, using the same wavelength to carry more than one traffic relation is of great help in reducing the total number of wavelengths required to solve the RWA problem. This offers an intuition on the reason why the S-RWA or ST-RWA solutions reduce the number of wavelengths required by more than 70%.

4 Conclusions

In this paper we extended the Lightpath concept to the Super-LightTree. Using a simple bit-level Time Division Multiplexing which can be implemented directly in the optical domain, the bandwidth of a wavelength is partitioned among several traffic flows. A LightTree is then used to build an optical tree rooted in the source node, that reaches all the destinations, by using Light-Splitters in each WR node.

The Super-LightTree approach modifies only the Routing and Wavelength Assignment problem. Numerical results show a gain of more that 70% in the number of wavelengths required to solve the RWA problem when using Super-LightTrees, making the RWA solution viable with today WDM technology. The use of Super-LightTrees allows to design networks with a larger connectivity degree, that is useful to reduce the average distance among nodes, and thus the number of electro-optic and opto-electronic conversions which traffic flow must face.

It also reduces the traffic congestion on logical links. In more details, as we shown in [9], it is possible to solve the RWA problem for the same LTD solution, by using optical transmitters that multiplex D channels on the same wavelength, and have data rate D time higher. In this case the number of wavelengths required is reduced. If instead we consider fixed the number of wavelength, then it is possible to design logical topologies with an increased number of channels, which produces a more connected logical topology, reducing the multihop traffic, the traffic congestion on links, and finally the number of electro/optical conversions.

We have compared the results obtained by allowing wavelength splitting in the optical domain to the solution in which no splitting is allowed, thus forcing the tree to degenerate into a linear path. The results presented shows that a small penalty is paid by removing the traffic splitting capability, which allows to use today optical technology which may lacks this splitting capability.

References

1. R. Ramaswami, K.N. Sivarajan, *Optical Networks: A Practical Perspective*, The Morgan Kaufmann Series in Networking, David Clark, Series Editor February 1998
2. Z. Zhang, A. Acampora. *A Heuristic Wavelength Assignment Algorithm for Multihop WDM Networks with Wavelength Routing and Wavelength Re-Use* , ACM/IEEE Transactions on Networking,.Vol.3, n.3, pp.281–288, June 1995.
3. R. Ramaswami, K.N. Sivarajan. *Routing and Wavelength Assignment in All-Optical Networks*, ACM/IEEE Transactions on Networking, Vol.3, n.5, pp.489-500, October 1995.
4. D. Banerjee, B. Mukhrjee. *A Pratical Appproach for Routing and Wavelength Assignment in Large Wavelength-Routed Optical Networks*, IEEE Journal in Selected Areas in Communications, Vol.14, n.5, pp.909-913, June 1996.
5. A. Mokhtar, M. Azizglu. *Adaptive Wavelength Routing in All-Optical Networks*, ACM/IEEE Transactions on Networking, Vol.6, n.2, pp. 197-206, April 1998.
6. E. Karasan , E. Ayanoglu. *Effects of Wavelength Routing and Selection Algorithms on Wavelength Conversion Gain in WM Optical Network*, ACM/IEEE Transactions on Networking, April 1998.
7. L. Li, A. K. Somani. *Dynamic Wavelength Routing using Congestion and Neighborhood Information*, ACM/IEEE Transactions on Networking, Vol.7, n.5, pp.779-788, October 1999.
8. D. Banerjee, B. Mukherjee. *Wavelength Routed Optical Networks: Linear Formulation, Resource Budgeting Tradeoffs, and a Reconfiguration Study*, ACM/IEEE Transactions on Networking, Vol.8, n.5, pp.598-607, October 2000.
9. M. Mellia, E. Leonardi, M. Feletig, R. Gaudino, F. Neri, *Exploiting OTDM technology in WDM networks*, IEEE Infocom 2002, New York, NY, USA, June 23-27 2002
10. K. Uchiyama, T. Morioka, *All-optical signal processing for 160 Gbit/s/channel OTDM/WDM systems*, Optical Fiber Communication Conference OFC 2001, paper ThH2, Los Angeles, CA, March 2001.
11. D. Chiaroni et al., *First demonstration of an asynchronous optical packet switching matrix prototype for Multi-Terabit-class routers/switches*, postdeadline ECOC 2001, Amsterdam, NL, Oct. 2001
12. L.H. Sahasrabuddhe, B. Mukherjee, *Light Trees: Optical Multicasting for Improved Performance in Wavelength Routed Networks*, IEEE Communications Magazine, Vol.37, n.2, pp.67-73, Feb. 1999
13. S. Ramanathan, “Multicast Tree Generation in Networks with Asymmetric Links”, *IEEE/ACM Transaction on Networking*, Vol.4, n.4, pp.558-568, August 1996.
14. K. Murakami and H. Kim, *Joint optimization of capacity and flow assignment for self-healing ATM networks*, IEEE ICC 1995, Seattle, WA, pp.216-220, June 1995.
15. N. Nagatsu, *Photonic network design issues and applications to the IP backbone*, IEEE Journal of Lightwave Technology, Vol.18, n.12, pp.2010-2018, December 2000.

DWDM for QoS Management in Optical Packet Switches

Franco Callegati, Walter Cerroni, Carla Raffaelli, and Paolo Zaffoni

D.E.I.S. - University of Bologna
Viale Risorgimento, 2 - 40136 Bologna, Italy
{fcallegati,wcerroni,craffaelli,pzaffoni}@deis.unibo.it,
http://www-tlc.deis.unibo.it

Abstract. This paper addresses the problem of quality of service management in optical packet switching over DWDM. Asynchronous, variable length packets are considered and algorithms able to offer congestion resolution and quality of service differentiation are presented, with reference to both connectionless and connection oriented network scenarios. The paper aims at showing that it is possible to guarantee differentiation of the quality of service among traffic classes, with very little buffering requirements.

1 Introduction

The very high capacity of DWDM optical fibers combined with all-optical packet switching (OPS) promises to be a powerful and flexible networking technology for implementation of future network infrastructure. In particular the DWDM environment offers new design parameters that could be exploited in switch design optimization.

Several projects, such as the UE funded KEOPS and DAVID [1] [2] [3], demonstrated the feasibility of switching matrixes able to switch all optical packet payloads, assuming synchronous switching operation to simplify the switching matrix design and providing optimal queuing performance. The drawbacks of this approach are the need of all optical synchronization, implying a substantial increase in hardware complexity [4], and inefficient interworking with network protocols using variable length packets, such as IP.

A solution adopting asynchronous and variable length optical packets would overcome these drawbacks [5], but arises the issue of congestion resolution. In particular optical queuing by means of fiber delay lines (FDLs) is not very effective because of the mismatch between the variable packet sizes and the fixed length of the FDLs [4] [6]. It has been shown, for instance in [7], that this limitation can be overcome by designing congestion resolution algorithm that combine the use of the time and wavelength domain, exploiting wavelength multiplexing on the output links in conjunction with some optical queuing with delay lines. These concepts may also be effectively extended to a network scenario adopting connection oriented operation [8], for instance based on MPLS [9].

M. Ajmone Marsan et al. (Eds.): QoS-IP 2003, LNCS 2601, pp. 447–459, 2003.

At the same time, when considering a transport network for the future Internet, quality of service (QoS) issues must be considered. In particular the support of a diff-serv paradigm [10] is mandatory to match the evolution of multimedia services over the Internet. In present day routers, QoS management is mainly performed by means of queuing priorities or scheduling algorithms based on the fair queuing paradigm [11]. In optical networks these methods are not applicable, because of the very limited queuing space and, most of all, because queuing by delay lines does not allow a new coming packet to overcome other packets already queued, a function that is necessary to implement conventional queuing priorities.

In this paper we show that by means of resource reservation, both in the time and in the wavelength domain, it is still possible to achieve QoS differentiation among different traffic classes, with a good degree of flexibility.

The paper is structured as follows. In section 2 an overview of the resource allocation algorithm used in a generic optical packet switch is provided to motivate the choice of the schemes used to provide differentiated QoS. In Section 3 the QoS management algorithms based on delay and wavelength reservation are presented and numerical results are provided in section 4. Finally in section 5 some conclusions are drawn.

2 Congestion Resolution Problem in OPS

The architecture of a generic all-optical packet switch is illustrated in Fig. 1. The figure is a general block diagram and is intended only to show the main logical building blocks of the switch architecture. The switch has N inputs and N outputs, each connected to a WDM fiber carrying W wavelengths. The wavelengths of each input fiber are split by an Input Unit (IU) making available at the switching matrix the single wavelengths of the bundles and are re-multiplexed together by an Output Unit (OU) for each output fiber. In between stays an intermediate stage where space switching and queuing are implemented by means of non-blocking Space Switches (SS) and by Fiber Delay Lines (FDL). Several arrangements of SS and FDL are possible, with feed forward and feed-back architectures, with advantages related to the adopted solution for hardware components. In this paper we do not focus on implementation issues, therefore we do consider a general architecture and just focus on its logical behavior. Finally the Switch Control Logic (SCL) takes all the decisions and configures the hardware in order to realize the proper switching actions.

In this section we review the functions to be performed by the SCL for congestion resolution, with particular emphasis on exploiting DWDM.

- The basic function is the forwarding algorithm, with header reading and address (or label) matching with the forwarding table to decide to which output fiber the packet should be sent.
- Having decided the output fiber, the SCL is supposed to choose one of the wavelengths of that fiber in order to properly control the OU. This function

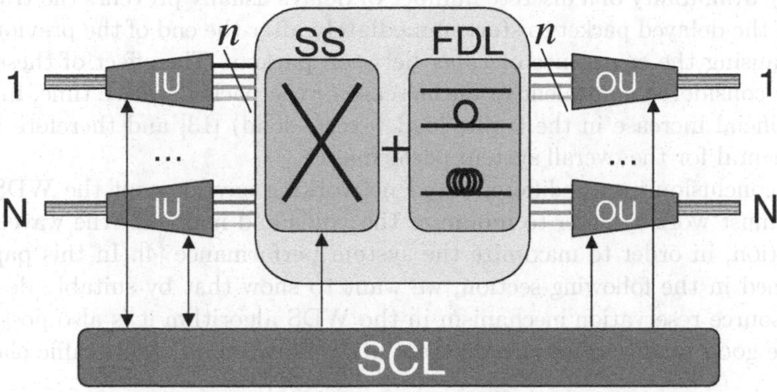

Fig. 1. Generic architecture of a WDM all-optical packet switch with FDL buffer

is routing independent, since the choice of the output fiber is enough to determine the network path.

– Selected the wavelength the question is whether it is immediately available or not. Indeed all the wavelengths of a given output fiber may be busy in transmitting other packets at a given instant in time. When this happens the SCL is supposed to decide whether the packet may be delayed by using the fiber delay lines or if it has to be dropped, because the required delay is not available.

The forwarding algorithm is independent of the wavelength and delay selection, while these latter actions are somewhat correlated, being the need to delay a packet related to the availability of the wavelength selected. This is what we call the Wavelength and Delay Selection (WDS) problem.

For packet i arriving at time t_i, after the output fiber is chosen by the forwarding component, the WDS algorithm chooses the wavelength $w \in [1 : W]$ and consequently the time t_{iw} when the packet will start being transmitted. If wavelength w is currently busy the packet has to be delayed using the FDLs and the choice of time t_{iw} is not free because the number of delays available is discrete. In the case of B FDLs only B delays D_j with $j \in [1 : B]$ are available. Therefore if the sample packet i needs to be delayed the only possible choices are $t_{iw} = t_i + D_j$, $j \in [1 : B]$. If t_{iw} should be greater than D_B the packet can not be accommodated in the buffer and is lost. To ease the discussion we assume that the available delays are linearly increasing with a fixed rate called D, that is the time unit in which the delays are measured. Therefore $D_j = k_j D$, with k_j an integer number. For the sake of simplicity we also assume a degenerate buffer [12] with $k_j = j$ but the ideas here presented are valid also for non degenerate buffers with different arrangements of k_j.

The availability of a discrete number of delays usually prevents the transmission of the delayed packet to start immediately after the end of the previous one, thus causing the emergence of *voids* between packets. The effect of these voids can be considered equivalent to an increase of the packet service time, meaning an artificial increase in the traffic load (excess load) [13] and therefore is very detrimental for the overall system performance.

In conclusion in a undifferentiated networking environment the WDS algorithm must work in order to minimize the voids and maximize the wavelengths utilization, in order to maximize the system performance [4]. In this paper, as explained in the following section, we want to show that by suitably designing the resource reservation mechanism in the WDS algorithm it is also possible to achieve good quality of service differentiation between different traffic classes.

3 Algorithms for QoS Management

In this section we propose some techniques to support different levels of QoS in both connectionless and connection oriented environment, adapting and developing a WDS algorithm that has proved to provide very good performance in term of packet loss probability [7] [8] [14]. The general model for QoS that is typically considered for core network environments is the Differentiated Services model proposed by IETF that aims at implementing different Per Hop Behaviors (PHBs) within each node to support different service classes [10]. The simplified environment we consider is characterized by two classes of service and provide QoS differentiation in terms of packet loss probability.

First of all we analyze the case of a connectionless network, for which the forwarding decision is required on a per packet basis and the same happens for the WDS. In future optical networks the packet arrival rate at the network nodes are going to be very high, because of the typically elevated bit rates. Therefore the packet header has to be processed in a very small time to avoid limiting the system throughput. For these reasons, it is easy to understand that the main issue is the trade-off between good performance and a reduced amount of computational complexity of WDS algorithms.

The basic idea adopted to design simple and efficient WDS algorithms, is that the algorithm works just on the basis of *an array of floating point variables* T_w, keeping track of when the future wavelength w will be free for transmission of new arriving packets. Obviously T_w is the time when the last bit of the last packet scheduled to wavelength w will leave the switching matrix. If $T_w = 0$ the wavelength is assumed not busy and a packet can be transmitted immediately.

The algorithm exploited in this paper for connectionless operations is called **Minimum length queue (MINL)** [7]. When a packet arrives at time t_i the algorithm:

1. searches for the set of wavelengths $\mathcal{F}$ such that $T_w < D_B \; \forall w \in \mathcal{F}$, if no such wavelength exist the packet is lost;
2. selects the shortest queue w to transmit the packet, that is $w \in \mathcal{F}$ such that $T_w < T_i$ for all $i \neq w$. In case two or more queues have the same length, that

which introduces the minimum void is chosen. This algorithm fully employs the WDM dimension. The complexity of SCL consists on comparing the values of delay and gap among all queues;
3. selects the delay D_j such that $t_i + D_j \geq T_w$ and then send the packet to FDL $j \in [1: B]$.

The further network scenario investigated in this paper is a connection-oriented scenario, based for instance on MPLS [9]. In order to design WDS algorithm in this network environment it is needed to take into account that packets are belonging to connections (the Label Switched Path or LSPs) and the switch forwarding table associates LSPs to output fibers.

In this case we implement the WDS algorithm following a policy that aims at realizing a trade-off between control complexity and performance as discussed in [8] [14]. The basic idea is to implement a dynamic algorithm in which packets belonging to the same LSP can be transmitted over more wavelengths. In fact, a static LSP to wavelength allocation allows to achieve minimum control complexity, since processing is required at LSP set up only, but it does not permit to fully exploit the wavelength domain to solve congestion [14]. The resulting dynamic WDS algorithms are called connection oriented WDS (CO-WDS).

According to the purpose of this paper, we do not address the routing issue, but present an example of CO-WDS algorithm that follow the previously explained principles. In particular, we refer to a CO-WDS algorithm, called **Minimum Queue Wavelength (MQWS)**, that is based on techniques similar to the above-mentioned connectionless approaches (MINL), but taking into account the connection-oriented nature of the MPLS traffic. This choice is made in order to have a benchmark to compare performances of the presented algorithms in both the investigated network scenarios. The algorithm follows the following strategy:

1. at LSP set up create a static LSP to wavelength allocation in the forwarding table, for instance packets belonging to LSP L_i will be forwarded to wavelength i;
2. when a new packet arrives, if $T_i \leq D_B$ send it to wavelength i, if $T_i > D_B$ proceed as follows (similar to the connectionless case);
3. look for the set of wavelengths $\mathcal{F}$ of the same fiber such that $T_w < D_B \; \forall w \in \mathcal{F}$, if no such wavelength exist the packet is lost;
4. choose the shortest waiting queue to transmit the packet, that is $w \in \mathcal{F}$ such that $T_w < T_i$ for all $i \neq w$. In presence of two or more queue with the same length, that which introduces the minimum void is selected;
5. select the delay D_j such that $t_i + D_j \geq T_w$, send the packet to FDL $j \in [1 : B]$ and modify the forwarding table to address the LSP packets to that wavelength from now on.

It is important to outline that the QoS management techniques in optical packet switches must be kept very simple due to the delay-oriented characteristics of sequential FDL buffers. In particular it is not possible to change the order of packets already fed in the delay lines, thus making pre-emption based techniques not applicable. Therefore mechanism based on a-priori access control of packets to the WDM buffers are necessary [15]. To this end we intend to exploit the above-mentioned WDS algorithm to differentiate the quality of service by differentiating the amount of choices given to the algorithm. We aim at applying some form of reservation to the resources managed through the WDS algorithm, that are the available wavelength and delay, in order to privilege one traffic class over the other. We have investigated both alternatives.

- *Threshold-based technique*: the reservation of resources is applied to the delay units, and a delay threshold T_{low} lower than the maximum D_B is defined. The WDS algorithm for low priority packets cannot use delays that are above threshold; therefore a low priority packet cannot be accepted if the current buffer delay is greater than the threshold T_{low}. Obviously high priority packets see the whole buffer capacity. This causes packets belonging to the two different classes to suffer different loss rate.
- *Wavelength-based technique*: the reservation of resources is applied to the wavelength domain. The WDS algorithm for high priority packets can send packets to all the W wavelengths of a fiber, while low priority packets use only a subset of W_{low} of them. The lower priority packets can access only a subset of the wavelength resources and in any case they share it with high priority packets. Lower priority packets are expected to suffer higher congestion and typically to experiment higher delays than higher priority ones.

These two concepts have been applied to the previously explained WDS algorithms in order to achieve service differentiation, leading to the algorithms named MINL-**D** and MQWS-**D** in the presence of the threshold-based technique, whereas MINL-**LIM** and MQWS-**LIM** with reference to the wavelength-based technique.

4 Numerical Results

In this section performance of the proposed algorithms is presented. The performance figures are the packet loss probability and the packet delay. The measures are obtained by means of an ad hoc, event-driven simulator, implementing an $N \times N$ optical packet switch. Packets incoming at each input wavelength are generated as outputs from an $\mathcal{M}/\mathcal{M}/1$ queue. Under such an assumption, it is guaranteed that the point process of packet arrivals is a Poisson process (i.e. random traffic) and that packets arriving on the same input wavelength do not overlap. The packet length is exponentially distributed, rounded up at byte scale with an average of 500 bytes and a minimum of 32 bytes (IP header).

The simulation set up is an optical packet switch with $N = 4$ fibers, each carrying $W = 16$ wavelengths, unless otherwise specified. The input load on each wavelength is assumed to be $\rho = 0.8$ and is uniformly distributed to the outputs. The packet loss probability has been evaluated simulating up to 1 billion packets, therefore values up to 10^{-6} have been obtained with good confidence.

Without loss of generality the value of the FDL time unit D is normalized to the average packet length. Therefore in the following, $D = 1$ means that the FDL time unit is equal to the average packet length, $D = 2$ that it is twice as much etc. In particular, analyzing the connection-oriented scenario, it is assumed that each input wavelength carries 3 different LSPs for a total of 192 incoming LSPs and the input load on each wavelength among the LSPs is equally divided. Moreover, in this environment the two classes of service are associated with LSPs at simulation start up.

4.1 Connectionless Scenario

Figure 2 plots the packet loss probability for the MINL-D algorithm for $B = 4$ and different values of T_{low}. The restricted exploitation of the buffer capacity for low priority packets leads to a high penalty in terms of packet loss probability for the low priority traffic, while a good behavior is shown for the high priority traffic. Moreover, performance of each class of service can be straightforwardly tuned by means of the threshold position of the buffer size. Figure 3 shows the delay distribution for low and high priority classes for $B = 4$ and for $T_{low} = 2$. In this figure, it can be seen that a high-usage of the zero-delay lines is made by both classes, while, as a consequence of the algorithm, highest delay lines are used only by high priority traffic and however it is clear how high priority traffic experiences the highest delay with an extremely limited probability, that is below a values equal to 10^{-6}. In fact, it is very important to note that, for the case of $B = 4$ here considered, the optical buffer is very short and the queuing delay is always very small, in particular when compared to propagation delays in a large geographical network. So, it is possible to affirm that the application of the threshold based scheme turns out to be an efficient technique to achieve QoS differentiation, with good values of performance and with a very slight processing effort.

In Fig. 4 the packet loss probability for the MING-LIM algorithm with different values of W_{low} and $B = 4$ is plotted. In this case, it is possible to note that a sensible differentiation in terms of packet loss is achievable. By exploiting the wavelength dimension, we are also able to tune very well the order of magnitude of packet loss probability of both classes of service, obtaining QoS differentiation with good flexibility. In comparison with Fig. 3 it is interesting to observe the delay distribution plotted in Fig. 5 for $W_{low} = 8$ and $B = 4$. It can be shown that differentiation is also achieved in terms of delay in the sense that high priority packets are mostly served with the minimum delay, while low priority packets typically suffer more congestion due to the limited wavelength resources.

4.2 Connection-Oriented Scenario

The packet loss probability for the MQWS-D algorithm is plotted in Fig. 6, for $B = 8$ and for different values of the threshold T_{low}. The behavior of these curves confirms that service differentiation can be achieved in a connection-oriented

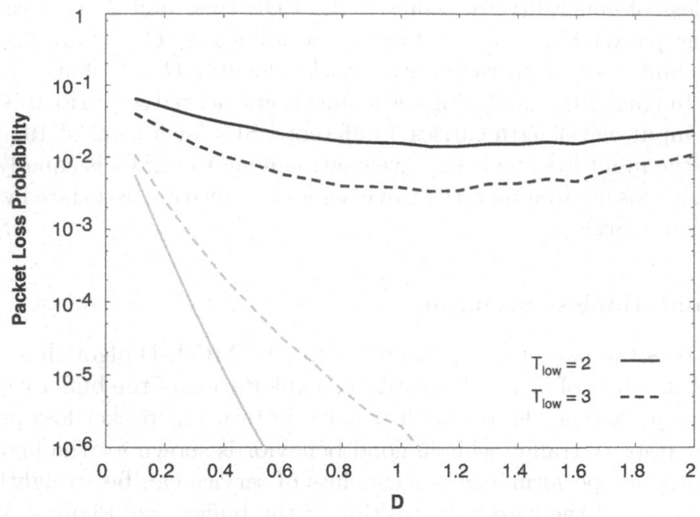

Fig. 2. Packet loss probability for MINL-D for both classes of services as a function of the delay unit D for a 4×4 switch with 16 wavelengths per fiber, 4 delay lines, uniform traffic and load 0.8, varying the value of the threshold T_{low} (the black curves refers to low priority class, while the gray ones to high priority class)

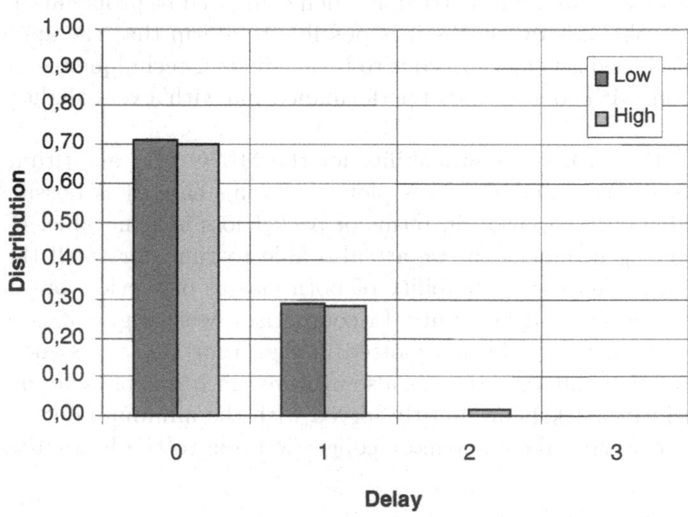

Fig. 3. Delay probability function for MINL-D for both classes of services for a 4×4 switch with 16 wavelengths per fiber, 4 delay lines, uniform traffic, load 0.8, for the delay unit $D = 1.5$ and $T_{low} = 2$

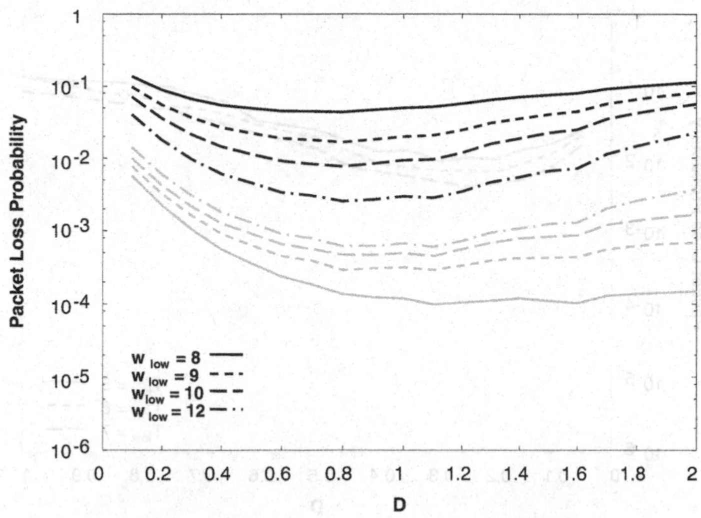

Fig. 4. Packet loss probability for MINL-LIM for both classes of services as a function of the delay unit D for a 4×4 switch with 16 wavelengths per fiber, 4 delay lines, uniform traffic and load 0.8, varying the number of wavelength available for low class (the black curves refers to low priority class, while the gray ones to high priority class)

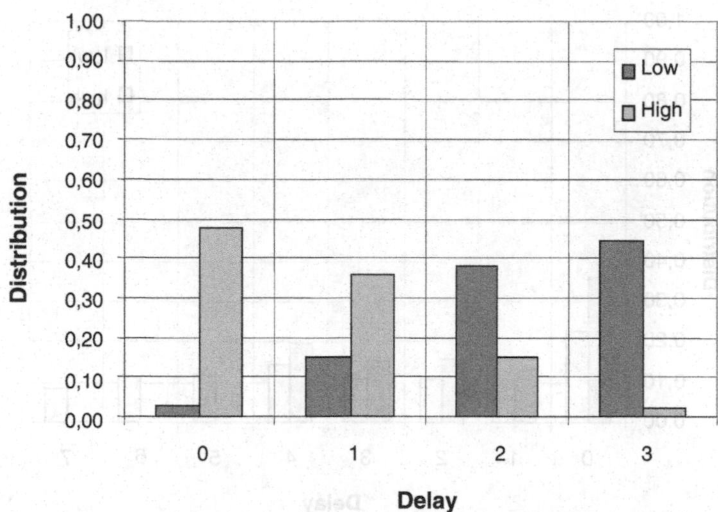

Fig. 5. Delay probability function for MINL-LIM for both classes of services for a 4×4 switch with 16 wavelengths per fiber, 4 delay lines, uniform traffic, load 0.8, for the delay unit $D = 1.3$ and $W_{low} = 8$

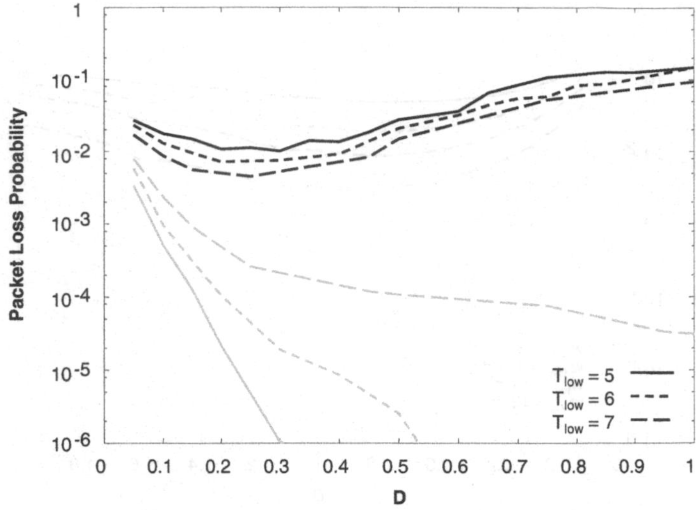

Fig. 6. Packet loss probability for MQWS-D for both classes of services as a function of the delay unit D for a 4×4 switch with 16 wavelengths per fiber, 4 delay lines, uniform traffic and load 0.8, varying the value of the threshold T_{low} (the black curves refers to low priority class, while the gray ones to high priority class)

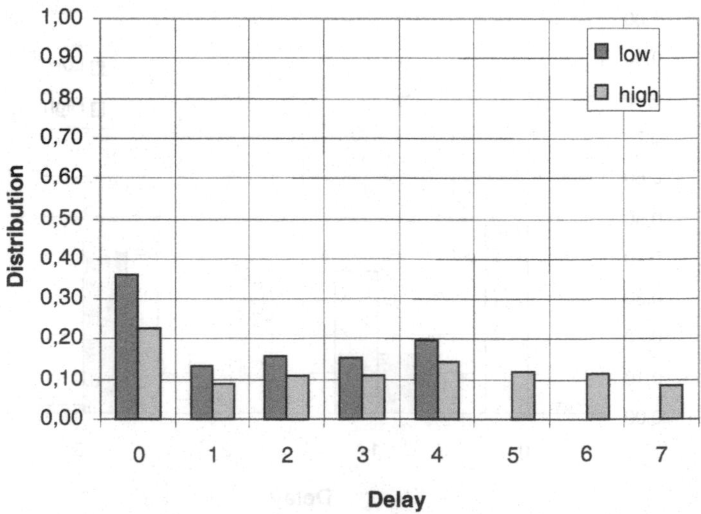

Fig. 7. Delay probability function for MQWS-D for both classes of services for a 4×4 switch with 16 wavelengths per fiber, 8 delay lines, uniform traffic, load 0.8, for the delay unit $D = 0.4$ and $T_{low} = 5$

scenario, exploiting the threshold-based technique and it is possible to note very well, how the packet loss rate of high priority traffic can benefit by varying the threshold position. Figure 7 plots the delay distribution for low and high priority classes for $B = 8$ and for $T_{low} = 5$. Both classes of service experience a high-usage of the lowest value of delay, while it is easy to observe the different exploitation of the buffer, in terms of delay distribution, by the two classes of service, as a consequence of the threshold value. On one side the delay for lower priority packets is limited by the threshold value while over this value the delay probability for high priority packets decreases with the delay value due to the lower usage of the buffer position related to high priority traffic only.

In Fig. 8 the performance of MQWS-LIM algorithm is presented for $B = 8$, varying the value of W_{low}. Likewise to the connectionless case, the figure shows that in the wavelength domain a sensible service differentiation can be achieved in terms of packet loss, tuning powerfully the performance of both classes of service. Figure 9 shows the delay distribution for low and high priority classes for $W_{low} = 8$ and $B = 8$ and for $T_{low} = 2$. The delay distribution exhibits the highest probability value for delay equal to zero for both classes while behaving in the same way for both classes. This is due to the fact that the two classes are characterized by the same load that results split almost equally on the two subsets of 8 wavelengths each. When queues are significantly not full, the algorithm tends to isolate the two subsets of wavelengths, devoting one to the lower priority packets plus a small fraction of high priority packets. This cause this subset to be loaded by a little more than half the total traffic and lead to a bit greater high delay probability with respect to high priority traffic. In fact, high priority LSPs find the wavelengths used also by lower priority LSPs less convenient in terms of queue occupancy and tend to use only their exclusive wavelengths.

5 Conclusions

To support quality of service in DWDM optical packet two different techniques have been considered and analyzed in connectionless and connection-oriented contexts. Both exhibit good differentiation and QoS tuning capabilities in the presence of two classes of service, although performance of the two classes are correlated. The threshold-based techniques achieve good discrimination between the two classes in terms of packets loss while naturally limiting the delay of the lower priority class that, on the other hand, suffers of the highest loss. The limitation in wavelength usage is more effective than the former for finer QoS tuning. Some effects of the algorithms on delay distribution are also shown and discussed.

It is important to outline that this study is related to a basic architecture, without considering implementation issues. For instance it is assumed that as many wavelength as needed can be used in the buffer, that is an idealistic assumption. Further study should consider the effect of more realistic architecture and the consequent conditioning on service class performance.

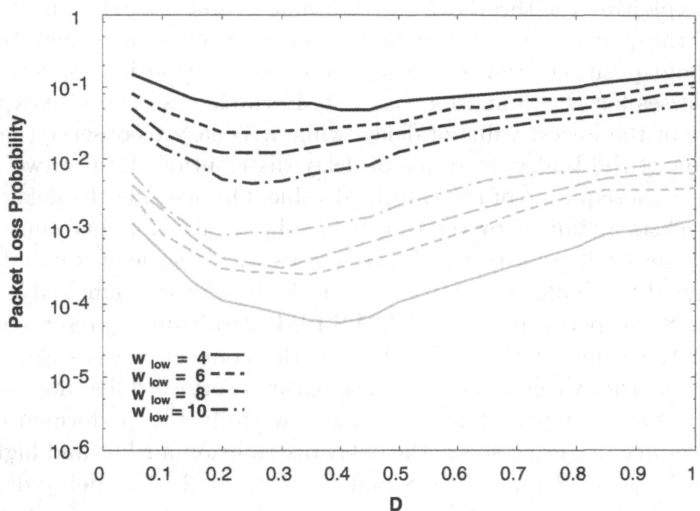

Fig. 8. Packet loss probability for MQWS-LIM for both classes of services as a function of the delay unit D for a 4×4 switch with 16 wavelengths per fiber, 8 delay lines, uniform traffic and load 0.8, varying the number of wavelength available for low class (the black curves refers to low priority class, while the gray ones to high priority class)

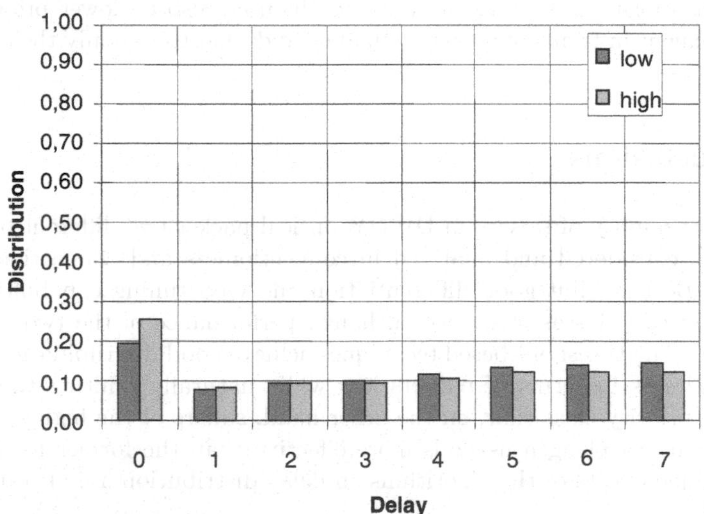

Fig. 9. Delay probability function for MQWS-LIM for both classes of services for a 4×4 switch with 16 wavelengths per fiber, 8 delay lines, uniform traffic, load 0.8, for the delay unit $D = 0.4$ and $W_{low} = 8$

References

1. P. Gambini et al., "Transparent optical packet switching: network architecture and demonstrators in the KEOPS project", *IEEE Journal on Selected Areas in Communications*, Invited paper, Vol. 16, No. 7, pp. 1245-1259, 1998.
2. C. Guillemot et al., "Transparent Optical Packet Switching: the European ACTS KEOPS project approach", IEEE/OSA Journal of Lightwave Technology, Vol. 16, No. 12, pp. 2117-2134, 1998.
3. `http://david.com.dtu.dk`
4. L. Tančevski, S. Yegnanarayanan, G. Castanon, L. Tamil, F. Masetti, T. McDermott, "Optical routing of asynchronous, variable length packets", *IEEEJournal on Selected Areas in Communications*, Vol. 18, No 10 , pp. 2084 -2093, 2000.
5. D. Chiaroni, et al., "First demonstration of an asynchronous optical packet switching matrix prototype for MultiTerabitclass routers/switches", *in Proceedings of 27th European Conference on Optical Communication* (ECOC 2001), Amsterdam, The Netherlands, October 2001.
6. F. Callegati, "Optical Buffers for Variable Length Packets", *IEEE Communications Letters*, Vol. 4, N.9, pp. 292-294, 2000.
7. F. Callegati, W. Cerroni, G. Corazza "Optimization of Wavelength Allocation in WDM Optical Buffers", *Optical Networks Magazine*, Vol. 2, No. 6, pp. 66-72, 2001.
8. F Callegati, W Cerroni, G. Corazza, C. Raffaelli, "Wavelength Multiplexing of MPLS Connections", *Proc. of ECOC 2001*, Amsterdam, October 2001.
9. E. Rosen, A. Viswanathan, R. Callon, "Multiprotocol Label Switching Architecture", *IETF RFC 3031*, January 2001.
10. S. Blake, D. Black, M. Carlson, E. Davies, Z. Wang, Z. W. Weiss, "An Architecture for Differentiated Services", *IETF RFC 2475*, December 1998.
11. D.C. Stephens, J.C.R. Bennett, H. Zhang, "Implementing scheduling algorithms in high-speed networks", *IEEE Journal on Selected Areas in Communications*, Vol. 17, No. 6, pp. 1145 -1158, June 1999.
12. L. Tančevski, L.S. Tamil, F. Callegati, "Non-Degenerate Buffers: a paradigm for Building Large OpticalMemories", *IEEE Photonic Technology Letters*, Vol. 11, No. 8, pp. 1072-1074, August 1999.
13. F. Callegati, "Approximate Modeling of Optical Buffers for Variable Length Packets", *Photonic Network Communications*, Vol. 3, No. 4, pp. 383-390, 2001.
14. F Callegati, W Cerroni, G. Corazza, C. Raffaelli, "MPLS over Optical Packet Switching", *2001 Tyrrenyan International Worksghop on Digital Communications*, IWDC 2001, September 17-20 Taormina, Italy.
15. F. Callegati, G. Corazza, C. Raffaelli, "Exploitation of DWDM for Optical Packet Switching with Quality of Service Guarantees", *IEEE Journal on Selected Areas in Communications*, Vol. 20, No. 1, pp. 190-201, Jan. 2002.

Space Division Architectures for Crosstalk Reduction in Optical Interconnection Networks

Andrea Borella, Giovanni Cancellieri, Dante Mantini

Dipartimento di Elettronica ed Automatica, Università di Ancona
Via Brecce Bianche, 60131 Ancona, Italy
Ph: +39.071.2204456, Fax: +39.071.2204835

Abstract. Omega space division architectures for photonic switching are investigated, with focus on optical crosstalk reduction and conflict avoidance. In such networks, many admissible permutations are not crosstalk free, due to the fact that different conflict-free paths are forced to share the same switching elements. To reduce those conflicts, a method operating in a multi-layer structure is proposed, that is based on a bipartite and two-colorable graph algorithm.

1 Introduction

Switching of data streams carried on optical signals is mainly performed today at electronic level, requiring O/E and E/O conversions before and after the switch. This function is usually called optical switching and is well established in SDH transport network, as well as in some layers of the access network. Switching of optical signals directly at optical level is possible, for example, by using MEMs, and has been widely tested in TDM transmissions and even in WDM-TDM transmissions [1]. In the latter case, the wavelengths are separated in front of the device by suitable optical demutiplexers, and then collected again at the device outputs.

Optical add-drop multiplexers (OADM) and cross-connects (XC) can be so arranged for traditional circuit switching [2]. When packet switching is considered, and it would be useful to switch optical packets or optical bursts directly, some additional problems arise, especially when self-routing functionality is requested. Strong research efforts are devoted to realize true photonic packet switches, where photon interactions in suitable optical materials are able to deviate the optical beam, in order to reach one of two (or more than two) output ports.

Among them, an interesting device called WRS (Wavelength Routing Switch) has been recently proposed [3,4]. This is a 1×2 photonic gate, exploiting the non-linear four-wave-mixing effect within a broad-area SOA (Semiconductor Optical Amplifier). Its fundamental role concerns the routing of an incoming WDM signal towards the selected output, interpreting the presence or absence of specific routing wavelenghts inside the spectrum of the same signal. The routing process is fully executed without requiring any optoelectric conversion.

M. Ajmone Marsan et al. (Eds.): QoS-IP 2003, LNCS 2601, pp. 460-470, 2003.

In a previous paper [5] we focused the attention on Omega networks, developing a possible architecture using WRS switching elements (SEs). The principal goal of our study has been the solution of crosstalk problem.
The procedures, developed for facing it optimizing routing processes and traffic engineering, are not limited to the choice of wavelength routing. On the contrary, they are general enough to cover also other possible signal domains. For example, MPλS is a particular method of forwarding packets at a high rate of speed. It combines the performance and speed of WDM physical layer with the flexibility of IP layer. In MPλS, wavelength-switched channels can be established instead of label-switched paths. Only the label needs to be processed in the network node; the packets can bypass the electronic conversion and processing and, thus, the network performance can be increased significantly.
A recently proposed optical CDM (Code Division Multiplexing) transmission format [6], which can be realized by means of reconfigurable arrayed-waveguide gratings, although completely different from the physical point of view, could replace WDM at all. CDM assures a two-dimensional domain, using time and wavelength simultaneously, to be exploited for crosstalk reduction and conflict avoidance efficiently.
However, both solutions described above require self-routing elements for burst optical switching.
This work is organized as follows. In Section 2 Omega networks are described, focusing the attention on optical crosstalk reduction. A space division approach is proposed to solve this problem, so that QoS in optical networks is improved. In Section 3 a more complex structure, based on the utilization of bipartition and two-colorability algorithms, is presented. Here we show the results of the simulation about the performances of the two methods. Finally, Section 4 concludes the paper.

2 Crosstalk in Omega Networks

Among the various multistage networks, proposed and studied as optical switching topologies, we have focused our attention on the structure of the Omega networks, which belong to the group of multistage cube-type networks (MCTN). However, this choice does not affect the generality of our study, since all MCTNs are topologically equivalent [7].
An N×N Omega network, with $N=2^n$, is composed by n stages (columns), each of them including a number of 2×2 switching elements (SEs) equal to N/2, so that the structure includes a number of SEs equal to $(N/2)\log_2 N$. The internal patterns follow different rules, which depend on the considered stage. In fact, the inter-stage connections adopt the shuffle-exchange scheme [8] and, similarly, the input ports of the network are also shuffled before reaching the first stage. On the contrary, the output ports of the last stage are directly connected to the network outputs. An example of a 8×8 topology is reported in Fig. 1, where inputs, outputs and SEs are numbered in a binary format.
Omega networks are fully accessible, that is every output can be reached from any input, and, by adopting the address bits of such output as routing tags, a self-routing

functionality can be obtained. In fact, the k-th bit (0≤k≤n-1) of the address is used to switch the SE at the k-th stage, following the rule that the upper port of the SE has to be selected if the bit equals 0 and the lower port otherwise. For example, in Fig. 1 is reported in dashed lines the path connecting input 001 to output 100. According to the latter binary sequence, the SE at the first stage switches to its lower port, whereas the other two SEs switch the path towards their upper one.

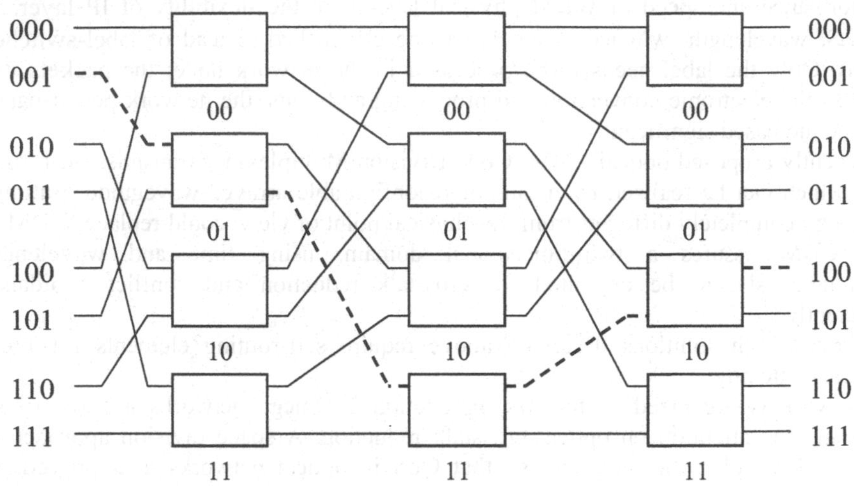

Fig. 1. 8×8 Omega network.

A N-permutation is composed by N different (i,j) input-output pairs, with 0≤i≤N-1 and 0≤j≤N-1, so that every input or output port is contained within only one of the above mentioned pairs. In practice, it represents the internal routing requests to be set within the switching network, once having avoided multiple connections addressed towards the same output port. Clearly, the number of possible permutations equals N!. Omega is a blocking network, since there are some permutations which imply a conflict within a SE. Let us consider, for example, a given permutation including the two input-output pairs (010→101) and (110→111). Even though they are addressed towards different outputs, both of them compete for the lower port within SE 10 at the first stage. On the contrary, a given permutation is said admissible if it implies N conflict-free paths within the switching fabric, that is to say, if it does not generate a block configuration of the switch.

Calling Ω the set including all the admissible permutations, the ratio N_Ω/N! equals the probability that a given permutation, in an N×N Omega network, is admissible. It can be underlined that Ω contains many permutations which are not crosstalk free (CF), since in many cases, different conflict-free paths are forced to cross the same SE, where crosstalk occurs.

Optical crosstalk effect is one of the main concerns as regards QoS in optical transport networks. In the context considered here, specific algorithms devoted to

identify suitable crosstalk-free permutations to be set in the Omega interconnection network can control such effect.
The RTDM (Reconfiguration with Time Division Multiplexing) approach [9] solves this problem by partitioning a permutation among different time slots, so that in each slot the subset of the considered partition does not imply any simultaneous crossing of a given SE by two different paths. Unfortunately RTDM, using a time multiplexing, needs optical buffering of data waiting for switching in subsequent slots. At the present state of the art, this is not always a practical solution, especially considering the length of the necessary photonic memories in an optical burst switching context [10].
Nevertheless, a similar method can be used to handle CF permutations, by adopting a reconfiguration technique operating according to space division multiplexing. In this way, through vertical replication of the considered Omega network, the original switching graph assumes a multi-layered architecture, where the dilated topology is equal at every overlapped level of the fabric [11]. The above mentioned permutation subsets are then allocated at different layers so that, once again, any SE is not simultaneously crossed by more than one path. An example of such structure is reported in Fig. 2, where a two-layered 4×4 Omega network is depicted.

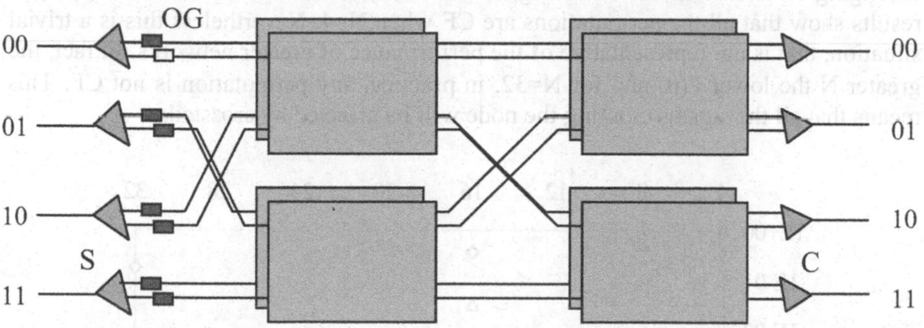

Fig. 2. Vertical expansion on two levels of 4×4 Omega network (OG: optical gate, S: 1×2 splitter, C: 2×1 combiner).

Even though the latter solution is characterized by a redundant structure (see Tab. 1), it allows a node crossing without delays due to temporarily buffering and, consequently, avoids the necessity of optical delay lines.

Tab. 1. Comparison between hardware requirements and throughput performance of time and spatial replication of a two-layered Omega networks.

architecture	switch	splitter	combiner	gate	buffer	throughput
RTDM	$(N/2)\log_2 N$	0	0	0	N	0.5
vertical expansion	$N\log_2 N$	N	N	N	0	1

Focusing our attention on a two-layers architecture, let us define θ as the set of N-permutations realizable with a two CF-mapping of an N×N Omega network. It can be demonstrated that for its size N_θ we have $N_\theta > N_\Omega$ but $\Omega \not\subset \theta$, which means that more CF permutations can be obtained at the expense of higher costs. Moreover, $N_\theta = N_{\Omega+1}$ [12], where class Ω+1 determines the set of admissible permutations of an Extra-Stage Omega network [13], that is obtained by adding one more stage of shuffle in front of an Omega. Since a single SE admits only two possible permutations, it follows that an N×N Omega network is characterized by $2^{N/2 \cdot \log_2 N}$ states. Furthermore, it can be observed that Omega networks admit a unique path from every input terminal to every output port. Consequently, every state of the network corresponds to a different permutation, so that $N_\Omega = 2^{N/2 \cdot \log_2 N} = N^{N/2}$. This is not true for the Extra-Stage Omega network, since it provides two different paths for every source-destination pair. For this reason, in this case, the value of N_θ cannot be computed analogously. We indicate $P(\theta) = N_\theta / N!$ as the probability that a given permutation is CF, in a two-layers N×N Omega network. Such probability can be estimated by simulation, verifying the percentage of randomly generated permutations belonging to θ. It is indicated in Fig. 3 case A, as a function of N. The numerical results show that all the permutations are CF when N=4. Nevertheless this is a trivial situation, that is not representative of the performance of greater networks. In fact, the greater N the lower P(θ) and for N=32, in practice, any permutation is not CF. This means that all the signals crossing the node will be affected by crosstalk.

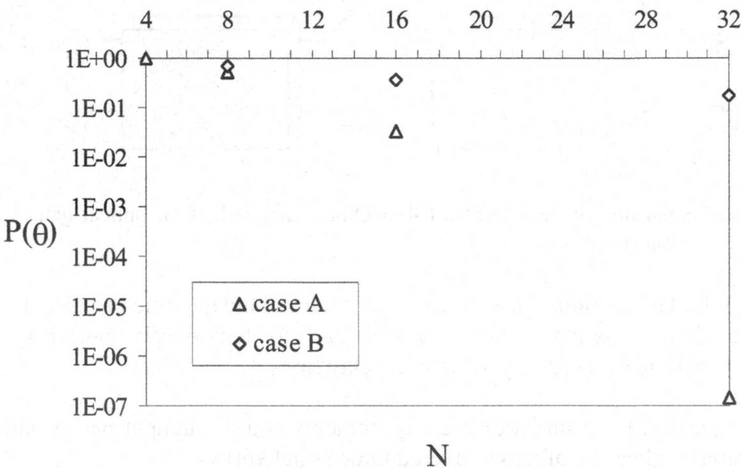

Fig. 3. Probability of a CF permutation in an two-layered N×N Omega network; case A) no deflections, case B) deflection routing.

In [5] we tried to recover partial CF permutations, by a progressive erasing of conflicting paths within a non-CF configuration. We introduced a simple mechanism which, once having verified that a given permutation contains conflicts (as regards

crosstalk), tries to transform it in a CF permutation. Re-routing has been obtained through a simple exchange of the output ports among conflicting paths. In other words, adopting the philosophy of deflection routing [8], those paths are rearranged and directed towards different outputs (with respect to those identified by the routing tables) to obtain a new CF setup. Assuming an effective interaction among the protocols of the MPλS stack [14], a prioritized management of the traffic can be developed, in order to force the deflection of lowest service classes first. Even though this operation does not respect the routing requests and causes a deflection of the considered paths towards different output ports, its effect in terms of crosstalk reduction is evident in Fig. 3 case B. There, $P(\theta)$ is greatly increased, with respect to the situation in which the paths are not modified.
However, this was only a first simple approach, which implies a minimum additional processing time, but generates a not negligible effect on the node performance. Here we propose a more involved algorithmic solution, detailed in the sequel.

3 Bipartite and Two-Colorable Omega Networks

In this paragraph we present the results obtained analyzing the performance of a more complex structure, where the considered topology is dilated on four space layers. Obviously, this solution increases complexity and cost but, through suitable routing algorithms, reduces crosstalk conflicts as well. Here we adopt a combined solution which is based on the integrated utilization of two algorithms.
The first approach is derived from [12], where a membership algorithm for θ class is presented. Through this technique, it is possible to verify if a given N-permutation π belongs to θ, so that it is CF in a two-layered Omega network. It works as explained in the sequel.
A transition sequence associated to the pair $(i,\pi(i))$, for $0 \leq i \leq N-1$, is the binary representation of the i-th node followed by its destination according permutation π. For example, if in the permutation π input 5 has to be connected to output 7, we have $(i,\pi(i))=101111$. Thus, it is evident that a given permutation can be represented as a transition matrix T, where every row is relative to a given input-output pair. For example, the permutation $(0\to0, 1\to4, 2\to7, 3\to3, 4\to5, 5\to1, 6\to2, 7\to6)$ of a 8×8 Omega interconnection network implies the matrix of Fig.4.
In Fig. 4 we have identified the so called optical windows too. Each of them is a sequence of n-1 adjacent columns and, in particular, the k-th one $OW_k(0 \leq k \leq n-1)$ includes the columns T[k+1, …, n+k-1]. In the considered example, where n=3 we have that every window contains only two columns. A string of bits $OW_k(i)$ placed on the same i-th row $(0 \leq i \leq N-1)$ of T and included in the same OW_k, individuates the k-th SE crossed by the path connecting input i with its destination port. See for example, for i=1, the row relative to $(1,\pi(1))=001100$. The three strings shadowed in Fig. 4 are $OW_0(1)=01$, $OW_1(1)=11$, $OW_2(1)=10$, which are exactly the three SEs that have to be crossed to connect input 1 with output 4 (dashed line in Fig. 1).

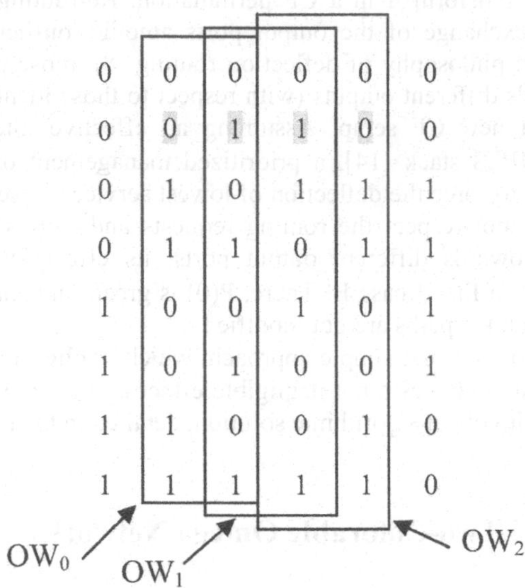

Fig. 4. An example of transition matrix for an 8×8 Omega network (OW: optical window).

For a given permutation, represented through its relative matrix T, a conflict graph G(V,E) includes the vertices $V=\{v_0, v_1,\ldots, v_{N-1}\}$, that coincide with the equally numbered input ports, and the edges E, that connect pairs of vertices (v_x, v_y) for which we have $OW_k(x) = OW_k(y)$, being $0 \le k \le n-1$ and $0 \le x,y \le N-1$. Therefore in such graph two vertices are adjacent and connected when they share a given SE on their path. Having assigned a color, black or white, to every vertex of the conflict graph, the latter is two-colorable when adjacent vertices are not of the same color. Now, an important feature has to be stressed: for a given N-permutation π, we have $\pi \in \theta$ if and only if the conflict graph G(V,E) is two-colorable [12]. This means that in this case we can partition the permutation on two different CF Omega layers. From a given π, the above mentioned graph verifies if the permutation belongs to class θ and, if so, creates the two groups of input-output pairs to be switched towards the two different layers. See for example the situation of Fig. 4. Applying the indicated procedure we obtain the conflict graph of Fig. 5. It is two-colorable and identifies the groups $\{v_0,$

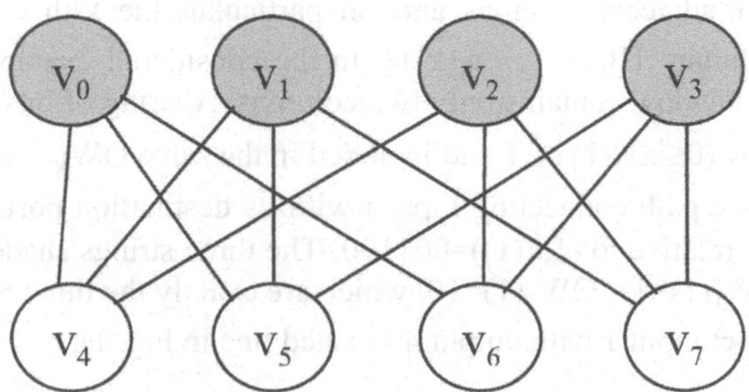

Fig. 5. Conflict graph relative to the transition matrix of Fig. 4.

$v_1, v_2, v_3\}$ and $\{v_4, v_5, v_6, v_7\}$. This means that at one layer of the Omega we will allocate the connections (0→0, 1→4, 2→7, 3→3), whereas at the other layer will be placed the connections (4→5, 5→1, 6→2, 7→6).

The correctness of this method is illustrated in Fig. 6, where the switching configurations are reported, evidencing the active edges only. In both cases a given SE is crossed by only one connection. Thus, every layer manages N/2×N/2 permutations only.

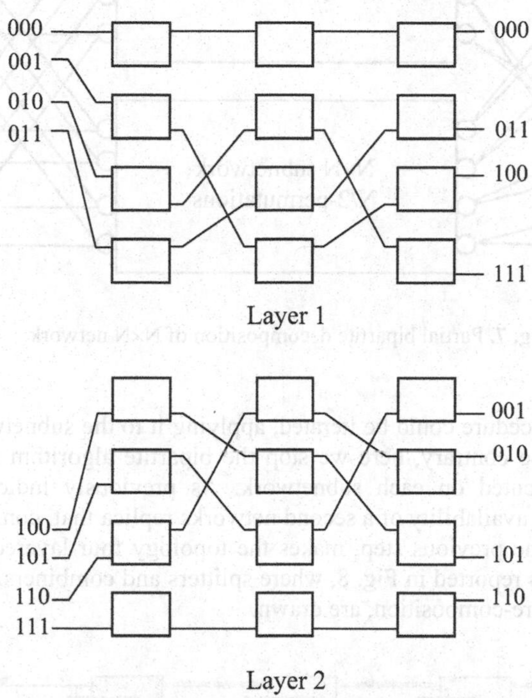

Fig. 6. Switching configurations for the two-colorable permutation of Fig. 4.

Using the illustrated procedure we can build a dilated Omega network, where the probability of obtaining a CF permutation is increased. Nevertheless, as stated above, our goal is the emphasizing of this approach, expanding the interconnection network on a multi-layered topology, to achieve higher CF performance. For this reason, before applying the two-colorable algorithm, a bipartite graph algorithm is used. For brevity sake, we report here just a description of its peculiar characteristics.

The bipartition of the original N×N Omega topology has been obtained here using an algorithm which guarantees CF conditions at the first and the last stage of the network only. That is to say, adopting the procedure proposed in [15], we split the incoming signal before they reach the first stage, so that at such stage no more than one port per SE is active. In addition, the splitting operation is executed taking into account the same problem for the last stage. In practice, the optical signals are addressed towards two overlapped N×N interconnection networks, each of them manipulating N/2-

permutations. The logical topology derives from the first step of the bipartite decomposition technique, which assigns an outgoing (incoming) degree equal to 2 to the inputs (outputs) and pairs them into 2-by-2 butterflies, as shown in Fig. 7.

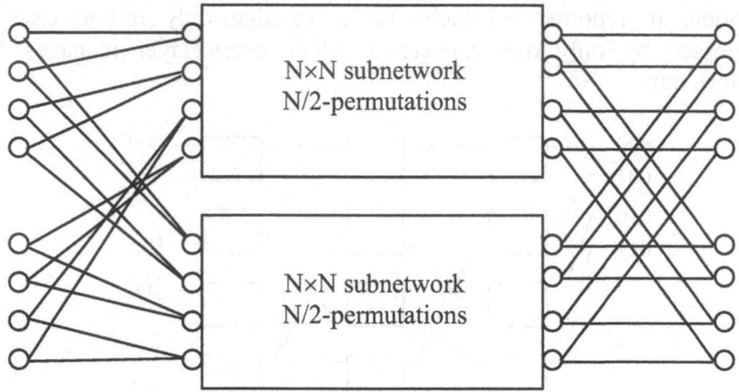

Fig. 7. Partial bipartite decomposition of N×N network.

In principle this procedure could be iterated, applying it to the subnetworks identified step by step. On the contrary, here we stop the bipartite algorithm to run the two-colorable one, executed on each subnetwork. As previously indicated, the latter algorithm needs the availability of a second networks replica that, considering the two ones requested at the previous step, makes the topology four-layered. The resulting physical topology is reported in Fig. 8, where splitters and combiners, responsible for signal partition and re-composition, are drawn.

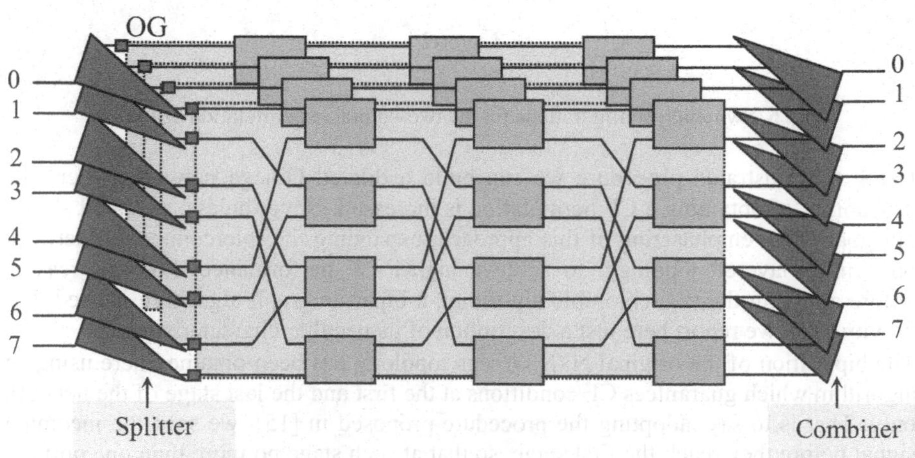

Fig. 8. Four-layered, bipartite and two-colorable 8×8 Omega network (OG: optical gate).

Now we can evaluate how this strategy is effective. In Fig. 9 the results obtained via simulation are plotted, having labeled by P(CF) the probability that a given permutation is CF in a two- or four-layered network. In fact, two curves relate to the case of two-layered N×N Omega networks, where the two-colorable algorithm have been used, with or without deflections of conflicting paths, whereas the third curve represents the performance of the combined bipartite and two-colorable algorithm. As a counterpart of an increased structural complexity, a new improvement is achieved. In fact, the third curve is always over the previous ones and the network is CF until N=16.

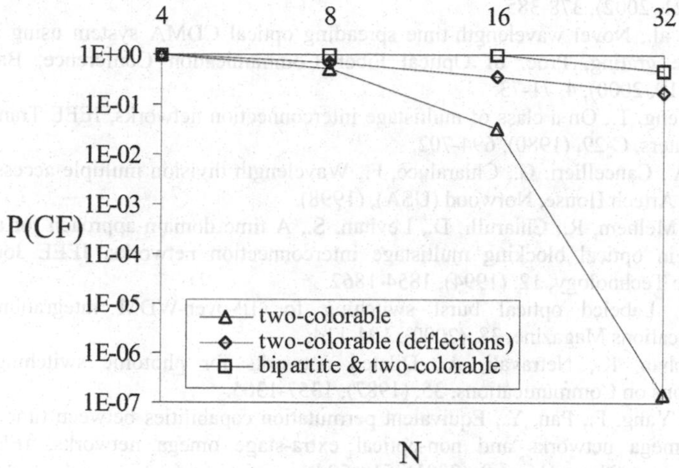

Fig. 9. Probability of a CF permutation in a dilated N×N Omega network.

For a deep comprehension of the proposed routing algorithm, we are developing a new and extended version of the two-colorable one presented above. It will works deeply in space division, as a membership algorithm for a physical topology dilated on four levels, whose architecture will be comparable with the solution presented here. This will be the subject of the next paper.

4 Conclusions

Omega networks have been investigated from the point of view of crosstalk reduction and conflict avoidance. Bipartite and two-colorable networks exhibit very good performance acting, in practice, as a 4-layered network.

The study has been developed for self-routing WDM transmission aver photonic switch elements, but the same model can be adopted also for a CDM transmission. The improvement of a combined time and wavelength domain gives us the opportunity of designing very innovative architectures.

References

[1] Barthel, J., Chuh, T., Optical switches enable dynamic optical add/drop modules, WDM Solutions, (2001), 93-98.

[2] Grusky, V., Starodubov, D., Feinberg, J., Wavelength-selective coupler and add-drop multiplexer using long-period fiber gratings, Proc. of Optical Fiber Communication Conference, Baltimora, (March 7-10, 2000), 4, 28-30.

[4] Zhu, D.X., et al., A novel all-optical switch: the wavelength recognizing switch, IEEE Photonics Technology Letters, 9, (1997), 1110-1112.

[5] Borella, A., Cancellieri, G., Optical burst switching in a WRS all-optical architecture, Proc. of 7th European Conference on Networks & Optical Communications, Darmstadt, (June 18-21, 2002), 378-385.

[6] Yu, K, et al., Novel wavelength-time spreading optical CDMA system using arrayed-waveguide grating, Proc. of Optical Fiber Communication Conference, Baltimora, (March 7-10, 2000), 4, 71-73.

[7] Wu, C., Feng, T., On a class of multistage interconnection networks, IEEE Transactions on Computers, C-29, (1980), 694-702.

[8] Borella, A., Cancellieri, G., Chiaraluce, F., Wavelength division multiple access optical networks, Artech House, Norwood (USA), (1998).

[9] Qiao, C., Melhem, R., Chiarulli, D., Levitan, S., A time domain approach for avoiding crosstalk in optical blocking multistage interconnection networks, IEEE Journal of Lightwave Technology, 12, (1994), 1854-1862.

[10] Qiao, C., Labeled optical burst switching for IP-over-WDM integration, IEEE Communications Magazine, 38, (2000), 104-114.

[11] Padmanabhan, K., Netravali, A., Dilated networks for photonic switching, IEEE Transactions on Communications, 35, (1987), 1357-1365.

[12] Shen, X., Yang, F., Pan, Y., Equivalent permutation capabilities between time division optical omega networks and non-optical extra-stage omega networks, IEEE/ACM Transactions on Networking, 9, (2001), 518-524.

[13] Adams III, G.B., Siegel, H.J., The extra stage cube: a fault-tolerant interconnection network for supersystems, IEEE Transactions on Computers, C-31 (1982), 443-454.

[14] Awduche, D., Utilizing multiprotocol lambda switching (MPLambdaS) to optimize optical connection management, Proc. of NMS-2001, Washington, (2001).

[15] Lea, C.T., Bipartite graph design principle for photonic switching systems, IEEE Transactions on Communications, 38, (1990), 529-538.

The RAMON Module: Architecture Framework and Performance Results

Aldo Roveri[1], Carla-Fabiana Chiasserini[2], Mauro Femminella[3], Tommaso Melodia[1], Giacomo Morabito[4], Michele Rossi[5], and Ilenia Tinnirello[6]

[1] University of Rome La Sapienza;
[2] Polytechnic of Turin
[3] University of Perugia
[4] University of Catania
[5] University of Ferrara
[6] University of Palermo

Abstract. A design study of a Re-configurable Access Module for Mobile Computing Applications is described. After a presentation of its cross-layered architecture, Control Parameters (CPs) of the module are introduced. The set of CPs both describes the functional state of the communication process in relation to the time-varying transport facilities and provides, as input of suitable Algorithms, the control information to re-configure the whole protocol stack for facing modified working conditions. The paper also presents the structure of the simulator realized to demonstrate the feasibility of the design guidelines and to evaluate reconfigurability performances.

1 Introduction

This paper presents a first insight in the design of a Re-configurable Access module for MObile computiNg applications (called RAMON in the sequel) and discusses some preliminary results of a feasibility study of this module which is being carried out within an Italian research program[1], with the participation of six academic research groups[2].

The framework of RAMON is given by mobile users in a TCP/IP environment (*mobile computing*). The mobility is allowed in different communication environments (*Reference Environment, RE*). These REs are identified in a WAN cellular system (e.g., UMTS or GPRS) and in a local area access system (e.g., IEEE 802.11 or Bluetooth).

In wireless communications, much effort will have to be put in the ability to re-configure all layers of the communication process to the time-varying transport facilities, in order to assure the respect of QoS *(Quality of Service)* requirements. The adaptability to different communication environments led us to overtake the OSI layered approach and to adopt the idea of *cross-layering*

[1] The RAMON project was partially funded by MIUR (*Italian Ministry of Research*).
[2] The address of the reference Author is: *roveri@infocom.uniroma1.it*

M. Ajmone Marsan et al. (Eds.): QoS-IP 2003, LNCS 2601, pp. 471–484, 2003.

design, which appears to be particularly appealing to enable dynamic optimization among all layers of the communication process. The idea has already been presented in [1], [2] as a design methodology for adaptive and multimedia networks. RAMON provides a first example of use of this methodology.

The remaining of the paper is organized as follows. In Section 2 we describe the RAMON Functional Architecture. Section 3 defines the relevant *Control Parameters* that constitute the basis of the re-configuration action and the main algorithms which perform it. Section 4 presents the overall RAMON simulator developed in the ns-2 framework [3] to demonstrate the feasibility of the design guidelines and to evaluate the reconfigurability performances, while Section 5 contains some samples of performance results.

2 The Architecture

RAMON allows the deployment of control algorithms which are independent of the specific RE considered. A *mobile station* (MS), equipped with RAMON, becomes a *virtual MS*, i.e. an MS which supports *abstract functionalities* common to all the REs. Such functionalities are obtained from the *specific functionalities* of each RE, by means of *translation entities*, as described below.

Figure 1 shows the RAMON functional architecture. It includes a *Common Control* plane (*CC-plane*) and an *Adaptation* plane (*A-plane*), which is divided in as many parts (*A/RE-i*) as the number of different REs involved. In the case of Figure 1 the number of different REs is let equal to two for the sake of simplicity. Below the A-plane, *Native Control* functionalities (*NC-plane*) for each RE (denoted as *NC/RE-i* for the *i*-th RE) are located. An RE-independent service development *API* (*Application Programming Interface*) is offered to the overlying Application layer.

CC-plane functions, which are also RE-independent, are grouped in three *functional sets*, according to the classical model adopted for wireless communications: i) *Radio Resource Control* (*RRC-c*); ii) *Session Control* (*SC-c*); iii) *Mobility Management* (*MM-c*).

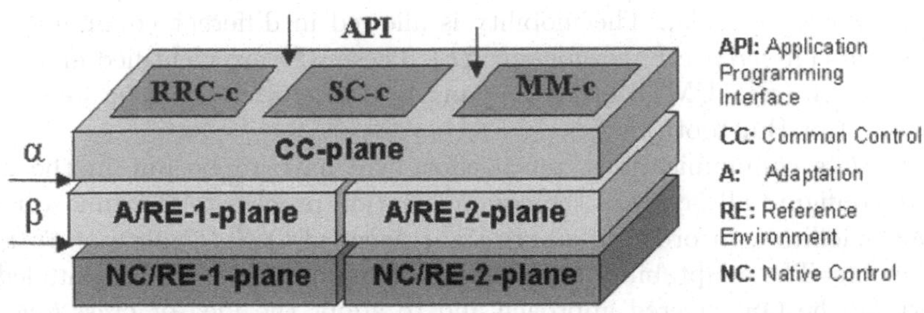

Fig. 1. Overall system architecture.

The A-plane translates: 1) *Primitives* exchanged between the CC-plane and the NC-plane for each RE; 2) *Control Parameters* (CP) data passed from the *Native User* plane (*NU-plane*) to the CC-plane.

Two different types of interfaces can be identified: i) the α *interface*, between the CC-plane and the A-plane; ii) the β *interface*, between the A-plane and the NC-plane of each RE. Figure 2 completes Figure 1. In its right part the CC-plane and the A-plane are depicted, with the relevant α and β interfaces. In its left part, which is concerned with two generic REs, the relevant NC-planes and NU-planes are shown.

Figure 2 helps us showing the way the SC-c, MM-c and RRC-c functionalities interact with the corresponding *SC-i*, *MM-i* and *RRC-i* (i=1,2) of the two REs through the translation performed by the A planes (A/RE-1, A/RE-2). The relations between the NC-planes and NU-planes are highlighted and particularly how the SC-i, MM-i and RRC-i functionalities are related with *Physical* (PHY-i), *Medium Access Control* (MAC-i) and *Radio Link Control* (RLC-i) functions is shown. Finally, *Applications*, *TCP/UDP* and *IP* layers, which are part of the NU-plane, directly interact with the CC-plane.

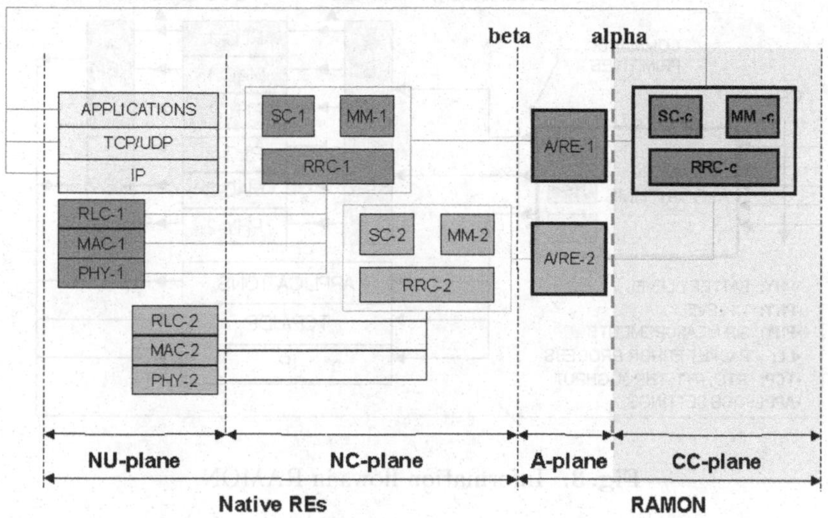

Fig. 2. Basic interactions among Control and User planes.

3 Reconfigurability Parameters

The main idea is to dynamically adapt the whole stack to the perceived QoS by means of adequate control actions. QoS measurements are passed to the RAMON entity by means of CP data.

Cross-layer interaction requires control information sharing among all layers of the protocol stack in order to achieve dynamic resource management. It may be performed in two directions: i) *upward information sharing*: upper layers parameters may be configured to adapt to the variable RE characteristics; ii) *downward information sharing*: MAC, RLC and PHY layer parameters can adapt to the state of Transport and Application layer parameters. Thus, e.g., the behavior of the TCP congestion window may be forced to adapt to the time-varying physical channel characteristics (upward), while lower layer parameters may be re-configured, with the aim of serving distinct applications with different QoS requirements (downward).

As shown in Figure 3, CP data flow from all layers of the NU-plane of each RE and are processed by *Algorithms* running in the CC-plane. The output of the Algorithms consists of *Command Primitives*, which are passed to all layers of the protocol stack. For lower layers, which are RE-dependent, these primitives have to be interpreted by the pertinent A-plane; for higher layers, which are typical of the Internet stack, the primitives can act without any mediation.

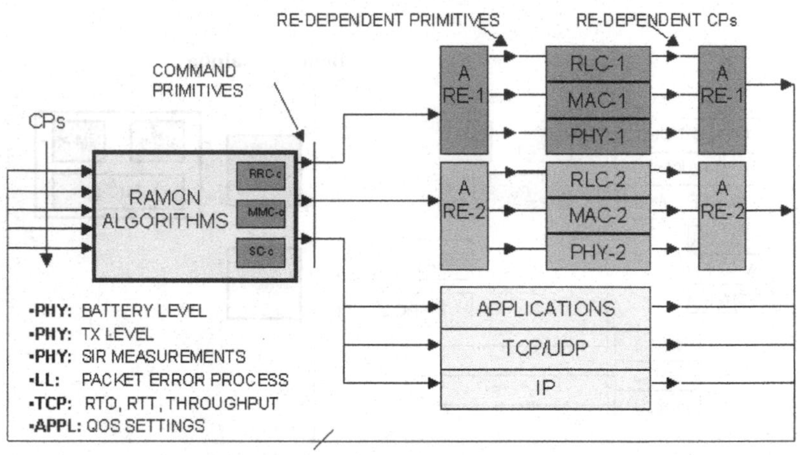

Fig. 3. Information flows in RAMON.

The effect of the Command Primitives, in concurrence with the variations of the RE transport characteristics, gives rise to modifications in the perceived QoS. These modifications cause corresponding change of the CP data. The CC-plane performs re-configuration actions only if they entail appreciable improvements in the QoS perceived.

The most important Algorithms adopted in the RAMON design are:

- *The Handover* Algorithm;
- *Session Control* Algorithm;

– *QoS & MM* Algorithm;
– RRC Algorithm for *Error Control;*
– RRC Algorithm for *Resource Sharing Control.*

The first two algorithms will be analyzed respectively in Sections 3.1 and 3.2; the other three will be briefly described in the following.

QoS and MM capabilities are carried out by means of the Mobile IP protocol, improved with an *Admission Control* function. This last is called *GRIP* (*Gauge and Gate Realistic Protocol*), is described in [7] and operates in a *Differentiated Services* framework. In addition, we assumed that a *micro-mobility strategy* [8] is adopted within the wireless domain: the scope is to hide local handoffs from Home Agent/Correspondent Nodes, thus reducing handoff latency.

To preserve information integrity of packet transmission over the radio channel while meeting the desired QoS and energy constraints, an algorithm for Error Control has been developed within the RRC functional set, which is suitable for a UMTS environment: it provides optimal operational conditions and parameter setting at the RLC layer. The performance results of the application of the algorithm to the UMTS RE are being published [10] and a module which implements it in the RAMON overall simulator has been developed.

An open, re-configurable scheduling algorithm called CHAOS (CHannel Adaptive Open Scheduling) has been defined, which is part of the RRC-c functional set. The Algorithm defines a scheduling strategy for efficient resource sharing control. It is adaptive to *traffic conditions* and *physical channel conditions.* Different CPs may be chosen to represent both variables. Traffic conditions may be represented by transmission buffer occupancy or queue age associated to packets in the transmission buffers; physical channel conditions may be expressed by averaged *Signal to Interfence Ratio* (SIR) and/or *Packet/Frame Error Ratio.* The basic CHAOS principles can be applied in a RAMON module residing in a VBS managing radio resources of different REs. The specific adaptations studied for UTRA-TDD and Bluetooth (whose description can be found in [11]) correspond to A-plane entities of the RAMON architecture.

3.1 Handover Algorithm

RAMON includes an abstract handover algorithm, running at the Virtual MS (VMS), based on the virtualization of the functions necessary for Mobility Management. This operation allows making handover services programmable and independent of the underlying technologies.

In a re-configurable wireless scenario, the selection of the "best" access point at any moment is not simply related to the physical channel quality, but implies also a technology choice as a trade-off between performance, cost, power consumptions, etc. Moreover, the roaming across different systems requires solving addressing and authentication problems, maintaining a low latency to prevent the adverse effects of TCP congestion control.

In order to develop a platform-independent handover algorithm, similarly to [4], [5] we have decoupled these problems. By hiding the implementation details of the MM algorithms from handover control systems, the handover decision can be managed separately from handover execution, so that the same detection mechanisms can interface with many different access networks.

The algorithm is based on a generic *Mobile Controlled Handover* (MCHO) style: handover decisions and operations are all done at the VMS. RAMON keeps track of a list of *Virtual Base Stations* (VBS) in its coverage area. The rationale is that some REs do not provide MCHO capabilities. Therefore, the entire RE is considered as a unique VBS. Conversely, for MCHO systems, a VBS coincides with a BS.

Periodically, for each VBS, the VMS collects CPs as QoS measures (possibly querying the VBS) and estimates cell load condition. According to the definition criterion of the best VBS, if the best serving VBS is not the one in use for a given period of time, the VMS hand-offs to it.

We can identify three main logical functional blocks: i) the *user profile specification*, which re-configures the decision metrics, according to the user requirements; ii) the *measurement system*, based on system-dependent available functions and signaling; iii) the *detection algorithm*, which compares abstract performance metrics computed on the basis of the measurements and of the user profile.

The cost of using a BS_n at a given time is function of the available *bandwidth* B_n, the related *power consumption* P_n, the *cost* C_n of the network the BS_n belongs to. While the last two parameters are easy to compute (mobile host battery life and service type cost), the bandwidth parameter is more complex. In order to account for the real available bandwidth, channel quality perceived by the MS and cell occupancy status have to be considered.

The cost function of the BS_n is then evaluated as:

$$f_n = w_b \cdot N(1/B_n) + w_p \cdot N(P_n) + w_c \cdot N(C_n) \tag{1}$$

where w_b, w_p and w_c are the weights of each parameter, which sum to 1, and $N(x)$ is the normalized version of the parameter x. The reciprocal of the bandwidth is considered in order to minimize the cost function of the best access BS.

Weights can be modified by the user at run-time, and can be set to zero for those parameters that are not of concern to the user: e.g., if high performance has to be pursued, we can assign $w_b = 1$, $w_p = 0$ and $w_c = 0$. Thus, by minimizing the cost function, we achieve load balancing across different REs.

3.2 Session Control Algorithm

Due to the low Mobile IP performance, the MS migrating from a RE to another may be unreacheable for time periods of the order of seconds, which considerably impacts the TCP operation. In particular, several consecutive RTOs (Retransmission Time Out) occur. This results in a very low value of the *Slow Start Threshold (ssthresh)* which, upon each RTO expiration, is updated as follows:

$$ssthresh = \max(cwnd/2,\ 2 \cdot MSS) \tag{2}$$

where *cwnd* denotes the current *Congestion Window.* Therefore, as the inter-RE handover is successfully completed, the sender immediately enters the *Congestion Avoidance* phase and *cwnd* increases slowly. This causes low throughput performance.

In order to solve this problem, the TCP sender implementation has been modified, still maintaining compatibility with standard TCP implementations. We observe that, after an inter-RE handover, the connection path may change dramatically. Consequently, what the TCP sender learned in terms of available bandwidth and estimations of the RTT (Round Trip Time) and RTO is not valid anymore. Based on the above observation, after the inter-RE handover is completed, the TCP sender resets its internal parameters (i.e., *ssthresh*, *Smoothed RTT*, and *RTO*), and enters into the so-called *Fast Learning* phase during which it rapidly estimates the available bandwidth and the RTT characterizing the connection in the new RE. Moreover, in order to avoid useless segment transmissions which would only result in power consumption, the TCP sender *stops* transmitting any data as the handover begins and *resumes* as the handover is successfully completed. The information about the beginning and the end of handovers is provided by command primitives.

The above Algorithm [6], which will be referred to as TCP-RAMON in the sequel, has been integrated in the overall RAMON simulator, and is effective when the MS, acting as a mobile host, is the TCP sender. Otherwise, a similar behavior can be obtained as follows. Before initiating the handover, the MS generates an ACK which informs the TCP sender that the *Receiver Window* (*rwnd*) is equal to 0. This results in a *freezing* of the TCP sender and avoids consecutive RTO expirations. When the handover is completed, the TCP receiver generates a new ACK with the original value of *rwnd*: this occurrence resumes the communication.

4 The Simulative Approach

In this section the RAMON simulator (Figure 4) is described with reference to UMTS-TDD and Bluetooth REs. An 802.11 implementation has also been developed. The simulator is based on the `ns` package [3] (ver. 2.1b7a). Specifically, modifications have been carried out to the Wireless Node object (*MobileNode* class). One protocol stack for every simulated RE is created in the same MobileNode object.

The RAMON module and the A-plane modules are interposed between Agents (RTAgent, MIP Agent, etc.) and radio access layers (LL, MAC, etc.). On the left part of Figure 4 the UMTS-TDD protocol stack is shown with its specific Adaptation plane (*A/RE-UMTS* in figure), where as the Bluetooth stack is on the right side with its A-plane (*A/RE-BT*). On top of the stacks the RAMON module is directly linked to the TCP (or UDP) layer, to the MIP layer (*Mobile IP*) and to the *NOAH* routing Agent (NOn AdHoc routing agent [13]). The UMTS-TDD protocol stack has been developed for this project. The Bluetooth protocol stack is derived from the `ns` *Bluehoc* simulator [12] and modified to

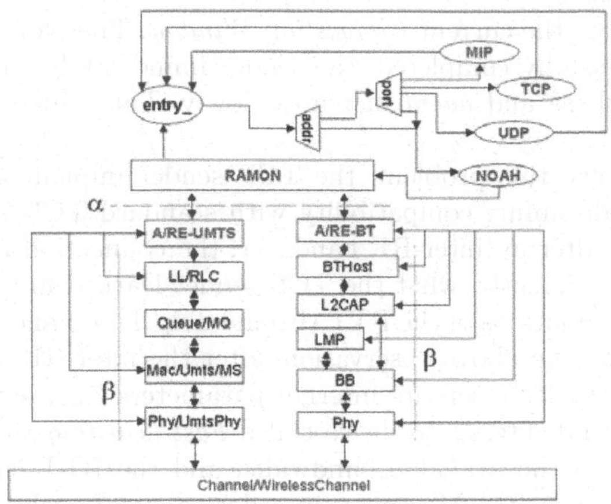

Fig. 4. Overall simulator architecture.

adapt the Bluetooth node object *BTnode* to the `ns` MobileNode object. The α interface in Figure 4 defines the messages exchanged between CC-plane and A-planes. The functions of the α interface can be related to the Command Primitives and the parameters they exploit to the CPs. A list of the most important of these functions follows:

- **`get(parameters_name)`**: gets the **`parameters_name`** parameter value.
- **`attach()`**: is used to create an *IP context* and to register the node to the *Foreign Agent* (*FA*).
- **`detach()`** deletes registration from the FA.
- **`monitor(RE)`**: returns a quality metric for the requested RE.
- **`set(parameters_name,value)`**: sets the **`value`** value to **`parameters_name`** parameter.
- **`send(packet,options)`**
- **`receive(options)`**

In particular, the "**`get`**" function is used to read the CPs from the A-planes for optimization purposes. CPs originated at different layers can be associated to different optimization tasks: i) physical layer CPs are related to power control and battery life saving; ii) link layer CPs are directly connected to link reliability and packet delay; iii) transport layer CPs are directly related to the QoS perceived from the user application in terms of packet delay and throughput. The "**`set`**" function is needed to pass some values to the A/RE-i modules: in fact, with this primitive, some CPs can be modified by the CC-plane. Both the "**`set`**" and the "**`get`**" functions are also defined for the interactions between the CC-plane and RE-independent upper layers protocols. The other functions defined above are Command Primitives (for the α interface).

Two more primitives (named respectively `start_handover_notification()` and `end_handover_notification()`) are defined for interaction between the CC-plane and the transport layer. The first one notifies the TCP about the beginning of the handover while the second one notifies the end of it.

The most important CPs passing through the α interface are briefly discussed in the following for the main layers interacting with RAMON.

Physical layer. CPs detectable at such layer are strongly dependent on the considered RE. A very useful CP is the *battery level* (`BAT_LEVEL`), used to maximize the MS life, e.g., by attaching the MS to the less power demanding RE. Moreover, *Transmitted Power* (`TX_POWER`) and *SIR (Signal to Interference Ratio)* (`SIR`) measurements, if available, are used to detect temporary link failures or high interference conditions: e.g., if the physical channel in a given instant is detected to be in an interference prone situation, the Link layer can be set in a "wait" status to avoid energy wastage.

Link layer. A description of the *error process* at such layer is significant in order to characterize the packet error process affecting the Application layer. Such a description can be achieved by translating the Link layer packet success/failure transfer process into a simple *Markov Model* [14], which is analytically tractable and RE independent; hence RE independent algorithms can easily be deployed which exploit such model.

The CPs for the models are: *loss probability* at the link layer (`P_LOSS_LINK`), *link layer packet size* (`LINK_PACKET_SIZE`), *average burst error length* at link layer (`BURST_ERR`) and *average link packet error probability* (`ERR_PR`). Other useful CPs are the *maximum number of retransmissions per packet* allowed in the link layer (`MAX_LINK_RETR`) and the *average number of retransmissions per correctly transmitted packet* (`AVG_LINK_RETR`); the last one is related to the energy spent per useful information bit, i.e., the amount of energy wasted due to Link layer retransmissions.

Transport layer. At this layer, the MS should be able to collect information regarding the packet *delay/error* process. When the TCP protocol is considered, in addition to the packet error rate, it communicates to RAMON the average and instantaneous *Throughput* value (`AVG_TH` and `ACT_TH`), the estimated *Round Trip Time* (`RTT`) and the actual `RTO`. These CPs are used to control the behavior of TCP state variables and to avoid their incorrect setting as packet errors occur. In particular, it is well known that TCP reacts to packet losses as if a congestion had occurred, i.e., by reducing the maximum flow that the sender is allowed to put into the network at a time. However, in a wireless environment this is not the correct way to proceed because errors affecting a wireless link are very often transient and do not represent at all a measure of network congestion. On the contrary they are a measure of the temporary link QoS. Moreover, performance measured at this level is directly related with the QoS perceived by the user, so it can be weighted in a cost function in order to derive an estimate of the actual QoS. Such an estimate is then compared against the required QoS and used by RAMON algorithms in their decisional tasks.

5 Performance Issues

The most meaningful simulated situations are:

1) Forced inter-RE Handover.
2) User-driven inter-RE Handover;

These scenarios are relevant in the RAMON context because they easily show how a Common Control Plane (RE-independent) can react to variations in the REs it handles, perceived by means of Control Parameters, for optimization purposes. In the first scenario the VMS loses connectivity and is forced to attach to another RE to maintain session continuity. In the second one the VMS is in the coverage area of two different REs (UMTS and 802.11) at the same time and chooses the best RE evaluating a cost function as described in Section 3.1.

5.1 Forced Handover

In this Section some simulation results relevant to an inter-RE forced handover are presented.

The scenario involves a UMTS BS, which is the MIP *Home Agent* (HA), and a Bluetooth BS, which has the role of *Foreign Agent* (FA). A Forced Handover between the former and the latter is simulated.

Figure 5 shows the temporal behaviour of the TCP sender *cwnd* when TCP NewReno is used and when the modified TCP-RAMON is used. When TCP NewReno is used, multiple RTOs expire during the handover and lead to the ssthresh to its minimum value. Thus, when the handover ends, the TCP enters in the *congestion avoidance* phase and thus its cwnd increases slowly.

When TCP-RAMON is used, as soon as the CC-plane detects a loss of connectivity, a `start_handover_notification()` primitive is issued to the TCP layer. This freezes the value of ssthresh to the one determined by the last congestion event occured. Accordingly, timeouts expiring during the handover do not impact the value of ssthresh. Once the `end_handover_notification()` is received the value of sstresh is set equal to the one determined by the last congestion event. Thus, TCP enters the slow start phase (it does not enter the congestion avoidance phase until cwnd reaches the ssthresh value). In these simulations, buffers are considered to be infinite and the initial ssthresh value has been set to 20 which is a typical value in several TCP implementations.

5.2 Handover Customization

In this Section we present some simulation results relevant to user-driven handovers. Our purpose is to demonstrate how user profile specifications affect handover trigger and detection phases. In fact, the optimization of different performance figures requires different choices in terms of RE attachment/detachment policy. For example, if user main goal is cost saving, connections to low-price access points have to be maintained as long as possible, even if better transmission

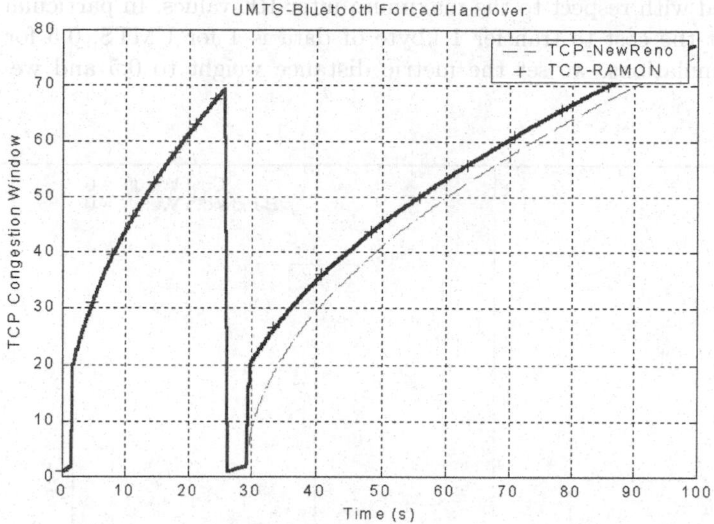

Fig. 5. UMTS-Bluetooth Forced Handover.

conditions (in terms of bandwidth or channel quality) towards other stations are available.

In our simulations we consider an area in which heterogeneous radio access technologies (namely UMTS and 802.11), experiencing different traffic conditions, overlap. In particular, we have considered a simple network topology: four 802.11 BSs (referred to as BS_2, BS_3, BS_4, BS_5) are placed at the vertices of a square and a UMTS BS (BS_1) is placed in the centre. The side of the square is set to 600m, while the coverage areas of the 802.11 and UMTS BSs are set respectively to 200m and 1000m. Channel rate in 802.11 cells is set to 2 Mb/s. 802.11 BSs have a different number of attached users: four mobile nodes are connected to BS_2 and BS_4, while nine mobile users are connected to BS_3 and BS_5.

In the simulation run, a RAMON mobile node, involved in a 6 Mbytes file transfer from the fixed network, moves clock-wise at 6m/s along the square starting from a vertex, for example from BS_2. Although during the simulation time RAMON node is covered by BS_1, handover to 802.11 BSs can be performed in order to save power or reduce cost. Nevertheless, an handover towards 802.11 BSs implies a lower amount of available bandwidth because of their load conditions. The required user trade-off is expressed by the decision metric (Section 3.1) settings. We compute such a metric considering distance, price and bandwidth offered by each BS. In order to make different system parameteres comparable, we normalize each parameter as follows: distance is expressed as the ratio between actual distance and maximum coverage radius, price and bandwidth are normalized with respect to the maximum inter-RE values. In particular, we assume that the cost to transfer 1 kbyte of data is 1 for UMTS, 0.5 for 802.11. In our simulations, we set the metric distance weight to 0.5 and we observe

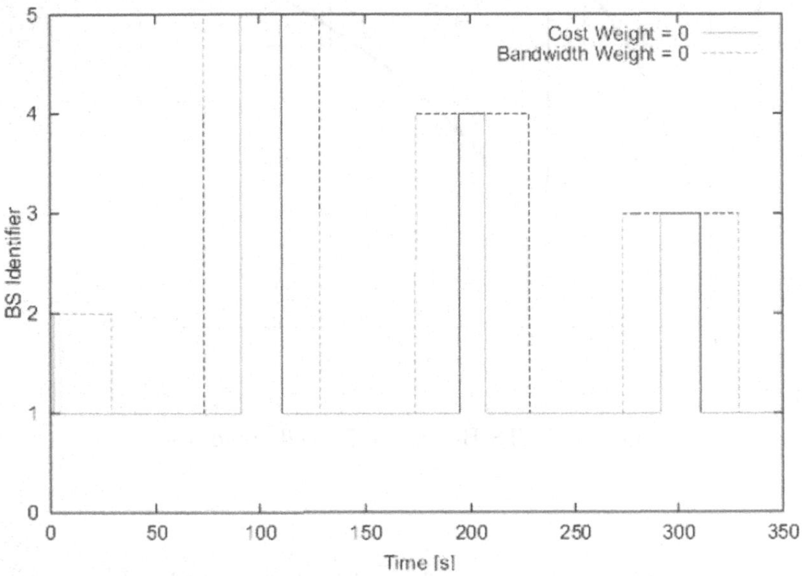

Fig. 6. Selected BS vs simulation time.

performance results varying the cost and bandwidth weights according to the relation: $w_b = 0.5 - w_c$. Consider preliminarly the extreme situations in which we want to minimize the file transfer-cost regardless of the transfer-time ($w_b = 0$) or, conversely, we want to minimize the transfer-time regardless of the cost ($w_c = 0$). Figure 6 shows the handover trigger and decision policies adopted in the cosidered extreme cases. We report the identifier of the selected BS versus the simulation time. In the w_b situation, connection to WLAN is kept as long as possible, since this access technology is the cheapest (handovers are triggered only when the RAMON node moves outside the coverage area of an 802.11 BS). In the w_c case, the RAMON node remains connected to UMTS for a greater time (switch to WLAN occurs only when the distance from the relevant BS is very small). Note that, since our metric accounts for both distance and load, the permanence time in 802.11 BSs is not constant and depends on the BS considered.

The time and the cost resulting for the considered 6 Mbytes file transfer is reported in Figure 7 versus the w_c setting. This figure shows that, by changing the weights of the cost function, the RAMON user can effectively configure its optimal cost/performance trade-off.

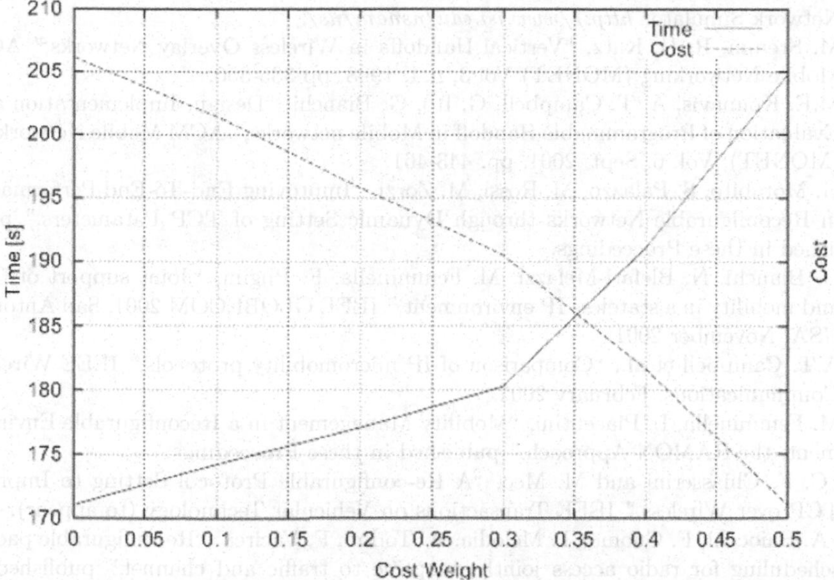

Fig. 7. Time and Cost Trade Off.

6 Conclusions

In this paper the guidelines adopted for the design of a module able to reconfigure different mobile communication environments for computing applications are introduced. The paper mostly concentrates on architectural issues and on control information flows. In particular, the Algorithms that constitute the processing core of the module are briefly described. The integrated RAMON simulator is described and some performance results are finally shown.

7 Acknowledgments

The authors would like to thank all the researchers of the Polytechnic of Turin and of the Universities of Catania, Ferrara, Palermo, Perugia and Rome "La Sapienza" who participated to the RAMON project and, notwithstanding their important contribution, do not appear as authors of this work.

References

1. Z.J. Haas, "Design Methodologies for Adaptive and Multimedia Networks," IEEE Communications Magazine., Vol. 39, no. 11, November 2001, pp 106-07.
2. T.S. Rappaport, A. Annamalai, R.M. Buehrer, W.H. Tranter, "Wireless Communications: Past Events and a future Perspective," IEEE Commun. Mag., Vol 39, no. 11, May 2002, pp 106-07.

3. Network Simulator *http://www.isi.edu/nsnam/ns/*.
4. M. Stemm, R. H. Katz, "Vertical Handoffs in Wireless Overlay Networks," ACM Mobile Networking (MONET) Vol 3, n 4, 1998, pp 335-350.
5. M.E. Kounavis, A. T. Campbell, G. Ito, G. Bianchi, "Design, Implementation and Evaluation of Programmable Handoff in Mobile networks," ACM Mobile Networking (MONET), Vol. 6, Sept. 2001, pp. 443-461.
6. G. Morabito, S. Palazzo, M. Rossi, M. Zorzi, "Improving End-To-End Performance in Reconfigurable Networks through Dynamic Setting of TCP Parameters," published in these Proceedings.
7. G. Bianchi, N. Blefari-Melazzi, M. Femminella, F. Pugini, "Joint support of QoS and mobility in a stateless IP environment," IEEE GLOBECOM 2001, San Antonio, USA, November 2001.
8. A.T. Campbell et al., "Comparison of IP micromobility protocols," IEEE Wireless Communications, February 2002.
9. M. Femminella, L. Piacentini, "Mobility Management in a Reconfigurable Environment: the RAMON Approach," published in these Proceedings.
10. C. F. Chiasserini and M. Meo, "A Re-configurable Protocol Setting to Improve TCP over Wireless," IEEE Transactions on Vehicular Technology (to appear).
11. A. Baiocchi, F. Cuomo, T. Melodia, A. Todini, F. Vacirca, "Reconfigurable packet scheduling for radio access jointly adaptive to traffic and channel," published in these Proceedings.
12. Bluehoc Simulator *http://www-124.ibm.com/developerworks/opensource/bluehoc/*.
13. J Widmer, "Network Simulations for a Mobile Network Architecture for Vehicles," *http://www.icsi.berkeley.edu/~widmer/mnav/ns-extension/*.
14. M. Zorzi, R.R. Rao, "Perspectives on the Impact of Error Statistics on Protocols for Wireless Networks," IEEE Personal Communications, vol. 6, Oct. 1999.

Reconfigurable Packet Scheduling for Radio Access Jointly Adaptive to Traffic and Channel

Andrea Baiocchi, Francesca Cuomo, Tommaso Melodia, Alfredo Todini, and Francesco Vacirca

University of Roma "La Sapienza", INFOCOM Dept., Via Eudossiana 18, 00184, Rome, Italy

Abstract. Adaptive packet scheduling for wireless data access systems, including channel state information based algorithms, is a quite well established topic. We develop here an original framework to define such an algorithm in a reconfigurable context, i.e. with the ability to exploit heterogeneous communication environments. The focus is to define platform independent algorithms that can be "adapted", by a specific software, to different environments, so that the core communication functions can be defined, modified and improved once for all. The specific communication function addressed in this work is packet scheduling. For better performance the packet scheduling design exploits the cross-layering approach, i.e. information and functions from different communication layers are used. The concept is proved with reference to the UMTS and Bluetooth technologies, as representatives of a cellular system and a local access one respectively.

1 Introduction

The importance of reconfigurability in wireless access interfaces (at least at the physical layer, e.g. software radio) is due to existence of various technologies spanning from personal area networks and wireless local area networks (e.g. IEEE 802.11, Bluetooth, Hiperlan/2) to cellular systems (e.g. GSM, GPRS, CDMA2000, UMTS, etc.).

The common meaning of reconfigurability [1] is the capability to provide a network element with different sets of communication functions and with the ability to switch among them so as to exploit any of a number of different wireless access environments that might be available at a given time and place. In the following we adopt a somewhat broader point of view: a system is reconfigurable if some communication functions and relevant algorithms are defined and implemented by abstracting from a specific wireless access. The specific wireless access is represented by models able to express the traffic load, the channel conditions and other communication parameters from a high level point of view. It becomes possible to design algorithms (e.g. packet scheduling) according to some general criteria independent of the specific technology.

A strictly related approach to high performance wireless access system is cross-layering [2] [3]. Optimization of data transfer and quality of service is best achieved by jointly considering functions and parameters traditionally belonging to different architectural layers (e.g. error control by means of hybrid ARQ/FEC techniques, power control, packet scheduling and radio resource sharing)

M. Ajmone Marsan et al. (Eds.): QoS-IP 2003, LNCS 2601, pp. 485–498, 2003.

The conception of communication algorithms and specifically of radio resource management algorithms is therefore "generalized" along two directions: reconfigurability, in order to exploit different wireless communication systems, and cross-layer design, to achieve a better efficiency and exploitation of the time-varying communication environment.

This work fits in a comprehensive effort to specify a reconfigurable wireless communication architecture for mobile computing applications, named *Re-configurable Access module for MObile computiNg applications* (RAMON in the sequel [1]). The framework of RAMON is a mobile computing environment adopting the TCP/IP protocol suite. Different communication environments (*Reference Environment*, RE) are provided to the mobile host; in the following, we consider as example network environments UMTS-TDD [4] and Bluetooth [5].

This paper focuses on the radio resource management functions, specifically on packet scheduling and resource sharing. There are some works on these subjects with specific reference to the wireless access. Papers [6]-[8] address wireless packet scheduling with emphasis on the fairness issue arising primarily from reuse of channel state information and the short term variability of channel condition that bring about unfairness at least in the short term if a high efficiency is targeted. Adaptation of the classic Weighted Fair Queuing schemes are investigated in [6] and [7]. Papers [9]-[12] deal with radio resource sharing for wireless packet access in a CDMA based air interface with mutual interference. Finally, papers [13] and [14] consider using channel state information for resource sharing in wireless packet access.

Some of these works take what can be viewed as a cross-layer approach, in that radio resource assignment is based on radio channel state for the competing links. This is also the approach considered in our research [15].

The novel points with respect to previous works are the reconfigurability framework and the specific scheduling algorithm with its adaptation to the considered REs (UMTS and Bluetooth). The first point has a conceptual value: from a large number of studies developed with reference to specific technologies, we abstract the major common ideas to define a general framework for wireless access thus simplify the design of efficient algorithms. The second point is the development of a packet scheduling scheme that is jointly adaptive to channel and traffic conditions. This scheme, named *CHannel Adaptive Open Scheduling* (CHAOS) fully exploits the abstract and general model available in the RAMON platform.

The rest of the work contains a brief description of the overall RAMON architecture (Section 2), aimed at stating our original context for the radio resource management algorithms. Section 3 defines the general, platform independent radio resource assignment algorithm, by exploiting RLC, MAC and physical layer information (cross-layer design). Section 4 describes in detail the adaptation modules to fit the general algorithm onto two specific REs, UMTS and Bluetooth. Section 5 shows some performance results. Final remarks and further work are addressed in Section 6.

2 Functional Architecture

RAMON allows the deployment of control algorithms independent of the considered RE. A *Mobile Station* (MS) or a *Base Station* (BS), equipped with the RAMON software module, supports abstract functionalities common to all the REs. Such functionalities are obtained from the specific operations of each RE, by means of translation entities, as described below.

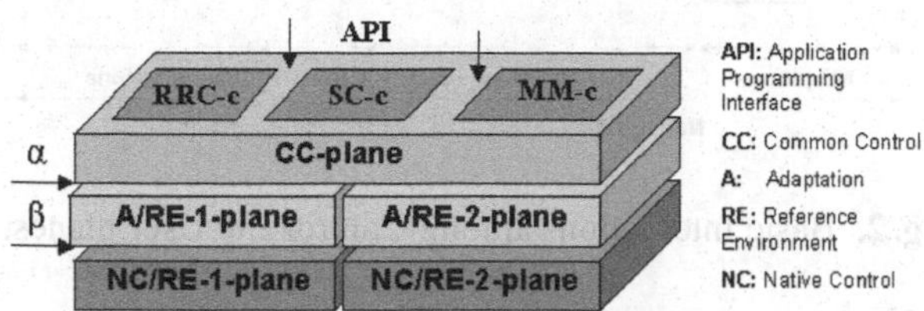

Fig. 1. Overall system architecture.

Figure 1 shows the RAMON functional architecture. It includes a *Common Control plane* (CC-plane) and an *Adaptation plane* (A-plane). *Native Control* functionalities (NC-plane) for each RE (denoted as NC/RE-i for the i-th RE) are located below the A-plane. A service development RE-independent API (*Application Programming Interface*) is offered to the overlying *Application Layer*. CC-plane functions, which are defined independently of the underlying RE, are grouped in three functional sets, according to the classical model adopted for wireless communications: i) *Radio Resource Control* (RRC-c); ii) *Session Control* (SC-c); iii) *Mobility Management* (MM-c) [1];

The A-plane, which is divided into as many parts (A/RE-i) as the number of different REs involved, translates: 1) Primitives exchanged between the CC-plane and the NC-plane for each RE; 2) *Control Parameters* (CP) data passed from the *Native User plane* (NU-plane) and the CC-plane. Two different types of interfaces can be identified: i) the α interface, between the CC-plane and the A-plane; ii) the β interface, between the A-plane and the NC-plane of each RE. Figure 2 completes Figure 1. The CC-plane and the A-plane with the relevant α and β interfaces are depicted on the right. The NC-planes and NU-planes are shown on the left. Figure 2 clarifies how the SC-c, MM-c and RRC-c functionalities interact with the corresponding SC-i, MM-i and RRC-i (i=1,2) of the two REs through the translation performed by the A planes (A/RE-1, A/RE-2). The relations between the NC-planes and NU-planes are highlighted and particularly how the SC-i, MM-i and RRC-i functionalities are related with Physical (PHY-i), *Medium Access Control* (MAC-i) and *Radio Link Control* (RLC-i) functions is shown. Finally, Applications, TCP/UDP and IP layers, which are part of the NU-plane, directly interact with the CC-plane. The main idea is to dynamically adapt the whole stack to the perceived link quality by means of adequate control actions. Quality measurements are passed to the RAMON entity by means of CP data.

[1] -c means in the common control plane.

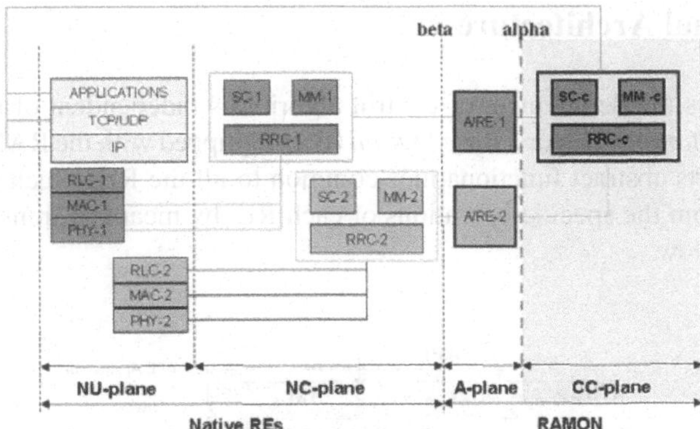

Fig. 2. Basic interactions among Control and User planes.

Cross-layer interaction requires control information sharing among all layers of the protocol stack in order to achieve dynamic resource management. It may be performed in two directions: i) *upward information sharing*: upper layers parameters may be configured to adapt to the variable RE characteristics; ii) *downward information sharing*: MAC, RLC and PHY layer parameters can adapt to the state of Transport and Application layer parameters.

3 Common Control Plane and Reconfigurable Algorithm

This Section is dedicated to describe the general CHAOS scheme in the framework of the RAMON architecture.

CHAOS is part of the RRC functional set of the CC-plane of the RAMON architecture. Thus it is defined independently of the underlying REs. The scheme sets a framework for designing scheduling strategies for efficient resource sharing control in a scenario constituted by a single cell served by a BS, based on an "abstract" representation of i) physical channel state; ii) traffic condition for each MS involved in the communication process. This representation is expressed by means of abstract quantized parameters, *CHANNEL STATE* and *TRAFFIC CONDITION* respectively. The system state can be described by four vectors of length N (where N is the number of MSs in the scenario):

1. CHANNEL_STATE for each MS (downlink);
2. CHANNEL_STATE for each MS (uplink);
3. TRAFFIC_CONDITION for each MS (downlink);
4. TRAFFIC_CONDITION for each MS (uplink).

Figure 3 shows the basic information flow, i.e., how the CHAOS entity running in the CC-plane gathers information about system state. For every MS, in both directions, the RRC-i (relative to the i-th RE) entity in the NC-plane of the BS collects information that will be used for the evaluation of the abstract parameters by the A-plane. The RRC-i

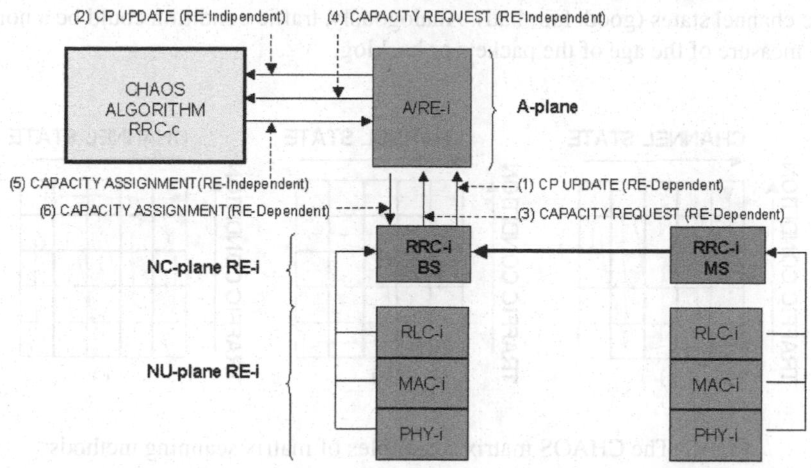

Fig. 3. Description of Information Flow from U-plane to CC-plane.

entity on the BS gets the BS-related information directly from the different layers of the NU-plane (i.e. the RLC-*i*, MAC-*i* and PHY-*i* layers). Information lying in the MS has to be exchanged between the two RRC peer entities (MS RRC-*i* and BS RRC-*i*) by means of the transport facilities offered by lower layers. As soon as new information describing the system state is acquired by the BS RRC-*i*, a *CP UPDATE indication primitive* (1), which is RE-dependent, is issued to the A-plane.

The mapping of these RE-dependent parameters into abstract parameters (CHANNEL STATE and TRAFFIC CONDITION), that can be used by the CHAOS algorithm, is up to the A-plane, which asynchronously communicates the abstract parameters to the CC-plane entity as soon as they are available, by means of an RE-independent *CP UPDATE primitive* (2).

The A/RE-*i* entity asynchronously issues RE-independent *CAPACITY ASSIGNMENT REQUEST primitives* (4) to the CHAOS entity in the CC-plane. These abstract requests are formulated by the A/RE-*i* plane on the basis of the RE-dependent explicit request coming from the RRC-*i* (RE-Dependent *CAPACITY REQUEST primitive* (3)) or on the basis of the CPs gathered when no explicit requests are issued by native system control mechanisms. These primitives contain the overall amount of capacity to be assigned. The CHAOS entity processes this request and returns an ordered vector of length N_T ($N_T \leq N$) whose element contains the MS *id* and the amount of capacity assigned to it through the RE-independent *CAPACITY ASSIGNMENT primitive* (5). The last step (6) consists in the A/RE-*i* entity translating the assignment into an RE-dependent *CAPACITY ASSIGNMENT primitive* command that can be issued to the BS NC-plane. This will result in the actual assignment with NC-plane specific mechanisms.

As a matter of example, the channel state could be a normalized measure of the distance between the target *Signal to Noise Ratio* (SNR) of the link and its actual SNR; or it could be derived from *Bit Error Rate* measurements. For instance, we could have three channel states (good, fair, bad). Analogously, traffic condition could be a normalized measure of the age of the packets or backlog.

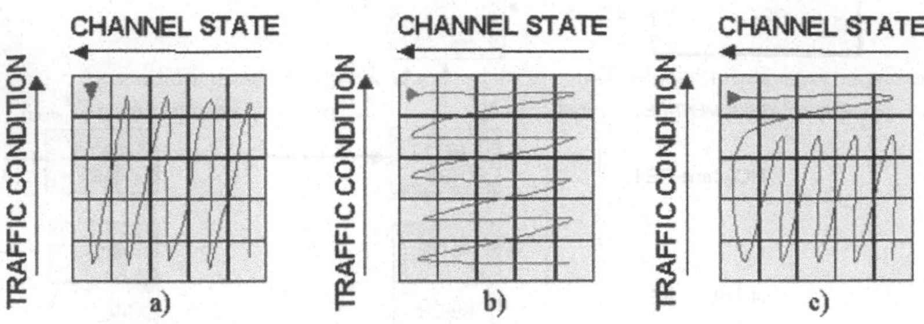

Fig. 4. The CHAOS matrix. Examples of matrix scanning methods

We will finish this section by describing how the CHAOS entity uses system state information to assign capacity to the different MSs. As soon as *CAPACITY REQUEST primitives* arrive from the A/RE-*i* plane, this information is arranged in a matrix. In Figure 4 the CHAOS matrix is shown with three different scanning methods. The horizontal dimension represents the channel state, and thus requests with a different value of CHANNEL STATE occupy different positions in the matrix. The vertical dimension is associated with traffic condition, thus requests with different values of TRAFFIC CONDITION are put in different vertical positions in the matrix. What the CHAOS entity obtains is thus a two-dimensionally ordered description of the requests coming from different MSs. User requests are served in the order defined by a predefined rule (scanning method). Each different rule results in a different scheduling discipline. In Figure 4(a) the matrix is scanned by column by giving priority to traffic state conditions; in Figure 4(b) it is scanned by row by giving priority to channel state conditions; a combination of the first two methods is shown in Figure 4(c).

Thus, CHAOS can be better understood as a class of algorithms, each qualified by a different service

discipline, which is somehow computed on the basis of the matrix.

4 Adaptation Planes and Reference Environments

In this Section the RE-dependent A-planes for the UMTS and Bluetooth systems are described. Native mechanisms and procedures are introduced and some design choices are explained. After a brief description of the capacity allocation procedures of the two REs (Subsections 4.1 and 4.3), we illustrate how to adapt these mechanisms to uniform them to the CHAOS paradigm (Subsections 4.2 and 4.4)).

4.1 UMTS-TDD

In UMTS-TDD air interface the time axis is slotted into *Time Slots* (TSs) and 15 TSs are grouped into a 10 ms frame. Within each time slot a set of variable spreading factor orthogonal codes are defined to spread QPSK modulated data symbols of up to 16 different users. The overall capacity of the radio interface is shared by common channels and data channels. In our study the *Random Access CHannel* (RACH) is allocated to the first TS of every frame and the *Forward Access CHannel* (FACH) and *Broadcast CHannel* (BCH) are allocated to the last one. We assume that transport blocks are 320 bit long, (including MAC/RLC overhead); a *Transmission Time Interval* (TTI) of $10\,ms$ and a convolutional coding with rate $1/2$ are used. This implies that four codes with spreading factor 16 are required to carry the coded transport block; the choice of this transport format implies that the minimum assignable capacity is $32\ kbit/s$ which will be referenced to in the sequel as *Resource Unit* (RU). From a modeling point of view, the entire transport resource of the radio interface (i.e. 60 RUs) is divided into common channels (8 RUs) and shared channels (52 RUs) for data transfer. Each slot can carry up to four data (plus associated signaling) transport blocks in each frame, if we require that the four RUs used by a transport block be all in the same time slot.

Different kind of logical channels are available for data transfer. The choice depends on the traffic type and on traffic volume. *Dedicated CHannels* (DCH) are well suited for *Real Time* services (RT), whereas common channels (RACH and FACH) or shared channels (*Uplink Shared CHannel* (USCH) and *Downlink Shared CHannel* (DSCH)) are suited for *Non Real Time* services (NRT). For mobile computing purposes (web-like and file transfer applications) the DSCH and the USCH are the most appropriate choices. These channels are allocated for a limited time and are released automatically. Figure 5 shows the data transfer procedure on USCH [16]. After scanning the RLC buffer, the MS sends a *Capacity Request* message to the UTRAN; this message contains a traffic measure (the RLC buffer length) and it is sent on the SHCCH (*SHared Control CHannel*), mapped on the RACH channel or on a USCH (if there is one already allocated to the MS). When the UTRAN RRC layer runs the *Scheduling Algorithm*, a resource allocation message (*Physical Shared Channel Allocation*) is sent to the MS on the SHCCH, which is mapped on the FACH or on the DSCH. At this point the MS can start data transfer. The allocation message allocates a number of RUs for a number t of frames (with $1 \leq t \leq t_{max}$). For this reason, it is not necessary to acknowledge the message. A similar procedure is employed for DSCH allocation.

4.2 UMTS-TDD Adaptation Plane

In this Section the interactions of the A/UMTS-plane with the UMTS-RRC entity and with RRC-c entity are described. On the one hand the UMTS-TDD Adaptation plane reads informations about channel and traffic conditions of MSs from the RRC-UMTS layer, processes these informations and informs the RRC-c plane about these updates, on the other hand it requests RRC-c plane for the resource allocation, maps it in a UMTS native language and passes it to the RRC-UMTS. To perform all these operations the plane must be aware of the UMTS resource allocation procedures (exchanged messages, parameters, etc.) and must interact with the RRC-UMTS entity to get all needed informations and to replace the RRC allocation algorithm. There are two basic procedures which can be better explained with two examples.

– *CP-UPDATE*. When the RRC-UMTS layer receives a new allocation request from a MS (see Figure 3 (3)), or when it receives new packets directed to a MS, it sends a

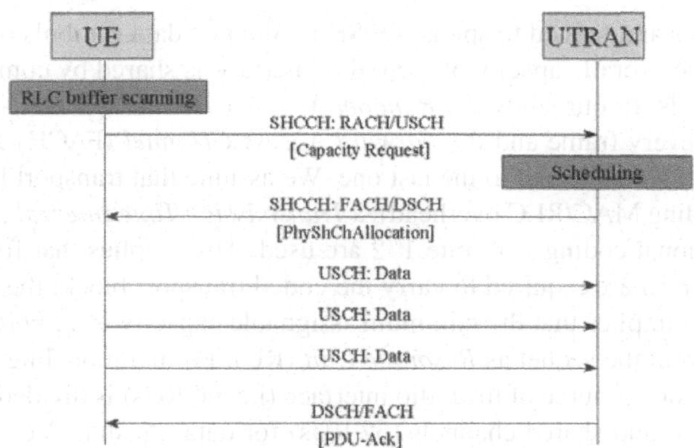

Fig. 5. USCH channel allocation procedure.

CP-UPDATE command to the RRC-c plane with the direction link (UL or DL), the mobile identity and the channel quality. The queue length indicated in the request is sent in a CAPACITY REQUEST (see Figure 3(4)). The RRC-c functional entity updates the matrix.

– *CAPACITY REQUEST*. Every 10 ms the A-plane sends a CAPACITY ASSIGNMENT REQUEST to the RRC-c entity indicating the unallocated capacity of the frame; the RRC-c scans the matrix and creates a vector with the MS identities, the capacity to allocate in one frame and the duration of the allocation. The A-plane translates it into the UMTS-TDD language (i.e. TSs, codes to assign and number of frames) and sends new allocations to the RRC-UMTS that will inform the involved MSs. As seen in the second example, the CAPACITY ASSIGNMENT REQUEST indication primitive is sent one time per frame and lets the RAMON module synchronize with the UMTS frame. The UPDATE messages, instead, are totally asynchronous and can be transmitted at every instant. From the RAMON point of view the whole UMTS-TDD system can be seen as shared capacity that varies in time but remains fixed in the frame.

4.3 Bluetooth MAC

In this paragraph the basic issues related to the MAC technique adopted in the Bluetooth (BT) technology are presented. Devices are organized into piconets, i.e., small networks which include up to 8 nodes. Different piconets share the same channel (2.4 GHz ISM band) by exploiting the FHSS (Frequency Hopping Spread Spectrum) technique, i.e., by following different hopping sequences.

In a piconet, one device has the role of master, all the other devices are slave. Time is slotted (625μs), the master controls the traffic and takes care of the centralized access control. Only master-slave communications are allowed. Odd slots in the TDD slotted structure are reserved for master transmission. The slave that receives a packet from the master is implicitly polled for the following even slot. Thus, the master controls the medium access by sending data to different slaves in even slots and implicitly authorizing the slave to transmit starting from the following slot. If the master has no data to send, it can authorize slaves transmission with an explicit POLL packet. Three data packet sizes are allowed: one slot, three slot, and five slot length.

4.4 Bluetooth Adaptation Layer

While the RRC-UMTS provides much information that can be easily adapted for the CHAOS scheduling, Bluetooth lacks mechanisms to retrieve physical layer (both in the UL and DL directions) and traffic state informations (for the UL direction). A great effort has been done to define a suitable A-plane[2] that would make up for the lack of native system informations. Moreover, due to the particular polling mechanism adopted in Bluetooth, master-slave pairs have to be jointly represented in a CHAOS matrix, since a slot assigned in the downlink direction implies at least a slot being assigned in the uplink. For every Master-Slave pair, the master keeps memory of the average Packet Error Rate (PER) on the link, by mediating on the last N_{tx} transmissions. This information is passed to the A-Plane which classifies the slaves into different classes of CHANNEL STATE according to their PER values.

Traffic condition for each link can be expressed by the amount of data to be transmitted in the L2CAP queue [17]. However, although the master node exactly knows queue-length for any of its slaves in the downlink direction (DLQ_i), uplink queue state must be somehow signaled by the slaves. We propose, as in [17], to use the flow bit in the Baseband packet header, which was intended in the Bluetooth Specifications [5] to convey flow information at the L2CAP level. In a way similar to [17], we use this bit to estimate the queue length of the slaves for the uplink direction, (ULQ_i). The master-slave pair, in the scheduling process, will be then represented by its *Virtual Queue*, defined as $VQ_i = ULQ_i + DLQ_i$. This parameter is then quantized by the A-plane and passed as TRAFFIC_CONDITION parameter to the CHAOS RRC-c entity. Due to the particular polling mechanism adopted in Bluetooth, the resource to be shared among the different M-S couples can not be easily described as capacity. The scheduler only decides which M-S couple has the right to transmit on the next available slot, but the amount of capacity assigned to each couple strongly depends also on other factors, such as the SAR (Segmentation & Reassembly), since choosing different packet sizes leads to different capacities being assigned.

An additional variable is defined in the A/BT-plane for each M-S couple, namely *Virtual Time*, to adapt the CHAOS framework to the polling mechanism. The Virtual Time assigned to the i-th M-S couple at step t is defined as:

$$V_t^i = V_{t-1}^i + \frac{l_{t-1}}{r_t^i} \tag{1}$$

where r_t^i represents the abstract capacity assigned by CHAOS to the i-th couple at step t and l_{t-1} is the duration of the last transmission of the M-S couple (which ranges from $625 \cdot 2\mu s$ if a 1 slot packets is used in both directions to $625 \cdot 10\mu s$ if both transmitters

[2] The authors would like to thank M. D. Marini for his precious work in adapting CHAOS to Bluetooth and for the simulation results presented in Section 5.2.

use a 5 slot packet). Thus, when CHAOS assigns high values of capacity, the *Virtual Time* increases slowly after each transmission, where as when it assigns high values of capacity it increases in a faster way. After one couple's transmission, the scheduler gives the right to transmit to the M-S couple with the minimum value of Virtual Time. This results in a higher number of chances to transmit being given to M-S couples CHAOS has assigned high capacity.

CHAOS assigns capacity on the basis of functions that can be computed on the matrix representing the system state, which somehow represent the scheduling discipline chosen. For the simulations presented in the following paragraph, a 5x5 matrix has been used to describe the system state. CHANNEL_STATE (C_S) and TRAFFIC_CONDITION (T_C) assume discrete values ranging from 1 to 5. The capacity assigned to the i-th couple is then calculated as:

$$r_t^i = \frac{T_C^2}{C_S^3} \tag{2}$$

which results in high capacity being assigned to couples with low values of CHANNEL_STATE (good channel) and high values of TRAFFIC_CONDITION (high traffic). However, channel state is given a higher exponent because the main target is avoiding potentially power-wasting transmissions when the state of the channel is bad.

5 Results

In this Section results about improvements that can be achieved with the CHAOS strategies are shown for the UMTS and for Bluetooth. Results have been obtained using an UMTS module for the ns network simulator [18] and extending the *Bluehoc* [19] functionalities. As said before it is important to notice that CHAOS constitutes a large class of algorithms and by modifying the matrix scanning methods different service disciplines (with different targets) can be easily achieved. This can be simply obtained by different scanning methods.

5.1 UMTS

In the simulated scenario N mobiles in a single cell send data through a gateway to N wired nodes. The entire data traffic is in the uplink direction. In the UMTS frame 9 time-slots have been statically allocated to the uplink (1 for the RACH and 8 for the USCHs), and the remaining 6 have been allocated to the downlink (1 for the FACH and BCH and 5 for the DSCH); thus, the maximum attainable throughput in the uplink direction is 1024 $kbit/s$. The traffic is generated by FTP traffic sources (one FTP agent per node) on a TCP transport layer.

Three matrix scanning methods have been used in simulations; the first one, named *Random*, scans the matrix randomly; the second one, named $CHAOS_1$, scans the matrix by giving priority to the oldest requests (see Figure 4(b)), and the last one, $CHAOS_2$, gives priority to the requests from users with better channel quality (see Figure 4(a)). In Figure 6 we can see that throughput is maximized by giving priority to

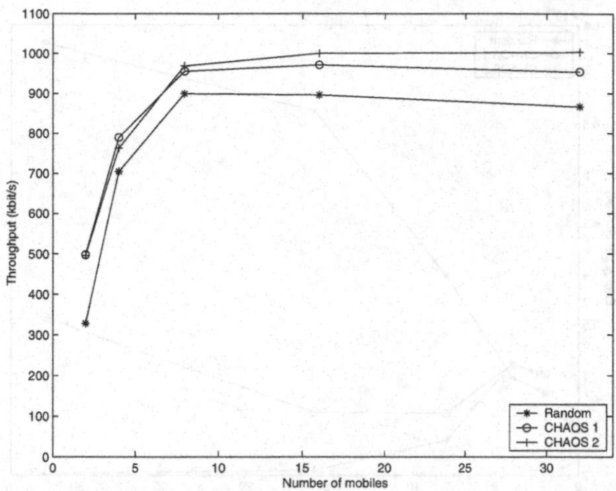

Fig. 6. FTP throughput vs. number of MSs varying matrix scanning methods.

"best channel" requests. This throughput gain is due to more efficient use of the radio interface: the adoption of a channel adaptive scheduling decreases the RLC-PDU error probability, leading to a better exploitation of the radio resource. This in turn reduces the need for packet retransmissions, as can be seen from Table 1. Figure 7 illustrates

Retransmissions	Algorithm	
	Random	CHAOS
0	95.676	99.888
1	2.765	0.1
2	0.98	0.011
3	0.363	0.00099
4	0.128	0.00013
5	0.05	0
≥ 6	0.068	0

Table 1. Percentual distribution of RLC PDU retransmissions

the gain in energy efficiency brought about by the use of a channel-adaptive packet scheduling algorithm. This result can be explained by observing that, with the CHAOS algorithm, mobiles tend to transmit more during the intervals in which they experience a high channel quality, when the transmitted power is reduced by the power control algorithm. Moreover, the decrease in the number of retransmissions also contributes to the increase in efficiency.

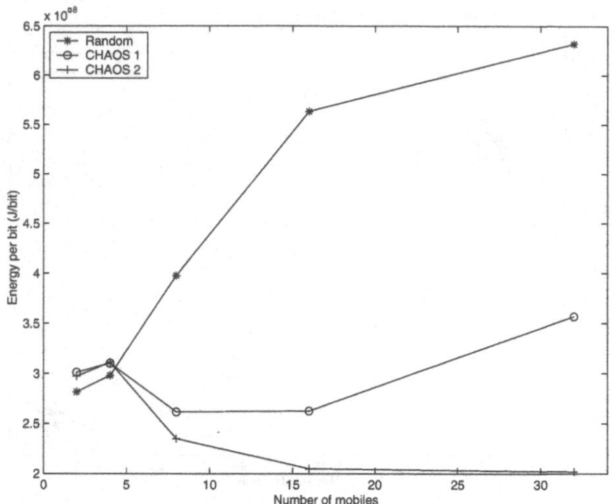

Fig. 7. Energy per bit vs. number of MSs varying matrix scanning methods.

5.2 Bluetooth

The simulation scenario involves a piconet with one master and two slaves. Two different CBR connections are supported in the downlink direction (from master to slave), and transport and network layer protocols are respectively UDP and IP. The channel is modeled as a two-state Markov Chain: in the BAD state the *Packet Error Rate* PER is very high (90%) where as in the GOOD state no errors occur on the channel. Residence time in each state is exponentially distributed with mean residence time equal to 5 seconds in the BAD state and 20 seconds in the GOOD state. This model tries to account for interference from co-located 802.11b devices. Figure 8 shows the throughput of the piconet with respect to the load offered to the piconet, normalized to the capacity of the piconet itself. CHAOS is shown to reach a better throughput with respect to a Deficit Round Robin (DDR) scheduler, especially as the load increases. Figure 9 shows the number of information bits received with respect to information bits transmitted obtained by varying the offered load to the piconet. This value can be interpreted as a measure of power efficiency since for low power devices (CLASS 2) no power control is used. It can be observed that while CHAOS constantly outperforms the DRR scheduler, higher values of power efficiency are obtained when offered load increases. This mainly happens because when much free capacity is available (which happens for lower values of offered load) useless retransmissions occur, which fail because the channel is in the BAD state.

6 Final Remarks and Further Work

In this paper a scheme for developing reconfigurable scheduling disciplines was presented. A modular architectural paradigm was introduced, which allows the deployment

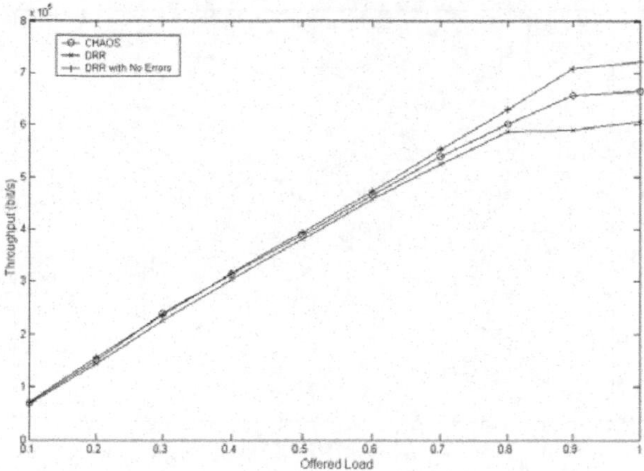

Fig. 8. Throughput vs Offered Load with different scheduling disciplines.

of general algorithms based on simple modules that abstract functionalities common to the wireless systems in use today. We exploited this architecture to develop scheduling algorithms adaptive to both traffic and channel quality; these algorithms are then applied to two different Reference Environments (Bluetooth and UMTS), and they are shown to achieve an efficient utilization of the radio interface. This occurs because the adoption of a cross layered architecture enhances native system functionalities. The adaptation plane algorithms for the two reference environments have been specified in detail. Results for each RE show how scheduling algorithms manage to reap the best from different REs.

References

1. Roveri A., Chiasserini C.F., Femminella M., Melodia T., Morabito G., Rossi M., Tinnirello I.: "The RAMON module: architecture framework and performance results," published in these Proceedings.
2. Haas Z.J.: "Design Methodologies for Adaptive and Multimedia Networks," IEEE Communications Magazine., Vol. 39, no. 11, November 2001.
3. Rappaport T.S., Annamalai A., Buehrer R.M., Tranter W.H.: "Wireless Communications: Past Events and a future Perspective," IEEE Commun. Mag., Vol 39, no. 11, May 2002.
4. Haardt M., Klein A., Kohen R., Oestreich S., Purat M., Sommer V., Ulrich T.: "The TDCDMA Based UTRA TDD Mode," IEEE Journal on Selected Areas in Communications, vol. 18, no. 8, August 2000.
5. Bluetooth Special Interest Group: "Specifications of the Bluetooth System, Volume 1, Core. Version 1.1, February 22 2001.
6. Bharghavan V., Lu S., Nandagopal T.: "Fair Queuing in Wireless Networks: Issues and Approaches," IEEE Personal Communications, vol. 6, no. 1, February 1999.
7. Lu S., Bharghavan V., Srikant R.: "Fair Scheduling in Wireless Packet Networks," IEEE/ACM Trans. on Networking, vol. 7, no. 4, August 1999.

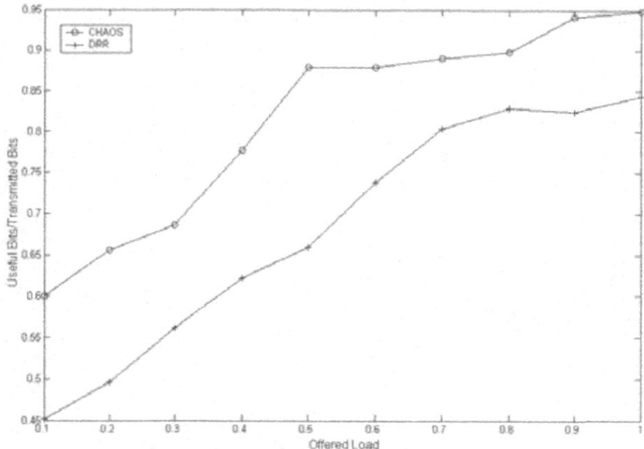

Fig. 9. Received Bits/Transmitted Bits vs Offered Load with different scheduling disciplines.

8. Nandagopal T., Kim T. E., Gao X., Bharghavan V.: "Achieving MAC Layer Fairness in Wireless Packet Networks," Proc. of ACM MOBICOM 2000, August 6-11 2000, Boston, Massachussets.
9. Liu Z., Karol M.J., El Zarki M., Eng K.Y.: "Channel access and interference issues in multicode DS-CDMA wireless packet (ATM) networks," ACM Wireless Networks, 2, 1996.
10. Alonso L., Agust R., Sallent O.: "A Near-Optimum MAC Protocol Based on the Distributed Queueing Random Access Protocol (DQRAP) for a CDMA Mobile Communication System," IEEE Journal on Selected Areas in Communications, vol. 18, no. 9, September 2000.
11. Elaoud M., Ramanathan P.: "Adaptive Allocation of CDMA Resources for Network-level QoS Assurances," Proc. of ACM MOBICOM 2000, August 6-11 2000, Boston, Massachussets.
12. Lee S. J., Lee H. W., Sung D. K.: "Capacity of Single-Code and Multicode DS-CDMA System Accommodating Multiclass Services," IEEE Transactions on Vehicular Technology, vol. 48, no. 7, March 1999.
13. Bhagwat P., Bhattacharya P., Krishna A., Tripathi S.K.: "Enhancing throughput over wireless LANs using channel state dependent packet scheduling," Proc. of IEEE INFOCOM 96, April 2-6 1996, San Francisco, California.
14. Fragouli C., Sivaraman V., Srivastava M.B.: "Controlled multimedia wireless link sharing via enhanced class-based queuing with channel state dependent packet scheduling," Proc. INFOCOM 98, April 1998, San Francisco, California.
15. Baiocchi A., Cuomo F., Martello C.: "Optimizing the radio resource utilization of multiaccess systems with a traffic-transmission quality adaptive packet scheduling," Computer Networks (Elsevier), Vol. 38, n. 2, February 2002.
16. 3GPP Technical Specification TS 25.331, V3.8.0 Release 99, "RRC Protocol Specification," September 2001.
17. Das A., Ghose A., Razdan A., Saran H., and Shorey R.: "Enhancing Performance of Asynchronous Data Traffic over the Bluetooth Wireless Ad-hoc Network,"IEEE INFOCOM 2001, Alaska, USA, April 2001.
18. ns - Network Simulator, UC Berkeley. Internet Site: http://www.isi.edu/nsnam/ns.
19. Bluehoc Simulator Site: http://www-124.ibm.com/developerworks/opensource/bluehoc/.

Mobility Management in a Reconfigurable Environment: The RAMON Approach

Mauro Femminella, Leonardo Piacentini

D.I.E.I., University of Perugia, Via G. Duranti 93, 06125 Perugia – Italy
{femminella, piacentini}@diei.unipg.it

Abstract. Reconfigurability is expected to play a critical role in the research area of Wireless/Mobile Communications, by increasing flexibility, reducing deployment as well as operation and maintenance costs, creating new business opportunities, facilitating enhancements to existing technology, service personalization, etc. On the long way leading to a complete Software Defined Radio terminal, in which a unique programmable hardware is on the fly configured via software download, the RAMON project places itself within a short/medium term time horizon. In this vision, a dual-mode terminal, enhanced with a common software module, is considered, with a massive use of Common Control Functions, limiting the use of system specific functionalities. The focus of the paper is to describe the mobility management developed in this project, providing details for both inter and intra reference environments mobility.

1 Introduction

The deployment of 2.5-3G architectures has given a great impulse to wireless networking. In addition to classical systems, able to provide wireless communications by means of cellular architectures (i.e., GSM, GPRS and, in the next future, UMTS), several technologies are emerging for the provision of wireless communications in a local area, such as 802.11 and HIPERLAN. In this framework, the Bluetooth (BT) technology [4][5] has the potential to provide access to Internet services, by extending the fixed network via the Wireless Personal Area Network (WPAN) paradigm.

The work has been carried out in the framework of a research project, named RAMON (Reconfigurable Access module for MObile computiNg Applications) [2]. One of the aims of the project is to define a reconfigurable radio terminal able to provide access to Internet services in private or public environments by means of the capability to move seamlessly between a wide area wireless network (UMTS) and a WPAN (BT). The role of the BT PAN is to complement the coverage provided by UMTS, for instance at home, or in specific public areas, such as airports, shops or offices, where, for instance, cheaper or gratuitous access could be available. In such a scenario, the emphasis is on the support of mobile computing applications. In this scenario, the terminal does not make use of classical Software Defined Radio concepts [1][3], but a short/medium term solution for reconfigurability is considered, i.e., a dual-mode mobile node, in which a set of Common Control Functions (CCF) is defined, independent from the underlying system, (Fig. 1). The system optimization is carried out trying to maximize the use of CCF rather than specific functionalities characteristic of each Reference Environments (RE).

M. Ajmone Marsan et al. (Eds.): QoS-IP 2003, LNCS 2601, pp. 499-512, 2003.

The focus of this paper is describing the mobility management approach developed in RAMON, where IP as common transport protocol has been chosen for its intrinsic capability of decoupling from underlying technology. In this way, integration and roaming between different access technology is greatly simplified. Moreover, the adoption of all-IP architectures eases the design, development and fruition by end-users of new services, often offered by third parties accessing the network infrastructure. In this scenario, the natural choice has been to adopt Mobile IP (MIP) as common platform to support mobility both inter and intra RE. Since UMTS is a complex and complete system, it has its own specific mobility management procedures [18][19]. However, in the perspective of a migration towards an all-IP architecture [12][13][14][15][16][17][20], we provide a promising scenario in which the IP protocol is used in the UMTS system both in core and access section, managing the mobility according to Mobile IP coupled with some micromobility strategy [11]. On the other side, the BT system does not have the same advanced procedures of the UMTS RE and can not offer the same sophisticated functionalities to manage directly the mobility from the RAMON module with "good" performance. This implies that, if the UMTS RE can provide the needed functionalities to the RAMON module with a very simple Adaptation Layer (AL), i.e., the AdaptU in Fig. 1, the BT RE needs a more complex one (the AdaptB in Fig. 1), in order to fill the gap in terms of functionalities. Here, we propose a mechanism operating at BT link layer to improve the overall performance in case of mobility.

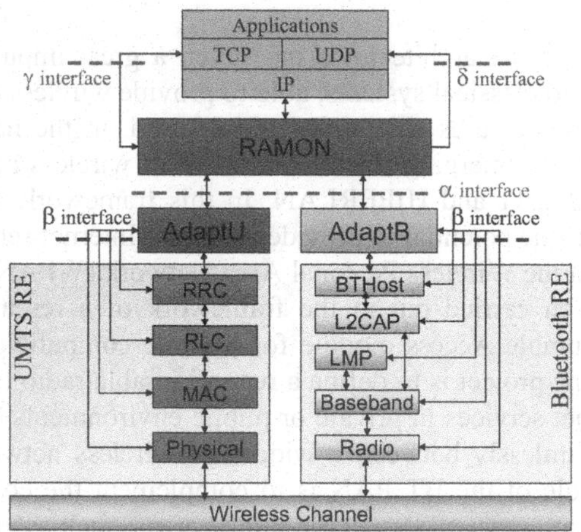

Fig. 1. RAMON protocol stack

The paper is organized as follows. In Section 2, we report the description of the approach used to manage logical mobility in the RAMON logic between different REs. In Section 3, we analyze a possible evolutionary scenario of the UMTS system towards an all-IP network. Section 4 is dedicated to the description of a novel approach to handle mobility at layer 2 in the BT system, in order to improve the delay performance in case of intra-RE handoffs. Finally, Section 5 reports our conclusions and future research directions.

2 The Mobility Management RAMON Logic

RAMON system architecture, illustrated in Fig. 1 is located upon the protocol stack of the different REs of a mobile node. RAMON can communicate with only one RE at time, chosen following a logic based on the QoS parameters configured by the user. The system is essentially divided into two parts:

- a basic layer that contains the logic implementing RAMON functionalities;
- two ALs that are specific for each RE, and present to the RAMON layer a common interface masquerading the differences between the REs.

RAMON and the ALs are sets of control entities that communicate between them (interface α). In addition, they communicate with protocol layers of the underlying REs (interface β), with the TCP agent (interface γ) and MIP agent (interface δ). From an object oriented point of view, RAMON architecture can be shown using an UML diagram as depicted in Fig. 2.

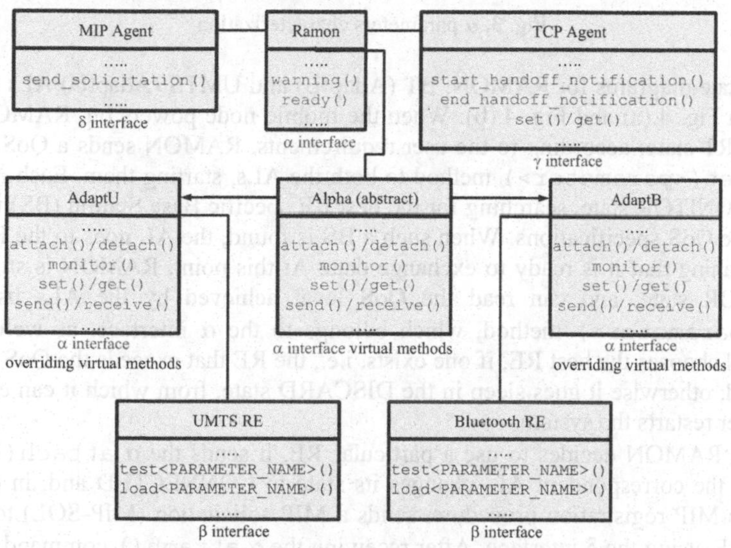

Fig. 2. UML diagram of the RAMON architecture

The β, γ and δ interface are, essentially, sets of public methods, while α is an abstract type containing RE common methods used by RAMON layer (i.e., by the CCF), which are inherited and overridden by the ALs, so becoming RE specific. Besides, the α interface is used for setting or getting the value assumed by several parameters belonging to RE specific protocols, making RAMON able to chose the best RE, to maintain QoS requirements or, if necessary, to perform a vertical handoff [22]. It is worth noting to distinguish RAMON-driven handoff from horizontal handoff, which is performed by AL exploiting RE functionalities, and is transparent to the RAMON logic. In fact, independently from the metric adopted by the RE, the QoS parameters are opportunely manipulated in order to provide an homogeneous scale to RAMON thanks to the α interface. As shown in Fig. 3, the user configures an optimal and a minimal threshold for each α quality parameter. The metric adopted by RAMON to weigh the goodness of each RE is such that the RE global quality reflects that of the worst parameter.

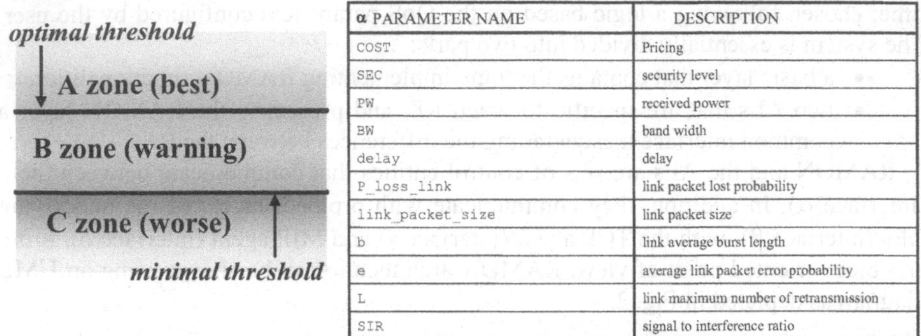

α PARAMETER NAME	DESCRIPTION
COST	Pricing
SEC	security level
PW	received power
BW	band width
delay	delay
P_loss_link	link packet lost probability
link_packet_size	link packet size
b	link average burst length
e	average link packet error probability
L	link maximum number of retransmission
SIR	signal to interference ratio

Fig. 3. α parameters characterization

The state diagrams for RAMON, BT (AdaptB) and UMTS (AdaptU) ALs are presented in Fig. 4.(a) and Fig. 4.(b). When the mobile node powers on, RAMON is in the START state; according to the user requirements, RAMON sends a QoS request by the `set(<parameter>)` method to both the ALs, starting them. Each AL goes to the MONITOR state, searching for the best RE specific Base Station (BS) in range fitting the QoS specifications. When such a BS is found, the AL goes to the READY state meaning that it is ready to exchange data. At this point, RAMON is still in the MONITOR state, and can read the QoS level achieved by the ALs using the `get(<parameter>)` method, which belongs to the α interface, as well as set. RAMON chooses the best RE, if one exists, i.e., the RE that exceeds the QoS optimal threshold; otherwise it goes sleep in the DISCARD state, from which it can exit only if the user restarts the system.

When RAMON decides to use a particular RE, it sends the α `attach()` command to the correspondent AL, changes its state to CONNECTED and, in order to speed up MIP registration procedure, sends a MIP solicitation (MIP-SOL) to the attached RE, using the δ interface. After receiving the α `attach()` command, the AL changes its state to CONNECTED and passes data and signaling packets to RAMON layer; anyway, the application starts sending data packets only after RAMON has intercepted a MIP registration reply (MIP-REG-REPLY) packet coming from the attached RE. At this point, the non connected AL (NC-AL) continuously tries to remain in the READY state exploiting the RE specific functionalities; when it finds a BS exceeding the minimal threshold, it calls the `ready()` method of the α interface from the RAMON side. Now, RAMON can consider the NC-AL READY or not, depending on the quality of the connected AL (C-AL), according to Fig. 5. In particular, if the C-AL is in the A zone, the NC-AL is considered READY if its quality is over the optimal threshold, while if the C-AL is in the B zone, the NC-AL is considered READY if the C-AL supplies a worse quality. Once READY, the quality of the AL is sampled every Tready_ seconds, and, if the above conditions are not satisfied, the AL returns in the MONITOR state looking for a better BS.

Since UMTS and BT have different connecting algorithms, their ALs must perform different procedures, even if identical from the RAMON point of view. So, the AdaptU, when not connected, can call the `ready()` method after a delay needed to find UMTS network [23]. In the BT RE, the AdaptB must wait for the reception of a L2CAP connection confirmation HCI event (`L2CA_ConnectCfm()`) by the below BT Host (BTHost) layer before calling the `ready()` function; this HCI event is the result of successful Inquiry and Paging procedures, that are started by the α interface `monitor()` function, and occur when the AdaptB is in the MONITOR state.

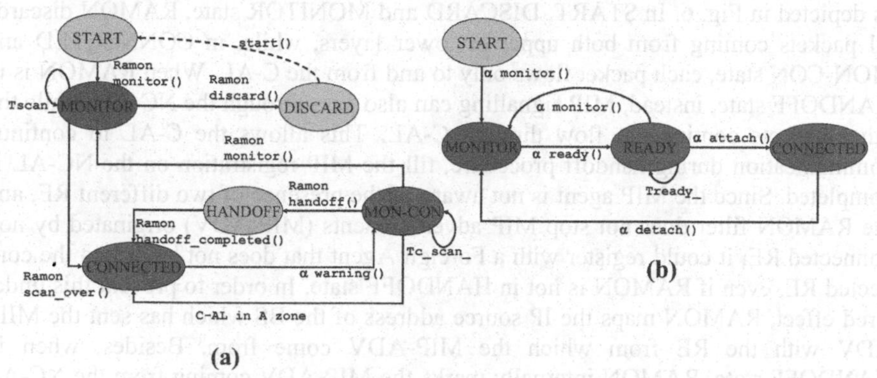

Fig. 4. (a) RAMON Layer state diagram; **(b)** Generic Adaptation Layer state diagram

	CONNECTED	MON-CON	MONITOR	HANDOFF	
C-AL (CONNECTED)	A	B	C	B	C
NC-AL (READY/MONITOR)	A (READY) B (MONITOR) C (MONITOR)	B (READY) if better than C-AL B (MONITOR) C (MONITOR)	C (MONITOR)	A (READY)	A/B (READY)

Fig. 5. Interaction between RAMON and Adaptation Layer state

The C-AL continuously checks the quality parameters of the communication channel, belonging to the RE it is attached to; in case these parameters go under the quality threshold, it notifies this event to RAMON, using the α method `warning()`. Once the `warning()` is received, RAMON changes its state to monitor-connected (MON-CON) and starts looking for a better RE, reading the quality achieved by the NC- AL at Tc_scan_ interval of time, or asynchronously when the AL calls the `ready()` method. As described in Fig. 5, if the non connected AL provides a QoS level greater than the optimal quality threshold, and the C-AL is in B zone, or, if NC-AL and C-AL are, respectively, in B and C zone, RAMON goes in the HANDOFF state, sending a MIP-SOL on the NC-AL, so starting MIP registration, and continues transmitting and receiving data on the C-AL. After receiving a MIP-REG-REPLY from the NC-AL, RAMON calls the α `detach()` method for the C-AL and the α `attach()` method for the NC-AL, thus returning CONNECTED. Of course, if in MON-CON state the quality of the C-AL rises over the threshold again, RAMON returns to CONNECTED state with the same AL at the next quality check. Finally, in case both C-AL and NC-AL go in C zone, and remain there for Tlink_supervision_timer_ seconds, RAMON assumes that a loss of connection has occurred, and the whole system goes in the MONITOR state.

It is worth nothing that, when RAMON is in MONITOR and HANDOFF state, it can freeze TCP transmission window sending TCP agent a γ interface `start_handoff_notification()` command to avoid TCP back-off algorithm and thus throughput starvation.

Moreover, during data transmission, RAMON layer also implements a packet filter as depicted in Fig. 6. In START, DISCARD and MONITOR state, RAMON discards all packets coming from both upper or lower layers, while, in CONNECTED and MON-CON state, each packet flows only to and from the C-AL. When RAMON is in HANDOFF state, instead, MIP signalling can also pass through the NC-AL, while the other packets continue to flow through C-AL. This allows the C-AL to continue communication during handoff procedure, till the MIP registration on the NC-AL is completed. Since the MIP agent is not aware of the presence of two different RE, and the RAMON filter does not stop MIP advertisements (MIP-ADV) originated by non connected RE, it could register with a Foreign Agent that does not belong to the connected RE, even if RAMON is not in HANDOFF state. In order to prevent this undesired effect, RAMON maps the IP source address of the BS which has sent the MIP-ADV with the RE from which the MIP-ADV come from. Besides, when in HANDOFF state, RAMON internally marks the MIP-ADV coming from the NC-AL with an high priority to force MIP agent to register with the BS that has send the advertisement.

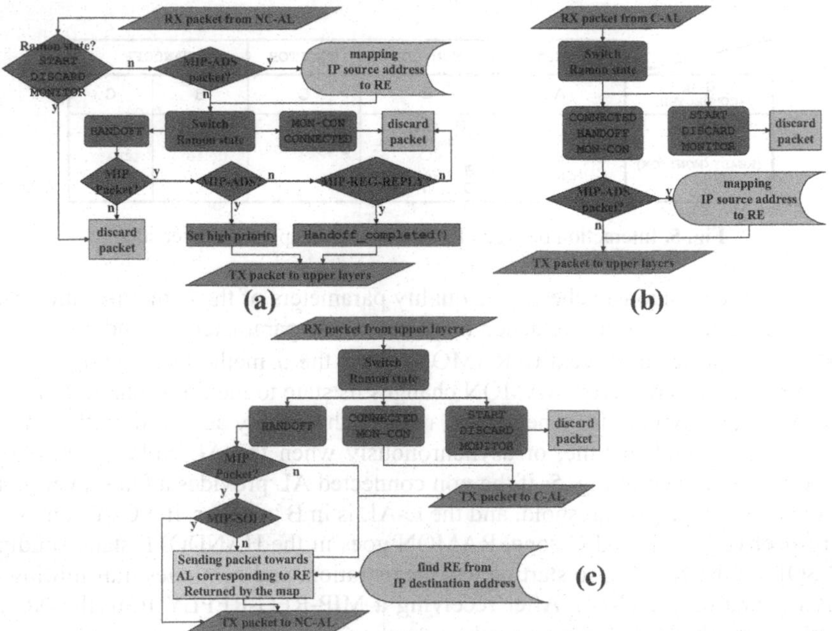

Fig. 6. (a) RAMON filter for packet from NC-AL; **(b)** RAMON filter for packet from C-AL; **(c)** RAMON filter for packet from upper layers

3 The UMTS Mobility Management

In this Section, we provide a description of the foreseen evolution of the UMTS system toward an all-IP network, both in core and access section. In the UMTS network architecture that likely will be firstly deployed, packet and circuit switched technologies will coexist, using IP only between the GPRS Support Nodes (i.e., SGSN and GGSN), to assure connectivity with the Internet world. In this scenario, the legacy GSM bearer network is enriched by GPRS/UMTS packet switched network.

A more advanced scenario is depicted in Fig. 7(a), where the IP protocol is employed not only between GGSN and SGSN, but also in the overall core network, including connections towards Radio Network Controllers (RNCs) (for complete survey of the network entities in Fig. 7(a), the reader is referred to [13][15]). It is worth nothing that the presence of the GERAN (GSM EDGE Radio Access Network) implies that IP is not fully deployed in the access network.

Finally, in Fig. 7(b), the all-IP UMTS Terrestrial Radio Access Network (UTRAN) is sketched. In this model, it is foreseen that RNCs and Node B are connected between an IP backbone, likely based on the DiffServ paradigm [21].

Given the scenario of an all-IP UMTS network, let us focus on the mobility management based on the MIP suite. MIP defines two entities to provide mobility support: a Home Agent (HA) and a Foreign Agent (FA). The mobile host has a permanent IP address registered at the HA, resident in its home network, while it is assigned to a FA on the basis of its current location. The FA has associated with it an IP address called care-of-address (CoA). Packets destined for a mobile host are intercepted by the HA and tunneled, using IP inside IP, to the FA using the CoA. The FA decapsulates the packets and forwards them directly to the mobile host. Therefore, the FA is the IP entity closest to the mobile host, i.e., the router directly attached to the BS (Node B). Improvements has been provided to this basic version of MIP in its implementation in IPv6 [9], that is supposed to be employed in the UMTS network.

However, even if MIP provides a good framework for allowing users to roam outside their home network without severe disruption of the running applications, it was not designed specifically to support wide-area wireless networking. In fact, it treats all handoffs uniformly (also those involving two close FAs), causing each time a new registration with the HA, with the consequent overhead, often implying delay and service disruption. In addition, it does not support paging, resulting in an not efficient power management at the mobile host. To solve these problems, a number of schemes known as micromobility strategies have been proposed in literature (for an excellent survey, see [11]). These approach can be classified into two distinct categories: hierarchical tunnel and routing update approaches. The former ones (e.g., Regional Tunnel Management and Hierarchical Mobile IP) are characterized by a tree-like structure of FAs, in which the traffic is encapsulated and then decapsulated at each FA on the tree until reaching the FA effectively handling the mobile host. In this way, due to the maintenance of location updated within the local network, signaling/delay due to updates at the HA are mostly reduced. The latter ones does not make use of tunnels within the local network, but maintains dedicated field in routing tables to route traffic to the mobile host, using directly its IP home address (Cellular IP, CIP [10]) or the CoA (Hawaii). Extra signaling generated by the mobile host is introduced only in the local network, in order to update routing tables, periodically or after handoffs.

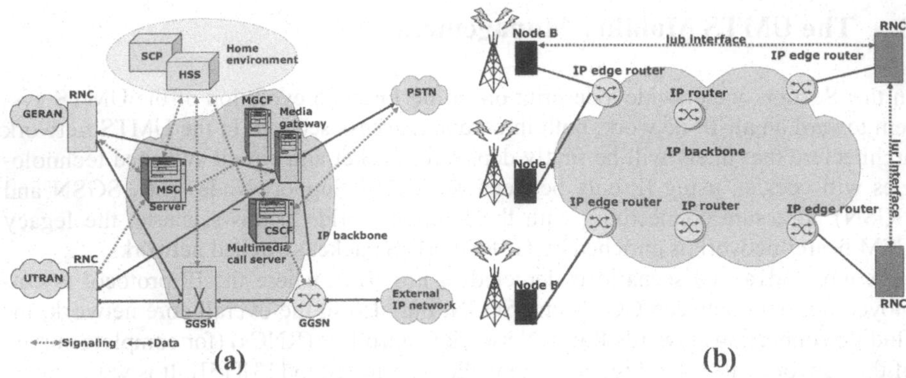

Fig. 7. (a) A simplified all-IP UMTS core network; **(b)** IP-based UMTS access network

It is worth noting that, using this all-IP approach both in core and access network, from a pure mobility management point of view, SGSNs are simply treated as core routers, GGSNs as edge routers, while RNCs (or, more precisely, the routers directly connected to Node B) act as access routers. Thus, the UMTS network can be seen as an IP domain employing an enhanced MIP approach for handling user mobility at layer 3, with a specific, efficient layer 2 technology [23], whose detailed operations are hidden to the RAMON logic by AdaptU.

4 The Bluetooth Mobility Management

The BT architectural model has been enriched by the Bluetooth Public Access IP (BLUEPAC IP) solution, which endows BT with handoff capability and layer 3 mobility management [6][7]. BLUEPAC IP exploits MIP [8], to manage the so-called "macro-mobility", i.e., the mobility among different domains, and CIP ([10][11]) to provide handoffs in a local area. This solution, even if elegant in its definition, presents some disadvantages in case of frequent handoffs, due to the time consuming procedures that a radio device uses to locate and to attach to a new BS. Reiterations of handoffs reduce the overall network performance, since a sort of "black out" of active transmissions occurs for each handoff.

In this Section, we propose an approach for the mobility management based on the intrinsic BT capability to form a number of piconets. In a piconet, a device assumes the role of master, while up to seven devices act as slaves. A device can participate to multiple piconets as slave while it can be master only in one piconet. Multiple piconets can coexist and overlap; a maximum of ten co-located piconets has been suggested, in order to assure a low level of interference. Fixed BT devices named Access Points assure the interworking with fixed networks (e.g., LAN and WAN).

A BT device can belong to different piconets also by keeping some of them in the so-called "low power modes". In such a modality, it does not participate actively (i.e., it does not transfer user data) in all the involved piconets; as a consequence, it consumes low power and bandwidth, but at the same time it is able to maintain synchronization with the masters. Our idea is to use low power modes [4] to maintain a sort of "back-up links" that can be used to speed up the handoffs. With this approach, a device maintains two layer 2 connections opened: one active, in which it sends/receives data, and one in a low power mode. By using low level information, such as field strength measurements, the device monitors the status of its connections and it is able to select the most suitable one, at AL level, performing a connection switch when needed. The time required to execute this operation is significantly reduced, when compared to a BLUEPAC IP handoff. Moreover, the number of times a BT device has to perform the classic BLUEPAC-IP handoff is highly reduced: in fact, this procedure shall be executed only when both the established links are unable to furnish a good level of connectivity.

Let us consider the state machine associated to a BT device (see [5]). In Standby mode, the BT device is waiting to join a piconet. In order to establish new connections, Inquiry and Page procedures are used:

- the Inquiry procedure enables a device to discover which other devices are in its radio range, and the relevant BT addresses and clocks;
- the Paging procedure is used to establish a new connection.

A device that establishes a connection, by means of the Page procedure, becomes automatically the master of the connection; later on, the devices can exchange roles. The paged device must be in scan mode in order to be found, and must answer to the inquiry messages with Frequency Hopping Synchronization (FHS) packets. In order to speed up the discovery process, we propose using a dedicated inquiry access code to find out Access Points (APs), instead of a Generic Inquiry Access Code (GIAC).

By assuming an error free environment and neglecting Synchronous Connection-Oriented (SCO) channels, since we focus on mobile computing applications, inquiry and page procedures last at most 10.24 s and 1.28 s, respectively. The inquiry procedure can be faster, for instance, when the inquiry sub-state is prematurely left, i.e., the BT Link Manager has received a sufficient number of answers from the APs. As regards the *page* procedure, its duration is a random variable uniformly distributed between 0 and 1.28 s, since a BT radio transceivers hops from one channel to another in a pseudo-random fashion. Thus, the expected mean value of the paging delay is 0.64 s.

Before entering in details with our BT layer 2 enhancement proposal, it is necessary to briefly describe the BLUEPAC-IP approach.

BLUEPAC IP is a novel solution proposed by University of Bonn in order to provide mobility support at IP level in public environments based on an infrastructured network architecture. A BLUEPAC network acts as a domain implementing MIP, in which a BT device can move in a transparent way with respect to the extern world by using CIP. In a typical topology of a BLUEPAC IP network, there is one main gateway that shields the internal network from the outer world, a number of internal routers supporting CIP and their leaves, represented by BT BSs. While for a detailed presentation of the BLUEPAC IP we refer to [6], here we are interested in individuating its pro and cons as regards the mobility management in BT, in order to fully justify and understand the solution we propose.

With BLUEPAC IP, the user profile is derived by a modification of the protocol stack of the so-called LAN profile (see [6]). Mobility is managed at IP layer by using the MIP/CIP approach, and at layer 2, by using the Link Supervision Time (LST). The latter approach is based on "hard" decisions, i.e., only when the layer 2 connection is lost and the relevant timer is expired, another connection is searched and set-up by means of an *inquiry* and a *page* procedure. As shown in [7] by means of simulations, the search procedure could be quite long (up to 10.24 s), and its effects at TCP level can be really disruptive for applications (up to 20 s of silence due to the back-off mechanism of TCP).

As a consequence, the handoff performance of this approach may be poor. The authors suggest to use smooth handoff techniques in order to fully exploit the capability of the reliable link layer of BT, able to limit the number of duplicated packets on the wireless link. However, even if smooth handoffs can surely improve BLUEPAC IP performance, our feeling is that layer 2 functionality should be exploited, if the handoff delay has to be appreciably reduced. On the other hand, the approach to use MIP/CIP techniques at layer 3 seems to be the right choice to enable a fully compatibility with the Internet world. In conclusion, it seems that an integrated approach based on both layer 2 and 3 could be the right choice.

4.1 The Layer 2 Handoff Procedure

As described in the previous Section, due to the *inquiry* and *paging* procedures, to the handoff detection delay, and to the TCP back-off algorithm, the overall handoff procedure of BLUEPAC IP may last a long time. Our solution is an integrated one with MIP and CIP that manage macro and micro-mobility, and with suitable enhancements at layer 2 to reduce the overall handoff delay.

We assume a LAN Access Usage Model and multiple BT devices that exploit an AP to connect to a wired CIP network. We also assume that APs have overlapping coverage areas. We recall that the handoff delay at layer 2 results from the sum of these contributions:

- handoff detection delay;
- search procedure delay;
- attach procedure delay.

To reduce the handoff detection delay, we may exploit the following layer 2 connection goodness indicators: Receiver Signal Strength Indicator (RSSI), Class of Device (CoD) measuring the resource utilization on the link, and Link_Quality [4]. However, the last parameter is vendor-dependent, so we neglect it and use only the former two.

In our solution, the BT device uses two layer 2 connections towards two different APs: a primary connection (PC), in active mode, which is actually used to transfer data, and a secondary connection (SC), in low power mode [1]. When the BT device loses the PC, it can switch to the SC, which is likely still available, by simply reactivating it from low power mode. The SC becomes a PC and another SC is searched for. When a new AP is found the BT device set up the SC with such AP; at this point the BT device is a master; thus it changes role performing a master-slave switch, becoming a slave. Obviously, this approach takes much less time (i.e., 2 ms) than looking for another AP, through inquiry and paging procedures.

[1] It is worth noting that the AdaptB hides this behavior to the RAMON layer, which sees only the PC. The term "connection" means that the terminal is attached to the WPAN, but RAMON and AdaptB are not necessarily in the state CONNECTED, as meant in Fig. 4.

In addition, to further decrease handoff delay, the BT device may transmit inquiry messages to APs in range. Such APs may answer with an FHS packet containing information about their hopping sequence, their clock and their load status. The BT device could store this information in a database (DB), in order to choose the best AP in its range (i.e., the one with the smaller value in the CoD field). This way of operation implies that successive handoffs need only a paging procedure, instead of both inquiry and paging ones (to connect to an AP whose data are in the DB and which is not actually supporting connections with the BT device). The DB information is handled via soft states, i.e., it must be periodically refreshed via an inquiry procedure. Note that, by using the CoD field, a sort of load balancing can be attained.

4.2 State Diagram

The state diagram in Fig. 8 describes the actions performed by a BT device, related to a handoff procedure. In the following, we will describe in details only the main state transitions. Once in a WPAN coverage area, the BT device can connect to an AP. If it has no data to transfer, it can sleep in Park mode, refreshing the CIP paging cache every T_B seconds. Otherwise, to connect to an AP, the BT device uses an inquiry and paging procedure. The address, clock and CoD information about discovered APs, received through FHS packets during inquiry procedure, are stored in a DB for future use. As described before, using the DB, the BT device need only the page procedure to connect to selected APs. If the BT device needs to transmit data, or receive a connection request, it can activate its parked connection or decide to connect to another AP.

Let us define the a mono-Connected state as a state in which the BT device is connected to a single AP only. Similarly a bi-Connected state is relevant to a BT device connected with two APs and a three-Connected state identifies a BT device with three simultaneous connections. To enter the mono-Connected state, the BT device has to perform the following actions: 1) ACL link set-up, 2) L2CAP connection oriented channel set-up and configuration, 3) CIP routing cache (CIP-RC) update, 4) MIP registration with the foreign agent. Of course, actions 3) and 4) occur only if the AdaptB is attempting to be connected. Once in the mono-Connected state, the BT device decides if using just one connection, thus managing mobility as in the BLUEPAC IP solution, or connecting to another AP, creating a SC.

In the bi-Connected state, the SC is set in Sniff mode, and the BT device measures the RSSI (RSSI_2) of this connection with a period of T_{sniff}. The PC monitors its received power too (RSSI_1), and, if RSSI_1<RSSI_2, then a switch between the two connections occurs by means of the soft Handoff state. This procedure guarantees that the PC has always a RSSI greater than the SC. The BT device perform this control only when the SC goes under the Golden Receive Power Range. If this happens, the BT device has to pass through a three-Connected state, with the PC and SC in Hold mode, in order to enable the paging procedure (the inquiry procedure has to be performed too, if the information in the DB is not refreshed and thus is not reliable anymore). Then, the BT device looks for an AP with a better RSSI, i.e., in the Golden Receive Power Range. If the BT device finds such an AP, it discards the previous SC and returns in the bi-Connected state passing through the soft Handoff state. If the BT device does not find such an AP it remains in the current state. Of course, if the

RSSI_1<0, the PC is changed in two steps: 1) the SC replaces the PC; 2) an alternative SC is searched for. Finally, if the BT device suddenly loses its PC, it has to activate the SC and return in the mono-Connected state passing through the hard Handoff state.

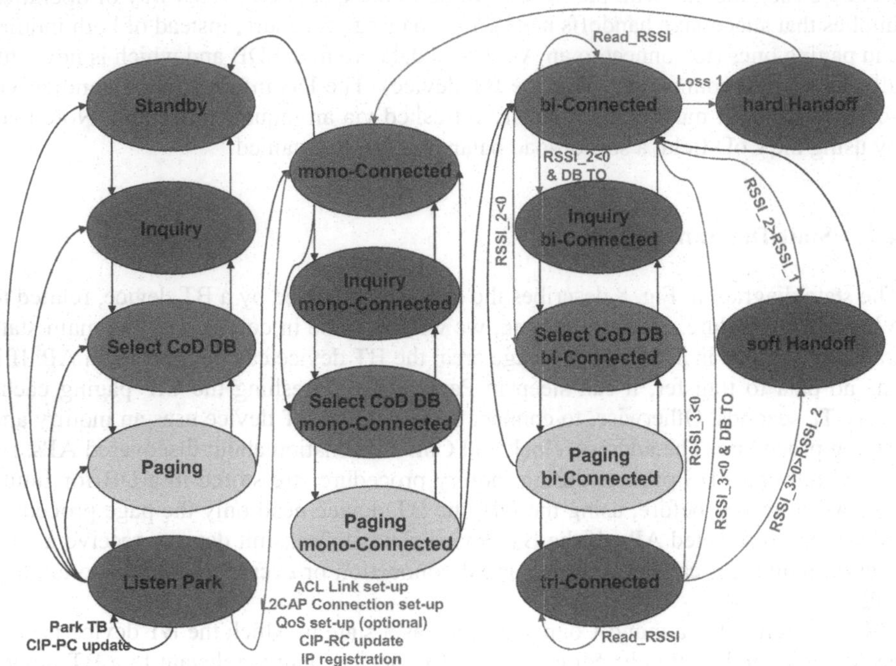

Fig. 8. State diagram of the BT mobility management procedure

4.3 Performance Analysis

The performance of our solution is strictly related to the values of the following parameters: T_{sniff} ,T_B, holdTO, the DB refresh period, the RSSI reading period, paging and inquiry periods. Of importance are also the topology of the network and how the BT device moves. A detailed study would be necessary to find optimal values for the above mentioned parameters and then to evaluate in details the performance of our approach. However, some preliminary considerations lead to a rough idea of the achievable performance limits:

- in the case of mono-Connected state, the performances is the same of BLUEPAC IP (hard Handoff);
- in the bi-Connected state:
  - in the best case, performing a handoff implies just a re-activation of the SC, which requires about 2 ms;
  - if the BT device has to page for another AP whose information is stored in its DB, the handoff delay is in average 0.64 s;

– in the worst case, the BT device has to perform an inquiry procedure; in this case, its performance is similar to the BLUEPAC one.

It is worth noting that the situations listed above occur with different probabilities, which are a function of the previous mentioned parameters. We stress also that the above possible values of handoff delays result from rough estimations, which do not take into account some overheads introduced by the procedure, like the impact of refreshing the DB, or the periodical reading of RSSI. Besides, we must underline that a piconet can contain at most 7 slaves, thus, using more than one connection for each BT device may result in a resource waste. To overcome the limit of 7 slaves in a piconet, it is possible to use Park instead of Sniff mode to support the SC. However, this way of operation seems to be implementation dependent. Finally, it has to be investigated the possibility that the increased workload on APs could make them the network bottlenecks, jeopardizing the benefits introduced by the proposed mechanism.

5 Conclusions

In this paper, we have investigated the mobility management in a reconfigurable system, based on a dual mode terminal (UMTS + BT), enhanced with a software logic (RAMON). We have described the mobility management approach in the new module, in charge of seamlessly switching from a RE to another, depending on the value of a given metric. Moreover, we have investigated the mobility management inside each RE. In particular, a typical evolutionary scenario for an all-IP UMTS architecture (core and access networks) has been provided, where terminal mobility is managed according to classical micromobility proposal. For what concerns BT, we have proposed to use a solution at IP layer, named BLUEPAC-IP, developed by the University of Bonn. In addition, we have enhanced such approach with a novel technique at link layer in order to decrease handoff latency, using the so called low power modes at the expense of an increased resource consumption.

Future research directions foresee the development of a simulator, in order to evaluate in details system performance. In addition, the RAMON logic can be naturally extended to support more than two REs. Moreover, the metric to compare RE quality may be implemented exploiting a unique cost function able to weigh all the RAMON α parameters with a set of coefficients configured by the user. Finally, another upgrade to RAMON can be the insertion of an hysteresis in the handoff algorithm, in order to prevent ping-pong effect; this can be made i) using a timer mechanism, ii) enhancing quality thresholds by extra values.

Acknowledgments

The authors would like to thank all their colleagues of the RAMON project.

References

1. J. Mitola III, ed., Special Issue on Software Radio, IEEE Communications Magazine, May 1995.
2. RAMON web page, http://conan.diei.unipg.it/RAMON.
3. J. M. Pereira, "Re-Defining Software (Defined) Radio: Re-Configurable Radio Systems and Networks," IEICE Transaction on Communications, vol. E83-B, N. 6, June 2000.
4. Bluetooth SIG, "Specification of the Bluetooth System–Version 1.1," Specification Volume I & II, February 2001.
5. J. Haartsen, M. Naghshineh, J. Inouye, "Bluetooth: Vision, Goals, and Architecture," Mobile Computing and Communications Review, ACM SIGMOBILE, vol. 1, N. 2.
6. M. Albrecht, M. Frank, P. Martini, M. Schetelig, A. Vilavaara, A. Wenzel, "IP Services over Bluetooth: Leading the Way to a New Mobility," LCN'99, Lowell, MA, October 1999.
7. S. Baatz, M. Frank, R. Gopffarth, D. Kassatkine, P. Martini, M. Schetelig, A. Vilavaara, "Handover Support for Mobility with IP over Bluetooth," LCN'00, Tampa, FL, November 2000.
8. C. Perkins, "IP Mobility Support," RFC 2002, October 1996.
9. D. B. Johnson, C. Perkins, "Mobility Support in IPv6," Internet Draft, draft-ietf-mobileip-ipv6-18.txt, June 2002, work in progress.
10. A. G. Valko, "Cellular IP - A New Approach to Internet Host Mobility," ACM Computer Communication Review, January 1999.
11. A.T. Campbell, J. Gomez, K. Sanghyo, C. Wan, Z. R. Turanyi, A. G. Valko, "Comparison of IP micromobility protocols," IEEE Wireless Communications, February 2002.
12. P. J. McCann, T. Hiller, "An Internet Infrastructure for Cellular CDMA Networks Using Mobile IP," IEEE Personal Communications Magazine, August 2000.
13. J. De Vriendt, P. Lainé, C. Lerounge, X. Xu, "Mobile Network Evolution: a Revolution on the Move," IEEE Communications Magazine, April 2002.
14. G. Patel, S. Dennett, "The 3GPP and 3GPP2 Movements Towards an All-IP Mobile Network," IEEE Personal Communications Magazine, August 2000.
15. L. Bos, S. Leroy, "Toward an All-IP-Based UMTS System Architecture," IEEE Network Magazine, January/February 2001.
16. R. Ramjee, T. F. La Porta, L. Salgarelli, S. Thuel, K. Varadhan, "IP-Based Access Network Infrastructure for Next-Generation Wireless Data Networks," IEEE Personal Communications Magazine, August 2000.
17. F. M. Chiussi, D. A. Khotmsky, S. Krishnan, "Mobility Management in Third-Generation All-IP Networks," IEEE Communications Magazine, September 2002.
18. 3GPP TS 23.002, "Network Architecture," rel. 5, v. 5.7.0, June 2002, http://www.3gpp.org
19. 3GPP TR 23.922, "Architecture for an All-IP Network," rel. 4, March 2001, http://www.3gpp.org.
20. D. R. Wisely, "The Challenges of an all IP Fixed and Mobile Telecommunication Network," IEEE PIMRC'00, London, UK, September 2000.
21. S. Blake, D. Black, M. Carlson, E. Davies, Z. Wang, W. Weiss, "An architecture for Differentiated Services," IETF RFC 2475.
22. Mark Stemm, Randy H.Katz, "Vertical handoffs in wireless overlay networks", ACM Mobile Networks and Applications", Volume 3, 1998, pp. 335-350.
23. Gessner, C.; Kohn, R.; Schniedenharn, J.; Sitte, A. "Layer 2 and layer 3 of UTRA-TDD",. 51st IEEE VTC 2000-Spring, Tokyo, Japan, 2000, Volume 2, pp. 1181-1185.

Improving End-to-End Performance in Reconfigurable Networks through Dynamic Setting of TCP Parameters

Giacomo Morabito[1], Sergio Palazzo[1], Michele Rossi[2], and Michele Zorzi[2]*

[1] Univeristá degli studi di Catania, Viale Andrea Doria n.6, I-95125, Italy
{gmorabi, spalazzo}@diit.unict.it

[2] Univeristá degli studi di Ferrara, via Saragat n.1, I-44100, Italy
{mrossi, mzorzi}@ing.unife.it

Abstract. In the next future, several wireless technologies will coexist and users will require to gain wireless access at any time through the most convenient technology using the same terminal. As a result, terminals will make handover between different access solutions, which we call inter-Reference Environment (RE) handovers. This involves the concept of reconfigurability: terminals must reconfigure their internal parameters to adapt their behavior to the new RE. However, in this scenarios there are more severe problems for TCP-based data services where compared to homogeneous wireless systems. In fact, the low Mobile IP performance causes long periods of terminal unreachability. Besides, when the reference environment changes the end-to-end path changes as well and what TCP learned in terms of round trip time and available bandwidth is not valid anymore. In this paper solutions to improve TCP performance in reconfigurable networks are introduced. The proposed solutions are based on appropriate setting of the maximum segment size (MSS) and modifications to TCP algorithm which are triggered on the following of an inter-RE handover. Performance results show that the proposed algorithms dramatically increase TCP performance.

1 Introduction

A mobile wireless terminal (MT) is typically affected by signal degradations, due to propagation phenomena. Moreover in Cellular Networks, due to inter-cell handovers, the user could be temporarily disconnected from the core network for non negligible periods of time. All these effects are misinterpreted by the Transmission Control Protocol (TCP) that reacts as a network congestion occurs. Accordingly, it updates its internal variables in order to reduce the amount of data to be put into the network. However, this is not the correct way to address the problem, because disconnections and signal degradations, in some particular cases, could be so substantial to cause the TCP to time out and to reduce its congestion window so that, once the link is restored a long period of

* This work has been supported by the RAMON project.

M. Ajmone Marsan et al. (Eds.): QoS-IP 2003, LNCS 2601, pp. 513–524, 2003.

time would be needed for TCP to re-set its internal variables to the correct value and entirely use the available bandwidth [1][2][3][4]. In the next future, several wireless technologies will coexist and users will require to gain wireless access at any time through the most convenient technology using the same terminal. This means that terminals will make handover between different access solutions, which we call inter-Reference Environment (RE) handovers. This involves the concept of reconfigurability: terminals must reconfigure their parameters to adapt their behavior to the new RE.

In this scenarios there are more severe problems for TCP-based data services where compared to homogeneous wireless systems. In fact, the MT could suffer from additional delays arising from mobile IP, and from the time needed to attach the MT to the new network, where with the term *attach* we mean the interaction between user and new network in order to allocate resources according to user requirements. Moreover, given that the connection path may completely change after an Inter-RE handover, what the TCP learned about the value of the round trip time and the available bandwidth is not valid anymore, which leads to further performance degradation [5].

Also note that at the MT the TCP interacts with lower layers, and in particular with the link layer (where low-level retransmissions are performed in order to reduce the residual IP packet error probability). For this reason, the size of each IP segment (MSS plus TCP and IP headers) must be chosen with care because, as will be shown in the following, it is directly related to the maximum achievable throughput. Hence, the MSS must be chosen accounting for both the channel state and the value of the link layer configuration parameters.

In this paper we introduce two new algorithms. The first one is aimed at finding the MSS value that leads to the maximum throughput, whereas the second algorithm has been introduced to handle temporary link failures and disconnections due to inter-RE handovers and to avoid wasting resources due to the erroneous settings, in such cases, of TCP internal variables.

The rest of the paper is organized as follows. In Sections 2 we report the algorithm used to choose the MSS leading to a maximization of the TCP throughput, in Section 3 the second algorithm is reported, i.e., the one used to handle link failure and disconnections. Finally, in Section 4 some conclusions are given.

2 Setting the Maximum Segment Size (MSS)

In order to present our algorithm and to prove its effectiveness, in this Section, we take as a reference scenario a third generation cellular system using W-CDMA as the channel access technique. We consider a single user that is transmitting data over a dedicated channel (DCH). The link layer is configured in Acknowledgment Mode (AM), i.e., retransmission of link layer packets (PDUs) is enabled. At the MT transport layer, TCP New-Reno [6] is used and the data traffic is considered continuous as in along FTP file transfer. TCP performance are plotted against the Frame Error Rate (FER), that is the frame error process seen after coding and interleaving.

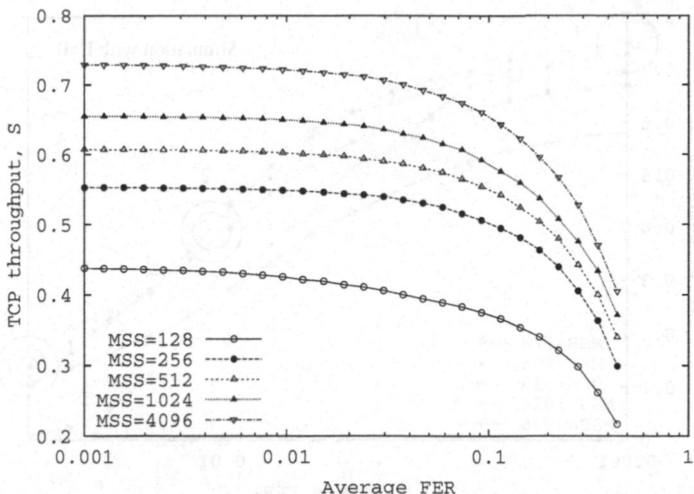

Fig. 1. TCP throughput as a function of the FER by varying MSS with $L = 10$.

In Figs 1-2, we report the TCP throughput considering a long FTP file transfer[1] as a function of the FER by varying the IP MSS and the maximum number of retransmissions performed at the link layer for each link layer PDU (L). From the observation of these figures, it appears evident that the choice of MSS depends on the value of L. In particular, when L is large (Fig. 1, $L = 10$), the curve with the maximum performance is the one where MSS is set at the maximum permitted value. In this case, in fact, the link layer is able to almost fully compensate the errors introduced by the wireless channel and the residual error probability is very low. In such a case, the best choice for MSS is the one that minimizes the amount of protocol overhead and so the one where the size of the IP payload is large with respect to the size of IP and TCP headers. The situation is reverted when the link layer does not perform retransmissions (Fig. 2, $L = 0$), or L is set to a small value. This is not uncommon and is justified, for example, by user requirements such as the need for a short interactive time. In this paper we consider that the choice of L is handled by algorithms placed at the MT based on delay and energy constraints. Furthermore, we assume that this value is

In the following we report the algorithm used in order to choose the MSS leading to the maximization of the TCP throughput, depending on both channel state and L. We consider a slotted time axis, where each slot corresponds to the time needed to transmit one full link layer PDU. The channel evolution is tracked by means of a Discrete Time two-state Markov Chain (DTMC). This simple model is characterized by two states, the Good state (0) and the Bad (1) state, and evolves slot-by-slot according to its transition probabilities. In

[1] In practice this is the throughput value reached at steady-state.

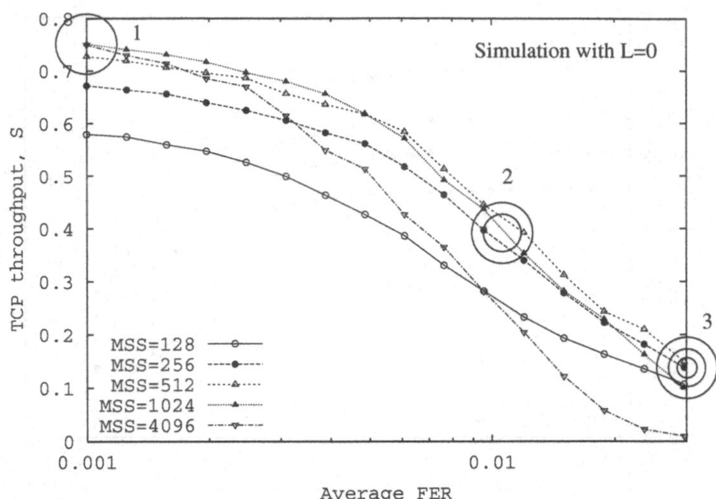

Fig. 2. TCP throughput as a function of the FER by varying MSS with $L = 0$.

particular, let k be the slot index and let $S_k \in \{0, 1\}$ be the channel state in slot k. The transition probabilities from slot k to slot $k + 1$ are then expressed as $p_{ij} = \text{Prob}\{S_{k+1} = j|S_k = i\}$, $i, j \in \{0, 1\}$. Note that, once the transition probabilities have been estimated, the steady-state channel error probability (ε) and the average length of a burst of erroneous PDUs (b) can be computed as $\varepsilon = p_{01}/(p_{01} + p_{10})$ and $b = 1/p_{10}$, respectively. Moreover, we consider that the number of slots elapsed between the beginning of the transmission of a given PDU and the reception of the corresponding ACK is equal to m slots, where m is an integer number. We refer to m as the link layer round trip delay.

In the next we present a simple algorithm to choose the MSS value leading to the maximum throughput for a TCP running over a link layer in presence of wireless channel errors. We first model two opposite effects: the overhead introduced by protocol headers (OH) and the residual TCP packet error probability (P_{res}). The normalized overhead (OH) per transmitted TCP packet can be written as follows:

$$OH = \frac{TCP_h + nPDU_h}{nPDU_l} \tag{1}$$

where $PDU_l = PDU_h + PDU_p$, PDU_h and PDU_p represent the length (in bits) of header and payload of one link layer PDU, respectively. With TCP_h we indicate the sum of IP and TCP headers occupancy. Moreover, n is the number of PDUs composing a full IP packet, $n = \lceil TCP_l/PDU_l \rceil$, where TCP_l is the total length of one TCP packet (including TCP and IP headers).

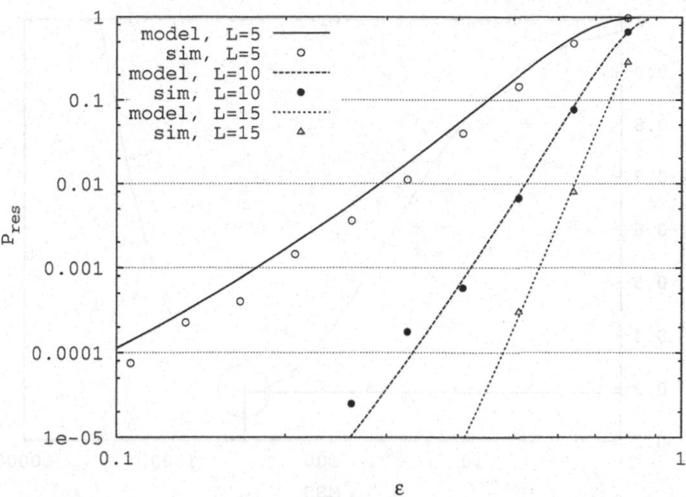

Fig. 3. P_{res}: comparison between analysis and simulation with $b = 5$, $m = 10$.

The residual IP packet error probability $P_{res} = P_{res}(\varepsilon, b, m, n, L)$, instead, can be modeled by means of the following formula:

$$P_{res}(\varepsilon, b, m, n, L) = 1 - \left[1 - \varepsilon p_{11}(m)^L\right]^n \quad (2)$$

where $p_{11}(m)$ is the DTMC m-step transition probability, i.e., the probability that $S_{k+m} = 1$ given that S_k was equal to 1. This probability is easily derived by the m-th power of the transition probability matrix [7]. Using the matrix function theory in [8], $p_{11}(m)$ can be easily written as:

$$\begin{cases} p_{11}(m) = 1 - \beta + \beta p_{11} \\ \text{where} \quad \beta = \dfrac{1 - [(p_{00} + p_{11}) - 1]^m}{2 - (p_{00} + p_{11})} \end{cases} \quad (3)$$

Moreover, $\varepsilon p_{11}(m)$ represents the mean probability that a given PDU is discarded by the LL, i.e., that both its first transmission and the following L retransmission attempts are in error. In Eq. (2), we adopt the following *independence* assumption: *each PDU composing a given TCP packet sees a correlated channel*, i.e., errors for that PDU are correctly tracked by means of the channel Markov model; *whereas the discarding processes of PDUs composing the same TCP packet are uncorrelated.* As an example, in Fig. 3 we compare P_{res} obtained by simulation against the one derived from our approximate analysis. In general, the approximation obtained from Eq. (2) is good as far as $b \leq m/2$.

In order to make an optimal choice of MSS, taking into account both the protocol overhead (OH) and the residual packet error probability (P_{res}), we define the following cost function:

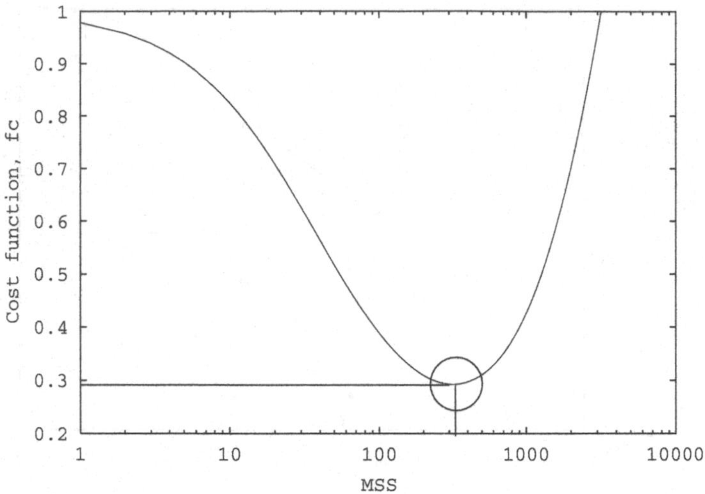

Fig. 4. f_c as a function of MSS for $\varepsilon = 0.1$, $b = 3$, $L = 0$ and $m = 10$.

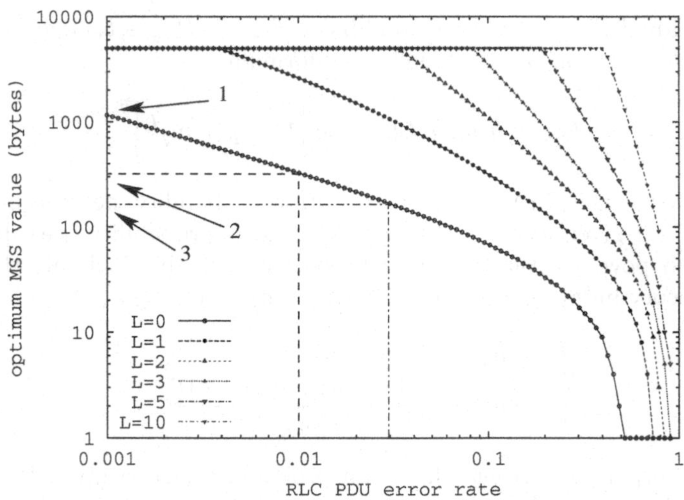

Fig. 5. Optimum MSS value as a function of ε by varying L for $b = 3$, $m = 40$.

$$f_c = w_1 P_{res} + w_2 OH \tag{4}$$

where the weights w_1 and w_2 are given by $w_1 = 1 + P_{res}$ and $w_2 = 1 - P_{res}$. This is justified by the fact that, when P_{res} is at a high value, more emphasis must be given on the first term, i.e., the TCP protocol performance, in this case, heavily depends on the residual packet error probability. On the contrary, when P_{res} is low, the TCP throughput is lightly affected by P_{res} and the most important factor to be minimized is the protocol overhead. To illustrate the behavior of f_C, in Fig. 4 we report the cost function as a function of MSS considering $\varepsilon = 0.1$, $b = 3$, $L = 0$ and $m = 10$. The MSS value to be used is the one which corresponds to the minimum of f_C.

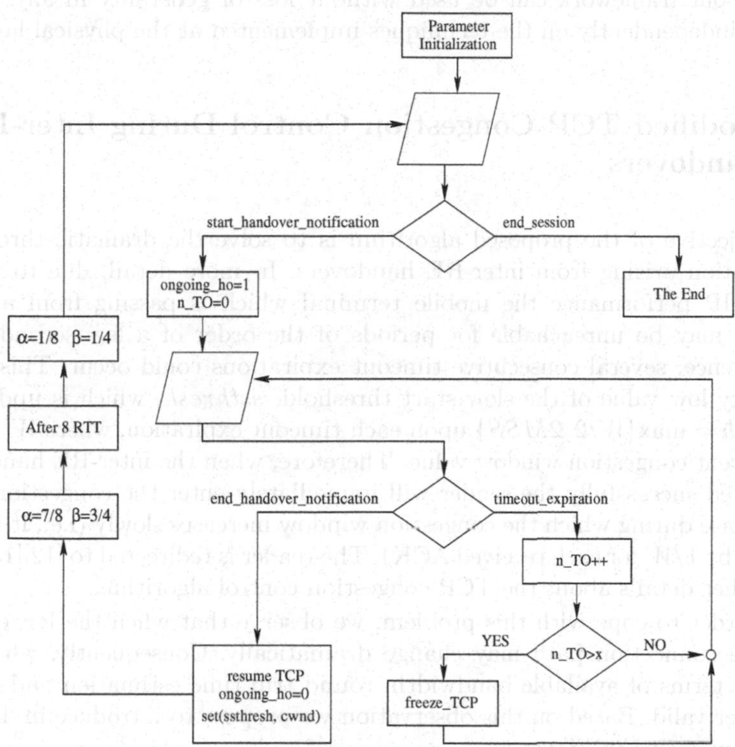

Fig. 6. Flow Chart of the Modified TCP Congestion Control algorithm.

In Fig. 5, we report the optimal MSS value as a function of ε, by varying L. This figure has been derived considering the same set of parameters (round trip delay, LL PDU length and TCP/IP header length) used to obtain Fig. 2. Note that the optimal MSS values indicated in Fig. 5 for the case $L = 0$ (labelled as 1, 2 and 3), correspond to the maximum throughput cases in Fig. 2. Note also that, in this particular case, the FER in Fig. 2 is well approximated by the PDU error rate (ε). This due to the fact that, for this particular user, the channel is characterized by an on/off behavior[2], in which whenever the frame contains errors, all the PDUs contained in the frame are tipically lost. Hence, in

[2] The selected user is characterized by a slow-fading (doppler frequency, $f_d = 6$ Hz). Therefore, off periods, that are due to user interference and fading, are characterized by a high correlation.

this case, the probability that a PDU get corrupted is well approximated by the probability that the frame contains errors.

Note that in our algorithm the channel state is tracked directly at the link layer (at the packet level), i.e., channel coding and interleaving processes are already included in our channel estimate (we not need to model them). In conclusion, our framework can be used without loss of generality in any wireless system independently on the techniques implemented at the physical layer.

3 Modified TCP Congestion Control During Inter-RE Handovers

The objective of the proposed algorithm is to solve the dramatic throughput degradation arising from inter-RE handovers. In more detail, due to the low Mobile IP performance the mobile terminal which is passing from a RE to another may be unreachable for periods of the order of a few seconds. As a consequence, several consecutive timeout expirations could occur. This results in a very low value of the slow start threshold, $ssthresh$, which is updated to $ssthresh = \max\{W/2, 2MSS\}$ upon each timeout expiration, where W denotes the current congestion window value. Therefore, when the inter-RE handover is completed successfully the sender will immediately enter the congestion avoidance phase during which the congestion window increases slowly (i.e., it is incremented by $1/W$ for each received ACK). The reader is redirected to [12][13][14][6] for further details about the TCP congestion control algorithm.

In order to cope with this problem, we observe that when the RE changes, also the connection path may change dramatically. Consequently, what TCP learn in terms of available bandwidth, round trip time estimation and RTO is no longer valid. Based on this observation we propose to introduce in TCP the algorithm shown in Figure 6.

The major functionalities of the new algorithm are triggered after an inter-RE handover. In more detail, when TCP receives notification of the beginning of a handover ($start\_handover\_notification$ message), a flag is set ($ongoing\_ho = 1$), and a counter is initialized ($n_{TO} = 0$). Every time a timeout expires, n_{TO} is increased by one. If n_{TO} is higher than a threshold, x (note that this is true in case of a long handover duration), TCP is frozen, i.e., the TCP sender cannot transmit any data. When a notification that the handover is completed arrives ($end\_handover\_notification$ message), the algorithm sets the new values of the TCP parameters and resets the values of RTT and RTO as explained in more detail below:

1. If TCP was frozen, then it is resumed, i.e., the sender can transmit segments again;
2. The flag is set to zero, ($ongoing\_ho = 0$);
3. $ssthresh$ is set to the original value it had before the beginning of the handover phase;

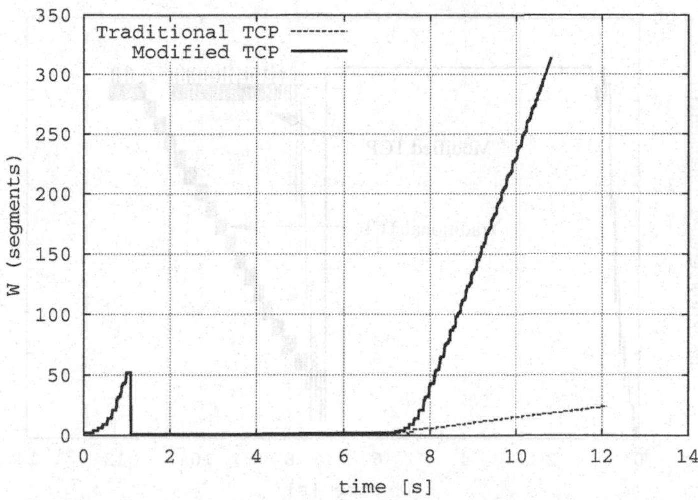

Fig. 7. Congestion window *Vs.* time, comparison between Traditional and Modified TCP.

4. W is set to MSS (one segment) such as in standard Slow Start;
5. A new RTT measurement R' is made.

In order to take into account that the connection path may have significantly changed, and accordingly the RTT, the TCP sender sets the values of α and β as $\alpha = 7/8$ and $\beta = 3/4$. In this way, more importance to the new values of RTT measurements is given, referred to the actual RE. The new values of α and β are maintained for a time equal to $8RTT$, then the algorithm sets the parameters to the original values ($\alpha = 1/8$ and $\beta = 1/4$). Note that α and β are the parameters used as weights of the new measurements in the RTT and RTO estimation:

$$\begin{cases} RTT = (1-\alpha)RTT_{old} + \alpha R' \\ RTTVAR = (1-\beta)RTTVAR_{old} + \beta|RTT - R'| \\ RTO = RTT + \max\{G, K \times RTTVAR\} \end{cases} \tag{5}$$

where R' is the last round trip time measurement, RTT is the current round trip time estimate, G is the clock granularity and K is a constant value usually set to 4 ($K = 4$).

The modified TCP algorithm has been implemented in the core of Linux operating system [15] and extensively studied in an emulated test-bed.

In Figures 7, 8 and 9 we report versus time the value of the congestion window, W, the number of unacknowledged data segments, and the number of segment correctly received, *acked*, respectively. In the above figures we have assumed that the inter-RE handover begun at time 1.7 sec and is completed at time 7.2 sec.

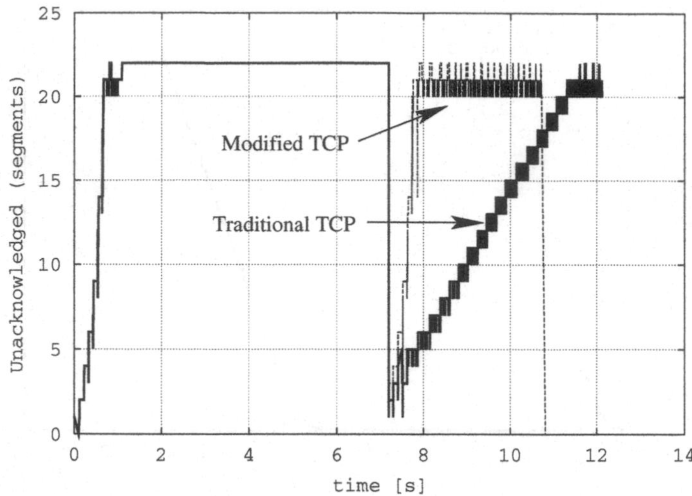

Fig. 8. Unacknowledged data segments *Vs.* time, comparison between Traditional and Modified TCP.

When the inter-RE handover occurs the congestion window W is equal to 50. Then the TCP timeout expires and the congestion window goes to 1. The inter-RE handover is completed at time 7.2. Since the value of the threshold is set to its initial value, which is infinity, the TCP sender enters the Slow Start phase and therefore, the congestion window, W, increases exponentially[3]. In Figure 8 we observe that between time 1.7 and 7.2 the number of unacknowledged segments is 21 which is the value set by the receiver window *awnd*. This is because during the handover the terminal is unreachable and therefore, no new ACKs arrive.

Similar behaviors have been obtained in all considered cases. For the sake of comparison, in the same figures we report the curves obtained by using traditional TCP implementations, i.e., TCP with FACK and SACK options. In Figure 7, we observe that after the inter-RE handover is successfully completed, the congestion window, W, for the traditional TCP increases much more slowly than in case of modified TCP. This is reflected in Figure 9 where we observe that modified TCP can deliver segments more rapidly than traditional TCP. In fact, the 1 Mbps file transfer considered in this case is completed by modified TCP in 10.9 secs and by traditional TCP in 12.2 sec.

Note that this last algorithm, as is, can be effectively used only when the MT is the sender (data packets are transmitted along the up-link connection, whereas on the down-link only TCP ACKs flow). In the down-link case, i.e.,

[3] Note that in real TCP implementations, there are no limitations for the congestion window, W. The bound on the transmission rate is given by the receiver window, *awnd*. In fact, a new segment can be transmitted if the number of unacknowledged segments is lower than $\min\{W, awnd\}$ as shown in Figure 8.

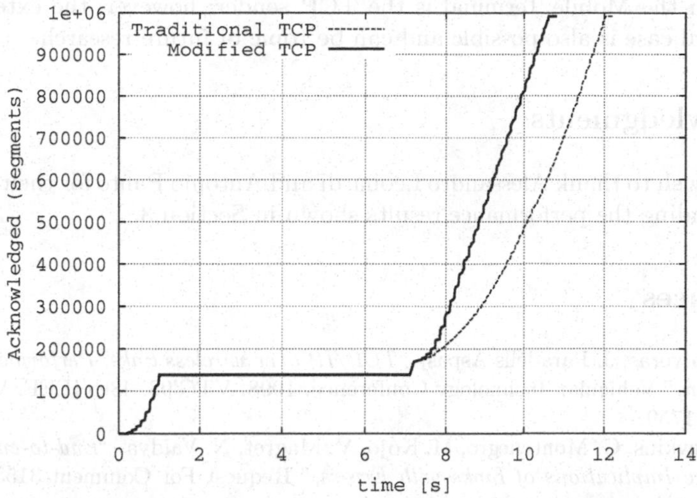

Fig. 9. Acknowledged data segments *Vs.* time, comparison between Traditional and Modified TCP.

when, for instance, the MT is receiving TCP data from a fixed host (FH) placed somewhere in the Internet, the FH should be informed about the link failure on the wireless part of the connection. In order to do that, at the MT serving BS some additional features must be implemented. In more detail, in the case of link failure the BS should send back to the FH a duplicate TCP ACK containing a *awnd* (receiver advertised window) field set to zero and, the MT, whenever the link is restored should send to the FH a new TCP ACK with the correct *awnd* value. In this case, if the first duplicate ACK received correctly by the FH, it will stop transmitting data (entering in a freeze status) avoiding unnecessary time outs during link failure periods. This strategy is similar to the one proposed in [16]. The performance investigation of this last strategy can be the topic of future research.

4 Conclusions

In this paper we addressed the performance problems of TCP in heterogeneous wireless systems and proposed some algorithms to improve it. In more detail, we proposed a methodology to dynamically set the TCP maximum segment size in order to minimize the residual packet error rate as a function of the channel state and of the link layer configuration. This solution leads to substantial throughput improvements and can be applied to both the up- and the down-link tranfer case.

We have also introduced some modifications to the TCP algorithm which are triggered on the following of an inter-RE handover in order to avoid the TCP time out problem due to link failures. This last technique can be effectively used only when the Mobile Terminal is the TCP sender; however, the extension to the general case is also possible and can be topic of future research.

Acknowledgments

Authors wish to thank Alessandro Leonardi and Antonio Pantó for their precious help in finding the performance results shown in Section 3.

References

1. A. Calveras, J. Paradells Aspas, *"TCP/IP over wireless links: Performance Evaluation,"* Vehicular Technology Conference, 1998. VTC'98. 48th IEEE. Vol. 3, pp: 1755-1759.
2. S. Dawkins, G. Montenegro, M. Kojo, V. Magret, N. Vaidya, *"End-to-end Performance Implications of Links with Errors,"* Request For Comment 3155. August 2001.
3. Lefevre, F., Vivier, G., *"Understanding TCP's behavior over wireless links,"* Symposium on Communications and Vehicular Technology, 2000, pp: 123-130.
4. G. Xylomenos, G.C. Polyzos, P. Mahonen, M. Saaranen, *"TCP performance issues over wireless links,"* IEEE Communications Magazine, Vol. 39, No. 4, pp: 52-58. April 2001.
5. S. Hadjiefthymiades, S. Papayiannis, L. Merakos, *"Using Path Prediction to Improve TCP Performance in Wireless/Mobile Communications,"* IEEE Comm. Mag., Vol. 40, No. 8, pp: 54-61. August 2002.
6. S. Floyd, T. Henderson, *"The New Reno Modification to TCP's Fast Recovery Algorithm,"* Request For Comment 2582. April 1999.
7. R. A. Howard, *"Dynamic Probabilistic Systems,"* New York: Wiley, 1971.
8. B. Friedman, *"Principles and Techniques of Applied Mathematic,"* New York: Dover Publications, Inc.
9. W. T. Chen, J.S. Lee, *"Some mechanisms to improve TCP/IP performance over wireless and mobile computing environment,"* Parallel and Distributed Systems, 2000. Proceedings. Seventh International Conference on, 2000, pp: 437-444.
10. H. Balakrishnan et al., *"A comparison of Mechanisms for Improving TCP Performance Over Wireless Links,"* Proc. ACM SIGCOMM. '96, pp: 256-269. August 1996.
11. K. Brown, S. Singh, *"M-TCP: TCP for Mobile Cellular Networks,"* Comp. Commun. Rev., Vol. 27, No. 5, pp: 19-43. October 1997.
12. W. R. Stevens, *"TCP/IP Illustrated, Vol. 1: The protocols,"* New York: Addison Wesley, 1994.
13. M. Allman, W. Richard Stevens, *"TCP congestion Control,"* Request For Comment 2581. April 1999.
14. W. Richard Stevens, *"TCP Slow Start, Congestion Avoidance, Fast Retransmit, and Fast Recovery Algorithm,"* Request For Comment 2001. January 1997.
15. `http://www.linux.org`
16. R. Ludwig, R.H. Katz, *"The Eifel Algorithm: Making TCP Robust Against Spurious Retransmissions,"* Comp. Commu. Rev., Vol. 30, No. 1, pp. 30-36. January 2000.

Adaptive MPEG-4 Video Streaming with Bandwidth Estimation

Alex Balk, Dario Maggiorini, Mario Gerla, and M.Y. Sanadidi

Network Research Laboratory, UCLA, Los Angeles CA 90024, USA

Abstract. The increasing popularity of streaming video is a cause for concern for the stability of the Internet because most streaming video content is currently delivered via UDP, without any end-to-end congestion control. Since the Internet relies on end systems implementing transmit rate regulation, there has recently been significant interest in congestion control mechanisms that are both fair to TCP and effective in delivering real-time streams. In this paper we design and implement a protocol that attempts to maximize the quality of real-time MPEG-4 video streams while simultaneously providing basic end-to-end congestion control. While several adaptive protocols have been proposed in the literature [20,27], the unique feature of our protocol, the Video Transport Protocol (VTP), is the use of receiver side bandwidth estimation. We deploy our protocol in a real network testbed and extensively study its behavior under varying link speeds and background traffic profiles using the FreeBSD Dummynet link emulator [23]. Our results show that VTP delivers consistent quality video in moderately congested networks and fairly shares bandwidth with TCP in all but a few extreme cases. We also describe some of the challenges in implementing an adaptive video streaming protocol.

1 Introduction

As the Internet continues to grow and mature, transmission of multimedia content is expected to increase and compose a large portion of the overall data traffic. Film and television distribution, digitized lectures, and distributed interactive gaming applications have only begun to be realized in today's Internet, but are rapidly gaining popularity. Audio and video streaming capabilities will play an ever-increasing role in the multimedia-rich Internet of the near future. Real-time streaming has wide applicability beyond the public Internet as well. In military and commercial wireless domains, virtual private networks, and corporate intra-nets, audio and video are becoming a commonplace supplements to text and still image graphics.

Currently, commercial programs such as RealPlayer [19] and Windows Media Player [16] provide the predominant amount of the streamed media in the Internet. The quality of the content delivered by these programs varies, but they are generally associated with low resolution, small frame size video. One reason these contemporary streaming platforms exhibit limited quality streaming

M. Ajmone Marsan et al. (Eds.): QoS-IP 2003, LNCS 2601, pp. 525–538, 2003.

is their inability to dynamically adapt to traffic conditions in the network during a streaming session. Although the aforementioned applications claim to be adaptive, there is no conclusive evidence as to what degree of adaptivity they employ as they are proprietary, closed software [20]. Their video streams are usually delivered via UDP with no transport layer congestion control. A large-scale increase in the amount of streaming audio/video traffic in the Internet over a framework devoid of end-to-end congestion control will not scale, and could potentially lead to congestion collapse.

UDP is the transport protocol of choice for video streaming platforms mainly because the fully reliable and strict in-order delivery semantics TCP do not suit the real-time nature of video transmission. Video streams are *loss tolerant* and *delay sensitive*. Retransmissions by TCP to ensure reliability introduce latency in the delivery of data to the application, which in turn leads to degradation of video image quality. Additionally, the steady state behavior of TCP involves the repeated halving and growth of its congestion window, following the well known Additive Increase/Multiplicative Decrease (AIMD) algorithm. Hence, the throughput observed by a TCP receiver oscillates under normal conditions. This presents another difficulty since video is usually streamed at a constant rate (VTP streams are actually piecewise-constant). In order to provide the best quality video with minimal buffering, a video stream receiver requires relatively stable and predictable throughput.

Our protocol, the Video Transport Protocol (VTP), is designed with the primary goal of adapting an outgoing video stream to the characteristics of the network path between sender and receiver. If it determines there is congestion, the VTP sender will reduce its sending rate and the video encoding rate to a level the network can accommodate. This enables VTP to deliver a larger portion of the overall video stream and to achieve inter-protocol fairness with competing TCP traffic. A secondary goal of VTP is the minimal use of network and end system resources. We make several trade-offs to limit processing overhead and buffering requirements in the receiver. In general, VTP follows a conservative design philosophy by sparingly using bandwidth and memory during the streaming session.

An important aspect of VTP is that it is completely end-to-end. VTP does not rely on QoS functionality in routers, random early drop (RED), other active queue management (AQM), or explicit congestion notification (ECN). It could potentially benefit from such network level facilities, but in this paper we focus only on the case of real-time streaming in a strictly best effort network. Possible interactions between VTP and QoS routers, AQM, or ECN is an area of future work.

VTP is implemented entirely in user space and designed around open video compression standards and codecs for which the source code is freely available. The functionality is split between two distinct components, each embodied in a separate software library with its own API. The components can be used together or separately, and are designed to be extensible. VTP sends packets using UDP, adding congestion control at the application layer.

This paper discusses presents the VTP design in Section 2. Section 3 covers the VTP experiments and results. The conclusion and a brief discussion of future work follow.

2 The Video Transport Protocol

A typical video streaming server sends video data by dividing each frame into fixed size packets and adding a header containing, for example, a sequence number, the time the packet was sent, and the relative play out time of the associated video frame. Upon receiving the necessary packets to reassemble a frame, the receiver buffers the compressed frame for decoding. The decompressed video data output from the decoder is then sent to the output device. If the decoder is given an incomplete frame due to packet loss during the transmission, it may decide to discard the frame. The mechanism used in the discarding decision is highly decoder-specific, but the resulting playback jitter is a universal effect. In MPEG-4 video, which we use in this paper, there are dependencies between independent or "key" frames and predicted frames. Discarding a key frame can severely effect the overall frame rate as errors will propagate to all frames predicted from the key frame.

The primary design goal of VTP is to adapt the outgoing video stream so that, in times of network congestion, less video data is sent into the network and consequently fewer packets are lost and fewer frames are discarded. VTP rests on the underlying assumption that the smooth and timely play out of consecutive frames is central to a human observer's perception of video quality. Although a decrease in the video bitrate noticeably produces images of coarser resolution, it is not nearly as detrimental to the perceived video quality as inconsistent, start-stop play out. VTP capitalizes on this idea by adjusting both the video bitrate and its sending rate during the streaming session. In order to tailor the video bitrate, VTP requires the same video sequence to be pre-encoded at several different compression levels. By switching between levels during the stream, VTP makes a fundamental trade-off by increasing the video compression in an effort to preserve a consistent frame rate at the client.

In addition to maintaining video quality, the other important factor for setting adaptivity as the main goal in the design is inter-protocol fairness. Unregulated network flows pose a risk to the stability and performance of the Internet in their tendency to overpower TCP connections that carry the large majority of traffic. While TCP halves its window in response to congestion, unconstrained flows are under no restrictions with respect to the amount of data they can have in the network at any time. VTP's adaptivity attempts to alleviate this problem by interacting fairly with any competing TCP flows.

The principal features of this design, each described in the following subsections, can be summarized as follows:

1. Communication between sender and receiver is a "closed loop," i.e. the receiver sends acknowledgments to the sender at regular intervals.

2. VTP is rate based; the bandwidth of the forward path is estimated and used by the sender to determine the sending rate.

2.1 Sender and Receiver Interaction

VTP follows a client/sever design where the client initiates a session by requesting a video stream from the server. Once several initialization steps are completed, the sender and receiver communicate in a closed loop, with the sender using the acknowledgments to determine the bandwidth and RTT estimates.

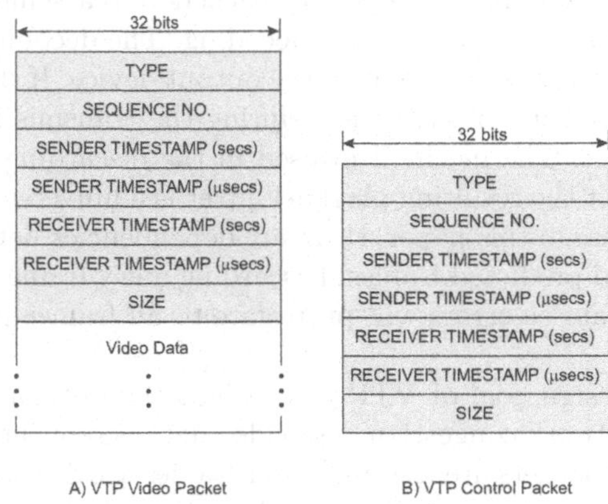

Fig. 1. VTP packet formats for a) video packets and b) control packets.

Figure 1 shows the VTP video header and acknowledgment or "control packet" formats. The symmetric design facilitates both bandwidth and RTT computation. The TYPE field is used by the sender to explicitly request a control packet from the receiver. For every k video packets sent, the sender will mark the TYPE field with an ack request, to which the receiver will respond with a control packet. The value of k is a server option that is configurable at run time by the user. The two timestamp fields for sender and receiver respectively are used for RTT measurement and bandwidth computation. VTP estimates the bandwidth available to it on the path and then calibrates its sending rate to the estimate, as detailed in the following paragraphs.

When the receiver receives a data packet with the TYPE field indicating it should send a control packet, it performs two simple operations. First, it copies the header of the video packet and writes its timestamp into the appropriate fields. Second, it writes the number of bytes received since the last control packet was sent into the SIZE field. The modified video packet header is then sent back to the sender as a control packet.

Upon receipt of the control packet, the sender extracts the value in the SIZE field and the receiver timestamp. The sender is able to compute the time delta between control packets *at the receiver* by keeping the value of one previous receiver timestamp in memory and subtracting it from the timestamp in the most recently received packet. The value of the SIZE field divided by this time delta is the rate currently being achieved by this stream. This rate is also the "admissible" rate since it is the rate at which data is getting through the path bottleneck. In essence, the measured rate is equal to the bandwidth available to the connection. Thus, it is input as a bandwidth sample into the bandwidth estimation algorithm described in the next section.

The sender uses its own timestamps to handle the RTT computation. When the sender sends a video packet with the TYPE field marked for acknowledgment, it remembers the sequence number. If the sequence number on the returning control packet matches the stored value (recall the receiver simply copies the header into the control packet, changing only its own timestamp and the SIZE field), the sender subtracts the sender timestamp in the control packet from the current time to get the RTT sample.

If either a data packet that was marked for acknowledgment or a control packet is lost, the sender notices a discrepancy in the sequence numbers of the arriving control packets. That is, the sequence numbers to not match those that the sender has recorded when sending out video packets with ack requests. In this case, the sender disregards the information in the control packets. Valid bandwidth or RTT samples are always taken from two consecutively arriving control packets.

2.2 Bandwidth Estimation and Rate Adjustment

Bandwidth estimation is an active area of research in its own right [1,4,5,13]. In this paper we provide only a brief summary following [5]. Recall from the previous section that the achieved rate sample b_i can be obtained by dividing the amount of data in the last k packets by the inter-arrival time between the current and $k-1$ previous packets. As a concrete example, suppose $k=4$ and four packets arrive at the receiver at times $t_1, ..., t_4$, each with $d_1, ..., d_4$ bytes of data respectively. The sum $\sum_{i=1}^{4} d_i$ is sent to the sender in the SIZE field of the control packet.

The sender, knowing t_1 from the last control packet and t_4 from the current control packet, computes

$$b_i = \frac{\sum_{i=1}^{4} d_i}{(t_4 - t_1)} \tag{1}$$

Exponentially averaging the samples using the formula

$$B_i = (\alpha)B_{i-1} + (1-\alpha)\left(\frac{b_i + b_{i-1}}{2}\right) \tag{2}$$

yields the bandwidth estimate B_i that is used by the sender to adjust the sending rate. The parameter α is a weighting factor that determines how much the two most recent samples should be weighed against the history of the bandwidth estimate. In experimental trials, it was determined that VTP performs best when α is a constant close to 1. Packet loss is reflected by a reduction in the achieved rate and thus in the bandwidth estimate. Since the bandwidth estimation formula takes into account losses due to both congestion and random errors, using an exponential average prevents a single packet drop due to a link error from causing a steep reduction in the estimate.

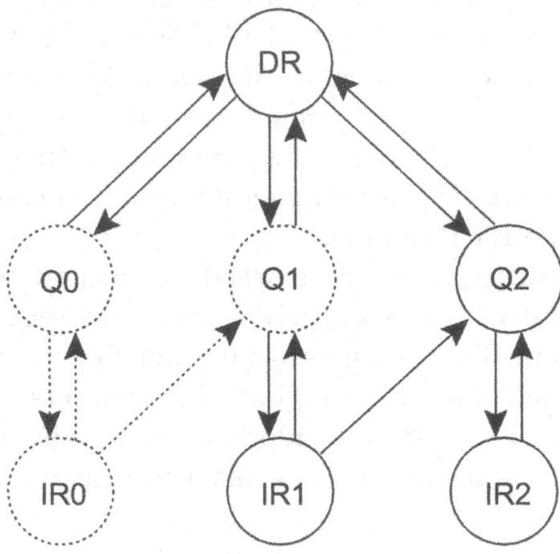

Fig. 2. VTP finite state machine with states and transitions involved in a video quality level increase represented with dashed lines.

Through the estimate of the connection bandwidth, the VTP sender gains considerable knowledge about the conditions of the path. The sender uses the estimate as input into an algorithm that determines how fast to send the data packets and which pre-encoded video to send. We describe the algorithm in terms of a finite state machine (FSM), shown in Figure 2. Assuming three video encoding levels, the states Q0, Q1, and Q2 each correspond to one distinct video level from which VTP can stream. We use three levels throughout this example for simplicity, but $n > 3$ levels are possible in general. Each of the IR states, IR0, IR1, and IR2, represent increase rate states, and DR represents the decrease rate state. In Figure 2, the states and transitions involved in a quality level increase are highlighted with dashed lines.

Starting in state Q0, a transition to IR0 is initiated by the reception of a bandwidth estimate that is equal to or greater than the current sending rate. Being in state Q0 only implies the VTP server is sending the lowest quality level, it says nothing about the sending rate. In state IR0, the server checks several conditions. First, it checks if the RTT timer has expired. If it has not, the server returns to Q0 without taking any action and awaits the next bandwidth estimate. If one RTT has passed, it remains in IR0 and investigates further. It next determines whether the sending rate is large enough to support the rate of the next highest level (level 1 in this case). If not, the server increases the sending rate by one packet size and returns to state Q0. If, on the other hand, the sending rate can accommodate the next quality level, the server checks the value of a variable we call "the heuristic."

The heuristic is meant to protect against over ambitiously increasing the video quality in response to instantaneous available bandwidth on the link that is short-lived and will not be able to sustain the higher bitrate stream. If the heuristic is satisfied, the server increases the sending rate by one packet size and transitions to state Q1. If the heuristic is not met, the server increases the rate by one packet and returns to state Q0. In normal operation, the server will cycle between states Q0 and IR0 while continually examining the RTT timer, the bandwidth estimate, and the heuristic, and adjusting the sending rate. When conditions permit, the transition to Q1 occurs. The process repeats itself for each of the quality levels.

In the current implementation the heuristic is an amount of time, measured in units of RTT, to wait before switching to the next higher level of video quality. Ideally, the heuristic would also take into account the receiver buffer conditions to ensure a video quality increase would not cause buffer overflow. Since the receiver is regularly relaying timestamp information to the sender, it would be expedient to notify the sender of the amount of buffer space available in the control packet. The sender would then be able to make the determination to raise the video quality with the assurance that both the network and the receiver can handle the data rate increase. [21] examines the factors that need to be taken into account in quality changing decisions in detail.

In a rate and quality decrease, the transition to DR is initiated when the server receives a bandwidth estimate less than its current sending rate. In DR, the server checks the reference rate of each constituent quality to find the highest one that can fit within the bandwidth estimate. The server sets its sending rate to the bandwidth estimate and transitions to the state corresponding to the video quality that can be supported. Unlike the state transitions to increase quality levels, the decrease happens immediately, with no cycles or waits on the RTT timer. This conservative behavior contributes greatly to the fairness properties of VTP discussed in Section 3.2.

As the FSM suggests, the selection of the encoding bitrates is important. VTP observes the rule that a particular video encoding level must be transmitted at a rate greater than or equal to its bitrate and will not send slower than the rate of the lowest quality encoding. This could potentially saturate the network and exacerbate congestion if the lowest video bitrate is frequently higher than the available bandwidth. Additionally, if the step size between each reference rate is large, more data buffering is required at the receiver. This follows from the fact that large step sizes lead to the condition where VTP is sending at a rate that is considerably higher than the video bitrate for long periods of time.

3 Experimental Evaluation

We implemented VTP on the Linux platform and performed extensive evaluations using the Dummynet link emulator [23]. We developed a technique to *smooth* the bandwidth required by the outgoing video stream and compute the client buffer requirement for specific pre-encoded video segments. The goals of our experiments were to assess inter-protocol fairness between VTP and TCP, and to evaluate the quality of the transmitted video played by the client. In this section we cover the results of our experimental evaluation.

3.1 Transmission Schedules for Variable Bitrate Video

In a constant bitrate (CBR) video source, the compression level is continuously adjusted to maintain the target bitrate of the overall video stream. This is beneficial for network transmission, but leads to varying video quality from frame to frame and can have an unpleasant effect on the viewer's perception. MPEG-4 preserves consistent quality by increasing the bitrate at times of high motion or detail, producing a variable bitrate (VBR) encoding. In some instances the bitrate can change dramatically during the course of a video clip. The amount of rate variability is codec-dependent. In this research we investigated three MPEG-4 video codecs: DivX 4.2 [6], FFmpeg 0.4.6 [10], and Microsoft MPEG-4 version 2 [16]. After several initial tests, the Microsoft codec was found to be inappropriate for VTP. This codec uses an algorithm that drops entire frames to achieve the desired compression level, conflicting with VTP's assumption of a similar frame pattern across the set of encodings.

Since it would be ineffective to transmit video data at uneven, bursty rates, we "smoothed" the VBR MPEG-4 video to develop a piecewise-constant sending rate profile. Figure 3 shows the results of applying a modified version of the PCRTT algorithm [15] to a 130 second sample of the movie "TRON" encoded with the DivX codec at three different levels of compression. The "QP range" in the figure represents the amount of compression applied by the codec: "QP" stands for "quantization parameters," where higher QPs imply more compression and the lower quality. The peak rate is reduced significantly, from more than 4 Mbps to around 1.6 Mbps in the transmission plan. Similarly, Figure 4 shows the smoothing algorithm applied to a 50 second sample of a trailer for the movie "Atlantis" produced with the FFmpeg codec for three different ranges of quantization parameters. These two sets of video sources are used throughout the experimental evaluation of VTP.

3.2 Fairness with TCP

The the first set of experiments was designed to quantitatively measure how much bandwidth TCP and VTP attain when competing directly with each other. We streamed both the "TRON" and "Atlantis" video sources in various scenarios differing mainly in link capacity and number of competing TCP connections. The experiments were performed using a relatively simple network topology in

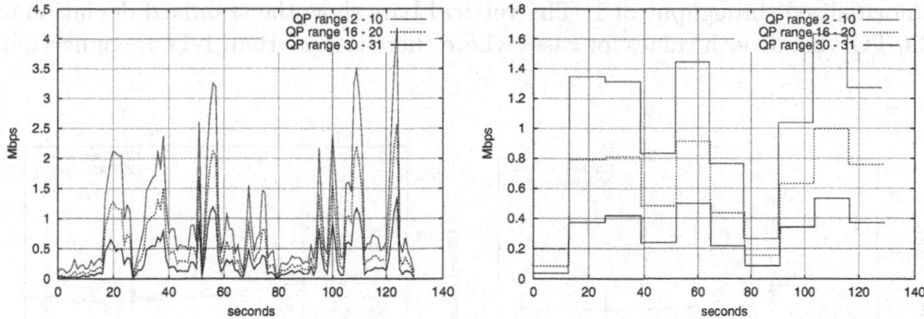

Fig. 3. Source bitrates (left) and sending rate profile (right) produced for "TRON."

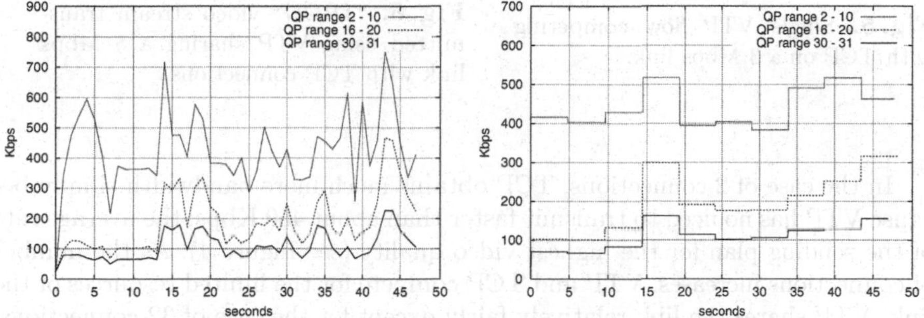

Fig. 4. Source bitrates (left) and sending rate profile (right) produced for "Atlantis."

which two independent LANs were connected through a PC running FreeBSD acting as a gateway. The Dummynet utility and the Iperf program[1] were used to vary the link capacity and generate background TCP traffic respectively. In this environment all packets arrive in order, so any gap in sequence numbers can immediately be interpreted by the receiver as packet loss.

Figure 5 presents the normalized throughput of VTP sending the "Atlantis" segment on a 3 Mbps, 10 ms RTT link with various numbers of TCP flows. Each column of data points represents a separate experiment where a single VTP flow and several TCP flows share the link. The x axis is labeled with total number of flows (e.g. the column labeled "16" is the result of one VTP and 15 TCP flows). The normalized throughput is computed by simply dividing the average bandwidth received by each flow by the fair share bandwidth value for each case. Perfect inter-protocol fairness would be exhibited by both VTP and TCP scoring

[1] http://dast.nlanr.net/Projects/Iperf/

a normalized throughput of 1. The vertical bars show the standard deviation of the TCP bandwidth values for cases where there is more than 1 TCP connection.

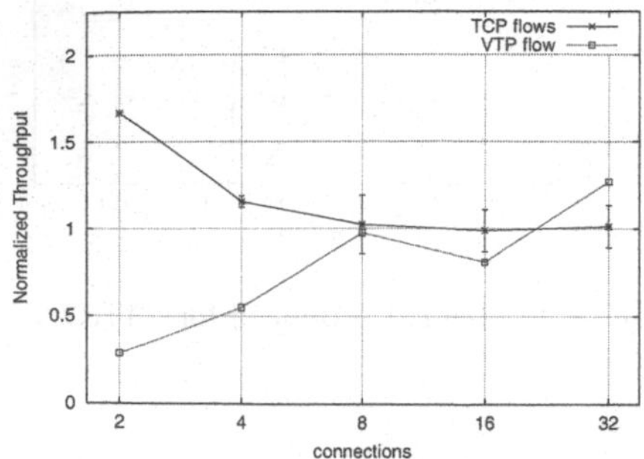

Fig. 5. Single VTP flow competing with TCP on a 3 Mbps link.

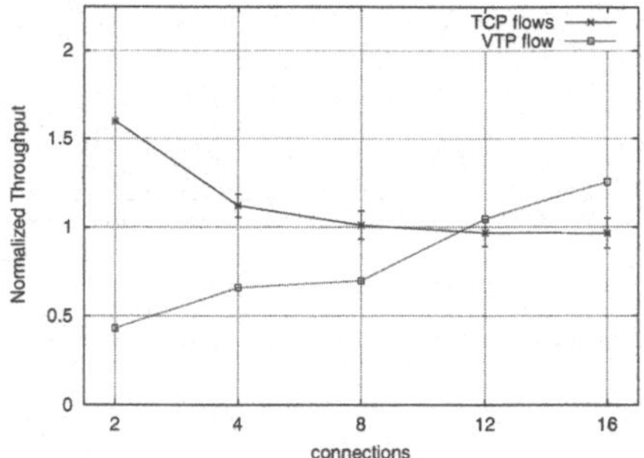

Fig. 6. "TRON" video stream transmitted using VTP sharing a 5 Mbps link with TCP connections.

In the case of 2 connections, TCP obtains much more bandwidth simply because VTP has no need to transmit faster than about 450 Kbps, the average rate of the sending plan for the highest video quality (see Figure 4). As the number of connections increases, VTP and TCP compete for the limited resources of the link. VTP shares the link relatively fairly except for the case of 32 connections. In this case, the fair share value is $3000/32 = 93.75$ Kbps, which is roughly three quarters of the rate of the lowest video quality according to Figure 4. Since VTP does not send slower than the rate of the transmission plan for the lowest video quality (about 125 Kbps according to Figure 4) it uses slightly more than the fair share value of the bandwidth. It is important to note that this unfairness is not an inherent limitation of VTP, but a circumstance of the relationship between the link capacity and the video encoding. The case where VTP shares the link with 7 TCP connections results in near perfect fairness.

In Figure 6, VTP sends the "TRON" video segment on a 5 Mbps, 10 ms RTT link against background TCP traffic. The "TRON" send plan requires significantly higher bitrates than "Atlantis," thus we set the link speed correspondingly higher. The "TRON" transmission plan also contains larger instantaneous jumps in send rate – as much as 1 Mbps for the highest video quality (see Figure 3). Both of these differences are a result of the dissimilar bitrate profiles produced by the DivX and FFmpeg codecs, as evident in Figures 3 and 4.

Figure 6 shows that VTP uses less than or equal to its fair share of bandwidth in all cases except that of 16 connections, where again the link limitation is reached. The figure verifies the "Atlantis" experiments: VTP behaves fairly, in some cases generously leaving bandwidth unused, if its bandwidth share allocation is at least enough to stream the lowest quality of video.

In summary, we have demonstrated that VTP uses network resources fairly when facing competition from the AIMD based congestion control of TCP. In lightly loaded networks, VTP uses only the bandwidth required to transmit at the rate of the highest quality video stream, the remaining bandwidth can be claimed by other connections. In environments of moderate congestion, VTP fairly shares the link as long as its fair share is at least the rate of the lowest quality video. Additionally, VTP's fairness properties are not codec specific, and that it is able to maintain stable sending rates when streaming source video with significantly different transmission plans.

3.3 Video Quality

In the evaluation of video quality delivered by VTP, we concentrated on two key parameters: the frame rate of the received video and the average values of the quantization parameters. We place a rather strict constraint on the player by configuring it to only display frames which are received completely intact, i.e., frames which have any errors due to packet loss are discarded. This bolsters the importance of the play out frame rate and magnifies the performance of VTP in terms of its key goal of providing a stable frame rate through quantization scale adjustment.

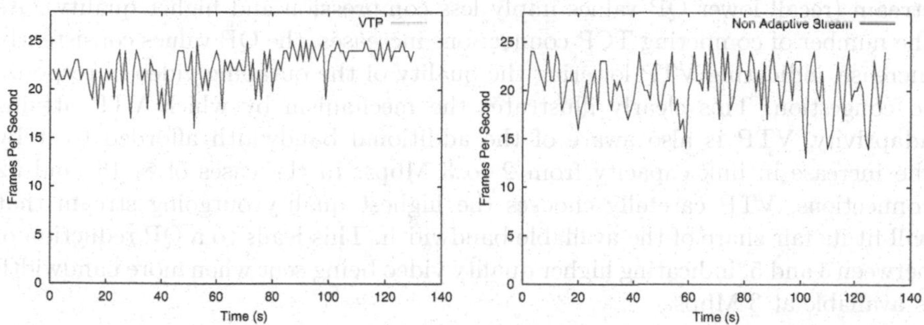

Fig. 7. Frame rate of received "TRON" stream with VTP and Non-Adaptive Streaming.

Figure 7 depicts a representative example of the advantage gained by VTP adaptivity. In this experiment, the conditions are those of the fourth case in Figure 6: 1 monitored flow (either VTP or non-adaptive streaming) sharing a 5 Mbps, 10 ms RTT link with 11 competing TCP connections. As the streaming session progresses, VTP discovers the fair share of available bandwidth and appropriately tunes to sending rate and video bitrate to avoid overflowing the router buffer. The resulting frame rate of the VTP stream stabilizes with time, while the frame rate of the non-adaptive stream increasingly oscillates toward the end of the segment, suffering from the effect of router packet drops.

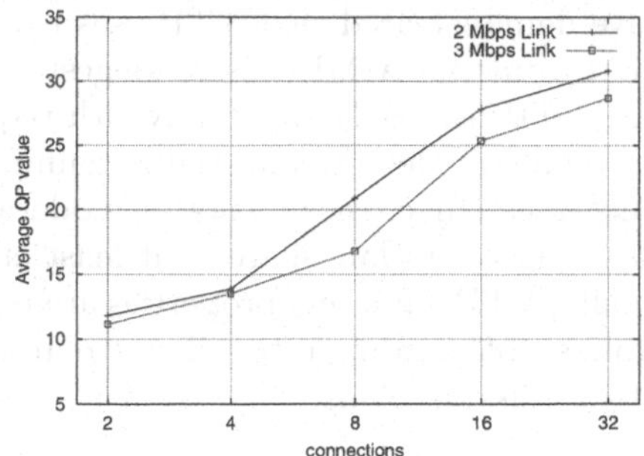

Fig. 8. Average values of quantization parameters of the delivered "Atlantis" stream.

In Figure 8 we present the average values of the QPs of the "Atlantis" segment throughout the duration of the session. We show the 3 Mbps case from the experiment in the previous section together with the case of a 2 Mbps link. The plot verifies that VTP adapts the outgoing video stream to fit the available network bandwidth. When there is little contention for the link, e.g. 2 and 4 total connections, VTP chooses video primarily from the high quality, high bitrate stream (recall lower QP values imply less compression and higher quality). As the number of competing TCP connections increases, the QP values consistently increase, indicating VTP lowering the quality of the outgoing video in response to congestion. This clearly illustrates the mechanism by which VTP attains adaptivity. VTP is also aware of the additional bandwidth afforded to it by the increase in link capacity from 2 to 3 Mbps. In the cases of 8, 16, and 32 connections, VTP carefully chooses the highest quality outgoing stream that will fit its fair share of the available bandwidth. This leads to a QP reduction of between 3 and 5, indicating higher quality video being sent when more bandwidth is available at 3 Mbps.

4 Conclusion

In this paper we designed, implemented and tested a new protocol to stream MPEG-4 compressed video in real-time. A distinct feature of VTP is the use of bandwidth estimation to adapt the sending rate and the video encoding in response to changes in network conditions. We developed VTP in accordance with open standards for video compression and file formats, and built a plug-in for a widely used video player to serve as the VTP receiver. We have made an effort to make VTP easily extensible.

VTP was evaluated in a controlled network environment under a variety of link speeds and background traffic. Experimental results show that VTP offers considerable gains over non-adaptive streaming in effective frame rate. To a large extent, VTP behaves fairly toward TCP when both protocols compete in a congested network. We found that VTP fairness toward TCP is vulnerable if the lowest video bitrate is higher than the average link fair share available to VTP. A priori knowledge of the general link capacity and typical network utilization can be extremely useful in the selection and configuration of the video sources for VTP. We believe this information is usually not difficult to obtain for administrators, and that a small amount of careful manual configuration is a reasonable price for the advantages of VTP.

For any streaming system to be fully useful, audio and video must be multiplexed into the data stream and synchronized during play out. A near term goal is to include the capability to adaptively stream audio and video in combination under the VTP protocol. We will also further investigate the effect of changing the k parameter, the number of packets used to compute a single bandwidth sample. We plan to implement an algorithm to dynamically adjust k during streaming to improve VTP's efficiency and fairness with TCP. Another advantage would be reducing the amount of manual user configuration required.

References

1. N. Aboobaker, D. Chanady, M. Gerla, and M. Y. Sanadidi, "Streaming Media Congestion Control using Bandwidth Estimation," In *Proceedings of MMNS '02*, October, 2002.
2. A. Augé and J. Aspas, "TCP/IP over Wireless Links: Performance Evaluation," In *Proceedings of IEEE 48th VTC '98*, May 1998.
3. D. Bansal and H. Balakrishnan, "Binomial Congestion Control Algorithms," In *Proceedings of INFOCOMM '01.* April 2001.
4. C. Casetti, M. Gerla, S. S. Lee, S. Mascolo, and M. Sanadidi, "TCP with Faster Recovery," In *Proceedings of MILCOM '00*, October 2000.
5. C. Casetti, M. Gerla, S. Mascolo, M. Y. Sanadidi, and R. Wang, "TCP Westwood: Bandwidth Estimation for Enhanced Transport over Wireless Links," In *Proceedings of ACM MOBICOM '01*, July 2001.
6. The DivX Networks home page. http://www.divxnetworks.com/
7. N. Feamster, D. Bansal, and H. Balakrishnan, "On the Interactions Between Layered Quality Adaptation and Congestion Control for Streaming Video," In *11th International Packet Video Workshop*, April 2001.
8. N. Feamster, *Adaptive Delivery of Real-Time Streaming Video.* Masters thesis, MIT Laboratory for Computer Science, May 2001.
9. W. Feng and J. Rexford, "Performance Evaluation of Smoothing Algorithms for Transmitting Variable Bit Rate Video," *IEEE Trans. on Multimedia*, 1(3):302-313, September 1999.
10. The FFmpeg homepage. http://ffmpeg.sourceforge.net/
11. S. Floyd, M. Handley, J. Padhye, and J. Widmer, "Equation-Based Congestion Control for Unicast Applications," In *Proceedings of ACM SIGCOMM '00*, August 2000.
12. International Organization for Standardization. *Overview of the MPEG-4 Standard*, December, 1999.
13. K. Lai and M Baker, "Measuring Link Bandwidths using a Deterministic Model of Packet Delay," In *Proceedings of ACM SIGCOMM '00*, August 2000.

14. X. Lu, R. Morando, and M. El Zarki, "Understanding Video Quality and its use in Encoding Control," In *12th International Packet Video Workshop*, April 2002.
15. J. McManus and K. Ross, "Video-on-Demand Over ATM: Constant-Rate Transmission and Transport," *IEEE Journal on Selected Areas in Communications*, 14(6):1087-1098, August 1996.
16. Microsoft Windows Media Player home page. http://www.microsoft.com/windows/windowsmedia/
17. The MPEG home page. http://mpeg.telecomitalialab.com/
18. J. Padhye, V. Firoio, D. Townsley, and J. Kurose, "Modeling TCP Throughput: A Simple Model and its Empirical Validation," In *Proceedings of ACM SIGCOMM '98*, September 1998.
19. The RealPlayer home page. http://www.real.com/
20. R. Rejaie, M. Handley, and D. Estrin, "RAP: An End-to-End Rate-Based Congestion Control Mechanism for Real-time Streams in the Internet," In *Proceedings of INFOCOMM '99*, March 1999.
21. R. Rejaie, M. Handley, and D. Estrin, "Layered Quality Adaptation for Internet Video Streaming," In *Proceedings of ACM SIGCOMM '99*, September 1999.
22. R. Rejaie, M. Handley, and D. Estrin, "Architectural Considerations for Playback of Quality Adaptive Video over the Internet," In *Proceedings of IEEE Conference on Networks*, September 2000.
23. L. Rizzo, "Dummynet and Forward Error Correction," In *Proceedings of Freenix '98*. June 1998.
24. D. Tan and A. Zahkor, "Real-time Internet Video Using Error Resilient Scalable Compression and TCP-friendly Transport Protocol," *IEEE Trans. on Multimedia*, 1(2):172-186, May 1999.
25. N. Wakamiya, M. Miyabayashi, M. Murata, and H. Miyahara, "MPEG-4 Video Transfer with TCP-friendly Rate Control," In *Proceedings of MMNS '01*. October 2001.
26. The xine video player home page. http://xine.sourceforge.net.
27. Q. Zhang, W. Zhe, and Y. Q. Zhang, "Resource Allocation for Multimedia Streaming Over the Internet," *IEEE Trans. on Multimedia*, 3(3):339-355, September 2001.

Dynamic Quality Adaptation Mechanisms for TCP-friendly MPEG-4 Video Transfer

Naoki Wakamiya, Masaki Miyabayashi, Masayuki Murata, and Hideo Miyahara

Graduate School of Information Science and Technology, Osaka University
1–3 Machikaneyama, Toyonaka, Osaka 560-8531, Japan
{wakamiya, miyabays, murata, miyahara}@ist.osaka-u.ac.jp

Abstract. When a considerable amount of UDP traffic is injected into the Internet by distributed multimedia applications, the Internet easily becomes congested. Consequently, the bandwidth available to TCP connections becomes limited and performance significantly deteriorates. In order that both multimedia applications and TCP-based ones can fairly co-exist on the Internet, it is becoming increasingly important to consider inter-protocol fairness.

In this paper, we first consider several issues related to TCP-friendly video transfer and propose several video-quality adjustment methods which accomplish high-quality, stable, and TCP-friendly FGS (Fine Granular Scalability) video transfer. Then, with consideration on the quality degradation caused by packet loss in video quality adaptation, we extend our method so that it is applicable to a lossy environment when it is used in cooperation with the FEC (Forward Error Correction) technique. Our mechanism adjusts the video quality in accordance with the TFRC (TCP-Friendly Rate Control) rate, the packet loss probability, and the resultant video quality, while avoiding undesirable video-quality degradation. Through simulation experiments using FGS video streams, we show that our proposed method can provide high-quality, stable and TCP-friendly video transfer even in the unstable and lossy Internet.

1 Introduction

Since the current Internet does not provide QoS (Quality of Service) guarantee mechanisms, each application chooses the preferable transport protocol to achieve the required performance. For example, traditional data applications such as `http`, `ftp`, and `telnet` employ TCP that accomplishes loss-free data transfer by means of window-based flow control and retransmission mechanisms. On the other hand, loss-tolerant real-time multimedia applications such as video conferencing or video streaming prefer UDP in order to avoid unacceptable delays caused by packet retransmissions. UDP is considered selfish and ill-behaved, because whereas TCP throttles its transmission rate to avoid network congestion, UDP does not have such control mechanisms.

It has been pointed out that selfish and greedy UDP traffic injected by multimedia applications easily dominates network bandwidth and drives the network into congestion. As a result, the bandwidth available to TCP connections can be said to be "oppressed" by UDP traffic and their performance deteriorates extremely. In recent years,

M. Ajmone Marsan et al. (Eds.): QoS-IP 2003, LNCS 2601, pp. 539–550, 2003.

several researches have focused on "TCP-friendly" rate control [1, 2, 3]. By "TCP-friendly" we mean that a non-TCP connection should receive the same share of bandwidth as that of a TCP connection if they traverse the same path [2]. A TCP-friendly system regulates its data sending rate according to the network conditions, typically expressed in terms of the round-trip-time (RTT) and the packet loss probability, to achieve the same throughput that a TCP connection would acquire on the same path. In particular, the TCP-Friendly Rate Control (TFRC) proposed in [3] has the feature of very smoothly adjusting the transmission rate while coping with network congestion.

If an application successfully adjusts its sending rate to the TFRC rate, TCP-friendly data transfer can be accomplished. However, TFRC itself does not consider the influence of the TCP-friendly rate control on the application-level performance. For example, the TFRC sender changes its sending rate at least once per RTT. Such a frequent rate control obviously affects the perceived video quality when a video application regulates the amount of coded video data by controlling video quality according to the TFRC rate. Thus, in this paper, we tackle issues like the appropriate choice of control interval, video quality regulation, and video rate adaptation.

We propose several rate control methods to accomplish TCP-friendly MPEG-4 video transfer with consideration of the application-level performance, i.e., video quality. In particular, in consideration of TCP-friendly MPEG-4 video transfer with a flexible rate control, we have employed the Fine Granular Scalability (FGS) [4, 5, 6] as the video coding algorithm to accomplish scalable rate adjustment, and the TFRC as the underlying rate control protocol.

Furthermore, taking into account the quality degradation caused by packet loss, we propose a video quality adjustment mechanism for stable, high-quality video transfer in a lossy environment that employs the FEC (Forward Error Correction) technique. Our mechanism adjusts the video quality in accordance with the TFRC rate, the packet loss probability, and the resultant video quality. Through simulation experiments, we show that our proposed method can provide high-quality, stable and TCP-friendly video transfer even in the unstable and lossy Internet.

The paper is organized as follows. In Section 2, we briefly introduce the FGS video coding algorithm and the TFRC mechanism. In Section 3, we consider several issues related to TCP-friendly video transfer and propose video-rate adjustment methods. In Section 4, we investigate the influence of packet loss on the perceived video quality. Then, we propose a dynamic quality adaptation mechanism for a lossy Internet environment. Finally, we summarize our paper and outline our future work in Section 5.

2 FGS and TFRC

In this section, we briefly introduce (1) the FGS (Fine Granular Scalability) algorithm [4, 5], which is excellent in regard to adaptation to the bandwidth variations among MPEG-4 video-coding standards, and (2) TFRC [3], which accomplishes fair-sharing of bandwidth among TCP and non-TCP connections.

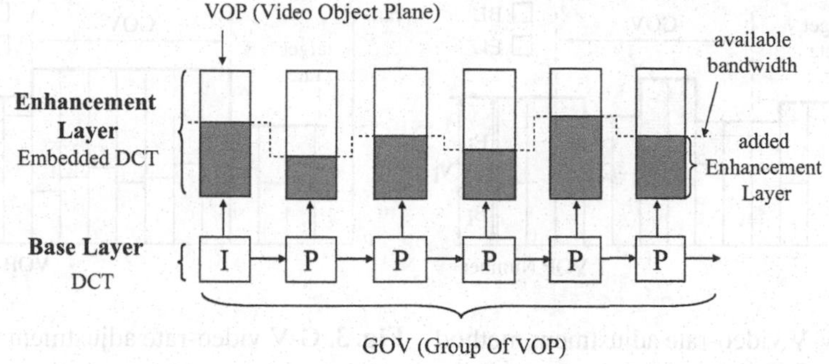

Fig. 1. Example of the FGS video structure

2.1 FGS Video Coding Algorithm

The FGS video coding algorithm [4, 5] is a compression method suitable for video streaming applications and is being incorporated into MPEG-4 standards. Figure 1 illustrates the basic structure of the FGS video stream. An FGS video stream is composed of a sequence of VOPs (Video Object Planes). The VOP is the basic unit of image data and is equivalent to the frame or picture of MPEG-1 and MPEG-2. A sequence beginning from an I-VOP is called a GOV (Group Of VOPs). An FGS video stream consists of two layers: the Base Layer (BL) and the Enhancement Layer (EL). The BL is generated using motion compensation and the DCT (Discrete Cosine Transform)-based conventional MPEG-4 coding algorithm, and it provides the minimum level of picture quality. The EL is generated from the BL data and the original frame. The embedded DCT method is employed for coding the EL, to obtain fine-granular scalable compression. By combining the BL and EL, one can enjoy higher quality video presentation.

Video quality depends on both the encoding parameters (quantizer scale, etc.) and the amount of supplemental EL data added. Even if only a little EL data is used in decoding a VOP, the perceived video quality will be improved. Losses of the BL data have a significant influence on perceived video quality because the BL is indispensable for decoding VOP.

2.2 TFRC: TCP-friendly Rate Control

TFRC [3] is a mechanism to ensure that a non-TCP connection behaves similarly to, but more stably than a TCP connection which traverses the same path. During a session, a sender transmits one or more control packets every RTT. On receiving the control packet the receiver returns feedback information required for calculating RTT and estimating the loss event rate p_{tfrc}. The sender then derives the estimated throughput of a TCP connection that competes for bandwidth on the path that the TFRC connection traverses. The estimated TCP throughput r_{TCP} is given as:

$$r_{TCP} \approx \frac{MTU}{RTT\sqrt{\frac{2p_{tfrc}}{3}} + T_0(3\sqrt{\frac{3p_{tfrc}}{8}})p_{tfrc}(1+32p_{tfrc}^2)}, \quad (1)$$

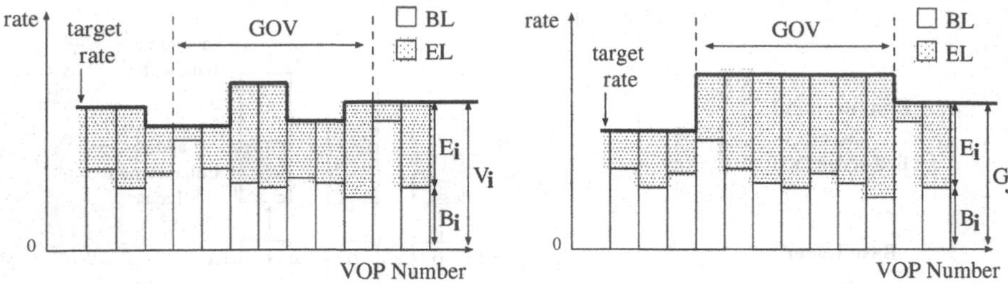

Fig. 2. V-V video-rate adjustment method **Fig. 3.** G-V video-rate adjustment method

where T_0 stands for retransmission timeout [1]. Finally, an application on TFRC adjusts its data rate to the estimated TCP throughput r_{TCP} (called "TFRC rate" in this paper) by means of, for example, video quality regulation. From now on, we call the estimated TCP throughput r_{TCP}, which determines the target rate of the application-level rate regulation, the "TFRC rate".

3 TCP-friendly MPEG-4 Video Transfer

In this section, we consider the several issues such as control interval, video rate adjustment, and BL rate violation and propose several variants of video-rate adjustment methods to accomplish TCP-friendly video transfer with consideration of the application-level performance, i.e., video quality.

3.1 Control Interval

The interval in which TFRC notifies the upper layer application of a new sending rate does not occur at the point in time when the application can change its data structure or amount. Considering the FGS video structure shown in Fig. 1, we propose a VOP-based method (V method) and a GOV-based method (G method). In the case of the V method, the target rate V_i of VOP$_i$ is defined as the TFRC rate at the beginning of VOP$_i$. Analogously, in the G-method, the target rate G_j of GOV$_j$ is defined as the TFRC rate at the beginning of GOV$_j$. They are illustrated in Figs. 2 through 4, where Fig. 2 corresponds to the V method, and Figs. 3 and 4 show the G method case.

A smaller interval leads to better agreement between the target rate and the TFRC rate determined by TFRC. As the instantaneous video rate can not fit the TFRC rate with a longer interval, TCP-friendly video transfer is performed when we observe the rate variation averaged over longer time. However, since video traffic is smoothed out by the underlying TFRC, a larger difference causes a longer buffering delay.

3.2 Video Rate Adjustment

Adjustment of the FGS video rate to the target rate is done by discarding a portion of the EL data. There are alternatives as to rate adjustment method, both VOP-based and GOV-based. We further consider two variants of the VOP-based adjustment, i.e., the V-V and

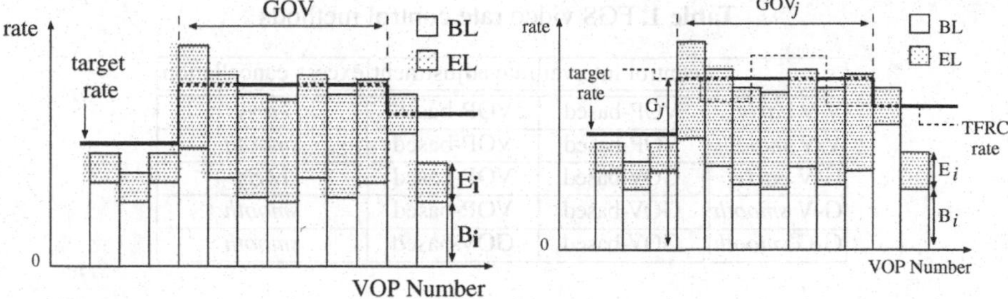

Fig. 4. G-G video-rate adjustment method

Fig. 5. G-G *smooth* video-rate adjustment method

G-V methods. In the case of the V-V method (Fig. 2), the target rate V_i of the VOP$_i$ is first determined by the V method from the TFRC rate, and the rate (or amount) of the additional EL data E_i is obtained by subtracting the BL data rate B_i from the target rate V_i. On the other hand, the G-V method (Fig. 3) first determines the target rate G_j of a GOV$_j$ by means of the G method, then applies the identical rate to all VOPs in the GOV ($V_i = G_j$, VOP$_i \in$ GOV$_j$). Then, the video rate adjustment is performed VOP by VOP as $E_i = G_j - B_i$ for a VOP$_i \in$ GOV$_j$ in the G-V method. In the GOV-based rate adjustment method, called as a G-G method (Fig. 4), the video rate averaged over GOV$_j$ matches the target rate G_j. The rate of the EL data added to each VOP in the GOV is given as $E_i = (NG_i - \sum_{VOP_k \in GOV_j} B_k)/N$ where N stands for the number of VOPs in a GOV and is identical among all VOPs in the GOV. The G-G method is supposed to achieve a smooth variation in video quality by equalizing the amount of supplemental EL data among VOPs, while the video rate may instantaneously exceed the target rate.

3.3 BL Rate Violation

IF the available bandwidth becomes smaller because of network congestion, the BL rate may occasionally exceed the target rate. Since the BL data are crucial for video decoding, they are always sent out and the excess is managed by reducing the EL rate of the following VOPs or GOVs. In the *smooth* method, the excess is divided and equally assigned to the rest of VOPs in the GOV, thus the average rate of several VOPs matches the target rate. An example of the "G-G *smooth* method" is illustrated in Fig. 5. The *early* method, to cancel the excess as fast as possible, assigns much of the excess to the VOP right after the greedy VOP, thus only a few VOPs are affected.

Table 1 summarizes the possible rate control methods obtained by combining the above-mentioned ones. We should note here that there is no G-G *early* method because the amount of EL data added to each VOP in the GOV must be identical in the "G-G" method.

Table 1. FGS video rate control methods

	control interval	rate adjustment	excess cancellation
V-V *early*	VOP-based	VOP-based	*early*
V-V *smooth*	VOP-based	VOP-based	*smooth*
G-V *early*	GOV-based	VOP-based	*early*
G-V *smooth*	GOV-based	VOP-based	*smooth*
G-G *smooth*	GOV-based	GOV-based	*smooth*

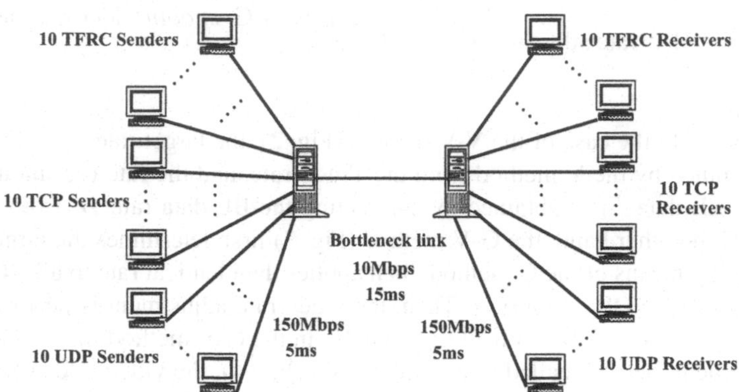

Fig. 6. Simulation network model

3.4 Simulation

In this subsection, we compare the five control methods proposed in the preceding section through simulation experiments. We use two MPEG-4 test sequences, "coastguard" and "akiyo". They are QCIF large (176×144 pixels), consisting of 300 frames. They are coded at 30 frames-per-second (30 fps) with a GOV structure of one I-VOP and 14 P-VOPs. No B-VOP is used, thus avoiding an inherent coding delay. We choose the quantization scale from 1 to 31 in order to investigate the effect of coding parameters on the coded video. The simulated network is depicted in Fig. 6, where ten TFRC connections, ten TCP connections and ten UDP connections compete for the bottleneck bandwidth. Link capacity and one-way delays are indicated beside the respective connections. We first conducted simulation experiments by assuming that persistent applications were using those connections. Then, our methods were simulated by assuming that MPEG-4 video streams were transferred over the TFRC connections whose rate variations had been obtained through the preceding experiments. Due to limited space, we only show results of the comparison between the V-V *smooth* and the G-G *smooth*" methods for "coastguard".

Figures 7 and 8, we show trajectories of target-rate variation and quality variation, respectively. We should note here again that the video traffic generated by our methods is smoothed by the underlying TFRC. Thus, the video traffic injected into networks

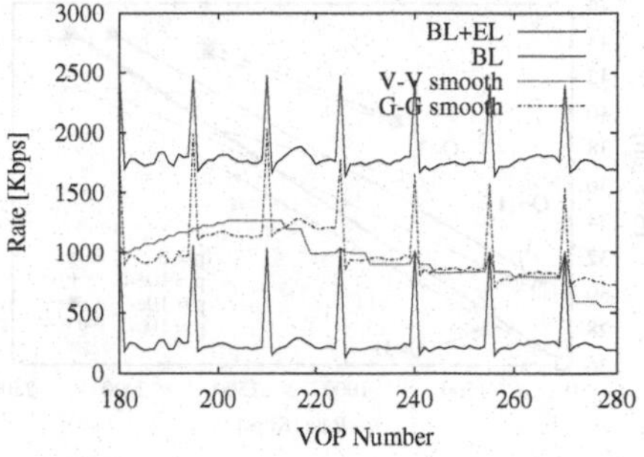

Fig. 7. Video rate (V-V *smooth* vs. G-G *smooth*)

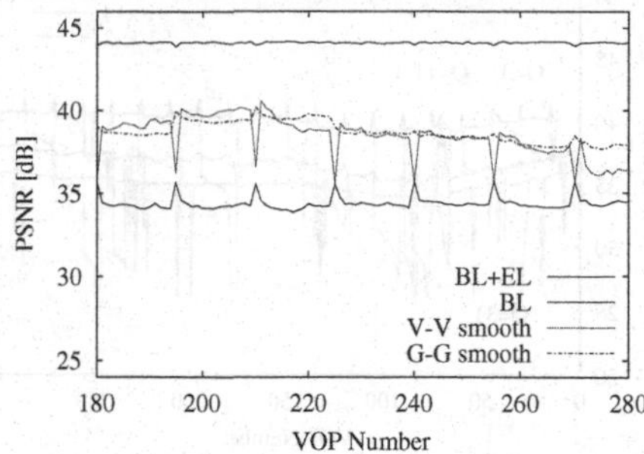

Fig. 8. Video quality (V-V *smooth* vs. G-G *smooth*)

becomes TCP-friendly. In these figures, we employ the quantizer scale of 3. "BL" and "BL+EL" correspond to the result of transmitting and decoding the BL data only and the BL with the entire EL data, respectively. Although it is not shown in the figure, the FGS video rate follows the TFRC rate in the V-V *smooth* method where the amount of EL data to be added is determined VOP by VOP. The step-wise variation of the VOP-based target rate is due to the TFRC rate variation. On the other hand, the variation of the FGS video rate in the G-G *smooth* method resembles that of the BL rate because an identical amount of EL data is added to each VOP. In addition, the variation of video quality in the G-G *smooth* method is more gradual than in the V-V *smooth* method. However, the instantaneous video rate of the G-G *smooth* method is not necessarily TCP-friendly and it may be necessary to introduce a smoothing delay to make the data sending rate TCP-friendly. Furthermore, without an appropriate estimator of BL rate variation, the G-G *smooth* method introduces one GOV-time delay because all of the BL rates B_i of the VOPs in the GOV must be known in advance, to determine the amount of EL data to be added to each VOP. Thus, the G-G *smooth* method is preferable when the video application emphasizes video quality while the V-V *smooth* method is faithful to the rate variation of TFRC sessions.

4 TCP-friendly MPEG-4 Video Transfer in a Lossy Environment

In the previous simulation, we did not take into account the quality degradation caused by packet loss since we wanted to investigate the ideal performance of the video-rate adjustment methods. However, once one or more packets are lost, the video quality deteriorates considerably. Especially when a packet that constitutes BL data is lost, the VOP and its succeeding VOPs are heavily affected.

In this section, we propose a video-quality adjustment mechanism is based on the G-G *smooth* method that is capable of high-quality video transfer in a lossy environment in cooperation with the FEC. We will start by investigating the basic characteristics of the G-G *smooth* method in a lossy environment.

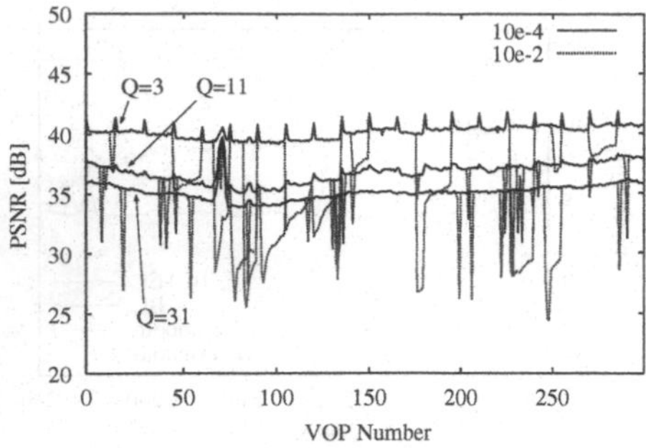

Fig. 9. FGS video quality ($p = 10^{-4}$ and 10^{-2}, rate = 1 Mbps)

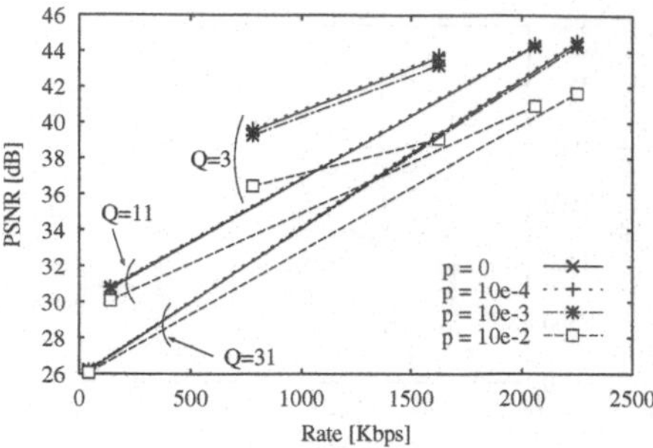

Fig. 10. Relationship between average rate and video quality

4.1 Investigation of Influence of Packet Loss on Video Quality

Figure 9 shows the results of experiments for "coastguard" transferred over a lossy connection with packet loss probability p of 10^{-4} or 10^{-2}. The video is coded at an average rate of 1 Mbps. The figure shows the video quality variation when only successfully received video data are decoded. In this figure, results for three quantizer scales, i.e., 3, 11, and 31, are depicted. In the experiments, each 1-Kbyte packet is examined and lost at the given packet loss probability.

For the packet loss probability of 10^{-4}, none of the about 1,860 packets was lost and no quality degradation was observed. As the packet loss probability increases, a video stream with a smaller quantizer scale begins to be affected. It is shown that the smaller the quantizer scale is, the higher the level of degradation is. The degradation of video quality is notable in the case of $Q = 3$ owing to the loss of BL data. In the cases of $Q = 11$ and 31, annoying spike-shaped degradations were observed. The subjective video qualities in terms of MOS (Mean Opinion Score) were 3.95, 3.88, 2.63 for video streams with of $Q = 3$, 11, and 31, respectively, at 10^{-2} of packet loss probability. From the result of MOS evaluation, we can see that users tend to feel uncomfortable when there are spike-shaped degradations caused by the loss of EL data (these were perceived as flickers). On the other hand, the degradations caused by the loss of BL data, which last for a while, were acceptable. Therefore, we conclude that we need to protect not only BL, but also the EL data, and that it is better to employ as small a quantizer scale as possible.

We also conducted experiments for the other settings of the video rate. All results are summarized in Fig. 10 in terms of average SNR, average video rate, and quantizer scale, for different levels of packet loss. Despite the trajectories of quality variation depicted in the figures, average SNR is higher when we choose a smaller quantizer scale for a given rate and packet loss probability.

However, the perceived video quality with such a smaller quantizer scale may suddenly fluctuate in a lossy environment. Therefore, for a stable-quality video transfer, we should employ a mechanism to minimize the influence of packet loss on video quality.

4.2 Dynamic Quality Adaptation Mechanism with Packet-Loss Protection for TCP-friendly MPEG-4 FGS Video Transfer

Our mechanism is based on the G-G *smooth* one, but it also reacts against packet loss and employs the FEC technique to protect video data from packet loss. The idea of Forward Error Correction (FEC) [7, 8] in a packet oriented video transmission scheme is to generate redundant packets at the sender, which can be used at the receiver to recover lost video data packets. When we send f redundant packets in addition to k information packets, all information packets can be reconstructed from successfully received packets as far as the number of lost packets is below f. We use Reed-Solomon (RS) codes for forward error correction in this paper.

To create an efficient control, we have to determine how many redundant packets should be added to the video packets in a lossy environment. Specifically, introducing redundancy suppresses the bandwidth that the server can use to transfer video data for a given TFRC rate. When the system is congested, the target rate given by the TFRC algorithm is decreased. At the same time, the sender should increase the number of redundant packets to achieve a higher level of protection against the higher loss probability. Consequently, the bandwidth left for the video data is greatly suppressed, and the video quality is expected to deteriorate.

To tackle the problem stated above, we determined the required amount of redundant packets for protection. Table 2 summarizes results of preliminary evaluations on the relationships among the average packet loss probability P_{ave}, the target packet loss probability P_{target}, and the resultant redundancy, for a video stream of average 1 Mbps. Each pair of values k and r_{FEC} corresponds to the number of redundant packets in a GOV and the bandwidth that FEC requires. In the experiment, the average number of packets admitted per one GOV time is 63 for video traffic of 1 Mbps. For example, the top and leftmost value of Table 2, i.e., 7, 112 Kbps, indicates that in order to improve the packet loss probability from $P_{ave} = 10^{-1}$ to $P_{target} = 10^{-2}$, we must send seven redundant packets out of 63 packets in one GOV, and then allocate 112 Kbps to the rate for FEC redundant data. The table also provides results of evaluations on the effect of the FEC protection in lossy environments. The pair in parentheses stands for the average SNR in terms of dB when FEC is employed to achieve the target probability P_{target} under a lossy condition of corresponding P_{ave} when the quantizer scale is 3 or 31. The corresponding average SNR which is obtained when no protection is applied, is indicated in the leftmost columns.

Although further results are omitted from the manuscript due to space limitations, we have conducted several experiments applying varying packet loss probability and target loss probability to sending FGS video data coded with various quantizer scales over a 1 Mbps session. We verified that independent of the actual packet loss probability and quantizer scale value, setting the target loss probability to 10^{-3} or 10^{-4} leads to the highest video quality.

The dynamic quality adaptation mechanism in cooperation with FEC is as follows.

1. Determine the TCP-friendly sending rate G_j by Eq. (1) at the beginning of GOV$_j$, based on the feedback information (RTT, p_{tfrc}) obtained by the TFRC algorithm.

Table 2. FEC redundancy packets and rate (Sending rate = 1 Mbps, 63 packets within one GOV)

P_{ave}	P_{target}			
	10^{-2}	10^{-3}	10^{-4}	10^{-5}
10^{-1} $(25.46, 26.83)$	7, 112 Kbps (36.27, 33.71)	10, 160 Kbps (39.02, 34.10)	13, 208 Kbps (39.59, 33.91)	15, 240 Kbps (39.64, 33.74)
10^{-2} $(36.74, 33.45)$		2, 32 Kbps (40.01, 34.80)	3, 48 Kbps (39.98, 34.83)	4, 64 Kbps (39.93, 34.73)
10^{-3} $(39.81, 34.81)$			1, 16 Kbps (39.98, 35.05)	2, 32 Kbps (40.08, 34.94)
10^{-4} $(40.22, 35.14)$				1, 16 Kbps (39.97, 35.05)

2. Derive the packet loss probability observed in the preceding interval, P_{j-1}, by dividing the number of lost packets by that of sent packets in the preceding GOV time.
3. Calculate the smoothed packet loss probability P_{ave} by applying the exponential moving average employed in TFRC as,

$$P_{ave} = \frac{\sum_{i=1}^{8} w_i P_{j-i}}{\sum_{i=1}^{8} w_i}. \tag{2}$$

   The weighting parameter w_i is, 1 for $1 \leq i \leq 4$ and $(9-i)/5$ for $4 < i \leq 8$.
4. Determine the redundant rate r_{FEC} to achieve the target loss probability P_{target}. r_{FEC} is determined as $G_j \times f/n$ where f and n are the number of redundant packets and the number of packets admitted in a GOV time. f is derived as;

$$f = \min_f \left\{ \sum_{i=f+1}^{n} \binom{n}{i} P_{ave}^i (1-P_{ave})^{n-i} \leq 1-(1-P_{target})^{n-f} \right\}. \tag{3}$$

5. Determine the sending rate R_{video} allocated to the video data as $G_j - r_{FEC}$ and apply the G-G *smooth* mechanism to R_{video}.
6. Generate redundant packets at the redundant rate r_{FEC} and send them with the video packets.

4.3 Simulation Results for a Lossy Environment

Figures 11 and 12 show simulation results of the method with the FEC technique and one without it, respectively. In the case of the original G-G *smooth* method, the quantizer scale of 5 is used during a session. The quantizer scale of 6 is applied in the with-FEC case since the actual video rate is suppressed by the FEC packets. Figure 11 shows the rate variation of the bandwidth allocated to the video data, R_{video}. Trajectories of perceived video quality under the lossy environment are shown in Fig. 12. In these figures, the target packet loss probability P_{target} is set to 10^{-3} or 10^{-4}.

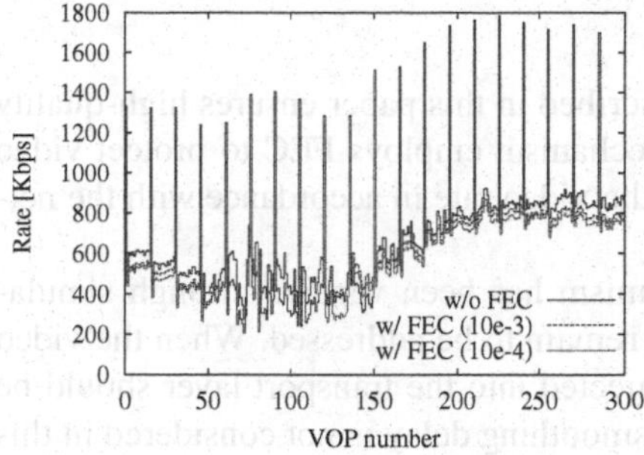

Fig. 11. Video rate variation (w/ FEC vs. w/o FEC)

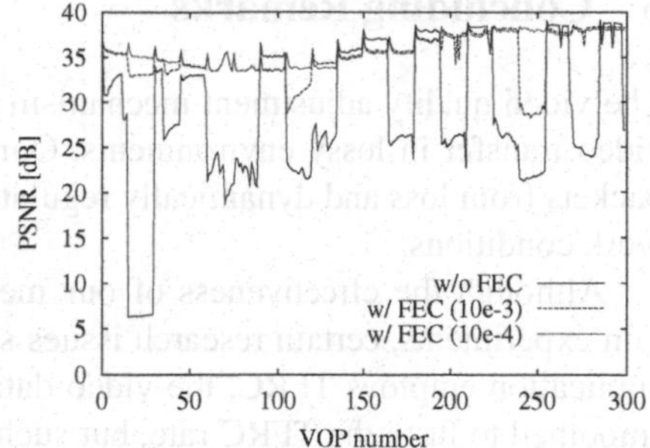

Fig. 12. Video quality variation (w/ FEC vs. w/o FEC)

Fig. 13. Displayed VOP 70

We can see that the FEC technique is effective in protecting the video quality from being degraded by packet loss. Although packets are lost at probability 10^{-2} to 10^{-1} in the network, the loss at the receiver is only 10^{-6} to 10^{-2}, with a help of FEC. However, quality degradation due to loss of BL data is noticeable when FEC is targeting the probability of 10^{-3}. This implies that the protection targeted at 10^{-3} is not effective enough against instantaneous but serious packet loss.

When we apply the target loss probability of 10^{-4}, the bandwidth available to the video data becomes obviously lower than that in the case of $P_{target} = 10^{-3}$ because of the increased FEC redundancy. As a result of increased redundancy, the video quality becomes lower for loss-free GOVs. However, the trajectories of quality are kept stable owing to the higher level of protection. Thus, on the basis of the above observations, we conclude that choosing the target packet loss probability of 10^{-4} can effectively protect the video packet from loss with the sacrifice of only a slight quality degradation, at most 1.5 dB in our experiments. Examples of images displayed on the monitor are shown in Fig. 13 for the method without FEC (left) and with FEC (right), respectively.

5 Concluding Remarks

The video quality adjustment mechanism described in this paper ensures high-quality video transfer in lossy environments. Our mechanism employs FEC to protect video packets from loss and dynamically regulates the video rate in accordance with the network conditions.

Although the effectiveness of our mechanism has been verified through simulation experiments, certain research issues still remain to be addressed. When the video application employs TFRC, the video data injected into the transport layer should be smoothed to fit to the TFRC rate, but such a smoothing delay is not considered in this paper. We should also consider improved mechanisms to change the quantizer scale dynamically in accordance with the packet loss probability, in order to attain real-time video transfer that gives high and stable video quality under both loss-free and lossy conditions. In addition, we only considered a random loss model; we should take into account bursty and correlated packet losses for which FEC cannot recover missing packets very well.

References

[1] Jitendra Padhye, Victor Firoiu, Don Towsley, and Jim Kurose, "Modeling TCP throughput: A simple model and its empirical validation," in *Proceedings of ACM SIGCOMM'98*, September 1998, vol. 28, pp. 303–314.

[2] Jitendra Padhye, Jim Kurose, Don Towsley, and Rajeev Koodli, "A model based TCP-friendly rate control protocol," in *Proceedings of NOSSDAV'99*, June 1999.

[3] Mark Handley, Jitendra Padhye, Sally Floyd, and Jorg Widmer, "TCP Friendly Rate Control (TFRC): Protocol Specification," Internet-Draft draft-ietf-tsvwg-tfrc-03.txt, work in progress, July 2001.

[4] Hayder Radha and Yingwei Chen, "Fine-granular-scalable video for packet networks," in *Proceedings of Packet Video'99*, April 1999.

[5] Hayder Radha, Mihaela van der Schaar, and Yingwei Chen, "The MPEG-4 fine-grained scalable video coding method for multimedia streaming over IP," *IEEE Transactions on Multimedia*, vol. 3, no. 1, pp. 53–68, March 2001.

[6] Text of ISO/IEC 14496-2, "MPEG-4 Video FGS v.4.0," Proposed Draft Amendment (PDAM), Noordwijkerhout, the Netherlands, March 2000.

[7] Uwe Horn, Klaus Stuhlmüller, M. Link, and B. Girod, "Robust Internet video transmission based on scalable coding and unequal error protection," *Image Communication*, vol. 15, no. 1–2, pp. 77–94, September 1999.

[8] Qian Zhang, Wenwu Zhu, and Ya-Qin Zhang, "Resource allocation for multimedia streaming over the Internet," *IEEE Transactions on Multimedia*, vol. 3, no. 3, pp. 339–355, September 2001.

Traffic Sensitive Active Queue Management for Improved Multimedia Streaming

Vishal Phirke, Mark Claypool, and Robert Kinicki

Worcester Polytechnic Institute, Worcester, MA 01609, USA
{claypool|rek}@cs.wpi.edu

Abstract. The Internet, which has traditionally supported throughput-sensitive applications such as email and file transfer, is increasingly supporting delay-sensitive multimedia applications such as interactive audio. These delay-sensitive applications would often rather sacrifice some throughput for lower delay. Unfortunately, the current Internet does not offer choices in the amount of delay or throughput an application receives, but instead provides monolithic best-effort service to all applications. This paper proposes and evaluates a new Active Queue Management (AQM) technique that employs source hints to provide service at network routers that is sensitive to the Quality of Service (QoS) expectations for a variety of applications. Applications indicate their delay or throughput sensitivity via a delay hint in their outgoing packets. The router, which we call RED-Boston, uses the delay hints to dynamically adjust the router to yield better delay performance for delay-sensitive applications and better throughput for throughput-sensitive applications. Using a new QoS metric, our simulations demonstrate that RED-Boston yields higher QoS than an adaptive version of RED for both throughput-sensitive flows and delay-sensitive flows. RED-Boston operates equally well in all traffic scenarios and fits the current best-effort Internet environment without requiring traffic monitoring.

1 Introduction

The Internet today carries traffic for applications with a wide range of delay and loss requirements. Traditional applications such as FTP and E-mail are primarily concerned with throughput, while Web traffic is moderately sensitive to delay as well as throughput. Emerging applications such as IP telephony, video conferencing and networked games have different requirements in terms of throughput and delay than these traditional applications. In particular, interactive multimedia applications, unlike traditional applications, have more stringent delay constraints than loss constraints. Moreover, with the use of repair techniques [3, 17, 20, 21] packet losses can be partially or fully concealed, enabling multimedia applications to operate over a wide range of losses, and leaving end-to-end delays as the major impediment to acceptable quality.

M. Ajmone Marsan et al. (Eds.): QoS-IP 2003, LNCS 2601, pp. 551–566, 2003.

Unfortunately, current Internet routers are not able to provide applications a choice in Quality of Service (QoS).[1] Due to the simplicity of the FIFO queuing mechanism, drop-tail buffers are the most widely used queuing scheme in Internet routers today. When drop-tail buffers overflow, newly arriving packets are dropped regardless of the application type of the arriving packet. To accommodate bursty traffic, drop-tail routers on the Internet backbone are over-provisioned with large FIFO buffers. When faced with persistent congestion, these drop-tail routers yield high delays for all flows passing through the bottle-necked router. This best-effort service provides no consideration for multimedia flows that can be severely affected by high delays.

Typical AQM mechanisms [8, 9, 6, 2, 15, 12] provide equal treatment to incoming traffic and tend to be tuned for high throughput without specific consideration for the traversing application's delay requirements. Other AQM mechanisms [25, 16, 4, 1, 18, 7] monitor the bandwidth use by unresponsive flows but ignore the willingness of interactive multimedia applications to accept reduced throughput if accompanied by reduced delay.

ABE [13] provides a queue management mechanism for low delay traffic. In ABE, delay-sensitive applications can sacrifice throughput for lower delays. However, ABE traffic classification is rigid in that applications are either delay-sensitive or throughput-sensitive without the applications themselves being able to choose relative degrees of sensitivity to throughput and delay.

DiffServ [11, 14] and other class-based approaches [22, 5] try to give differentiated service to traffic aggregates. However, they require complicated mechanisms to negotiate service level agreements. In addition, DiffServ architectures require traffic monitors, markers, traffic shapers, classifiers and droppers to enable components to work together.

We present a new AQM approach, called *RED-Boston*, for IP networks that provides QoS service for applications with a variety of delay and throughput requirements, such as file transfer, Web browsing and interactive audio. Unlike approaches that provide fixed classes of service, each application sending traffic into the RED-Boston router chooses a customized delay-throughput trade-off based on its own requirements. The service is still best-effort in that it requires no additional policing mechanisms, charging mechanisms or usage control. Under the proposed service, RED-Boston routers operate equally well under scenarios with only traditional traffic and operate better under scenarios with mixed or multimedia only traffic.

In our approach, applications mark each packet with a suggested delay, referred to as a *delay hint*, indicating the relative importance of delay versus throughput. At a congested router, the queue manager, which we have named *RED-Boston*[2], uses the delay hints of all packets to calculate an average delay

[1] Throughout this work, we use the term QoS to refer explicitly to *delay* and *throughput* provided by the network.

[2] Similar to various versions of TCP, which are named after cities in Nevada, we have named our version of the RED [8] active queue management technique after a city in Massachusetts.

hint and a target average queue size chosen to best fit the aggregate traffic. RED-Boston inserts packets in the outbound queue based on their delay hint relative to the average delay hint. Packets with a delay hint lower than the average delay hint are inserted towards the front of the queue, while being dropped with a higher than average drop probability. Conversely, packets with a delay hint higher than the average delay hint are inserted towards the back of the queue, while being dropped with a lower than average drop probability. Thus, using a low delay hint, delay-sensitive applications experience reduced end-to-end delay, while receiving the lower throughput resulting from an increased drop rate. Similarly, using a high delay hint, throughput-sensitive applications receive an increased throughput resulting from a decreased drop rate, while having an increased end-to-end delay.

We evaluate RED-Boston via simulation under a wide variety of traffic mixes. The results show that RED-Boston achieves better overall application performance for both delay-sensitive multimedia applications and throughput-sensitive applications. As traffic mix changes from mostly throughput-sensitive to mostly delay-sensitive, RED-Boston gives consistently better performance than Adaptive RED [9].

This paper is organized as follows: Section 2 explains the RED-Boston mechanism and provides an example of how an application may use the RED-Boston service; Section 3 describes simulation experiments used to evaluate RED-Boston; Section 4 analyzes the simulation results; and Section 5 contains conclusions and suggests possible future work.

2 RED-Boston

In RED-Boston, applications mark each outgoing packet with a delay hint that suggests the relative importance of delay versus throughput for the packet. Delay hints are explained in Section 2.1. The router mechanism itself is explained in Section 2.2. The router uses the delay hints to calculate a target average queue size as explained in Section 2.2. Packets experience a drop probability based on their delay hint and average drop probability, as explained in Section 2.2, while being inserted in the outgoing queue based on their delay hint relative to the average delay hint, as explained in Section 2.2. Lastly, the space and processing requirements for RED-Boston are discussed in Section 2.3.

2.1 Delay Hints

In RED-Boston, application end-hosts indicate their sensitivity to queuing delay as source hints in the IP packet header that we refer to as *delay hints*. The delay hint is not an absolute bound on queuing delay, but is rather a suggestion to Internet routers as to the relative importance of delay versus throughput.

RED-Boston does not guarantee delay based on the delay hint, but the delay hint facilitates the router in providing an appropriate service. At the RED-Boston router, a relatively low delay hint indicates an application's desire for lower delay and lower throughput while a higher delay hint suggests the application prefers higher throughput even if this implies higher delays. Applications such as FTP or Email that want to minimize overall transfer time without significant concern over individual packet delays would choose a high delay hint. Applications such as multimedia streaming that seek to minimize end-to-end delay could choose to send a low delay hint. However, under congested network conditions these applications may discover that the low delay hint also yields unacceptably low throughput. Applications could then raise their delay hints until measured throughput reached acceptable levels. Thus, sources are able to vary their delay hints in any manner that increases their user-defined QoS. Packets that carry no delay hints are handled using a default delay hint that corresponds to the router's target queue size.

Figure 1 presents a simple illustration of how a delay-sensitive TCP-friendly application might use delay hints. An interactive videoconference running over a company T1 link competes with n TCP flows for the bandwidth. The videoconference requires a minimum data rate, R, of about 384 Kbps[3] to insure acceptable video quality. This rate is depicted by the horizontal dotted line in Figure 1. Under these conditions, the videoconference is free to choose the source hint. The curved dashed lines in Figure 1 depict the approximate bandwidth an application would receive for source hints of 20 ms, 40 ms, 60 ms and 80 ms (reading left to right). The higher the source hint, the more bandwidth the videoconference will receive, but the higher the delay. The lower the source hint, the less bandwidth the videoconference will receive, but the lower the delay. The "best" source hint for the application depends upon the perceived quality, pq, for each throughput and delay combination. For this simplified example, we assume that pq is 0 when the data rate, R', is less than R and increases inversely with the delay for $R' > R$. In other words, once the videoconference obtains its minimum data rate, the greatest benefit to perceived quality comes from lower delay. Under these conditions, the vertical lines in Figure 1 indicate regions A, B, C, D and E where the videoconference would choose different delay hints. In region A, the required minimum bandwidth can be obtained by using the lowest delay hint, 20 ms. As the T1 link becomes more congested, the videoconference would choose source hints of 40 ms and 60 ms for regions B and C respectively. In region D, the videoconference is forced to use the largest source hint, 80 ms, to obtain the minimum bandwidth. This is the same source hint as is likely used by the competing TCP flows. In region E, even with a large source hint, the available bandwidth drops below the minimum acceptable rate, in which case the videoconference user would probably terminate the connection. Note that this example is greatly simplified and we would expect perceived quality functions that trade off throughput and latency in a more sophisticated manner would be used.

[3] A typical minimum data rate for an H.261 videoconference.

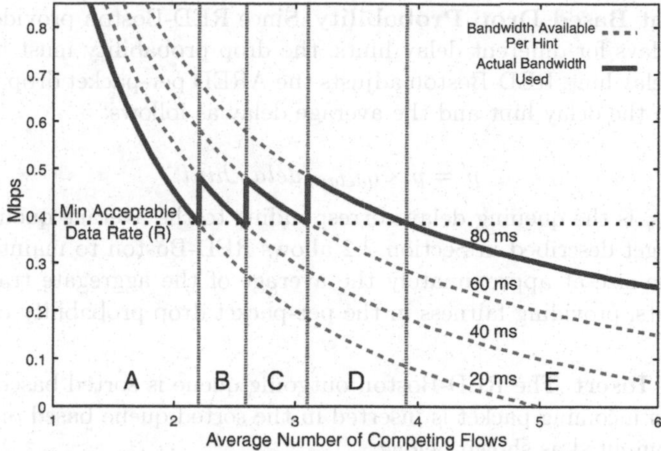

Fig. 1. Possible Delay Hint Strategy for an Interactive Multimedia Application

2.2 Mechanism

Arriving delay hints affect three aspects of the RED-Boston router mechanism. First, the target average queue size moves to adjust the average queue size as described in Section 2.2. Second, the delay hint determines the drop probability, where higher delay hints mean lower drop probabilities and vice versa, as explained in Section 2.2. Third, the delay hint also determines where the incoming packet is placed in the outgoing queue as described in Section 2.2.

Moving Target Adaptive RED [9] extends RED [8] by keeping the average queuing delay q_{ave} within a target range. In ARED, the targeted average queue size is independent of traffic QoS requirements. We argue that the targeted average queue size should be based on the incoming traffic's requirements. RED-Boston extends ARED by providing a moving target queue size based on an exponentially weighted moving average of the delay hints of the incoming packets, as shown below:

$$target = (1 - w_t)target + w_t C \times delay\_hint/(packet\_size \times 8) \qquad (1)$$

where C is the router's outgoing link capacity and w_t is the weight for calculating the exponentially weighted moving average (set to $w_t = 0.02$ in our implementation). The weighted average allows the queue target to stabilize even if the individual packets arriving from different flows have different delay hints. When overall traffic changes from throughput-sensitive to delay-sensitive, the target reflects the average queuing delay requirements of the incoming packets.

Delay Hint Based Drop Probability Since RED-Boston provides different queuing delays for different delay hints, the drop probability must be adjusted based on delay hint. RED-Boston adjusts the ARED per-packet drop probability p' based on the delay hint and the average delay as follows:

$$p' = p \times q_{delay} / delay\_hint \tag{2}$$

where q_{delay} is the queuing delay corresponding to the average queue size. The moving target described in Section 2.2 allows RED-Boston to maintain the average queue size at approximately the average of the aggregate traffic's delay requirements, providing fairness in the per-packet drop probability calculation.

Weighted Insert The RED-Boston outgoing queue is sorted based on packet weights. An incoming packet is inserted in the sorted queue based on its weight which is computed as shown below:

$$weight = arrival\ time + delay\_hint \tag{3}$$

Arrival time is used as an aging mechanism in this calculation to avoid packet starvation. Under normal conditions a flow's delay hints will not vary and differences in arrival times will prevent packet reordering. Since packets are served from the front of the sorted queue, the packet with the earliest deadline will be served first. Note the delay hint is not an upper bound on queuing delay for the packet but rather a relative hint such that packets with lower delay hints experience lower delays than packets with higher delay hints. However, our results show that on average, packets receive a queuing delay close to the delay hint specified.

Thus, RED-Boston's *weighted insert* provides delay-sensitive packets with a relatively lower queuing delay. On the other hand, RED-Boston's *delay hint based drop probability* compensates by providing delay tolerant packets with a lower drop probability and hence higher throughput.

Algorithm Figure 2 summarizes the RED-Boston algorithm. Each packet contains a delay hint, as described in Section 2.1. For each incoming packet, the delay hint is used to update the moving target for average queue size, kept by the router, as described in Section 2.2. If there is extreme congestion, indicated by the queue average being above max_{th}, the packet is dropped. If there is congestion, as indicated by the queue average being between min_{th} and max_{th}, the drop probability for the packet is computed based on the queue parameters, the *delay_hint*, and the average queuing delay, *delay*, as described in Section 2.2. If the packet is not dropped, it is inserted in the queue based on a weighting of its arrival time and delay hint, as described in Section 2.2.

2.3 Overhead

To use the QoS services provided by RED-Boston, IP packets have to be labeled with delay hints. Based on discussion in [26], there are from 4 to 17 bits in the

```
on receiving packet pkt:
    // Calculate moving target, (see Section 2.2)
    target = (1 − w_t) target + w_t C × pkt.delay_hint/(pkt.size × 8)
    if (q_avg>=max_th) then

        dropPacket(pkt, 1)

    elseif (q_avg>=min_th) then

        // Calc drop prob p, based on RED (see [8])
        p=calcDropP(q_avg,min_th,max_th,max_p)

        //Calc delay hint based drop prob, p' (see Section 2.2)
        p'=p×(q_delay/delay_hint)

        if (!dropPacket(pkt,p')) then
        // Insert in the queue based on weight, (see Section 2.2)
        weight=arrival time+pkt.delay
        insertPacket(pkt,weight)

Every 500 msec: // Default in [9]
    // Adjust max_p so q_ave hits target, (see [9] or [23])
    max_p= adaptMaxP(max_p,target,q_avg)
```

Fig. 2. RED-Boston Algorithm

IP header that may be available to carry delay hints. Using milliseconds as the units for delay hints, 10 bits covers queuing delays from 0 to 1023 ms. Since 10 ms granularity is more than sufficient for most applications, the number of bits needed can be reduced to 7.

The labeling of delay hints itself can be done either by end hosts, most likely the applications, or by edge routers at the ingress of a network. As the example in Section 2.1 shows, the labeling of packets by applications gives them flexibility to adjust the hints based on measured levels of throughput and delay to maximize the application-specific QoS. However, if the edge routers label hints, legacy applications can be supported without change.

The RED-Boston router has to read the labels from the incoming packets, an overhead similar to that in other approaches such as ABE [13], AF [11] and CBT [22]. RED-Boston calculates the queue average target, a calculation similar to the average queue calculation in ARED.

RED-Boston's weighted insert mechanism is slightly more complex than ABE's duplicate scheduling with deadlines mechanism, but provides much more flexibility. The weighted insert mechanism can be implemented using a probabilistic data structure such as skip lists [24], giving complexity $O(log(n))$.

The overall complexity of RED-Boston is much less than IntServ approaches and is comparable to class based approaches, while providing better QoS service than typical class based approaches. Unlike DiffServ approaches, RED-Boston does not require negotiation of service level agreements and does not required traffic monitoring and shaping, and integrates more easily into the current best-effort Internet environment.

3 Experiments

This section discusses the experimental methods and NS-2 simulation details associated with our comparison study of the performance of RED-Boston versus ARED. The NS-2 simulator [19] provides the ability to simulate drop-tail, RED and ARED routers. Additionally, NS-2 includes code to simulate both TCP protocols and TCP-friendly rate controlled protocols such as TFRC [10]. RED-Boston was simulated by extending the ARED implementation with the *moving target*, *weighted insert* and *delay hint based drop probability* algorithms as explained in Section 2 and, TCP NewReno and TFRC were modified to send delay hints.

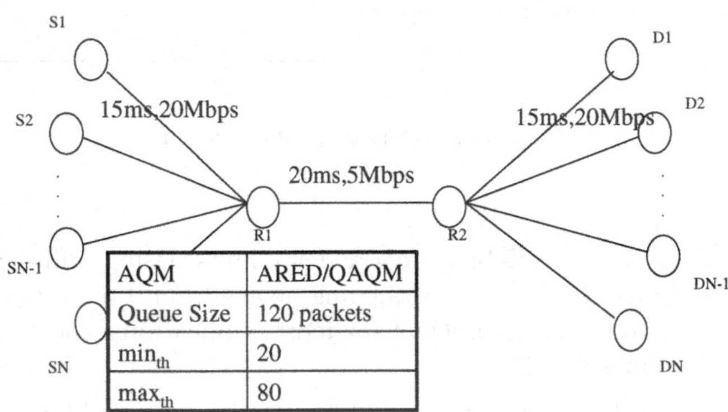

Fig. 3. Network Topology

The generic network topology used for all experiments is shown in Figure 3. S1-SN are traffic sources and D1-DN are destinations. The S-D pairs were varied to provide traffic mixes that included different mixes of TCP NewReno and TFRC flows (see Section 4), where the TFRC flows are meant to represent TCP-friendly multimedia flows. All sources send fixed-length 1000-byte packets. All links connecting sources to router R1 and all links connecting destinations to router R2 have a 20 Mbps capacity and a 15 ms delay. With the bandwidth and delay of the bottleneck link going from R1 to R2 set at 5 Mbps and 20 ms respectively, this topology is such that packet drops only occur at R1, where all measurements are made. For both RED-Boston and ARED, the queue size at the congested router is 120 packets and min_{th} and max_{th} are set to 20 packets and 80 packets respectively. All simulated flows start at time 0 sec and end at time 100 sec. Performance measurements are taken during the stable interval between 20 sec and 80 sec to avoid transient conditions from startup and stopping.

To ascertain overall router performance, we measure link utilization, average queue size and drop rate. Moreover, since application performance realistically involves tradeoffs between throughput and delay, we believe analyzing either alone is not sufficient. We propose a new metric, *QoS*, that provides a quantitative performance metric designed to capture to some degree the nature of the throughput-delay tradeoff. The correct choice of QoS function depends upon the the application's sensitivity to delay and throughput. As illustrated in Section 2.1, an application may dynamically adjust its delay hints to current network conditions in attempt to improve its QoS. To include a wide spectrum of individual application types that can select their own tailored QoS objective, we provide a QoS measure that is generic while permitting customized combinations of delay and throughput measures for each individual application:

$$QoS = T^{\alpha}/D^{\beta} \tag{4}$$

where $\alpha + \beta = 1$. While the choices of α and β will depend upon the throughput and delay tolerances of the specific applications themselves, the expectation is that the relativity of the delay hints from RED-Boston enabled sources will vary accordingly. Note our new metric, QoS, is in fact a traffic-sensitive variant of the power performance metric.

In the experiments discussed in the next section, $\alpha = 1$, is used for throughput-sensitive flows (such as Email or file transfer), since these flows can tolerate delay, and QoS for them depends only upon throughput. We use $\alpha = 0.5$ and $\beta = 0.5$ for delay-sensitive flows (such as an interactive videoconference), since these flows require low delay, but also moderate amounts of throughput. For medium throughput and medium delay-sensitive flows (such as some HTTP flows for Web browsing), we would use $0.5 < \alpha < 1$ and $0 < \beta < 0.5$.

4 Analysis

This section presents results analysis of one set of NS-2 simulations designed to compare RED-Boston's performance against ARED. We focus on changing the percentages of delay-sensitive and throughput-sensitive flows in the incoming traffic mix. Results from a second set of simulations consisting of incoming traffic with a range of throughput and delay requirements is presented in [23], but is not presented here due to lack of space.

The objective of the simulations is to evaluate RED-Boston's robustness as the incoming traffic mix at the router varies from mostly throughput-sensitive traffic to mostly delay-sensitive traffic. Figure 4 provides details on the traffic mixes used for the first five simulations.

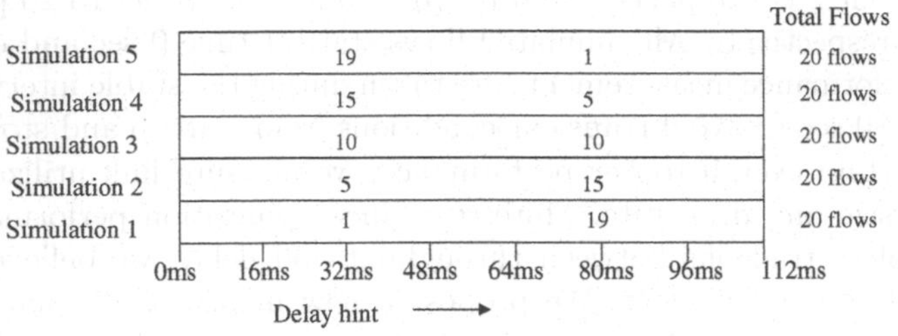

Fig. 4. Simulation Traffic Mix

Each simulation ran 20 flows using the network topology shown in Figure 3. Throughput-sensitive flows were represented by TCP flows carrying a delay hint of 80 ms, which corresponds to a queue size of 50 packets at the bottlenecked router R1. Delay-sensitive flows were represented by TFRC flows carrying a delay hint of 32 ms which corresponds to a queue size of 20 packets at R1. The X-axis for all the graphs in this section indicate the changing traffic mix corresponding to the five different simulations described in Figure 4.

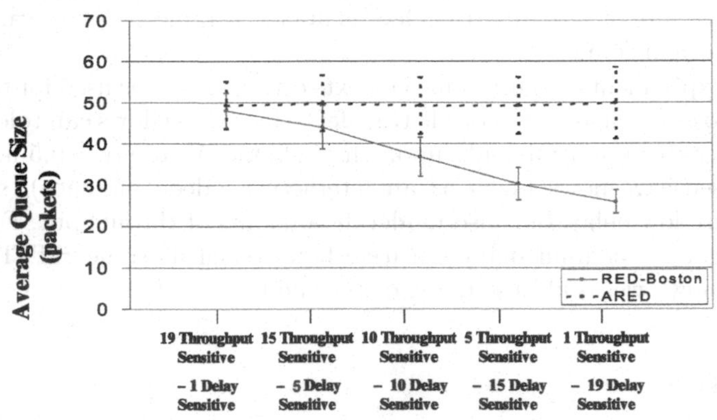

Fig. 5. Average Queue Size

Figure 5 shows the average queue size in packets for RED-Boston and ARED as the traffic mix varies. While the average queue size remains constant across all the ARED simulations, the moving target mechanism in RED-Boston allows the average queue size to adjust to the delay hint distributions caused by changes in the incoming traffic's QoS requirements. Thus, when traffic is mostly throughput-sensitive, the average queue size is higher to increase the overall throughput. Correspondingly, when traffic is mostly delay-sensitive, the average queue size becomes smaller to reduce the overall queuing delays. The RED-Boston average queue sizes in these five experiments reflects the average requirements of incoming traffic. On the other hand, Adaptive RED's target range is insensitive to the incoming traffic's requirements.

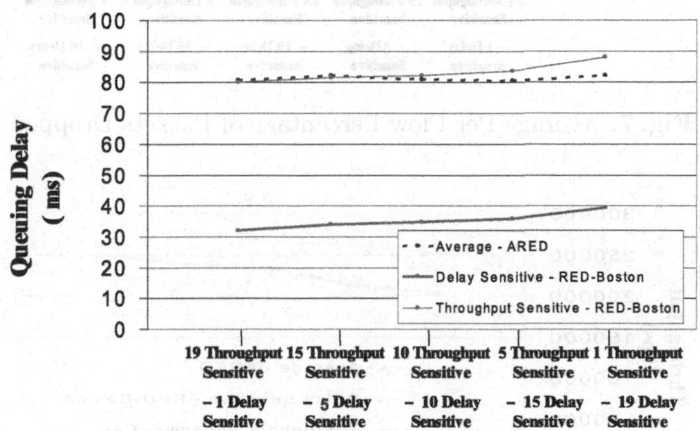

Fig. 6. Queuing Delays

Figure 6 shows the queuing delays experienced by the throughput-sensitive and the delay-sensitive flows. For ARED, graphing only the average delay curve is necessary because the throughput-sensitive and delay-sensitive flows receive identical treatment from ARED despite having different QoS requirements. Figure 6 indicates that RED-Boston gives lower queuing delays to delay-sensitive flows as compared to throughput-sensitive flows. Moreover, when the delay-sensitive flows represent a small percentage of the incoming traffic, the queuing delays experienced by the delay-sensitive flows is close to the 32 ms delay hint. However, as the percentage of delay-sensitive flows increases in the incoming traffic mix, the average queuing delay for these flows increases slightly. This is caused by many delay-sensitive flows competing with each other for lower delays. Conversely, when most of the incoming flows are throughput-sensitive, their average queuing delay is near the 80 ms delay hint. As the percentage of delay-sensitive flows increases in the incoming traffic, the average queuing delay for throughput-sensitive flows increases slightly. This increase happens because RED-Boston lets the delay-sensitive flows cut in the queue ahead of throughput-sensitive flows to get lower delays. However, from Figure 6, it is clear that the aging component of the RED-Boston packet weight calculation prevents starvation of throughput-sensitive flows.

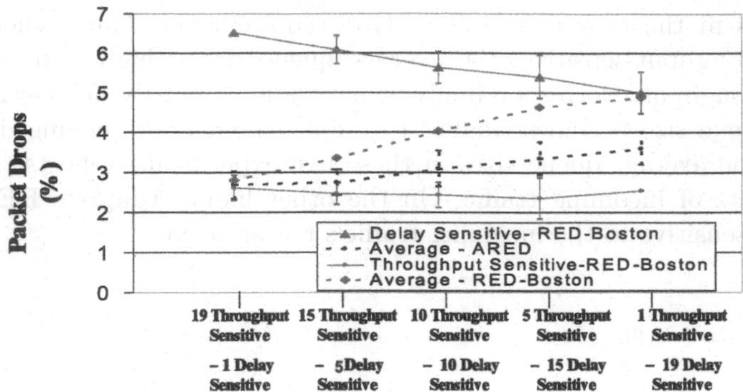

Fig. 7. Average Per Flow Percentage of Packets Dropped

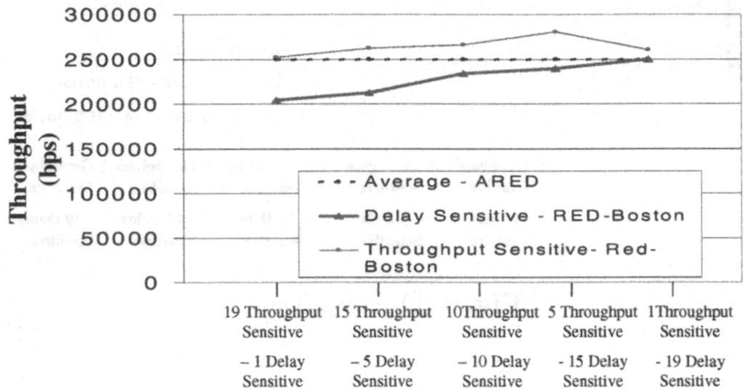

Fig. 8. Average Per Flow Throughput

Figure 7 shows the average percentage of packets dropped per flow in RED-Boston and ARED. Again only one curve is needed for ARED, since ARED treats all flows equally regardless of QoS requirements. For RED-Boston, Figure 7 does separate out the average per flow percentage of packets dropped for throughput-sensitive flows and delay-sensitive flows. The delay sensitive flow competing with 19 throughput-sensitive flows has to pay a high drop rate for the low delay service it receives because it is inserted far below the average queue size determined by the throughput sensitive flows. As the fraction of delay-sensitive flows increases, the average per flow drop rates for the delay-sensitive flows decreases. This is because as the number of delay sensitive flows increases, RED-Boston's moving target brings the average queue size down to give a lower delay service and the corresponding drop rate is shared equally by all the delay-sensitive flows. For throughput-sensitive flows, the RED-Boston drop rate is consistently lower than that of ARED and is almost constant for all simulations irrespective of number of delay-sensitive flows.

Figure 8 shows the average per flow throughput for ARED and average per flow throughput for throughput-sensitive flows and delay-sensitive flows for RED-Boston. The ARED throughput is nearly constant for all simulations and is the same for all flows. In RED-Boston, delay-sensitive flows get lower throughput than throughput-sensitive flows in exchange for low delay service. When there are 19 throughput-sensitive and one delay-sensitive flow, the delay-sensitive flow sacrifices some throughput for lower delay. This additional throughput sacrificed by the one delay-sensitive flow is equally shared by 19 throughput-sensitive flows. However, as the number of delay-sensitive flows increases, their throughput penalty decreases since the average queue size is reduced. The throughput sacrificed by delay-sensitive flows is shared equally by the existing throughput-sensitive flows in each case. In practice, as described in Section 2.1, if the reduced throughput given to a delay-sensitive flow falls below an acceptable threshold for that flow, the flow can send a larger delay hint if it is able to trade higher delay for higher throughput.

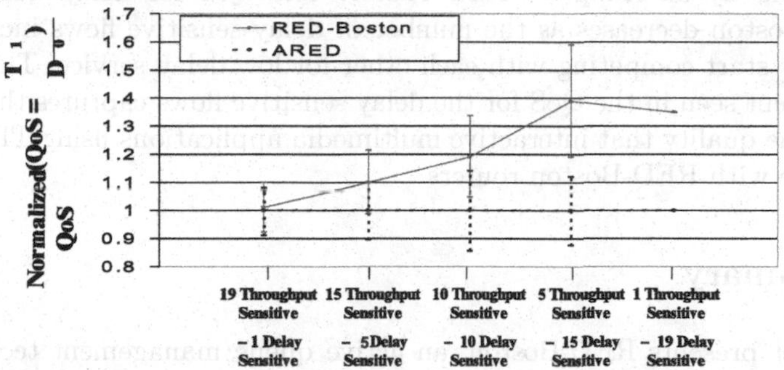

Fig. 9. Normalized QoS for Throughput-Sensitive Flows

Figure 9 presents the QoS for throughput-sensitive flows with a RED-Boston router normalized with respect to the QoS for throughput-sensitive flows traversing an ARED router. As described in Section 3, QoS for throughput-sensitive flows is defined as $QoS = T^1/D^0$. The RED-Boston throughput-sensitive flows record consistently higher QoS than the equivalent ARED flows. Furthermore, the RED-Boston QoS advantage grows as the percentage of delay-sensitive flows in the traffic mix increases. When delay-sensitive flows dominate the mix, more flows are willing to give up throughput for lower delay and as a result the throughput sensitive flows experience higher QoS even though they are in the minority.

Paralleling Figure 9, Figure 10 graphs the QoS for delay-sensitive flows in RED-Boston normalized with respect to the QoS for delay-sensitive flows in ARED. As described in Section 3, the specific QoS used for delay-sensitive flows is $QoS = T^{0.5}/D^{0.5}$. Using this metric, RED-Boston yields a QoS improvement

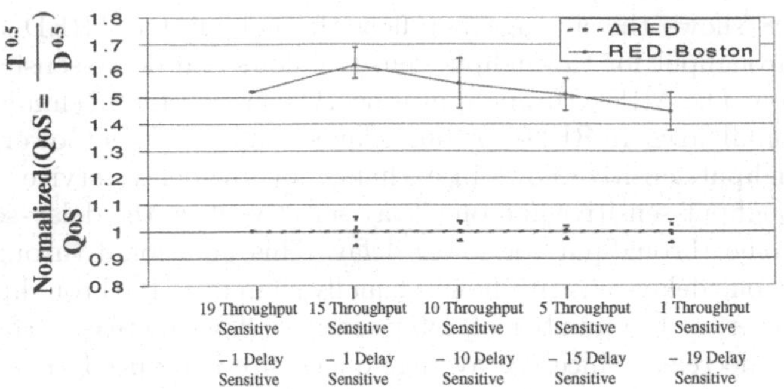

Fig. 10. Normalized QoS for Delay Sensitive Flows

(on average) for delay-sensitive flows more than 50 percent over delay-sensitive flows served by an Adaptive RED router. The QoS for delay-sensitive flows in RED-Boston decreases as the number of delay-sensitive flows increases, because they start competing with each other for low delay service. The resulting improvement seen in the QoS for the delay sensitive flows captures the improvement in the quality that interactive multimedia applications using TFRC would experience with RED-Boston routers.

5 Summary

This paper presents RED-Boston, an active queue management technique designed to provide better QoS support to all applications under the current best-effort Internet environment. With RED-Boston, applications include a delay hint in their packets as an indication of their preference for either lower queuing delay or higher throughput at the router. RED-Boston gives a relatively lower delay service to delay-sensitive flows and relatively higher throughput to throughput-sensitive flows. Thus, not only can low-bandwidth multimedia flows, such as interactive audio, obtain the low delays they need to support interactive communication, but delay-tolerant non-interactive flows, such as file downloading, can obtain the bandwidth they require to support efficient network transfer.

RED-Boston does not provide the delay and throughput guarantees provided by IntServ approaches, but nor does RED-Boston have the inherent scalability difficulties required by such per-flow reservations. RED-Boston does not predefine classes of service as do DiffServ approaches, but instead provides a continuum of service classes. The service afforded by RED-Boston does not require additional monitoring or charging of either flows that take advantage of the low delay service or flows that take advantage of the high throughput service.

In evaluating RED-Boston with varying percentages of delay-sensitive and throughput-sensitive flows, we find that RED-Boston is able to adapt the overall router queue parameters to better suit the current traffic mix. Additionally,

RED-Boston provides differentiated services by providing lower delay for the delay-sensitive flows and higher-throughput for throughput-sensitive flows. This improves QoS support for all flows in RED-Boston as compared to ARED.

Currently, RED-Boston expects flows to be responsive to network congestion in a TCP-friendly fashion. Since RED-Boston provides a higher drop rate for flows with a low delay hint, unresponsive flows should not gain more of a bandwidth advantage under RED-Boston than they do under current Internet environments. Still, an extension to RED-Boston would be to enhance it with a rate-based active queue management technique such as CSFQ [25], so that the dropping probability can be calculated based on a delay hint and the flow rate, helping ensure bandwidth fairness for unresponsive traffic.

Another extension to RED-Boston could be interaction between cascaded RED-Boston routers. When dequeuing a packet, RED-Boston could be modified to update the cumulative amount of time the packet has waited in router queues, thus providing additional information for trading of delay and throughput for RED-Boston routers downstream.

References

[1] A. Clerget and W. Dabbous. Tag-based Unified Fairness. In *Proceedings of IEEE INFOCOM*, April 2001.

[2] S. Athuraliya, V. H. Li, S. H. Low, and Q. Yin. REM: Active Queue Management. *IEEE Network*, May 2001.

[3] J-C. Bolot, S. Fosse-Parisis, and D. Towsley. Adaptive FEC-Based Error Control for Internet Telephony. In *Proceedings of IEEE INFOCOM*, March 1999.

[4] Z. Cao, Z. Wang, and E. Zegura. Rainbow Fair Queuing: Fair Bandwidth Sharing Without Per-Flow State. In *Proceedings of IEEE INFOCOMM*, March 2000.

[5] Jae Chung and Mark Claypool. Dynamic-CBT and ChIPS - Router Support for Improved Multimedia Performance on the Internet. In *Proceedings of the ACM Multimedia Conference*, November 2000.

[6] W. Feng, D. Kandlur, D. Saha, and K. Shin. Blue: An Alternative Approach To Active Queue Management. In *Proceedings of the Workshop on Network and Operating Systems Support for Digital Audio and Video (NOSSDAV)*, June 2001.

[7] W. Feng, D. Kandlur, D. Saha, and K. Shin. Stochastic Fair Blue: A Queue Management Algorithm for Enforcing Fairness. In *Proceedings of IEEE INFOCOM*, April 2001.

[8] S. Floyd and V. Jacobson. Random Early Detection Gateways for Congestion Avoidance. *IEEE/ACM Transactions on Networking*, August 1993.

[9] Sally Floyd, Ramakrishna Gummadi, and Scott Shenker. Adaptive RED: An Algorithm for Increasing the Robustness of RED's Active Queue Management. Under submission, 2001. http://www.icir.org/floyd/papers/adaptiveRed.pdf.

[10] Sally Floyd, Mark Handley, Jitendra Padhye, and Jorg Widmer. Equation-Based Congestion Control for Unicast Applications. In *Proceedings of ACM SIGCOMM Conference*, pages 45 – 58, 2000.

[11] J. Heinanen, F. Baker, W. Weiss, and J. Wroclawski. Assured Forwarding PHB Group. *IETF Request for Comments (RFC) 2597*, June 1999.

[12] C. Hollot, V. Misra, D. Towsley, and W. Gong. On Designing Improved Controllers for AQM Routers Supporting TCP Flows. In *Proceedings of IEEE INFOCOMM*, April 2001.

[13] P. Hurley, M. Kara, J. Le Boudec, and P. Thiran. ABE: Providing a Low Delay within Best Effort. *IEEE Network Magazine*, May/June 2001.
[14] V. Jacobson, K. Nichols, and K. Poduri. Expedited Forwarding PHB Group. *IETF Request for Comments (RFC) 2598*, June 1999.
[15] S. Kunniyur and R. Srikant. Analysis and Design of an Adaptive Virtual Queue. In *Proceedings of ACM SIGCOMM*, August 2001.
[16] D. Lin and R. Morris. Dynamics of Random Early Detection. In *Proceedings of ACM SIGCOMM Conference*, September 1997.
[17] Yanlin Liu and Mark Claypool. Using Redundancy to Repair Video Damaged by Network Data Loss. In *Proceedings of IS&T/SPIE/ACM Multimedia Computing and Networking (MMCN)*, January 2000.
[18] Debasis Mitra, Keith Stanley, Rong Pan, Balaji Prabhakar, and Konstantinos Psounis. CHOKE, A Stateless Active Queue Management Scheme for Approximating Fair Bandwidth Allocation. In *Proceedings of IEEE INFOCOMM*, March 2000.
[19] Universiy of California Berkeley. The Network Simulator - ns-2. Interent site `http://www.isi.edu/nsnam/ns/`.
[20] C. Padhye, K. Christensen, and W. Moreno. A New Adaptive FEC Loss Control Algorithm for Voice Over IP Applications. In *Proceedings of IEEE International Performance, Computing and Communication Conference*, February 2000.
[21] K. Park and W. Wang. QoS-Sensitive Transport of Real-Time MPEG Video Using Adaptive Forward Error Correction. In *Proceedings of IEEE Multimedia Systems*, pages 426 – 432, June 1999.
[22] Mark Parris, Kevin Jeffay, and F. Smith. Lightweight Active Router-Queue Management for Multimedia Networking. In *Proceedings of Multimedia Computing and Networking (MMCN), SPIE Proceedings Series*, January 1999.
[23] Vishal Phirke, Mark Claypool, and Robert Kinicki. Traffic Sensitve Active Queue Management for Improved Multimedia Streaming. Technical Report WPI-CS-TR-02-10, Worcester Polytechnic Institute, April 2002.
[24] William Pugh. Skip Lists: A Probabilistic Alternative to Balalnced Trees. *Communications of the ACM*, 33(6):668–676, June 1990.
[25] Ion Stoica, Scott Shenker, and Hui Zhang. *Core*-Stateless Fair Queueing: Achieving Approximately Fair Bandwidth Allocations in High Speed Networks. In *Proceedings of ACM SIGCOMM Conference*, September 1998.
[26] Ion Stoica and Hui Zhang. Providing Guaranteed Services Without Per Flow Management. In *Proceedings of ACM SIGCOMM Conference*, September 1999.

A QoS Providing Multimedia Ad Hoc Wireless LAN with Granular OFDM-CDMA Channel

Hyunho Yang[1] and Kiseon Kim[2]

[1] Dept. of Computer and Information, Suncheon Cheongam College,
224-9, Deogweol-dong, Suncheon-si, Jeonnam, 540-743, Republic of Korea
hhyang@scjc.ac.kr
[2] Dept. of Information and Communications,
Kwangju Institute of Science and Technology (K-JIST)
1 Oryong-dong, Puk-gu, Kwangju, 500-712, Republic of Korea
kskim@kjist.ac.kr

Abstract. A QoS providing distributed resource management scheme for multimedia Ad Hoc Wireless LANs(AWLANs) based on orthogonal frequency division multiplexing-code division multiple access(OFDM-CDMA) is presented. This scheme implements distributed resource management with granular OFDM-CDMA channel architecture to support multimedia services with QoS provisions. The performance evaluation result for broadband wireless access(BWA)-type physical layer supporting MPEG traffic sources shows that when the number of nodes is 50 and the mean session arrival rate is less than 0.1, we can get the blocking rate of 10^{-2} and the QoS loss probability of 10^{-1} when there is no control packet loss.

1 Introduction

Provision of QoS has been widely discussed in wireline and wireless networks. Also, multiple access systems for ad hoc networks need a carefully designed MAC protocol, to accommodate QoS requirements with flexibility and, at the same time, keep complexity limited. To this end, we choose a demand assignment MAC protocol with distributed admission control between the nodes in order to provide QoS to multimedia ad hoc wireless LANs.

Multimedia traffic requires a large bandwidth, at data rates greater than tens of megabits per second. Also, with a rather challenging radio channel in local area environments for multiple access, signal reception can be impaired significantly due to multipath fading, delay spread interference etc. OFDM-CDMA is a robust modulation/multiple access technique to encompass all channel impairment for high data rate multimedia, which were proposed in [1]-[3]. In addition, by virtue of the channel granularity, OFDM-CDMA has high flexibility in bandwidth assignment for multimedia traffic.

Further, to allocate network resources for multimedia traffic more efficiently, the resource management scheme is typically provided so that the QoS requirements are met for respective services. Conventionally, the wireless LANs are configured as a centralized, where the access point is responsible for the management

M. Ajmone Marsan et al. (Eds.): QoS-IP 2003, LNCS 2601, pp. 567–580, 2003.

of network resources, as in 802.11[4] and HYPERLAN[5]. There is another type of configuration, namely a distributed, by which ad hoc wireless LAN(AWLAN) can compose peer-to-peer networks. It is noteworthy that distributed systems must offer a carefully designed decentralized MAC protocol to accommodate QoS requirements with flexibility and limited complexity.

In [6] and [7] there were proposed MAC protocols using OFDM-CDMA technique for wireless local area network(WLAN) and wireless local loop(WLL) respectively. Both of them introduced and evaluated the applicability of OFDM-CDMA as a centralized, not distributed, MAC protocol for multimedia wireless networks. They also provide QoS guarantees by fair scheduling algorithm on the IP layer rather than the MAC layer.

In [8], they proposed a distributed resource management scheme for CDMA-based WLAN, so called distributed resource negotiation protocol(DRNP). They also evaluated this scheme by three resource allocation criteria; minimum power, maximum signal to interference ratio(SIR) or maximum transmission rate. Despite of its well-defined procedure, DRNP scheme is rather a complicated job to measure and manage the minimum transmission power, the maximum SIR or the maximum allowed data rate of each nodes.

We propose a new resource management scheme, namely distributed channel allocation protocol (DCAP), which basically adopts DRNP without measuring the available resources, based on OFDM-CDMA technique. This scheme utilizes a granular OFDM-CDMA channel architecture which consists of uniform size allocation units so called time slot-code(TC) pair. Subsequently, in this scheme, resources are allocated and managed by the TC unit rather than by laborious measurement of resources such as maximum allowed transmission power, interference level or data rate for each session. This scheme makes the node architecture less complex than that of the original DRNP and the system more robust for high data rate multimedia. Moreover, using the resource allocation list(RAL) at each node to monitor and prohibit new sessions which deteriorate QoS of ongoing sessions, we can implement the distributed resource management mechanism with QoS guaranteed for multimedia AWLANs.

In the following Section 2, we described overall system architecture of ad hoc wireless local area network(AWLAN) and physical and logical channel architecture based on OFDM-CDMA. In Section 3, the detail functional elements and system flow of the proposed distributed channel allocation protocol(DCAP) was given. There follows performance evaluation with typical realtime multimedia traffic source and discussions on the results in Section 4, and finally in Section 5 we concluded our work.

2 System Architecture

2.1 Overall Architecture

The ad hoc wireless LAN (AWLAN) architecture considered in this paper includes homogeneous nodes each serving as an independent control station as well

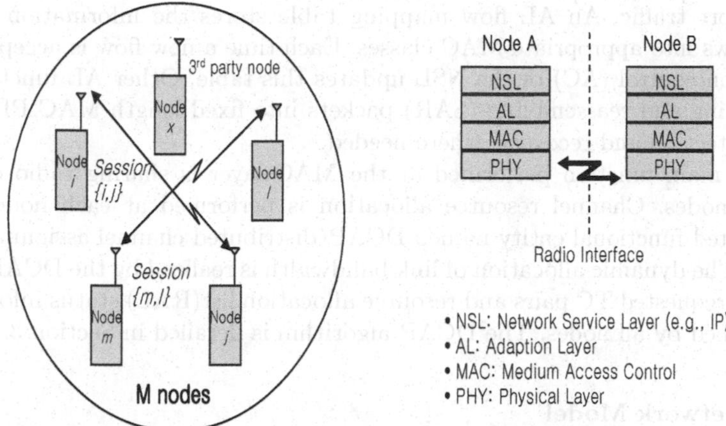

Fig. 1. System Architecture of Ad Hoc Wirleless LAN (a)Overall Network (b)Protocol Architecture

as an end user terminal. (see Fig. 1(a)). The AWLAN exploits the OFDM-CDMA (orthogonal frequency division multiplexing-code division multiple access) technique [1]-[3] which provides protection against fading, peak-average power ratio reduction capabilities, and high flexibility in bandwidth assignment. Duplexing can be managed dynamically to provide tight tracking of traffic asymmetry, by sharing the available pool of codes (code division duplex). All nodes provides the functionality to control the wireless interface, i.e., the logical functions related to the data link layer and the network control and the transmission functions. In Fig. 1(b) the user plane protocol architecture of the AWLAN is shown. Three layers can be distinguished: i) network service layer(NSL); ii) adaptation layer (AL); iii) MAC layer; iv) physical layer.

A network service layer (NSL) stands on the top of the AWLAN protocol stack. The NSL corresponds to classical network functions (addressing, routing) plus traffic handling functions (packet flow description and classification, admission control, traffic policing and/or shaping). A major example of NSL is the IP layer enhanced with QoS handling capabilities (e.g., IntServ or DiffServ paradigms [9]-[11]). Another example is ATM, where QoS is pursued by means of the standard ATM traffic control framework based on the traffic categories CBR, VBR, ABR, and UBR [12]. AWLAN is conceived to give support to such NSLs; therefore, it is assumed that basic traffic control functions are placed in the NSL.

The main task of the adaptation layer(NL) is mapping NSL traffic classes into MAC service classes. NSL traffic classes do not need to have a one-to-one correspondence with MAC classes, provided that support for the targeted QoS can be maintained. So, to keep things simple, two service classes have been assumed in the MAC layer-one for guaranteed bandwidth traffic and one for best effort traffic. An AL flow mapping table stores the information to map NSL flows into appropriate MAC classes. Each time a new flow is accepted, the admission control (AC) of the NSL updates this table. Other AL functions are segmenting and reassembling (SAR) packets into fixed length MAC_PDUs and error detection and recovery, where needed.

The main function performed at the MAC layer is sharing radio capacity among nodes. Channel resource allocation is performed at each nodes, by a distributed functional entity named DCAP(distributed channel assignment protocol). The dynamic allocation of link bandwidth is realized by the DCAP on the basis of requested TC pairs and resource allocation list(RAL) status information maintained by all nodes. The DCAP algorithm is detailed in Section 3.

2.2 Network Model

In our proposed scheme, we consider a single-hop ad hoc OFDM-CDMA based multimedia wireless LAN with m nodes. M is defined as the set of nodes in the network

$$M = \{1, 2, 3, M, m\} \tag{1}$$

The nodes are assumed to be distributed randomly. Also, it is assumed that their positions are either fixed or slowly varying. Let $M^t \subset M$, denote the set of nodes that are currently transmitting. Then, $M^c = (M^t)^c$ is the set of nodes monitoring the common control channel(CCCH). Obviously, $M^t \cup M^c = \phi$. Moreover, $M^r \subseteq M^c$, defines the set of nodes that are currently receiving data frames. A session between nodes, $i \in M^t$, and $j \in M^r$, is denoted by $\{i, j\}$. Also, $M^a = M^t \cup M^r$, is defined as the set of currently active nodes. Further, it is assumed that a transmitter can communicate with only a single receiver at a time and vice versa.

Each node is equipped with two receivers. One receiver is always synchronized to the common code allocated to common control channel(CCCH), while the second receiver is synchronized to the unique receiver code combination for data channel(DCH) allocated to each node. Each channel structures are described in the following subsection 2.3. A node cannot transmit and receive simultaneously. A node that is not transmitting, monitors the control channel CCCH. Moreover, the dual receiver architecture allows a node to receive control and data messages simultaneously.

2.3 Channel Structure

Physical Channel Orthogonal frequency division multiplexing(OFDM) has drawn a lot of attention in the field of radio communications as an effective way to combat frequency selectivity in hostile radio channels. In OFDM, the entire channel is divided into many narrow subchannels transmitted in parallel. Concerning bandwidth granularity, to obtain a rate of h data symbols per symbol time (with $h< K$), h codes (h subcarriers) in the same symbol interval are used. The number of OFDM subcarriers comes from a compromise of several factors: increasing the number of subcarriers improves multipath robustness, reduces the guard interval overhead, and increases the flexibility in bandwidth assignment; on the other hand, it increases phase noise sensitivity and makes baseband processing(i.e. FFT) more complex[7]. It is well known that conventional OFDMs have a drawback that the envelope power of the transmitted signal fluctuates widely, e.g. a K-subcarriers OFDM system has peak-average power ratio (PAPR) of K. As an alternative, OFDM-CDMA systems can spread the data symbols across the frequency domain and employ OFDM modulation for conveying each spread data symbol, subsequently, this high PAPR can be mitigated by appropriate selection of the spreading codes and moderate amplifier backoff[13].

In this paper, we adopt the OFDM-CDMA as a physical channel architecture, for typical broadband wireless access(BWA) applications, such as 802.11a[4], HIPERLAN Type2[5] and 802.16[14], to testify the applicability of our new channel resource management scheme.

We assume a channel bandwidth of 110MHz, the intercarrier spacing is 212 kHz. Thus, the Fourier period of the OFDM-CDMA symbol is equal to 4.71 μs and ISI is prevented in typical AWLAN channels at millimeter-wave frequencies. Hence symbol duration results equal to 5.15μs, where an overhead of 8.5 percent is used to provide protection against multipath. Typical OFDM implementations sacrifice some of the carriers to create a guard band between adjacent channels. Accordingly, we decided to use 400 out of 512 subcarriers, corresponding to a guard band of 11.8 MHz at each side. Therefore, if one adopts robust QPSK modulation, an aggregated bit rate of 155 Mb/s is obtained. For a system designed with the above described parameters and adopting an MMSE equalization criterion, BER of less than 10^{-3} can be obtained for an E_b/N_0 of 8 dB, and a special efficiency of about 1.4 b/s/Hz.

Logical Channel The logical structure of our proposed scheme is structured into frames lasting T_{FRAME} and consisted two types of logical channels; i) data channel(DCH) and ii) common control channel(CCCH). Time

For the DCH, we assume the radio capacity is structured as $K-1$ out of K orthogonal codes that can be used simultaneously. One code is reserved for CCCH. Each code is regarded as a channel used in a time-division multiple-access fashion; A time slot of a code channel, referred as time slot-code(TC) pair, carries a MAC protocol data unit(MAC_PDU), i.e., the duration of a time slot is equal to the duration carrying a MAC_ PDU($T_{SLOT} = T_{MAC\_ PDU}$). Further, a time frame is consisted of N time slots. Consequently a frame can be filled by $(K-1) * N$ MAC-PDU's. All nodes are assumed to be synchronized properly and the capacity assignment is performed frame by frame. Each node can transmit on several time slots in a frame, distributing the MAC-PDU's on a given number of codes.

The CCCH is assumed to have a mini-slotted structure with each mini-slot duration(T_{MINI_SLOT}) is equals to the transmission time of a control message and is much shorter than T_{SLOT}. This code channel is shared by all nodes to exchange the resource allocation requests and answers. The data rate on the CCCH is fixed. All control messages are assumed to be of the same size and transmitted with the same power, i.e., no power control is used. Moreover, it is assumed that control messages are transmitted with a high enough transmission power so that they can be received by all nodes M^r. Since the CCCH is a single channel, it is vulnerable to collisions. Due to the relatively light traffic on the CCCH, a simple multiple access scheme such as slotted ALOHA is proposed. In addition, due to the critical nature of the control messages it is assumed that some form of forward error correction (FEC) is used for messages on the CCCH.

3 Distributed Channel Allocation Protocol (DCAP)

In this section we describe the proposed scheme, distributed channel allocation protocol(DCAP). It is different from DRNP [8] in the following three points

- *Resource allocation policy;* In DCAP resources are allocated by the assignment of fixed size allocation unit(i.e., TC pair) to the communicating node pair, while in DRNP, physical parameters, e.g., minimum transmission power, maximum sustainable interference(MSI), maximum data rate, for each session condition should be measured before resource allocation and calculated results are assigned.
- *Contents of RAL;* Both in DCAP and DRNP, each node have resource allocation list(RAL), described below, by which resource allocation information of all sessions around are maintained. Source/destination addresses, estimated session duration are common to both schemes. In addition, in DCAP, allocated TC Pairs are maintained while, in DRNP, path loss and MSI are maintained instead.
- *RTS message contents;* In request to send(RTS) message, defined in the following, source/destination address, message type('RTS'), estimated session duration are common exchanged information in both schemes. However, the resource requesting information of DCAP differs from that of DRNP, i.e., in DRNP, the number of required TC pairs and proposed TC list are carried in the RTS, while in DRNP, detailed physical parameter values such as minimum SNR, maximum data rate and transmission power level are transferred by the RTS message.

3.1 Resource Allocation List (RAL)

All the nodes, as mentioned in[8], have the resource allocation list(RAL), which is an extension of the network allocation vector (NAV) used in 802.11 WLAN's. Each node maintains a database encoding its knowledge about other ongoing sessions in the network. Let $\{i,j\}$ be a currently active session from a node i to a node j, and l be a third party node with respect to $\{i,j\}$ that can track the CCCH. A third party node l encodes information about the session in its database as a record containing the following information

– the source (i) and destination (j) addresses for $\{i,j\}$
– the allocated TC(Time slot-Code) pairs for the session $\{i,j\}$
– the estimated duration for the $\{i,j\}$ session, τ_{ij}.

RAL of node l contains the same set of records for every active session in the network.

3.2 Control Message Formats

Control messages are broadcast on CCCH and are used to setup/tear down sessions, etc. The following convention is used in the presentation of the control message formats.

– Physical layer entities such as synchronization and error detection/correction fields have been excluded. Only the data structures specific to the DRNP are presented.
– Each message is prefixed by the source and destination addresses.
– Each message contains a unique entry identifying the message type and is denoted by the respective message name in the presentation of message formats that follows.

Whenever a session is to be established, a session negotiation procedure is performed. At first the request to send(RTS) message is issued by a transmitter i to another node j. Its format is

$$RTS_{ij} = \{i, j, RTS, N^r_{ij}, \{(t,c)_1, ..., (t,c)_n\}, \tau_{ij}\}$$

where N^r_{ij} is the requested number of TC pairs, $(t,c)_n$ are the TC pair candidates for the session $\{i,j\}$ and τ_{ij} is the estimated duration of the session.

The clear to send(CTS) message is issued by node j, the recipient of the RTS message, if it can support the QoS requested by node i. The format of CTS is

$$CTS_{ij} = \{j, i, CTS, N^a_{ij}, \{(t,c)_1, ..., (t,c)_n\}\}$$

where N^a_{ij} is the number of allocated TC pairs.
The primary reject (PREJ) is issued by the receiver j, if it is unable to support the QoS requested by the transmitter i. The format of PREJ is

$$PREJ_{ji} = \{j, i, PREJ, N^r_{ij}, \{(t,c)_1, ..., (t,c)_n\}\}.$$

The end of session request(ESR) message

$$ESR_{ij} = \{i, j, ESR\}$$

is used by the transmitter i to signify the end of a session. The end of session acknowledgement(ESA) message is sent by the receiver node j, in response to the ESR. Its format is

$$ESA_{ji} = \{j, i, ESA\}.$$

Every 3rd party node hearing ESRij or ESAji updates their RAL to release the allocated TC pairs for reuse by other modes.

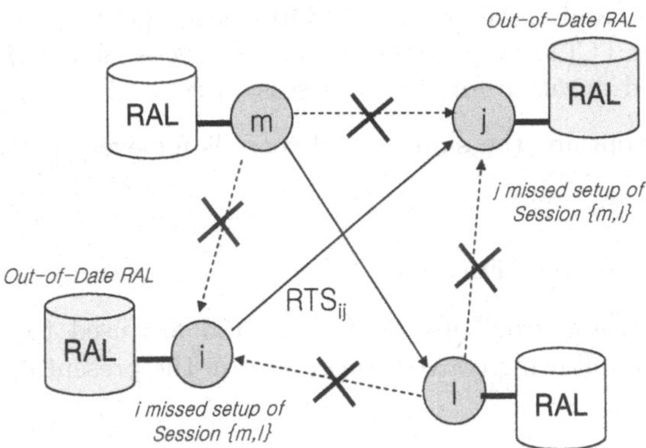

Fig. 2. Out of Date RALs in i and j Leads Incorrect Channel Allocation During Session Negotiation

Secondary Reject(SREJ) As a node is essentially deaf while transmitting, even if no other form of message loss occurs, a node's RAL will not be updated. Out-of-date RAL's on both sides can cause an invalid TC pair allocation, which would degrade the QoS of the currently active sessions in the network. This scenario is illustrated in Fig. 2, where i has out of date information. When it tries to establish a session with j, a busy TC pair could be requested by RTS_{ij}. The receiver j has no way of validating TC pair's current state and may allocate requested TC pair to a new session. The secondary reject (SREJ) mechanism allows third-party receivers to interrupt the setup of sessions that violate their QoS guarantees. The receiver of each third party nodes(l) tracks the CCCH while it is receiving data. If the allocation of TC pairs for the session $\{i, j\}$ causes QoS violation for an existing session, node l issues a SREJ message

$$SREJ_{li} = \{l, i, SREJ, \{(t,c)_1, ..., (t,c)_n\}\}$$

to the transmitter. The transmitter i waits for the SREJ message for certain amount of time (τ_{DATA}), i.e., several minislots of CCCH, before it starts transmitting. If no SREJ message is received in the given time interval, it starts transmission or regards the session as dropped.

In principle, the probability that a SREJ message is lost can be minimized by making τ_{DATA} as large as possible. Realistically, the gain in terms of QoS guarantees is far outweighed by the overhead through delay caused by excessively large values for τ_{DATA}. Ideally, the timeout period should be varied with network load, although this again might entail too much computational overhead, although it does form an interesting avenue for research. The effect of message loss involving SREJ is investigated later.

Update Resource Allocation List(UPD_RAL) UPD_RAL message is broadcast by all third party receivers, $l \in K_{ij}^r$ when the CTS_{ij} message is received. This

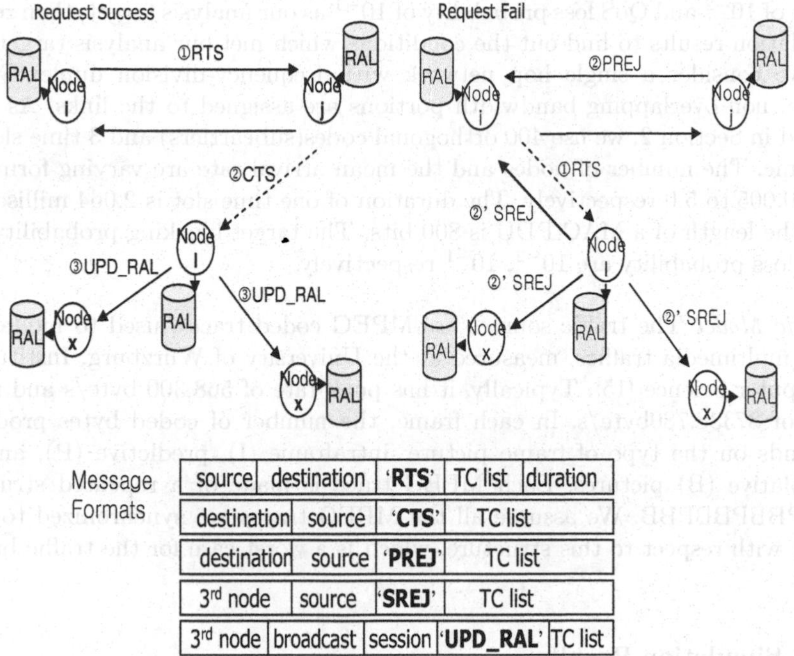

Fig. 3. Schematic Diagram of Distributed Channel Allocation Procedure for Session Negotiation (a)Request Success (b)Request Fail

message updates the RAL due to the addition of $\{i, j\}$ to the network. Its format is

$$UPD\_RAL_{lx} = \{l, x, \{m, l\}, UPD\_RAL, \{(t, c)_1, ..., (t, c)_n\}\}$$

where x is a network specific broadcast address. Since UPD_RAL messages are transmitted by one or more third party receivers, they are particularly vulnerable to loss through collision. Thus the same random slot mechanism used for SREJ is applicable here as well. In fact, UPD_RAL and SREJ messages are extremely similar, i.e., they are both used to update third party RAL's. Thus, in the current incarnation of DCAP, the time window is shared by both SREJ and UPD_RAL messages.

Fig.3 shows the schematic diagram of the session negotiation for two cases, i.e., request successful and request fail.

4 Performance Evaluation

4.1 Simulation Model

Channel Model The behavior of the proposed MAC protocol, with reference to key targets such as availability i.e., the blocking probability, and QoS guarantees(QoS loss probability), has been investigated. We choose the blocking probability of 10^{-2} and QoS loss probability of 10^{-1} as our analysis target, then review simulation results to find out the conditions which met our analysis targets.

We consider a single hop network with frequency-division duplex(FDD); hence, non overlapping bandwidth portions are assigned to the links. As mentioned in Section 2, we use 400 orthogonal codes(subcarriers) and 3 time slots in a frame. The number of nodes and the mean arrival rate are varying form 5 to 500, 0.005 to 5.0 respectively. The duration of one time slot is 2.064 millisecond and the length of a MAC_PDU is 800 bits. The target blocking probability and QoS loss probability are 10^{-2}, 10^{-1} respectively.

Traffic Model The traffic sources are MPEG coded traces, used to model real time multimedia traffics, measured at the University of Würzburg, Institute of Computer Science [15]. Typically, it has peak rate of 568,300 byte/s and mean rate of 87338.780byte/s. In each frame, the number of coded bytes produced depends on the type of frame picture: intraframe (I), predictive (P), and interpolative (B) pictures. Each MPEG trace is made of a repeated structure IBBPBBPBBPBB. We assume all the MPEG traces are synchronized to each other with respect to this structure, which is a worst case for the traffic burstiness.

4.2 Simulation Results

Using the above models, simulations had been done to analyze the performance of DCAP according to the variation of the number of nodes and mean arrival rate, for an idealized network where no message loss occurs[8]. Our simulation results show overall trends from which we can get conceptual insight for performance behavior of proposed scheme.

Ideal Mode The blocking probability is essentially a throughput measure and represents sessions deemed lost due to the receiver unavailable.

When the number of nodes is small the destination nodes are more probable to be engaged in a session. Thus the availability of destination node is low, hence, as shown in the Fig.4, blocking probability is comparatively high. As the number of nodes increases, blocking probability decreases because of the reason described above.

Intuitively, blocking probability is also influenced by network load, e.g., arrival rate. The figure also shows the relations. As shown in the figure, in the light load condition e.g., less than 1 arrival/sec, blocking probability increases rapidly as the network load increases

Our analysis target, i.e., blocking probability of 10^{-2}, achieved at the mean arrival rate below 0.1 and the number of nodes is greater than 50. When the number of nodes is less than 50, our target satisfied only for light load conditions say, inter-arrival time greater than 10 second. On the other hand , in the heavier load condition i.e. arrival rate of greater than 0.1, our target achieved only for the larger number of the nodes.

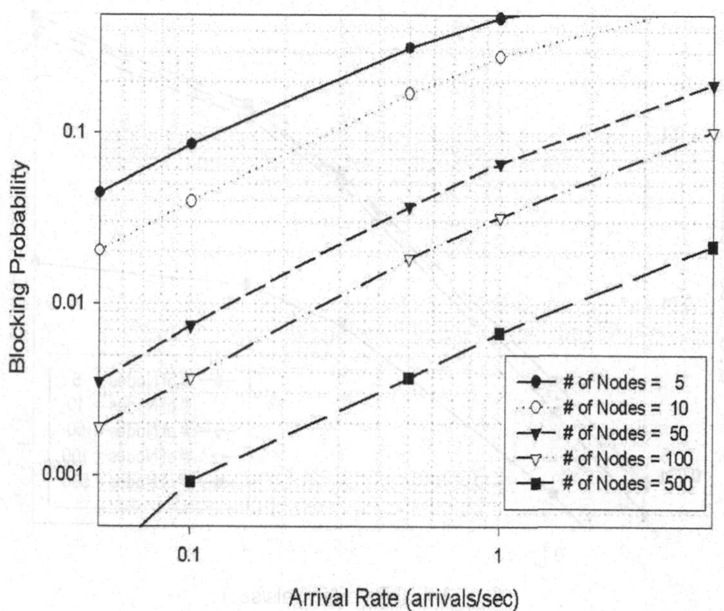

Fig. 4. Blocking probability distribution for varying mean arrival rate.

Real Mode We now turn our attention to investigating the effect of message loss on DCAP's ability to maintain QoS guarantees. Unless stated otherwise, τ_{DATA} is set to sufficiently long and the backoff count set to zero. The loss of QoS(QoSLoss) probability parameter denotes the percentage of sessions that are successfully set up but lose their QoS guarantees because some other nodes interfere and try to use the same TC already used by ongoing sessions. The deaf transmitter mode can be thought of as the best case real mode scenario. In this case, the only messages lost are due to a transmitter's inability to receive control messages. This implies that nodes, while transmitting, have no knowledge of sessions that arrive or leave. Thus when these transmitters issue RTS messages for future session setups, they advertise an out of date TC pair matrix usage information. Without the SREJ mechanism, this causes loss of QoS for already active sessions.

As mentioned above, when the number of nodes is small, the probability of being busy for the destination node is high. Even though the total blocking probability is high for this case, QoS loss probability is relatively low because a third party node has no chance to interfere the session negotiation.

When the mean arrival rate increases, the number of concurrently allocated TC pairs also increases. Consequently, for a given set of TC pairs being selected by the transmitter, the probability that a TC from the set being already used by

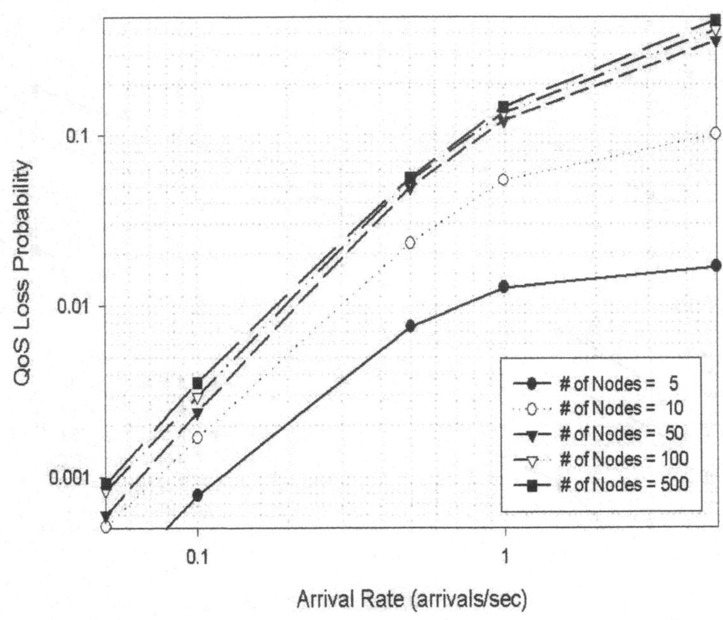

Fig. 5. QoS loss probability distribution for varying mean arrival rate.

some other nodes increases rapidly . This is evident from the Fig.5. When the mean arrival rate is below 1.0, the QoS loss probability increases rapidly with increasing the mean arrival rate.

As shown in the figure, our analysis target, the QoS loss probability of 10^{-1}, met for the mean arrival rate less than 1.0 with independent of the number of nodes. For the arrival rate greater than 1.0 given target satisfied only for small number of the nodes, e.g., less than 50. At the large values of the number of nodes and the mean arrival rate, the QoS loss probability value approaches around 50%.

In summary, our proposed scheme best performs, say, satisfies both targets; blocking probability and QoS loss probability, at relatively low load condition, arrival rate of less than 0.1, and the number of nodes about 50.

It is intuitively evident that our proposed scheme, with granular OFDM-CDMA channel architecture consists of uniform sized allocation units, can implement simplified resource allocation and management scheme. This fact makes the node architecture less complex than DRNP and the system more robust in the time-varying radio environment. Moreover, from the results of the simulations e.g. blocking probability, QoS loss, it seems to be possible to guarantee QoS of ongoing sessions in the reasonable network load conditions.

5 Conclusions

DCAP was designed especially for the allocation and management of resources in multimedia AWLAN environments. In this work, we proposed a MAC protocol on the top of an OFDM-CDMA physical channel. Further, we combined this with distributed resource allocation scheme to implement QoS guaranteeing channel resource management especially for multimedia AWLAN.

We investigated the performance of DCAP, in terms of network wide metrics such as blocking probability and QoS loss probability. The performance evaluation result for BWA-type physical layers supporting MPEG traffic sources shows that, when the number of nodes is about 50 and mean session arrival rate below 0.1, we can get the blocking probability of 10^{-2} and the QoS loss probability of 10^{-1}. Further, blocking those users that violate the QoS constrains of the currently active sessions, the protocol fulfills the role of a call admission control (CAC) mechanism for AWLAN. It is completely distributed and adaptive to changing network conditions.

It is evident that, with some performance tuning, e.g., fair scheduling, this new scheme can be a good choice of multiple access and resource management protocol for AWLAN especially for broad wireless access(BWA) framework.

Acknowledgement

This work is supported in part by BK21 Project, MOE and ERC-UFON, MOST, Republic of Korea.

References

1. S. Hara and R. Prasad, "Overview of multicarrier CDMA," *IEEE Commun. Mag.*, Dec. 1997, Vol.35 No.12 pp. 126-133
2. H. Rohling, K. Bruninghaus, and R. Grunheid, "Comparison of multiple access schemes for an OFDM downlink system," *in Multi-carrier Spread Spectrum, K. Fazel and G. P. Fettweis, Eds. Norwell, MA: Kluwer Academic*, 1997, pp. 23-30.
3. S. Kaiser and K. Fazel, "Spread-spectrum multi-carrier multiple-access system for mobile communications," *in Multi-Carrier Spread Spectrum, K. Fazel and G. P. Fettweis, Eds. Norwell, MA: Kluwer Academic*, 1997, pp. 49-56.
4. IEEE Computer Society LAN/MAN Standards Committee, Wireless LAN Medium Access Control(MAC) and Physical Layer(PHY) Specifications: High-speed Physical Layer in the 5 GHZ Band, *IEEE Std. 802.11a-1999.*
5. European Telecommunications Standards Institute(ETSI), Broadband Radio Access Network(BRAN); HIPERLAN Type 2: Physical(PHY) layer, *ETSI TS 101 475 V1.1.1 (2000-4).*
6. F. Cuomo, A. Baiocchi and R. Cautelier, "A MAC protocol for a wireless LAN based on OFDM-CDMA," *IEEE Communication Magazine*, Sep. 2000, Vol.38 No.9, pp.152-159.
7. A. Baiocchi, F. Cuomo and S. Bolognesi, "IP QoS delivery in a broadband wireless local loop: MAC protocol definition and performance evaluation," *IEEE JSAC*, Sep. 2000, Vol.18 No.9, pp.1608-1622.

8. S. Lal and E.S. Sousa, "Distributed resource allocation for DS-CDMA based multimedia ad hoc wireless LAN's," *IEEE JSAC*, Oct. 1991, Vol.9 No.8, pp.1265-1279.
9. X. Xiao and L. M. Ni, "Internet QoS: A big picture," *IEEE Network*, pp. 8-18, Mar.-Apr. 1999.
10. P. P. White, "RSVP and integrated services in the Internet: A tutorial," *IEEE Commun. Mag.*, pp. 100-106, May 1997.
11. D. Black, S. Blake, M. Carlson, E. Davies, Z. Wang, and W. Weiss, "An architecture for differentiated services," RFC 2475, Dec. 1998.
12. J. W. Roberts, U. Mocci, and J. Virtamo, Eds., Broadband Network Tele-traffic: Final Report of Action COST 242: Springer-Verlag, 1996.
13. B. J. Choi, E. L. Kuan and L.Hanzo, "Crest-factor study of MC-CDMA and OFDM," *IEEE VTC*, Amsterdam, The Netherlands, 1999, pp.233-237.
14. [online], http://grouper.ieee.org/groups/802/16/index.html.
15. O. Rose, "Statistical properties of MPEG video traffic and their impact on traffic modeling in ATM systems," *Proc. 20th Annu. Conf. Local Computer Networks, Minneapolis*, Oct. 15-18, 1995, pp.397-406.

SIP Originated Dynamic Resource Configuration in DiffServ Networks: SIP / COPS / Traffic Control Mechanisms

Stefano Giordano [1], Marco Listanti [2], Fabio Mustacchio [1], Saverio Niccolini [1], Stefano Salsano [3], Luca Veltri [4]

[1] UNIPI – Università di Pisa, Italy,
{s.giordano, f.mustacchio, s.niccolini}@iet.unipi.it

[2] INFOCOM – Università di Roma "La Sapienza", Italy
marco@infocom.uniroma1.it

[3] DIE – Università di Roma "Tor Vergata", Italy
stefano.salsano@uniroma2.it

[4] Università di Parma, Italy
luca.veltri@unipr.it

Abstract. Voice, video and multimedia sessions are applications sensitive to the QoS provided by the underlying IP network. Therefore a lot of interest is currently devoted to the interaction of application level protocols with the QoS mechanism in IP networks. Among them SIP is currently having a lot of attention as a protocol for session signaling over the Internet. This work will describe an enhancement to SIP protocol for the interworking with a QoS enabled IP network. The proposed mechanism is simple and it fully preserves backward compatibility and interoperability with current SIP applications. Moreover the paper describes the application of this mechanism to a particular QoS enabled IP network, which implements DiffServ as transport mechanisms (the DiffServ mechanisms are obtained by means of Traffic Control functionalities with the TCAPI software libraries) and modified COPS clients for resource admission control. A test-bed implementation on Linux PCs of the proposed solutions is finally described.

1 Introduction

There is much interest in the Internet community about how to automate the process of resource allocation within a QoS (Quality of Service) network and the related topics constitute an area of active research today.

Looking at the standardization effort in the area of IP QoS the two main approaches that have been proposed in the IETF are the Integrated Services (Intserv) model and the Differentiated Services (Diffserv) model. Additional proposals consider a combination of the two approaches. In addition, the MPLS (Multi Protocol Label Switching) technology is going to play an important role in this field, for example as transport backbone for Diffserv. A very good introduction to IP QoS topics can be found in [2], [3].

M. Ajmone Marsan et al. (Eds.): QoS-IP 2003, LNCS 2601, pp. 581-591, 2003.

The DiffServ architecture has the potentiality, with its Per Hop Behaviors (PHBs), to differentiate the QoS in a scalable mode; unfortunately such architecture is still utilized in a static manner; the providers are configuring the network resources statically with a capacity planning study without a time-dependent optimization. Since the amount of traffic offered to the network is intrinsically variable with time there is the possibility of underutilizing the resources or overloading the network.

The IntServ, on the other hand, is a more dynamic architecture and is more suited to solve time-variant configuration issues but it has shown its weakness when dealing with scalability problems.

In the DiffServ framework, even if a static SLA (Service Level Agreement) is possible to define nowadays, it is difficult for a client to rely on such SLAs for several reasons:

- the maintenance of the information about all the user introduces a scalability issues;
- the user could change the call parameters during the call, paying for unused resource;
- static SLAs that well suite to the user demand are quite difficult to define.

For those reasons we are dealing with dynamic resource allocation in a QoS domain. In particular our work is focused in providing a scalable architecture for dynamic resource allocation / configuration when an access network (which may be QoS aware or not) request transport to a "*QoS enabled*" network.

In such an area of research we are focusing in providing QoS to the multimedia applications (e.g. IP Telephony, Videoconferencing, etc.) based on the SIP (Session Initiation Protocol) protocol as application level protocol.

SIP [1] is currently having a lot of attention within IETF (Internet Engineering Task Force) and it is seems to be the more promising candidate as call setup signaling for the present day and future IP based telephony services. It could even be a real competitor to the Plain Old Telephone Service (PSTN).

For the realization of this scenario, there is the obvious need to provide a good speech quality. This quality in turn depends on the Quality of Service delivered by the IP network. Reservation and/or admission control mechanisms could be needed to get QoS from the IP network. Unfortunately, at present day, there is not a clear picture about the "elected" mechanism for QoS provisioning in IP network, as much research and standardization effort is ongoing in this area. The interaction of these QoS mechanisms with the call setup procedures (i.e. SIP) is therefore a very hot topic. There is a work, an Internet Draft of the SIP IETF Working Group [4], which deals with the interaction between SIP and resource management for QoS.

Our goal is to enable seamless inter-operation between the application level protocol, the policy/resource management protocol and the router configuration.

To accommodate this aim we propose, in this work, a very simple solution that is based on an enhancement of the SIP protocol to convey QoS related information. The solution preserves backward compatibility with current SIP applications and it decouples as much as possible the SIP signaling from the handling of QoS. Moreover the solution foresees the use of COPS (extended with QoS handler features) protocol as policy/bandwidth control and the use of Linux Traffic Control (TC) to build the QoS mechanisms.

The rest of the paper is organized as follows. Section 2 gives an overview of the proposed mechanisms for dynamic resource allocation explaining the rationale and overall requirements. Those mechanisms are detailed in Section 3, 4 and 5. Moreover Section 6 deals with the software modules and the test-bed as we implemented in the framework of the Nebula project. Finally we give our conclusions.

2 Dynamic Resource Configuration: Mechanisms Overview

This section introduces the overview of our proposed solution in order to make the reader able to fully understand the rationale of the proposed solution.

In order to not delegate the user terminal to be aware of the network QoS model, it should not originate the resource reservation with an "ad hoc" protocol (e.g. RSVP, Resource ReserVation Protocol); in our belief such a solution limits the scalability of the architecture and needs a capillary control on every user terminal.

However triggering the resource reservation is mandatory in order to take advantage from the flexibility of a dynamic Service Level Agreement (SLA) scheme. In our work we propose to use the SIP protocol itself as resource reservation triggering protocol.

Hence in the following sections we define the generic mechanism in the SIP protocol and we present our implemented interaction with a specific QoS mechanism based on COPS (Common Open Policy Service) [5] for admission control in a DiffServ network. The basic idea is that Admission Control entities running on the network borders (e.g. in the Edge Routers, ER) dialogues with external QoS clients (the Q-SIP servers) and with Bandwidth Broker (BB) in the "*QoS enabled*" network. The Admission Control entities use a variant of the COPS protocol, called COPS-DRA (COPS-DiffServ Resource Allocation) to dialogue with the QoS clients and with the BB.

The solution foresees the enhancement of the SIP proxy server to handle QoS aspects. In the following, the enhanced SIP server will be called Q-SIP server (QoS enabled SIP server). All the QoS aspects can be covered by the Q-SIP servers in the originating and terminating sides. This is also justified by the fact that in a DiffServ QoS scenario there will be servers dedicated to policy control, accounting and billing aspects. Hence, a solution based on SIP servers is really suited to this QoS scenario.

The SIP signalling protocol, opportunely extended in order to support a QoS enabled resource reservation, is managed by SIP proxy servers which, in turn, originate, by means of COPS-DRA protocol, the resource reservation request to the DiffServ network (to the local or to the remote Bandwidth Broker as needed). The managing of such information is done at application level by means of SIP-COPS cooperation. In order to have the correct resource reservation and configuration the QoS enabled SIP servers have to cooperate with a COPS-DRA client. The first COPS-DRA client encountered in the resource reservation process is located in the same machine as the SIP proxy server. The COPS-DRA client asks to a COPS-DRA server located in the Edge Router (ER) of the DiffServ network for the resource reservation. The decision to ask for resources to the remote BB is taken with attention to the local resource availability; a mix of the so-called COPS configuration and outsourcing model is preferred in order to take advantage from the scalability of the former and the high control provided by the latter.

Once the resource reservation is accepted the configuration phase takes place by means of kernel configuration. In our solution a Traffic Control Server is used configure the Linux kernel; it takes the input from the COPS client located in the ER and configures the DiffServ classes. Once the reservation process ends the Linux PCs are properly configured in order to forward the user packets with the requested QoS (the necessary parameters are extracted from the Session Description Protocol, SDP, in the SIP messages).

3 Q-SIP Signaling Mechanism

This section describes the signaling mechanisms used by the proposed SIP based reservation architecture (Q-SIP); further details are given in [8].

The proposed solution makes possible to use existing SIP clients with no enhancements or modifications. It is possible to interact with no problem with other parties that do not intend or are not able to use QoS. Moreover backward compatibility with standardized SIP protocol is preserved.

The IP phones/terminals are located on the access networks; standard SIP clients can be used, set with an explicit SIP proxying configuration. When a call setup is initiated, the caller SIP client starts a SIP call session through the SIP proxy server. If a Q-SIP server is encountered, this can start a QoS session interacting with a remote Q-SIP server and with the QoS providers for the backbone network (i.e. the access ERs). Fig. 1 shows the reference architecture.

According to the direction of the call, the two Q-SIP servers are named caller-side Q-SIP server and callee-side Q-SIP server.

As far as the reservation procedure is concerned, two different models are possible: i) unidirectional reservations and ii) bi-directional reservations. The choice between the two models can be done on the basis of a pre-configured mode or through the exchange of specific parameters (*qos-mode* parameters) between the Q-SIP servers during the call setup phase. We are now going to detail the uni-directional model because the bi-directional one is easily evinced from the previous.

With reference to Fig. 1 (see also Fig. 2), the call setup starts with a standard SIP *INVITE* message sent by the caller to the local Q-SIP server (i.e. caller-side Q-SIP server). The message carries the callee URI in the SIP header and the session specification within the body Session Description Protocol (SDP) (media, codecs, source ports, ecc). The Q-SIP server is seen by the caller as a standard SIP proxy server. The Q-SIP server, based on the caller id and on session information, decides whether a QoS session has to be started or not. If a QoS session is required/opportune, the server inserts the necessary descriptors within the *INVITE* message and forwards it towards the callee. The *INVITE* messages can be relayed by both standard SIP proxy servers and Q-SIP servers until they reach the callee-side Q-SIP server and then the invited

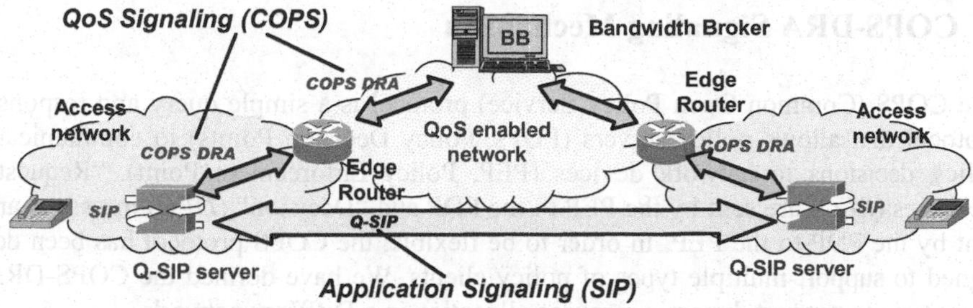

Fig. 1. Reference scenario for the proposed QoS architecture

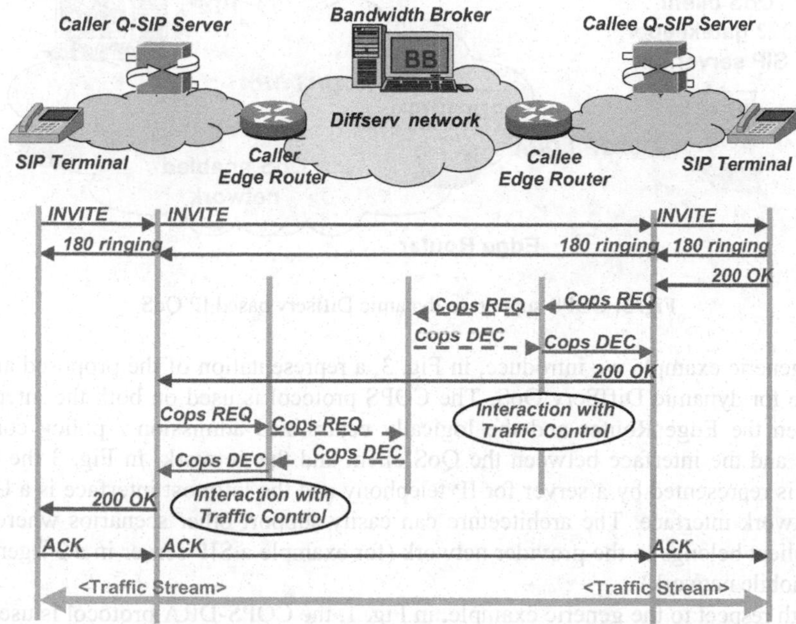

Fig. 2. Q-SIP call signaling flow - QoS enabled model

callee. When the callee responds with a *200 OK* message, it is passed back to the callee-side Q-SIP server. At this point, the callee-side Q-SIP server can request a QoS reservation to the ER on the callee access network (i.e. the QoS provider for the callee). Subsequently, the *200 OK* response, opportunely extended by the callee-side Q-SIP server, is forwarded back to the caller, via standard SIP servers and via the caller-side Q-SIP server. When the caller-side Q-SIP server receives the *200 OK* message, it performs QoS reservation with the ER on the caller access network (i.e. the QoS provider for the caller).

It is important to note that the proposed architecture keeps the compatibility with standard SIP clients and standard SIP servers. All the information needed by the Q-SIP servers to perform the QoS session setup is inserted within the SIP messages in such a way that non Q-SIP aware agents can transparently manage the messages.

4 COPS-DRA Signaling Mechanism

The COPS (Common Open Policy Service) protocol is a simple query and response protocol that allows policy servers (PDPs, Policy Decision Points) to communicate policy decisions to network devices (PEP, Policy Enforcement Point). "Request" messages (*REQ*) are sent by the PEP to the PDP and "Decision" (*DEC*) messages are sent by the PDP to the PEP. In order to be flexible, the COPS protocol has been designed to support multiple types of policy clients. We have defined the COPS-DRA client type to support dynamic resource allocation in a DiffServ network.

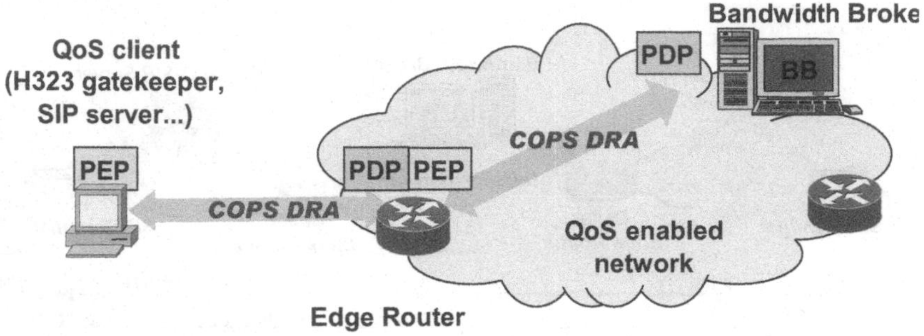

Fig. 3. COPS support to dynamic Diffserv based IP QoS

As a generic example we introduce, in Fig. 3, a representation of the proposed architecture for dynamic DiffServ QoS. The COPS protocol is used on both the interface between the Edge Router and the logically centralized admission / policy control server and the interface between the QoS client and the network. In Fig. 3 the QoS client is represented by a server for IP telephony and the leftmost interface is a User-to-Network interface. The architecture can easily support other scenarios where the QoS client belongs to the provider network (for example a SIP server in a 3rd generation mobile network).

With respect to the generic example, in Fig. 1, the COPS-DRA protocol is used on two different interfaces.

First, it is used as a generic signaling mechanism between the user of a "*QoS enabled*" network and the QoS provider. In our case the Q-SIP proxy server plays the role of QoS user and will implement a COPS-DRA client, while the Edge Router plays the role of QoS provider and will implement a COPS-DRA server. On this interface, COPS-DRA provides the means to transport: the scope and amount of reservation, the type of requested service and the flow identification. The second interface where the COPS-DRA is applied is between the Edge Router and the Bandwidth Broker, in order to perform the resource allocation procedures. A flexible and scalable model for resource allocation is implemented. A set of resources can be allocated in advance by the Bandwidth Broker to the Edge Router in order to accommodate future request (according to the so-called COPS *provisioning* model). The amount of this "aggregated" allocation can also be modified with time. Moreover, specific requests can be sent by the Edge Router to allocate resources for a given flow (according to the so-called COPS *outsourcing* model). The set of Edge Routers and the Bandwidth Broker realize a sort of distributed bandwidth broker in a DiffServ network.

The generic COPS-DRA architecture is better described in [7], the protocol details can be found in [6]. In Fig. 2 we provide an example of the message exchange between Q-SIP server, ER and BB. The first COPS message is originated by the COPS Client co-located with Callee Q-SIP Server once it receives the *200 OK* message from the Callee SIP Terminal. The dotted lines in Fig. 2 are optional messages needed if the outsourcing model is adopted and an outsourced *REQ* message is sent to the BB. Once the *200 OK* message arrives to the Caller Q-SIP Server an analogous message exchange (*REQ-DEC*) is performed (since we are detailing an uni-directional model).

5 Traffic Control and COPS DRA Interaction Mechanism

In this section we detail how we perform the resource allocation on routers. Routers have been implemented on Linux PCs where the kernel provides the traffic control software functions allowing the building of a DiffServ router.

The Linux operating system part responsible for both bandwidth sharing and packet scheduling is called Traffic Control (TC) [11]. The main goal of the TC is to manage packets queued on the outgoing interface with respect to the configured rules for the outgoing traffic. To achieve this goal the components used are: queuing disciplines, classes, filters and policing functions.

Filters are needed to mark packets on the ingress interface with a *tcindex* value (a kernel level packet specific value); it is used to forward the packets properly on the egress interface. Filtering rules could be performed on the following fields of the TCP/IP header: IP source address, IP destination address, source port, destination port, protocol, type of service.

Policing functions are used in order to keep the traffic profile under the negotiated SLA level.

Queuing disciplines and classes allow to implement PHBs in the DiffServ network.

The module responsible for resource allocation is named TC server and is located in the ER. The COPS-DRA client, located in the ER itself, acting as a TC client issues commands on the TC Server by means of "ad-hoc" messages detailed later on.

TC server has been built as a distinct module to preserve the modularity of the architecture and to realize a flexible element that could work also with different resource allocation protocol (e.g. SNMP, etc.).

Communications between TC server and client use a simple proprietary protocol. The messages sent are TLV (Type-Length-Value) messages. The design foresees five message types:

- *REQUEST* (to set-up filter and bandwidth allocation for a given flow);
- *RELEASE* (to delete structures set by the *REQUEST* message);
- *MODIFY* (to modify the bandwidth allocation for a specific flow);

- *ACK* (to notify the correct reception of the message);
- *NACK* (to notify an error in the reception).

REQUEST messages brings information about the flow identifier, the service class, the amount of reservation and the filter set-up information.

The operational model of this message exchange is the following: when the COPS client receives, from the COPS server running on the Bandwidth Broker, a specific request for a given flow, it, in turn, sends a *REQUEST* message to the TC server to configure the kernel parameters negotiated by means of COPS-DRA protocol. Then, TC server replies to the TC client with an *ACK* or *NACK* message to notify whether the message is received correctly or not. The operational model is the same for the *RELEASE* and the *MODIFY* messages.

6 Q-SIP / COPS-DRA / Traffic Control Testbed

The proposed architecture has been implemented in a test-bed composed of a set of Linux PCs. The DiffServ components of the test-bed have already been discussed in [9]. The overall picture of the test-bed is described in Fig. 4.

The Q-SIP proxy servers have been implemented on a Linux PCs based on RedHat7.1 distribution. The Q-SIP server is developed in Java (running on Sun JDK 1.2.2 virtual machine) while the COPS DRA and TC clients/servers are developed in C. The TC modules are based on the TCAPI [12] a library that allows the dynamic configuration of filters and scheduling mechanisms. The internal architecture of the test bed elements is shown in Fig. 5.The source code of the Q-SIP server is available under the GNU license at the URL in [8]. Note that also the COPS-DRA and TC server source code is available under the GNU license. The publicly available "Ubiquity SIP User Agent" version 2.0.10 [10] has been used as SIP terminals, running on Win98 PCs.

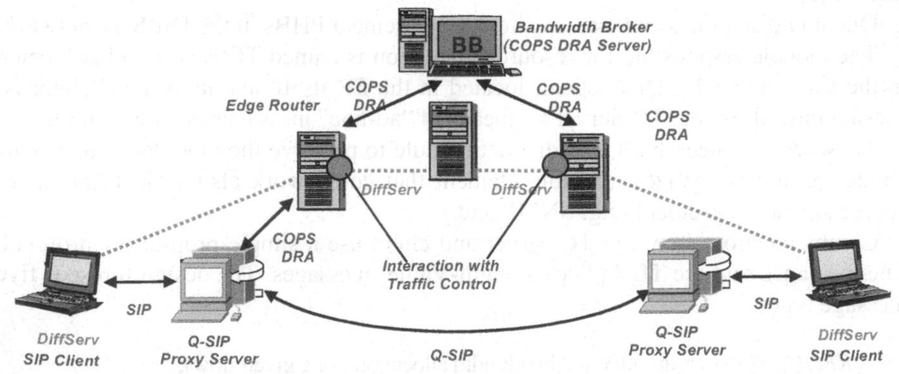

Fig. 4. Overall test bed scenario

The Q-SIP server has a modular architecture, in order to be able to handle different QoS mechanisms. As shown in Fig. 5, there is a JAVA module called Generic Q-SIP Protocol Handler, which is independent of the underlying QoS mechanism. This module dialogues through a JAVA interface with a QoS-specific Interface module (realized in JAVA as well), which is specific of the underlying QoS model. In the picture, the COPS-DRA specific module is shown, which interacts through a socket interface to the COPS-DRA client process, realized in C. The Edge Routers, that act as QoS Access Points, include a COPS DRA server that communicate through a socket interface with a process implementing the Local Decision Server and the COPS DRA client. This process communicates through another socket interface to the TC server that is able to configure the traffic control mechanisms provided by the Linux kernel. Communications between TC server and Linux kernel are made through a netlink socket. The PDP/BB is composed by a COPS DRA server and a Decision Server, that interact through a socket based interface.

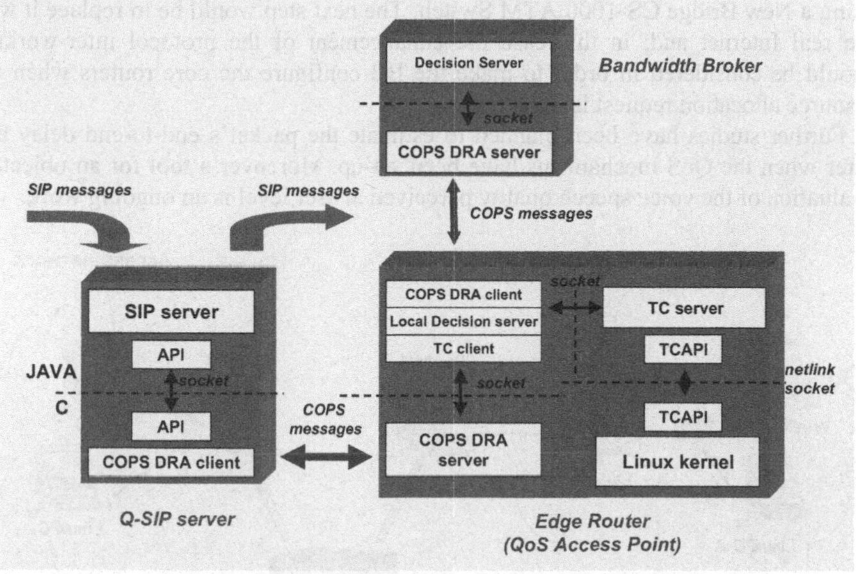

Fig. 5. Q-SIP server, ER, and BB internal architectures

6.1 Testbed in the NEBULA Project

The test-bed, developed in the Nebula project [13], has been shown "up and running" during the GTTI [14] annual meeting in Trieste in June. The overall picture of the test-bed is described in Fig. 6.

Some tests have been performed on this test-bed to verify the speech quality received at application level in both "*no-QoS*" and "*QoS enabled*" scenarios. In the "*no-QoS*" test we have observed that the quality perceived is good in condition of low traffic but degrades quickly when the background traffic causes the link congestion.

In the "*QoS enabled*" test even if an heavy background traffic is present the quality perceived is good and not affected by the traffic volume itself.

The access network A is constituted by a Win98 PC that plays the role of the caller SIP Client and by two Linux PCs. Linux PC A acts as the background traffic generator in both scenarios and, in the "*QoS enabled*" one as the Q-SIP server too. Linux PC B is a router that connects the access network to the backbone link in both scenarios and, in the "*QoS enabled*" one plays the role of the COPS client too. The access network B is constituted by two computers, a Win98 PC and a Linux one directly connected to a router on the backbone link. The Win98 PC is the SIP called client, while the Linux PC acts as the background traffic receiver and, as it concerns the "*QoS enabled*" scenario as both Q-SIP server and COPS client. The router A and B are two Linux PCs on which we have implemented the TC mechanisms. In order to reduce the topology complexity we have decided to make the Router A playing the role of the BB too. The backbone link is emulated by means of an ATM connection using a New Bridge CS-1000 ATM Switch. The next step would be to replace it with the real Internet and, in this case the enhancement of the protocol inter-working should be considered in order to make the BB configure the core routers when the resource allocation request is accepted.

Further studies have been planned to estimate the packet's end-to-end delay and jitter when the QoS mechanisms have been set-up. Moreover a tool for an objective evaluation of the voice speech quality perceived at user level is an ongoing work.

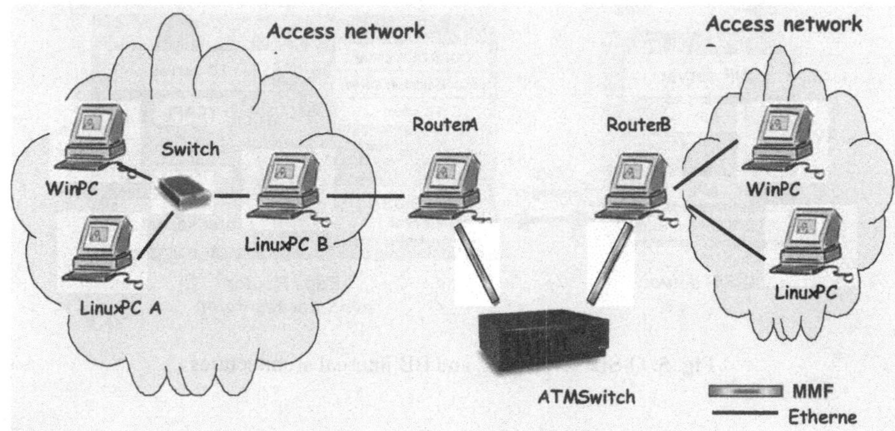

Fig. 6. Overall Nebula test-bed

7 Conclusions

In this paper we have proposed an architecture for the interaction of SIP protocol with QoS mechanisms in a *"QoS enabled"* network and for the interaction of those mechanisms with the Linux Traffic Control in order to obtain a Dynamic Resource Allocation / Configuration in a DiffServ network. The enhancements to SIP protocol that

that have been given are basically independent of the specific "*QoS enabled*" network. Obviously the mechanism should be specialized for each specific case. In particular the interaction with a Diffserv network using a refined version of COPS (COPS-DRA) for admission control has been described in this paper. The DiffServ implementation was achieved by means of Linux PCs dynamically configured by means of Traffic Control and TCAPI functionalities. A possible deployment scenario based on a Q-SIP proxy server is proposed, having the advantage that "legacy" SIP user application can be fully reused. We note also that the solution is fully backward compatible with current SIP based equipment that does not support QoS, allowing a smooth migration. The test bed implementation of the proposed solution and the tests detailing, including the internal architecture of the software modules, has been described.

Acknowledgements

The authors would like to thank Donald Papalilo and Enzo Sangregorio for their work in support of the specifications and of the test-bed implementation.

References

1. J. Rosenberg, H. Schulzrinne, G. Camarillo, A. Johnston, J. Peterson, R. Sparks, M. Handley, E. Schooler, " SIP: Session Initiation Protocol", IETF RFC 3261, June 2002.
2. X. Xiao, L.M. Ni "Internet QoS: A Big Picture", IEEE Networks, March 1999
3. W. Zhao, D. Olshefski and H. Schulzrinne, "Internet Quality of Service: an Overview" Columbia University, New York, New York, Technical Report CUCS-003-00, Feb. 2000.
4. G. Camarillo (Ed.), et al. "Integration of Resource Management and SIP", IETF RFC 3312, October 2002, Work in Progress.
5. D. Durham, Ed., J. Boyle, R. Cohen, S. Herzog, R. Rajan, A. Sastry, The COPS (Common Open Policy Service) Protocol, IETF RFC 2748, January 2000.
6. S. Salsano "COPS usage for Diffserv Resource Allocation (COPS-DRA)", <draft-salsano-cops-dra-00.txt>, September 2001, Work in Progress, http://www.coritel.it/projects/cops-bb
7. S. Salsano, L.Veltri, "QoS Control by means of COPS to support SIP based applications", IEEE Network, March/April 2002
8. L. Veltri, S. Salsano, D. Papalilo, "QoS Support for SIP based Applications in Diffserv Networks", <draft-veltri-sip-qsip-01.txt>, October 2002, Work in Progress, http://www.coritel.it/projects/qsip
9. W.Almesberger, S.Giordano, R. Mameli, S. Salsano, F.Salvatore "A prototype implementation for Intserv operation over Diffserv Networks", IEEE Globecom 2000, S. Francisco, December 2000.
10. "SIP User Agent", Ubiquity Software Corporation, http://www.ubiquity.net.
11. B.Hubert "Linux Advanced Routing & Traffic Control HOWTO" http://lartc.org/howto.
12. D.P.Olshefski, TCAPI Project, http://oss.software.ibm.com/pub/tcapi
13. Nebula Project Home Page, http://nebula.deis.unibo.it/
14. GTTI Home Page, http://www.gtti.cnit.it/

Virtual Flow Deviation: Dynamic Routing of Bandwidth Guaranteed Connections

Antonio Capone, Luigi Fratta, and Fabio Martignon

DEI, Politecnico di Milano, Piazza L. da Vinci 32, 20133 Milan, Italy
{capone, fratta, martignon}@elet.polimi.it

Abstract. Finding a path in the network for a new incoming connection able to guarantee some quality parameters such as bandwidth and delay is the task of QoS routing techniques developed for new IP networks based on label forwarding.

In this paper we focus on the routing of bandwidth guaranteed flows in a dynamic scenario where new connection requests arrive at the network edge nodes. When more than a path satisfying the bandwidth demand exists, the selection of the path is done in order to minimize the blocking probability of future requests.

We propose a new routing algorithm named Virtual Flow Deviation (VFD) which exploits the information of the ingress and egress nodes of the network and the traffic statistics. We show that this new algorithm allows to reduce remarkably the blocking probability in most scenarios with respect to previously proposed schemes.

1 Introduction

The development of the Internet has been impressive in the last years. The upcoming high-speed networks are expected to support a wide variety of communication-intensive real-time multimedia applications. However, the current Internet architecture offers mainly a best-effort service and does not meet the requirements of future integrated services networks that will be designed to carry heterogeneous data traffic [1]. In order to offer guaranteed end-to-end performance (as bounded delay, jitter or loss rate), it's necessary to introduce some sort of resource reservation mechanism in the Internet. With classical IP routing, however, when the resources are not available on the shortest path the connection request must be rejected even if they are available on alternative paths.

Many Quality of Service routing algorithms have been recently proposed [1, 2, 3, 4]. With new label based forwarding mechanisms, such as MPLS (Multi Protocol Label Switching) [5], per flow path selection is possible and Quality of Service parameters can be taken into account by the routing algorithm. The notion of Quality of Service (QoS) has been introduced to capture the qualitatively and quantitatively defined performance contract between the service provider and the user applications. The QoS requirement of a connection can be given as a set of link constraints. Such contraints can be expressed, for instance,

M. Ajmone Marsan et al. (Eds.): QoS-IP 2003, LNCS 2601, pp. 592–605, 2003.

as bandwidth constraints specifying that the path selected for the connection of the requesting user has sufficient bandwidth to meet the connection requirement.

The goal of QoS routing algorithms is twofold:

– satisfying the QoS requirements for every admitted connection;
– achieving global efficiency in resource utilization.

In this paper, we focus on the problem of QoS routing. First of all, we review some of the proposed QoS routing algorithms, such as the Min-Hop Algorithm (MH) [6], the Widest Shortest Path Algorithm (WSP) [7] and the Minimum Interference Routing Algorithm (MIRA) [8]. We'll describe in some detail MIRA, which has somewhat different features compared to other algorithms, as it takes explicitly into account the topology disposition of the ingress and egress points of the network, i.e. those routers through which the traffic enters into and exits form the network.

We analyze and compare their performance based on results obtained under a variety of simulated scenarios. We observe that these algorithms are unable to achieve good performance when the traffic statistics at each ingress point are different (for instance, when an ingress node offers to the network a traffic significantly higher than other nodes). Furthermore, all the proposed algorithms lack to consider traffic statistics which can be easily measured at each ingress node.

To overcome these limitations, we propose a new QoS routing algorithm, called Virtual Flow Deviation (VFD), which keeps into account the information of the ingress and egress nodes of the network and the traffic statistics. More precisely, VFD exploits the knowledge of the disposition of the ingress/egress nodes of the network, and uses the statistics information about the traffic offered to the network through each ingress point in order to forecast future connection arrivals. For every connection request, VFD creates a set of *virtual calls* based on the observed traffic statistics. These virtual calls represent the calls which are likely to request resources to the network in the immediate future, and will thus interfere with the current one. In order to improve the global resource utilization, VFD routes the current call together with the virtual calls using the Flow Deviation method [9].

In order to assess the effectiveness of the proposed scheme, we analyze the performance of VFD under a variety of scenarios, and we compare it with that achieved by existing routing algorithms.

The paper is structured as follows: in Section 2 we address the QoS routing problem and some existing routing algorithms. In Section 3 we introduce the Virtual Flow Deviation algorithm. In Section 4 we discuss and compare the performance of these algorithms under a variety of simulated scenarios. Finally, Section 5 concludes the paper.

2 QoS Routing

The concept of Quality of Service (QoS) indicates the performance trading between the Internet Service Provider and the hosts' applications. The QoS constraints parameters of a single connection can be specified in terms of:

- minimum guaranteed bandwidth;
- maximum tolerable delay and/or jitter;
- maximum tolerable loss rate.

The main goal of a QoS routing technique is to determine a path able to guarantee the constraints requested by the incoming connection and to reject as few connections as possible.

Let's model a network as a graph (N, A), where the nodes N represent routers and Arcs A represent communication links, as shown in Figure 1.

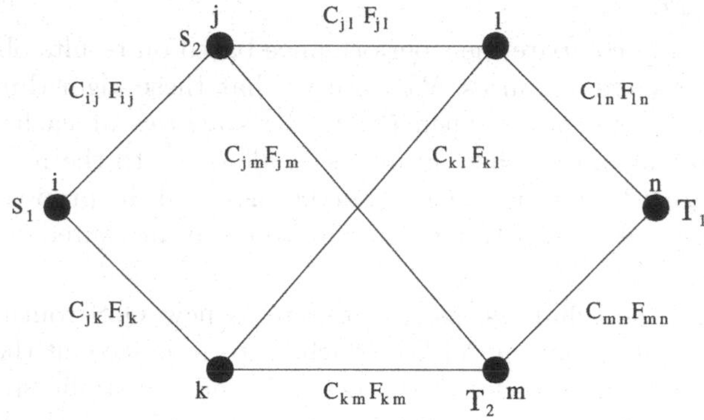

Fig. 1. QoS Network State

The traffic enters the network at ingress nodes S_i and exits at egress nodes T_i. Each single connection requires a path from S_i to T_i. Each link (i, j) has associated some parameters such as the capacity C_{ij} and the actual flow F_{ij}. The residual bandwidth is defined as $R_{ij} = C_{ij} - F_{ij}$. A new connection can be routed only over links with R_{ij} greater or equal to the requested bandwidth.

Referring to a new connection with requested bandwidth d_k, a link is defined as *feasible* if $R_{ij} \geq d_k$. A connection can be accepted if at least one path between S_i and T_i exists in the feasible network. The minimum R_{ij} over a path defines the maximum residual bandwidth of that path.

In the following we review some of the algorithms available in the literature.

Min-Hop Algorithm

The Min-Hop Algorithm (MHA) [6] routes an incoming connection along the path which reaches the destination node using the minimum number of feasible links.

This scheme, based on the Dijkstra algorithm, is simple and computationally efficient. However, MHA can build up bottlenecks in the network, as it tends to overload some links leaving others underutilized. The cost given to each link, in fact, remains unvaried and independent of the current link load and therefore MHA tends to use the same paths until saturation is reached before switching to other paths with underutilized links.

Widest Shortest Path Algorithm

The Widest Shortest Path Algorithm (WSP), proposed in [7], is an improvement of the Min-Hop algorithm, as it attempts to load-balance the network traffic. In fact, WSP choses the feasible min-hop path with the maximum residual bandwidth, thus discouraging the use of already heavily loaded links.

However, WSP still has the same drawbacks as MHA since the path selection is performed among the shortest feasible paths.

Minimum Interference Routing Algorithm

The Minimum Interference Routing Algorithm (MIRA), proposed in [8], explicitly takes into account the location of the ingress and egress routers. The key idea of MIRA is to route an incoming connection over a path which least interferes with possible future requests.

Specifically, an incoming connection request between (S_i, T_i) is routed with the goal of maximizing an objective function which is either the minimum maximum-flow (maxflow) of all *other* ingress-egress pairs or a weighted sum of maxflows in the latter case, where weights α_{ST} assigned to each ST pair reflect the "importance" of the flow.

In order to achieve an on-line routing algorithm, MIRA keeps an updated list of the *critical* links, i.e. the links whose use by the incoming call diminishes the maxflow between other pairs.

When a new call has to be routed between the sorce/destination pair (S_i, T_i), MIRA determines the set L_{ST} of the critical links for all the source/destination pairs (S_j, T_j) *other* than (S_i, T_i). The weight w of each link l is then set according to the equation $w(l) = \sum_{(S,T):l \in L_{ST}} \alpha_{ST}$, and the route which causes the minimum interference to other source/destination pairs is selected.

In spite of its more sophisticated functions, MIRA still has the following limitations whose effect will be shown in the discussion of numerical results:

- MIRA discourages the use of critical links based only on the number of other S-T pairs which could use them, without verifying if these S-T pairs actually use these links. Evidently, if one of these other S-T pairs introduces a low traffic in the network, the *criticality* of the links which diminish its maxflow is far less important than that of S-T pairs which produce a large amount of traffic. As a consequence, MIRA preserves the use of certain links which remain underutilized, thus causing a suboptimal use of the network.

To overcome this limitation, it has been proposed to maximize a weighted sum of the source/destination maxflows. However, in [8] the weights are chosen offline and do not adapt to changes in network traffic. Hence this solution does not provide the flexibility necessary to an on-line routing scheme.
- In its on-line implementation, MIRA sets the link weights almost in a static way according only to their level of criticality. In fact, the only event which can cause the redistribution of new weights is the saturation of some links, similarly to the Min-Hop algorithm.
- While chosing a path for an incoming request, MIRA does not take into account how the new call will affect the future requests of the *same* ingress/ egress pair (auto-interference).

3 Virtual Flow Deviation

In the previous Sections we have summarized the features and the limitations of the existing QoS routing algorithms. In this section we propose a new technique, called Virtual Flow Deviation (VFD), which aims to overcome these limitations by exploiting all the information available which has been underevaluated in the other routing algorithms.

To better describe the current state of the network and to forecast its future one can add to the topological information on the location of ingress/egress pairs, used by MIRA, the traffic statistics obtained by measuring the load offered to the network at each source node. This information plays a key role in deciding how to route incoming requests to prevent network congestion.

Exploiting the knowledge of the offered traffic, we can forecast how many new connections will probably be generated at each S-T pair in the immediate future. These new calls, which are likely to be offered to the network, will interfere with the current call to be routed, and they should thus be considered in the routing process.

In order to take into account the future traffic offered to the network, VFD thus routes not only the real call, but also a certain number of *virtual* calls, which represent an estimate (based on measured traffic statistics) of the connection requests that will probably interfere with the current, real call. The number of these virtual calls, as well as the origin and the bandwidth requested should reflect as closely as possible the real future network conditions. Hence, we determine these parameters based on the past traffic statistics of the various ingress/egress pairs, as we'll explain in detail in the next subsection. With this mechanism, we can take into account both the interference and the auto-interference produced by the current and future calls between every S-T pair.

All the information concerning network topology and estimated offered load must be used to produce a path selection which uses at the best the network resources and minimizes the number of rejected calls. Such a path selection is performed in VFD by the Flow Deviation algorithm, which allows to determine the optimal routing of all the flows entering the network through all the different source/destination pairs.

Before describing in detail the VFD algorithm, we give an high-level scheme of its functionalities.

3.1 The Virtual Calls

Each call offered to the network, either real or virtual, can be represented with the notation (S, T, d), where S and T are the source and destination nodes of the call, respectively, and d is its bandwidth requirement. The determination of these three parameters for each virtual call is quite critical, as they must reflect as closer as possible the real evolution of the network. More precisely, we have to determine how many virtual calls should be generated, their source/destination pairs, and their bandwidth request. In this process, we can easily measure and distribute to each S-T pair the two following parameters:

- the average traffic (λ_{S_i,T_i}) offered by the i-th S-T pair, defined as the average number of connections entering the network through the node S_i in an interval Δt
- the probability density distribution of the bandwidth required at each S-T pair, which can be estimated as the ratio between the number n_b of calls which have requested b bandwidth units and the total number N of calls considered for the estimation.

Note that, for simplicity of exposition, we have considered bandwidth requests which are multiple of a given bandwidth unit. However, the algorithm works as well with bandwidth requirements which can assume any real value.

If we define the total average load offered to the network, Λ, as:

$$\Lambda = \sum_{\forall \text{ pairs } S_i - T_i} \lambda_{S_i,T_i}$$

we can evaluate the probability P_{S_i,T_i} to receive a call between the node pair (S_i, T_i) as $P_{S_i,T_i} = \frac{\lambda_{S_i,T_i}}{\Lambda}$ while the probability P_{b_i} to have a request of b bandwidth units at the i-th source node, is estimated by $P_{b_i} = \frac{n_b}{N}$.

The parameters (S_i, T_i, d_i), which completely determine the virtual calls, are generated by extracting random values according to the probability density functions P_{S_i,T_i} and P_{d_i} derived as described above. The virtual calls are routed together with the real call, represented by (S_R, T_R, d_R), using the Flow Deviation algorithm, in order to ensure an optimal flow assignment.

To determine the number N_v of virtual calls which must be generated we have considered two different approaches listed below.

The first one takes into account the variations in the total load offered to the network, and estimates the average number $\overline{N}$ of calls routed (active calls) in the network over the past T seconds. When the new call request arrives, $N_v = \lfloor(\overline{N} - N_A)\rfloor$ virtual calls are generated, where N_A is the current number of active calls. Note that if $N_A > \overline{N}$, no virtual call is generated.

The second approach is based on the maximum number of active calls which can be routed in the network, N_{max}, and the number of virtual calls is given by $N_v = \lfloor (N_{max} - N_A) \rfloor$. The underlying assumption in this approach is to consider the network operating close to its saturation. This condition, which stresses the effectiveness of the routing algorithm, is useful when comparing performance. All the numerical results presented in Section 4 have been derived by implementing this second approach.

3.2 The Virtual Flow Deviation Algorithm

The VFD algorithm operation is described in the flow diagram of Figure 2.

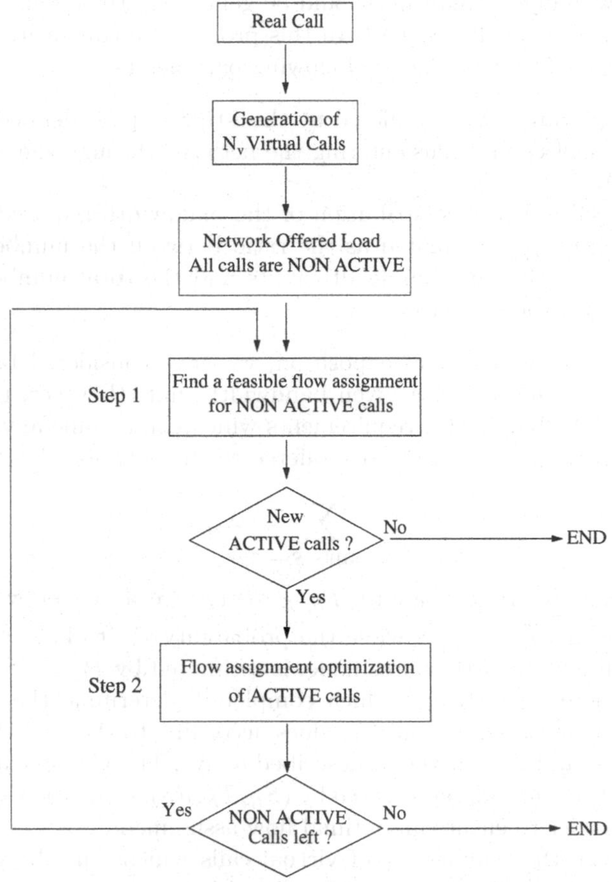

Fig. 2. The Virtual Flow Deviation algorithm

Upon a new call request the process for generating N_v associated virtual calls, as described in the previous section, is activated. The real call and the virtual calls are then offered to the network. The procedure to route the new traffic operates in two steps.

In the first step an initial feasible flow assignment is obtained. Calls are routed one by one starting from the real call. A call can be either defined as ACTIVE, if a feasible path has been found, or NON ACTIVE otherwise. This step is repeated until all calls have been considered. The procedure stops if the real call cannot be routed.

In step two the routing of all ACTIVE calls is optimized using a slightly modified version of the Flow Deviation Method, described in [9, 10], in which the flows cannot be splitted.

Then step one is repeated for the NON ACTIVE calls. If at least one NON ACTIVE call is declared ACTIVE the step two is repeated and the procedure is iterated until either all calls are ACTIVE or step one does not define any new call as ACTIVE.

At the end of the procedure the real call has been routed on an optimal path considering an expected future evolution of the network traffic load.

The feasible flow assignment is obtained in step one by using the Shortest Path Algorithm (Dijkstra) applied to the network whose links weights reflect the actual channel utilization. More specifically, for each link a weight $w_{ij} = \frac{1}{C_{ij}-F_{ij}}$ is assigned and updated at each iteration.

A more formal description of VFD is given by the pseudo-code in Table 1. The description of a connection has been enriched by a flag to identify ACTIVE and NON ACTIVE connections.

4 Numerical Results

In this section we compare the performance of the Virtual Flow Deviation algorithm with that of the Min-Hop Algorithm and MIRA referring to three different network scenarios in order to cover a wide range of possible environments. The performance function we consider is the percentage of rejected calls versus the average total load offered to the network.

The first scenario we consider is illustrated in Figure 3. In this network the links are unidirectional with capacity equal to 120 bandwidth units. The network traffic, offered through the source nodes S_1, S_2 and S_3, is unbalanced as the traffic offered by sources S_2 and S_3 is four times that offered by S_1. Each connection requires a bandwidth uniformly distributed between 1 and 3 units.

In this simple topology only one path is available to route connections between S_1-T_1 and S_3-T_3, while connections S_2-T_2 can choose between two different paths.

This case evidentiates the main limitation of MIRA that does not consider the information about the total load offered to the network. Since the links (1,2),(2,3) and (8,9) are critical for S_2-T_2, the route selected by MIRA follows the path with the minimum number of critical links (5-8-9-6 in the example). Unfortunately this interferes with the path (7-8-9-10) that carries the high load of S_3-T_3. This choice will penalize the performance as shown in Figure 4.

VFD achieves the best performance since it exploits the information on the unbalanced load. The performance of MHA and MIRA are exactly the same. In

```
for ( ∀ connection (S_k, T_k, d_k, flag_k))
    flag_k = NON ACTIVE
end for

do
    for (∀ connection (S_k, T_k, d_k, flag_k = NON ACTIVE)
        for (∀ link l_ij)
            weight assignment:
            w_ij = 1/(C_ij − F_ij) if F_ij < C_ij
            w_ij = ∞ if F_ij = C_ij
        end for
        execution of Dijkstra Shortest Path algorithm:
        if (∃ a path between S_k and T_k with bandwidth d_k)
            update F_ij and memorize the path
            flag_k = ACTIVE
        end if
    end for

    for ( ∀ connection (S_k, T_k, d_k, flag_k = ACTIVE))
        execution of the Flow Deviation method
    end for

while (in the last iteration at least one flag_k has been set to ACTIVE)
```

Table 1. Pseudo-code specification of *Step 1* and *Step 2* introduced in Figure 2

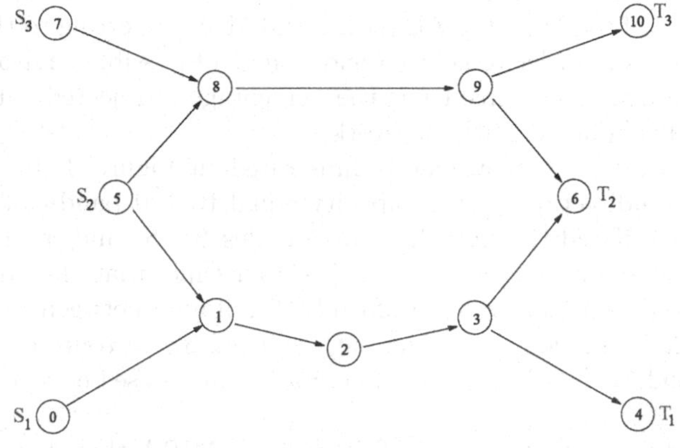

Fig. 3. Network topology with unbalanced offered load: the source/destination pairs S_2-T_2 and S_3-T_3 offer to the network a traffic load which is four times higher than that offered by the pair S_1-T_1

fact MIRA operates for the connections between S_2-T_2 the same path selection of MHA, since the path (5-8-9-6) is shorter than (5-1-2-3-6).

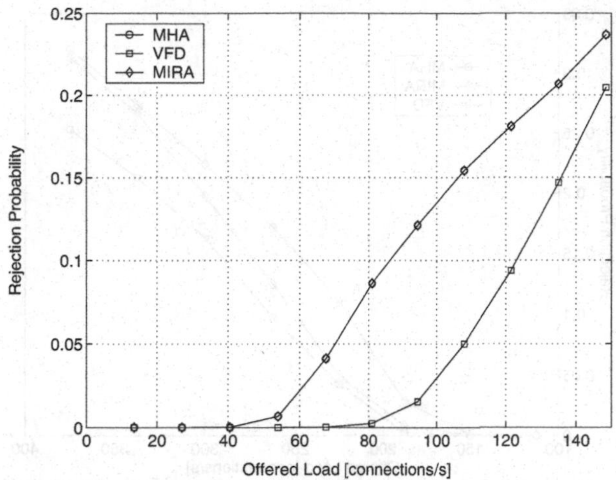

Fig. 4. Connection rejection probability versus the average total offered load to the network of Figure 3

The second network considered is shown in Figure 5 where a balanced traffic is offered at S_1 and S_2. All links have the same capacity (120 bandwidth units) and are bidirectional.

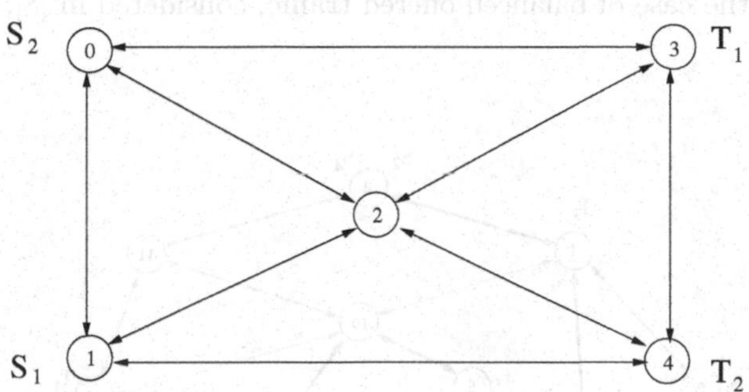

Fig. 5. Network Topology with a large number of critical links

The critical links identified by MIRA are (0,1),(0,2),(0,3),(1,4),(2,4),(3,4) for connections S_2-T_2 and (1,0),(1,2),(1,4),(0,3),(2,3),(4,3) for connections S_1-T_1. This leads to have only the path (1-2-3) available for connections S_1-T_1 and the path (0,2,4) for connections S_2-T_2.

This is a very limiting way of operation that penalizes MIRA. As shown in Figure 6, VFD can reach a more balanced routing using all the available paths with no limitation.

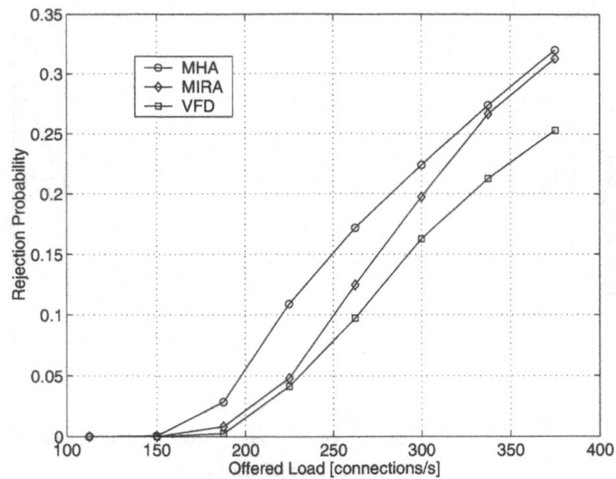

Fig. 6. Connection rejection probability versus the average total offered load to the network of Figure 5

As third scenario we have considered the network shown in Figure 7 that was proposed in [8] and represents a more realistic scenario. All links are bidirectional. Those marked by heavy solid lines have a capacity of 480 bandwidth units while the others have a capacity equal to 120 bandwidth units. The performance for the case of balanced offered traffic, considered in [8], are shown in Figure 8.

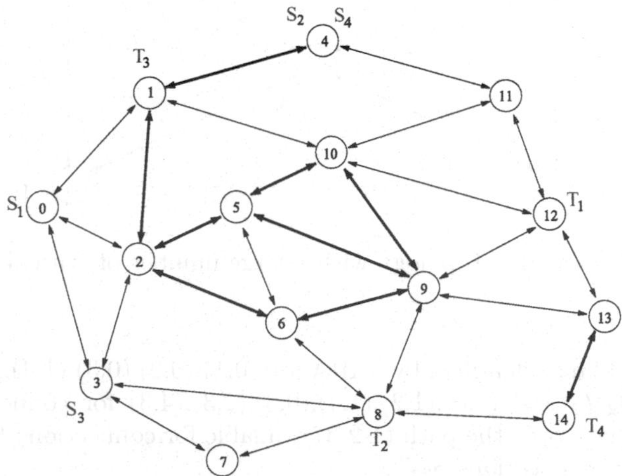

Fig. 7. Network Topology with a large number of nodes, links, and source/destination pairs

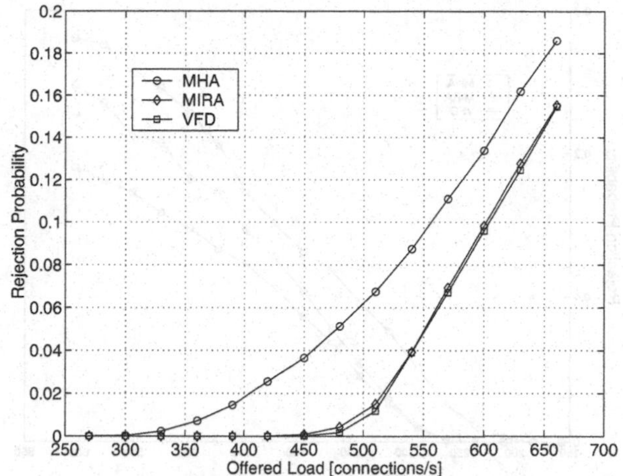

Fig. 8. Connection rejection probability versus the average total offered load to the network of Figure 7

VFD and MIRA achieve almost the same performance. However, VFD presents some advantages at low rejection probabilities since it starts rejecting connections at an offered load 10% higher than MIRA. We have measured that a rejection probability of 10^{-4} is reached at an offered load of 420 connections/s by MIRA as opposed to 450 connections/s for VFD.

If we consider on the same topology an unbalanced load where for instance traffic S_1-T_1 is four times the traffic of the other sources, the improvement in the performance obtained by VFD is much more significant. The curves shown in Figure 9 confirm that the unbalanced situations are more demanding on network resources with respect to the balanced case as the rejection probability for the same given offered load is much higher. In these more critical network operations VFD has proved to be more effective providing improvements of the order of 20%.

5 Conclusions

We have proposed a new QoS routing scheme, called Virtual Flow Deviation (VFD), which exploits the informations about the ingress and egress nodes of the network and the traffic statistics.

As a key innovation with respect to existing QoS routing algorithms, VFD performs a path selection based not only on the current state of the network, but also on an estimate of its future evolution. This goal is achieved by routing a set of *virtual* calls together with the current call using the Flow Deviation algorithm to ensure an optimal disposition of all the flows. Virtual calls represent an estimate (based on traffic statistics measured at each ingress node) of the calls which are likely to be offered to the network during the current connection lifetime, thus interfering with it.

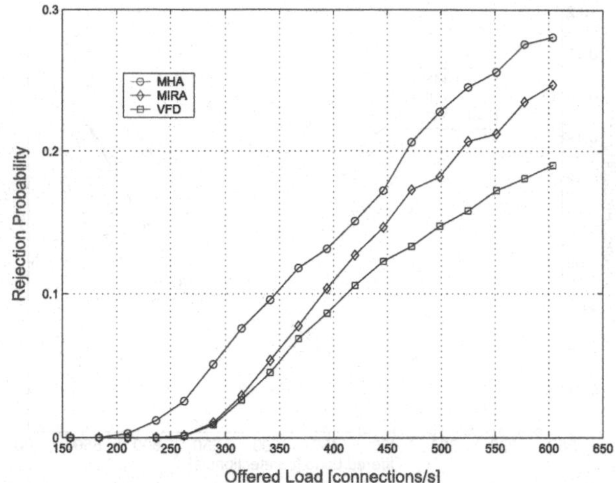

Fig. 9. Connection rejection probability versus the average total offered load to the network of Figure 7, where the traffic between S_1-T_1 is four times higher than the traffic produced by the other pairs

VFD allows to achieve lower connection rejection rates than existing algorithms, expecially in more critical network operations whith unbalanced traffic offered at the ingress nodes.

We have shown that this new algorithm allows to reduce remarkably the blocking probability in most scenarios with respect to previously proposed schemes.

References

[1] S. Chen and K. Nahrstedt. An Overview of Quality-of-Service Routing for the Next Generation High-Speed Networks: Problems and Solutions, 1998.

[2] Z.Wang and J.Crowcroft. QoS Routing for Supporting Resource Reservation. In *IEEE JSAC*, Sept.1996.

[3] A.Orda. Routing with End to End QoS Guarantees in Broadband Networks. In *IEEE INFOCOM'98*, Mar.1998.

[4] B.Awerbuch et al. Throughput Competitive On Line Routing. In *34th Annual Symp. Foundations of Computer Science*, Palo Alto, CA, Nov.1993.

[5] E.Rosen, A.Viswanathan, and R.Callon. Multiprotocol Label Switching Architecture. In *RFC 3031*, January 2001.

[6] D.O.Awduche, L.Berger, D.Gain, T.Li, G.Swallow, and V.Srinivasan. Extensions to RSVP for LSP Tunnels. In *Internet Draft draft-ietf-mpls-rsvp-lsp-tunnel-04.txt*, September 1999.

[7] R.Guerin, D.Williams, and A.Orda. QoS Routing Mechanisms and OSPF Extensions. In *Proceedings of Globecom*, 1997.

[8] Murali S. Kodialam and T. V. Lakshman. Minimum Interference Routing with Applications to MPLS Traffic Engineering. In *Proceedings of INFOCOM (2)*, pages 884–893, 2000.

[9] L.Fratta, M.Gerla, and L.Kleinrock. The Flow Deviation Method: An Approach to Store-and-forward Network Design. In *Networks 3*, pages 97–133, 1973.
[10] D.Bertsekas and R.Gallager. *Data Networks*. Prentice-Hall, 1987.

Design and Implementation of a Test Bed for QoS Trials

Giorgio Calarco[1], Roberto Maccaferri[1], Giovanni Pau[2], and Carla Raffaelli[1]

[1] D.E.I.S. - University of Bologna
Viale Risorgimento, 2 - 40136 Bologna, Italy
{gcalarco,rmaccaferri,craffaelli}@deis.unibo.it,
http://www-tlc.deis.unibo.it
[2] Department of Computer Science
The Henry Samueli School of Engineering and Applied Science
Campus Box 951596, UCLA, Los Angeles, California 90095-1596, USA
gpau@cs.ucla.edu

Abstract. This paper describes the design, implementation and testing of a test bed supporting flow-based classification functions for multi-service traffic. Protocol and statistical analysis of application flows is performed in the edge routers to provide EF treatment to multimedia traffic without any user signaling. These functions take advantage of the Linux Traffic Control environment and implement SLA management and traffic statistics collection. Sample measurement performed on the test bed shows the effectiveness and feasibility of the proposed solution.

1 Introduction

Quality of Service (QoS) is the capability of a network to forward packets in different ways by grouping them into traffic categories called classes. Several different solutions to the QoS problem have been devised: ATM (Asynchronous Transfer Mode), RSVP (Resource ReSerVation Protocol) [1], the Integrated Services [2] and the Differentiated Services [3] architectures are examples of complementary and interoperable approaches addressing different needs. The Differentiated Services architecture, that is considered here, supports a scalable solution to QoS in IP networks being it based on few fundamental concepts and components: the identification of the packet QoS class through a code point and the differentiated treatment of that packet within a diffserv node as Per Hop Behaviour (PHB). Two main PHBs have been standardised so far: the Expedited Forwarding PHB [4] - for the support of services requiring time guarantees - and the Assured Forwarding PHB [5] - for packet treatment according to three types of drop precedence. PHBs are identified through a 6 bits label, called Differentiated Services Code Point (DSCP) which is placed into the Diffserv Field of the IP header.

In order to permit the end-to-end QoS management a hierarchical, two-tier [6] architecture was also proposed. This model defines the inter- and intra-domain resource allocation, needed to achieve the end-to-end QoS support. The approach

M. Ajmone Marsan et al. (Eds.): QoS-IP 2003, LNCS 2601, pp. 606–618, 2003.

requires the interaction between the RSVP signalling in the stub networks and the bandwidth brokers within the diffserv domains [7]. So the user application should be RSVP capable in order to take benefit from traffic differentiation.

In this paper a new approach to end-to-end QoS is proposed to allow QoS unaware users to access network QoS capabilities in a "plug and play" fashion. The basic idea behind it is a DSCP marking of the traffic based on a content-oriented micro flow classification. The flow classification is made by the edge routers by looking at the traffic aggregate generated within the stub network. The proposed classification process is developed for real time traffic and considers both protocol and statistical analysis of stub network traffic. It is implemented in a Linux environment using Traffic Control utilities.

The paper is organized as follows. In section 2 interactive multimedia traffic characteristics are analysed; section 3 describes the test bed characteristics; section 4 introduces the real time classifier architecture; section 5 focuses on real time classifier functionality and performance evaluation. The paper is concluded with section 6 that summarizes the main achievements and focuses on some ideas for future work.

2 Interactive Multimedia Traffic

From the overall performance point of view interactive multimedia applications are more resistant to packet loss than to high end-to-end delay or jitter when they are transmitted across IP networks. TCP flow control mechanisms assure the correctness of TCP streams but the delay introduced by the retransmission of lost packets creates a bigger damage than the loss itself, if this is reasonable small (i.e. 10%).[8] . Typically these applications are based on the UDP protocol [9]. The impossibility for current best effort IP networks to assure a better service to real-time applications has lead to the development of special protocols. The main contributors on this direction are the IETF and ITU. To address the previous problems, the IETF Audio-Video Transport and Multiparty Multimedia Session Control working groups have developed RTP/RTCP [10] protocols for the transport of real-time content, and RTSP [11] protocol optimised for multimedia streaming. The benefits introduced by these protocols, together with the need for a common standard base, has given RTP the role of standard protocol for the transport of real-time contents over the Internet. The RTP protocol is being used by the most common interactive multimedia applications covering both the commercial and the scientific community as shown in table 1. This means that the use of RTP by an application is a sufficient condition to classify the transmitted data as real-time data. The key concept, behind our classification and marking scheme, is to try to recognise RTP as the protocol used above layer four, typically above UDP.

As regards specific delay requirements, the ITU [12] studies the transmission delay constraints for PSTN. Three different classes of delay that satisfy most of the applications have been identified for connections with adequately controlled echo [12]. In order to keep the end-to-end delay as low as possible, it is better to transmit the audio stream as a bigger number of small packets, instead of a smaller number of big packets [13]. There are different reasons that justify this choice. Smaller packets are more unlikely to be fragmented or dropped due to buffer management problems and moreover the loss of one packet introduces a very limited source of noise at the receiver side.

Table 1. Most commonly used interactive multimedia tools

	RTP/RTCP	RTSP	RVSP	Audio Video
Netmeeting	Yes	No	Yes	AV
Vic	Yes	No	Yes	V
Rat	Yes	No	No	A
Real Server	Yes(live)	Yes	No	AV

The packet size depends also on audio and protocol aspects. The audio part depends in its turn on the codec frame size and bandwidth while the protocol one is related to the use of different headers (i.e. IP, UDP, RTP). In the case of very limited transmission bandwidth (i.e. analog dial-up modems) the transmission of different audio frame within the same IP packet is required in order to limit the protocol overhead [14]. Given a set of codecs it is possible to estimate the typical mean packet frequency and size for audio conference applications over IP networks in order to keep the end-to-end delay in the range of few hundreds of milliseconds as specified in ITU-T G.114 to assure acceptable quality to delay-aware users [12]. The lower bound for this packet frequency can be considered about 10 packets per second. This value can be drawn from table 2 where the values of packet size and frequency are shown for G.723.1 codec, assuming 24 byte audio frames generated every 20 ms.

Table 2. Rate and packet sizes for G.723.1 codec with 24 byte frames every 20 ms

IP Packets per second	Coding Delay (msec)	Audio payload size (byte)
1	1000	1200
5	200	240
10	100	120

These considerations about delay, with minor differences, can be applied also for interactive video conference applications and videophone connections. Both packet size and frequency values are considered in the classification process.

3 Test Bed Layout and Configuration

In order to perform quality of service trials, a local test bed was designed to emulate functions of a real network environment that offers real time services with quality of service guarantees. Figure 1 shows the test bed layout, which consists of five Intel i810-board systems connected through a Layer2-switch and equipped with a 1Ghz- Pentium III processor and a 256MB bank of RAM. An Internet link is also provided for geographical connection and testing. The edge router (called Alfa) is based on the popular Linux operating system. In detail, a 2.2.19 version of this kernel was used as a developing platform and partially modified to satisfy our aims. It represents the core of the test bed, since it performs the quality of service functions. To this end, it takes advantage of classification, SLA, and bandwidth management utilities, eventually by the interaction with a bandwidth broker (called DeisBB), connected through the switch. 100 Mbit/s Ethernet cards are used. Nevertheless the output links of the edge router are forced to work at 10 Mbit/s to be more easily saturated by test traffic.

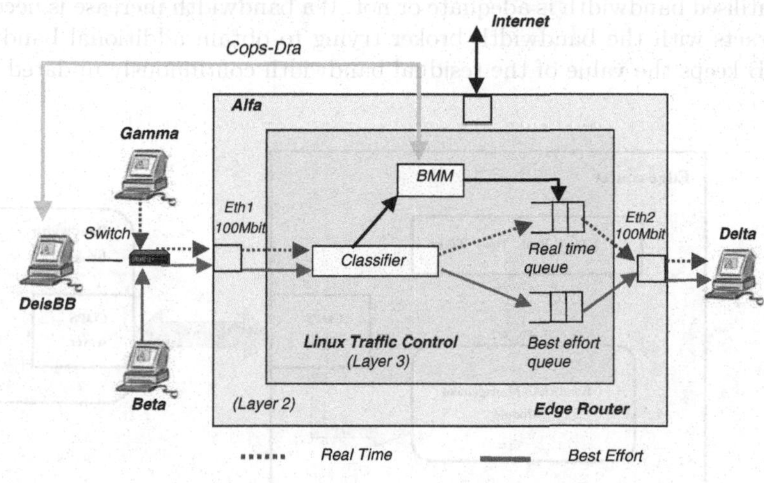

Fig. 1. Functional diagram of the test layout

The other three computers (called Beta, Gamma and Delta) are dual-boot systems having Linux and Windows 2000 installed; they are exclusively utilised for traffic generation and analysis. In particular, Rude, a traffic generator [17] was installed on Beta and Gamma and used for injecting three distinct flows of traffic into the input port of the router. Specifically, these are a real time flow, a non real time flow (both at 64 Kbit/s), and a best effort flow (at 16 Mbit/s, thus sufficient to saturate alone the output port of the router). Access to the router by Beta and Gamma is obtained through the 10/100 Mbit/s Ethernet switch. A traffic receiver, Crude [17], is set up on Delta: it collects information about the packets coming from the output interface of the router, helping us to verify if the real time flow had been correctly treated. Other applications were also useful for generating the real time flow and evaluating how the system can significantly improve the quality of the communication under a human perspective. Examples of these are the popular Microsoft "NetMeeting", GnomeMeeting and RAT. The traffic control can be configured via the "tc" command, a user-space application which interacts with the Linux kernel to create various objects as queues, classes, SLAs, etc and to initialise them. A graphical front-end interface was also released for easiness of use. The Linux Traffic Control queuing discipline chosen for service differentiation is here based on the CBQ (Class Based Queuing) algorithm. Its configuration assumes two classes, with 1 Mbit/s and 9 Mbit/s rates, respectively, each with a 100 packet-long FIFO buffer attached. Figure 2 shows the basic software architecture of the system. The Bandwidth Management Module (BMM), is a Unix bash script and interacts with both the Linux Traffic Control and the Bandwidth Broker through the COPS client/server protocol (Common Open Policy Service) [18]. Figure 2 shows the architecture of software for bandwidth management. By measuring the real time traffic flows, the BMM decides if the utilised bandwidth is adequate or not. If a bandwidth increase is necessary, it interacts with the bandwidth broker trying to obtain additional bandwidth. The BB keeps the value of the residual bandwidth continuously updated.

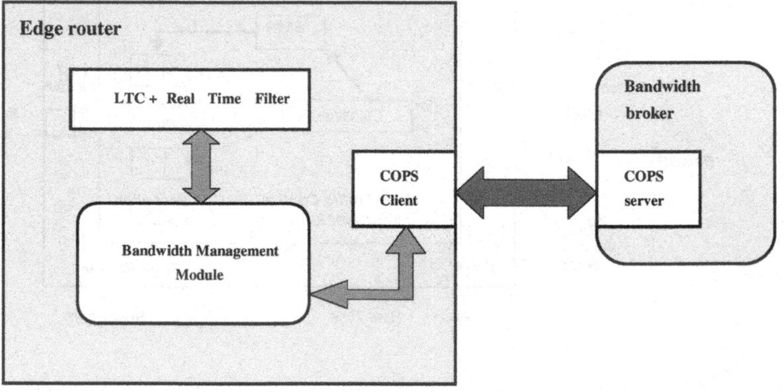

Fig. 2. Software architecture

4 The Real Time Classifier

In this section the approach to allow QoS unaware users of multimedia tools to take benefit of network QoS is described. The new functionality is introduced into the edge router in relation to the network scenario of figure 3, although it can be even introduced within user equipments. The new feature consists in a classifier that, according to a given set of SLAs, performs both protocol and statistical analysis on the traffic incoming from the stub network. The new functionality and its prototype implementation are called Real Time Classifier (RTC). RTC is designed for interactive multimedia applications and, at this moment, it is able to recognize and mark that kind of traffic. In terms of diffserv PHB, RTC marks the traffic recognized as belonging to real-time multimedia streams as EF, setting the IP packet DSCP field. The number of packets necessary used for classification can be chosen independently for each classification algorithm with the aim to optimise the classification delay and failure rate trade-off. RTC is actually composed by the following four logical units:

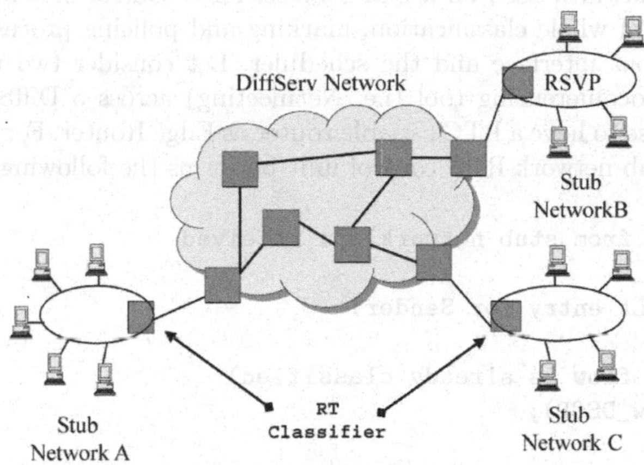

Fig. 3. RTC classifier functionality on the DiffServ Architecture

- Classifier: performs traffic classification on protocol and statistical basis;
- Marker: marks the packets according to the classifier policy;
- Meter: meters the incoming traffic;
- Control Unit: manages the SLAs and performs supervision of other units actions.

The control unit has in charge the management of SLA's and the supervision of the whole classification process. For our purposes a 7th-tuple as shown in the table 3 defines a SLA.

The ID parameter is the unique SLA identifier. The fields IP and MASK are used to identify the host/network belonging to the SLA. The BW parameter is the bandwidth allowed for the considered SLA. The policy parameter can take different values according with the policy adopted for the out-profile traffic of the considered SLA. In particular at the moment the following values are allowed:

Table 3. SLA format

ID	IP	Mask	BW	Shared	DSCP	Policy

OK when the out-profile traffic is forwarded as in-profile and no actions is taken; DROP when the out-profile traffic is discarded.

Finally, "shared" is used to specify the degree of fairness to among flows belonging to the same SLA. The value of the field ranges between 0 and 100, and represents the percentage of the bandwidth used on a FCFS basis, with zero meaning that all the bandwidth is equally split between the flows and 100 meaning all bandwidth used on a FCFS basis. RTC control unit has in charge the supervision of whole classification, marking and policing process as a filter between the input interface and the scheduler. Let consider two user A, and B, using a videoconferencing tool (i.e. Netmeeting) across a DiffServ capable network. Suppose to have a RTC capable router as Edge Router. For each Packet coming from stub network RTC control unit performs the following algorithm:

```
when (packet  from stub network) is received
{
 If (exists SLA entry for Sender)
 {
   if (packet flow is already classified)
     mark(Flow_DSCP);
   else
    Classify;

   if (Meter(packet,SLA)==out-profile)
     Policy(SLA policy);
 }
 else{
      if (source==untrasted) mark(Best Effort);
     }
 Forward(scheduler); }
```

The "classify" procedure tries a classification of the incoming packet using both application protocol header and statistical information. Once sufficient information has been collected, the flow is classified and its socket tuple is recorded in the hashing table of classified flows. All its subsequent packets are recognized and marked accordingly by matching the value of the tuple.
The RTC classifier is the first functional unit encountered by the traffic entering the Differentiated Services domain from a stub network. RTC is integrated in Linux Kernel QoS Mechanisms named Linux Trafic Control, so once classified and marked a packet can be forwarded using a Linux QoS scheduler.

RTC classifies traffic using both statistical and protocol analysis. The protocol analysis is based on the RTP header characteristics; in particular there are a few parameters within the RTP header keeping constant their values during the whole session. The RTP header has a minimum length of 96 bits and 42 of them remain constants during the whole session.
The protocol-based classification algorithm can be tuned changing the size of the population used to classify. This tuning affects both the precision of the results and the time needed to obtain them. The presence of the RTP protocol in the analysed flow is considered a sufficient, but not necessary, condition to classify the flow as an interactive multimedia stream.
The statistical classification algorithm adds classification capability in the case of real time applications that are not RTP compliant. It takes into account the flow rate and the packet size as main parameters for the classification process. As described in section II, the flow rate depends on the codec used and bandwidth. Typically each IP packet contains one single audio frame. On the other hand when the introduced overhead becomes an issue two or more audio frames are grouped in one IP packet. The number of audio frames per packet is kept as small as possible in relation to the line bandwidth. For example, the codec G.723.1 [15] produces audio frames of 30 ms (33 audio frames per second) and they are generally transmitted one per IP packet if the link bandwidth is enough, but considering a 14.4 kbs modem, they are grouped in three audio frames per IP packet in order to satisfy the bandwidth constraints (table 4).

Table 4. Flow rate and coding delay

	Modem 14.4	LAN
Packet size (Bytes)	100 (3 audio frames)	52 (1 audio frame)
Packet/s	11	33
Bit/s	8800	13728
Overhead	28%	53%
Coding Delay	90 ms	30 ms

Taking into account all the parameters a flow is supposed to be interactive multimedia flow if the number of packet per second will be greater than 10 packets per second. This is the lower bound to keep the delay in the constraints defined by G.114 [15]. In this scenario, in order to obtain an acceptable delay for end users, an application has to encode the audio using more than 10 packets per second. Finally RTC marker unit writes the DSCP value in the DSFIELD of the classified packets according to the classification results.
RTC meter unit, in conjunction with the policy unit, performs the enforcement of the traffic profiles.

5 Experimental Evaluation

The RTC tool has been implemented hacking a Linux kernel (release 2.2.19) and the command TC included in the package iproute2 [16] in order to realize a new filter (called RTC) of the Linux Traffic Control. Several tests have been done in order to evaluate classification effectiveness and the amount of resources, in terms of memory and time per classified flow, required for the classification process. The classification process is tested using both the RTP-based and statistical approaches. Three different traffic flows have been generated by Rude [17] in order to saturate the 10 Mbit/s router output link: a real time flow and a non real time flow, both at 64 Kbit/s, and a best effort flow at 16 Mbit/s. Access to the router by Beta and Gamma is obtained through a 100 Mbit/s Ethernet switch.

The Linux Traffic Control queuing discipline for service differentiation is here based on the CBQ (Class Based Queuing). Its configuration assumes two classes, with 1 Mbit/s and 9 Mbit/s rates, respectively, each with 100 packet FIFO queues attached. The main performance figures of interest are the bandwidth used by each flow, the packet loss rate and the time jitter. Jitter is defined with reference to figure 4 as J_tx - J_rx.

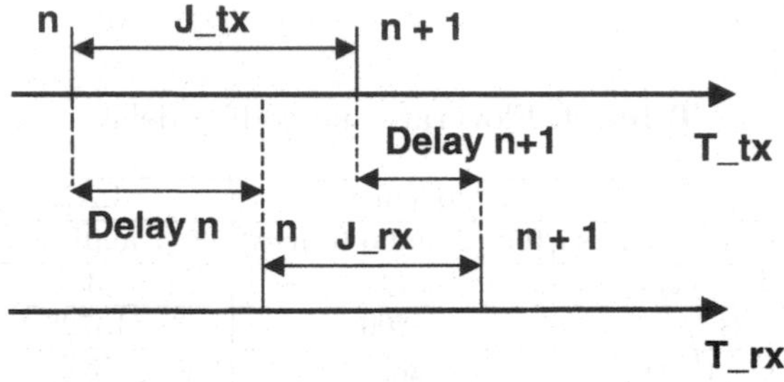

Fig. 4. Main quantities for jitter evaluation

Figure 5 shows the bandwidth usage for real time and non real time traffic during congestion. It is evident that the real time traffic after the classification process has recognized this kind of traffic, obtains the required bandwidth of 64 Kbit/s even if saturation is present. The bandwidth used by non-real time traffic is on the other hand not stable. Some losses are present for real time traffic due to the layer 2 transmission buffer overflow, where both EF and BE traffics are considered in the same way . This losses can be eliminated by reducing the upstream CBQ queue size for the BE class, thus limiting the BE traffic offered to the transmission queue. The packet lost in time are represented in figure 6 where is evident the different behaviour of the two flows. The packet loss

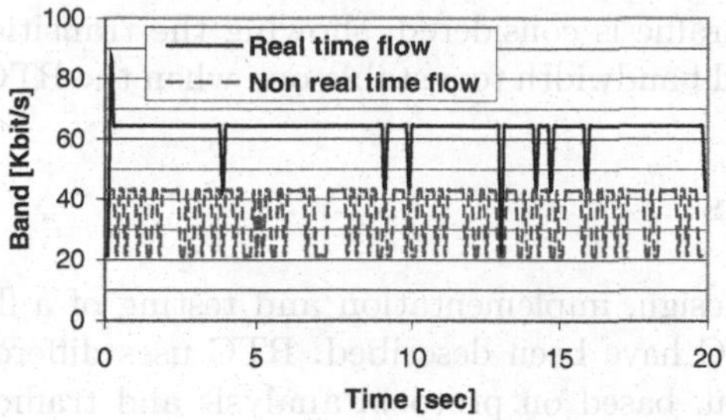

Fig. 5. Link bandwidth usage as a function of time for real time and non real time 64 Kbit/s flows in the presence of a 16 Mbit/s best effort flow over a 10 Mbit/s link.

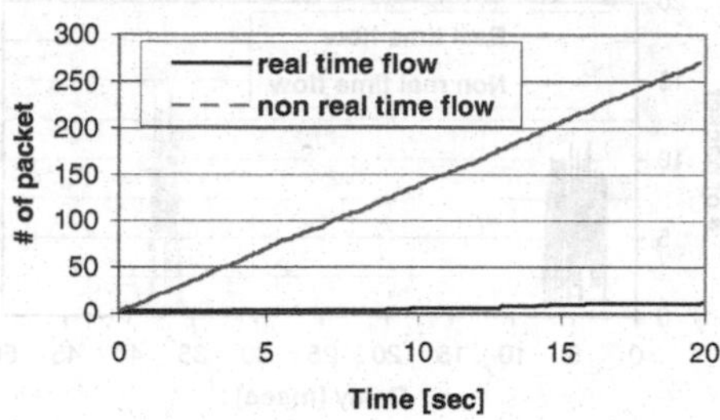

Fig. 6. Packets lost during test for real time and non real time traffic, both at 64 Kbit/s, in the presence of a 16 Mbit/s best effort flow over a 10 Mbit/s link.

rate for real time traffic has been calculated to be lower than 2 %. The analysis of transmission delay is presented in figure 7 and 8. Figure 7 shows the percentage of packets at different delay for the to 64 Kbit/s flows. The real time flow has a delay limited at 5 ms, much lower than the delay for the non real time flow. Figure 8 evidences the effect of the RTC after the time necessary for recognizing the real time flow. At the beginning the two flows are dealt in the same way, then, after less then 10 ms, when the classification procedure is completed, the different behaviour in terms of delay is evident. Figure 9 shows the temporal jitter as previously defined for real time and non real time traffic. The values of the jitter for real time traffic are acceptable being within 5 ms after the flow has been classified. As regards the jitter for non real time traffic, it is limited only because almost all-non real time traffic is lost and the small percentage of packets that enters the queue typically finds it full.

A temporal analysis focused only on real time traffic has been also performed in the presence of best effort traffic at 16 Mbit/s with the insertion at a given time of the RTC function. In figure 10 the time behaviour for the bandwidth used by real time traffic is considered, showing the transition from a situation with not guaranteed bandwidth to a stable one when the RTC function is active.

6 Conclusions

In this paper the design, implementation and testing of a flow based real time classifier called RTC have been described. RTC uses different methodology to perform its function, based on protocol analysis and traffic patterns. Being it flow-oriented, it can perform functions such as SLAs management and bandwidth usage measurement. This can be fruitfully used in dynamic bandwidth management (i.e. bandwidth broker support). The results show effectiveness of RTC in recognizing real time flows and in guaranteeing bandwidth as limited delays for these kinds of applications.

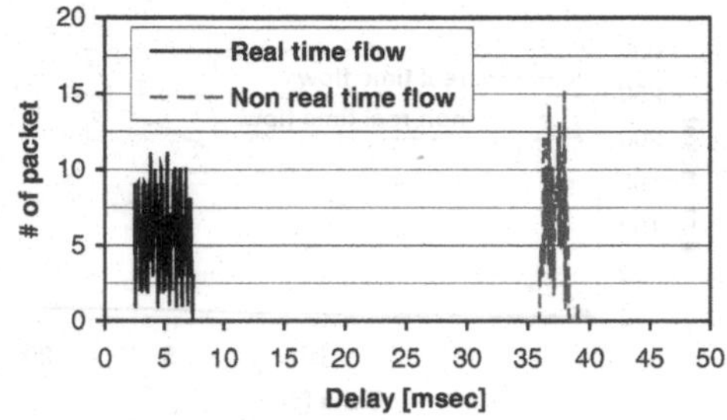

Fig. 7. Distribution of delay for real time and non real time traffic.

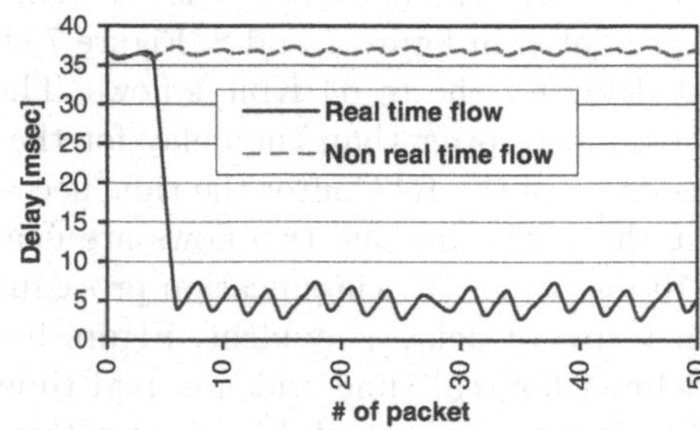

Fig. 8. Packet delay during test for real time and non real time traffic.

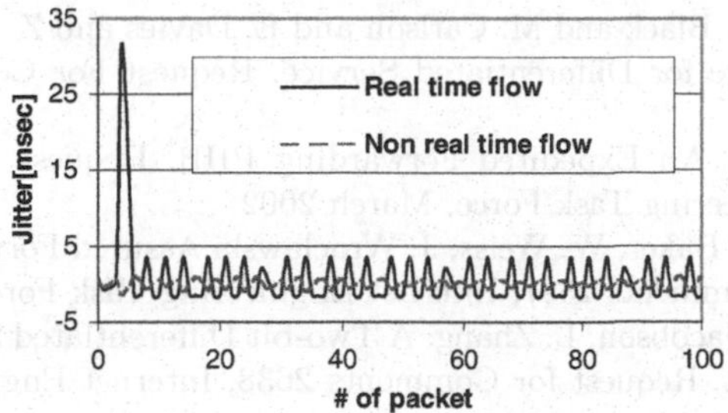

Fig. 9. Time Jitter for real time and non real time traffic.

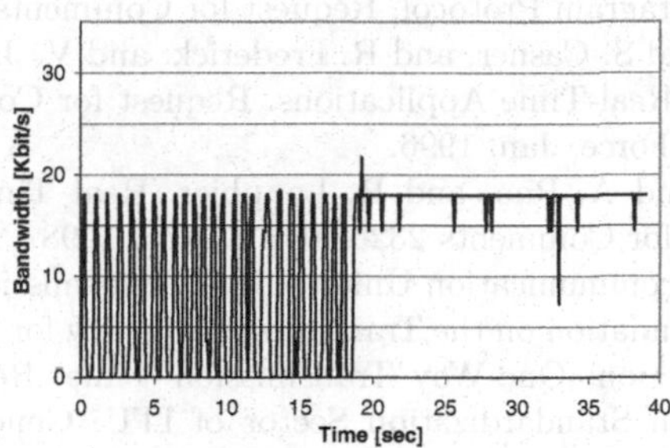

Fig. 10. Effect of the introduction of the RTC function on the behaviour of real time traffic.

7 Acknowledgements

The authors wish to thank Dr. Christian Benvenuti for the technical support in the development of the classification procedure.

References

[1] R. Braden Ed. and L. Zhang and S. Berson and S. Herzog and S. Jamin. Resource ReSerVation Protocol RSVP) Version 1 Functional Specification. Request For Comment 2205, IETF, 1997.

[2] Opens H.323 group. Codec Bandwidth and Latency Calculations. Url-http://www.openh323.org/bandwidth.html, June 2000. R. Braden, and D. Clark, and S. Shenker, Integrated Services in the Internet Architecture: an Overview. Request for Comments 1633, Internet Engineering Task Force, June 1994.

[3] S. Blake and D. Black and M. Carlson and E. Davies and Z. Wang and W. Weiss. An Architecture for Differentiated Service. Request For Comment 2475, IETF, Dec. 1998.
[4] B. Davie et al. An Expedited Forwarding PHB. Request for Comments 3246, Internet Engineering Task Force, March 2002.
[5] J. Heinanen, F. Baker, W. Weiss, J. Wroclawski.Assured Forwarding PHB Group. Request for Comments 2597, Internet Engineering Task Force, June 1999.
[6] K. Nichols, V. Jacobson, L. Zhang. A Two-bit Differentiated Services Architecture for the Internet. Request for Comments 2638, Internet Engineering Task Force,. July 1999.
[7] Andreas Terzis and Jun Ogawa and Sonia Tsui and Lan Wang and Lixia Zhang. A Prototype Implementation of the Two-Tier Architecture for Differentiated Services. In RTAS99, Vancouver, Canada, 1999.
[8] D. Su and J. Srivastava, and Jey-Hsin Yao. Investigating factors influencing QoS of Internet phone. In Proc. of IEEE International Conference on Multimedia Computing and System, pages 308-313,June 1999.
[9] J. Postel. User Datagram Protocol. Request for Comments 768, IETF, Aug 1980.
[10] H. Schulzrinne, and S. Casner, and R. Frederick, and V. Jacobson. RTP:a Transport Protocol for Real-Time Applications. Request for Comments 1889, Internet Engineering Task Force, Jan. 1996.
[11] H. Schulzrinne, and A. Rao, and R. Lanphier. Real Time Streaming Protocol (RTSP). Request for Comments 2326, IETF, Apr. 1998.
[12] International Telecommunication Union (ITU). Transmission Systems and Media, General Recommendation on the Transmission Quality for an Entire International Telephone Connection; One-Way Transmission Time. Recommendation G.114, Telecommunication Standardization Sector of ITU, Geneva, Switzerland, Mar. 1993
[13] Bolot, Jean-Chrysostome and Garcia, Andres Vega. Control Mechanisms for Packet Audio in the Internet. In INFOCOM, San Fransisco, California, Mar. 1996.
[14] Opens H.323 group. Codec Bandwidth and Latency Calculations. Url-http://www.openh323.org/bandwidth.html, June 2000.
[15] http://www.itu.int
[16] ftp://ftp.sunet.se/pub/Linux/ip-routing/.
[17] http://www.atm.tut.fi/rude/.
[18] R. Mameli, S. Salsano, "Use of COPS for Intserv Operations over Diffserv: Architectural Issues, Protocol Design and Test-bed implementation", ICC 2001, Helsinky.

A Linux-Based Testbed for Multicast Sessions Set-Up in Diff-Serv Networks*

Elena Pagani, Matteo Pelati, and Gian Paolo Rossi

Computer Science Dept., Università degli Studi di Milano
via Comelico 39, I-20135 Milano, Italy
pagani@dsi.unimi.it, matteo@dolce.it, rossi@dsi.unimi.it

Abstract. In this work, we describe the implementation of an architecture to perform admission control and traffic management for multicast sessions in Diff-Serv networks. The *Bandwidth Broker* functionalities are carried out by the *Call Admission Multicast Protocol* (CAMP). CAMP performs the set-up of a multicast RTP session. It supports dynamic changes in the group membership. The implementation has been performed in a testbed network based on the Linux platform; we discuss the measurement results obtained by performing experiments with the testbed.

1 Introduction

In the last few years, many distributed applications have been deployed, that require some sort of *Quality-of-Service* (QoS) to the network protocols for the transmission of their data, e.g. in terms of available bandwidth or limited end-to-end delay. Currently, those applications can only exploit the *best effort* service supported by the IP protocol, which is unable to provide adequate QoS guarantees. Many research efforts are carried out to design novel network protocols and devices that are *QoS-aware*, that is, that can appropriately manage different classes of traffic. The Diff-Serv framework [1] proposed by IETF is scalable and can support QoS with only limited changes to the current Internet structure, by charging the most part of the control overhead on the domains boundary routers. The Diff-Serv model includes *Bandwidth Broker* (BB) agents [2] that perform the functionalities of admission control, resource reservation and system configuration. A BB exists for each administrative domain; the BBs cooperate to support inter-domain sessions. The two approaches proposed in the literature to perform call admission control are either the *passive* measurement policy or the *active* measurement policy [3]. The active measurement approach allows a more accurate estimate of the resource availability than the passive approach, and it is more lightweight because routers are stateless. The solutions for admission control so far proposed in the literature [3, 4, 5, 6] only consider unicast sessions,

* This work was supported by the MURST under Contract no.MM09265173 "Techniques for end-to-end Quality-of-Service control in multi-domain IP networks".

M. Ajmone Marsan et al. (Eds.): QoS-IP 2003, LNCS 2601, pp. 619–633, 2003.

although many applications are multicast in nature. On the other hand, multicast support in the Diff-Serv model is difficult [7]. Moreover, the behaviours of the existing solutions in realistic environments are often not sufficiently analyzed.

In this work, we describe the implementation of an architecture to perform admission control and traffic management for multicast sessions in Diff-Serv networks. The admission control service is provided by the *Call Admission Multicast Protocol* (CAMP), that carries out the *Bandwidth Broker* functionalities. CAMP performs the set-up of a multicast RTP session, and it supports dynamic changes in the group membership. The system exploits an active measurement approach with dropping of the `probe` packets as the congestion signal. The implementation has been performed in a testbed network based on the Linux platform; we discuss the measurement results obtained by performing experiments with the testbed.

The paper is structured as follows: in Section 2, we describe the testbed architecture, and we give an overview of the admission control mechanism. In Section 3, we discuss in detail the testbed implementation on the Linux platform. In Section 4, we present the experimental results. Section 5 concludes the work.

2 Testbed Architecture

In figure 1, we show the architecture of the implemented system. At the highest level, the most part of the modules are implemented in both the source and the destinations; the shadowed modules only exist at the destinations. The modules at the network and the data link layers are instantiated in all the routers belonging to the multicast routing tree. At the server side, the traffic generator can either produce a dummy data stream having a Constant Bit Rate, or it can read the data (1) from a MPEG file. The traffic generator is linked (2) to the RTP/RTCP library [10], and the admission control module (CAMP). The RTP library is used to support the appropriate playback of the data traffic at the destinations (10). The RTCP library is exploited to exchange information about the profile of the traffic generated by the source, and the quality of the traffic received at the destinations. It may use either UDP or TCP; in section 2.1 we specify how the transport service is chosen. At the receiver side status information is maintained about the profile of the traffic generated by the source, to decide about the session acceptance. This information is available via the RTCP source report; it is used by the admission control module. To support multicast communications, we use IGMP [11] as a membership service, and PIM-SM [12] as a multicast routing protocol. They cooperate to maintain the multicast routing table (Multicast Forwarding Cache, MFC) (4). To notice the dynamic changes in the multicast group membership, thus performing the appropriate admission control over the new multicast tree branches, the CAMP part at the kernel level implements a hook (5) to PIM-SM. Both data (7) and probing (6) multicast traffic is routed by PIM-SM. At the data link layer, packets are managed according to a *priority* packet scheduling policy: the data packets belonging to sessions requiring QoS have the highest priority. The priority level is determined by the

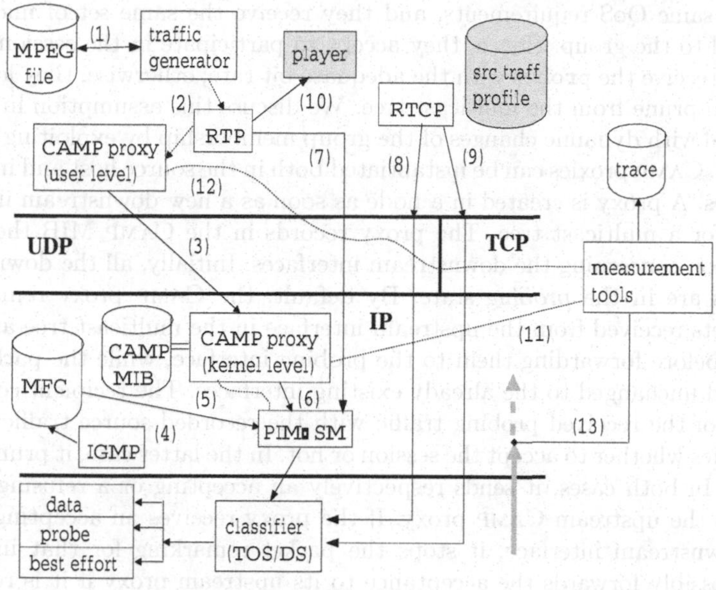

Fig. 1. Functional architecture of the multicast testbed.

value of the TOS/DS byte carried in the IP header [13]. We deployed ad hoc measurement tools that trace the traffic generation and delivery. At the source side, they are linked to the CAMP part at the kernel level, to monitor the packet transmission (11). At the destination side, they are linked to the CAMP part at the user level (12), that receives the packets; moreover, they trace the packet arrivals immediately above the network interface (13).

2.1 The CAMP Protocol

In this section, we outline the admission control mechanism; for more details, the interested readers may refer to [8, 9]. The admission control procedure is *receiver-driven*: each receiver autonomously decides whether it can receive the data traffic with the adequate QoS level or not, that is, whether to participate in the multicast session or not. The *Call Admission Multicast Protocol* (CAMP) operates on-demand; it performs the admission control for multicast sessions requiring bandwidth guarantees, thus supporting the *Premium Service* (EF PHB [14]). CAMP adopts an active measurement approach, with dropping of the `probe` packets as the congestion signal. The `probe` packets are marked with a lower priority than the QoS data packets, but a higher priority than the best effort packets. This way, they can drain the available bandwidth at the expenses of the best effort traffic, while they do not affect the established QoS sessions. For the sake of simplicity, throughout this work we hypothesize that all the recipients have the same QoS requirements, and they receive the same set of microflows addressed to the group. Hence, they accept to participate in the session only if they can receive the `probes` with the adequate bit rate; otherwise, they leave the group and prune from the multicast tree. We discuss this assumption in section 5. We deal with dynamic changes of the group membership by exploiting a *proxy* approach. CAMP proxies can be instantiated both in the source host and in the in-tree nodes. A proxy is created in a node as soon as a new downstream interface appears for a multicast tree. The proxy records in the CAMP MIB the status information concerning the downstream interfaces. Initially, all the downstream interfaces are in the *probing* state. By default, the CAMP proxy remarks all the packets received from the upstream interface in the multicast tree as `probe` packets, before forwarding them to the probing interface, while the packets are forwarded unchanged to the already existing interfaces. The recipient compares the QoS of the received probing traffic with the recorded source traffic profile, and decides whether to accept the session or not. In the latter case, it prunes from the tree. In both cases, it sends respectively an accepting or a refusing RTCP report to the upstream CAMP proxy. If the proxy receives an accepting report for a downstream interface, it stops the packet remarking for that interface, and it possibly forwards the acceptance to its upstream proxy if it is receiving remarked packets in turn. The interface state is changed from *probing* to *data*. When all the *probing* interfaces of a proxy either disappear as a consequence of a pruning recipient or are marked as *data* interfaces, the proxy turns off and all the status information it recorded in the CAMP MIB can be deleted. The proxy approach allows to perform the admission control for dynamically joining destinations, and to immediately forward the data to recipients joining during the transmission without negatively affecting other established sessions. It also allows to hide the membership changes to the source: in the initialization phase, the CAMP source generates `probe` packets until an accepting report is received. At that point, the source switches *without discontinuity* to the transmission of the data packets, while the proxies terminate the admission control procedure for the remaining destinations. To guarantee the termination of the session set-up phase, the decision reports cannot be lost. While the usual RTCP reports can be sent exploiting the UDP transport service (link 8 in fig.1), the decision RTCP reports are sent via TCP connections (link 9 in fig.1). The decision is included in an application-dependent field. In order to expedite the switch to the data transmission phase, the decision reports are generated using the *early RTCP feedback* mechanism [15].

3 Testbed Implementation

We deployed a prototype version of the described architecture in the framework of the NS-2 simulation package: the obtained performance measures were promising [8]. To evaluate the behaviours of our approach in a realistic environment, we implemented the architecture in a testbed network based on the Linux platform. In this section we supply the technical details on the implementation of the system modules shown in figure 1. As the IGMP and PIM-SM implementations, we used those available with the Linux kernel.

TRAFFIC GENERATOR. In order to deploy our streaming server, we evaluated the performance of two existing traffic generators: MGEN [16] and RUDE [17]. Our primary concern was the time granularity at which network packets were sent over the network. The obtained results suggested us to implement our streaming server using the same technique adopted by RUDE. We performed a test by generating bursts of 10 packets, of size 1024B each, every 10 ms, which is the tick value, that is, the minimum time granularity achievable on an x86 platform. As it is shown in Figure 2(a), we are able to transmit data with a packet interval of about 10 ms (continuous line). Hence, the performance measured at the clients (section 4) is not affected by burstiness in the generated traffic. This

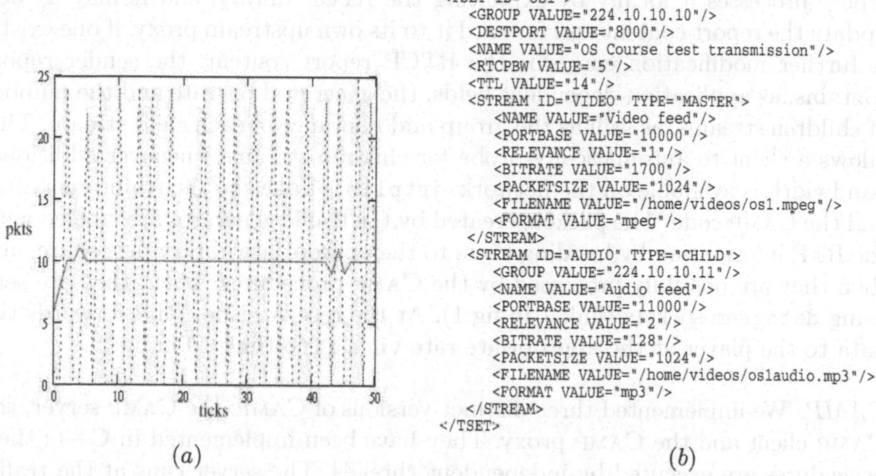
25
20
15
10
5
0
pkts
0
10
20
30
40
50
ticks
(a)
<TSET ID="OS1">
<GROUP VALUE="224.10.10.10"/>
<DESTPORT VALUE="8000"/>
<NAME VALUE="OS Course test transmission"/>
<RTCPBW VALUE="5"/>
<TTL VALUE="14"/>
<STREAM ID="VIDEO" TYPE="MASTER">
<NAME VALUE="Video feed"/>
<PORTBASE VALUE="10000"/>
<RELEVANCE VALUE="1"/>
<BITRATE VALUE="1700"/>
<PACKETSIZE VALUE="1024"/>
<FILENAME VALUE="/home/videos/os1.mpeg"/>
<FORMAT VALUE="mpeg"/>
</STREAM>
<STREAM ID="AUDIO" TYPE="CHILD">
<GROUP VALUE="224.10.10.11"/>
<NAME VALUE="Audio feed"/>
<PORTBASE VALUE="11000"/>
<RELEVANCE VALUE="2"/>
<BITRATE VALUE="128"/>
<PACKETSIZE VALUE="1024"/>
<FILENAME VALUE="/home/videos/os1audio.mp3"/>
<FORMAT VALUE="mp3"/>
</STREAM>
</TSET>
(b)

Fig. 2. (a) Profiles of the packet flow produced by the traffic generator and by MGEN. (b) Example of the parameters specified in the sender-side configuration file.

granularity was obtained by using a busy-waiting technique instead of timers, and by scheduling every send operation using the `gettimeofday()` function. By contrast, the same traffic load generated by MGEN yields the transmission of bursts of 25 packets every 30 ms (dashed line), thus showing a sawtooth behaviour. The streaming server integrates the RTP/RTCP library and interfaces directly to the transport layer via sockets. The server can generate multiple sessions concurrently. Each session is composed by one or more streams, each one of them sent to a specific multicast group. With this approach, multiple channels can be included in a session (e.g., audio, video, slides), and compatibility is offered with future layered video codecs. The streams that are to be generated are described in a configuration file, which is interpreted by the traffic generator. A server-side configuration file is shown in figure 2(*b*): in that example, a two-streams transmission is set-up, involving a master stream of 1700 Kbps and a child audio stream of 128 Kbps. In the example, data are read from files. The generator can also produce synthetic CBR traffic.

RTP/RTCP. As the implementation of RTP/RTCP, we use the `jrtplib` library [18], appropriately modified according to our needs. In particular, we modified the library so that the reports can be sent either multicast or unicast. Multicast reports are sent accordingly to the standard specification, that is, they are sent via UDP (datagram sockets) to all the multicast group members (link 8 in fig.1). The unicast addressing is used for the decision reports. Unicast reports are sent upon request of the CAMP proxy, and their content is specified by the proxy. The unicast decision reports are sent using TCP (stream sockets, link 9 in fig.1); each recipient sends the decision to the upstream proxy. A proxy receiving a unicast report processes it locally by exploiting the RTCP library, and it may as well update the report content and forward it to its own upstream proxy, if one exists. A further modification concerns the RTCP report content: the sender report contains, as application-dependent fields, the generated bit rate and the number of children streams, as well as the group and relevance of each child stream. This allows a client to automatically probe for children streams whenever additional bandwidth is available on the network. `jrtplib` is linked to the traffic generator and the CAMP code. The packets created by the traffic generator are labeled with the RTP information by handing them to the appropriate library procedure, and then they are possibily processed by the CAMP procedures. Then, they are sent using `datagram` sockets (link 7 in fig.1). At the receiver side, RTP forwards the data to the player at the appropriate rate via a `fifo` (link 10 in fig.1).

CAMP. We implemented three distinct versions of CAMP: the CAMP server, the CAMP client and the CAMP proxy. They have been implemented in C++; their procedures are executed by independent threads. The server runs at the traffic source, while the client runs in each group member. The server involves both the traffic generator and the CAMP proxy, detailed below.

The CAMP client only records the multicast group it belongs to, and the address of the upstream proxy throughout the set-up phase. The client directly reads data from the network via datagram sockets and integrates with the RTP/RTCP library. The client is able to receive multiple RTP streams. RTP and RTCP data are automatically de-multiplexed by the `jrtplib` library: RTP streams are then forwarded to an external program (an `mpeg` player, for example) using `fifos`, while RTCP information is collected by the CAMP module of the client to create stream profiles. These profiles store information such as the expected and actual reception rate, jitter, delay, and other QoS metrics. The client uses this information whenever a probing session must be accepted or refused. In order to accept a probing session, the client invokes the `jrtplib` to create a new RTCP report and sends it to the upstream proxy using TCP. In addition, once a session has been accepted, if additional children streams exist, the client will try to join them in order of their relevance and, if enough bandwidth is available, accept them as well.

The CAMP proxy is executed in the in-tree routers. It maintains the status information of the currently controlled multicast output interfaces, recorded in the CAMP MIB shown in figure 1, and performs the packet remarking. Moreover, the CAMP proxy substitutes the IP source address of every remarked multicast packet with the address of its outgoing interface, so that the downstream recipient (or downstream proxy) can appropriately address the decision reports.

In figure 3(*a*), we show the modules implemented for the CAMP proxy; in figure 3(*b*), we represent the operations performed by the proxy. The modules

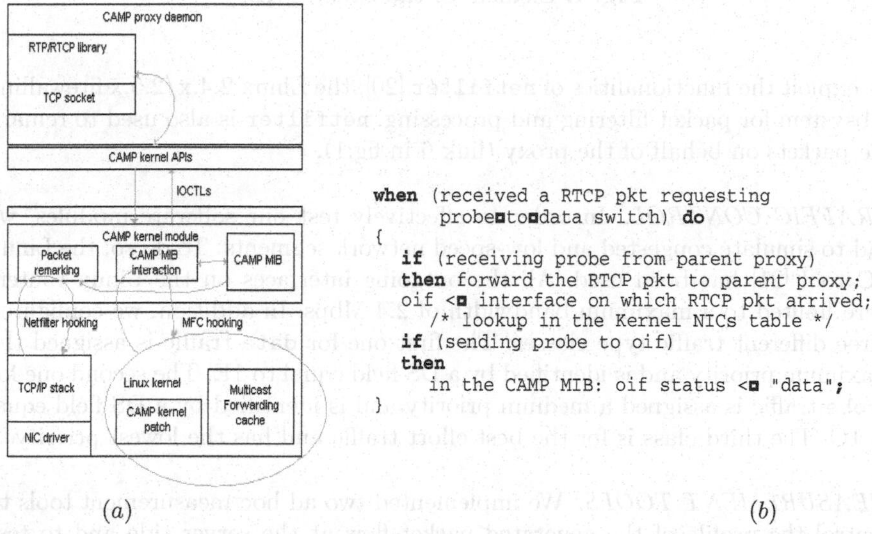
CAMP proxy daemon
RTP/RTCP library
TCP socket
CAMP kernel APIs
IOCTLs
CAMP kernel module
Packet remarking
CAMP MIB interaction
CAMP MIB
Netfilter hooking
MFC hooking
TCP/IP stack
Linux kernel
Multicast Forwarding cache
CAMP kernel patch
NIC driver
when (received a RTCP pkt requesting
probe□to□data switch) do
{
if (receiving probe from parent proxy)
then forward the RTCP pkt to parent proxy;
oif <□ interface on which RTCP pkt arrived;
/* lookup in the Kernel NICs table */
if (sending probe to oif)
then
in the CAMP MIB: oif status <□ "data";
}
(a)
(b)

Fig. 3. (*a*) Layout of the CAMP proxy. (*b*) Pseudo-code of the CAMP proxy operations.

communicate via `ioctl` (link 3 in fig.1): the CAMP user space daemon handles the RTCP unicast messages, while the kernel level module manages both the reports and the data packets. We deployed a kernel patch that implements a hooking mechanism to monitor the MFC and receive a notification whenever a downstream multicast interface is added or removed; the interested readers may refer to [19] for technical details about the kernel hook implementation. Upon the reception of a decision report, the daemon must decide whether the report must be further forwarded upstream or not. If the packets received for the multicast group indicated in the report are `probes`, then the report is forwarded upstream after the appropriate update of the destination address. The daemon cooperates with the kernel module to retrieve the network interface from which the report has been received. The information for the report management are recorded in the CAMP MIB, which is composed by entries having the structure shown in figure 4. The NIC to which the report is addressed is compared against the `dev` field of the CAMP MIB. For the matching entry, the `mark` field is changed from `probe` to `data`. To support the appropriate management of the decision reports,

```
typedef struct _PROBING_TRAFFIC_INFO {
    unsigned char mark; // downstream interface status
    unsigned long mcastAddress; // multicast IP address
    unsigned long srcIpAddress; // downstream interface IP address
    struct net_device *dev; // kernel list for attached NICs
} PROBING_TRAFFIC_INFO, *PPROBING_TRAFFIC_INFO;
```

Fig. 4. Element of the CAMP MIB.

we exploit the functionalities of `netfilter` [20]: the Linux 2.4.x/2.5.x firewalling subsystem for packet filtering and processing. `netfilter` is also used to remark the packets on behalf of the proxy (link 6 in fig.1).

TRAFFIC CONTROL. In order to effectively test our software modules, we had to simulate congested and low-speed network segments. To do so, the Linux TC tool [21] has been used. All the outgoing interfaces on the Linux routers were limited to a maximum bandwidth of 2.4 Mbps. In addition, we configured three different traffic type classes. The first one for `data` traffic is assigned the maximum priority and is identified by a DS field equal to 1E. The second one for `probe` traffic is assigned a medium priority and is identified by a DS field equal to 1C. The third class is for the best effort traffic and has the lowest priority.

MEASUREMENT TOOLS. We implemented two ad hoc measurement tools to control the profile of the generated packet flow at the server side and to test the reception quality at the client side. The former tool is a user level thread implemented into the client software. Instead of delivering all RTP packets to the external player program, the packets are just logged to an external file. The log file contains remote SSRCs, reception rate, jitter and intra-packet delay. The latter tool has been written to be integrated into the kernel. It is a kernel module that hooks all outgoing packets addressed to a particular multicast address. We used it to trace the packet generation process. We have been able to achieve a great precision (10 ms) by monitoring the system wide clock counter (the `jiffies` global variable of the Linux kernel) whenever network packets were about to hit the network wire. The trace files produced by the measurement tools can be used as input to graphic tools to plot the performance measures.

4 Experimental Results

Our test network is based on four machines running RedHat Linux with kernel 2.4.18. In figure 5(a), we show the layout of our network, while in figure 5(b), we

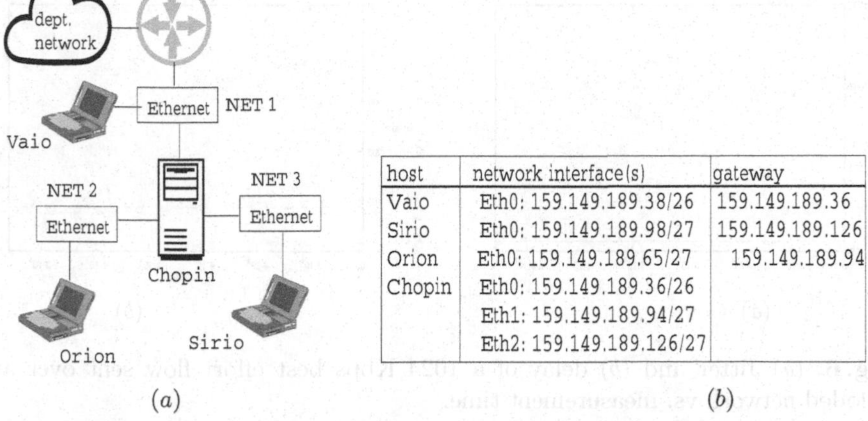

host	network interface(s)	gateway
Vaio	Eth0: 159.149.189.38/26	159.149.189.36
Sirio	Eth0: 159.149.189.98/27	159.149.189.126
Orion	Eth0: 159.149.189.65/27	159.149.189.94
Chopin	Eth0: 159.149.189.36/26	
	Eth1: 159.149.189.94/27	
	Eth2: 159.149.189.126/27	

Fig. 5. (*a*) Layout of the test network. (*b*) Host configuration.

report the network configurations of the hosts. `Chopin` acts as the router among three different subnets. In order to perform multicast routing, PIMd has been installed on `Chopin` and multicast routing support has been enabled in the kernel. `Vaio` has been primarily used to generate traffic. `Sirio` and `Orion` have been used to receive the incoming data generated by `Vaio`. The client running on those two hosts has been deployed ad hoc to support CAMP. We performed experiments with the clients joining at different times during the data transmission: the proxy approach worked properly.

In all the discussed experiments, we generated CBR data traffic with packet size 1024B. As the performance indexes we considered the bit rate received at the destinations, the intra-packet delay, estimated by the measurement tools as explained in section 3, and the jitter. The jitter is the variance of the packet inter-arrival time; it is a measure of the regularity with which the flow is received at the destinations. The jitter is evaluated by `jrtplib` according to the RTP standard. We tried to measure the overhead due to the packet processing performed by the CAMP proxy: the overhead is negligible, comparable to that involved by a firewall. In figure 6, we show the jitter and the delay measured by performing an experiment with one 1024 Kbps flow sent over a link of 2.4 Mbps, without any packet prioritization. The measures are reported against the measurement time; all the described experiments last around 120 sec. This experiment aims at validating our system. The received bit rate equals the sending rate, while the jitter is negligible and the intra-packet delay behaviour confirms that the traffic is generated with a regular profile. We initially performed experiments without enabling the admission control procedure, nor the packet prioritization. All the packets fairly share the available bandwidth, according to a stochastic fair queuing packet scheduling policy. We injected two flows of 1.7 Mbps over a channel having 2.4 Mbps capacity. The second flow starts around 60 sec. after the first flow. As it can be observed in figure 7, the average rate obtained by each

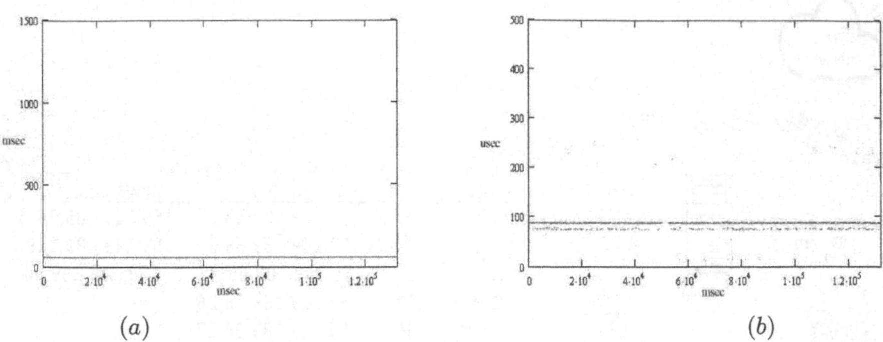

(a) (b)

Fig. 6. (a) Jitter and (b) delay of a 1024 Kbps best effort flow sent over an unloded network vs. measurement time.

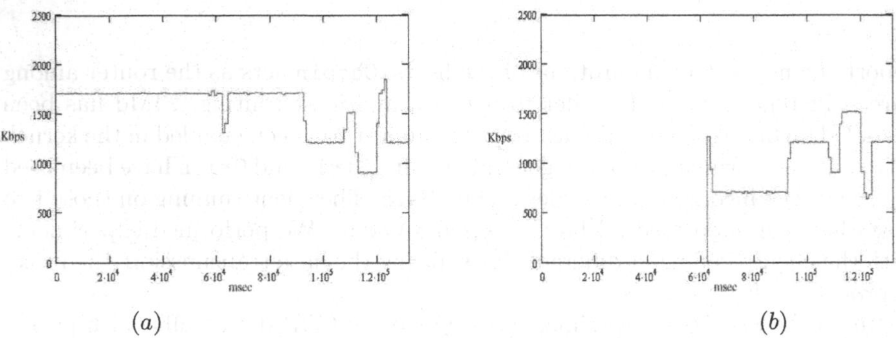

(a) (b)

Fig. 7. Received bit rate for (a) the first and (b) the second flow, vs. measurement time.

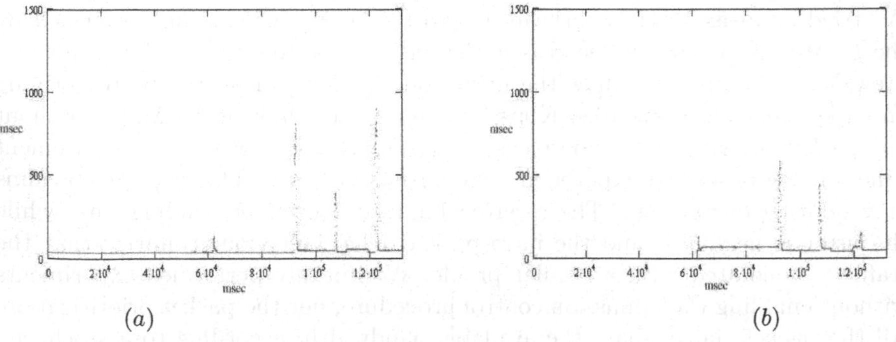

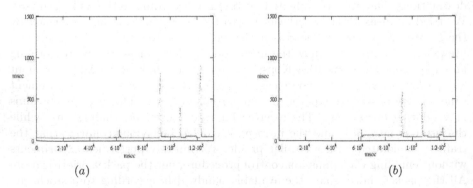

(a) (b)

Fig. 8. Instantaneous jitter for (a) the first and (b) the second flow, vs. measurement time.

flow is roughly half of the link capacity. As queuing delays cannot be predicted, the traffic behaviour is not regular, and this negatively affects the measured jitter (figure 8) and intra-packet delay (figure 9). Under this system configuration, QoS sessions cannot receive any guarantee in terms of available bandwidth, jitter or delay. We tried to evaluate to what extent the traffic prioritization succeeds in protecting the QoS traffic at the expenses of the best effort traffic. We generated two flows of 1.7 Mbps; the former one is classified as best effort traffic, while the packets of the latter one are marked as `data` packets. As expected, the QoS data is received at the correct rate, and it shows a negligible jitter and a constant delay, comparable with the measures shown in figure 6. By contrast, the best effort traffic can only use the remaining bandwidth (figure 10(a)). Moreover, the best effort packets suffer transient and unpredictable queuing delays. As a

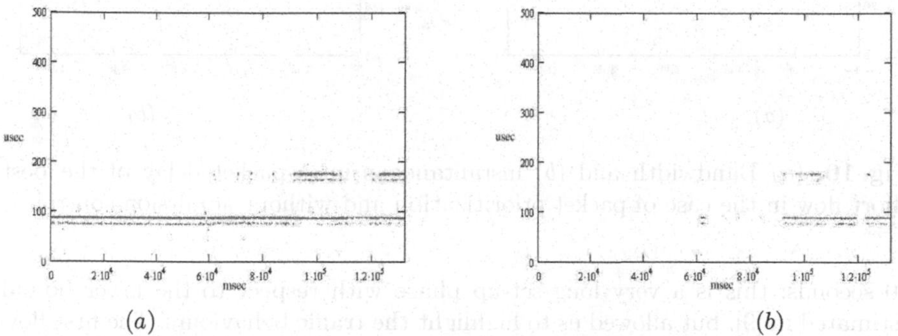

Fig. 9. Instantaneous intra-packet delay for (a) the first and (b) the second flow, vs. measurement time.

consequence, its profile at the recipient is bursty (figure 10(b)). The jitter shows a sawtooth behaviour. Similar results are achieved starting the QoS flow 60 sec. after the best effort flow: the QoS packets gain the required bandwidth at the expenses of the best effort traffic, whose performance degrade.

The priority packet scheduling allows to protect the QoS traffic from the best effort traffic, but, it cannot guarantee the required QoS in the case the network is congested by QoS sessions. We performed an experiment by generating two 1.7 Mbps flows both having a `probe` priority; the flows start at the same time. Although our network was underutilized during the experiments, some background traffic was possibly present. With the packet prioritization, the effects of that background traffic are negligible and the two flows have a regular pattern (figure 11). But, they compete for the available bandwidth, that is fairly shared among them, so that the recipients observe a degradation in the quality of both flows. As a consequence of the contention for the use of the available resources, the packets of both flows suffer queuing delay, that in our measures is around 10 msec, yielding a highly variable jitter.

A priority packet scheduling policy alone cannot guarantee that the multimedia sessions receive the needed amount of resources. To this purpose, the network congestion must be avoided by allowing that in the network is only injected the traffic that can be appropriately handled. We generated two flows with bit rate 1.7 Mbps as before, this time activating the CAMP mechanisms. The second flow starts 45 sec. after the first one. The probing phase lasts about

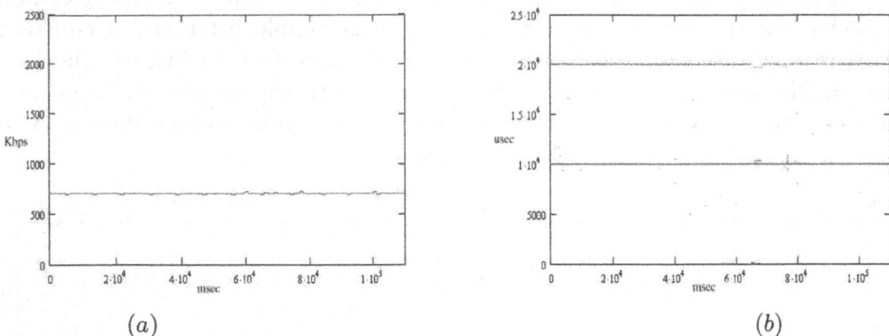

(*a*) (*b*)

Fig. 10. (*a*) Bandwidth and (*b*) instantaneous intra-packet delay of the best effort flow in the case of packet prioritization and without admission control.

30 seconds: this is a very long set-up phase with respect to the lower bound estimated in [9], but allowed us to highlight the traffic behaviours. The first flow does not compete with any other `probe` or QoS traffic, and as a consequence it succeeds to complete the admission control and the packets are transmitted as `data`, with the highest priority. When the second flow starts, the transmission of the first flow is already ongoing. Thanks to the higher priority, the traffic profile of the first flow is not negatively affected by the second flow (figure 12(*a*)). The `probe` packets of the second flow can drain at most 0.7 Mbps of available bandwidth (figure 12(*b*)); they are enqueued and suffer a queuing delay of about 10 msec. As a consequence, the recipients refuse the second session. By activating the CAMP procedures and repeating the same experiment with two flows that start contemporarily, the recipients refuse both the flows because their arrival bit rates are unacceptable. The measured performance is comparable with that shown in figure 11. Hence, the admission policy is conservative, and it can lead to poor network utilization. Anyway, in those cases, the traffic sources can wait a random time before re-trying to perform the admission control, so that a greater success probability is guaranteed.

5 Concluding Remarks

In this paper we describe the implementation of a Linux-based testbed for the set-up and management of multicast sessions requiring QoS in a Diff-Serv environments; the source code of the testbed is available [22]. The CAMP protocol

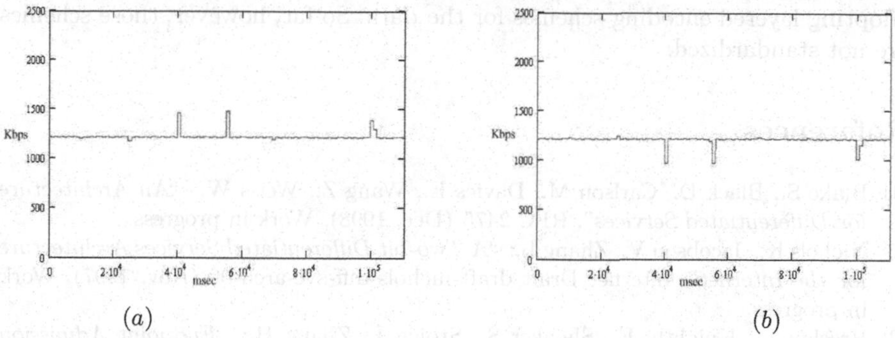

Fig. 11. Received bit rate for (*a*) the first and (*b*) the second QoS flow, vs. measurement time.

has been exploited to implement the Bandwidth Broker functionalities. Experiments have been performed to evaluate the system behaviour under different traffic and network conditions. The implemented architecture is able to provide bandwidth guarantees. The established sessions are not affected by the concurrent best effort traffic or the probing traffic for the set-up of new sessions. The admission control mechanism is effective, although possibly conservative in the case several sessions start simultaneously. The adopted approach does not impose drastical changes to the current network architecture, and charges a low processing overhead on the routers. As a consequence, it can effectively represent a first step in the deployment of the QoS-aware Internet structure. As a future work, we plan to extend our testbed with mechanisms for traffic policing, to support multiple service levels by guaranteeing the separation among different classes of traffic. Moreover, we are designing mechanisms to support heterogeneous receivers with different QoS requirements. This could be easily achieved by

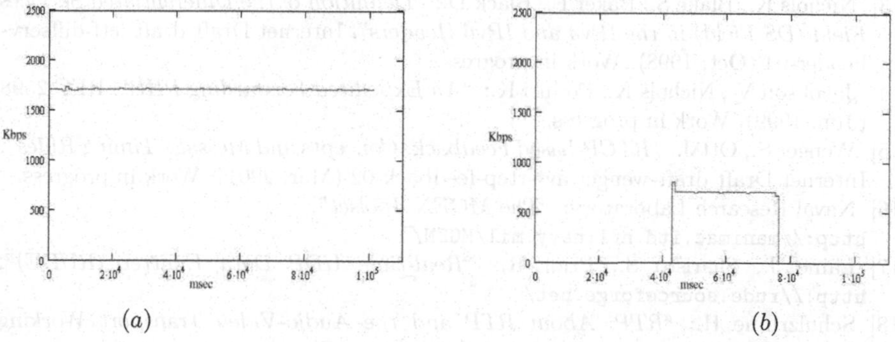

Fig. 12. Received bit rate for (*a*) the first and (*b*) the second QoS flow, vs. measurement time.

adopting layered encoding schemes for the data. So far, however, those schemes are not standardized.

References

[1] Blake S., Black D., Carlson M., Davies E., Wang Z., Weiss W.: *"An Architecture for Differentiated Services"*. RFC 2475 (Dec. 1998). Work in progress.
[2] Nichols K., Jacobson V., Zhang L.: *"A Two-bit Differentiated Services Architecture for the Internet"*. Internet Draft draft-nichols-diff-svc-arch-00 (Nov. 1997). Work in progress.
[3] Breslau L., Knightly E., Shenker S., Stoica I., Zhang H.: *"Endpoint Admission Control: Architectural Issues and Performance"*. Proc. SIGCOMM'00 (2000) 57-69.
[4] Räisänen V.: *"Measurement-Based IP Transport Resource Manager Demonstrator"*. Proc. International Conference on Networking, LNCS Vol. 2094. Springer (Jul. 2001) 127-136.
[5] Lai K., Baker M.: *"Measuring Link Bandwidths Using a Deterministic Model of Packet Delay"*. Proc. SIGCOMM'00 (2000) 283-294.
[6] Elek V., Karlsson G., Rönngren R.: *"Admission Control Based on End-to-End Measurements"* Proc. INFOCOM'00 (2000).
[7] Bless R., Wehrle K.: *"IP Multicast in Differentiated Services Networks"*. Internet Draft draft-bless-diffserv-multicast-01 (Nov. 2000). Work in progress.
[8] Pagani E., Rossi G.P.: *"Measurement-Based Admission Control for Dynamic Multicast Groups in Diff-Serv Networks"*. Proc. 2nd Intl. IFIP Networking Conference, Lecture Notes in Computer Science, Vol. 2345. Springer (May 2002) 1184-1189.
[9] Pagani E., Rossi G.P.: *"Distributed Bandwidth Broker for QoS Multicast Traffic"*. Proc. 22nd IEEE Intl. Conf. on Distributed Computing Systems - ICDCS'02 (Jul. 2002) 319-326.
[10] Schulzrinne H., Casner S., Frederick R., Jacobson V.: *"RTP: A Transport Protocol for Real-Time Applications"*. RFC 1889 (Jan. 1996). Work in progress.
[11] Cain B., Deering S., Thyagarajan A.: *"Internet Group Management Protocol, Version 3"*. Internet Draft draft-ietf-idmr-igmp-v3-01 (Feb. 1999). Work in progress.
[12] Estrin D. et al.: *"Protocol Independent Multicast - Sparse Mode (PIM-SM): Protocol Specification"*. RFC 2362 (Jun. 1998). Work in progress.
[13] Nichols K., Blake S., Baker F., Black D.: *"Definition of the Differentiated Services Field (DS Field) in the IPv4 and IPv6 Headers"*. Internet Draft draft-ietf-diffserv-header-04 (Oct. 1998). Work in progress.
[14] Jacobson V., Nichols K., Poduri K.: *"An Expedited Forwarding PHB"*. RFC 2598 (Jun. 1999). Work in progress.
[15] Wenger S., Ott J.: *"RTCP-based Feedback: Concepts and Message Timing Rules"*. Internet Draft draft-wenger-avt-rtcp-feedback-02 (Mar. 2001). Work in progress.
[16] Naval Research Laboratory: *"The MGEN Toolset"*. `http://manimac.itd.nrl.navy.mil/MGEN/`
[17] Laine J., Saaristo S., Prior R.: *"Real-time UDP Data Emitter (RUDE)"*. `http://rude.sourceforge.net/`
[18] Schulzrinne H.: *"RTP: About RTP and the Audio-Video Transport Working Group"*. `http://www.cs.columbia.edu/~hgs/rtp/`
[19] Pelati M.: *"Multicast Routing Code in the Linux Kernel"*. Linux Journal. SSC Publications (Nov. 2002) 16-21.

[20] Netfilter Team: *"The netfilter/iptables project"*. http://www.netfilter.org/
[21] *"Differentiated Services on Linux"*. http://diffserv.sourceforge.net/
[22] http://homes.dsi.unimi.it/~pagae/NPTLab/Software.

Light-Trails: A Solution to IP Centric Communication in the Optical Domain

Imrich Chlamtac and Ashwin Gumaste

Center for Advance Telecommunications Systems and Services
The University of Texas at Dallas, TX 75083, USA
chlamtac@utdallas.edu

Abstract. We propose a solution for implementing a conceptual framework for IP centric communication in the optical domain. The solution, termed Light-trails, is a combination of node architecture and protocol for realizing efficient optical communications from IP bursts to dynamic lightpaths. It is a paradigm shift from conventional optical communication modes, in supporting amongst others, very fast optical connection set up and tear down for burst of lightpaths communication, dynamic and highly bandwidth efficient sub-lambda provisioning. Light-trails also provide a first solution to optical multicasting, a key element for many of the emerging services that motivate the need for optical capacity. Contrary to existing proposals for IP type communication in the optical domain light-trail node architecture also presents the first practically implementable solution to enable optical transport with mature technology, non stringent optical switching requirements, and presenting a much more cost effective alternative to electronics.

I Introduction

All optical circuits each on a separate wavelength called lightpaths [1], represented the first major method for optical communication and have today have graduated into practical circuit based solutions. The lightpath granularity between a source and destination node pair is a whole wavelength, and no wavelength multiplexing between multiple nodes along the lightpath is allowed. Thus unless the lightpath pipe can be filled up by efficient aggregation and traffic engineering, its utilization will be low and no multicasting is possible. In contrast IP traffic is characterized by burstiness and high variability, requiring bandwidth on demand, and sub-lambda wavelengths allocation. Currently, there is no optical solution which utilizes efficiently the bandwidth offered by a single wavelength between multiple users. Opto-electronic solutions such as Gigabit Ethernet and Resilient Packet Rings (RPR) have been proposed for providing optical and opto-electronic sub-lambda granularity respectively. Access and metro Gigabit Ethernet is an end-to-end solution on a lightpath creating an information highway between a source and destination node. Gigabit Ethernet solution does not provide for sub-lambda type traffic between a stream of users, as it creates data flows that are inaccessible to intermediate nodes. RPR on the other hand allows a stream of consecutive nodes to communicate in a downstream direction by sharing the capacity of an optical wavelength by dropping

M. Ajmone Marsan et al. (Eds.): QoS-IP 2003, LNCS 2601, pp. 634-644, 2003.

and electronically processing the optical signal at each node. As a result RPR faces several limitations. The dropping of optical signal at every node, creates a need for high cost and high performance electronics at each network element. Further RPR is a slotted solution, involving non trivial issues of synchronization further limiting scalability, as well as classical bandwidth utilization problems as fixed size slots are not a good match for the demands of variable length Ethernet type packets in bursts of IP traffic. Current versions of RPR have a speed restriction of 2.5 Gbps.

Optical burst switching (OBS) is a recent paradigm proposed for creating burst level communication at the optical layer. OBS is therefore in principle a natural solution for providing support to IP centric communication optically in its native mode. However, current OBS solutions require high speed switches to create a practically allowable ratio of burst length to setup time, and hence reasonable utilization. Moreover, OBS solutions are based on pre-allocation of resources to avoid contention between competing bursts. From the time a resource is requested and until it is allocated, resources, such as wavelengths, are blocked. Given the large ratio between propagation delays and switching, relative to the duration of IP burst packets the network may become severely underutilized. Individual wavelengths remain underutilized due to gaps between individual packets of a burst. That means conventional OBS, while providing burst level granularity, cannot sustain this granularity at reasonable loads. Lastly, there is an absence of a solution allowing multicasting for lightpath or bursts, a centerpiece for most of the "bandwidth killer applications" as envisioned today. The proposed light-trail architecture comes to resolve these various problems, through a combination of a new node architecture based on existent optical technology and an effective novel provisioning protocol.

II Light-Trails

The concept of light-trails is proposed to enable IP centric communication at the optical layer. This concept consists of an architecture and a protocol that allows dynamic the opening of an optical path, or "trail" of length t, between any chosen source and destination node, while allowing optical communication (access) to all the nodes en route to the destination without the need for optical switch reconfiguration at individual nodes. With the principle of access to the all optical path by any node on the trail, a light-trail solution offers full optical connectivity between up to $\binom{t}{2}$ number of connections which can share the wavelength in time domain leading to dynamic and self regulating wavelength multiplexing. Thus the light-trails architecture can provide high utilization, low access delays and multicasting without the need for fast optical switching.

In order to demonstrate the light-trails concept in its simplest form we consider a 2-unidirectional fiber rings of N nodes. A light-trail of 't' nodes is a sequence of nodes using a wavelength λ_i. In a light-trail (N_1, ... N_t) node we call the first node, N_1, the *convener node*, and the last node N_t, the *end node.* A light-trail can be viewed as an optical bus between the convener and end nodes, with the characteristic that intermediate nodes can also access this bus, contrary to lightpath or burst level path provisioned under conventional architectures.

Shown in Fig. 1 is a proposed node configuration for realizing light-trails. On each of the two uni-directional fibers, there is a full de-multiplex section that de-multiplexes a composite DWDM signal and feeds individual channels to a local access section. The local access section for each wavelength (channel), (Fig. 2) consists of two passive couplers separated by an optical shutter. The first coupler is called the drop coupler (DC) (for dropping the signal) while the second coupler is called the add-coupler (AC) (for adding a local signal). The optical shutter is a fast ON/OFF optical switch typically using Mach Zehnder Interferometer technology on Lithium Niobate substrates [3]. A light-trail is set up between two nodes, by configuring the optical shutters on the desired wavelength at the convener and end nodes (in the OFF position) as well as by configuring the optical shutters (in the ON position) at each of the intermediate nodes. Within a light-trail we thus obtain a unidirectional communication in the direction of convener node to end node.

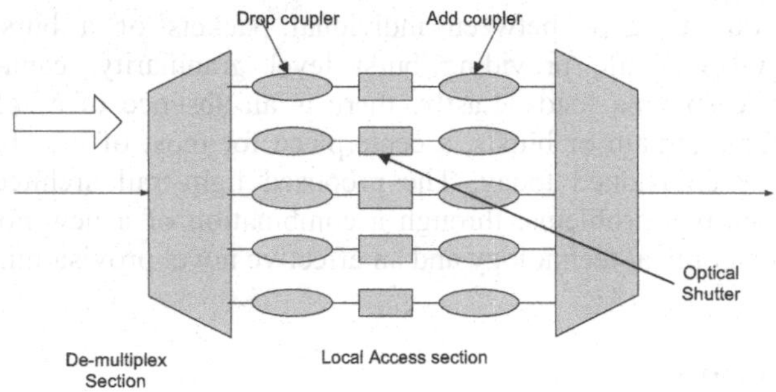

Fig.1 Light trail node configuration.

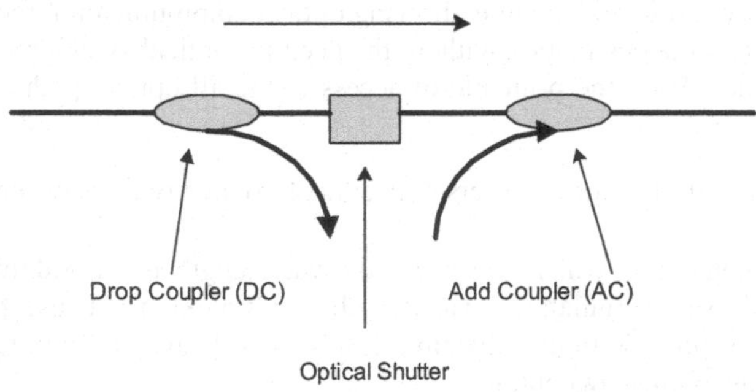

Fig. 2. Local access section of the Light-trail architecture. The top arrow represents the direction of communication.

It is easy to see that in this configuration in a light-trail of 't' nodes connection between any source destination pair within the light-trail can be established without requiring ON/OFF switching. Therefore, dynamic and fast lightpaths or bursts provisioning can occur without the need for fast ON/OFF switches. Specifically, if the optical shutter separating the two couplers is ON then the system represents a drop and continue function. On the other hand if the optical shutter is OFF wavelength reuse is obtained by virtue of spatial diversity. The same structure therefore also supports optical multicasting. Lastly, as specified above, the network elements can be built from available on-the-shelf technology.

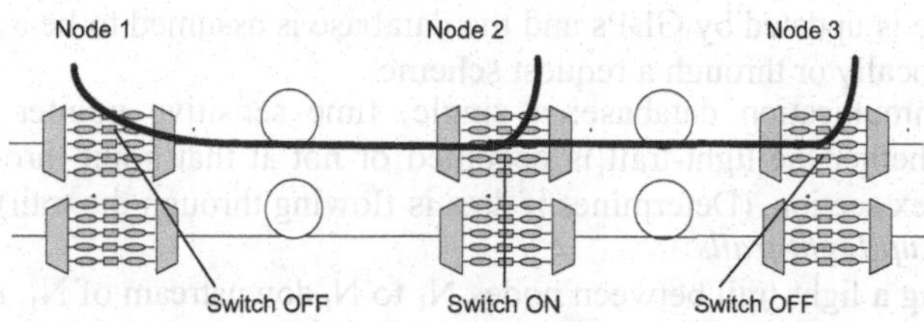

Fig. 3. Three nodes in a light-trail architecture.

The proposed configuration of the node shown in Fig. 3 supports a bi-directional light-trails based ring.

The Protocol

To establish light-trails and to provide optical connections within light-trails we define the following protocol. For obtaining the signaling required to set up light-trails and for optical connections management within light-trails we assume there exists an out of band communication channel called optical service channel (OSC), which is dropped and processed at each node. The OSC carries information about all the light-trails in the network, and is responsible for provisioning connections within light-trails by providing a method for signaling.

In light trail communication OSC control packets are sent ahead of the data using an offset which is a function of propagation delay from ingress to egress node as well as the control packet processing time at each node.)

Five types of control packets are defined:

Setup packets(SP): Are control packets used to set up light-trails. They contain the ingress, egress node information as well as the wavelength on which the light-trail is to be established. Setup packets are deciphered at each intermediate node en route adding the intermediate node to the light-trail.

Communication control packets (CCP): Are used to set up connections within a light-trail. CCPs carry information about the ingress node/egress node(s), light-trail number (a unique number which identifies the light-trail based on convener node, end node and wavelength used).

Dimensioning packets (DP): Are used by the convener node to dimension light-trails, i.e. allowing new nodes to become members, or eliminating nodes that are no longer active participants in the light-trail

Global Broadcast Packets (GBP): At regular intervals, on the OSC each light-trail sends GBPs throughout the network appraising all the nodes (and hence EMS's) of existence of themselves, and member nodes.

ACK packets: Are used when a light-trail is set up, sent by the end node and ratified by intermediate nodes to the convener node, indicating the acceptance of the request to set up a light-trail.

Light-trail database: The network management system (NMS) for light-trails contains the necessary information regarding the present light-trails in the network. This database is updated by GBPs and the database is assumed to be available to each node either locally or through a request scheme.

Local communication database: a single, time sensitive pointer at each node, indicating whether the light-trail is occupied or not at that point through the node's local multiplex section. (Determines if data is flowing through the trail).

A. Setting up Light-trails:

For creating a light-trail between nodes N_1 to N_t downstream of N_1, node N_1 selects an available wavelength λ_t, based on the light-trails database information. Node N_1 sends a control packet (SP) through the OSC requesting opening of this optical connection to N_t through the intermediate nodes N_2, N_3, ... N_{t-1}. Node N_t upon receiving the SP from N_1 replies through the OSC in the fiber (in the opposite direction) with an ACK. This ACK is validated by intermediate nodes also. If node N_t cannot allow the light-trail to be established, it indicates so. Nodes N_2, N_3, ... N_{t-1} upon receiving the control packet (SP), switch their optical shutters on the selected wavelength in ON position while nodes N_1 and N_t keep the shutter in OFF position. In this way a light-trail is created whose member nodes are N_1, N_2, ... N_t enabling downstream communication between them.

B. Communication within Light-trails: Setting up optical connections

If node N_j desires to send data (to set up dynamic lightpath or burst) to downstream node N_k, both nodes of the same light-trail, it proceeds as follows: To avoid conflict of usage of same light-trail by an upstream node within the same trail, the initiating node N_j determines availability of the light-trail by examining the "local communication database", and finds out the occupancy of the light-trail at its own port (multiplex section). If no upstream node is using this light-trail for communication then N_j can initiate communication to node N_k by sending a control packet (CCP) to be followed by the data after an offset interval. Nodes N_{j+1}, ... N_{k-1}, ...N_t now become aware of this communication. Note that after sending a connection set up control packet, node N_j is guaranteed successful connection to downstream node or multiple destination nodes (as in case of multicasting), since all the destination nodes can detect the data as a part of the optical power is split through the multiplex section of each node locally.

If a connection in a light-trail is in progress, and an upstream node wants to start another connection. Then the previous connection (assume started by a downstream node) has to be called off to facilitate the new connection initiated by the upstream node. Such kind of connection overlapping can happen in light-trails and several solutions are possible to deal with "overlapping" connections. Node N_j which is upstream of node N_q can access the optical path, even when N_q is transmitting information. If such does happen, then N_j requests for opening of an optical connection in the light-trail. Node N_q upon realizing the possibility of conflict inhibits its data and allows the data from node N_j to pass through. In this event, the downstream node (N_q) may either send its data after the upstream node (N_j) has finished its data transmission or send the data on another light-trail. However for delay-sensitive applications, it may not always be possible to inhibit the local transmission. In that case, the downstream node (N_q) may switch its optical shutter (for that light-trail) in the OFF position and collect the data from the upstream node (N_j) (through its drop coupler). It may then either buffer this data or send it on another light-trail usually involving O-E-O from the initial light-trail to the new light-trail. In this process of re-transmission (either over time or over different light-trail) the incumbent node does not restraint its data flow into the light-trail.

C. Dimensioning light-trails: Expanding and Contracting

As seen earlier light-trails can efficiently support dynamic optical communication, such as bursty IP centric traffic as they do not require the resetting of optical switches along the light-trail for connections within the trail. If a connection cannot use an existing light-trail a new trail needs to be established, a procedure which takes additional time. It is therefore desirable, from this point of view, to have the longest light-trail possible. However, since only one connection is active per trail at a time, light-trails that are too long would lead to underutilization. If light-trails are too short, optical communication deteriorates to that of a standard optical architecture. Therefore, light-trails may require dimensioning (expanding or contracting) to optimize performance in terms of utilization, set up times, etc. The most efficient way of managing the light-trail length is naturally a self regulating procedure in which a light-trail is extended or shortened so it covers all active connections (bursts) that use the trail at any given time. Specifically the left most active trail will define the left most (convener node) of the trail, while the rightmost node of the existing connection defined the trail's end node. Since connections (bursts) come and go dynamically, to obtain this type of light-trail management it is necessary to have a distributed mechanism to move the trail's end nodes dynamically. This can be done using the following mechanism.

A node N_a upstream of node N_1 and not part of a trail LT_1={ N_1, N_2, ..., N_t }, may request communication to a node in LT_1. The convener node by virtue of its dominant status may allow N_a to join the trail. It does so, it shifts the convener status to node N_a and the new trail LT_{NEW}= {N_a, N_{a+1}, ... N_1, ...N_t }is formed and this information is broadcast through the network by GBP. The act of expanding a light-trail is similar to setting up a light-trail though we use dimensioning packets (DP) as control mechanisms for this purpose.

Similarly, if the end node, through the local communication database learns it is no longer recipient of information, it may us a DP to the convener request to relive itself (or the group) from the light-trail. In that case, the first node from the end node in the reverse direction which is still an active member of the light-trail now becomes the end-node of the light-trail and configures its shutter accordingly.

Over time light-trails thus expand and contract depending on demands of traffic, allowing burst communication within optical connections (without switching) to occur, intuitively, for most of the traffic most of the time. On re-configuring a light-trail (expanding or contracting) the new light-trail information is broadcast throughout the entire network to facilitate nodes to learn about pre-set optical paths that can guarantee seamless communication of lightpaths as well as bursts.

III Light-Trail Evaluation

As pointed out in the Introduction there are important benefits to light-trail communication in provisioning wavelength sub-lambda allocation through time based wavelength multiplexing, supporting variable length bursts and variable length packets within a burst, allowing very fast optical connection set up, being technologically feasible, and providing multicasting. In this section we provide a basic evaluation of light-trails by a comparison to conventional burst switching and show how the basic mechanism of delay between control and data packet sending can be used to add quality of service to the network.

A. Provisioning Time in Light-Trails for Optical (Bursts) Connections

For efficient burst type communication the ratio of burst set up time to duration is a key parameter for effective burst switching. Burst transport algorithms due to the constraints on switching speed and uncertainty of resource availability due to the distributed nature of a network, have relied on pre-allocation of resources to create an end-to-end optical path. JET[2] is a leading burst transport algorithm based on pre-allocation of resources ahead in time, using an out of band signaling approach. In light-trails because the optical connection does not need to reset switches, one can expect an advantage in provisioning time. Shown in Fig. 4 is the provisioning time for light-trails, including connections established within a light-trail and those that require establishing a new trail, For both light-trail and JET burst switching we consider the provisioning time to be a function of the hop length, switch configuration time and control packet processing time Optical bursts are generated by multiplexing different classes of traffic (namely voice and data). In the simulation study we assume Poisson and Pareto distributions for burst aggregation. Scheduling policy for bursts that are delay sensitive is shown in [5]. In the simulation we use line rate of 1Gbps and bursts of 22 ms in length (average). In the simulation propagation delays are taken to be for 20 km links between consecutive nodes to emulate a typical metro area. Control packets are 20 kb in length and we assume a 1 GHz. processor at each node to process the dropped control packets. For a collection of simulated rings of sizes varying from 10~16 nodes and 40 wavelengths, we have the speed of control channel to be 51 Mb/s to avoid collision and to guarantee control packets to nodes as desired with a probability 0.999. Fig. 4 shows a significant benefit in the provisioning times for light-trail communication. Quantitatively we see that even if a contemporary fast switch [3] having configuration time of 0. 1 ms is assumed for JET we still see an approximate 1000% decrease in provisioning time for single connection using light-trails. If we further assume that the light-trail is already set up and burst transport is the act of communication (creating connections) within a light-trail we observe results such that there is on an average an order of two advantage in provisioning as compared to the provisioning using JET, showing the importance of optimized light-trail length. This validates the light-trails architecture as a method for providing high bandwidth on demand to end-users on a real time basis.

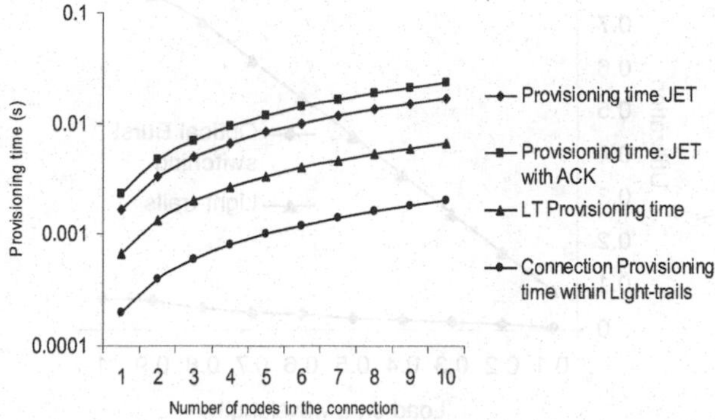

Fig.4. Comparison of provisioning times for conventional burst based communication algorithm JET and for Light-trails.

B. Network Utilization Benefits of Light-Trails

A major advantage of using light-trails for burst transport as compared to classical optical burst switching solution is the benefit observed in network utilization. In classical optical burst switching, the data burst is sent upon the successful reservation of bandwidth in the path. That is, for every burst to be broadcast, a control packet has to be sent, and switches have to be configured. This procedure is cumbersome and lengthy. By requiring an exclusive wavelength path to be set up for each burst, long voids are created within the channel since there is no utilization of the channel when the control packet is in transit, or when switches are being configured. In contrast, for a light-trails solution, we do not have to configure any switches. This leads to excellent provisioning times as seen in the previous sub-section. Moreover this also leads to better utilization of the system. Shown in Fig. 5 is a comparison of utilization of a single wavelength for different loads for both light-trails as well as classical burst switching. We assume 1Gbps channel in both cases. Bursts are aggregated according to algorithm in [5]. The average length of the path (for both light-trails and OBS) is 5 hops. Propagation delays are assumed to be for 20 km per hop, and processing at each node done by a 1 Ghz processor (processing for control packet is 1.25 micro seconds) Control packets were 20 kb maximum length. Load is computed stochastically in Erlangs. Utilization is defined as the ratio of capacity used over time for actual data transmission to the total capacity. Switching time for classical OBS switches are 0.1 ms which are very fast by today's standards [3]. We observe that utilization in OBS is severely degraded as compared to that in light-trails. On an average the utilization of light-trails is a single order of magnitude better than that seen in OBS under similar conditions.

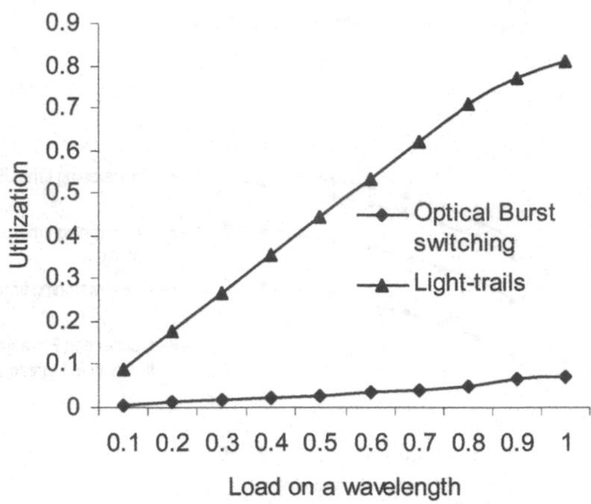

Fig. 5. Utilization of a wavelength using Light-trails and compared to classical optical burst switching.

C. QoS in Light-Trails

Quality of Service (QoS) is an important parameter for amalgamating different service types in a single network. The most important QoS parameter is delay on account of its relevance to voice and video traffic. End-to-end delay for bursts is a very critical, to ensure bandwidth savvy and high quality multimedia transfers. Since provisioning bursts over light-trails is much faster, easier and simpler as compared to optical burst switching, this means, light-trails is a natural candidate for providing high QoS. In [5] we proposed a policy to aggregate bursts and schedule them for delay sensitive applications. The policy for aggregating bursts is based on the time out of the first delay sensitive packet in the aggregating buffer. In Fig. 6 we observe average delay per packet within a burst, for light-trails with QoS policy and without QoS policy.

QoS policy: In light-trails we implement delay based QoS. The policy works as follows: Given a certain delay bound depending on the type of traffic, say T seconds. The control packet reserving/setting up the communication is sent after aggregating the burst for T-k-d seconds. Where d is the offset delay between the control packet and the data and k is the delay, reserved for retransmission in case the burst is blocked.

Aggregation process of bursts is a multiplexing operation of video, voice and data packets. Video files are created as a self similar arrival process with Hurst parameter 0.75 to depict the pictorial heavy tailed ness. Voice is a Poisson arrival process and data is a Pareto process of high burstiness. QoS delay levels are accorded to voice and video packets. In the simulation we have the distribution of voice packets as 21.8 % while that of video is19.7% and data packets is58.5%. The packets are maximum 1500 bytes long and bursts are aggregated by coalescing multiple packets. Buffers are maximum 100 Mb in length. In Fig. 5 we observe the average delay experienced by packets, in light-trails for providing QoS and for schemes where QoS is not provided (control packets are not scheduled to ensure end-to-end delay within a bound). We observe that as utilization increases the end-to-end delay experienced by packets within a burst also increases. For utilization of 0.4 and 0.8 we see that the delay is lesser than the QoS cut off delay of 4.5 x 10 $^{-6}$. The average burst size is 23 Mb and variance is 76 Mb. Load is measured in Erlangs

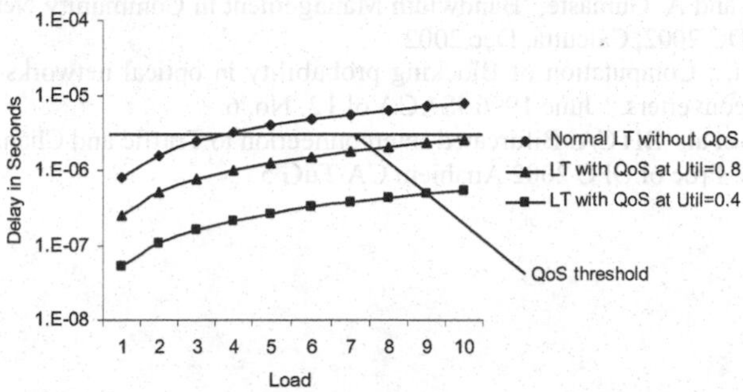

Fig. 5 Average delay for packets using light-trails with a QoS policy and without a QoS policy.

IV Conclusion

In this paper we have proposed the concept of light-trails which creates a framework for IP centric communication in the optical domain. By creating multi-point accessible information highways we showed that Light-trail architecture can provide an efficient platform for dynamic provisioning of optical connections on a per-burst time scale providing sub lambda multiplexing in the time domain and supporting variable size packet transitions leading to multiple befits not only in utilization but in simplicity of node design and reduced constrains on the optical network in term so buffer management and node synchronization.. The practicality of building a light-trail architecture for all optical communication can also be expected to provide significant cost advantage, scalability in speed and

Lastly, we demonstrated light-trails as a platform for QoS applications. QoS sensitive applications, previously requiring dedicated connections can now be dynamically configured due to ease of provisioning and guaranteed delivery using light-trails.

References

[1] I. Chlamtac, A. Ganz, and G. Karmi, 'Lightpath Communications A Novel Approach to High Speed Optical WANs' *IEEE Trans. on Commun.* Vol. 40 No. 7 July 1992. pp 1152

[2] M. Yoo, C. Qiao and S. Dikshit, ' Optical Burst Switching for Service Differentiation in the Next-Generation Optical Internet,' *IEEE Communications Mag.* Feb. 2001 pp 98-104

[3] A. Gumaste and T. Antony 'DWDM Networks Design and Engineering Solutions,' *McMillan Publishers.*
[4] M. Bouda et. al. ' Tunable AOTF switches' *Proc of OFC* Baltimore 2000
[5] A. Gumaste and J. Jue, 'Burst aggregation strategies based on according QoS for IP traffic,' First International Conference on Optical Communications and Networks, Singapore Nov. 2002
[6] R. Ramaswami and G. Sasaki, Limited Wavelength Conversion in Optical WDM ring networks', IEEE/ACM Trans. On Networking, Vol 3 No 5 Oct. 1995 pp 1152-1162
[7] I Chlamtac and A. Gumaste, 'Bandwidth Management in Community Networks,' Key note address, IWDC 2002, Calcutta, Dec 2002
[8] B. Humblet, ; Computation of Blocking probability in optical networks with and without wavelength converters, ' June 1996 JSAC Vol 12. No. 6.
[9] A.Gumaste et al, 'BITCA: Bifurcated Interconnection to Traffic and Channel Assignment in Metro Rings' Proc of OFC 2002 Anahiem CA TuG 5

End-to-End Bandwidth Estimation for Congestion Control in Packet Networks

Luigi Alfredo Grieco[1] and Saverio Mascolo[2]

[1] Dipartimento d' Ingegneria dell' Innovazione, Università di Lecce
Via Monteroni, 73100 Lecce, Italy
alfredo.grieco@unile.it

[2] Dipartimento di Elettrotecnica ed Elettronica, Politecnico di Bari
Via Orabona 4, 70125 Bari, Italy
mascolo@poliba.it

Abstract. Today TCP/IP congestion control implements the *additive increase/multiplicative decrease* (AIMD) paradigm to probe network capacity and obtain a "rough" but robust measurement of the best effort available bandwidth. Westwood TCP proposes an *additive increase/adaptive decrease* paradigm that adaptively sets the transmission rate at the end of the probing phase to match the bandwidth used at the time of congestion, which is the definition of best-effort available bandwidth in a connectionless packet network. This paper addresses the challenging issue of estimating the best-effort bandwidth available for a TCP/IP connection by properly counting and filtering the flow of acknowledgments packets using discrete-time filters. We show that in order to implement a low-pass filter in packet networks it is necessary to implement an anti ACK compression algorithm, which plays the role of a classic anti-aliasing filter. Moreover, a comparison of time-invariant and time-varying discrete filters to be used after the anti-aliasing algorithm is developed.

1 Introduction

The today dominant Internet is a global packet network that implements resource sharing through statistical multiplexing. Packets are delivered hop by hop by a connectionless network layer that employs store and forward switching using *first in first out* (FIFO) queuing. The stability of the Internet and in particular the prevention of congestion requires that flows use some form of end-to-end congestion control to adapt the input rate to the available bandwidth [1,2,4,5]. Thus the goal of obtaining an end-to-end estimate of the bandwidth available for a TCP connection is crucial in order to design more efficient and fair congestion control algorithms.

Today TCP congestion control was introduced in the late eighties [1,2] and follows the additive increase/multiplicative decrease probing paradigm (AIMD) [6]. The additive phase linearly increases the transmission rate until the network capacity is hit and the sender becomes aware of congestion via the reception of duplicate acknowledgments (DUPACKs) or the expiration of a timeout. Then the sender reacts to light congestion (i.e. 3 DUPACKs) by halving the congestion window (fast recovery) and sending again the missing packet (fast retransmit), and to heavy

M. Ajmone Marsan et al. (Eds.): QoS-IP 2003, LNCS 2601, pp. 645-658, 2003.

congestion (i.e. timeout) by reducing the congestion window to one (multiplicative decrease phase). After a timeout or at the beginning of the connection, the TCP enters a faster increasing phase that is exponential and aims at grabbing the best-effort available bandwidth faster.

The AIMD paradigm can be viewed as an endless series of cycles, having a linear or exponential increasing phase and a multiplicative decreasing phase, which aim at continuously probing the network capacity to obtain a "rough" but robust measurement of the best effort bandwidth that is available for a TCP connection. As reported in [1], "This mechanism can insure that network capacity is not exceeded, but it cannot insure fair sharing of that capacity". In fact, it has been shown that the throughput of a TCP connection is proportional to the inverse of its round trip time [13], which favors shorter connections. Furthermore, today TCP is not well suited for wireless links since losses due to unreliable radio channels are misinterpreted as a symptom of congestion thus leading to an undue reduction of the transmission rate and low utilization of a wireless path. As a consequence, today TCP requires supplementary link layer protocols such as reliable link-layer or split-connections approach to efficiently operate over wireless links [20].

Westwood TCP proposes to improve fairness and efficiency of congestion control by introducing an innovative end-to-end bandwidth estimation algorithm. In particular, Westwood TCP leaves unchanged the probing phase of classic TCP but it substitutes the multiplicative decrease phase with an adaptive decrease phase, which sets the control windows by taking into account the bandwidth estimate. It has been shown that Westwood TCP increases the TCP throughput over wireless links, increases the fairness in bandwidth allocation w.r.t. Reno TCP and is friendly towards Reno TCP [12,19].

It is worth remarking that the end-to-end bandwidth estimation mechanism proposed in Westwood TCP, and that we are going to investigate in this paper, is built on the standard probing mechanism of today TCP, that is, it is based on probing the network capacity using the slow-start and congestion avoidance phases. A congestion episode at the end of a probing phase points out that all the best-effort available bandwidth has been grabbed. Therefore, an estimate of the used bandwidth at the end of a probing phase is, by definition, an estimate of the available best effort bandwidth in a statistically multiplexed packet network. The latter observation is valuable to be kept in mind because the used bandwidth exhibits high variability with time in packet networks due to statistical multiplexing, whereas the available best-effort bandwidth might not.

The end-to-end bandwidth estimation algorithm is a critical issue. The basic idea reported in [12] is to get an estimate of the used bandwidth by counting and low-pass filtering the flow of ACK packets during the data transfer. In particular, when an ACK arrives it can be argued that a certain amount of data has been delivered since the previous ACK arrival time and a sample of the used bandwidth can be computed. Bandwidth samples must be low-pass filtered to obtain the low frequency components of the used bandwidth because congestion is only due to the low-frequency components of the traffic [9]. Low-pass filtering using discrete-time filter is a challenging goal due to the fact that ACK packets do not arrive at constant sampling intervals since packets networks are asynchronous systems. In particular, ACK packets experience congestion along the backward path they traverse and arrive bunched. The latter phenomena, known as ACK compression [14], provokes considerable bandwidth overestimate that is disruptive of the adaptive decrease mechanism and leads to connection starvation.

The disruptive effects of ACK compression on the bandwidth estimation algorithm were not evident in the original paper on TCP Westwood because the phenomena of ACK compression was negligible in the considered scenarios. They were shown in [19] along with a new filtering technique and a mathematical model for the long-term Westwood TCP throughput.

This paper is entirely devoted to the challenging issue of end-to-end bandwidth estimation in packet networks. In particular it is shown that ACK compression generates aliased bandwidth samples that lead to greatly overestimate the used bandwidth. Therefore, it is necessary to implement an anti ACK compression algorithm, which plays the role of a classic anti-aliasing filter, before using any discrete-time filter in packet networks. Time-invariant and time-varying discrete filters are compared. Simulation results show that it is possible to obtain an estimate of available bandwidth which enhances the performance of TCP congestion control.

The paper is organized as follows: Section 2 summarizes the related work; Section 3 describes the Westwood TCP congestion control algorithm; Section 4 focuses on time-varying and time-invariant discrete-time filters for low-pass filtering and bandwidth estimation; Section 5 introduces the anti-aliasing filter and test time-invariant and time-varying discrete-time filters using the ns-2 simulator [15]; finally, Section 6 draws the conclusions.

2 Related Work

TCP Vegas is the first significant example of congestion control algorithm that partially departs from the AIMD paradigm by proposing two estimates: (a) the expected connection rate $cwnd/RTT_{min}$ and (b) the actual connection rate $cwnd/RTT$, where RTT is the round trip time and RTT_{min} is the minimum measured RTT [11]. When the *difference* between the expected and the actual rate is less than a threshold $\alpha>0$, the *cwnd* is additively increased. When the *difference* is greater than a threshold $\beta>\alpha$ then the *cwnd* is additively decreased. When the difference is between α and β, *cwnd* is kept constant [11]. Vegas can be viewed as the first attempt to use a bandwidth estimation scheme to improve the internet congestion control. However, it should be noted that Vegas employs the actual sending rate $cwnd/RTT$ rather than the rate of the returning ACKs to infer congestion; the sending rate $cwnd/RTT$ is a measure of bandwidth that is based on the number of sent packets (*cwnd*) and not on the number of acknowledged packets. As a consequence, it does not take into account that a fraction of sent packets could be lost and could not correspond to an actual bandwidth capacity, that is, the Vegas actual rate overestimates the used bandwidth.

The first attempt to exploit ACK packets for bandwidth estimation is the packet pair (PP) algorithm, which tries to infer the bottleneck available bandwidth at the starting of a connection by measuring the interarrival time between the ACKs of two packets that are sent back to back [21]. Hoe proposes a refined PP method for estimating the available bandwidth in order to properly initialize the *ssthresh* [22]: the bandwidth is calculated by using the least-square estimation on the reception time of three ACKs corresponding to three closely-spaced packets. Allman and Paxson evaluate the PP techniques and show that in practice they perform less well than expected [23]. Lai and Baker propose an evolution of the PP algorithm for measuring the link bandwidth in FIFO-queuing networks [16]. The method consumes less network bandwidth while maintaining approximately the same accuracy of other methods, which is poor for paths longer than few hops. Jain and Dovrolis proposes to use streams of probing packets to measure the end-to-end available bandwidth, which is defined as the maximum rate that the path can provide to a flow, without reducing the rate of the rest of the traffic [7]. Finally, they focus on the relationship between the available bandwidth in a path they measure and the throughput of a persistent TCP connection. They show that the averaged throughput of a TCP connection is about 20-30% more than the available bandwidth measured by their tool due to the fact that the TCP probing mechanism gets more bandwidth than what was previously available in the path, grabbing part of the throughput of other connections. A similar technique has been proposed in [3]. It uses sequences of packet pairs at increasing rates and estimates the available bandwidth by comparing input and output rates of different packet pairs. Estimating the available bandwidth at the beginning of a TCP connection is a different and much more difficult task than measuring the actual rate a TCP connection is achieving during the data transfer as it is done by Westwood TCP [12]. In particular, Westwood TCP low-pass filters the flow of returning ACKs to get an estimate of the bandwidth a TCP connection is using. However, the filter proposed in [12] does not work properly in the presence of ACK compression because of aliasing. In particular, when in the presence of ACK compression, the estimation algorithm described in [12] causes an overestimate of the bandwidth that is disruptive of the Westwood adaptive decrease mechanism and leads to starvation and fairness disruption [19].

3 Westwood TCP Congestion Control

TCP-W is a sender-side only modification of the TCP stack. It proposes an end-to-end bandwidth estimation algorithm that is built on the standard probing mechanism of today TCP. It implements slow-start and congestion avoidance phases such as classic Reno TCP to probe the network but, after congestion, it employs the estimate of the best-effort available bandwidth B to properly set the congestion window and the slow-start threshold. In particular, when a TCP-W sender receives 3 DUPACKs, it sets both *ssthresh* and *cwnd* equal to $max[2,(B*RTT_{min})/seg\_size]$, where RTT_{min} is the minimum measured RTT and *seg_size* is the size of the sent segments. On the other hand, when a timeout expires, *ssthresh* is set as in the previous case whereas *cwnd* is set equal to 1 segment.

It is important to notice that the setting $ssthresh=(B*RTT_{min})/seg\_size$ provides a slow-start threshold that follows exactly the best-effort available bandwidth as it is computed by the TCP Westwood sender (RTT_{min}/seg_size is a scale factor). Therefore, the effective ability of Westwood TCP to track the available bandwidth can be tested by plotting the *ssthresh*.

The *adaptive decrease* mechanism improves the stability of the standard TCP *multiplicative decrease* algorithm since it can ensure that the congestion window is reduced enough in the presence of heavy congestion and not too much in the presence of light congestion or losses not due to congestion, such as in the case of radio links. Clearly, a key requirement in order to have a properly working adaptive decrease mechanism is that the bandwidth estimate is correct.

To conclude this background section we also notice that the adaptive setting of the control windows increases the fair allocation of available bandwidth to different TCP flows. This can be intuitively explained by considering that the window setting of TCP Westwood tracks the estimated bandwidth so that, if the estimate is a good measurement of the fair share, then the fairness is improved (for a mathematical proof of this results see [19]). Moreover, the setting $cwnd{=}B{*}RTT_{min}$ sustains a transmission rate $cwnd/RTT{=}B{*}RTT_{min}/RTT$ that is less than the bandwidth B estimated at the time of congestion: as a consequence, the considered TCP flow clears out its path backlog after a congestion episode thus leaving room for coexisting flows, and improving statistical multiplexing and fairness.

4 Bandwidth Estimation Algorithms

This section focuses on the issue of estimating the used bandwidth by counting ACK packets and by filtering the information they convey.

Fig. 1 depicts the end-to-end sender-based bandwidth estimation framework. It shows a Westwood TCP sender injecting data segments into the Internet and receiving ACKs from the receiver. When an ACK is received at time t_k, it means that a certain amount of data d_k has been received by the TCP receiver. In particular, on ACK reception, the following *sample* of bandwidth used by the TCP connection can be computed:

$$b_k = \frac{d_k}{t_k - t_{k-1}} = \frac{d_k}{\Delta_k} \tag{1}$$

where t_{k-1} is the time the previous ACK was received and $\Delta_k = t_k - t_{k-1}$ is the last interarrival time. Since congestion is due to low-frequency components of used bandwidth, samples (1) must be low-pass filtered by using a discrete-time filter. The latter is a delicate task because samples (1) contain high frequency components due to the fact that ACK packets can come back to the sender bunched as well as equally spaced. As a consequence, any discrete-time low-pass filter will remarkably overestimate the available bandwidth due to the phenomena of aliasing.

In this section we will first show through simulations that both time-invariant and time-varying discrete-time filters fail to estimate the used bandwidth due to aliasing effects. Then, we will introduce an anti ACK-compression algorithm that plays the classic role of an anti-aliasing filter in packet networks [10].

We consider the following three types of discrete-time low-pass filters, which are all obtained by discretizing a first-order low-pass continuous filter:

– A time-varying filter obtained by discretizing the continuous filter using a Zero Order Holder (ZOH) [10];
– A time-invariant filter obtained using a ZOH;
– A time-varying filter obtained by discretizing the continuous filter using the bilinear approximation [10].

In order to compare the three filters above, we employ the topology in Fig. 2.

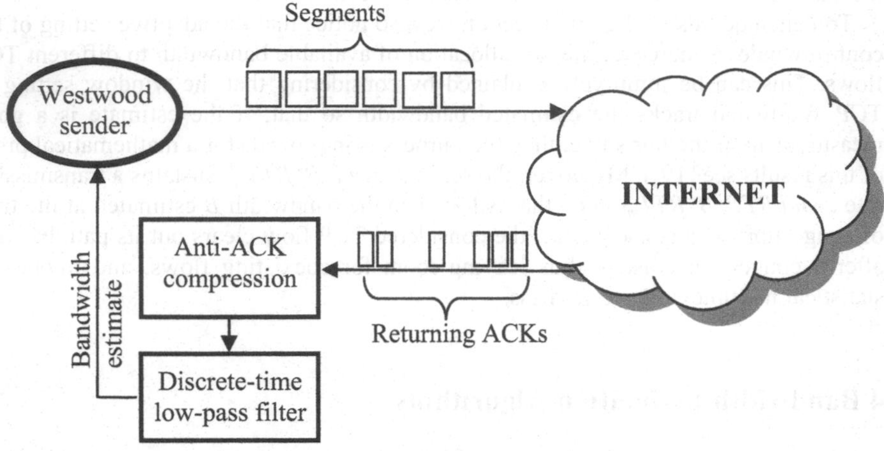

Fig. 1. Bandwidth estimation scheme.

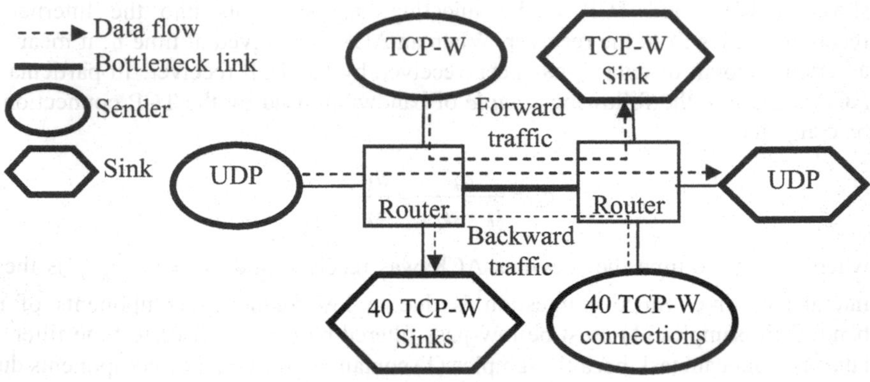

Fig. 2. Topology for testing the discrete-time filters.

It consists of a duplex 5Mbps bottleneck link shared by one persistent TCP Westwood (TCP-W) source implementing the New Reno feature [17], and one ON/OFF constant bit rate UDP source transmitting data at 1Mbps during the ON period that lasts 100s and is silent during the OFF period that lasts 100s too. On the reverse path, 40 TCP-W sources implementing the New Reno option provoke congestion along the TCP Westwood ACK path and excite ACK compression. Packets are 1500 Bytes long whereas ACKs are 40 Bytes long. The bottleneck buffer size is set equal to the link capacity times the round trip time of the forward TCP connection whose value is 250ms. All TCP sinks implement the delayed ACK option [2]. The initial congestion window has been set equal to 3 segments [18]. In the first set of simulations, the UDP source is turned OFF to test various estimation algorithms in the presence of constant available bandwidth environment, then it is turned ON to investigate the estimate behavior in the presence of time varying available bandwidth.

4.1 A Time-Varying Discrete Time Filter Using a ZOH

The first filter we consider is obtained discretizing a first order low-pass continuous filter with time constant τ by using a zero order holder (ZOH). The discretization procedure is described below:

A sample of used bandwidth (1) is considered as an impulse, which arrives every time t_k the sender receives an ACK. The impulsive sample is interpolated via a zero order holder, which generates a piecewise constant signal. The continuous piecewise constant signal is filtered by a first order single pole low-pass continuous filter with time constant τ. The output of the time-continuous filter is sampled at t_k times. Such steps lead to the following discrete time filter:

$$\hat{b}_k = \alpha_k \cdot \hat{b}_{k-1} + (1-\alpha_k) \cdot b_k \qquad (2)$$

where $\alpha_k = e^{-\Delta_k / \tau}$. The Infinite Impulse Response filter (2) is identical to the one proposed in [8] to estimate the arrival rate of a flow at network edges.

Fig. 3 (a) plots the used bandwidth samples and Fig. 3 (b) the used bandwidth estimate obtained using a TCP-W sender implementing the filter (2). It is worth noticing that, in this case, time-varying coefficients counteract the non-uniform sampling time and mitigates the effects of ACK compression. In order to illustrate high frequency components due to ACK compression, Fig. 3 (a) shows bandwidth samples feeding the filter (2). It should be noticed that bandwidth samples exhibit a peak value equal to 375 Mbps. Such a value is equal to the returning link capacity (5Mbps) times the ratio between the segments size (1500Bytes) and the ACKs size (40Bytes), times two to take into account that each ACK acknowledges two segments because of the delayed ACK option [2].

Even though filter (2) mitigates ACK compression, it does not work for every value of τ due to aliasing effects. In fact, Fig. 4 shows the over estimate that is obtained by using a filter with time constant τ=0.1s.

4.2 A Discrete Time Invariant Filter

By considering a constant interarrival time $\Delta = \Delta_k$, the filter (2) gives the following time-invariant filter:

$$\hat{b}_k = \alpha \cdot \hat{b}_{k-1} + (1-\alpha) \cdot b_k \qquad (3)$$

where

$$\alpha = e^{-\Delta / \tau} \qquad (4)$$

Fig. 5 (a) plots samples (1) of the used bandwidth and Fig. 5 (b) plots the output of the filter (3) with $\alpha = 0.9$ when fed by these samples. Fig. 5 (b) shows that the filter overestimates the available bandwidth up to more than 10 times. The reason is that bandwidth samples of Fig. 5 (a) contain aliased frequency components due to ACK compression.

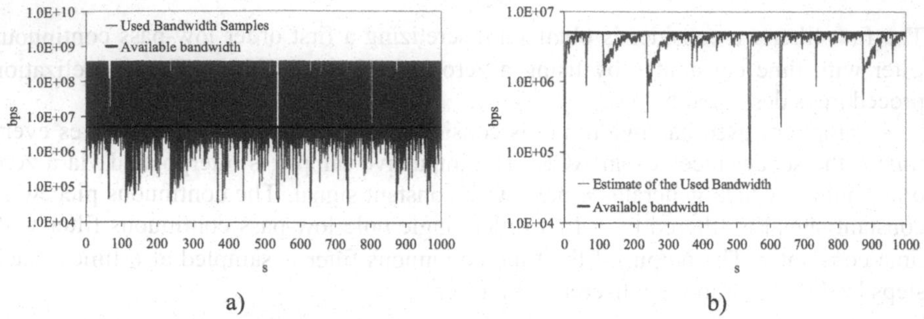

Fig. 3. Used Bandwidth: (a) Samples; (b) Output of the filter (2).

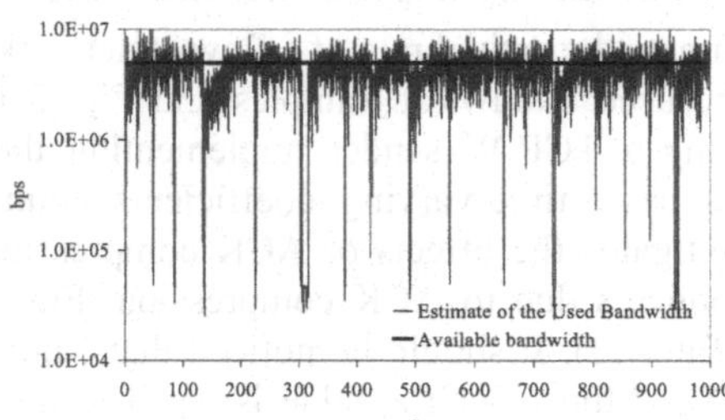

Fig. 4. Estimate of the used bandwidth using the filter (2) with τ=0.1s.

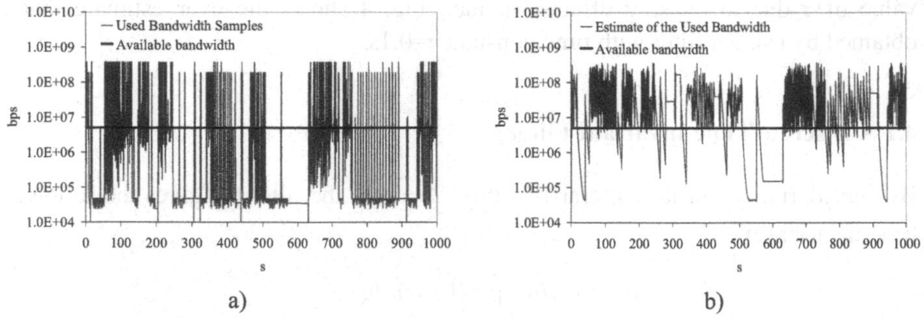

Fig. 5. Used bandwidth: (a) Samples; (b) Output of the filter (3).

4.3 A Time Varying Filter Using the Bilinear Transformation

As in the previous section, a sample of used bandwidth (1) is computed every time t_k the sender receives an ACK. By assuming, for the time being, that inter-arrival time between samples is uniform and equal to Δ, the continuous first order low-pass filter is discretized using the bilinear transformation [10], which leads to the following discrete-time filter:

$$\hat{b}_k = \frac{2\tau - \Delta}{2\tau + \Delta}\hat{b}_{k-1} + \Delta\frac{b_k + b_{k-1}}{2\tau + \Delta} \quad (5)$$

4.3.1 The Filter of Westwood TCP

To take into account that the inter-arrival time of bandwidth samples is not uniform, the following time varying form of the filter (5) has been employed in Westwood TCP [12]:

$$\hat{b}_k = \frac{2\tau - \Delta_k}{2\tau + \Delta_k}\hat{b}_{k-1} + \Delta_k\frac{b_k + b_{k-1}}{2\tau + \Delta_k} \quad (6)$$

where Δ_k is the inter-arrival time between the (*k*-1)-*th* sample and the *k-th* sample. In order to satisfy the Nyquist-Shannon sampling Theorem, the inter-arrival time Δ_k must be less than $\tau/2$. Therefore if it happens that $\Delta_k > \tau/2$, then the filter is fed by N virtual samples with inter-arrival time $\tau/2$ and amplitude equal to 0, where $N = \text{integer of }(2 \cdot \Delta_k / \tau_f)$. Moreover, an additional samples b_k feeds the filter with inter-arrival time equal to $\Delta_k - N \cdot \tau/2$. In the following, we set $\tau = 1s$ unless otherwise specified.

Fig. 6 reports the used bandwidth samples and the estimated used bandwidth obtained using the filter (6). As it can be viewed, the filter greatly overestimates the bandwidth because of ACK compression that generates aliased samples. In the original Westwood TCP paper [12] this phenomena was not evident because the ACK compression was weak in the considered scenarios.

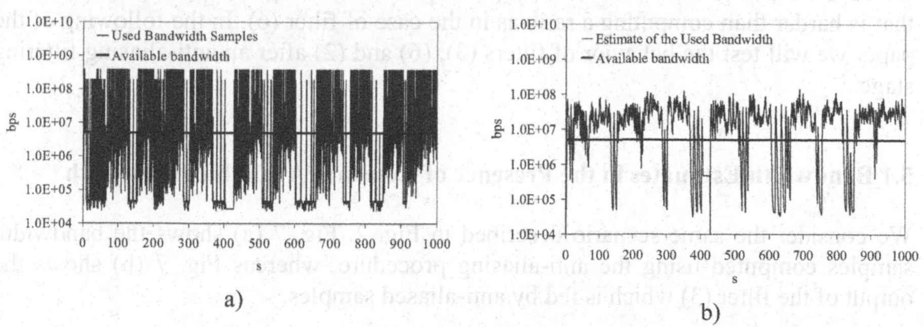

Fig. 6. Used Bandwidth: (a) Samples; (b) Output of the filter (6).

5 Anti Aliasing Filter in Packet Networks

We have seen that ACK compression generates bandwidth samples containing aliased frequency components that cannot be low-pass filtered using a digital filter. Therefore, similarly to the case of filtering an analog signal, which requires the analog signal be pre-filtered by an anti-aliasing filter before using a discrete-time filter, also in packet networks it is necessary to implement a sort of anti-aliasing filter. The basic idea to implement an anti-aliasing filter in packet networks is to group the ACKs received in a sufficiently large time interval T and compute a unique corresponding bandwidth sample. This operation has the effect of smoothing single ACKs, which corresponds to filter out high frequency components. In particular, ACKs received during the last *RTT* are grouped into a unique bandwidth sample that is computed by considering the total data acknowledged over the *RTT*. To show that the proposed algorithm avoids ACK compression effects, we consider the same scenario investigated in the previous simulations. It is worth noting that the anti-aliasing algorithm generates one bandwidth sample for each *RTT* (that is the sampling time Δ is equal to *RTT*). Assuming a constant *RTT*, the α coefficient in filter (3) can be set accordingly to (4). The difference between the time-invariant filter (3) and the time-varying filters (2) and (6) is that filter (3) employs a constant coefficient α, whereas filters (2) and (6) dynamically adjust their coefficients by taking into account the last samples interarrival time. The latter feature is important to provide proper filtering also in scenarios with large *RTT* variance.

In order to employ the time-varying filter (6), it is still necessary to satisfy the Nyquist-Shannon sampling theorem, that is, it must be $\Delta_k \le \tau/2$ [10]. To be conservative, we assume $\Delta_k \le \tau_f/4$. Thus, if it happens that $\Delta_k > \tau/4$ then we interpolate and re-sample by creating $N = \text{integer}(4 \cdot \Delta_k / \tau)$ virtual samples b_k that arrive with the inter-arrival time $\tau/4$ and one more sample b_k arriving with inter-arrival time $\Delta T = \Delta_k - N \cdot \tau/4$.

It should be noticed that filter (2) already implements interpolation and resampling through the ZOH that keeps constant the signal during the last interarrival time. However, the coefficient of the filter (2) requires the computation of an exponential that is harder than computing a ratio as in the case of filter (6). In the following of the paper we will test the behavior of filters (3), (6) and (2) after an anti-aliasing filtering stage.

5.1 Bandwidth Estimates in the Presence of Constant Available Bandwidth

We consider the same scenario described in Fig. 2. Fig. 7 (a) shows the bandwidth samples computed using the anti-aliasing procedure, whereas Fig. 7 (b) shows the output of the filter (3) which is fed by anti-aliased samples.

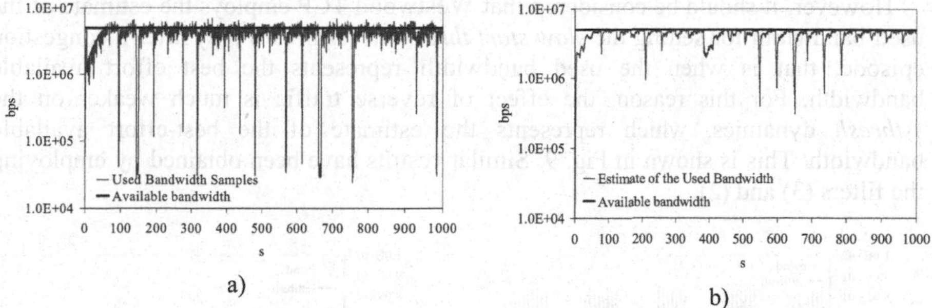

Fig. 7. Used bandwidth: (a) Anti-aliased samples; (b) Output of the filter (3).

A comparison of Fig. 7 (a) and Fig. 3 (a) shows that now there is no overestimate of the bandwidth samples due to aliasing. Once the bandwidth samples are obtained as shown in Fig. 7 (a), a further stage of discrete-time filtering extracts the low frequency components of the used bandwidth as it is shown in Fig. 7 (b). Analogous results have been obtained with filters (6) and (2).

5.2 Bandwidth Estimates in the Presence of Time Varying Available Bandwidth

In this section we investigate the behavior of the three low pass filters introduced above in the presence of a time-varying available bandwidth when bandwidth samples are properly prefiltered using the anti-aliasing procedure. To the purpose, the scenario in Fig. 2 has been considered with the UDP source turned ON. The *RTT* of the considered TCP connection is 100ms. At first, the reverse traffic has been turned OFF to observe the output of various filters without disturbances, and then the reverse traffic has been turned ON to test the filters in a realistic scenario with ACK compression.

Figs. 8 (a) and (b) report the estimate of the used bandwidth obtained using the filter (6) with and without reverse traffic, respectively. As it can be viewed, the reverse traffic affects the estimate of the used bandwidth because of ACK congestion.

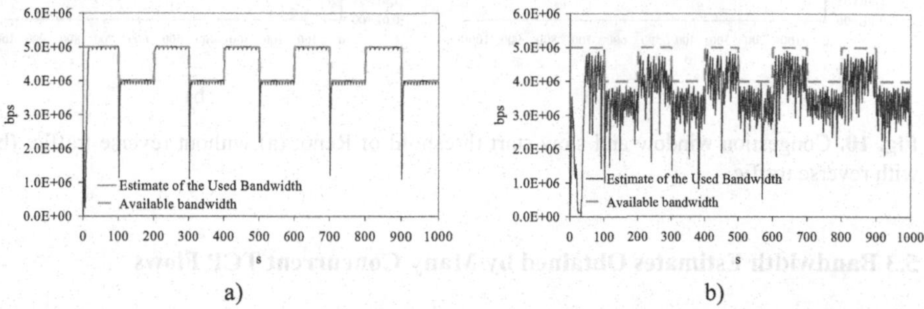

Fig. 8. Estimate of the used bandwidth with filter (6): (a) without reverse traffic; (b) with reverse traffic.

However, it should be considered that Westwood TCP employs the estimate of the used bandwidth for setting the *slow start threshold* (*ssthresh*) only after a congestion episode, that is when the used bandwidth represents the best effort available bandwidth. For this reason, the effect of reverse traffic is much weaker on the *ssthresh* dynamics, which represents the estimate of the best-effort available bandwidth. This is shown in Fig. 9. Similar results have been obtained by employing the filters (3) and (2).

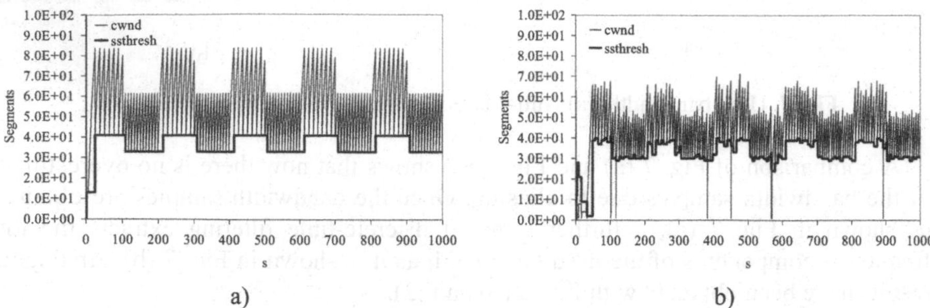

Fig. 9. Congestion window and slow start threshold with filter (6): (a) without reverse traffic; (b) with reverse traffic.

It is interesting to compare the *ssthresh* of Westwood TCP w.r.t the *ssthresh* of Reno TCP, which is plotted in Fig. 10. The comparison points out that the *ssthresh* of Westwood is larger than the one of Reno. This proves that the proposed bandwidth estimate enhances the ability of TCP congestion control to match the available bandwidth, which provides better utilization of network bandwidth.

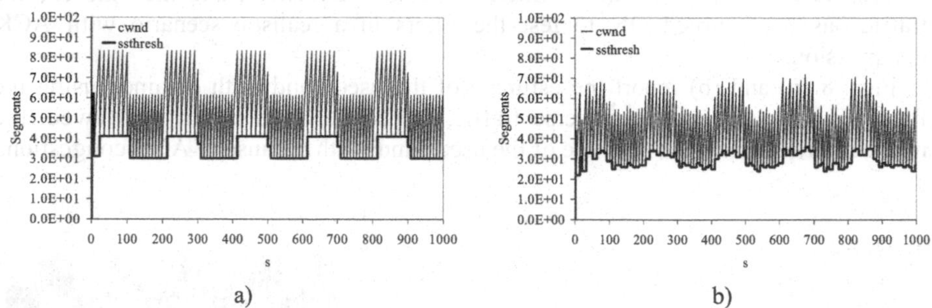

Fig. 10. Congestion window and slow start threshold of Reno: (a) without reverse traffic; (b) with reverse traffic.

5.3 Bandwidth Estimates Obtained by Many Concurrent TCP Flows

In this section, we investigate the bandwidth estimates of many Westwood TCP flows competing for the same bottleneck capacity. We would like that, given a link capacity C shared by N flows each flow obtains an estimate of the used bandwidth oscillating around the fair-share C/N and a *ssthresh* which is close to the available bandwidth.

This would prove that the end-to-end bandwidth estimation algorithm can improve the fairness of TCP congestion control in bandwidth allocation. We consider a single bottleneck scenario similar to the one depicted in Fig. 2, where a 10Mbps FIFO bottleneck is shared by 10 TCP-W persistent flows with *RTTs* ranging uniformly from 25ms to 250ms. Figs. 11 (a) and (b) show the estimates of the used bandwidth and of the best effort available bandwidth $ssthresh*Seg\_size/RTT_{min}$, respectively, obtained by the 10 TCP-W connections using the filter (6). Similar results have been obtained for filters (2) and (3). Fig. 11(a) shows that the used bandwidth estimates oscillate around the fair-share *C/N*, which is denoted by the dashed line. Furthermore Fig. 11(b) shows that the estimates of the best effort available bandwidth are less oscillating and closer to the fair share *C/N*. Similar results have been obtained by using the filters (3) and (2).

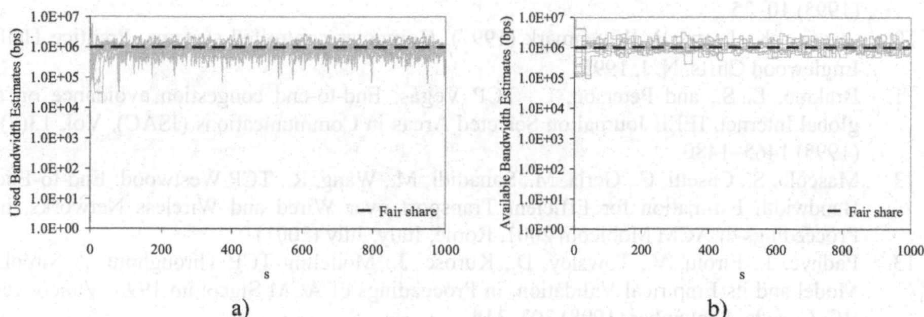

Fig. 11. Bandwidth estimates of 10 Westwood flows implementing the filter (6): (a) Used Bandwidth; (b) Available Bandwidth.

6 Conclusions

This work has focused on end-to-end bandwidth estimation schemes to be used for TCP congestion control. It has been shown that it is necessary to implement an anti-aliasing filter before using discrete-time filters in packet networks. Simulation results have shown that anti-aliasing plus low-pass discrete time filtering provide a reliable estimate of the used bandwidth that, when coupled with a probing congestion control algorithm, gives also a reliable estimate of the best-effort available bandwidth. This estimate enhances the efficiency of TCP congestion control [12,19].

References

1. Jacobson, V.: Congestion avoidance and control, in Proceedings of ACM Sigcomm '88, Stanford CA, August (1988) 314–329.
2. Allman, M., Paxson, V. and Stevens W.R.: TCP congestion control, RFC 2581, April 1999.
3. Melander, B., Bjorkman, M. and Gunningberg, P.: A New End-to-End Probing and Analysis Method for Estimating Bandwidth Bottlenecks, in Proceedings of Global Internet Symposium, 2000.

4. Clark, D.: The design philosophy of the DARPA Internet protocols, in Proceedings of ACM Sigcomm'88, Stanford CA, August (1988) 106–114.
5. Floyd, S., Fall, K.: Promoting the use of end-to-end congestion control in the Internet. IEEE/ACM Transactions on Networking, Vol. 7(4), (1999), 458–472.
6. Dah-Ming Chiu, Jain, R.: Analysis of the increase and decrease algorithms for congestion avoidance in computer networks. Computer Networks and ISDN Systems, Vol. 17(1), (1989) 1–14.
7. Jain, M., Dovrolis, C.: End to End Available Bandwidth: Measurement Methodology, Dynamics, and Relation with TCP Throughput, in Proceedings of ACM Sigcomm 2002.
8. Stoica, I., Shenker, S. and Zhang, H.: Core-Stateless Fair Queueing: Achieving Approximately Fair Bandwidth Allocations in High Speed Networks, in Proceedings of ACM Sigcomm '98, Vancouver, Canada, August (1998) 118–130.
9. Li, S. Q., and Hwang, C.: Link Capacity Allocation and Network Control by Filtered Input Rate in High speed Networks, IEEE/ACM Transaction on Networking, Vol. 3(1), (1995) 10–25.
10. *Aström* , K. J. and B. Wittenmark (1997). Computer controlled systems, Prentice Hall, Englewood Cliffs, N. J, 1995.
11. Brakmo, L. S., and Peterson, L.: TCP Vegas: End-to-end congestion avoidance on a global Internet. IEEE Journal on Selected Areas in Communications (JSAC), Vol. 13(8), (1995) 1465–1480.
12. Mascolo, S., Casetti, C., Gerla, M., Sanadidi, M., Wang, R.: TCP Westwood: End-to-End Bandwidth Estimation for Efficient Transport over Wired and Wireless Networks, in Proceedings of ACM Mobicom 2001, Rome, Italy, July (2001).
13. Padhye, J., Firoiu, V., Towsley, D., Kurose, J.: Modeling TCP Throughput: A Simple Model and its Empirical Validation, in Proceedings of ACM Sigcomm 1998, Vancouver BC, Canada, September (1998) 303–314.
14. Mogul, J. C.: Observing TCP dynamics in real networks, in Proceedings of ACM Sigcomm 1992, 305–317.
15. Ns-2 network simulator (ver 2). LBL, URL: http://www-mash.cs.berkeley.edu/ns.
16. Lai, K. and Baker, M.: Measuring Link Bandwidths Using a Deterministic Model of Packet Delay, in Proceedings of ACM Sigcomm 2000, Stockholm, Sweden, August (2000) 283 –294.
17. Floyd, S., Henderson, T.: NewReno Modification to TCP's Fast Recovery, RFC 2582, April 1999.
18. Allman, M., Floyd, S., Partridge, C.: Increasing initial TCP's initial window, RFC 2414, September 1998.
19. Grieco, L. A., and Mascolo, S.: Westwood TCP and easy RED to improve Fairness in High Speed Networks, in Proceedings of IFIP/IEEE Seventh International Workshop on Protocols For High-Speed Networks, PfHSN02, Berlin, Germany, April (2002) 130–146.
20. Chaskar, H.M., Lakshman, T.V. and Madhow, U.: TCP Over Wireless with Link Level Error Control: Analysis and Design Methodology, IEEE/ACM Transactions on Networking, Vol. 7(5), (1999) 605–615.
21. Keshav, S.: A Control-theoretic Approach to Flow Control, in Proceedings of ACM Sigcomm 1991, Zurich, Switzerland, September (1991) 3–6.
22. Hoe, J. C.: Improving the Start-up Behavior of a Congestion Control Scheme for TCP, in Proceedings of ACM Sigcomm'96, Palo Alto, CA, August (1996) 270–280.
23. Allman, M. and Paxson, V.: On Estimating End-to-End Network Path Properties, in Proceedings of ACM Sigcomm 1999, Cambridge, Massachusetts, August (1999) 263–276.

Priority-Based Internet Access Control for Fairness Improvement and Abuse Reduction*

Tsung-Ching Lin [1], Yeali S. Sun [2], Shi-Chung Chang [1], Shao-I Chu [1], Yi-Ting Chou [1], Mei-Wen Li [3]

[1] Department of Electrical Engineering, National Taiwan University
Taipei, Taiwan
{tclin, scchang, shaoi, eddy} @ac.ee.ntu.edu.tw
http://www.ee.ntu.edu.tw

[2] Department of Information Management, National Taiwan University
Taipei, Taiwan
sunny@im.ntu.edu.tw

[3] Computer and Information Network Center, National Taiwan University
Taipei, Taiwan
mli@ccms.ntu.edu.tw

Abstract. In this paper, we exploit a prioritized-service architecture and apply a priority based traffic control scheme to reduce abusive Internet access, improve fairness among users and study users' behavior under a prioritized service for Internet access. The Internet access by dormitory users of National Taiwan University (NTU) serves as a conveyer problem. There are two classes of service. The regular class has a volume quota for each user, which is designed to meet majority users' essential demands while limiting abusive usage of quality service. The custody class is lower in service priority with no volume quota limitation. Our mathematical models for design and analysis include individual and aggregate user demand models and a network performance model. The priority control scheme is implemented over an existing NTU dormitory network with additions of a QoS router, a meter reading server, an accounting server and a user interface server. The control leads to 48.9% reduction in average packet drop rate, 42.2% improvement in a fairness measure, reduction of abusive Internet access by 57.82% and 145.09% Internet access increase in majority users. Such results are quite consistent with predictions of our mathematical models.

1. Introduction

In a best effort and free of charge Internet access environment, there often exists abusive and unfair usage of network resources. Abusive usage of limited resources by a small number of users may lead to the poor network performance such as high packet loss rate or long delay. All the other users have to suffer from the

* This work was supported in part by the National Science Council of the Republic of China under Grants NSC88-2215-E-002-026, NSC 89-2215-E-002-037 and NSC-90-2213-E-002-078.

M. Ajmone Marsan et al. (Eds.): QoS-IP 2003, LNCS 2601, pp. 659-671, 2003.

consequences. It is obviously unfair. Thus, how to prevent users from abuse and achieve fairness among all users are important issues for our study. In the literature of network management, the most commonly used fairness [1,2] is the max-min fairness, which tries to forbid the ignorance of small users. Such fairness may result in network inefficiency. Therefore, Kelly [3] advocated the proportional fairness instead. This kind of fairness aims at maximizing the overall utility of rate allocations assuming that the traffic of Internet is elastic and the utility of each traffic flow is logarithmic.

To limit network usage by a user, the volume-based quota has been a straightforward common practice adopted for the management of a free of charge network environment [4]. Each user is allocated a fixed number of bytes (volume) of usage in a period of time. Once the quota is used up, a user is refrained from using the network service until the next period. The quota scheme sets a hard constraint to each user, prevents users from abusing network resources and is fair. Nevertheless, it is not flexible to allow excessive usage when the network resource is lowly utilized, i.e., disadvantageous for network efficiency.

Prioritization of service provides a soft way of reducing abusive and unfair usage. In the development of Internet technology, a differentiated-service architecture, called "DiffServ" [5], has recently been proposed by IETF. There is a class of qualitative services under DiffServ, where a service with a higher priority always gets a better quality of service than that with a lower priority. A high priority is given to users' essential needs while other needs are served in a low priority. One's unimportant or excessive usage in low priority will not affect the essential or important usage in high priority.

In this paper, we combine the ideas of both the volume quota limitation and the priority scheme into a priority-based scheme to control Internet access traffic. Internet access by dormitory users of NTU campus networks serves as a conveyer problem of our study. How the priority and quota should be assigned to users' network service demands and how the priority and quota settings may affect network resource allocation, performance, quality of service to users and user behaviors are the challenging research issues in this paper.

The remainder of this paper is organized as follows. In Section II, we describe the issues of unfair and abusive Internet access by NTU dormitory network users. In Section III, we propose a mathematical model for analysis and design of the priority-based Internet access control scheme, system implementation is described as well. In Section IV, we present experimental results to validate our design. Finally, Section V concludes the paper.

2. Unfair and Abusive Internet Access

2.1 Network Configuration without Prioritized Service

Fig. 1 depicts the configuration of the dormitory network in NTU campus without the priority-based control. There are fourteen dormitories and approximately 5400 network users. Each user is provided with a UTP Ethernet 10/100Mbps access point and assigned an Internet IP address. The dormitory network is connected to NTU campus networks and Taiwan Academic Networks (TANet) /Internet via an Alcatel router and a Cisco 7513 router. The access link is a 100Mbps Ethernet link. However, to help alleviating the frequent congestion on the International link between TANET and the Internet, NTU has set a policy to restrict the outbound Internet traffic of its dormitories to a maximum rate of 54Mbps.

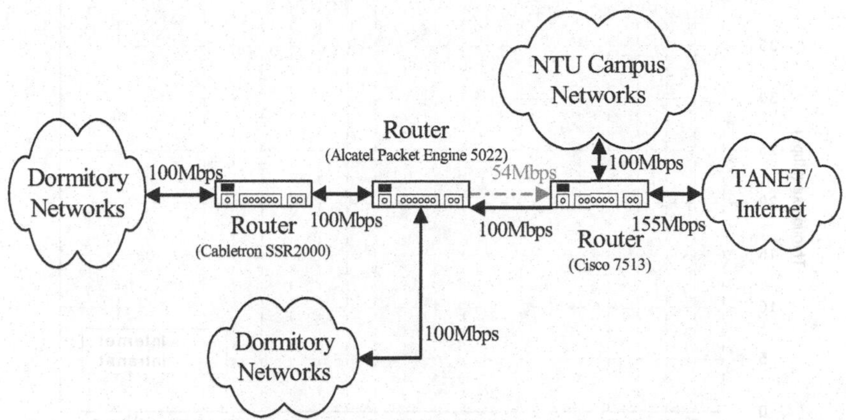

Fig. 1. Original Internet Access Architecture from Dormitory users

A network management tool, Netflow™ [6], is installed in NTU campus networks to monitor all packet flows passing the Cisco 7513 router. Currently, Netflow collects, every 10 minutes, MIB information items about each flow and calculates many statistics such as the top heavy users and application share by volume. Under the current environment, the volume per IP used is easily obtained from Netflow.

2.2 Traffic Characteristics

National Taiwan University is a national university and funded by the government. The Internet access service to the dormitory users is unlimited and almost free of charge (a flat charge of less than 9 US$/person-semester). We define the traffic for Internet access as that is transmitted from dorm networks to the Internet, and the traffic for intranet access as that is transmitted from dorm networks to other NTU campus networks. Our analysis (Fig. 2) shows that the peak hours for Internet access and intranet access are 5AM and 5PM respectively. The Internet access traffic (36.58Mbps) is more than two times of that for intranet access (17.18Mbps). However, under the current campus network environment, Internet access is bandwidth tight while there is excess bandwidth for intranet communications. Almost all the Internet access traffic (99.88%) is TCP traffic.

The drop rate of the traffic at the 54Mbps link is measured every 5 minutes. The drop ratio is defined as the drop rate divided by the link bandwidth. The 54Mbps link is considered congested once the drop ratio is more than 0.1. Abusive usage naturally exists and leads to network congestion. Representative patterns of the total outbound load and the drop rate at Cisco router 7513 during a day are shown as Fig. 3, where the congestion lasts over 17 hours, from 11am to 4am. The average daily network utilization is 99.57% with an average packet drop ratio of 0.11. Such network service quality is definitely unsatisfactory.

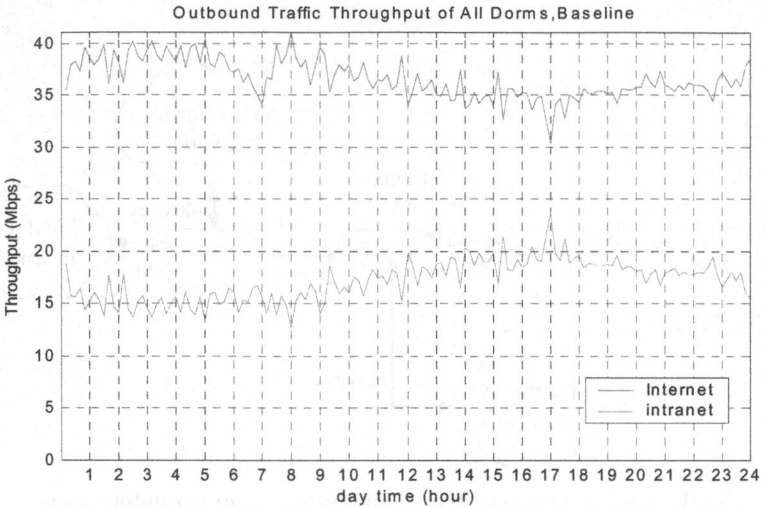

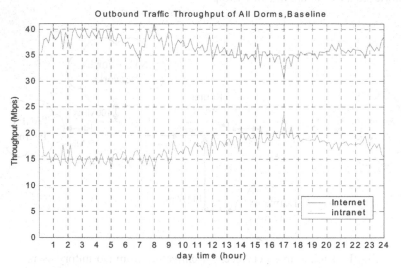

Fig. 2. Internet and Intranet Traffic Components in Outbound Traffic

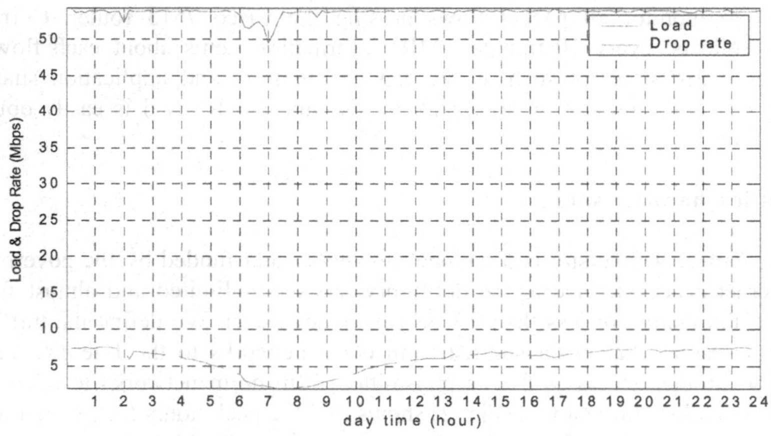

Fig. 3. Drop rate and throughput of outbound traffic

2.3 User Characteristics

The Internet access is both abusive and unfair. Fig. 4 gives a dormitory Internet access profile of daily volume per IP. It clearly shows a three-peak profile. Empirical data further shows that the daily usage for Internet access is 81.7 Mbytes per user in average, but with a large standard deviation, 471.83MB, and a coefficient of variation of 5.78. The top 2% heavy users, whose daily average usage is more than 1GB, create

66% of the traffic load. However, the lightest 90% users only create 3 % of the traffic load. It is unfair that all users suffer from the same level of network congestion. In this paper, we define the abuse index as the total volume (bytes) transmitted by the top 2% heavy users and the unfairness index as coefficient of variation (variance/mean) of users' daily volume usage.

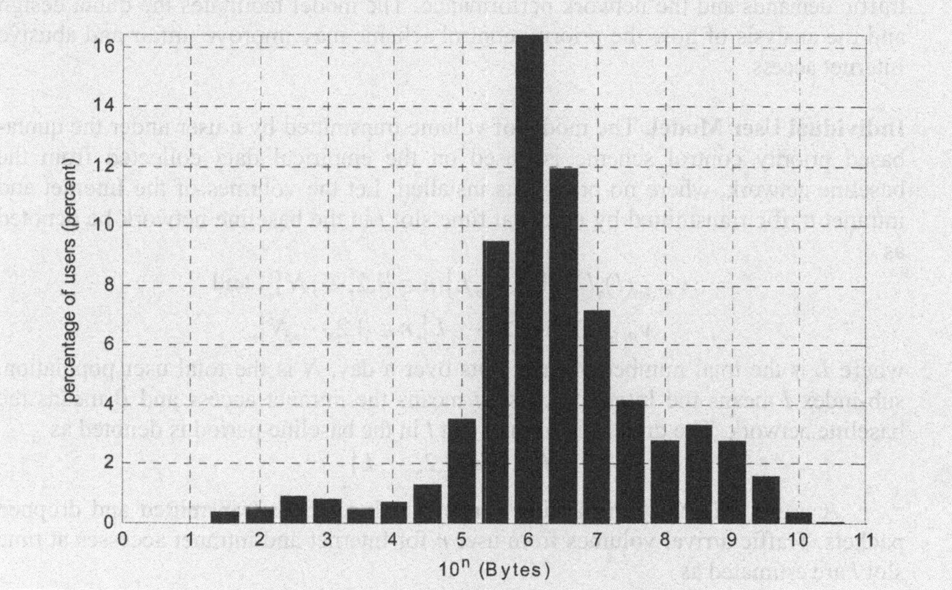

Fig. 4. The load distribution and user clusters

3. Priority Control System Design

3.1 Policy Design

As seen in many Internet access networks, performance issues mainly arise from overuse and congestion externalities, which are often factors of human behavior. On account of limited network resources, for a network manager, the goal is to assure the benefit of most users, i.e. to maximize total user utility [7]. One way to alleviate the problem of poor performance caused by abusive Internet access is to take some measures to ensure that each user is guaranteed with a basic service quota - a fair share of resource and access. The philosophy is not to prevent heavy users from using the network when their usage exceeds the quota but to make sure their over-usage is under control, not affecting other users' basic service.

Our prioritized service plan consists of two service classes: regular and custody. The regular service has a higher priority than the custody service on data transmission. There is a volume quota for each user's regular service, which is designed to meet majority users' essential demands while limiting abusive usage of quality service. The custody class has no quota limitation, which allows the heavy users to access the Internet at a lower quality.

3.2 Mathematical Model for Priority Control System

We now construct a mathematical model of the relationship among the quota, user traffic demands and the network performance. The model facilitates the quota design and the analysis of how the priority control scheme may improve unfair and abusive Internet access.

Individual User Model. The model of volume transmitted by a user under the quota-based priority control scheme is based on the empirical data collected from the baseline network, where no priority is installed. Let the volumes of the Internet and intranet traffic transmitted by user n at time slot l in the baseline network be denoted as

$$v_{BI,\ln}(l), l \in \{1,2,\cdots,L\}, n \in \{1,2,\cdots,N\} \text{ , and}$$
$$v_{Bi,\ln}(l), l \in \{1,2,\cdots,L\}, n \in \{1,2,\cdots,N\},$$

where L is the total number of time slots over a day, N is the total user population, sub-index I means the Internet access, i means the intranet access and B means the baseline network. The drop ratio at time slot l in the baseline period is denoted as

$$d_{B\_l}, l \in \{1,2,\cdots,L\}.$$

A user's original transmission demand includes both transmitted and dropped packets. Traffic arrival volumes from user n for Internet and intranet accesses at time slot l are estimated as

$$\hat{a}_{BI\_\ln} = v_{BI\_\ln} \times (1 + d_{B\_l}), l \in \{1,2,\cdots,L\}, n \in \{1,2,\cdots,N\}, \text{ and}$$
$$\hat{a}_{Bi\_\ln} = v_{Bi\_\ln} \times (1 + d_{B\_l}), l \in \{1,2,\cdots,L\}, n \in \{1,2,\cdots,N\}.$$

Let the volume quota be H. If the cumulative regular Internet traffic volume of a user exceeds the quota H, the subsequent Internet access of the user will be downgraded to the custody service only. For the regular Internet access, the traffic arrival volume from user n at time slot l is denoted as $\hat{a}_{HI\_\ln}$ and is estimated as,

$$\hat{a}_{HI\_\ln} = \begin{cases} \hat{a}_{BI\_\ln}, \text{if } \sum_{j=1}^{l-1} \hat{v}_{HI\_jn} < H \\ 0, \text{otherwise} \end{cases},$$
$$l \in \{1,2,\cdots,L\}, n \in \{1,2,\cdots,N\}$$

where $\hat{v}_{IH,\ln}$ is the regular Internet traffic volume successfully transmitted by user n at time slot l.

Thus, the total traffic arrival volume from user n at time slot l is

$$\hat{a}_{H\_\ln} = \hat{a}_{HI\_\ln} + \hat{a}_{Bi\_\ln}, l \in \{1,2,\cdots,L\}, n \in \{1,2,\cdots,N\},$$

and the total traffic arrival volume from all users at time slot l is calculated as

$$\hat{A}_{H\_l} = \sum_{n=1}^{N} \hat{a}_{H\_\ln}, l \in \{1,2,\cdots,L\},$$

the estimate of the traffic arrival rate at time slot l is calculated as

$$\hat{X}_{H\_l} = \frac{\hat{A}_{H\_l}}{T}, l \in \{1,2,\cdots,L\}, \tag{1}$$

where T is the length of each time slot.

Once the estimated traffic arrival rate (i.e. the offered load) exceeds the bottleneck bandwidth B, i.e., $\hat{X}_{H\_l} \geq B$, we consider that congestion occurs. The drop ratio is predicted as

$$\hat{d}_{H\_l} = \begin{cases} (\hat{X}_{H\_l} - B)/B, \text{if } \hat{X}_{H\_l} > B \\ 0, \text{otherwise} \end{cases}, l \in \{1,2,\cdots,L\}. \tag{2}$$

It is assumed that the regular Internet access suffers from the same drop ratio. Hence, the regular traffic volumes successfully transmitted by user n for Internet and intranet accesses are estimated as

$$\hat{v}_{HI\_\ln} = \begin{cases} \hat{a}_{HI\_\ln} \times \left(1 - \hat{d}_{H\_l}\right), \text{if } \sum_{j=1}^{l-1} \hat{v}_{HI\_jn} < H \\ 0, \text{otherwise} \end{cases}, \tag{3}$$

and

$$\hat{v}_{Hi\_\ln} = \hat{a}_{Hi\_\ln} \times (1 - \hat{d}_{H\_l}), l \in \{1,2,\cdots,L\}, n \in \{1,2,\cdots,N\}.$$

In summary, we have developed a model to predict, under a given quota H, every user's behavior and network performance within a time slot, including per user's volume usage (3), the aggregated traffic rate (1) and the drop ratio (2). Predictions of all time slots can therefore be obtained.

Aggregate User Model. Based on subsection 3.2.1, the means and the standard deviations of the regular traffic volumes transmitted by users for Internet and intranet access are calculated as

For Internet access

$$\hat{\bar{v}}_{HI} = \frac{\sum_{n=1}^{N} \hat{v}_{HI\_n}}{N}, \quad \hat{S}_{HI} = \sqrt{\frac{\sum_{n=1}^{N} (\hat{v}_{HI\_n} - \hat{\bar{v}}_{HI})^2}{N-1}}, \text{ and}$$

For Intranet access

$$\hat{\bar{v}}_{Hi} = \frac{\sum_{n=1}^{N} \hat{v}_{Hi\_n}}{N}, \quad \hat{S}_{Hi} = \sqrt{\frac{\sum_{n=1}^{N} (\hat{v}_{Hi\_n} - \hat{\bar{v}}_{Hi})^2}{N-1}}$$

where $\hat{v}_{HI\_n} = \sum_{l=1}^{L} \hat{v}_{HI\_\ln}$ and $\hat{v}_{Hi\_n} = \sum_{l=1}^{L} \hat{v}_{Hi\_\ln}$.

As a result, we have the relationship between the quota and the aggregate user behavior (mean and standard deviation of transmitted volume).

Network Performance Model. We have calculated the drop rate and users' usage for all time slots. The daily mean of the drop ratio is then

$$\hat{d}_H = \frac{\sum_{l=1}^{L} \hat{d}_{H\_l}}{L}, l \in \{1,2,\cdots,L\}.$$

The daily means of the regular traffic rate for Internet and intranet accesses are

$$\hat{R}_{IH} = \frac{\sum_{n=1}^{N} \hat{v}_{IH\_n}}{L}, \text{ and}$$

$$\hat{R}_{iH} = \frac{\sum_{n=1}^{N} \hat{v}_{iH\_n}}{L}.$$

The network utilization is $\hat{u}_H = \frac{\hat{R}_{IH} + \hat{R}_{iH}}{B \times T_{day}}$,where T_{day} is the length of one day.

3.3 Prediction of Control Effects

We apply the user behavior and network performance models to predict control effect of our proposed scheme. We construct a numerical experiment with the following parameters: the quota H is given as 2GB and 1GB to satisfy at least 99% and 98% of users in the baseline network respectively. The link bandwidth B is 54Mbps, the total user population is 5355. Following the above models, we predict controlled user behavior of Internet access as listed in Table 1. Results show that a 42.2% reduction of unfairness and a 62.32% reduction of abusive usage as compared to baseline network behavior.

Table 1. Numerical results for Internet Access

Information \ Treatment	Baseline	2GB	1GB
Standard deviation (MB)	449.03	282.66	186.49
Mean volume (MB/user)	73.78	68.93	52.97
Unfairness index; Coefficient of Variation	6.09	4.10	3.52
Abuse index; Usage of top 2% user (GB)	261.71	174.30	98.60

3.4 System Implementation

In this section, we design a network system architecture to realize the control policy proposed in section 3.1. The whole design is depicted in Fig. 5, where the infrastructure consists of: (1) QoS router (2) metering router, (3) meter reading server, (4) accounting server, (5) user interface server. All these devices but the metering router do not exist in the baseline network and need new design and development.

We adopt the QoS router developed by Sun et al [7] to provide prioritized services. The classifier and priority queueing scheduler of the router is specialized to the two–class service provisioning. The meter reading server is based on Netflow™, which collects traffic statistics of each IP from the metering router. The accounting server is developed in Perl[9] and C-shell. Service class of metered traffic is determined at the accounting server by checking if the quota of IP is exceeded. These two servers jointly implement the quota control function. In addition, we construct a user interface server (http://ntunm.ntu.edu.tw), through which users can check with the accounting server their available quota, transmitted volume, ranking of all users, and the current service class.

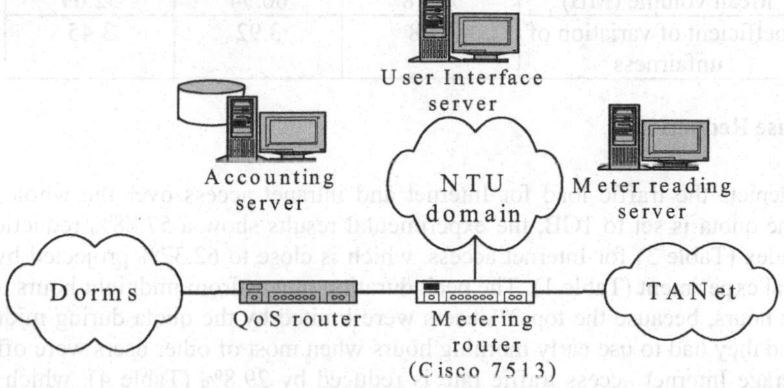

Fig. 5. System Architecture

4. Experiment Results

To assess the effectiveness of our priority-based scheme and to investigate users' response to the two-priority service, we design and conduct a set of experiments on the priority-based Internet access control system just described in Section 3. We announced the priority control experiment to dormitory users in the fall semester of 1999. The users then had two months of trial period to familiarize with the experiment and feedback their comments for system tuning. In this section, we considered, as the baseline data set, the network data of a representative but arbitrarily selected week of 2/22/'00 to 2/28/'00. In that week, the network is without priority or quota control. The daily quota H is our control variable, there were two levels of H experimented, 2GB and 1GB, each set as the daily quota over one week from 3/5/'00 to 3/23/'00. The quota was renewed at the time of least traffic, i.e. 6AM, each day. Performance statistics of our study include the daily volume used by each user, the drop rate, and the traffic load over a day, etc.

4.1 Fairness Improvement

Table 2 shows the mean and standard deviation of individual user's daily volume for Internet access under different *H* levels. It can be clearly observed that the average and standard deviation of users' Internet access traffic are both reduced. Compared with the baseline statistics, the coefficients of variation of unfairness are reduced by 35.53% and 43.3% for the two *H* levels respectively which is close to 32.68 and 42.2% projected by our numerical experiment (Table 1).

Table 2. Comparison information of Internet Access

Treatment / Information	Baseline	2GB	1GB
Standard deviation (MB)	449.03	239.14	179.53
Mean volume (MB)	73.78	60.94	52.09
Coefficient of variation of unfairness	6.08	3.92	3.45

4.2 Abuse Reduction

Fig. 6 depicts the traffic load for Internet and intranet access over the whole day. When the quota is set to 1GB, the experimental results show a 57.88% reduction in abuse index (Table 3) for Internet access, which is close to 62.32% projected by our numerical experiment (Table 1). The peak duration moves from midnight hours to the morning hours, because the top 2% users were limited by the quota during midnight hours and they had to use early morning hours when most of other users were offline. The average Internet access traffic rate is reduced by 29.8% (Table 4), which may imply the reduction of unnecessary access of abusive users. This is supported by Fig. 7, which shows a decrease in the numbers of users whose usage is more than 1GB/day. Furthermore since heavy users had the same quota in using the regular class just like everybody else, the majority of users must be benefited and their network usage were encouraged as indicated by Table 3.

Table 3. Comparisons among different users of Internet Access

Quota / Total transmitted volume	Baseline	2GB	1GB
Abuse index; Top 2% users (GB)	261.71	161.5624	110.24
Lower 90% users (MB)	9.98	24.46	15.97

Table 4. Comparisons of load of Internet Access

	Baseline	2GB	1GB
Throughput (Mbps)	32.89	25.85	23.09

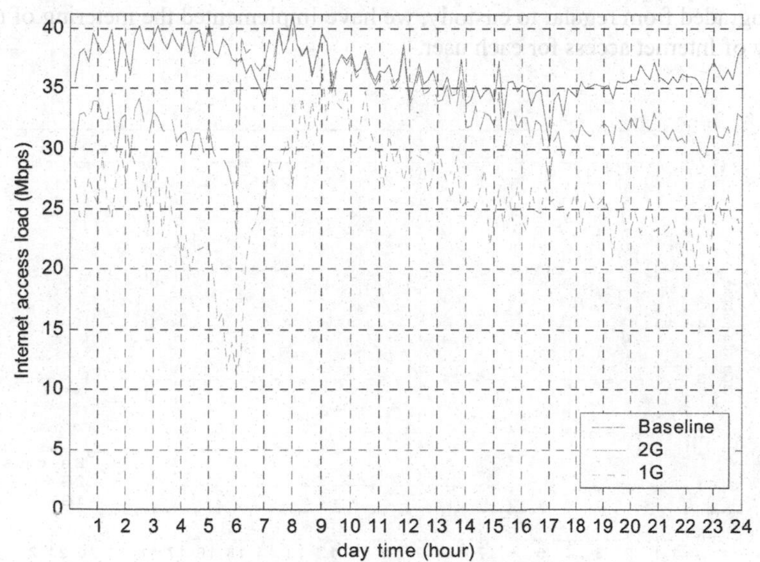

Fig. 6. The Internet access load over a day

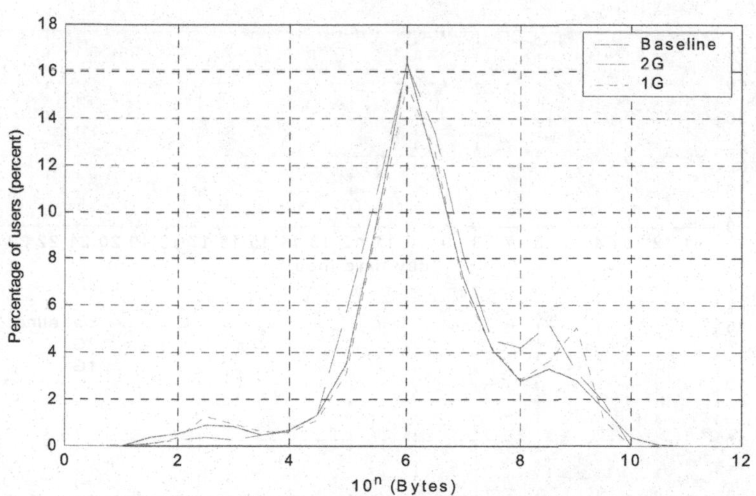

Fig. 7. The comparisons of Internet usage distribution

4.3 Network Performance Improvement

Following the reduction of daily Internet access volume, the network congestion is soothed too. Fig. 8 and Table 5 indicate a 48.97% reduction of the drop rate when the quota is set to 1GB. Fig. 9 indicates a 9.9% reduction of bandwidth utilization. To identify the performance seen by each user when the user's service class is downgraded from regular to custody, we have implemented the metering of round trip delay of Internet access for each user.

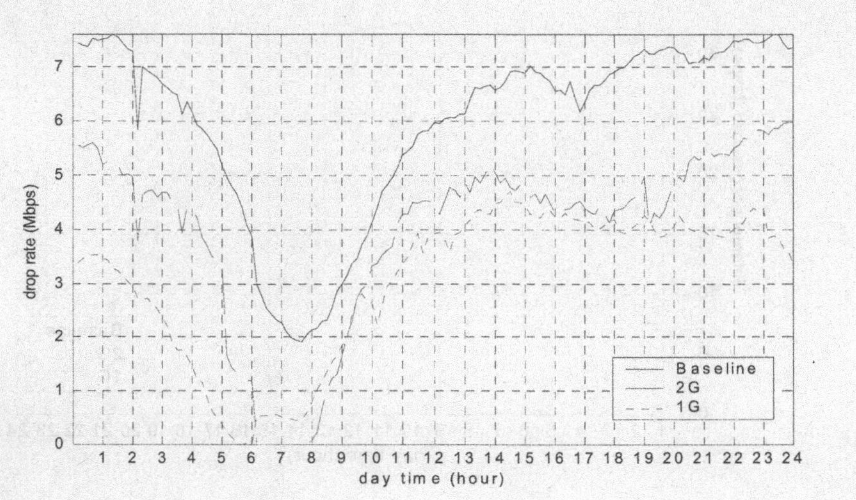

Fig. 8. The comparisons of drop rates between Baseline, 2G and 1G

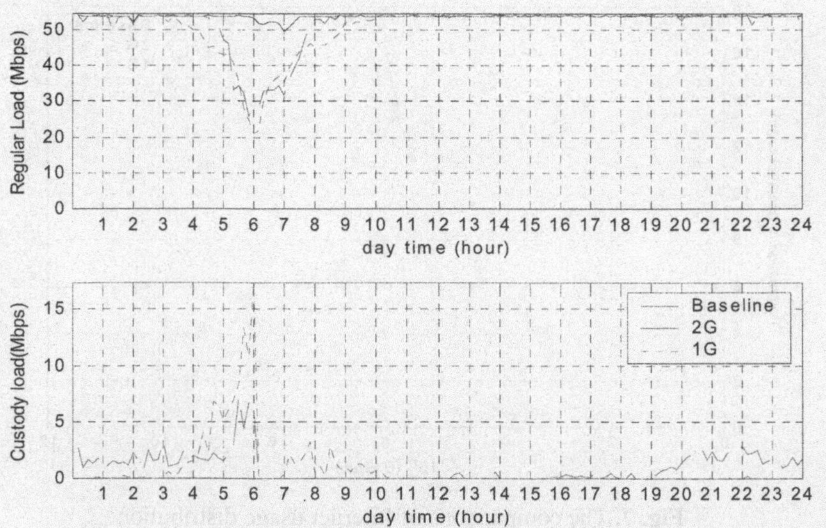

Fig. 9. The comparisons of drop rates between Baseline, 2G and 1G

Table 5. Comparisons of drop rate

	Baseline	H=2GB	H=1GB
Drop Rate (Mbps)	5.95	3.98	3.04

5. Conclusions

In this paper, a priority-based Internet access control scheme has been proposed to alleviate the problem of unfair and abusive utilization for Internet access by an intra-network. We have built simple user behavior and traffic models for the design and analysis of the control scheme. We have also established the necessary experimentation environment over the dormitory networks of NTU. Experiments with dormitory users demonstrate a significant reduction of abusive usage, improvement in fairness and alleviation of the congestion. Models of users' behavior under the prioritized service are also validated. We are now constructing detailed models to better predict traffic generated by users that fits empirical data.

Acknowledgements

The authors would like to thank Yu-Qun Pan, Guang-Wei Li, Yu-Mei Shao, Yin-Ren Chien, Zhi-Xiu Lin and Yu-Quan Li for very valuable discusions and technical supports at various stages of this work.

References

1. L.Massoulie, J. Roberts: "Bandwidth sharing: objectives and algorithms, *INFOCOM '99, Eighteenth Annual Joint Conference of the IEEE Computer and Communications Societies. Proceedings. IEEE*, vol. 3 (1999) 1395-1403
2. Y. Le Boudec: Rate adaptation, congestion control and fairness: a tutorial, http://www.statslab.cam.ac.uk/~frank/pf/
3. F. Kelly, A. Maulloo, and D. Tan: Rate control for communication networks: Shadow price proportional fairness and stability, *Journal of Operations Research Society*, vol. 49 (1998) 237–252
4. S.-T. Cheng, C.-M. Chen and I.-R. Chen: Dynamic quota-based admission control with sub-rating in multimedia servers, *Multimedia Systems*, vol. 8, no. 2 (2000) 83-91
5. S. Blake, D. Black, M. Carlson, E. Davies, Z. Wang, W. Weiss: An Architecture for Differentiated Services, IETF RFC 2475, http://www.ietf.org/rfc/rfc2475.txt (1998)
6. http://www.cisco.com/univercd/cc/td/doc/product/rtrmgmt/nfc/nfc_3_0/nfc_ug/nfcover.htm
7. Yeali S. Sun and J.-F. Lee: Policy-based QoS Management in NBEN – Differentiated Services Provisioning, *TANET'2000* (2000).
8. http://www.cisco.com/warp/public/732/Tech/netflow.
9. http://www.perl.com

A Probing Approach for Effective Distributed Resource Reservation

Lihua Yuan[1], Chen-Khong Tham[1], and Akkihebbal L. Ananda[2]

[1] Department of Electrical and Computer Engineering
National University of Singapore, Singapore 119260
{dcsylh, eletck}@nus.edu.sg
[2] Center for Internet Research, School of Computing
National University of Singapore, Singapore 119260
ananda@comp.nus.edu.sg

Abstract. Resource reservation is a essential component in providing QoS guarantees to distributed multimedia applications that run over the internetwork. Early reservation systems, including both immediate reservation and advance reservation, have taken an "all-or-nothing" approach in which both QoS and temporal parameters of requests are inflexible. This paper describes a new probe-based adaptive reservation approach using *Probing Requests* that exploits the potential flexibility in reservation requests to increase resource utilization and reduce rejection rates. Compared to other approaches that support flexibility, the probe-based approach is more efficient in finding alternatives, and it causes less signaling overhead and incurs a lower computational load on the often busy resource providers. The *Probing Request* mechanism has been implemented as part of our inter-domain bandwidth broker, can also be applied to other resource reservation schemes.

Key words: Flexible Resource Reservation, Adaptive Reservation, Resource Probing, Scalable QoS Architecture

1 Introduction

Multimedia applications like video conferencing, Video on Demand (VoD) and IP Telephony are becoming increasingly popular. These applications need their particular QoS requirements to be met in order to work properly. In order to meet these QoS requirements, the availability of certain amounts of different resources (bandwidth, CPU time, buffer space etc.) from the various involved parties (the network, sender and receiver) needs to be assured during the lifetime of the application. In both the earlier Integrated Services model and the recent Differentiated Services model, resource reservation is crucial in ensuring this resource availability.

Early works [1, 2, 3] on resource reservation mostly focused on reservations for immediate usage (known as "immediate reservation") which lasts for an unspecified amount of time. However, for applications that need a large amount of

M. Ajmone Marsan et al. (Eds.): QoS-IP 2003, LNCS 2601, pp. 672–688, 2003.

resources or that involve multiple resource users or providers, the probability of a successful immediate reservation might be unacceptably low. It is also undesirable for applications that involve many parties to be informed of the reservation failure at the last moment. Therefore, the capability of making reservation in advance is both desirable and necessary. In various advance reservation schemes [3, 4, 5, 6], in addition to the QoS parameters, requestors also indicate the desired temporal parameters, i.e. *start time* and *duration*, to resource providers. Providers have to ensure all the QoS and temporal parameters can be met in order to admit an advance reservation. In other words, resource providers have to ensure the availability of a certain amount of resources over the entire requested period.

In early works on reservation systems, including both immediate reservation and advance reservation, requestors either get exactly what is requested or nothing at all. This "all-or-nothing" approach implies that the QoS parameters and temporal parameters are all hard and inflexible[1]. However, not all QoS parameters and temporal parameters are hard and inflexible. Some reservations could be flexible in the temporal domain parameters (start time, duration). For example, instead of accepting a reservation failure, a VoD user may be willing to wait for an extra 5 minutes, or a video conference may be willing to shorten itself a bit. Other reservations could be flexible with their QoS requirements, including almost every aspect of QoS, e.g. throughput, delay, jitter and loss etc. For example, a VoD user might be willing to watch in gray-scale mode or on a smaller screen size if bandwidth is not enough to support the full-color large-screen mode. In another example, although IP Telephony applications prefer a one-way delay less than 200ms and zero packet loss to perform satisfactorily, users can still communicate as long as the delay does not exceed 450ms [7], and most encoder/decoder can still function if the packet loss is below 30% [8].

In brief, many applications or users are flexible in certain aspects. If the reservation system can help to find out *suitable alternatives* for an originally rejected reservation, the requestors' chance of success is increased, and the providers' resources are better utilized. Naturally, one would expect higher resource utilization and lower rejection rate in a reservation system that supports flexibility. In our approach, requestors use a "probing request" (P_r) which has the requestor's flexibility information to probe the network for possible support. On its way toward the destination, each resource provider check the P_r against its own resource availability and reduces P_r accordingly. When P_r reaches its destination, the resource availability information will be discovered.

Section 2 reviews related works on resource reservation. Section 3 presents the probe-based algorithm in detail. Section 4 studies the performance of the probe-based reservation system. Effective in improving resource utilization and reducing rejection rate, the probe-based algorithm is also faster in making reser-

[1] It is possible for users to find an alternative reservation by using the "trial-and-error" approach, which we discuss later. But for individual requests, there is no explicit flexibility support.

vation and incurs a lower signaling and computational overheads, as compared to other approaches. The last section presents the conclusion and future work.

2 Related Work

Ferrari and the Tenet group [4] were among the first to realise the need for advance reservation for real-time connections. The design and mechanism were proposed as part of the Tenet Real-Time Protocol Suite 2 [3]. In [4], resources are dynamically partitioned into "immediate partition" and "advance partition". For "advance partition", reservation information was kept in a variable-length "interval table" and reservation change was only allowed to take place at the edge of a "time granule". They noted that when rejecting a request, the provider should notify the client of "what changes to what parameters (including the start time and the duration)" might be acceptable.

Wolf *et. al* [9] proposed a general model for advance reservation. Their later work [5] provided a detailed study on the temporal sequence of advance reservation events and a state transition diagram. Similar to the Tenet approach, resources for ReRA and Non-ReRA[2] reservation were partitioned with a moving boundary [5]. The time axis was divided into slices.

Greenberg *et. al* [10], by using a 2-D Markov-chain model, mathematically proved that resource sharing between Immediate Reservation (IR) and Book-Ahead (BA) calls can bring more revenue through higher resource utilization than partitioning the bandwidth. Following their study, IR and BA calls share the same pool of resources in our bandwidth broker (BB) system [11], and the only difference between them is that IR calls have a book-ahead time equals to zero.

Schelèn *et. al* [12, 13] proposed an agent-based reservation architecture. In [13], Schelèn proposed a binary search tree (BST) data structure to support continuous time reservations and a segment tree structure to support slotted time reservations. He compared the performance of these two data structures in terms of its memory usage and admission decision speed. Our study in [11] shows that continuous time reservation outperforms slotted time reservation in terms of resource utilization and rejection rate. Therefore, our implementation of the probing request algorithm implements an AVL-tree, which is an enhanced BST where the height of the two subtrees of a node differs at most by one [14], as the baseline data structure.

The above mentioned works, together with many others [1, 2, 6], have *No Explicit Flexibility Support* (NEFS). If an application or user is flexible with certain parameters, it has to use a "trial-and-error" approach to find a suitable

[2] Wolf *et. al* [9] named reservation in advance as "Resource Reservation in Advance (ReRA)" and reservation for immediate usage as "Non-ReRA". Greenberg *et. al* [10] called the former "Book-Ahead" (BA) calls and the latter "Instantaneous-Reservation" (IR) calls. Schelèn *et. al* called them "Advance Reservation" and "Immediate Reservation" respectively. Despite the difference in names, their definitions are the same. This paper follows the naming convention used by Schelèn *et. al*

set of reservation characteristics. However, since requestors generally have no information about the downstream resource availability, it is often difficult for requestors to find suitable reservation characteristics even if they themselves are flexible enough. The best that the requestors can do is to continuously try based on guesswork and hope that the request will be accepted. It could take the requestor many trials to find a suitable alternative, if one could be found at all. Not only that, repeated trials also introduces undesirable extra signaling and computation overheads.

The QBone [15] group proposed an inter-domain bandwidth broker (BB) architecture [16], along with some initial work on the Simple Inter-domain Bandwidth Broker Signaling (SIBBS) [17] message format and the state machines. Similar to Tenet, SIBBS proposed that BB should find another acceptable set of reservation characteristics and attach it to the Resource Allocation Acknowledgment (RAA) that is being returned to the requestor when rejecting a Resource Allocation Request (RAR). The requestor can then, based on the proposal, make a retry that is guaranteed to succeed at the BB which made the proposal.

Hafid and Bochmann *et. al* [18] proposed a negotiation approach with future reservation (NAFUR) to support distributed multimedia applications. In NAFUR, the client indicates the desired QoS parameters and duration in the "ServiceInq" and the QoS manager returns "best" proposals based on the available service projection (ASP). The requestor can choose among the proposals and seal the reservation with a corresponding "ServiceReq" operation. We categorize the Tenet proposal, QBone and NAFUR as "proposal-based" approaches since they all rely on providers to make proposals if the initial reservation characteristics cannot be met. Since the resource provider making proposals usually does not have sufficient information on the requestor's flexibility, preference and other resource providers' resource availability, the proposal could be unacceptable to the requestor or other involved resource providers. If a proposal is not accepted in the end, it merely causes unnecessary signaling and computational overheads. Furthermore, if a proposal from one provider is rejected by another, requestors would still need multiple rounds of signaling to get a reservation done.

Chen and Lee [19] proposed a flexible service model for advance reservation. This model introduced a flexible interval for start time for advance reservations. The resource manager uses the time adaptation approach to determine the exact start time that incurs the minimum cost. Then the manager informs the client about this exact start time in a "Admission Confirmation" message. Through request scheduling, the system shifts some peak period workload to off-peak periods so as to achieve a better acceptance ratio. While this approach limits the flexibility to start time alone, our probing approach takes the flexibility of other parameters (duration, bandwidth, etc) into consideration. Furthermore, Chen and Lee's model let the manager, which do not have the requestor's preference information, make the final decision. Instead, our approach let either the requestors or destination requestors make the final decision.

3 The Probe-Based Mechanism

This section presents the probe-based mechanism in detail. The probing mechanism is based on a new *probing request* message, and it can be divided into three stages, namely the probing initiation stage, resource discovery stage and reservation confirmation stage.

3.1 Probing Request

The *probing request* (PR) is largely the same as a reservation request, with the only difference being that the value of the flexible requirement (e.g. start time, duration etc.) is now replaced by a *probing range* (P_r) field. If a normal reservation request R is represented as:

$$R = (T^{st}, D, Q) \tag{1}$$

where T^{st} is the start time of the reservation, D is the duration of the reservation and Q is the QoS requirement[3], then a probing request looking for possible start time range from T^{st}_{min} to T^{st}_{max} would be represented as:

$$\begin{aligned} PR &= (P^{st}_r, D, Q) \\ \text{with } P^{st}_r &= (T^{st}_{min}, T^{st}_{max}) \end{aligned} \tag{2}$$

and a probing request looking for a possible duration range from D_{min} to D_{max} can be represented as:

$$\begin{aligned} PR &= (T^{st}, P^{D}_r, Q) \\ \text{with } P^{D}_r &= (D_{min}, D_{max}) \end{aligned} \tag{3}$$

Similarly, a probing request for a certain QoS parameter[4] is:

$$\begin{aligned} PR &= (T^{st}, D, P^{Q}_r) \\ \text{with } P^{Q}_r &= (Q_{min}, Q_{max}) \end{aligned} \tag{4}$$

The *Probing Range* is the flexible requirement range in which the requestor is probing the network for possible support. It can also be interpreted as the acceptable range of the flexible parameter of the requestor. Therefore, resource providers receiving a PR will know which parameter is flexible and what is its flexible range.

[3] Q can in turn be represented as a $< B, D, J, L >$ where B, D, J and L represent bandwidth, delay, jitter and loss respectively.

[4] Depending on which QoS parameter is actually flexible, P^{Q}_r should be $< B_{min}, B_{max}, D, J, L >$ or $< B, D_{min}, D_{max}, J, L >$ and so on. In this paper, we summarize them under one notion to avoid the lengthy representation.

3.2 Probing Initiation

The probing mechanism is initiated by the requestor choosing what is the flexible requirement and the flexible range. For example, a VoD user may decide that the start time is flexible and it can vary from 9 : 00am to 10 : 00am. Thus, the flexible parameter is the start time and its flexible range (hence the probing range) is $P_r^{st} = (Ts_{min}, Ts_{max}) = (9:00\text{am}, 10:00\text{am})$. The requestor then constructs a probing request message based on this P_r, and other inflexible requirements are set in the same way as a normal reservation request. The probing request PR is then sent to the downstream BB to initiate a probing session.

3.3 Resource Discovery

Each resource provider receiving the probing request checks this request against its own resource availability. If the provider is flexible enough to support the request with any start time in the P_r^{st}, the probing request is forwarded downstream with unmodified P_r^{st}. If the provider can only support the request with a subset of P_r^{st}, where $P_{r'}^{st} \subseteq P_r^{st}$, the P_r^{st} of the probing request is reduced to $P_{r'}^{st}$ and forwarded down. Otherwise, if $P_{r'}^{st} = \emptyset$, the probing request is considered to have failed and a failed "Probing Reply" message is sent back to the requestor. This process is repeated at every involved BB until the probing request reaches the destination. If the probing request reaches the destination with a non-empty $P_{r'}^{st}$ (which may be different from the original P_r^{st}), the range that can be supported by the network would have been discovered.

For example, if provider A receives the PR described in Sec.3.2, it will check whether it can support Premium Service bandwidth 2Mbps lasting for 2 hours with any start time withing the range of 9 : 00am to 10 : 00am. If the answer is yes, PR is forwarded downstream with P_r^{st} unmodified. Otherwise, if provider A finds that it can only support $P_{r'}^{st} = (9:30\text{am}, 10:00\text{am})$, the probing range is modified accordingly and the modified PR is forwarded downstream. In the worst case, if provider A finds it cannot support the request with any start time from 9 : 00am to 10 : 00am, a failed "Probing Reply" message is sent back to the requestor. The probing request could reach the destination with $P_{r'}^{st} = (9:40\text{am}, 9:45\text{am})$, this information indicates that the network can only support start time range from 9 : 40am to 9 : 45am.
[Notes:] If a probing request has indicated itself as a "reservation probe", all resources successfully probed in the P_r will be reserved for a short period of time until a "Probe Holding Timer" expires. This will guarantee resource availability before the follow-up action is taken. If a probing request has indicated itself as "reservation probe" but did not make the reservation, a penalty could be imposed on the requestor. A probing request can also indicate itself as a "discovery probe", meaning it only serve as a dynamic resource discovery protocol and the requestor is not affirmative about a follow-up reservation even if the probing succeeds. In this case, no resource will be reserved for the probing request.

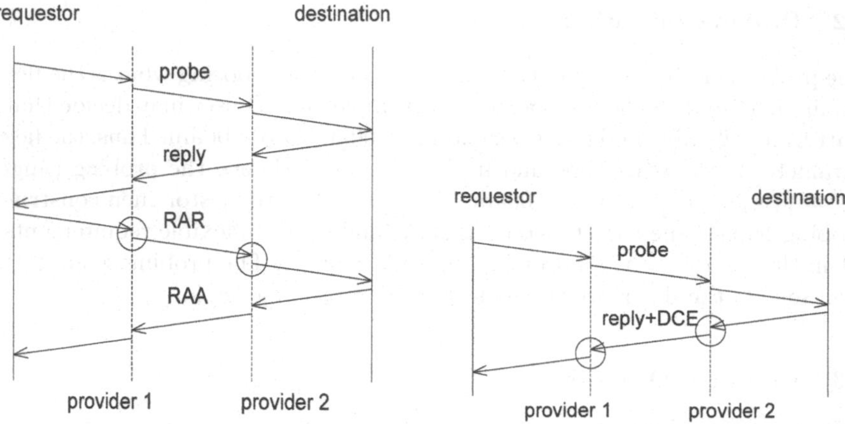

Fig. 1. Probe with Follow-up Reservation

Fig. 2. Probe with DCE

3.4 Reservation Confirmation

At this stage, a range of reservation characteristics acceptable to both requestor and the network have been discovered and reached the destination. The system now needs to choose among the available choices and finalize the reservation. Figures 1 and 2 depict two methods we have developed to finalize the reservation. We explain them in detail below.

Probe with Follow-Up Reservation. If the destination cannot confirm the reservation, the information discovered in "resource discovery" stage would be passed back to the requestor in a "Probing Reply" message. The requestor would make the final choice according to the reduced P_r and send a corresponding reservation request. If the original "Probing Request" is a "Reservation Probe", providers would have reserved all the successfully requested resources in the probing process. Therefore, upon receiving a follow-up reservation request, only the chosen resources will be kept and other resources will be released. In this case, the reservation request is guaranteed to succeed unless the "Probing Holding Timer" had expired. If the original *probing request* is a "Discovery Probe", no resource has been reserved. Therefore, the follow-up reservation request is treated in the same way as a new reservation request and is subject to failure.

Probe with Destination Confirmation Extension (Probe+DCE) If, however, the destination can decide what is the desirable reservation characteristics, the Destination Confirmation Extension (DCE) will be used. In DCE, the destination will choose from the P_r for a final reservation characteristics. The chosen characteristics will then be put into the "Probing Reply" message with the DCE flag set. Providers receiving a "Probing Reply" message with

DCE flag will then release the unrequested resources and keep only the chosen resources. When the "Probing Reply" reaches the requestor, the reservation has been successfully made.

Compared to probe with follow-up reservation, probe with DCE is faster in getting reservation done, as illustrated in Fig.2. It also incurs less signaling overheads and releases faster the resources which are not needed. However, if the destination does not have the requestor's preference information, DCE may end up making sub-optimal reservations. Therefore, if the destination does not have enough information about user's preference, yet it still wants to avoid sub-optimal reservations, the destination should avoid confirming the reservation and leave the final confirmation to the requestor.

4 Performance Studies and Discussions

This section studies the performance of the probe-based reservation system by comparing it with systems with no explicit flexibility support (NEFS) and systems using a proposal-based approach. For any reservation system, the primary performance metrics are resource utilization and rejection rate. Time-To-Reservation (TTR), which indicates the responsiveness of the system, and signaling and computational overheads, which determines the implementation cost and scalability, are also important metrics.

4.1 Simulation Implementation and Topology

The resource reservation system is implemented in NS-2 [20]. The resource requestors and providers are implemented as bandwidth brokers ("Application/BB") that run over the "SimpleTcp" agent. Bandwith broker with request generators ("Application/RARGen") attached can generate both reservation requests and probing requests. Resource requestors and providers can run in *NEFS* mode which simulates a system that does not support flexibility. In this mode, requestors will simply give up after a rejection. Requestors can also run in *NEFS-R* mode, in which requestors will modify its reservation characteristics and retry if being rejected. This is an implicit way of supporting flexibility, and it does not need any support from the network. The *probe* and *probe-dce* modes implement the probing algorithm with follow-up reservation and with DCE respectively. As a comparison, we have also implemented a *proposal* mode using the proposal-based approach proposed by a few [4, 16, 18]. In the *proposal* mode, a provider, when rejecting a reservation request, also make a proposal about a suitable set of reservation characteristics. The requestor can then retry based on the proposal. In both *NEFS-R* and *proposal* mode, requestors will not retry more than 10 rounds for any single request.

Figure 3 depicts the simulation topology. S1 to S5 are requestors that send reservation requests to D1 to D5 correspondingly. M1 to M4 are the resource providers. All the links have a 100ms delay and 10Mbps bandwidth. Requests from S1 need to compete with requests from other four requestors for resources.

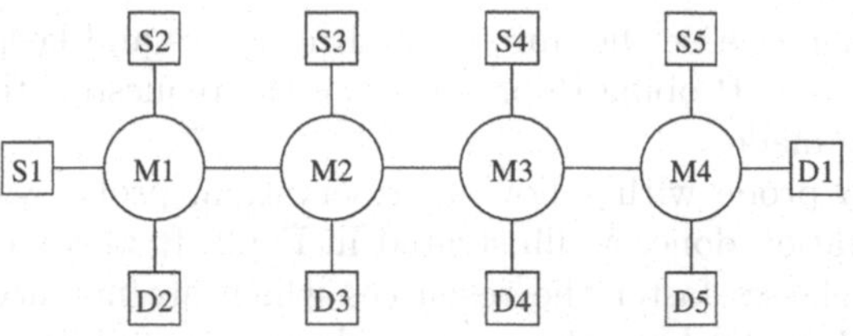

Fig. 3. Simulation Topology

These requests need to survive independent admission tests from M1 to M4 to before they can get a successful reservation. This is a close simulation of the inter-domain resource reservation scenario. All the five requestors send requests according to a Poisson process at rate λ, with different random number generator seeds. The book-ahead time of the requests are all uniformly distributed from 0 to 100 units and the amount of resources required and the duration length are both normally distributed with the mean set to 10 units and the variation equals to 3 units. Resource available at each provider is 100.

4.2 Resource Utilization and Rejection Rate

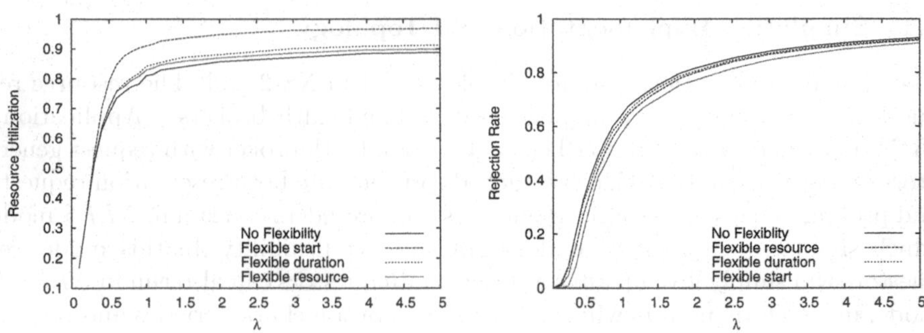

Fig. 4. Resource Utilization

Fig. 5. Rejection Rate

Figures 4 and 5 show that supporting flexibility in resource reservation can effectively improve resource utilization and reduce rejection rate. In the simulation, we assume that all users' flexible range for start time is at most 20 units later than the original desired start time. The flexible range for duration and bandwidth are assumed to be at least 50% and 10% of the original respectively. In real environment, users' flexible ranges might be very different from each other. Therefore, the actual improvement varies depend on the flexible range.

The comparisons show that all three mechanisms that supports flexibility, *NEFS-R*, *proposal* and *probe*, get similar improvements on resource utilization and rejection rate. This is reasonable because it is the *flexibility* itself, instead of the flexibility support mechanism, that improves the resource utilization and rejection rate. However, the probe-based mechanism is more efficient in supporting flexibility, as we will show through the following comparison on Time-To-Reserve, signaling overheads and computational load.

4.3 Time-To-Reserve (TTR)

Time-to-reservation (TTR) is defined as the time interval start from the moment that the requestor sends out a reservation request until the results is confirmed. Since the results can be either success or fail, we further define time-to-reservation-success (TTRS) and time-to-reservation-failure (TTRF) to be the time to get a positive result and negative result respectively. Ideally, a reservation system should not only admit every possible request, but also admit them fast. If it is impossible to support a request, the system should also reject it fast.

System with No Explicit Flexibility Support (NEFS). Assuming a traditional NEFS system without user or application initiated retry, every request will reach its conclusion after one rounds of signaling. A successful reservation needs to survive every admission test on the path until it reaches the destination and come back. Therefore, TTRS normally equals to the round-trip-time. On the other hand, a rejected request could be rejected by any single provider on the path. Consequently, a rejected request do not need to travel to the destination to know its fate. Therefore, TTRF would be smaller than the round-trip-time.

But even in system without explicit flexibility support, requestors can still decide whether they want to try another reservation with modified reservation characteristics or simply give up. In fact, if retried reservations have a reasonable probability of success, requestors will be motivated to retry by the incentive of potential successful reservation. However, since requestors generally have no information about the resource availability of downstream resource providers, it is often difficult for requestors to find suitable reservation characteristics even if they themselves are flexible enough. The best that the requestors can do is to continuously try based on guesswork and hope that the reservation will be accepted. Therefore, in NEFS with user or application retry (NEFS-R), a request might reach its success after several rounds of trial-and-error. The eventual failure could only be reached after the requestor gives up, either because of the requestor has tried every acceptable reservation or simply lost the patience. In this case, the many rounds of signaling significantly increase TTRS and TTRF, as shown in Fig.8. and Fig.9. In addition, every round of signaling adds to the system extra signaling overheads and computational burden.

System Using Proposal-Based Approach. In proposal-based approaches, providers try to guide to requestors on what could be a feasible reservation by making "proposals". However, when a resource provider is making a proposal, it only considers the characteristics indicated in the request and the the resource

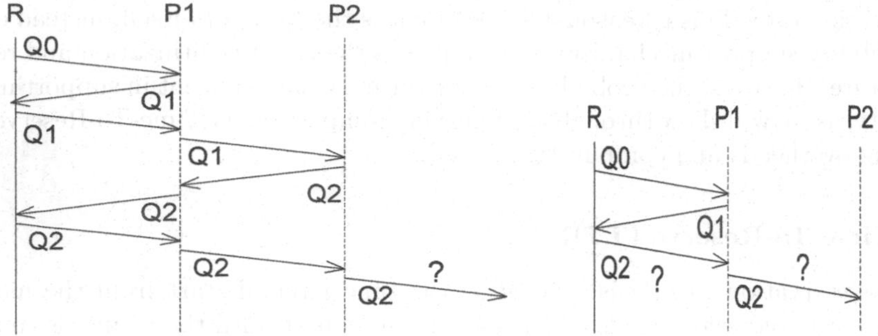

Fig. 6. Rejected by Another Provider

Fig. 7. Rejected by Requestor

usage information of the provider itself. Without considering other providers' resource availability and requestors' limitation, the proposal from one provider could be rejected by another provider on the path or the requestor. If a proposal is rejected, the requestor would have to return to the "trial-and-error" approach or give up. Again, multiple rounds of signaling brings more signaling and computational overheads and increases TTRS and TTRF.

Figure 6 depicts a typical scenario where proposal from one provider rejected by another leading to multiple rounds of signaling. Reqeustor R initially sends a request Q_0. Provider P_1 receiving this request cannot satisfy Q_0 and proposes Q_1, say, with another start time. R receives proposal Q_1 and request accordingly. However, it is rejected by provider P_2 and another round of proposal/re-negotiation begins. Actually, Q_2 could also be rejected at provider P_1 or other downstream providers. Requestor R may eventually get a successful reservation after many rounds of proposal/re-negotiation process. But, it is also possible that after several rounds, one provider has to reject the request and cannot give any proposal, possibly because it is fully reserved. In the worst case, if the resource availability at different providers does not overlap at all, the requestor will never get a success despite the different proposals coming from the different providers.

Figure 7 depicts another scenario that proposals unacceptable to the requestor leads to multiple rounds of signaling. Same as the previous scenario, provider P_1 propose Q_1 after the original request Q_0 is rejected. However, Q_1 is unacceptable the requestor R, and therefore, R has to use the trial-and-error approach. A proposal might be unacceptable to requestor for two reasons. Firstly, it could be the proposed change of the flexible parameter is outside the acceptable range. For example, the proposed start time is 2 hours later while the user can only wait for 30 minutes. Another possibility is that the provider is assuming a flexible parameter in making the proposal but that parameter is not flexible to the requestor at all. For example, if a video conference is very strict about the start time but is flexible on quality (bandwidth etc.), then the proposal for a later start time will not be suitable to the requestor.

In both the two scenarios described previously, proposal-based approach cannot effectively eliminate the need for multiple rounds of signaling. Same to the NEFS-R system, the consequent increase in TTR value, signaling and computational overheads are undesirable.

In [18], Hafid *et. al* proposed for one central QoS manager to have the usage information of other QoS components so that it can consolidate the resource usage information and make a proposal which will be acceptable to all involved components. This consolidated proposal can eliminate multiple rounds of signaling if the request only involves resources from a single domain and the central manager has the information of every components. For reservations span multiple administrative domains, making a consolidated proposal will be difficult since sharing resource availability information across multiple domains is not realistic. Also, this approach needs a central manager to perform the consolidation. This is unscalable if a large number of proposals need to be offered. In addition, this consolidation algorithm does not consider requestors' limit. Consequently, it can not eliminate the multiple rounds of signaling described in the second scenario.

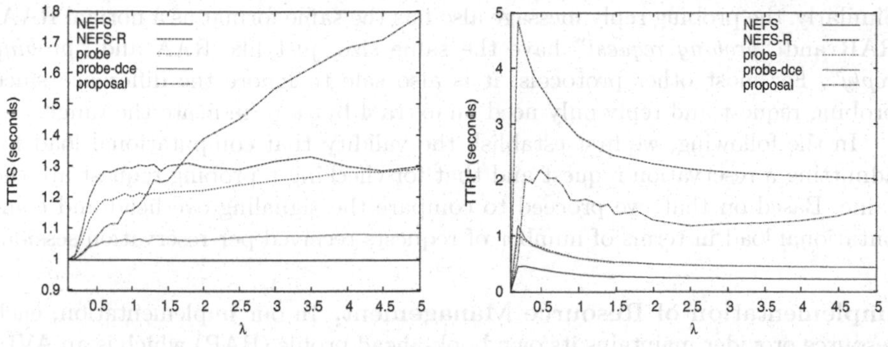

Fig. 8. Time To Reservation Success **Fig. 9.** Time To Reservation Failure

Comparison of TTR. Compared to NEFS-R and proposal-based systems, our probing approach effectively eliminates the need for multiple rounds of signaling. If the probe is successful, the requestor can choose the most preferable characteristics. Otherwise, since the whole flexible range has been searched, requestors have no incentive to retry the same flexible range. Figure 8 shows that the probe-based mechanism setup reservation faster. Destination confirmation extension (DCE) can further increase the reservation setup speed by saving from the follow-up reservation. Figure 9 shows that the probe-based mechanism can also effectively reduce TTRF. Requestors, after a failed probe, would be convinced that there is no possibility for a successful reservation and therefore, will not perform meaningless retry.

4.4 Signaling Overheads and Computational Load

Computational load for a reservation system depends not only on how many reservation requests the system needs to process, but also on how efficient the resource management and admission control algorithm are. For reservation systems that support flexibility, the computational load also depends on the complexity of making a proposal or checking a probing request and the number of proposals made or probing request checked. Here, we discuss our implementation of resource management, which determines the complexity of admitting a new request, making a proposal or checking a probing request. We show that in our implementation, the complexities of the three are almost the same. Therefore, the computational load can be compared on the basis of number of requests per reservation session, including both reservation requests and probing requests, received at resource providers.

Signaling overheads can be compared in a similar fashion if the difference between the size of probing request and that of normal reservation request can be ignored. In fact, in our implementation, both probing request and normal reservation request use the SIBBS [17] RAR message format, the only difference being that probing request use the originally undefined "experimental" field. Similarly, the probing reply message also has the same format as a normal RAA. RAR and "*probing request*" have the same size, just like RAA and "*probing reply*". For most other protocols, it is also safe to ignore the difference since probing request and reply only need an extra 4 bytes to indicate the range.

In the following, we first establish the validity that computational load for admitting a reservation request and that for checking a probing request are the same. Based on that, we proceed to compare the signaling overhead and computational load in terms of number of requests received per reservation session.

Implementation of Resource Management. In our implementation, each resource provider maintains its own book-ahead profile (BAP) which is an AVL-tree [14], a variant of the binary search tree (BST), sorted by time. For each reservation request, one start time entry (SAT) and one stop time entry (SOT) are created and inserted into the tree. The SAT entry presents an increment of resource usage (a positive delta) and the SOT entry presents a corresponding decrement (a negative delta). In every entry, the sum of the left subtree is also kept which indicates the total resource reservation before this entry. Therefore, an in-order traversal from the start time to the stop time in the tree can effectively find out the resource usage information.

This approach is based on the BST-based approach proposed by Schelèn [13]. But an AVL-tree offers additional advantage that it is always close to balance. A normal BST will degenerate and lose its $O(log(n))$ search property if the order of inserting entries is not fully random. If entries are pre-sorted, a binary search tree will degenerate to a linked list, which will require $O(n)$ search. In advance reservation systems, although incoming requests are not exactly pre-sorted, they are inclined to have a *sorted order*. Consequently, a normal BST will lose its optimality, and the height of the tree (which determines the number of comparisons needed to locate an entry) will increase quickly. A regular "rebalance" will be needed for normal BST to keep it close to the optimal state. As a comparison, "an AVL-tree will never be more than about 50% taller than the corresponding optimally balanced tree" [21]. Actually, the height of the two subtrees of a node differs at most by one [14]. This ensures that our book-ahead profile (BAP) based on AVL-tree has a good performance even in the worst case condition. We have used the GNU *libavl* version 2.0 [21, 22] implementation for AVL-tree as the base data structure.

Complexity of Single Requests. As discussed in [13], admission of a new request requires an "insert" operation for the start time entry (SAT) and a travel from the SAT to the appropriate location of the stop time entry (SOT). The "insert" operation has a complexity of $O(log(n))$ where n is the number of entries in the BAP. The travel has a complexity of $O(m)$ where m is the number of entries falling in between SAT and SOT.

The processing for probing requests works very much the same as admission of new reservation requests. After receiving a probing request P_r, the provider constructs one earliest possible start time entry (SAT') and another latest possible stop time entry (SOT'). The insertion of SAT' and traveling to SOT' have the same complexity as normal admission process. In the case of start time probing, the travel will have a larger complexity of $O(m+n)$ where m is the number of entries that fall in between T_{max} and $T_{max} + Dur$ and n is the number of entries that falls in between T_{min} and T_{max}.

Therefore, we conclude that the complexities of admitting a normal request or a probing request are the same except for the start time probing. Even for start time probing, these complexities are still very close if the duration of the request is significantly larger than the flexible start time range.

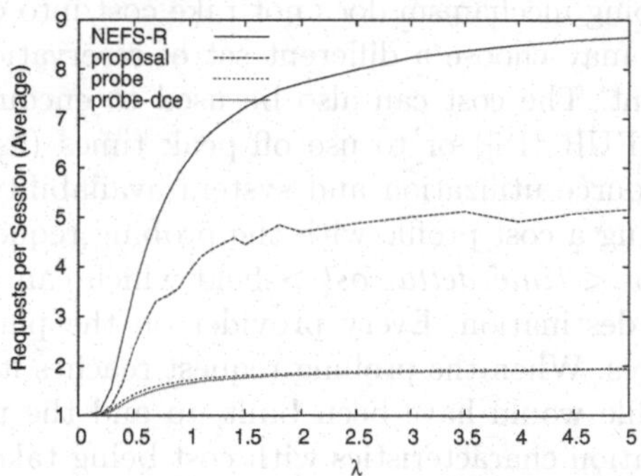

Fig. 10. Number of Requests Per Session

Comparison of Number of Requests per Session. Figure 10 presents the average number of requests received from S1 at M1 per session, which corresponds to the computational load and signaling overheads per session. Figure 10 shows that*NEFS-R* is inefficient since requestors retry blindly. The proposal-based approach reduce the number of requests, since, by making proposals, the reservation system is guiding the retry requests. However, without the resource availability information of other involved providers and the requestors' flexibility, the guidance itself is made blindly. The probe-based approach, by effectively discovering the resource availability in one round of probing, reduces the signaling overheads and computational load to a minimum.

5 Conclusion and Future Work

In this paper, we have proposed a probe-based approach that increases the system resource utilization and reduces the rejection rate by exploiting the flexibility in reservation request. Compared to the provider-initiated proposal-based approach, the requestor-initiated probe-based approach is more efficient in finding a suitable set of reservation characteristics. We have implemented this algorithm as part of our work on BB, and it should be relative easy to applied the same approach to other resource reservation systems.

The probing request mechanism currently supports the flexibility of only one parameter. It is possible that some requests are flexible with multiple parameters at the same time. For example, some VoD users may be willing to wait longer and are also willing to accept a lower quality. In the current probe-based system, such users would have to probe for one parameter at a time. Allowing probing for multiple parameters at the same time would further improve the efficiency of the reservation system. However, in a system with multi-dimensional flexibilities, one flexible parameters may conflict with other flexible parameters. Therefore, a conflict resolution mechanism needs to be developed to handle the potential conflicts. One possibility is to apply the knowledge in two-dimensional space filling algorithms in our proposed scheme.

The current probing mechanism does not take cost into consideration. However, the requestors may choose a different set of reservation characteristics if the costs are different. The cost can also be used to encourage users to share resources (as in NAFUR [18]) or to use off-peak times (as in [19]). This will in turn improve resource utilization and system availability. We are currently working on associating a cost profile with the probing request. The cost profile is essentially a list of $< time, delta\_cost >$ field which can be concatenated on its way toward the destination. Every provider on the path adds to the cost profile its own charges. When the probing request reaches its destination, a full end-to-end cost profile would have been built up and the requestor can select the preferred reservation characteristics with cost being taken into account.

References

[1] L. Delgrossi and L. Berger Eds., "Internet Stream Protocol Version 2 (ST2) Protocol Specification - Version ST2+," IETF Request for Comments 1819, Aug 1995.

[2] R. Braden Ed., L. Zhang, S. Berson, S. Herzog, and S. Jamin, "Resource ReSerVation Protocol (RSVP) – Version 1 Functional Specification," IETF Request for Comments 2205, September 1997.
[3] A. Banerjea, D. Ferrari, B. A. Mah, M. Moran, D. C. Verma, and H. Zhang, "The Tenet Real-Time Protocol Suite: Design, Implementation, and Experiences," *IEEE/ACM Transactions on Networking*, vol. 4, no. 1, pp. 1–10, Feb 1996.
[4] D. Ferrari, A. Gupta, and G. Ventre, "Distributed Advance Reservation of Real-Time Connections," in *Fifth International Workshop on Network and Operating System Support for Digital Audio and Video*, Durham, NH, USA, Apr 1995.
[5] L. C. Wolf and R. Steinmetz, "Concepts for Resource Reservation in Advance," Special Issue of *Journal of Multimedia Tools and Applications, The State of the Art in Multimedia Computing*, May 1997.
[6] W. Reinhardt, "Advance Resource Reservation and its Impact on Reservation Protocols," in *Proceedings of Broadband Island'95*, Dublin, Ireland, September 1995.
[7] A. Percy, "Understanding Latency in IP Telephony," Tech. Rep., Brooktrout Technology, Feb 1999.
[8] A. Bouch, M. A. Sasse, and H. de Meer, "Of packets and people: A User-centered Approach to Quality of Service," in *Proceeding of 8th International Workshop on Quality of Service*, Jun 2000, pp. 189–197.
[9] L. C. Wolf, L. Delgrossi, R. Steinmetz, S. Schaller, and H. Wittig, "Issues of reserving resources in advance," in *Proceedings of 5th International Workshop on Network and Operating Systems Support for Digital Audio and Video (NOSSDAV'95)*, Durham, NH, Apr 1995, pp. 27–37, 18–21.
[10] A. G. Greenberg, R. Srikant, and W. Whitt, "Resource Sharing for Book-Ahead and Instantaneous-Request Calls," *IEEE/ACM Transactions on Networking*, vol. 7, no. 1, Feb 1999.
[11] L. Yuan, "Bandwidth broker for advance reservation in diffserv networks," M.S. thesis, National University of Singapore, 2002, Available at http://bronco.sytes.net/research/my/master-thesis.pdf.
[12] Olov Schelèn and Stephen Pink, "Resource Sharing in Advance Reservation Agents," *Journal of High Speed Networks, Special issue on Multimedia Networking*, vol. 7, no. 3-4, 1998.
[13] O. Schelèn, A. Nilsson, J. Norrgård, and S. Pink, "Performance of QoS Agents for Provisioning Network Resources," in *Proceedings of IFIP Seventh International Workshop on QoS (IWQoS)*, Jun 1999.
[14] D. E. Knuth, *The Art of Computer Programming, Volume 3: Sorting and Searching*, Addison-Wesley, 2nd edition, 1998.
[15] B. Teitelbaum, "QBone Architecture Version 1.0," Tech. Rep., Internet2 QoS Working Group, Aug 1999.
[16] B. Teitelbaum and P. Chimento, "QBone Bandwidth Broker Architecture — Work in Progress," Tech. Rep., Internet2 QoS Working Group, Jun 2000.
[17] QBone Working Group, "QBone Signaling Design Team Charter," http://qbone.internet2.edu/bb.
[18] A. Hafid, G. von Bochmann, and R. Dssouli, "A Quality of Service Negotiation Approach with Future Reservations (NAFUR): A Detailed Study," *Computer Networks and ISDN Systems*, vol. 30, pp. 777–794, 1998.
[19] Y. T. Chen and K. H. Lee, "A Flexible Service Model for Advance Reservation," *Computer Networks*, vol. 37, pp. 251–262, 2001.

[20] NS, "The Network Simulator - ns-2," http://www.isi.edu/nsnam/ns.
[21] B. Pfaff, "An Introduction to Binary Search Trees and Balanced Trees," http://www.msu.edu/user/pfaffben/avl/.
[22] B. Pfaff, "GNU libavl," http://www.msu.edu/user/pfaffben/avl/.

Dynamic Adaptation of Virtual Network Capacity for Deterministic Service Guarantees

Stephan Recker[1], Heinz Lüdiger[1], and Walter Geisselhardt[2]

[1] IMST GmbH, Information and Communications Systems,
Carl-Friedrich-Gauß-Str. 2, 47475 Kamp-Lintfort, Germany
{recker,luediger}@imst.de

[2] Gerhard-Mercator-University, Institute of Data Processing,
Bismarckstr. 81, 47057 Duisburg, Germany
gd@uni-duisburg.de

Abstract. We consider the case of a virtual data network comprised of a set of end-to-end virtual leased lines. Our work aims at tackling the aspect of appropriate dynamic dimensioning of end-to-end paths subject to minimizing the consumed resources under the constraint of providing deterministic QoS guarantees. In particular, we propose a novel application of fuzzy logic control in order to dynamically adjust the resources assigned to one path to variations of traffic traversing that path. We design the required fuzzy sets and the respective rule base of the controller and evaluate its asymptotic stability and performance. As controller input a measure of the traffic variation shall be averaged over a variable size window. We design a second fuzzy logic controller in order to adapt this averaging window, such that significant traffic variations are captured while the measurement overhead during periods of nearly time invariant traffic characteristics is reduced.

1 Introduction

The role of virtual networks comprised of a set of virtual leased lines may significantly change in future data networks. While in today's packet-switching networks we find virtual network structures most commonly to interconnect non-adjacent branches of a certain institution, the administrative instances of a future virtual network may act as capacity wholesaler, i.e. network transport resources are booked over medium to long periods and in turn partitioned for sale over medium to short periods [1]. The reason for the potential increase of the technical and economic significance of the role of a capacity wholesaler lies within the inevitable evolution from a single service to a multi service environment, where services are distinguished by the QoS perceived by the respective packets.

1.1 Virtual Domains and User-Centric Quality of Service

Quality of Service is defined in [2] as "collective effect of service performances which determine the degree of satisfaction of a user of the service" and [3] distinguishes between QoS and *Network Performance* (NP). NP parameters ultimately

M. Ajmone Marsan et al. (Eds.): QoS-IP 2003, LNCS 2601, pp. 689–703, 2003.

determine the user observed QoS, but do not necessarily describe it in a way meaningful to users. In our work we generally distinguish between *user-centric QoS* (UC-QoS) and *network-centric QoS* (NC-QoS). The former is stipulated by the customer demanding for a network transport service, either from a set of network specific QoS classes or arbitrarily defined, and can be denoted as time invariant, while the latter is depending on the current network state and thus may vary over time. In context of [3], NP parameters associated with UC-QoS are guaranteed and time invariant, while NP parameters of one NC-QoS class depend on the current state of the network and are meaningful in comparison with those of other NC-QoS classes, i.e. relative service differentiation among NC-QoS classes.

Most commonly a user contracts to one provider, which is in turn handling all issues related to the entire end-to-end service. This is denoted in [4] as *One-stop Responsibility* concept. In order to achieve end-to-end connectivity, nested bi-lateral *Service Level Agreements* (SLAs) are frequently used, since multi-lateral SLAs only rarely work. Nested SLAs correspond to the case where the access provider in the retail market is subscribed to another network service provider providing one stop responsibility. This concept is also applicable for provision of end-to-end connectivity for NC-QoS classes, where one nested SLA construct is established for each NC-QoS class. But UC-QoS classes provided by one ISP are in general not compliant with any UC-QoS class provided by other ISPs. Therefore for each UC-QoS class one ISP needs to maintain bi-lateral SLAs with all destination ISPs and potential intermediate providers. This clearly leads to the problem of appropriate selection of the spatial and qualitative granularity, i.e. how many UC-QoS classes are provided to which destinations. On the scale of nation-wide or even world-wide network infrastructures, end-to-end provision of a broad range of UC-QoS service seems to be questionable. And even if technically possible, the establishment of all required SLAs for all user and network-centric QoS classes can be considered as a non-negligible market entry barrier.

Dealing with the question, whether we can avoid the obligation of one ISP[1] to ensure end-to-end connectivity for all QoS classes to all global destinations, we want to consider domains specialized in a restricted set of QoS classes and providing services only in specific areas. Such domains may not necessarily deploy own physical network infrastructure. Consequently, we generally include the case of renting capacity from other ISPs and denote such domains as being *virtual*. The operator of a virtual domain shall have full control over the capacity provided by a virtual link, i.e. the capacity can be partitioned arbitrarily to forward data of different flows.

This in fact adds a hierarchical layer of virtual domains on top of existing administrative domains already existing today. The task of providing UC-QoS classes can now be spread over two layers, where the qualitative and spatial granularity can be adapted to the domain dimensions. In particular, large scale

[1] ISPs in this context are assumed to possess physical and geographically limited network infrastructure.

physical domains have to provide capacity only in terms of a *Virtual Leased Line* service on a long term basis. Virtual domains partition the rented resources to provide user-centric QoS on a short term basis and thus act as capacity wholesaler. Apart from the technical aspect of spreading complexity over hierarchical layers, there is also the economic aspect of risk sharing, where the virtual domains permanently book resources from physical domains for sale at their own risk.

The set of virtual domains providing specialized services in restricted geographical areas[2] may appear as set of patches, which have to be composed depending on one particular service demand characterized by the required QoS class and the source and destination geographic position. The challenging task of composing a particular end-to-end services from service segments provided by virtual domains is one objective of the IST project *Whyless.com*, which explicitly deals with a resource trading concept based on deployment of at least one brokerage instance. This brokerage instance, which has superior knowledge of the physical topology and all existing domains[3] and which provides maximum fairness to both, requesting parties and service providing domains, shall be capable to determine a feasible end-to-end arrangement with minimum cost. The brokerage instance itself can be seen as new player, as it is not necessarily bound to any network operator or application service provider. This player could run his/her own business by charging a fee for composition of end-to-end services, where the above mentioned fairness shall be ensured by mandatory deployment of algorithms certified by a trusted entity. Consequently also a number of brokerage instances in competition may exist. For further information on the whyless.com resource trading concept we refer to [5] and assume in the remainder of this paper that the individual transport services provided by specialized virtual domains can be composed to meet any end-to-end service demands.

1.2 Resource Management of Virtual Domains

The attractive feature of virtual domains to be exploited is that the booked resources can be adjusted on shorter time scales than physical resources can be adapted. The rate of resource adjustments, i.e. service contract adaptations, shall have a negotiated fixed upper bound. A virtual domain shall be comprised of end-to-end paths between domain boundaries, where a sufficient number of individual flows is aggregated into each of these paths. We exclusively deal with the case of providing deterministic guarantees for UC-QoS. Since the duration of user traffic flows is generally not aligned with the minimum time between two possible resource adjustments we need to decouple resource adaptations from birth-death events of traffic flows. Simplistic resource management schemes either use over-provisioning (e.g. peak rate allocation) or over-booking for that purpose. Stochastic measurements for resource adaptation, such as effective bandwidths

[2] Without inter-domain SLAs to achieve end-to-end service provision.

[3] Existing domains shall be those domains participating in this resource brokerage concept.

[6], can be a powerful tool to combine at least stochastic QoS guarantees and efficient resource utilization. In addition to accounting for traffic characteristics, network state information has been proposed to be used in the process of appropriate capacity dimensioning, e.g. Bouillet et al. [7] discuss off-line and on-line DiffServ SLA crafting and adaptation depending on the network load along the current route.

In our approach we deploy network calculus for deterministic bounds on network performance in combination with a novel application of *Fuzzy Logic Control* to determine the appropriate amount of capacity to be adjusted. By additionally considering the variability of traffic and using available information about the network state in terms of traffic load we assess important indicators of increased risk of QoS violations and of network congestion respectively. Fuzzy logic is employed as an appropriate tool to derive a decision from this set of input variables, while being robust and flexible to accomodate also for individual operator strategies. Thus in case of high variation of incoming traffic with the current network performance being close to the guaranteed bounds higher capacity could be allocated, while in situations of low traffic variations but high traffic load in the network the assigned capacity could be limited to a reasonable minimum.

Our approach intends to be an alternative to resource allocation based on stochastic traffic measurement or conservative allocation proportional to previously defined or measured peak rates. One advantage of our approach is the more easily configurable trade-off of more deterministic performance guarantees for less efficient resource utilization, which can even be varied over time. This attractrive feature comes at the expense of a more complex design and implementation phase and the lack of a general rule of thumb applicable in all networks for all operators. The required information about the variation of the traversing traffic shall be obtained by means of on-line measurements[4]. Furthermore we anticipate availability of state information about all network links, which is a reasonable assumption in light of already existing link state routing protocols [8].

1.3 Outline

In the following section we briefly introduce the theory for QoS performance prediction of an end-to-end path to be performed in the head-end nodes. In section 3 we present the designed controller, including plant model and configuration parameters. In particular, we describe the fuzzy membership functions and the rule base in terms of surface plots. The stability and performance evaluation is described afterwards in section 4, where we present the results of eight experiments that intend to illustrate characteristic properties of the designed controller. Finally we conclude our work and give an outlook on further activities.

---

[4] Per aggregate measurements are usually not considered as unacceptable implementation burden.

2 Path Performance Assessment

For dimensioning of end-to-end paths at the head-end nodes of the path, the evaluation and prediction of the performance perceived by packets traversing the path is essential. In the specific case of leaky bucket[5] characterized traffic the (ρ, σ) calculus can be deployed [9, 10] to assess QoS performance. We will use the more generic notions of arbitrarily shaped *Arrival Curves* and *Service Curves* to characterize incoming traffic and the service guaranteed by a forwarding node. A minimum (maximum) service curve $\bar{f}$ $(\hat{f})$ indicates the minimum guaranteed (maximum possible) number of bits served over time. The arrival curve $\hat{A}$ is an upper bound of an associated arrival process such that $A(t) - A(s) \leq \hat{A}(t-s)$ for all $t \geq 0$. For details and background information on service and arrival curves we refer to [11, 10, 12] and introduce only the basic operations and theorems deployed in the following sections.

A legacy forwarding node in packet switching networks can be modeled by means of an *universal f-server* which guarantees a minimum *service curve* $\bar{f}$:

Theorem 1. *The output sequence S of an universal f-server guaranteeing a minimum service curve $\bar{f}$ for an input with constraint function $\hat{A}$ satisfies*

$$S(t) \geq (A \star \bar{f})(t) = \min_{0 \leq s \leq t} \{A(s) + \bar{f}(t-s)\} \tag{1}$$

with $\star$ denoting the min-plus convolution.

The concatenation of n f_i-servers for an arbitrary arrival process is in fact an f-server for the arrival process with service curve

$$f(t) = (f_1 \star f_2 \star ... \star f_n)(t) = \bigotimes_{i=1}^{n} f_i \tag{2}$$

For performance evaluation it is important to use sub-additive arrival curves but it is also equally important to use the minimum arrival curve [10]. According to [12] the following definition describes one way to obtain a valid arrival curve.

Theorem 2. *Given an arrival function $A(t)$ of any arrival process A with $t \geq 0$ the function $A \oslash A = \max_{s \in \mathcal{R}} \{A(t+s) - A(s)\}$ is an arrival curve, i.e traffic constraint function of arrival process A, which is sub-additive. The operator $\oslash$ shall denote the min-plus deconvolution.*

Provided an appropriate arrival curve and the minimum service curve we can calculate the end-to-end performance of a path:

Theorem 3. *Consider a server for any traffic with an arrival curve A. Let $S(t)$ be the traffic leaving that server, $q(t) = A(t) - S(t)$ be the queue length at the server at time t and $d(t) = \inf \{d \geq 0 : S(t+d) \geq A(t)\}$ the virtual delay of the*

---

[5] A leaky bucket characterized traffic is said to be constrained by $\hat{A}(t) = \rho t + \sigma$

last packet that arrives at time t. Suppose $A(t)$ is upper constrained by $\hat{A}(t)$. Then the queue length and the virtual delay are respectively bounded by

$$q(t) \leq \max_{0 \leq s \leq t} \left\{ \hat{A}(s) - \bar{f}(s) \right\} = \left(\hat{A} \oslash \bar{f} \right)(0) \tag{3}$$

$$d(t) \leq \inf \left\{ d \geq 0 : \hat{A}(t) \leq \bar{f}(s+d), s = 1, ..., t \right\} \tag{4}$$

3 Controller Design

The basis for any controller design is a sufficient analysis of the system including the definition of the controlled object, which is frequently called *Process* or *Plant*, and an appropriate selection of the type of controller to be deployed. In traditional controller design an explicit mathematical model of the process in terms of *State Transition Equations* is required [13]. A process model built from expert system knowledge or from open loop measurements is typically used to determine such equations. When faced with time varying systems, as communication networks generally are, adaptive control mechanisms need to be deployed. This could be, for example, a recursive estimation of the parameters of the plant model. The difficulty of accurate system modelling and dynamic adaptation suggest deployment of *Fuzzy Logic Control* [14][15]. The applicability of Fuzzy Logic Control in communication networks has been been discussed in previous studies, where the intuitive design of the rule base can be considered as main attractor [16, 17]. This comes at the expense of an inevitable need for iterative controller tuning to achieve stability and sufficient performance characteristics.

The purpose of the controller to be deployed is to adapt the resources reserved for one end-to-end traffic aggregate to traffic variations such that the QoS requirements are met. We therefore identify the QoS provided to one end-to-end traffic aggregate predicted at the ingress node based on the theory introduced in section 2 as the controlled object. Given a maximum tolerable delay D_{req} we can determine the minimum required service curve $\bar{f}$ for a certain arrival curve $\hat{A}$ using (4). In turn we can compute the minimum buffer size to be provided to achieve zero loss from (3). For each UC-QoS class the maximum tolerable delay is defined in the service establishment process and therefore we need a measure of the current arrival curve $\hat{A}(t)$ for appropriate resource adaptation.

The arrival curve of a certain traffic class shall be computed from h measurements of the number of bits arrived of this traffic class, where the sampling rate r_h shall be fixed. Using the measured time discrete arrival process $A(n)$, $0 \leq n < h$ and theorem 2 we can compute the discrete arrival curve $\hat{A}(n)$. After one measurement window of length $h \cdot 1/r_h$ a new arrival curve shall be computed, such that we obtain a sequence of arrival curves $\hat{A}_m(n)$. At the end of each measurement window m a minimum required service curve $\bar{f}_m(n) = \hat{A}_m(n - D_{\text{req}})$ can be computed, or in other words, the current maximum delay is given by

$$d(m) = \inf \left\{ d \geq 0 : \hat{A}_m(n) \leq \bar{f}_{m-1}(s+d), s = 1, ..., n \right\} \tag{5}$$

Thus at discrete times $m \cdot h \cdot 1/r_h$ we obtain the arrival curve $\hat{A}_m(n)$ and the maximum delay $d(m)$ that packets perceived in the previous measurement window. Our objective is to track traffic variations and, with respect to end-to-end delay performance, the horizontal deviation between two subsequent arrival curves suitably indicates the delay impact of the most recent traffic variation:

$$d_{\text{dev}}(m) = \inf\left\{\tau \geq 0 : \hat{A}_{m-1}(n) \leq \hat{A}_m(s+\tau), s = 1, ..., n\right\} \quad (6)$$

Before we define the controller output we want to discuss the dynamics of the controlled object in terms of potential bounds on the system response time. The time needed for signaling during the resource adaptation process, which is clearly topology dependent, obviously provides a lower bound on the system response time. But operators may even want to perform resource adaptations less frequently than with a maximum rate given by the system response time. We assume that operators may first collect k samples of the exhibited maximum delay and arrival curve before triggering resource adaptation. Please note that the service curves $f_m(n)$ remain unchanged over the interval $[t, t + k \cdot h \cdot 1/r_h]$. In order to compute a single input figure from these k samples of the exhibited delay we deploy the *exponential moving average* algorithm recursively defined by:

$$d^*(m) = (1-\alpha)\, d(m) + \alpha\, d^*(m-1) \quad (7)$$

Figure 1 depicts the relations between the number of traffic arrival measurements h, the number of averaged delay and arrival curve deviation samples and the resource adaptation times.

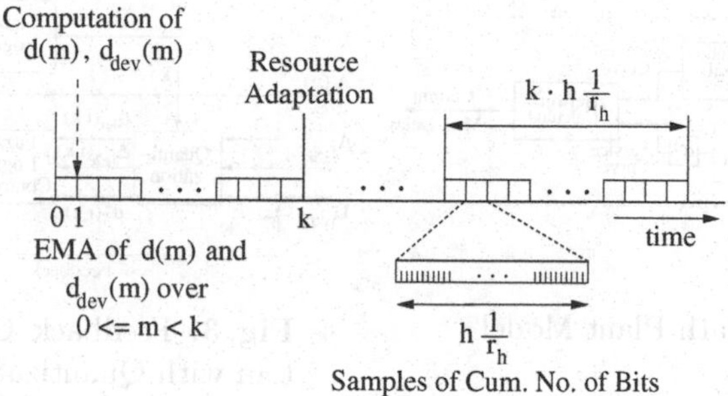

Fig. 1. Time Scales of Measurements

Independently of traffic variations and delay performance another objective shall be to minimize the interference on other traffic flows, which can be achieved by minimizing resources assigned along a path. This objective becomes more important when the network load increases and the competition for link resources becomes more severe. We assume that link state information is distributed regularly, e.g. by a QoS extended link state routing protocol [18]. From the distributed link state information we derive the maximum available service curve $\hat{f}_i$ for link i and compute the amount of available, unreserved resources along the entire path $\hat{f}^m_{\mathrm{E2E}}(t)$ using (2). Whenever $\hat{f}^m_{\mathrm{E2E}}(t)$ is small or even approaches zero the traffic load along that path is high. As indicator for high traffic load situations with respect to end-to-end delay we will take the horizontal deviation between the current available service curve $\hat{f}^m_{\mathrm{E2E}}(t)$ and the initial available service curve $\hat{f}^0_{\mathrm{E2E}}(t)$:

$$\Delta_{\hat{f}_{\mathrm{E2E}}}(m) = \inf\left\{\tau \geq 0 : \hat{f}^m_{\mathrm{E2E}}(t) \leq \hat{f}^0_{\mathrm{E2E}}(s+\tau), s = 1, \ldots, t\right\} \tag{8}$$

where large positive (negative) $\Delta_{\hat{f}_{\mathrm{E2E}}}(m)$ indicate high (low) traffic load compared to the initial conditions.

By using $d_{\mathrm{dev}}(m)$ instead of $\hat{A}(m)$ we need to discuss, which arrival curve shall be deployed to determine the new required minimum service curve using equation (4). Here we deploy the output of the Fuzzy Logic Controller. By using the averaged $d_{\mathrm{dev}}(m)$, $\Delta_{\hat{f}_{\mathrm{E2E}}}(m)$ and $d(m)$, the controller shall compute a delay additive *out* for the originally requested maximum delay D_{req}, such that the $(m+1)^{\mathrm{th}}$ service curve is determined by:

$$\bar{f}_{m+1}(n) = \hat{A}_0(n - (D_{\mathrm{req}} - out)) \tag{9}$$

For simplicity reasons we assume that buffers will always be dimensioned accordingly to achieve zero loss for any arrival-service curve pair $(\hat{A}_0, \bar{f}_{m+1})$. Figure 2 depicts the pertaining plant model of QoS performance prediction for traffic traversing a certain path and figure 3 illustrates the feedback control system.

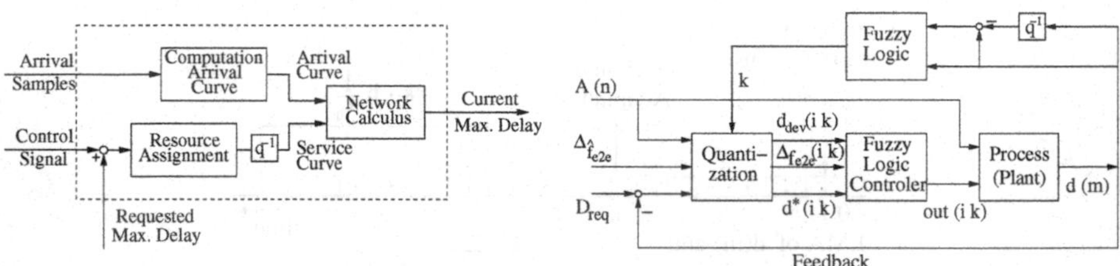

Fig. 2. Path Plant Model

Fig. 3. Feedback Control System with Quantization Control

The process of computing the exponential averages shall be denoted as *Quantization*. With respect to deterministic service provision we have to recognize that by computing averages we introduce some statistical properties depending on the averaging method. By appropriately choosing parameter α of the exponential moving average algorithm in (7) we can put more or less emphasis on single peaks in the available sample set. In [19] a dynamic adaptation of α depending on the system characteristics has been proposed. In highly dynamic situations more weight is given to peaks in the sample set and thus the average better reflects high variation peaks. Alternatively, one can fix α and decrease the averaging window. In context of our control problem a decrease of the averaging window (decrease k) is equivalent with increasing the resource adaptation frequency. An upper bound of this frequency shall be given by the lower bound on the system response time[6] When either $d(m)$ approaches or exceeds the requested bound D_{req} or the delay variation $d(m) - d(m-1)$ exceeds a certain threshold, the service curve shall be adapted as fast as possible, and thus the averaging window k shall be decreased. In order to adapt k appropriately we deploy a second fuzzy logic block as depicted in figure 3 with $d(m)$ and $\Delta_{\text{delay}}(m) = d(m) - d(m-1)$ as input values.

We will briefly discuss the characteristics of fuzzy control with respect to the most general *Mamdami Controller* [20]. A *Fuzzification* block converts each piece of crisp input data to degrees of membership by lookup in the according *membership functions.* Each crisp input variable must be within a certain range, which is called *Universe of Discourse.* For the crisp input signals and for the output we need to define the universe of discourse, the *Term Set* and the membership functions in order to obtain the associated fuzzy variables. The rule base contains the control rules in a linguistic form. The *Inference Engine* processes the input with respect to the rule base. The *Aggregation* checks the degree of fulfillment or firing strength of one particular rule, while the *Accumulation* computes a final conclusion from the accumulated conclusions of all fired rules. In the *Defuzzification* block the concluded outcome of the inference engine, which is a fuzzy set, is converted to a crisp output. For connectives, aggregation and implication we deploy only max and min operations and for de-fuzzification the *Center of Gravity* method.

Figures 4, 5, 7 and 8 depict our straight forward selection of term sets and membership functions based on trapezoid functions for all inputs and the output of the controller. All crisp input variables shall first be normalized by the required maximum delay for the respective traffic class D_{req}. The term set of the current delay contains the linguistic variables *Very Low* (VL), *Low* (L), *Good* (G), *High* (H) and *Very High* (VH). The high-pass filtered deviation of the arrival curves has a similar term set, where the linguistic variable *Good* is substituted by *Medium* (M). The linguistic variables of the normalized available service curve along the path are defined as *Much Less* (ML), *Less* (L), *Similar* (S), *More* (M) and *Much More* (MM), while for the output we defined the term set *Maximum Decrease* (MD), *Decrease* (D), *Decrease Slightly* (DS), *UnChanged* (UC), *Increase Slightly* (IS), *Increase* (I) and *Maximum Increase* (MI).

Our objective shall be to achieve a delay of 75% of the targeted delay and therefore the membership functions of $D_{\text{E2E}}(m)/D_{\text{req}}$ have been symmetrically defined around 0.75. With the fuzzy variables defined by these term sets we have intuitively defined a rule base in linguistic form, e.g.

$$\texttt{if } \frac{\Delta_{\text{arrival}}(m)}{D_{\text{req}}} \texttt{ is H and } \frac{\Delta_{f_{\text{E2E}}}(i)}{D_{\text{req}}} \texttt{ is M and } \frac{D_{\text{E2E}}(m)}{D_{\text{req}}} \texttt{ is H then output is I .}$$

[6] Minimum assignment duration required by the the physical network operator providing resources for virtual domain.

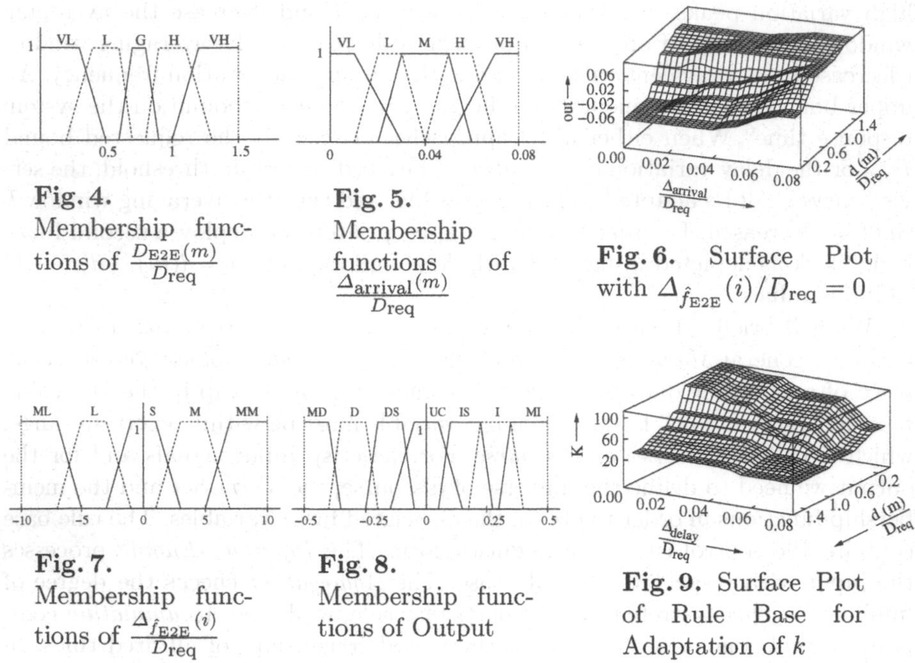

Fig. 4. Membership functions of $\frac{D_{\mathrm{E2E}}(m)}{D_{\mathrm{req}}}$

Fig. 5. Membership functions of $\frac{\Delta_{\mathrm{arrival}}(m)}{D_{\mathrm{req}}}$

Fig. 6. Surface Plot with $\Delta_{\hat{f}_{\mathrm{E2E}}}(i)/D_{\mathrm{req}} = 0$

Fig. 7. Membership functions of $\frac{\Delta_{\hat{f}_{\mathrm{E2E}}}(i)}{D_{\mathrm{req}}}$

Fig. 8. Membership functions of Output

Fig. 9. Surface Plot of Rule Base for Adaptation of k

We abstain from depicting the entire rule base and exemplarily show in figure 6 the control surface with $\Delta_{\hat{f}_{\mathrm{E2E}}}(i)/D_{\mathrm{req}}$ fixed at zero. The surface illustrates well the increasing output when either the arrival curve variation or the current delay increase. Clearly more weight is put on reacting when the current delay dangerously approaches the threshold of D_{req}. The fuzzy logic block for the adaptation of the averaging window k uses the membership function depicted in figure 4; for $\Delta_{\mathrm{delay}}(m)/D_{\mathrm{req}}$ the membership function applied before to $\Delta_{\mathrm{arrival}}(m)/D_{\mathrm{req}}$ depicted in figure 5 and the rule base represented by the surface plot in figure 9. We set $10 \leq k \leq 120$. Whenever the end-to-end delay is close or above D_{req} or when the delay variation is high the window size is decreased for faster reaction.

4 Stability and Performance Assessment

In order to assess the performance and stability of the fuzzy logic controller we generate representative artificial input to the controller with clearly defined expected results. The evaluation process is organized in eight experiments characterized by a given arrival curve deviation $d_{\mathrm{dev}}(m)$ and capacity deviation $\Delta_{\hat{f}_{\mathrm{E2E}}}(m)$. Table 1 summarizes the input signal shape for all experiments, where the arrival curve impulses are emulated by an artificial decrease of the service curve $\bar{f}^{m}_{\mathrm{E2E}}(n)$. The utilized sinusoidal shapes for $d_{\mathrm{dev}}(m)/D_{\mathrm{req}}$ and $\Delta_{\hat{f}_{\mathrm{E2E}}}(m)$ are depicted in the lowermost diagram of figure 12.

In all experiments we set $h = 100$ where a sample of the arrival curve is taken every 10 ms. As assessment criteria we use the normalized predicted delay per-

Table 1. Experiments

No.	$d_{\text{dev}}(m)/D_{\text{req}}$	$\Delta_{\hat{f}_{\text{E2E}}}(m)$
1	Zero with Periodic Impulses	Zero
2	Zero with Periodic Impulses	Sinusoidal
3	Normally Distributed with Periodic Impulses	Zero
4	Normally Distributed with Periodic Impulses	Sinusoidal
5	Sinusoidal	Zero
6	Sinusoidal	Sinusoidal
7	Sinusoidal with high frequent normally distributed variations	Zero
8	Sinusoidal with high frequent normally distributed variations	Sinusoidal

ceived by packets traversing the path $d_{\text{dev}}(m)/D_{\text{req}}$, the service curve deviation given by:

$$\Delta_{\bar{f}_{\text{E2E}}}(m) = \inf\left\{\tau \geq 0 : \bar{f}^{m}_{\text{E2E}}(n) \leq \bar{f}^{0}_{\text{E2E}}(s+\tau), s = 1, \ldots, n\right\} \qquad (10)$$

and the averaging window size k, which are shown from top to bottom in the diagrams of figure 10 to figure 13.

Stability of fuzzy systems is a somewhat open question. Nonlinear systems, such as fuzzy control systems, are said to be asymptotically stable, if they converge to the equilibrium when the system starts close to it [14]. For the designed controller we therefore have to show that both the computed delay and the assigned service curve reach an equilibrium state. In experiments one and two the controller decreases upon detection of the impulse the averaging window k and thus increases the resource adaptation frequency. As a result the service curve is adapted quickly. As the only arrival curve variation value known by the controller is the extremely high impulse the service curve is in fact over dimensioned, which is indicated by the equilibrium delay $0.10 \cdot D_{\text{req}}$. In case of sinusoidal capacity variation (experiment two) the assigned service curve varies synchronously. In presence of normally distributed arrival curve variations in addition to impulses we can see a better approximation of an average delay of $0.75 \cdot D_{\text{req}}$. Intuitively this can be explained by the additional smaller (normally distributed) variations of the arrival curve deviation which weaken the dominance of the extreme impulses. The second experiment demonstrates the adaptation of the averaging window k depending on the current delay and delay variation. In the first four experiments an equilibrium state synchronously changing with capacity variations has been reached.

So far we have verified that the controller appropriately reacts on impulses and that it is also stable in presence of random input variations. The last four experiments aim at validating the responsiveness on periodic and strong input variations expressed by a sinusoidal shape of $d_{\text{dev}}(m)/D_{\text{req}}$ with high amplitude.

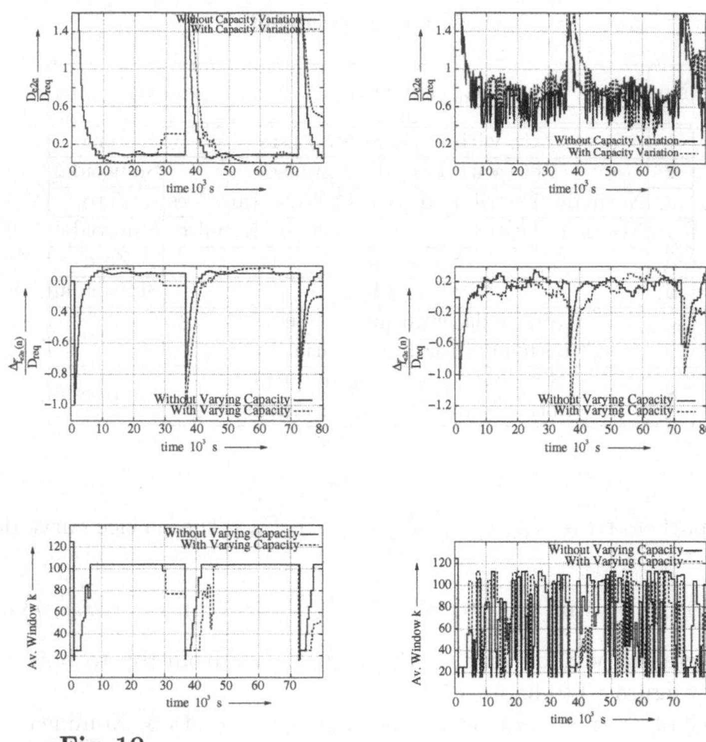

Fig. 10. Experiments One and Two

Fig. 11. Experiments Three and Four

In experiment five we see the sinusoidal peaks of the predicted delay approximately upper bounded by D_{req}. The averaging window again adjusts well to delay, i.e. when $d_{\mathrm{dev}}(m)/D_{\mathrm{req}}$ is low, k is maximal. At the strongly increasing or decreasing slopes of $d_{\mathrm{dev}}(m)/D_{\mathrm{req}}$ k is decreased to improve the responsiveness. Experiment six illustrates the impact of capacity variations, i.e. at contemporary minima of $\Delta_{\hat{f}_{\mathrm{E2E}}}(m)$ and of the slope of $d_{\mathrm{dev}}(m)/D_{\mathrm{req}}$ the service curve is significantly decreased in order to free unnecessarily assigned resources when the overall available capacity is low. This comes at the expense of a delay bound violation at the next increase of the arrival curve deviation. In presence of highly frequent random variations this behavior generally does not change. As observed with the impulse input, k varies more and due to the random peaks the delay bound D_{req} is slightly exceeded. This may be an indicator for the need for a more conservative policy, e.g. a target of 60 % of D_{req}.

Summarizing, the designed controller is stable under the conditions considered above. It shows reasonable sensitivity to random delay variations without becoming instable. Mainly due to the dynamic adaptation of the capacity update frequency $1/k$ the responsiveness is appropriate over the timescales under consideration. While we do not have to worry about delay bound variations caused by extreme impulses the slight but frequently occurring violations caused by random input values suggest to go for a more conservative target value.

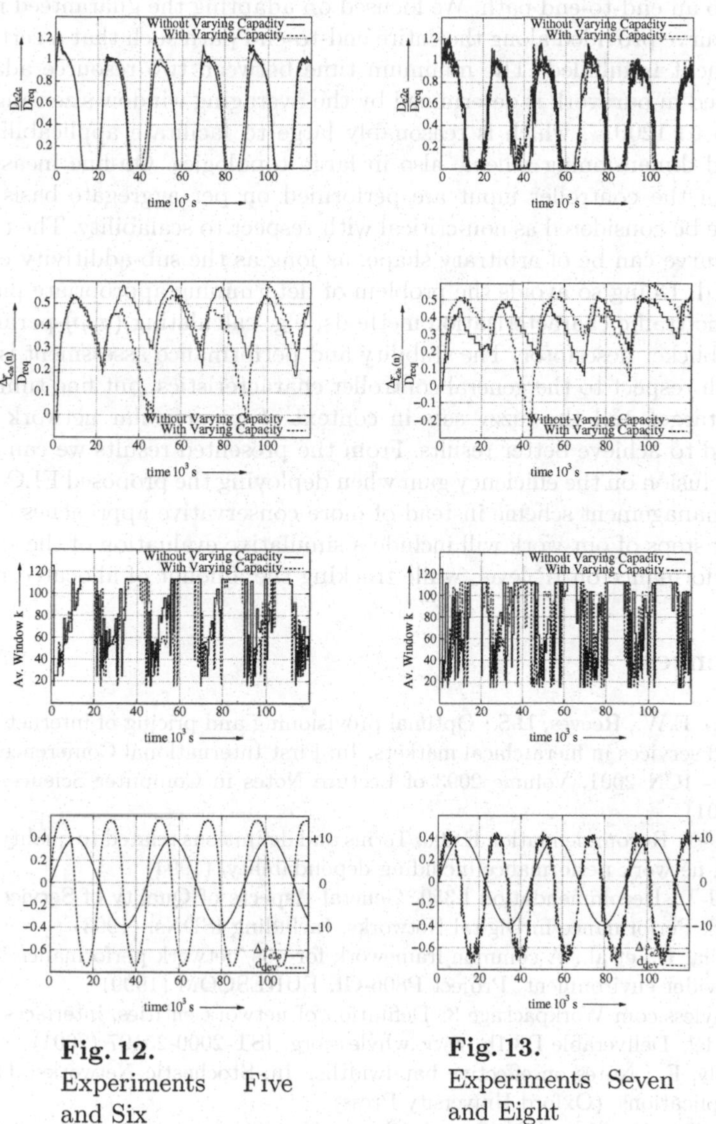

Fig. 12. Experiments Five and Six

Fig. 13. Experiments Seven and Eight

5 Conclusion and Outlook

We presented a dynamic resource management scheme for virtual domains, which act as capacity wholesalers. In this scheme we exploit one attractive feature: network resources maintained by virtual domains can be adjusted on significantly shorter time scales than physical capacities can be changed. In particular, we presented a fuzzy logic controller that dynamically adapts network resources assigned to an end-to-end path. We focused on adapting the guaranteed minimum service curve provided along the entire end-to-end path such that a certain delay requirement is fulfilled. The minimum time between two resource adaptations considered in our work is determined by the averaging window size ranging from 100 s up to 1200 s , which is reasonably large to facilitate applicability of the proposed dimensioning scheme also in large topologies. On-line measurements to obtain the controller input are performed on per aggregate basis and can therefore be considered as non-critical with respect to scalability. The calculated arrival curve can be of arbitrary shape, as long as the sub-additivity constraint is fulfilled. Doing so avoids the problem of determining appropriate parameters of specific traffic characterization methods, e.g. calculating (σ, ρ)-parameters of a leaky bucket descriptor. The stability and performance assessment is satisfactory with respect to the general controller characteristics, but fine tuning of the control target and the fuzzy sets in context of a particular network topology is needed to achieve better results. From the presented results we can not draw any conclusion on the efficiency gain when deploying the proposed FLC-based resource management scheme instead of more conservative approaches. Therefore the next steps of our work will include a simulative evaluation of the end-to-end QoS performance on IP level, while tracking the amount of allocated resources.

References

[1] Fulp, E.W., Reeves, D.S.: Optimal provisioning and pricing of internet differentiated services in hierarchical markets. In: First International Conference Networking - ICN 2001. Volume 2093 of Lecture Notes in Computer Science., Springer (2001)

[2] ITU-T: Recommendation E.800, Terms and definitions related to quality of service and network performance including dependability. (1994)

[3] ITU-T: Recommendation I.350, General Aspects of Quality of Service and Network Performance in Digital Networks, Including ISDNs. (1993)

[4] Jutila, U., et al.: A common framework for qos/network performance in a multiprovider environment. Project P806-GI, EURESCOM (1999)

[5] Whyless.com Workpackage 3: Definition of network entities, interfaces and early model. Deliverable D3.1b, www.whyless.org, IST-2000-25197 (2001)

[6] Kelly, F.: Notes on effective bandwidths. In: Stochastic Networks: Theory and Applications, (Oxford University Press)

[7] Bouillet, E., Mitra, D., Ramakrishnan, K.: The structure and management of service level agreements in networks. IEEE Journal on Selected Areas in Communications **20** (2002) 691–699

[8] Bitar, N., et al.: Traffic engineering extensions to OSPF. Work in progress (July 2001)

[9] Parekh, A., Gallager, R.: A generalised processor sharing approach to flow control in integrated services networks: The multiple node case. IEEE/ACM: Transactions on Networking **2** (1996) 344–357

[10] Chang, C.S.: Performance Guarantees in Communication Networks. Springer (2000) Telecommunication Networks and Computer Systems Series.

[11] Cruz, R.L.: Quality of service guarantees in virtual circuit switched networks. IEEE Journal on Selected Areas in Communications **13** (1995) 1048–1056

[12] Boudec, J.L.: Application of network calculus to guaranteed service networks. IEEE Trans on Information theory **3** (May 1998)

[13] Lygeros, J., Godbole, D., Coleman, C.: Model based fuzzy logic control. Technical Report ERLM94, UCB (1994)

[14] Driankov, D., Hellendoorn, H., Reinfrank, M.: An Introduction to Fuzzy Control. Springer Verlag, Heidelberg (1993)

[15] Berenji, H.R.: Fuzzy logic controllers. In: An Introduction to Fuzzy Logic Applications and Intelligent Systems, Boston, MA, (Kluwer Academic Publisher)

[16] Tsang, D.H.K., Bensaou, B., Lam, S.T.C.: Fuzzy-based rate control for real-time MPEG Video. IEEE-FS **6** (1998) 504

[17] Bensaou, B., Lam, S., Chu, H., Tsang, D.: Estimation of the cell loss ratio in atm networks with a fuzzy system and application to measurement-based call admission control. IEEE/ACM Transactions on Networking **5** (1997) 572–584

[18] Apostolopoulos, G., Guerin, R., Kamat, S., Orda, A., Przygienda, T., Williams, D.: QoS routing mechanisms and OSPF extensions. RFC, Internet Engineering Task Force (1997)

[19] Burgstahler, L., Neubauer, M.: New modifications of the exponential moving average algorithm for bandwidth estimation. In: Proceedings of IP SPecialist Seminar ITC, Wuerzburg, Germany (2002)

[20] Mamdami, E., Assilian, S.: Applications of fuzzy algorithms for control of simple dynamic plant. Proc. Institute of Electronic Engineering (1974) 1585–1588

Towards RSVP Version 2

Rosella Greco, Luca Delgrossi, and Marcus Brunner

Networks Laboratories, NEC Europe Ltd.,
Adenauerplatz 6, D-69115 Heidelberg, Germany
{greco,luca,brunner}@ccrle.nec.de

Abstract. The main elements of a communications system that is able to provide QoS over the Internet are a scheduling policy and a signaling protocol. This paper is concerned with the design of the reservation protocol. The most widely used reservation protocol is RSVP which we take as a basis of this work. RSVP has been criticized mainly because of its complexity and poor scalability. This paper presents the first steps towards the definition of a new version of RSVP, which we call RSVPv2. The goal of RSVPv2 is to provide a more "light" approach, that can help improve handling reservations in the network by means of a simplified behaviour.

1 Introduction and Motivation

A key to the success in the provision of QoS over the Internet is the ability to schedule resources along the communications paths according to the reservations issued by the users. As a result of reservation signaling, the nodes along the communications paths should contain sufficient information to detect packets belonging to specific data flows and be prepared to provide them with an adequate quality of service.

A wide variety of applications needs to be served. To mention only a few of them, some applications may be targeted to large multicast scenarios with a high number of receivers involved, whereas others may be targeted to peer-to-peer communications or to more advanced mobile scenarios, where the users are allowed to move across different zones as they produce and consume time-sensitive data.

The main elements of a communications system that is able to provide QoS over the Internet are a scheduling policy and a signaling protocol. This paper is not concerned with scheduling policies (we assume that an effective policy is in place, as for instance DiffServ [1]) and it focuses primarily on the design of the reservation protocol. Although many signaling protocols have been proposed in the past ten years, it is felt that a conclusive solution has not been reached yet. This is confirmed by ongoing work at the IETF to define requirements for QoS signaling protocols [2]. The goal of the IETF NSIS Working Group is to investigate into new QoS signaling functions both to local access networks and end-to-end.

M. Ajmone Marsan et al. (Eds.): QoS-IP 2003, LNCS 2601, pp. 704–716, 2003.

Currently, the most widely used reservation protocol is RSVP [3]. RSVP has a large number of implementations in place and it has been standardized by the IETF in September 1997. For these reasons, we take RSVP as the basis of this work. However, since its standardization, RSVP has been criticized mainly because of its complexity and poor scalability. The protocol has been designed for large multicast scenarios and for applications as the distribution of audio and video contents over IP multicast, as it is done with the MBONE [4]. Although RSVP responds well to the needs why it has been designed, it is felt to be too complex to be used in many other scenarios, that are perhaps more restricted but nevertheless useful for a wide range of applications.

This paper presents the first steps towards the definition of a new version of RSVP, which we call RSVPv2. The goal of RSVPv2 is to provide a more "light" approach, that can help improve handling reservations in the network by means of a simplified behavior. In particular, the new protocol should be able to be more efficient in terms of reservations setup time. We also introduce into the protocol a series of new mechanisms that extend it to make it able to handle mobility. It is a first experiment to open up the road for applications based on mobile IP that need an adequate provision of quality of service. RSVPv2 is not intended to replace RSVP but, on the contrary, to coexist with it in a dual stack architecture, so that both reservation protocols are available to applications with different needs.

In the rest of this paper, we present in details a possible design of RSVPv2, including the new features for mobility. We conclude the paper with results from a number of simulations targeted to validate the effectiveness of the protocol design.

2 Related Works on Signaling Protocols

Some protocols have already been proposed to provide end-to-end QoS over Internet, like ST-II [5], DRP [6], Boomerang [7], SSR [8], Yessir [9] and RSVP [3]. These protocols can be divided in two categories according to the approach used to provide QoS over Internet: *per-packet* or *per-flow*. With a per-packet approach (e.g. used by SSR and DRP), all the information necessary to obtain differentiated service is carried inside each data packet and routers do not need to maintain any state; with a per-flow approach (e.g. used by ST-II, Yessir, Boomerang and RSVP), resources for data packets are reserved in advance using specific messages sent between a source and its destination.

The per-packet approach has advantages like limiting the protocol overhead and not saving per flows information along the path. However, we preferred to focus this work on the per-flow approach, considering that separation between data and signaling offers a more flexible and clean solution, and that, with a per-packet approach, strict guarantees on QoS cannot be provided since the amount of traffic generated is not bounded.

Among the already existing protocols using the per flow-approach, Boomerang seems to be not very extensible or adaptable to different scenarios because it uses ICMP messages as protocol messages and a reservation scheme based only on the IP address. Yessir can be used only by applications that send their data with the Real Time Transport protocol (RTP). In fact, the protocol uses RTP signaling messages to send the reservation requests.

3 Towards RSVP Version 2

3.1 General Features

The main features of RSVPv2 have been designed by evaluating pros and cons and trying to learn as much as possible from RSVP, other protocols and from the critics moved to them.

Unicast. RSVPv2 has been thought as a *unicast* protocol. Even if we consider multicast as an important feature of a signaling protocol, we focused, as a first step in the protocol design, on simplicity and a lightweight approach that can fit very well in many scenarios. We think that multicast support can be regarded as an extension to be added to the protocol. The idea is to have an underlying common structure that can be specialized adding different functionalities required by particular scenarios.

Sender-Oriented. We propose a *sender-oriented* protocol. This approach has been chosen considering that, in this way, the complexity of the protocol can be reduced. In fact, the cost of sending advertisements, processing them in each aware router along the path is eliminated, together with the necessity of symmetric routing (from source to destination and vice versa). Besides, the time to obtain a reservation decreases. This approach does not give the receiver the possibility to choose the traffic characteristic. However, from the experience with RSVP version 1, it has been seen that in most of the cases the receiver will simply request whatever traffic bandwidth the sender has indicated. The receiver-oriented approach is helpful in multicast scenario where there is the need to accommodate large groups membership and heterogeneous receiver requirements. As previously mentioned, RSVPv2 has been thought for unicast scenarios and as a minimal protocol where multicast support can be provided by RSVP version 1 running dual stack with RSVPv2 or by extension of the protocol itself.

Soft State. RSVP version 1 is a *soft state* protocol, which means that information about the flows are temporary saved along the path. Soft state gives adaptivity to route changes during the lifetime of a reservation and increases the protocol robustness to loss of messages. Besides, unexpected loss of connectivity from an end-point will simply lead to timed out states after some time along the path. These characteristics seem to be very important in every scenario where a QoS signaling protocol can be used.

Soft state, on the other hand, introduces the issue of refreshing the information stored along the path. For RSVPv2, we propose to make the end points of the reservation responsible for sending end-to-end refresh messages and in particular the source. In this way, the effort done by intermediate nodes is reduced to the forwarding of the refresh message as opposed to issuing such messages upon expiration of the correspondent soft state lifetime.
Besides, even if the soft state approach does not require the existence of any mechanism for the explicit end of a reservation, this feature can be useful to avoid keeping resources allocated when they are not needed any longer.
RSVPv2 supports the explicit end of a reservation coming from the source that requested the reservation.

Reservation and Error Notification. After sending a reservation request, it is desirable to receive an answer about its result. RSVPv2 has been designed including a reservation acknowledgment sent from the destination back to the source in the case of a successful reservation. This acknowledgment can be used by the source as an indication that the reservation has been completed along the path. A source may decide whether to wait for this acknowledgment or not before sending the data.
In the same way, it is desirable to receive information in case of failure of a reservation in a node along the network. RSVPv2 sends an error notification from the node where the reservation failed back to the source. This message may contain information on the state of the resources at the node where the failure occurred and it can be used by the source to perform a new reservation request with different parameters.

Service Specification. RSVP version 1 can provide one of the following types of service: Guaranteed Load and Integrated Service. RSVPv2 wants, on the other hand, to be as general as possible as far as offering QoS is concerned. The idea is that the protocol should be able to carry specifications for a large number of services in order to be easily used in different scenarios.

Transport Mechanism. There are basically two levels which can be considered to transport signaling packets. These are transport and network level. Transport level is used, for example, by Yessir to send its signalling. At network level, signaling information can be included into ICMP echo-request/echo-reply messages (this is the approach used by the Boomerang protocol) or sent as IP datagram with a specified protocol number. IP messages are easily handled by routers. On the other hand, this requires the ability to perform 'raw' network I/O and many host systems may not support this feature. For RSVPv2, all messages are sent as raw IP packets, following the same approach used in the original RSVP and considering the fact that encapsulation can be used for end systems which don't support raw I/O.

Flow Identifier and Reservation Identifier. To keep things separate as much as possible, RSVPv2 uses different information to identify a flow and a reservation. In other words, each RSVP message carries a Reservation Identifier to address the reservation and a Flow Identifier that allows to match data with the reservation previously set. Benefits from that can be seen in many cases like the mobile scenario and sending refresh messages.

4 Outline of the Protocol Operations in Some Scenarios

To explain how the protocol works, some possible scenarios are described. In the first scenario, many implementation details are given in order to give a full overview of the protocol. In the others, only a number of specific features are pointed out.

4.1 Reservations between Two End-Systems

The scenario presented here consists of two end-systems placed in different part of the network that want to request a reservation before transmitting the data. The goal of the protocol is to carry a service request from the data source to the destination. This operation is carried out in two step. As a first step, the source sends a reservation (RESV) message to the destination. If the message reaches the receiver, than an acknowledgment (ACK) is sent back from the receiver to the source as a confirmation of the reservation set up. If an error occurs along the path, then, the RESV message is not further forwarded and an error message (NACK) is returned back to the source. While processing a RESV message, each aware router along the path saves state and sets an appropriate reservation for the flow. The format of a RESV message is represented in Figure 1.

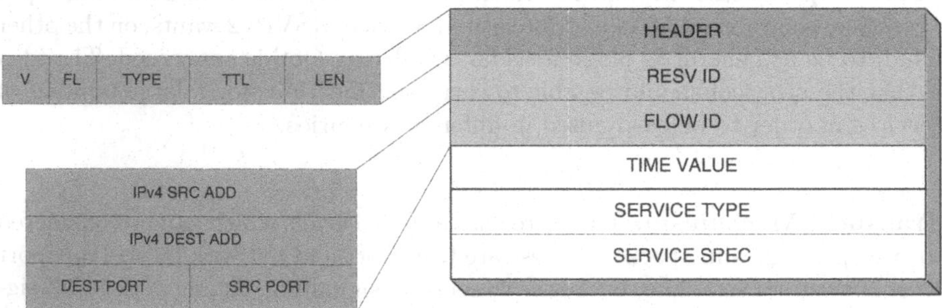

Fig. 1. RSVPv2 RESV Message

The structure of RSVPv2 messages consists of a common part, included in all the packet types, containing the header and the Identifiers. It follows then a certain number of fields that carry the specific information required by each message type.

RSVPv2, following its philosophy of being a light weight protocol, attempts to minimize the amount of state information stored in each router along the path. In particular, reducing the amount of information stored in each node can help the overall scalability of the system. The state information allows us to uniquely identify a flow and a reservation. Each flow is represented by the IP source and destination address and by the TCP/UDP source and the destination port numbers that are used to distinguish among different flows originated by the same host (e.g., audio and video applications).

After the Flow Id and the Reservation Id, a time information is saved for each entry. The time saved is the arrival time of the RSVP packet plus the time value carried in the Time field. In this way, the estimated expiring time is calculated and saved.

To sum up, for a reservation using IPv4 addressing, only 16 Bytes are saved along the data path.

As a first stage in the implementation, one new service type has been coded: Assured Bandwidth. It gives the possibility to the source to ask for a certain amount of bandwidth, leaving all the other parameters (delay, loss rate, jitter, etc.) unspecified although in a reasonable range. The service specification carried in the message is not saved in the RSVP state but it is directly passed to the traffic control. If an error occurs or resources are not available, an error messages (NACK) is immediately generated.

As previously pointed out, the soft state approach requires to set and manage a number of timers for the RSVP state saved along the path. In order to keep the protocol simple, we propose a solution that reduces the number of timers simplifying its managing even if the deallocation of timed out resources can be slightly penalized. RSVP version 1 sets a timer for each state saved. This fact adds complexity and increases scalability problems in the core network due to the management of these of timers. RSVPv2 uses just one timer for each RSVP list of entries. This timer periodically scans the list of the saved states and erases all the timed out entries.

In order to reduce the impact of refresh messages in the overhead of the protocol, the application responsible for the reservation request, can specify, as a time value, the whole expected duration of the data transmission without sending any refresh. In case of an earlier end, an explicit tear down of the reservation can be performed using an appropriate TearDown message (RTEAR) included among the protocol messages. This solution is not suggested when there are frequent route changes.

4.2 Signaling of Service between Mobile Hosts

The mobile scenario has to be analyzed separating two possible cases: mobile sender and mobile receiver.

Mobile Sender. In this case, RSVPv2 tries to increase the re-use of resources and the probability to obtain a reservation while moving from one place (IP

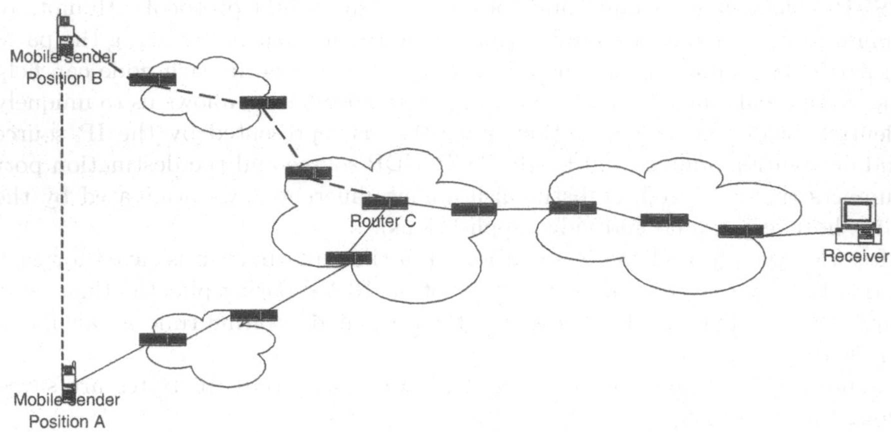

Fig. 2. Mobile scenario

address A) to another (IP address B). Referring to Figure 2, we assume that a mobile user starts a reservation from its home site (A). The reservation is sent through all the path till the destination according normal routing. The reservation is identified by the Flow Id and the Reservation Id as always. When the mobile sender moves to B (which can be everywhere and we are not making any assumptions on that), it tries to set a new reservation from B. The Reservation Id remains unchanged as the source moves. This new reservation can be sent through a certain numbers of routers which are in common with the previous reservation. If this is the case, when a new reservation request comes in a router where a reservation with the same reservation Id already exists, the flow Id is modified to reflect the change of the sender IP address and the same resources allocated for the old reservation are reused. At the merging point (C), a message is sent back to inform the source that, from there on, the protocol tries to exploit the old reservation.
Had no Reservation Id field be used by the protocol, as with RSVP version 1, then the reservation from B would have been seen as completely new because the sender has a different IP address.
This solution allows to reuse resources and increases the probability of obtaining a reservation while moving but, on the other hand, it raises the problem of using a globally unique Reservation Id which has no easy solution.

Mobile Receiver. If the mobile end-system is the receiver of the data, the following approach is possible with RSVP version 2. Let us consider Figure 2 again, assuming that a receiver host located in zone A moves to zone B. The proposed solution is to use a tunnel between the home agent located in A to the foreign agent located in B and to make a separate and independent reservation across the tunnel. When a reservation request arrives at the home

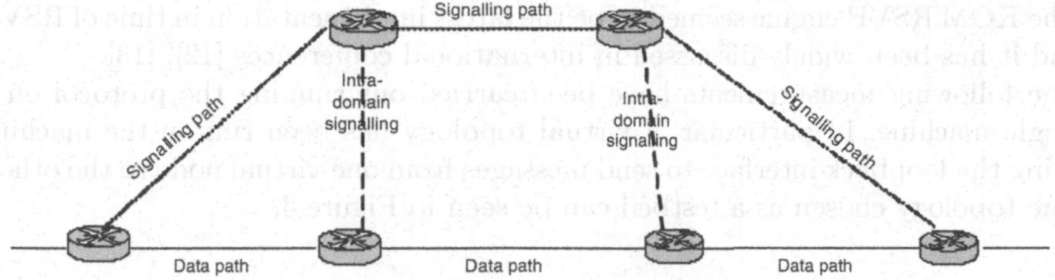

Fig. 3. Out of band signaling

agent, this agent is responsible for the creation of a new reservation to the foreign agent and for forwarding signaling and data messages along the tunnel. The creation of the tunnel may introduce additional delay to the delivery time which means that mobile users can experience temporary disruptions of QoS. Since tunnel management is separated from end to end reservation, it is possible to make tunnel reservation in advance or in a proprietary way limiting this service disruption.

4.3 Signaling between Networks with Centralized Management of Resources

In a heterogeneous network as Internet is, there can be situations where signaling should follow a different path from the one followed by data. In specific, a possible scenario are DiffServ networks [10] where admission control and all the management of resource inside the domain is done by Bandwidth Broker (BB) [11]. In this case, a RESV message should be addressed at IP level to the BB (whose address should be known by the source). When this message arrives at the BB, it is processed and resources are reserved in the appropriate routers accordingly intra-domain decisions. From there, the RESV message should be sent to the BB of the next domain or to the destination according to its routing table. After reserving resources inside a domain, a notification may be sent back to the ingress router to notify the result of the reservation inside the domain. The situation described can be seen in Figure 3. Benefits of this mechanism are that it allows an independent management of resources inside a domain and it separates data to signaling path.

5 Test and Results

In order to verify and quantify the design of the protocol, an implementation in C++ has been developed and this implementation has been used to make some measurements. The implementation has been developed modifying the KOM RSVP Engine developed at the University of Darmstadt. The KOM RSVP engine has been taken as a starting point of this work mainly in order to have a running version of the original RSVP protocol and to be able to make a fair comparison.

The KOM RSVP engine seemed to be the latest implementation in time of RSVP and it has been widely discussed in international conferences [12], [13].
The following measurements have been carried out running the protocol on a single machine. In particular, a virtual topology has been run on the machine using the loopback interface to send messages from one virtual node to the other. The topology chosen as a testbed can be seen in Figure 4.

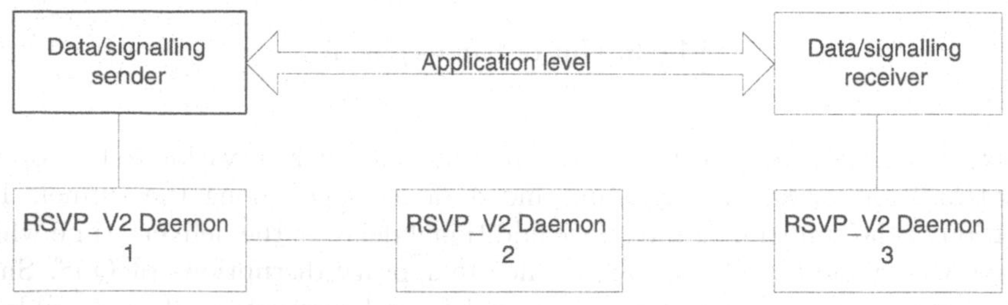

Fig. 4. Virtual topology

5.1 Set Up Time

The *set up time* is an important parameter to measure in a resource reservation protocol. With the term set up time, it is meant the time needed to receive a feedback of a reservation request. In other words, it is the time a source has to wait before sending data with the guarantee of receiving the request QoS. In the specific case of the RSVPv2 protocol, the set up time has been computed as a difference between the time when the confirmation (ACK) of the reservation arrived and the time when the reservation (RESV) message has been sent.
This set up time has been compared to an as much as possible equivalent set up time in the original version of RSVP. For the comparison, it has been considered the difference between the time when a reservation (RESV) message is received and the time a PATH message is sent. This means that the measurement are done at the source of the data flows responsible to send the first signaling packet (PATH message). This means also that time needed to send a reservation confirmation (RESV CONF) to the receiver is not taken into account.
The set up time is not constant but depends on the number of reservation already set in the network. This is mainly due to the increased number of checks that should be performed before accepting a new reservation in each router along the communication path. For this reason, we measured the set up time with different numbers of reservations already in place. Since the measurements have been taken on a single machine, the number of reservations (flows) could not exceed 20000 without overloading the CPU and running out of memory.
The result can be seen in Figure 5. The variability of the curve is justified by the multiprocess nature of the simulation. In fact, the measurements are taken

more than once, but since we run the tests on a single machine, a larger number of measurements is needed to obtain a good average.

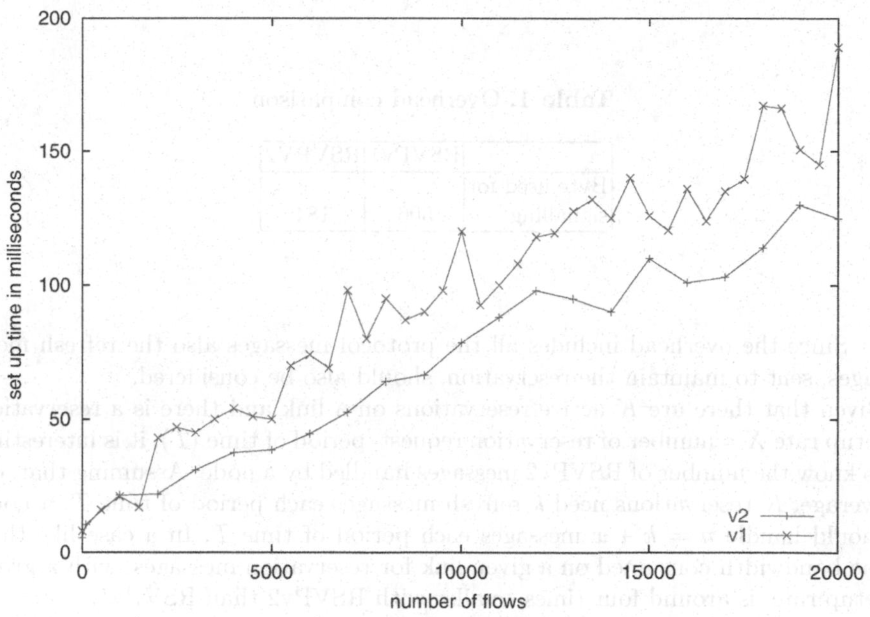

Fig. 5. Set up time

5.2 Overhead

Another important parameter to consider evaluating the protocol is the overhead. The protocol overhead is determined by three factors: the number of messages sent, the size of these messages, and the refresh frequencies of both path and reservation messages. In this specific case, we make a comparison between the number of messages and their size used to set up and tear down a reservation with RSVP version 1 and RSVP version 2.

This comparison can be interesting in a scenario where the bandwidth is expensive (e.g. wireless environment) or there are often congestions. In these cases, having a protocol with a large overhead instead of one with a smaller one can make the difference or at least not make the situation worse than it already is. In the comparison, the whole message exchange to obtain a reservation has been considered. For RSVP version 1 it has been considered the path message, the reservation message, the reservation confirmation and tear down message, while for RSVPv2 are considered the reservation, the ACK message and the reservation tear down. In the messages themselves, all parameters for setting up a reservation (like AdSpec object, Flow spec object, Filter spec object in RSVP) have not taken into account. Policy data and Security object has not been considered neither, in order to make a comparison as fair as possible between the two protocols. The Table 1 shows the overhead of the two protocols in number of Byte.

Table 1. Overhead comparison

	RSVPv1	RSVPV2
Byte used for signalling	506	184

Since the overhead includes all the protocol messages also the refresh messages, sent to maintain the reservation, should also be considered.
Given that there are K active reservations on a link and there is a reservation setup rate X =number of reservation request/period of time (T), it is interesting to know the number of RSVPv2 messages handled by a node. Assuming that, on average, K reservations need k refresh messages each period of time T, a node should handle $n = k + x$ messages each period of time T. In a case like this, the bandwidth consumed on a given link for reservation messages, with a given setup rate, is around four times smaller with RSVPv2 than RSVPv1.
In all these simulations we needed to fix many parameters, otherwise the simulations were getting too extensive and not presentable in the paper.

5.3 Memory Usage

Another interesting parameter to evaluate is the memory used by the protocol. In order not to give absolute value difficult to interpret, it has been compared the percentage of memory used by RSVPV2 and RSVP running 5000 flows. The results can be seen in table 2.

Table 2. Memory usage

	RSVP %Mem	RSVPV2 %Mem
Daemon1	12.2	3.7
Daemon2	10.5	2.8
Daemon3	12.2	3.7

These results finds explanation in the fact that RSVP version 1 saves information for path and reservation messages, including routing information like next and

previous hop address. On the other hand, RSVPv2 tries to limit the number of information to be stored along the path in order to obtain a better scalability and reduce the impact of the protocol on routers.
Since RSVPv2 is supposed to be 'light', these results should be considered as a first confirmation of working into the right direction, leaving optimization towards better performance for further studies.

6 Conclusions and Future Works

This paper proposes a first step towards a new reservation protocol able to suit the largest number of present and future scenarios and to operate within different technologies. The idea is to have a basic, network architecture independent, signaling structure for exchanging requests where extra features can be build on. The protocol will be tested on a real network topology and other and maybe more significant measurements will be done in that environment to better study the protocol performances.
Integration of the protocol with DiffServ and MPLS is foreseen, together with the integration of security functions not yet included in the protocol at the moment.

References

[1] F. Baker K. Nichols, S. Blake and D. Black, "Definition of the Differentiated Service Field (DS field) in the IPv4 and IPv6 Headers," RFC 2474, December 1998.

[2] M. Brunner et al, "Requirements for QoS signalling protocols," draft-brunner-nsis-req-03.txt, July 2002.

[3] R. Yavatkar D. Durham, "Inside the Internet's Resource reSerVation Protocol," Wiley Computer Publishing, 1999.

[4] B. Quinn K. Almeroth, "IP multicast applications: Challenges and Solutions," RFC 3170, September 2001.

[5] L. Delgrossi and L. Berger, "Internet Stream Protocol Version 2 (ST2), Protocol specification - Version ST2+," RFC 1819, August 1995.

[6] J. Crowcroft P. White, "A case For Dynamic Sender-Based Reservations in the Internet," *The Journal of High Speed Networks*, 1998.

[7] G. Fehér et al., "Boomerang - A simple Protocol for Resource Reservation in IP networks," in *IEEE Workshop on QoS Support for Real-Time Internet Application*, Vancouver, Cananda, June 1999.

[8] T. Ferrari W. Almesberger, J. Le Boudec, "Scalable Resource Reservation for the Internet," in *Proceeding of IWQoS'98*, Napa, CA, May 1998.

[9] P. Pan and H. Schulzrinne, "YESSIR: A simple reservation mechanism for the internet," in *Procedings of International Workshop on Network and Operating System Support for Digital Audio and Video (NOSSDAV)*, July 1998.

[10] M. Blake et al., "An Architecture for Differentiated Services," RFC 2475, December 1998.

[11] V.Jacobson L.Zhang, K.Nichols, "A two-bit differentiated services architecture for the Internet," RFC 2638, July 1999.
[12] R. Steinmetz M. Karsten, J. Schmitt, "Implementation and Evaluation of the KOM RSVP Engine," in *Proceeding of the 20th annual Joint conference of the IEEE Computer and Communications Societies (INFOCOM'2001)*, April 2001.
[13] M. Karsten, "Design and Implementation of RSVP based on Object-Relationships," Shortened version appears in Proceedings of Networking, March 2000.

Analysis of SIP, RSVP, and COPS Interoperability

Csaba Király, Zsolt Pándi, Tien Van Do

Department of Telecommunications, Budapest University of Technology and Economics, Pf. 91., 1521 Budapest, Hungary
{cskiraly, pandi, do}@hit.bme.hu
http://www.hit.bme.hu/

Abstract. The All-IP network concept with end-to-end QoS provisioning has received particular attention in 3GPP recently. The UMTS proposals, however, have not yet solved some protocol interoperability issues. This paper analyzes the IP Multimedia Subsystem from the aspect of call control, resource reservation and network policing interoperability from the viewpoint of implementations. More specifically, the experiences based on a prototype implementation of the IMS based on SIP, RSVP and COPS are analyzed and conclusions are drawn to support the standardization process, as well as future implementations. The considered architecture is general and can be applied also to fixed IP networks.

1 Introduction

Mobile communications, as well as mobile telephony, is one of the hottest areas in the networking world. The number of mobile users has grown steadily throughout the recent years, which indicates the global success of second generation mobile networks such as GSM, CDMA and TDMA. The future of mobile communications is just being defined in the framework of the 3rd Generation Partnership Project (3GPP) which organizes worldwide research on the wireless standardization. The standardization of the radio access network of Universal Mobile Telecommunications System has already been finished, but that of the core network is still in progress. In the evolving standards some trends can be perceived, which represent the intentions of the standardization body [1], [2].

At the present 3GPP is aiming at the All-IP network concept as a final target. However, one of the most serious issues that still need to be resolved is the end-to-end QoS (Quality of Service) provision. To provide QoS an appropriate networking technology must be selected with appropriate signaling protocols and mechanisms which implement the following major functions: call control, QoS architecture for resource reservation and network policing.

3GPP standards regarding the IP Multimedia Subsystem (IMS) [1], [2] define an architecture which consists of the following components.

Firstly, for call control and signaling the Session Initiation Protocol (SIP) is proposed. SIP is used in various multimedia services and originates from the IP world, that is, it incorporates concepts and design patterns characteristic of well-known pro-

M. Ajmone Marsan et al. (Eds.): QoS-IP 2003, LNCS 2601, pp. 717-728, 2003.

tocols applied in the Internet, such as the Hypertext Transfer Protocol (HTTP) [3]. Recently the role of SIP has gained strength due to its flexibility and scalability. This protocol could be a means of implementing the Intelligent Network concept and is certainly an important step towards NGNs [4].

Secondly, as far as QoS architecture for resource reservation is regarded, the Internet Engineering Task Force (IETF) has elaborated two fundamental service architectures for QoS provisioning in IP networks: the Integrated Services (IntServ) architecture and the Differentiated Services (DiffServ) architecture ([5], [6] and [7]). DiffServ provides QoS for aggregate flows; therefore, it is primarily intended for use in the core network domain due to its scalability. On the other hand, the flow based IntServ provides QoS for each flow on a separate basis through the Resource Reservation Protocol (RSVP) at the expense of higher administrative costs; consequently, it is only suitable for the access network domain. The interworking of the two architectures has already been demonstrated by for example the ELISA and MQOS projects [8]; moreover, IETF recommendations also exist [9]. 3GPP standards propose the use of both architectures.

Finally, network policy control is inevitable in QoS provision. This is generally related to authentication, authorization and accounting (AAA), as well as resource management and call admission control. Service requests of users must be either accepted or refused based on several policy rules. Then this decision must be executed and adhered to at the appropriate network devices. The Common Open Policy Service (COPS) is a promising candidate protocol for communication on policy decisions between the so called Policy Decision Points (PDPs) and Policy Enforcement Points (PEPs) [10]. It has been defined so that it could be applied with RSVP, as well [11]. 3GPP standards propose COPS for the communication between PDPs and PEPs.

Although the functional components of the IMS and their operation have already been defined to a certain extent, the necessary interoperability of call control, resource reservation and network policing has not yet been covered. Therefore, this paper aims at the analysis of the IMS architecture focusing on service provision and protocol interoperability. To facilitate the analysis a prototype implementation was prepared. Based on the experiences gathered from the implementation important conclusions can be drawn regarding issues that still need further clarification and recommendations can be given to help the standardization process and future implementations. Moreover, as the same problems arise in fixed IP networks, our results can be applied in that context, as well.

Obviously, the implementation of additional functions, such as authentication and accounting is also inevitable for service provision; however, these remain out of the scope of the present paper.

The rest of the paper is organized as follows. Section 2 gives a short overview of the IMS architecture, while Section 3 briefly considers some issues related to the prototype implementation. Section 4 discusses call setup signaling, then Sections 5 and 6 deal with the Policy Control Function and SIP Proxy Server entities, respectively. Section 7 analyzes the requirements for a communication protocol between these two entities and Section 8 concludes the paper.

2 Short Overview of the IMS Architecture

In this section only the most important functional entities of the IMS are covered. For a more detailed description the reader is referred to [12].

A general scenario for the application layer signaling in the IMS architecture is illustrated in Fig. 1. According to [2] the most important functional elements in the IMS are the Gateway GPRS Support Node (GGSN), the Proxy-Call Session Control Function (P-CSCF) and the User Equipment (UE). Within the P-CSCF there are two fundamental functional elements: a local SIP proxy and a Policy Control Function (PCF).

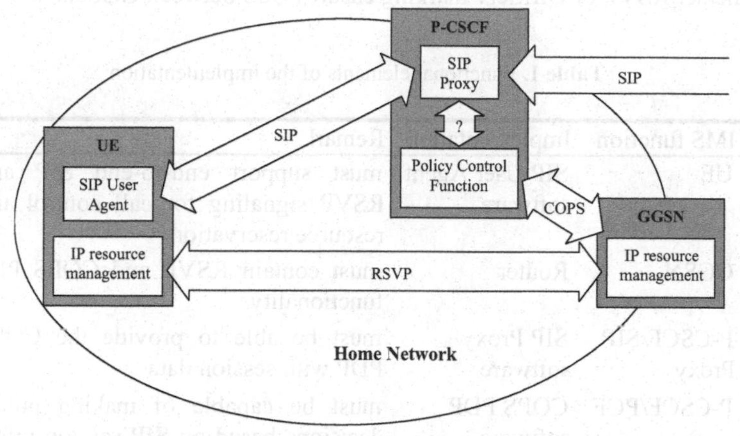

Fig. 1. Application layer signaling scenario in the IMS

[2] discusses several possible scenarios where these functional elements have different capabilities. For example UEs may support RSVP signaling or DiffServ edge functions, but they do not have to have IP bearer service management functionality at all.

Assuming a scenario where UEs are RSVP-capable, that is, RSVP signaling is end-to-end and where GGSNs are not transparent forwarders of RSVP messages (scenario 4 in Annex A of [2]) the components of the architecture must have the following functionality. The *IP resource management function in the UE* is responsible for QoS requests using a suitable protocol (e.g. RSVP), whereas the *IP resource management function in the GGSN* must contain IP policy enforcement and DiffServ edge functionality. The *PCF* in the P-CSCF communicates with the GGSN through the Go interface, which is used for transmitting policing related data and policy decisions between the two entities. COPS is proposed for use in the Go interface. The P-CSCF also contains *SIP Proxy* functionality to be able to track current SIP calls and thus make appropriate policy decisions about resource reservation requests. However, the interface between the local SIP Proxy and the PCF is still undefined in the standard.

3 Adopted Approach: Prototype Implementation

In order to be able to analyze the IMS we had to implement a SIP User Agent (UA), a SIP Proxy, and a COPS Policy Decision Point (PDP) software with the appropriate functionalities listed in Table 1. These requirements were derived from [1] and [2], and will be detailed later on. As the implementation of the GGSN a commercially available and widespread IP router was used.

In the GGSN the DiffServ functionality was not used due to the following reasons. Firstly, in the focus of interest there are the requirements for the Home Network and not the IP cloud beyond the GGSN. Moreover, [8] already demonstrated how RSVP might be used over a DiffServ domain; therefore, from our point of view it is irrelevant whether RSVP or DiffServ marking ensures QoS between GGSNs.

Table 1. Functional elements of the implementation

IMS function	Implementation	Remark
UE	SIP User Agent software	must support end-to-end SIP and RSVP signaling for call control and resource reservation
GGSN	Router	must contain RSVP and COPS PEP functionality
P-CSCF/SIP Proxy	SIP Proxy software	must be able to provide the COPS PDP with session data
P-CSCF/PCF	COPS PDP software	must be capable of making policy decisions based on SIP session information and a priori configuration data

During the software design and development several problems were faced regarding the interoperability issues of the applied protocols. In the rest of the paper these points will be covered. Firstly, we will overview the actions that must be taken during service provision, especially at call setup, as this is the most difficult process in the IMS. Other processes related to QoS provisioning can be handled in a fairly straightforward manner, therefore their discussion will be omitted. After discussing the call setup all the requirements derived from the functionalities listed in Table 1 will be analyzed for each system component that belongs to the service provider, except for the GGSN, as the role of this entity is adequately defined, and its implementation does not entail interoperability issues for the investigated protocols.

4 Call Setup Signaling Scenario

Due to the separation of the bearer and control planes the call setup process is a difficult process with multiple entities, which may fail due to the following reasons:

1. Either the caller or the callee is not authorized to make the call.
2. Either the caller or the callee tries to make an illegal step during call setup, for example issues a reservation request for too much resources.
3. The authorized reservation request fails due to the lack of resources (supposing that QoS assured operation mode is implemented).

Moreover, "ghost rings" must be avoided, that is, the device of the callee must not ring if the call cannot be set up for any reason. Fig. 2 demonstrates a successful call setup process until the phone of the callee begins to ring. Dark (light) background shading marks the RSVP and COPS messages associated with the media stream sent by the caller (callee).

The functional elements interact during call setup as follows:

1. A UE initiates the call setup, the UEs negotiate and agree in the media streams, the set of applicable codecs for each stream and the necessity of QoS with an SDP offer/answer pair [13] in the INVITE and 183 messages[1].
2. Both on the caller and callee side the SIP Proxies forward the SDP information to the PCFs.
3. For both users the appropriate PCF decides whether the user is authorized and which billing category the call belongs to.
4. The RSVP messages sent by UEs for each media stream arrive at the GGSN, which requests a decision using COPS from the PCF. The PCF then examines for each request if the RSVP parameters conform to the parameters negotiated via SDP. If they do, the resource allocation will be permitted, and rejected otherwise.

Finally, the UEs notice the success or failure of their RSVP requests. A request might be rejected by the PCF (due to policy reasons) or CAC at the GGSN (due to the lack of resources). Depending on the result of this preliminary negotiation the call setup may proceed by sending Ringing messages.

At the time when the presented work was in progress the current version of the SIP RFC [14] did not contain any recommendations regarding the integration of resource reservation management in the SIP protocol state machine. Neither work that was under progress in the IETF [15] nor the new proposed standard published recently [16] contained such recommendations. However, the topic was addressed in other IETF drafts [17] and [18], which were being developed continuously. The available draft version at the time of the specification of the architecture recommended the use of COMET and PRACK messages for this purpose. Hence we adhered to these specifications. However, out of these two messages the first was replaced by the UPDATE message used for the same purpose here, while the second one became a proposed standard [19]. Even though the UPDATE message replaced COMET and its function, the analysis of the architecture using COMET will not loose its relevance, as the basic functionality of the two alternatives does not differ.

As far as SDP usage is regarded, we adhered to the SDP offer/answer model of [13] (published later as an RFC [20]) and [18].

---

[1] The offer/answer can also be in 183 and PRACK when media streams are defined by the callee.

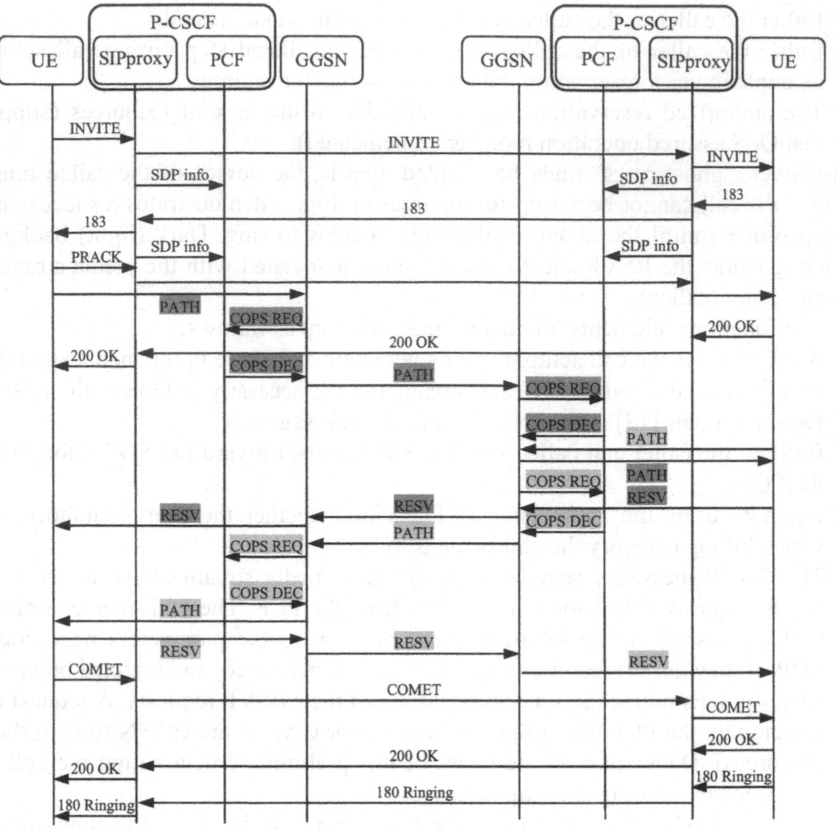

Fig. 2. Message sequence diagram of a successful call setup

5 Policy Control Function

As it was already mentioned, the PCF collects all the necessary call parameters from SIP and RSVP signaling and decides whether the resource reservation request of the user may proceed. For making the policy decision the PCF must use a Policy Information Base (PIB), that contains identification information and service contract details for each user (user profile). Moreover, the PCF must have a predefined set of decision rules, and another database describing resource requirements of different codecs in terms of RSVP parameters.

5.1 Input Information for the Decision

When defining the decision process, the starting point was the fact that the information contained in RSVP messages is insufficient for deciding whether the caller (callee) is authorized for requesting the indicated resources. Transmitting resource management related information in SDP parameters embedded in SIP messages is more conformant to the business model of service providers.

Various pieces of information might be used for authorization and accounting depending on the business model, which leads to a decision mechanism that is far more general and flexible than decision making based only on RSVP parameters. The most important features of a general decision mechanism are as follows:

- integrated management of multiple media streams used in the same session,
- distinction based on the locations of the other party (same network/foreign network),
- distinguished callers or callees (e.g. customer service, emergency calls),
- distinction between different media types (e.g. audio, video, whiteboard),
- distinction between codec performance levels (e.g. high and low quality video conference),
- constraints on the number of simultaneous calls.

It is important to note, however, that although these aspects might be involved in the decision process, some of them imply the conformance of user terminals, which cannot be controlled by the service provider. For example, SDP parameters contain the codec information, but it is not reasonable to check whether user terminals apply one of the negotiated codecs, as this would mean huge additional load for GGSNs. We do not deem the violation of this conformance requirement to be a serious business risk, as this fraud would require the modification of user terminals.

5.2 Decomposition of the Decision Process

The decision process can be decomposed to the following stages:

1. The PCF checks whether the service contract permits the call for the user according to the conditions mentioned above, and determines which billing category the call belongs to.
2. The PCF checks whether the resource reservation request is reasonable based on call parameters.

The first stage can be carried out based on the contents of the SDP offer/answer and the user profile, and a so-called RSVP Envelope can be determined for each direction of each media stream, which describes the maximal allowed resource reservation. When the COPS request arrives later on[2] and the final decision has to be made, this RSVP Envelope, which in fact contains the necessary information in a contracted form, will provide the boundaries for the comparison.

This two-stage process simplifies the definition of decision rules, and, although unsupported by COPS, facilitates pushing decision information to the GGSN, resulting in a reduced call setup time. Even though this two-stage process is not equivalent

---

[2] If the PRACK message carries SDP information, the COPS REQ might arrive at the PCF earlier than the SDP information forwarded by the SIP proxy due to the independence of paths taken by SIP and RSVP messages. In this case, the decision must be postponed until SDP information arrives.

to a single decision made when all the necessary information is available, it is still flexible enough to support adequate differentiation.

5.3 RSVP Envelopes

Annex C of [2] demonstrates through an example how to determine the amount of resources that the user is allowed to reserve, that is, the RSVP Envelope. This is intended to prohibit unauthorized resource reservations. The envelope may introduce constraints on the following parameters based on the SDP description:

- FilterSpec parameters:
  - source IP address
  - destination IP address
  - protocol ID
  - destination port(s)
- FlowSpec parameters:
  - mean rate (r)
  - peak rate (p)
  - bucket depth (b)
  - minimum policed unit (m)
  - maximum packet size (M)

Determining the IP addresses based on the SDP description guarantees that the reservation can only be accepted between the caller and the callee. Port numbers depend on media type and the applied transport protocol according to [21], and this RFC only requires that they could be determined by an algorithm using the number contained in the SDP parameters. The PCF, therefore, must be aware of these media and transport dependent calculation methods, as well.

FlowSpec parameters can be calculated based on the set of codecs determined by SDP request and answer. The SDP information only describes the set of codecs supported by both parties, not the currently applied codec. In addition to this, the applied codec may be selected arbitrarily from this set, and can be changed during the call; consequently, the envelope must fit all of the indicated codecs. The RSVP parameters necessary for an appropriate reservation can be determined for each codec based on their parameters like frame size and frame duration. From the aspect of implementation it is more straightforward to store RSVP parameters associated with each codec in the PCF database. The envelope can then be calculated by finding the maximum (r, p, b, M) and the minimum (m) of the RSVP parameters associated with the codecs in the set.

This calculation method might significantly overestimate real resource use (consider the case when the set of codecs contains a G.711 and a GSM FR codec, and all users apply the latter), but the user will have no interest in issuing a too large reservation request if it costs more money. However, if accounting is based on the media type (like telephone quality audio) and not on the actual codec in use, the only reason why a user should choose a lower bandwidth codec is the slightly higher probability of the successful reservation.[3]

6 SIP Proxy Server

The SIP proxy is primarily applied in the PCF to extract SIP session and SDP codec data from SIP call flows. This additional function is a relatively simple extension of the standardized SIP proxy function set.

[15] defines two types of SIP proxy server behavior, namely transaction stateful and transaction stateless proxies. The former should track transaction state (but not necessarily call state) whereas the latter does not keep state information. Since SIP and SDP specifications have been changing rapidly during the last year, we recommend the use of a stateless proxy for the following reasons:

- The proxy does not generate SIP messages, so the SIP signaling can be end-to-end.
- A stateless proxy can simply forward messages without following transactions and dependencies between messages, thus implementation and future adoption according to the non-final standard is more easy.
- Better scalability can be achieved with the stateless design.

However, there are some counterarguments to consider:

- The proxy can only perform syntactic check on SIP messages, but it cannot force the correct order of SIP messages. This check can be implemented in the PCF but it violates our goal to separate responsibilities, since the PCF itself does not take part in SIP signaling.
- The proxy can not filter out repeated (retransmitted) SIP messages. Retransmitted SIP messages, however, should be identical, so in case of repeated SIP messages the stateless proxy sends repeated messages to the PCF as well, which can easily be recognized.
- The stateless proxy cannot support advanced functions like forking proxy mode. This restriction can be relaxed if necessary, since another general-purpose SIP Proxy can be chained after the stateless proxy.
- Implementing a SIP authentication scheme stronger than "HTTP Basic" authentication can be complicated.

A stateful proxy does not have these problems, however, its implementation and maintenance requires far more efforts than that of a stateless proxy, due to the fact that changes in SIP and SDP standards must be followed.

[3] Unfortunately, some RSVP implementations do not support the modification of RSVP reservation parameters without teardown, which, in fact, forces the users to issue larger reservation requests in order to ensure that during the call they will be able to change to a codec that needs more resources without loosing the QoS guarantee.

7 Protocol between the SIP Proxy and the PCF

As it was previously noted, the communication protocol between the PCF and the SIP Proxy is not yet standardized, therefore we had to elaborate its details. In this paper we restrict ourselves to review some considerations and decisions without describing the protocol syntax.

The PCF has to receive enough information from the SIP Proxy to

- identify the user,
- recognize protocol messages belonging to the same session,
- calculate RSVP Envelopes based on media and codec information in the SDP offer/answer pair,
- couple COPS requests with the corresponding RSVP Envelope.

7.1 Protocol Messages and Statefulness of the SIP Proxy

Assuming that the SIP Proxy is stateless, it cannot gather information about the INVITE transaction or call state, thus a message must be sent to the PCF every time a SIP message contains any information necessary for the decision. These SIP messages are INVITE, 183, PRACK, COMET, 200 messages for call setup and BYE, CANCEL for ending the session.

When communicating with the PCF the protocol messages sent by the SIP proxy must contain data from the *to*, *from*, *call-id*, and *cseq* fields of the SIP message to facilitate user and call identification and authentication. At the same time, RSVP Envelope calculation requires the transmission of SDP body copied from the SIP message, as well. The latter provides enough information also to match COPS requests and RSVP Envelopes. To extract this information from the forwarded SIP messages the SIP Proxy has to do only minimal processing on SIP messages and it can handle SDP data transparently, which may simplify the proxy implementation even further.

A stateful proxy, on the other hand, may gather all the SIP and SDP information necessary for the decision and forward it in a single message. However, this delays the decision process significantly, as the PCF can only consult the PIB when it receives the message from the SIP Proxy. Therefore, it is advantageous to forward different pieces of information as soon as they become available.

7.2 Transport Protocol

The transport protocol used for the transmission of the Proxy-PCF protocol messages must be reliable and it should be non-blocking to allow for the easy implementation of a stateless proxy. A persistent connection should be set up between the SIP Proxy and the PCF, thus the TCP protocol is a straightforward choice for this purpose. If, however, the Proxy and the PCF entities are co-located in the same network device, or they are components of the same software, then any other means of non-blocking and reliable data transfer might be suitable.

7.3 Enhancements via Feedback

The protocol messages mentioned so far are all uni-directional (the SIP Proxy notifies the PCF). We think that all the necessary functionality can be implemented with this uni-directional protocol, although bi-directional communication would facilitate:

- the modification of SDP and SIP parameters in view of the user profile,
- user notification via SIP in the following cases:
  - The SIP session could be ended in case of an RSVP error, which is of particular importance if the error is between the UE and the GGSN, and the UE loses the connection with the GGSN (which may happen in mobile networks).
  - Authentication parameters could be forwarded to the UE.
  - Information calculated from the user profile and the SDP could be forwarded to the UE.
  - If the authentication fails the UE could be notified instantly, as opposed to the present situation, where it will only be notified about rejected PATH messages. (The latter results in superfluous delay, unnecessarily reserved resources and the fact that the other end will also begin to set up the RSVP session.)

Nevertheless, those opting for utilizing the advantages of bidirectional communication must be aware of the fact, that interacting in the SIP session this way breaks the end-to-end nature of SIP signaling and it is not conformant to the current SIP standard.

8 Conclusions

This paper aimed at the analysis of the IP Multimedia Subsystem of third generation mobile networks from the aspect of call control, resource reservation and network policing protocol interoperability issues related to service provisioning.

The investigation was carried out through discussing some considerations based on a prototype implementation of the IMS developed by the authors beforehand. We analyzed several implementation options for the two most important functional entities of the system, the Policy Control Function and the SIP Proxy.

We proposed a two-stage decision process at the PCF, which simplifies the definition of decision rules, while remains flexible at the same time. We also showed that a stateless SIP Proxy capable of communicating with the PCF is easy to implement and that any existing proxies can be enhanced with this function via proxy chaining. Finally, we overviewed the necessary characteristics of a candidate protocol between these two functional entities.

Although the aim was to analyze the IMS of UMTS, the results presented in this paper are general enough to be applied for service provisioning in fixed IP networks, as well.

9 Acknowledgements

This work was supported by NOKIA Hungary. The authors would like to express their gratitude to György Wolfner of NOKIA for the valuable discussions. We would like to thank also Dóra Erős of NOKIA for her continous support during the course of this work.

References

1. 3GPP TS 23.228 (V5.5.0): IP Multimedia Subsystem (IMS). June 2002
2. 3GPP TS 23.207 (V5.4.0): End-to-End QoS Concept and Architecture. June 2002
3. Schulzrinne, H., Rosenberg, J.: The Session Initiation Protocol: Internet Centric Signaling. IEEE Communications Magazine, October 2000, p. 134
4. Canal, G., Cuda, A.: Why SIP will Pave the way towards NGN. In Proceedings of the 7 th ITU International Conference on Intelligence in Networks, October 2001
5. Braden, R. et al.: Resource ReSerVation Protocol (RSVP) - Version 1 Functional Specification. RFC 2205, September 1997
6. Herzog, S.: RSVP Extensions for Policy Control. RFC 2750, January 2000
7. Blake, S. et al.: An Architecture for Differentiated Service, RFC 2475, January 2000
8. Detti, A. et al.: Supporting RSVP in a Differentiated Service Domain: an Architectural Framework and a Scalability Analysis. International Conference on Communications, Vancouver (Canada), June 1999
9. Bernet, Y. et al.: A Framework for Integrated Services Operation over Diffserv Networks. RFC 2998, November 2000
10. Durham, D. et al.: The COPS (Common Open Policy Service) Protocol. RFC 2748, January 2000
11. Herzog, S. et al.: COPS Usage for RSVP. RFC 2749, January 2000
12. Lin, Y., Pang, A., Haung, Y., Chlamtac, I.: An All-IP Approach for UMTS Third-Generation Mobile Networks. IEEE Network, vol. 16, no. 5, September 2002
13. Rosenberg, J. et al.: An Offer/Answer Model with SDP. draft-rosenberg-mmusic-sdp-offer-answer-00 (formerly RFC2543 Appendix B), October 2001
14. Handley, M. et al.: SIP: Session Initiation Protocol. RFC 2543, March 1999
15. Handley, M. et al.: SIP: Session Initiation Protocol. draft-ietf-sip-rfc2543bis-05, October 2001
16. Rosenberg, J. et al.: SIP: Session Initiation Protocol. RFC 3261 (formerly draft-ietf-sip-rfc2543bis-09), June 2002
17. Camarillo, G. et al.: Integration of Resource Management and SIP. draft-ietf-sip-manyfolks-resource-02, February 2001
18. Johnston, A. et al.: SIP Call Flow Examples. draft-ietf-sip-call-flows-05, June 2001
19. Rosenberg, J. et al.: Reliability of Provisional Responses in the Session Initiation Protocol (SIP). RFC 3262, June 2002
20. Rosenberg, J. et al.: An Offer/Answer Model with the Session Description Protocol (SDP). RFC 3264 (formerly draft-rosenberg-mmusic-sdp-offer-answer-00), June 2002
21. Handley, M. et al.: SDP: Session Description Protocol. RFC 2327, April 1998

Simulation Study of Aggregate Flow Control to Improve QoS in a Differentiated Services Network*

Sergio Herrería-Alonso, Andrés Suárez-González, Manuel Fernández-Veiga, Raúl F. Rodríguez-Rubio, and Cándido López-García

ETSE de Telecomunicación
Universidade de Vigo
36211 Vigo, SPAIN
sha@det.uvigo.es

Abstract. The Differentiated Services architecture is a simple and scalable approach to provide Quality of Service (QoS) in IP Networks. Several studies have shown that the number of microflows in aggregates, the round trip time (RTT) or the mean packet size are key factors in the throughput of aggregates obtained using this architecture. In this paper, we examine the behaviour of one of the techniques suggested to improve fairness in a Diffserv network: the Aggregate Flow Control mechanism. We also propose two alternatives in the control overlay of this scheme and compare them with the original approach. Simulation results indicate that our proposed modifications improve throughput assurance and fairness requirements.

1 Introduction

The Differentiated Services [1] architecture is one of the recent proposals to address QoS issues in IP networks. This architecture relies on packet tagging and lightweight router support to provide assured services that extend beyond best effort. The differentiated service is obtained through traffic conditioning at the edge of the network and simple differentiated forwarding mechanisms at the core.

One of the packet-handling schemes introduced by this architecture is the *Assured Forwarding* (AF) [2] *Per Hop Behavior* (PHB). The basis of the AF PHB is to differentiate packets by marking them, based on conformance to their target throughputs. Subscribed traffic profiles for customers must be maintained at the edge of the network. The *Time Sliding Window Three Color Marker* (TSWTCM) [3] is one of the packet marking algorithms proposed to work with AF. In this algorithm, two target rates are defined: *Committed Information Rate* (CIR) and *Peak Information Rate* (PIR). The aggregated traffic is monitored. When the measured traffic is below its CIR, packets are marked with the lowest

* This work was supported by the project TIC2000-1126 of the Plan Nacional de Investigación Científica, Desarrollo e Innovación Tecnológica.

M. Ajmone Marsan et al. (Eds.): QoS-IP 2003, LNCS 2601, pp. 729–741, 2003.

drop precedence, *AFx1*. If the measured traffic exceeds its CIR but falls below its PIR, packets are marked with a higher drop precedence, *AFx2*. Finally, when traffic exceeds its PIR, packets are marked with the highest drop precedence, *AFx3*.

At the core of the network, the different drop probabilities can be achieved with the RIO (RED with In/Out) [4] scheme, an active queue management technique with three different sets of RED parameters, one for each of the drop precedence markings.

The two major requirements of assured services are [7]:

- **Throughput assurance**: each aggregate should receive its subscribed target rate on average.
- **Fairness**: in under-subscribed networks[1], the extra (unsubscribed) bandwidth should be evenly shared among aggregates.

Unfortunately, unfairness has been found in sharing the extra bandwidth among aggregates with different RTTs, different mean packet sizes or different number of microflows in the aggregate [5,6]. Many smart packet marking mechanisms have been proposed to overcome these fairness problems [9,10,11,12] but, either they are unable to mitigate fully key unfairness factors or their deployment is really complicated. Another approach is to address these issues by enhanced RIO queue management algorithms [7,8]. However, these solutions may not be scalable since they depend on state information at the core of the network.

Aggregate Flow Control (AFC) [13] is an edge-to-edge control mechanism that, combined with Diffserv traffic conditioning, addresses assurance issues for AF-based services. The AFC mechanism is based on some TCP control connections associated with each customer traffic aggregate. In this paper, we propose two modifications in the control overlay of the AFC scheme which improve throughput assurance and facilitate the deployment of this technique.

The rest of the paper is organized as follows. In Section 2 we give a brief overview of the operation of AFC (a complete description may be found at [13]). Two new kinds of control flows are presented in Section 3. Section 4 presents and compares the performance of the original AFC scheme and the two modifications derived from it under different simulation scenarios. We end the paper with concluding remarks in Section 5.

2 Aggregate Flow Control

AFC [13] manages the customer traffic aggregates in a controlled, fair manner. This scheme regulates the flow of the aggregated customer traffic into the core of the network. Customer aggregates that are exceeding their committed rate and causing congestion at the core are throttled at the edge of the network.

AFC works in the following manner: (a) Several control TCP connections are associated with each customer aggregate between two network edges; (b)

[1] Under-subscription occurs when aggregate demand does not exceed network capacity.

Control TCP packets are injected into the network to detect congestion along the path of the aggregated data (control and data packets must follow the same path between the edges); (c) Congestion control is enforced on the aggregated traffic at the ingress edge router based on control packets drops.

User packets are classified as belonging to a particular customer aggregate and queued according to that classification. User packets can only be forwarded when credit for the corresponding aggregate is available. When a packet is sent, the credit is decremented by the size of the packet. For every *vmss* (virtual maximum segment size) user bytes transmitted, a control packet is generated and sent. This increments the credit by *vmss* bytes. The user and control packets are metered, policed, and marked in the same manner of any standard Diffserv traffic conditioner. Therefore, this scheme does not require any changes to the Diffserv module.

The loss of control packets in the network slows down the transmission rate of control packets. Since control packets regulate data packets, the loss of control packets will slow the transmission rate of user data.

3 New Proposed TCP Control Flows

In [13] the authors propose the use of four TCP NewReno [14] control flows for each aggregate. If each aggregate was regulated by a single control flow, the congestion control would be too harsh: a control packet loss would cause the entire aggregate to halve its sending rate.

Since this is the only reason to use more than an unique control flow per aggregate, we suggest and study two interesting alternatives as control flows: binomial congestion control algorithms [16] and TCP Vegas implementation [17].

3.1 Binomial Congestion Control Algorithms

Binomial algorithms [16] are nonlinear congestion control algorithms that generalize the AIMD (Additive-Increase/Multiplicative-Decrease) rules in the following manner:

$$\text{I: } w_{t+R} = w_t + \alpha / w_t^k$$
$$\text{D: } w_{t+\delta} = w_t - \beta w_t^l$$

where w_t refers to the window size at time t, R the RTT of the flow, α and β are constants ($\alpha > 0$, $0 < \beta < 1$), and k and l are the parameters of the binomial algorithm.

For $k = 0$, $l = 1$, we get AIMD. The binomial algorithm SQRT ($k = l = 0.5$) is significantly less aggressive than the AIMD while still being TCP-friendly[2], so we evaluate the use of a single SQRT control flow for each aggregate (AFC-BI).

[2] TCP-friendly condition: $k + l = 1$ and $l \leq 1$

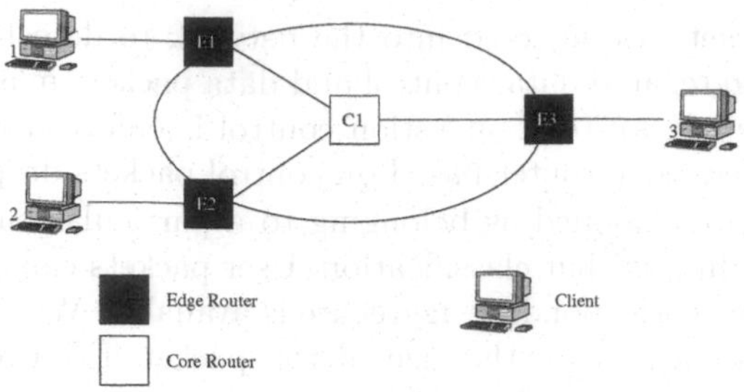

Fig. 1. Simulation topology

3.2 TCP Vegas

TCP Vegas [17] is an implementation of TCP that employs some techniques to increase throughput and decrease losses. One of these techniques permits a proactive congestion detection and avoidance (Reno is reactive). This mechanism can detect the incipient stages of congestion before losses occur.

Vegas calculates the expected throughput and the current actual throughput. Then Vegas compares these throughputs and adjusts the window accordingly. Let $\mathit{Diff} = \mathit{Expected} - \mathit{Actual}$. Vegas defines two thresholds $\alpha, \beta, \alpha \leq \beta$. When $\mathit{Diff} < \alpha$, Vegas increases the congestion window linearly each RTT. If $\mathit{Diff} > \beta$, Vegas decreases the congestion window linearly each RTT. When $\alpha < \mathit{Diff} < \beta$, the congestion window remains unchanged.

This proactive behaviour makes TCP Vegas interesting for its use as control flow (AFC-VE). Also in this case, employing more than one control TCP Vegas flow per aggregate would have no meaning.

4 Simulation Results

We have implemented the AFC mechanism [13] in the ns-2 simulator [15]. We employ the same network topology (Fig. 1) and configuration parameters used in [13]. Each link in the topology has a 5 Mbps capacity and a 1 ms delay. We consider aggregate 1 runs between client 1 and client 3; and aggregate 2 runs between clients 2 and 3.

The marking scheme used in the edge routers is TSWTCM. Each edge node uses a single-level RED queue configured with the parameters listed in Table 1.

Table 1. Edge RED parameters

Size	min_{th}	max_{th}	max_p	w_q
100 pkts	30 pkts	60 pkts	0.02	0.002

Table 2. Core RED parameters

Precedence	min_{th} (pkts)	max_{th} (pkts)	max_p
AFx1	200	240	0.02
AFx2	160	200	0.06
AFx3	40	160	0.12

The core node implements the RIO scheme: three sets of RED thresholds are maintained, one for each drop precedence (Table 2). The physical queue is limited to 250 packets and w_q equals 0.002.

Customer TCP flows use the NewReno implementation. The original AFC scheme employs four TCP NewReno control flows per aggregate. The AFC-BI scheme only uses one binomial control flow per aggregate with parameters $k = l = 0.5$ (SQRT binomial congestion control algorithm). The AFC-VE scheme employs one control TCP Vegas flow per aggregate with parameters $\alpha = \beta = 1$.

The packet size of customer aggregates is set to 1514 bytes. Control flows send 40-byte packets. A *vmss* of 1514 bytes is used in the original AFC and AFC-BI tests, while a *vmss* of 2×1514 bytes is used with AFC-VE. Therefore the overhead introduced due to control packets is 2.6% in original AFC and AFC-BI. Under the AFC-VE case the bandwidth overhead is reduced to 1.3%.

Each simulation scenario is repeated 10 times, and from them a 95% confidence interval (CI) for the mean value of the measured parameter is computed. We let simulations run for 5 seconds before AFC scheme starts to work. Then we simulate 60 seconds for each run.

4.1 Impact of Number of Microflows

The number of microflows in competing aggregates is a key factor in the throughput obtained using Assured Services. In this scenario, there are two sets of aggregated TCP flows. Each aggregate has the same target: a 0.5 Mbps CIR and a 1.0 Mbps PIR. Aggregate 1 contains 10 TCP flows, while the number of TCP flows in aggregate 2 varies from 5 to 25.

In an under-subscribed Diffserv network, the aggregate with a larger number of TCP flows obtains a greater share of the excess bandwidth. However, under AFC each aggregate obtains an equal amount of bandwidth. We obtain similar results with the proposed control flows (see Fig. 2).

It is worth emphasizing that, with TCP Vegas control flows, the sharing of the excess bandwidth is completely fair. Also, as it happens in almost all the tests, the AFC-BI scheme has wider CIs than the other two. Thus, even though it has a suitable behaviour in average, not so fair behaviour in each one of the simulations can be expected.

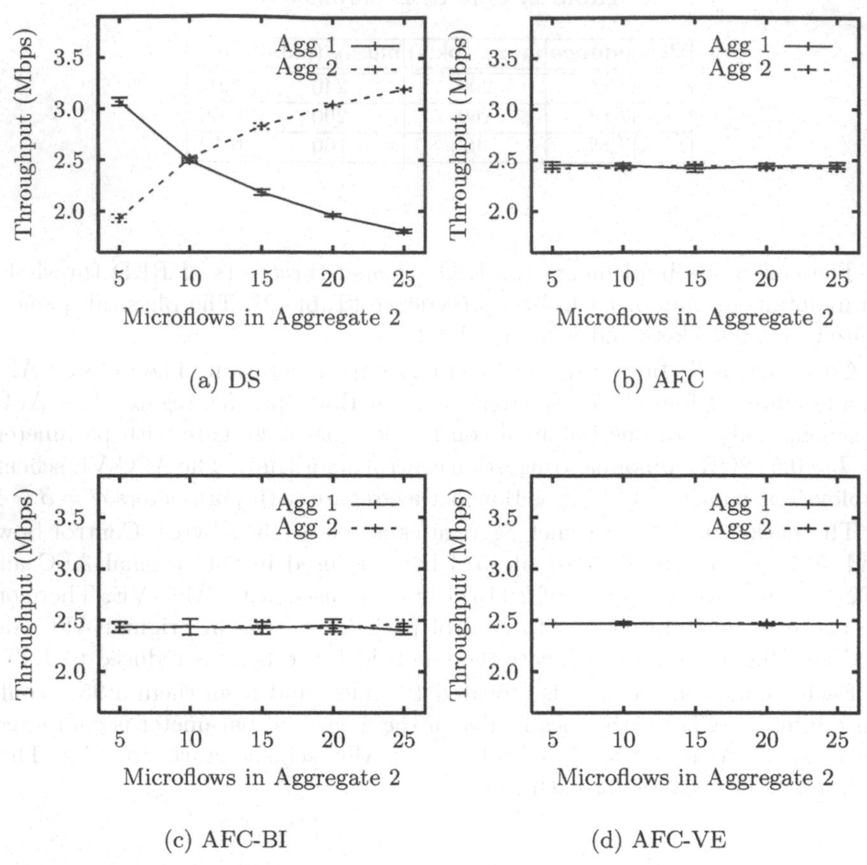

Fig. 2. Impact of number of microflows

4.2 TCP/UDP Interaction

It is important to protect responsive TCP flows from non-responsive UDP flows since this unresponsive traffic may impact the TCP traffic in an adverse manner. In this scenario, the two aggregates have the same target: a 1.0 Mbps CIR and a 2.0 Mbps PIR. Aggregate 1 contains 10 TCP flows, while aggregate 2 has an UDP flow with a sending rate increasing from 1 Mbps to 5 Mbps.

Under the Diffserv case, as the UDP rate increases, the amount of bandwidth obtained by the TCP aggregate decreases. However, with AFC, the bandwidth is shared in a TCP-friendly manner. The results obtained with our proposed control flows are suitable too (see Fig. 3).

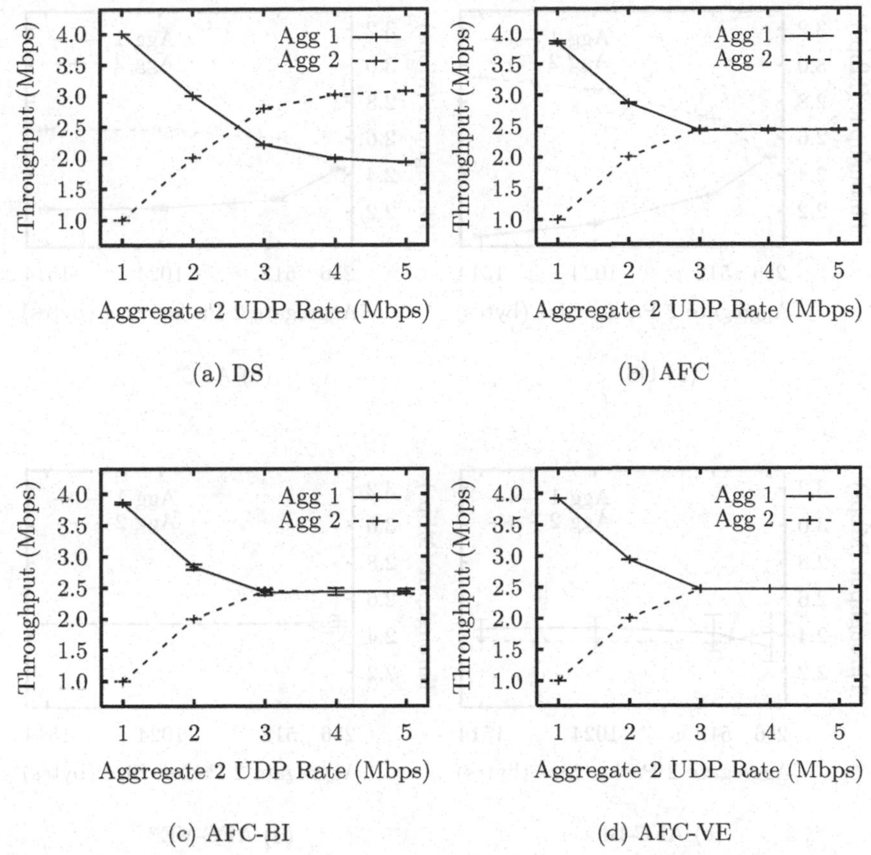

Fig. 3. TCP/UDP interaction

4.3 Impact of Packet Size

Fairness is also desired between aggregates of different packet sizes. In this scenario, aggregate 1 packet size is 256 bytes, while aggregate 2 packet size increases from 256 to 1514 bytes. Each aggregate comprises 10 TCP flows, and has a 1.0 Mbps CIR and a 2.0 Mbps PIR.

Through Diffserv, the aggregate that is sending larger packets consumes more of the available bandwidth. With AFC, the results do not seem to improve. However, with the proposed control flows, the sharing of excess bandwidth can be made insensitive to the packet sizes (see Fig. 4).

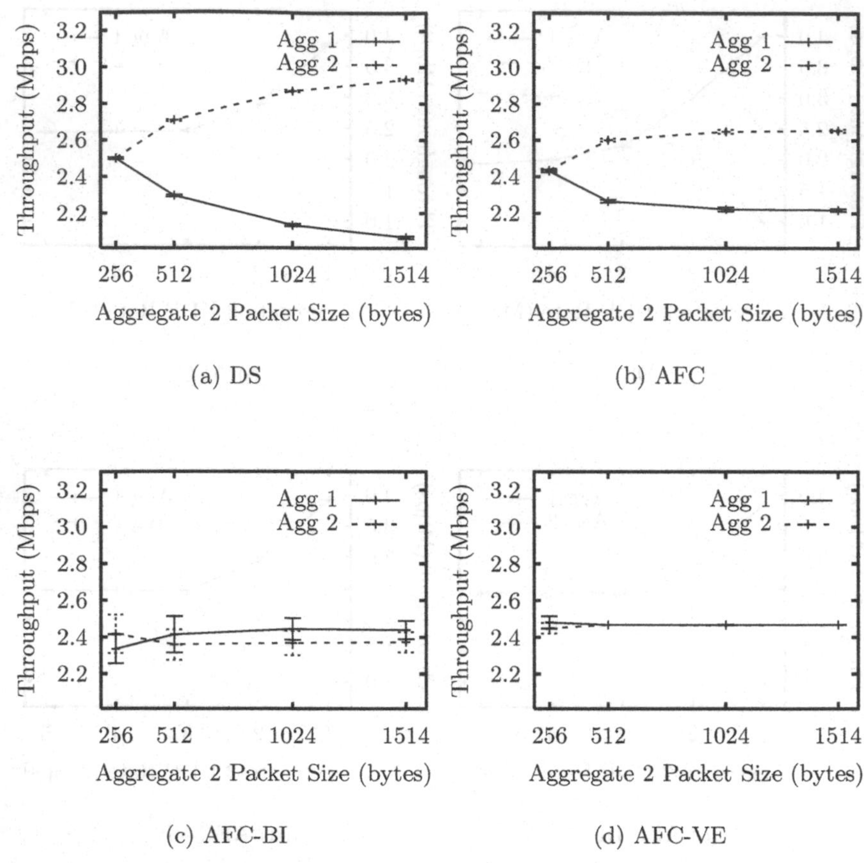

Fig. 4. Impact of packet size

4.4 Impact of Packet Size/VMSS Relationship

So far we have been working with a packet size of 1514 bytes. This value is equal to the selected *vmss* value in the original and binomial AFC cases. The objective of this experiment is to study what happens if we decrease the packet size of both aggregates.

In this scenario, aggregate 1 contains 10 TCP flows, while aggregate 2 comprises 5 TCP flows. Each aggregate has the same target: a 0.5 Mbps CIR and a 1.0 Mbps PIR. The packet size of both aggregates varies from 200 to 1514 bytes.

Figure 5 shows that, under the original AFC, as packet size decreases (differs of *vmss*), the disparity in achieved bandwidth increases. Nevertheless, with AFC-VE, the throughput among aggregates is equalized.

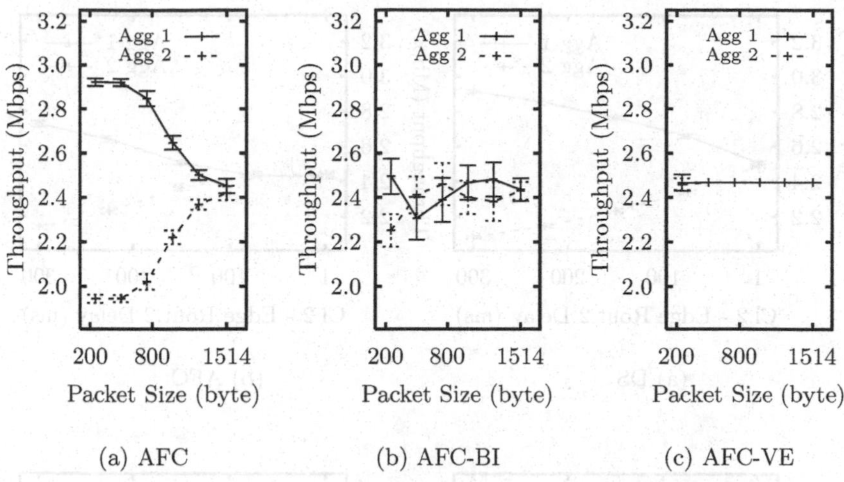

Fig. 5. Impact of packet size / *vmss* relationship

4.5 Impact of RTT

The objective of this test is to study the behaviour of AFC when aggregates have different RTTs. So the delay of access links is changed in order to vary RTTs of customer aggregates.

In this scenario, the link that joins client 1 with edge router 1 has a 1 ms delay, while the link that joins client 2 with edge router 2 has a delay increasing from 1 ms to 300 ms. Each aggregate comprises 10 TCP flows, and has a 1.0 Mbps CIR and a 2.0 Mbps PIR.

Under the Diffserv case, the aggregates with different RTTs cannot achieve a fair share of extra bandwidth. With the original AFC, the results do not improve too much. We obtain better results with both proposed control flows (see Fig. 6).

4.6 Protection of Short TCP Flows

Short-lived TCP flows should be protected under congestion because most interactive web traffic is in short flows and should be transferred faster. In this experiment, each aggregate has a 1 Mbps CIR and a 2 Mbps PIR. Aggregate 2 contains a base traffic of long-lived flows plus a series of short TCP flows. A short flow (transaction) consists of a single packet request from client 3 to client 2, followed by a 16 Kbytes response.

We studied the behaviour of standard Diffserv and AFC under different scenarios. Transactions per second is the metric used. Table 3 lists the results of the five tests performed.

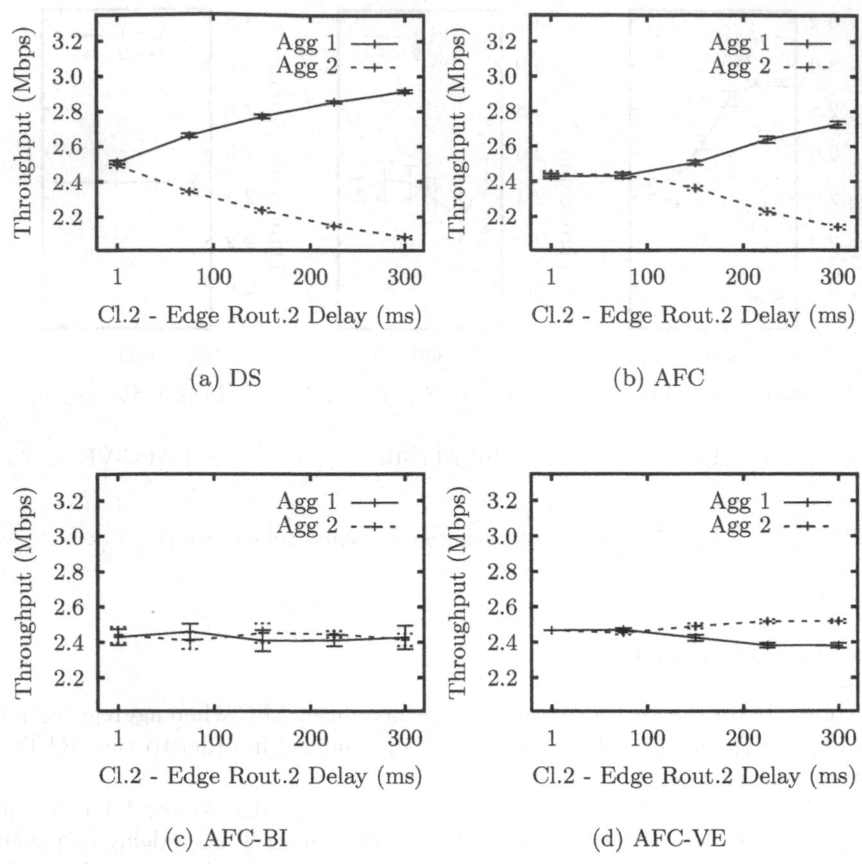

Fig. 6. Impact of RTT

Both of our proposed control mechanisms increase the transaction rate significantly when short flows mix with UDP traffic, especially AFC-VE. With TCP traffic, the transaction rate decreases slightly.

4.7 Average Core Queue Size

AFC pushes congestion from the shared core of a Diffserv network onto the edges whose ingress traffic is causing congestion. Thus the queues at the core of the network remain small due to the congestion management performed at the edge routers.

In this experiment, we metered the average core queue size under four different scenarios. Each aggregate has a 1 Mbps CIR and a 2 Mbps PIR. Table 4 shows the results obtained.

Table 3. Protection of short TCP flows (in Trans/sec)

Agg 1	Agg 2	Diffserv	AFC	AFC-BI	AFC-VE
4M UDP	2M UDP	0.40 ± 0.03	1.58 ± 0.06	2.08 ± 0.09	4.92 ± 0.03
4M UDP	10 TCP	0.33 ± 0.02	0.37 ± 0.03	0.19 ± 0.01	0.20 ± 0.01
10 TCP	2M UDP	0.85 ± 0.03	1.53 ± 0.06	2.12 ± 0.09	5.05 ± 0.05
10 TCP	10 TCP	0.43 ± 0.03	0.36 ± 0.02	0.21 ± 0.01	0.20 ± 0.01
10 TCP	-	1.46 ± 0.14	2.24 ± 0.31	3.35 ± 0.01	9.28 ± 0.01

Table 4. Average core queue size (in packets)

Agg 1	Agg 2	Diffserv	AFC	AFC-BI	AFC-VE
4M UDP	2M UDP	160 ± 1	92 ± 3	63 ± 1	5.7 ± 0.1
4M UDP	10 TCP	162 ± 1	98 ± 3	67 ± 4	8.1 ± 0.1
10 TCP	2M UDP	89 ± 2	89 ± 3	61 ± 1	5.9 ± 0.1
10 TCP	10 TCP	102 ± 4	98 ± 1	72 ± 4	7.6 ± 0.1

The core queue size under AFC-BI is smaller than under original AFC but the smallest core queue size is obtained with AFC-VE.

4.8 Multiple Congestion Points

The AFC scheme can be applicable to more complex scenarios. In this experiment we use AFC with a network topology which adds two more congestion points (see Fig. 7). Aggregates run from clients 1, 2, 3 and 4 to client 5. The four aggregates are sent at a time. Thus, the links are shared by between two and four aggregates. Each aggregate comprises 10 TCP flows and has a 1 Mbps CIR and a 2 Mbps PIR. The results of this test are shown in Table 5. All the AFC schemes share the bandwidth in a fair manner.

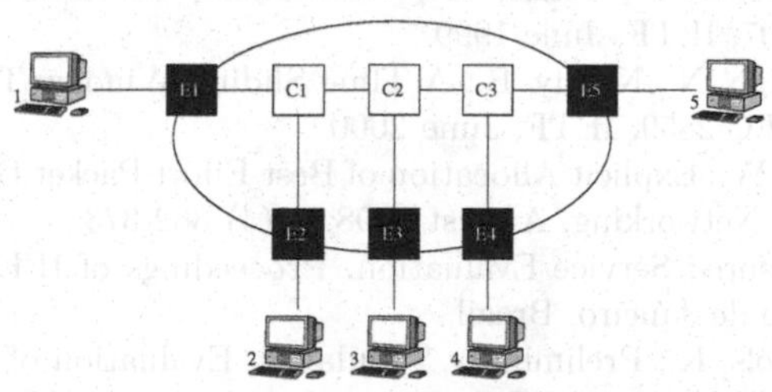

Fig. 7. Multiple congestion points topology

Table 5. Bandwidth distribution (Mbps)

Aggregate	AFC	AFC-BI	AFC-VE
Agg. 1-5	1.23 ± 0.02	1.19 ± 0.01	1.26 ± 0.05
Agg. 2-5	1.23 ± 0.01	1.20 ± 0.01	1.27 ± 0.04
Agg. 3-5	1.22 ± 0.01	1.22 ± 0.01	1.18 ± 0.04
Agg. 4-5	1.22 ± 0.03	1.26 ± 0.01	1.23 ± 0.03

5 Conclusions

The AFC scheme is a very interesting technique to improve bandwidth assurance of differentiated services. On the one hand this solution is scalable because the congestion control is performed on the aggregate level. On the other hand this scheme permits an incremental deployment because it does not require any special support at the core of the network. If the AFC traffic coexists with other uncontrolled Diffserv traffic, a separate queue could be used to allocate the AFC traffic in order to avoid its unfair treatment.

The deployment of this scheme can be made even simpler if we only use one control flow per customer aggregate. We propose binomial congestion control algorithms and TCP Vegas implementation as suitable control flows. AFC with the proposed control flows not only simplifies deployment but also improves fairness in the bandwidth sharing among aggregates with different packet sizes, different RTTs or small *packet size / vmss* ratios. In addition, the transaction rate for short-lived TCP flows under congestion is increased while average core queue size is greatly decreased, especially with TCP Vegas control flows.

References

1. Blake, S., Black, D., Carlson, M., Davis, E., Wang, Z., Weiss, W.: An Architecture for Differentiated Services. RFC 2475, IETF, December 1998
2. Heinanen, J., Baker, F., Weiss, W., Wroclawski, J.: Assured Forwarding PHB Group. RFC 2597, IETF, June 1999
3. Fang, W., Seddigh, N., Nandy, B.: A Time Sliding Window Three Colour Marker (TSWTCM). RFC 2859, IETF, June 2000
4. Clark, D., Fang, W.: Explicit Allocation of Best Effort Packet Delivery. IEEE/ACM Transactions on Networking, August 1998, **6** (4) 362-373
5. Rezende, J.: Assured Service Evaluation. Proceedings of IEEE GLOBECOM'99, March 1999, Rio de Janeiro, Brazil
6. Ibanez, J., Nichols, K.: Preliminary Simulation Evaluation of an Assured Service. Internet Draft, draft-ibanez-diffserv-assured-eval-00.txt, IETF, August 1998.
7. Lin, W., Zheng, R., Hou, J.: How to Make Assured Services More Assured. Proceedings of ICNP, October 1999, Toronto, Canada
8. Ling, S., Hou, J.: An Active Queue Management Scheme for Internet Congestion Control and Its Application to Differentiated Services. Proceedings of IEEE ICCCN, October 2000, Las Vegas, NV

9. Feng, W., Kandlur, D., Saha, D., Shin, K.: Adaptive Packet Marking for Maintaining End-to-end Throughput in a Differentiated-services Internet. IEEE/ACM Transactions on Networking, October 1999, **7** (5) 685-697
10. Nandy, B., Seddigh, N., Pieda, P., Ethridge, J.: Intelligent Traffic Conditioners for Assured Forwarding Based Differentiated Services Networks. NetWorld+Interop 2000 Engineers Conference, May 2000, Paris, France
11. Alves, I., Rezende, J., Moraes, L.: Evaluating Fairness in the Aggregated Traffic Marking. Proceedings of IEEE GLOBECOM'00, November 2000, San Francisco, CA
12. El-Gendy, M., Shin, K.: Equation-Based Packet Marking for Assured Forwarding Services. Proceedings of IEEE INFOCOM, 2002, New York, NY
13. Nandy, B., Ethridge, J., Lakas, A., Chapman, A.: Aggregate Flow Control: Improving Assurances for Differentiated Services Network. Proceedings of IEEE INFOCOM, 2001, Anchorage, AK
14. Fall, K., Floyd, S.: Simulation-based Comparison of Tahoe, Reno, and Sack TCP. Computer Communication Review, July 1996, **26** 5-21
15. The Network Simulator, ns-2. Version ns-2.1b9a, July 2002. URL http://www.isi.edu/nsnam/ns/
16. Bansal, D., Balakrishnan, H.: Binomial Congestion Control Algorithms. Proceedings of IEEE INFOCOM, 2001, Anchorage, AK
17. Brakmo, L. S., O'Malley, S. W., Peterson, L. L.: TCP Vegas: New Techniques for Congestion Detection and Avoidance. Proceedings of the ACM SIGCOMM, 1994, London, UK

Quality of Service Multicasting over Differentiated Services Networks*

Giuseppe Bianchi[1], Nicola Blefari-Melazzi[2], Giuliano Bonafede[1], and Emiliano Tintinelli[2]

[1] University of Palermo, D.I.E, Viale delle Scienze, 90128 Palermo, Italy
giuseppe.bianchi@tti.unipa.it

[2] University of Roma Tor Vergata, D.I.E, Via del Politecnico 1, 00133 Roma, Italy
blefari@uniroma2.it

Abstract. This paper proposes a solution to support real-time multicast traffic with Quality of Service (QoS) constraints over Differentiated Services (DiffServ) IP networks. Our solution allows multicast users to dynamically join and leave the multicast tree. Moreover, it allows a multicast user which has negotiated a best-effort session to upgrade to a QoS-enabled session. Our solution is backward compatible with the Protocol Independent Multicast (PIM) scheme. It combines two ideas. First, resource availability along a new QoS path is verified via a probe-based approach. Second, QoS is maintained by marking replicated packets with a special DSCP value, before forwarding them on the QoS path.

1 Introduction

The Internet is witnessing the emergence of several Web-based real-time multimedia and multicast applications, such as video-conferencing, staggered multimedia information retrieval, etc [ALM00]. Unfortunately, the current best-effort Internet does not offer Quality of Service (QoS) guarantees to effectively support unicast streaming services, let alone multicast ones.

Several proposals have appeared in the literature in the area of QoS multicast. Most of the work concerns the development of multicast QoS routing protocols [KPP93, CNS00, YFB02], i.e., protocols that select multicast paths under QoS constraints. Conversely, the issue of endowing current multicast protocols with resource reservation and admission control mechanisms has been generally confined to be a somehow straightforward extension of related unicast protocols. For example, [FIT96] proposes a reservation mechanism for multicast traffic based on a reference Internet model very similar to that proposed in the Integrated Services approach. In addition, the RSVP protocol specification, version 1, [R2205] has been devised to efficiently support multicast traffic.

We argue that novel problems arise in QoS multicast when the reference unicast QoS architecture is not based on explicit and stateful resource reservation

* This research is supported in part by the European Union in the frame of the IST project ICEBERGS

M. Ajmone Marsan et al. (Eds.): QoS-IP 2003, LNCS 2601, pp. 742–755, 2003.

protocols. This is specifically the case for the Differentiated Services (DiffServ) framework [R2474, R2475]. DiffServ is a current trend in the Internet community for the development of a scalable QoS architecture not burdened with the complex task of reservation states creation and maintenance. The potential problems and the complexity of supporting multicast in a DiffServ environment are sketched in [R2475] and, in greater details, in [BW01, BW02, SM01, YM02] and references therein contained. In particular, [BW01, BW02] highlight the following issues.

- First, dynamic addition of new members to a multicast group can adversely affect existing other traffic, if resources are not explicitly reserved after each join, since replicated packets get the same Differentiated Services Code Point (DSCP) of the original packet and thus enjoy the relevant treatment consuming un-reserved resources.
- Second, resources should be reserved separately for each multicast tree associated to a given sender, to allow simultaneous sending by multiple sources, with QoS constraints.
- Finally, it appears difficult to support heterogeneous multicast groups, i.e., groups in which different users have different necessities.

More into details, as regards the last point, participants who can cope with a best-effort service should coexist with participants needing specified and different levels of QoS assurances, so that the same multicast group can deliver differentiated services. For instance, a user could browse a multicast multimedia session in best-effort mode and then decide to switch to a QoS mode (eventually by paying for it).

In this paper, we propose a QoS multicast solution for DiffServ IP networks. The goal of our proposed solution is i) to provide flexible QoS support with respect to heterogeneous multicast groups, and ii) to maintain compatibility with currently deployed multicast protocols, with specific reference to PIM (Protocol Independent Multicast [R2362, PIMV2]).

The rest of this paper is organized as follows. The basic motivations and directions of our proposed approach are outlined in Section 2. Our solution is illustrated in Sections 3, which shows the basic protocol operation, and in Section 4, which focuses on the operation necessary for a QoS user to join the multicast tree, and thus to check and reserve resource availability. For convenience of the reader, sub-section 4.1 is devoted to briefly review a unicast DiffServ QoS solution, formerly proposed in [BB01, BBF02] by some of the authors of this paper, and used as a basic brick for our multicast solution. Conclusions are drawn in Section 5.

2 Motivation and Directions

The design requirements for our proposed solution were:

- QoS should be envisioned as an incremental addition to existing multicast protocols devised for Best-Effort traffic - in other terms, the QoS multicast solution should be backward compatible;

– we aim at devising an overlay solution, i.e., whose applicability over the DiffServ network should be done with minimal modification to the DiffServ router operation and to the underlying routing protocols;
– the solution should be flexible with respect to heterogeneity of users within a group, and of distribution trees (source-based or core-based).

Because of these design requirements, we have selected PIM (Protocol Independent Multicast), in the Sparse Mode (SM) version [R2362, PIMV2], as the reference multicast routing protocol. The reason is that PIM is independent on the underlying unicast routing protocols, and it allows several different configurations of the multicast distribution tree (unique group shared distribution tree based on a central Core Router; different distribution trees per specific source sending to a group; coexistence of shared and source-specific trees for the same group).

To provide QoS multicast over DiffServ, two important issues need to be considered. The first issue is how to differentiate the service level supported on different paths inside the multicast distribution tree. The second one is how to provide an admission control function in order to support dynamic joins, from QoS-enabled users. In both cases, recall that no per-flow reservation state can be employed in the DiffServ routers, whose role, as specified in the DiffServ architecture [R2475], is simply to apply different forwarding disciplines to packet aggregates, based on their DSCP value.

To solve the first issue, Bless and Wherle [BW01] suggest to add an additional field in the Multicast Routing Table, which specifies the DSCP(s) to be used for each output link. It is thus possible to specify whether the next hop is QoS-enabled (i.e., mark packets with a specified DSCP, corresponding to a negotiated QoS level), or not (i.e., mark packets with the Best-Effort DSCP). This solution allows heterogeneous users to share the same multicast distribution tree, as well as it allows deployment of several different QoS levels in the network.

As discussed in section 3, our solution inherits from [BW01] this marking strategy. However, this handy tool is not sufficient by itself to guarantee that resources are available to QoS multicast sessions. To solve this problem, [BW01] relies on an unspecified management procedure (e.g., a centralized entity such as a Bandwidth Broker), which performs an admission control test, and lists the following requirements:

– "there must be a mechanism for DiffServ nodes to inform a management entity about the join request of a new sub-tree";
– "a mechanism must be supplied for instructing a router to suitably change (and update) the DSCP value in the related multicast routing table entry. This mechanism may be also incorporated into an existing multicast routing protocol as an extension."

In our proposal, we do not rely on the existence of a centralized control entity. Instead, we combine the marking strategy proposed in [BW01] with a DiffServ-compliant admission control function, based on data-plane operation (forwarding/dropping selected packets), which allows to control the traffic load within each given DSCP class. This function, named GRIP (Gauge & Gate Reservation with Independent Probing), has been proposed by some of the authors of this paper in [BB01, BBF02]. For convenience of the reader, it will be briefly reviewed in Section 4.1. GRIP can provide QoS guarantees by means of stateless DiffServ-compliant procedures and without modifying the basic DiffServ router operation. Thus, one of the aims of Section 4 is to show how this admission control mechanism can be adapted to operate within the PIM-SM framework.

3 QoS Multicast Forwarding

This Section shows the basic protocol operation. More details on the joining operation of a QoS user to the multicast distribution tree will be given in Section 4. For convenience of presentation, we focus on the example network scenario illustrated in Figure 1.

Our discussion will assume PIM-SM as the multicast protocol of reference, being the particularization to PIM-SSM (Source-Specific Multicast) rather straightforward. In the basic operation of PIM-SM, a router, called Rendez-vous Point (RP), is used, for all traffic sources S, as the root of the distribution tree for the multicast group. Destination hosts H_i avail themselves of Designated Routers (DR_i) on their LAN, which act on behalf of those hosts as far as the PIM-SM protocol is concerned. In other words, the DR manages, on one side, all local group management information (e.g., via the IGMP protocol), and, on the other side, it emits PIM join/prune messages towards the RP. We assume that all routers are DiffServ capable. In addition, we assume that while routers A, B and C in Figure 1 support PIM, intermediate DiffServ routers, indicated with the label R, are transparent to the multicast protocol.

Multicast packets are replicated at each multicast router (these routers are the nodes of the multicast tree), and are delivered to the multicast destinations according to the routing information provided by the PIM protocol operation (see [PIMV2]) and stored in a table called Multicast Routing Table, hereafter referred to as MRT.

Following [BW01], we also assume that each multicast routing table stores, for each router output interface (oif), an additional entry. This entry represents the Differentiated Services Code Point (DSCP) value, which is used to mark replicated packets that are forwarded along the considered interface.

Figure 1 shows, in the leftmost column, labeled "Only $H_1 + H_2$", the MRT states for routers A, B, and C, in the assumption that only hosts H_1 (QoS-enabled) and H_2 (Best-Effort) are members of the multicast group. In Figure 1, we use the label BE to denote a Best-Effort service, (i.e., DSCP 000000) and the label QoS to denote a generic QoS-aware service. For the sake of simplicity, here we will present our proposal assuming that we want to offer a single QoS level, in addition to the Best-Effort service. However, we remark that extension to multiple QoS levels and typologies is rather simple.

Let us first focus on the case "only $H_1 + H_2$". By looking at Figure 1, we see that the MRT of router A marks packets directed to host H_1 (interface 2)

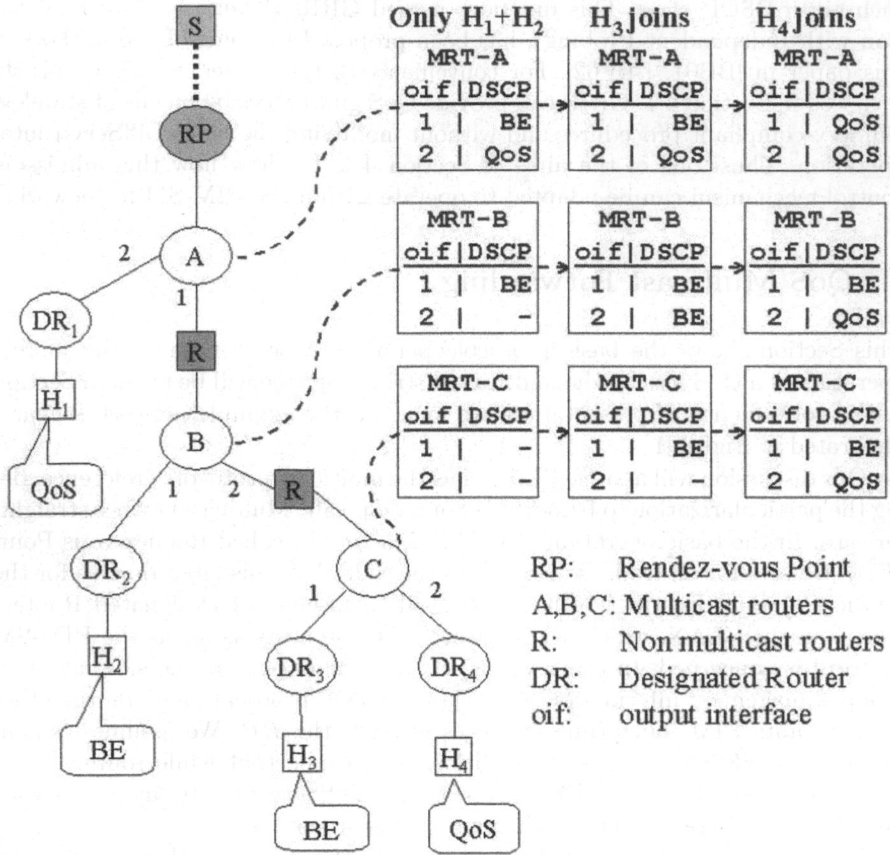

Fig. 1. Example Network Scenario

with a QoS marking, while it labels packets forwarded to interface 1 as BE. In the MRT of router B, clearly, only the interface 1 is active, while router C has no multicast state, at the moment.

Let us now discuss what happens when, first the Best-Effort host H_3, and then the QoS host H_4, join the multicast group. In the case of H_3, we adopt a standard PIM Join operation. The Designated Router of H_3, DR_3, sends a PIM $Join(*, G)$ message towards the RP. When a multicast router receives this message, it adds a new entry to its MRT, with a BE value, for the associated DSCP marking field. This state will indicate that future datagrams destined for group G must be marked as BE, and forwarded on the interface where the join message came from. The $Join(*, G)$ message is regenerated upstream along what we define as the "Branching Path". This is the path from the requesting DR to the first router that already has a $Join(*, G)$ state for that group, which is called

"Branching Router". Eventually the Branching Router could coincide with the RP itself. In the specific case of Figure 1, the Branching Path goes from DR_3 to router B, which is the Branching Router, since it already has a $(*, G)$ state created by the *Join* message coming from host H_2, by way of DR_2. The second column of Figure 1, labeled "H_3 joins", shows the modified MRTs in routers A, B and C after a successful join of the BE host H_3.

In the case of QoS-host H_4, there are two fundamental differences. The first difference is the extent of the Branching Path. If the Designated Router DR_4 were to send a standard $Join(*, G)$ message, the Branching Router would have been router C, since it already has a $(*, G)$ state. Therefore, when a QoS host wants to join the group, it uses a modified[1] *Join* message, hereafter referred to as $Join(*, G, QoS)$ message, which explicitly carries the QoS level the host is requesting. This $Join(*, G, QoS)$ message travels upstream until it either reaches the RP, or it reaches a router with a $(*, G, QoS)$ state (i.e., a state that instructs a router to send QoS marked traffic over at least one of its interfaces). We define "QoS Branching Path" the path from the Designated Router to the first router that already has a $Join(*, G, QoS)$ state for that group. We name such a router as "QoS Branching Router" (note that it may be the *RP* itself). In Figure 1, the QoS Branching Router for host H_4 results to be router A. The rightmost column of Figure 1, labeled "H_4 joins", shows the modified MRTs in routers A, B and C after a successful *join*[2] of the QoS host H_4.

The second difference in the QoS-host operation is that, unlike the best effort case, a simple $Join(*, G, QoS)$ message is not sufficient. In order to provide a service with QoS requirements, there is the need to check the availability of network resources along the QoS Branching Path. This requires supplementary operation (and related message exchange), after the delivery of the $Join(*, G, QoS)$ message. The detailed operation needed to establish QoS on a new QoS branching path is described in the next section.

Finally, note that a given host may upgrade its service level. For example, an host may first send a $Join(*, G)$ message to preview the multicast group, and then upgrade to QoS by sending a $Join(*, G, QoS)$. This second message is regenerated upstream along the QoS Branching Path, and, along its way, it modifies the state of the crossed multicast router(s).

[1] The new PIM messages considered in this paper, i.e., $Join(*, G, QoS)$, $Confirm(*, G, QoS)$, $Prune(*, G, QoS)$, can be built upon the PIM message format specifications, as extensions. In fact, PIM defines only 8 different packets, while a 4-bit packet type field (i.e., 16 possibilities) is available.

[2] Note that now the distribution tree has a QoS path from *RP* to DR_4. Therefore, hosts H_2 and H_3 will receive a BE service only on the last hop of their path. In other words, the fact that some members of the multicast groups require QoS, indirectly brings benefits also to Best-Effort users (even if they are not paying for such benefits...).

4 Establishing QoS

This section focuses on the detailed operation necessary for a QoS user joining the multicast tree to establish QoS on the new QoS Branching Path. With reference to the example discussed in Figure 1, in our presentation we focus on the establishment of QoS on the QoS Branching Path from router A as a starting point (the network path from the RP to router A is already QoS enabled, because of QoS user H_1), to the designated router DR_4 as a destination point[3].

Our proposal consists in adapting, to the multicast case, a DiffServ compatible stateless admission control operation based on pure data-plane operation, called GRIP (Gauge & Gate Reservation with Independent Probing), proposed in [BB01] with unicast traffic in mind. For convenience of the reader, the basic principles of GRIP are briefly reviewed in section 4.1. The reader interested in additional technical details can refer to [BB01, BBF02]. The adaptation of GRIP to multicast is tackled in sections 4.2 and 4.3.

4.1 Review of the Basic, Unicast, GRIP Operation

Provisioning of QoS in DiffServ networks is frequently assumed to be accomplished via static over-provisioning (i.e., over-dimensioning of core network links with respect to the expected offered traffic). When dynamic provisioning of DiffServ domains is considered, the traditional approach is to rely on centralized control entities, often referred to as Bandwidth Brokers (BB). A BB is an agent that has sufficient knowledge of resource availability and network topology to make admission control decisions. Border routers use control protocols to interact with this agent. The existence of BBs has been assumed in [BW01, YM02], to manage multicast traffic handling strategies.

We have shown in [BB01, BBF02] that per-flow admission control can be supported on stateless DiffServ domains, i.e., with no need of managing per-flow states within each core router. We recall that, in the DiffServ approach, core routers cannot support any explicit signaling protocol (this would imply parsing and interpreting higher layer information contained in signaling packets payload). Instead, their unique duty is to implement low-layer mechanisms devised to forward/drop packets, and apply packet scheduling mechanisms, on the basis of the DSCP value marked on each IP packet header.

Our admission control procedure, named GRIP (Gauge & Gate Reservation with Independent Probing), is based on the idea of using "implicit signaling", i.e. GRIP uses pure data-plane operation (packet dropping/forwarding) to convey, at the network borders, the information that the network is congested and a new flow cannot be accepted. A GRIP router is a plain DiffServ router, which handles a number of aggregate classes. A packet belongs to a class on the basis of its

[3] We assume that QoS on the last hop $DR_4 \rightarrow H_4$ is provided by some layer-2 mechanism, or, in other words, that when the host H_4 wants to join the multicast tree with QoS, a local admission control function on the last-hop segment has been already performed with positive answer.

DSCP value. A GRIP router supports at least three different DSCP values: Best Effort (BE) packets, QoS information packets, and QoS probing packets. The following description will assume that only one QoS level and one traffic class is supported (i.e., only one information and one probing DSCP label in addition to BE). Some additional considerations about how GRIP can be extended to to support several levels of QoS and different traffic classes can be found at the end of this section.

A *measurement module* in each GRIP module is in charge of taking a smoothed and filtered measure of the load offered by information packets. It will soon be clear that this is a measure of the aggregate accepted traffic. On the basis of these traffic measurements, and according to a suitable *Decision Criterion*, the measurement module drives an *ACCEPT/REJECT* switch. When the switch is in the ACCEPT state, incoming probing packets are forwarded to the output interface. Conversely, probing packets are dropped when the switch is in the REJECT state. In other words, the router acts as a gate for packets labeled as probing, where the gate is opened or closed on the basis of traffic estimates taken on the aggregate accepted load (hence the Gauge&Gate in the acronym GRIP).

As extensively shown in [BB01], the described operation is not only compatible with DiffServ, but it is already supported in the specification of the DiffServ Assured Forwarding Per-Hop Behavior (AF-PHB) [R2597].

Admission control support via implicit signaling arises when the above described operation (localized and independent within each core router) is combined with a suitable end-point operation. When a user terminal requests a unicast connection for the considered QoS traffic class, the source node transmits a packet whose DSCP is marked as probing. Meanwhile, the source node activates a probing phase timeout, lasting for a reasonable time. If no response is received from the destination node before the timeout expiration, the source node enforces rejection of the connection setup attempt. Otherwise, if a *feedback packet* is received in time, the connection is accepted, and control is given back to the user application, which starts a data phase, simply consisting in the transmission of data packets, whose DSCP is marked as *information*.

In order for a call setup procedure to succeed, the probing packet needs to find all the routers along the path in the ACCEPT state (if the probing packet encounters a router in the REJECT state, it gets discarded; hence it does not reach the destination, no feedback packet will be relayed back, and the call will be blocked as soon as the probing phase timeout expires).

It remains to understand which level of QoS provisioning can be offered by the GRIP operation. Since the decision criterion is locally taken by each router, on the basis of filtered and smoothed measurements taken on the aggregate accepted load (i.e., throughput), GRIP effectiveness is comparable to Measurement Based Admission Control (MBAC) schemes. In fact, GRIP's measurement module can make use of state-of-the art MBAC mechanisms (e.g., [BJS00] and references therein contained), for which it has been shown that a target QoS performance can be obtained by simply controlling throughput. An example of a very simple Decision Criterion is to switch from ACCEPT to REJECT state when the aggregate load measurement exceeds a given threshold, and conversely. A more specific example of a Decision Criterion is proposed in [BBF02], where we show that, with suitable additional assumptions, it is possible to provide as much as hard performance guarantees.

In the above description, as well as in the rest of this paper, for simplicity of presentation, we have assumed that only one QoS level and one traffic class is supported, i.e. the network provides only one information and one probing DSCP label. Indeed, we remark that the GRIP router operation can be extended. To support several levels of QoS and different traffic classes it is necessary: i) to add a new pair of DSCP values (for information and probing packets) for each additional QoS class, and ii) to suitably configuring the (standard) DiffServ scheduling rules among the supported QoS classes to reciprocally protect each QoS class from congestion eventually arising on other QoS classes. This is perfectly coherent with the DiffServ paradigm, as different QoS classes are handled in a differentiated way. It is worth to note that different DSCP pairs can also be used to manage heterogeneous traffic classes (where a traffic class is defined as a set of sources with equal or very similar traffic parameters, e.g., peak rate, sustainable rate, etc.). In this case, different traffic classes are handled by separate GRIP modules, each implementing a suitably engineered measurement module and decision criterion.

4.2 Adaptation of Unicast GRIP to Multicast

Figure 2 illustrates the message exchange involved in the set-up of a QoS path. The scenario is the same considered in the previous Figure 1. We focus on the case of a new QoS user H_4 which is willing to join the multicast tree with QoS requirements. As discussed in section 3, the QoS Branching Router for H_4, i.e. the first router in the multicast tree with a QoS state, results to be router A. For comparison purposes, Figure 2 also illustrates the message exchange (i.e. a simple $Join(*, G)$ message) involved in a previous set-up of a best effort path for host H_3.

Our solution consists in running GRIP considering the network router A as a starting point and the designated router DR_4 as a destination point. When the QoS Branching Router A first receives a $Join(*, G, QoS)$ message, it sends a GRIP Probe down on the multicast tree. The probe is a normal multicast packet. The correct routing of the Probe packet is achieved by suitably modifying the PIM state machine running at each multicast router. As described in section 4.3, each multicast router along the QoS branching path forwards the Probe packet along the output interfaces from which a former $Join(*, G, QoS)$ message has arrived (typically only one interface, unless concurrent $Join(*, G, QoS)$ message arrive from different end nodes).

Moreover, as reviewed above, the Probe packet is distinguished from information packets via a special DSCP marking. Specifically, if QoS packets are marked with DSCP value, say, X_1, GRIP Probes are marked with a different DSCP value, say, X_2, which is logically associated to X_1. For each QoS level, a different pair of DSCPs should be reserved. All network routers (i.e., both

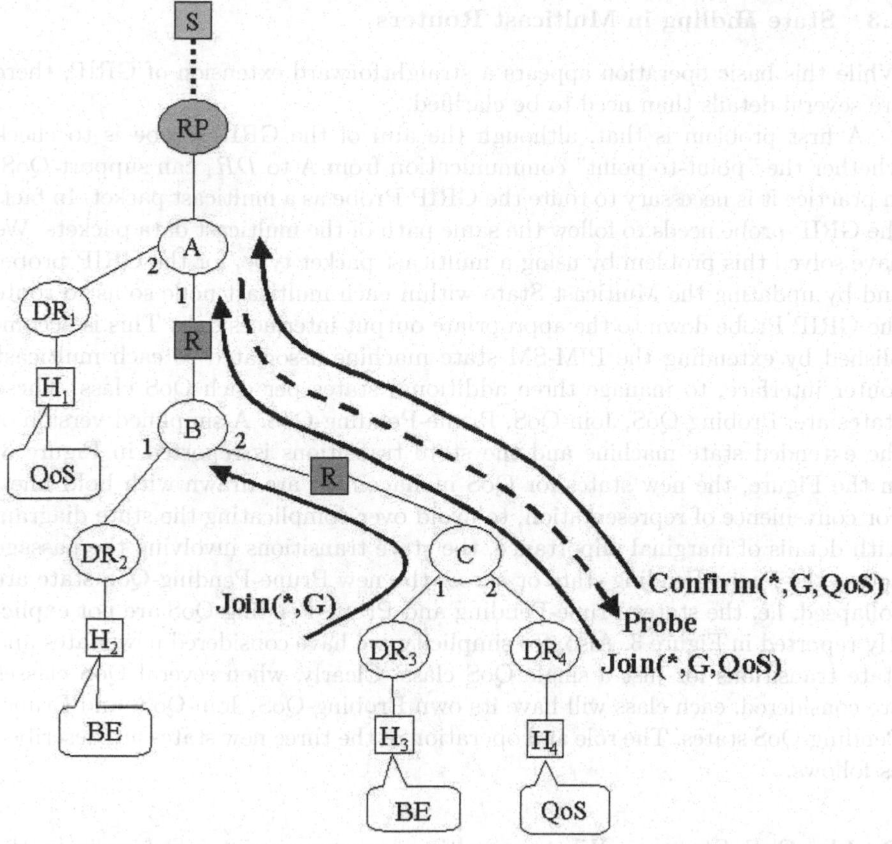

Fig. 2. Join(*,G,QoS), Probe and Confirm(*,G,QoS) message routing

multicast-capable routers and non-multicast capable routers) support a DiffServ Per-Hop-Behavior (i.e., a data-plane mechanism) which consists in letting packets marked as X_2 go through only, for instance, if the load run-time measured on X_1-marked packets is lower than a given threshold.

The result of this operation is that a GRIP Probe is received at the destination DR_4 only if all the routers along the path are found in an non-congested state, defined with a suitable criterion (see e.g., [BBF02]). When the GRIP Probe arrives at DR_4, a $Confirm(*, G, QoS)$ message is sent back as a feedback packet, to notify the sender node A that all the routers along the QoS Branching Path can accept a QoS connection. When the $Confirm(*, G, QoS)$ message is finally received at router A, the multicast data can be forwarded on the new path. As discussed in the previous Section, subsequent replicated packets are marked according to the DSCP value specified in the MRT.

4.3 State Hhdling in Multicast Routers

While this basic operation appears a straightforward extension of GRIP, there are several details than need to be clarified.

A first problem is that, although the aim of the GRIP Probe is to check whether the "point-to-point" communication from A to DR_4 can support QoS, in practice it is necessary to route the GRIP Probe as a multicast packet. In fact, the GRIP probe needs to follow the same path of the multicast data packets. We have solved this problem by using a multicast packet type, for the GRIP probe, and by updating the Multicast State within each multicast node so as to route the GRIP Probe down to the appropriate output interfaces only. This is accomplished by extending the PIM-SM state machine associated to each multicast router interface, to manage three additional states per each QoS class. These states are: Probing-QoS, Join-QoS, Prune-Pending-QoS. A simplified version of the extended state machine and the state transitions is reported in Figure 3. In the Figure, the new states for QoS management are drawn with bold lines. For convenience of representation, to avoid over-complicating the state diagram with details of marginal importance, the state transitions involving the passage across the Prune-Pending state or across the new Prune-Pending-QoS state are collapsed, i.e. the states Prune-Pending and Prune-Pending-QoS are not explicitly reported in Figure 3. Also, for simplicity, we have considered new states and state transitions for just a single QoS class. Clearly, when several QoS classes are considered, each class will have its own Probing-QoS, Join-QoS, and Prune-Pending-QoS states. The role and operation of the three new states are described as follows.

ProbingQoS State - When a multicast router output interface is in the Probing-QoS state, it means that a $Join(*, G, QoS)$ message has been received, but no $Confirm(*, G, QoS)$ message has been received yet. When in the Probing-QoS state, packets are forwarded according to the following rules:

- Packets labeled as information packets are replicated and forwarded according to the existing MRT. The MRT is updated only when a $Confirm(*, G, QoS)$ message is received. With reference to the example of Figures 1 and 2, while host H_4 is joining the multicast group (i.e., before a $Confirm(*, G, QoS)$ has arrived), routers A and B continue to operate according to the previous MRT (central column in Figure 1), i.e., they replicate and forward packets labeled as BE, while router C does not replicate multicast data packets on the output interface 2.
- Probes are recognized based on their special DSCP value. They are forwarded only toward interfaces whose state is Probing-QoS, and only if traffic measurements taken on the aggregate traffic indicate that no congestion is occurring on the considered interface. As illustrated in Figure 2, a Probe packet is generated at router A and forwarded only on interface 1, while the same Probe packet is forwarded only on interface 2 at both routers B and C.

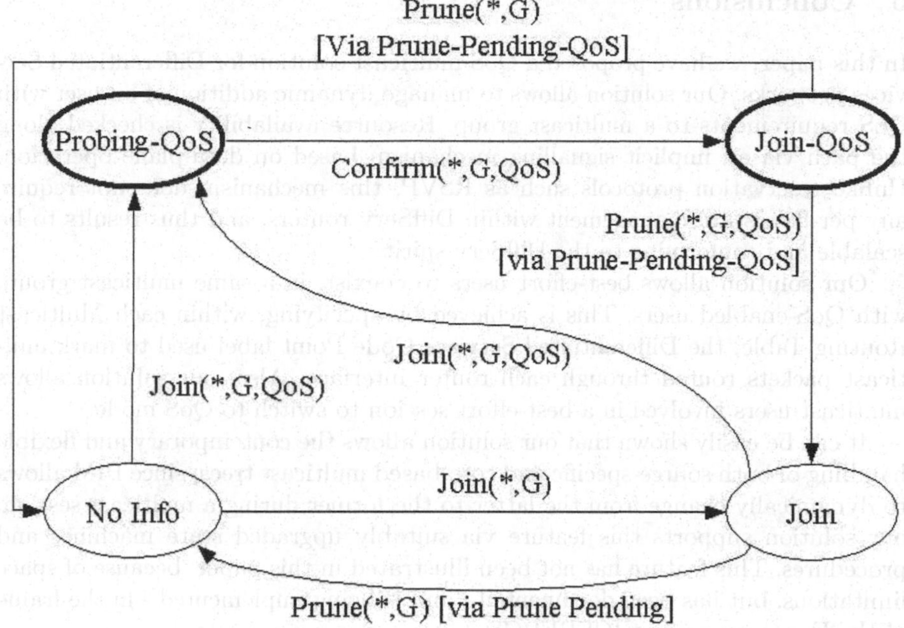

Fig. 3. PIM-SM extended state machine

JoinQoS State - This state is activated after a confirmation message, and it implies that the router output interface is QoS-enabled. When in the Join-QoS state, the MRT is set as described in Section 3.

PrunePendingQoS - This state has the same role of the standard PIM-SM Prune-Pending state. Entrance in the Prune-Pending-QoS state occurs when a new prune message, called $Prune(*, G, QoS)$, is received while the router is in the Join-QoS state (in PIM, entrance in the Prune-Pending state occurs when a standard $Prune(*, G)$ message is received while in the Join state). The system remains in this state until a timeout expires, or a $Join(*, G, QoS)$ is received to restore ("refresh") the Join-QoS state.

In fact, we recall that multicast states are "soft" i.e., as long as a group G is active on a router interface, the router will periodically receive $Join(*, G)$ messages in order to refresh the relevant states. When information transfer for this group is no longer desired, the requiring entity should stop sending joins for that group, so that the relevant states on the routers along the Branching Path expire. In alternative, an explicit PIM $Prune(*, G)$ messages can be sent towards the RP for that multicast group.

5 Conclusions

In this paper, we have proposed a QoS multicast solution for Differentiated Services networks. Our solution allows to manage dynamic addition of an user with QoS requirements to a multicast group. Resource availability is checked along the path via an implicit signalling mechanism based on data-plane operation. Unlike reservation protocols such as RSVP, this mechanism does not require any per-flow state management within DiffServ routers, and thus results to be scalable and conforming to the DiffServ spirit.

Our solution allows best-effort users to coexist, in a same multicast group, with QoS-enabled users. This is achieved by specifying, within each Multicast Routing Table, the Differentiated Services Code Point label used to mark multicast packets routed through each router interface. Also, our solution allows multicast users involved in a best-effort session to switch to QoS mode.

It can be easily shown that our solution allows the contemporary and flexible handling of both source-specific and core-based multicast trees; since PIM allows to dynamically change from the latter to the former during a multicast session, our solution supports this feature via suitably upgraded state machines and procedures. This feature has not been illustrated in this paper, because of space limitations, but has been documented - and is being implemented - in the frame of the European project ICEBERGS.

Finally, regarding performance evaluation, a preliminary performance investigation has been carried out in [BBB03]. In such work, we have quantified the benefits indirectly brought to Best-Effort users when near-by members of the multicast groups require QoS. From the network efficiency point of view this is a definite advantage, but from the market point of view this might be a potential problem.

References

[ALM00] K. G. Almeroth, "The Evolution of Multicast: from the MBone to Inter-Domain Multicast to Internet2 Deployment", IEEE Network, January/February 2000

[BB01] G. Bianchi, N. Blefari-Melazzi: "Admission Control over Assured Forwarding PHBs: a Way to Provide Service Accuracy in a DiffServ Framework", IEEE Globecom 2001, San Antonio, Texas, USA, 25-29 November 2001.

[BBB03] G. Bianchi, N. Blefari-Melazzi, G. Bonafede, E. Tintinelli: "QUASIMODO: QUAlity of ServIce-aware Multicasting Over DiffServ and Overlay networks", to appear in IEEE Networks, Special Issue on: "Multicasting: an enabling technology"; January-February 2003.

[BBF02] G. Bianchi, N. Blefari-Melazzi, M. Femminella: "Per-flow QoS Support over a Stateless DiffServ Domain", Computer Networks, Elsevier, special issue on "Towards a New Internet Architecture", Vol. 40, Issue 1, September, 2002, pp. 73 - 87.

[BJS00] L. Breslau, S. Jamin, S. Schenker: "Comments on the performance of measurement-based admission control algorithms", IEEE Infocom 2000, Tel-Aviv, March 2000.

[BW01] R. Bless, K. Wehrle: "Group communication in differentiated services networks", First IEEE/ACM International Symposium on Cluster Computing and the Grid, 2001, pp. 618-625.

[BW02] R. Bless, K. Wehrle: "IP Multicast in Differentiated Services Networks", Internet Draft, draft-bless-diffserv-multicast-03.txt, March 2002, work in progress.

[CNS00] S. Chen, K. Nahrstedt, Y. Shavitt, "A QoS-aware multicast routing protocol", IEEE JSAC Vol. 18, No. 12, Dec. 2000.

[FIT96] V. Firoiu, D. Towsley, "Call admission and resource reservation for multicast sessions", IEEE INFOCOM 1996, pp. 94-101.

[KPP93] V. P. Kompella, J. C. Pasquale, G. C. Polyzos, "Multicast routing for multimedia communication", IEEE/ACM Transactions on Networking, Vol. 1, No. 3, pp. 286-292, June 1993.

[PIMV2] B. Fenner, M. Handley, H. Holbrook, I. Kouvelas: "Protocol Independent Multicast - Sparse Mode (PIM-SM) Protocol Specification (Revised)", draft-ietf-pim-sm-v2-new-05.txt, Internet-Draft, work in progress, March 2002.

[R2205] R. Braden, L. Zhang, S. Berson, S. Herzog, S. Jamin, "Resource ReSerVation Protocol (RSVP) - Version 1 Functional Specification", RFC 2205, Sep. 1997

[R2362] D. Estrin, D. Farinacci, A. Helmy, D. Thaler, S. Deering, M. Handley, V. Jacobson, C. Liu, P. Sharma, L. Wei: "Protocol Independent Multicast-Sparse Mode (PIM-SM): Protocol Specification", RFC 2362, June 1998.

[R2474] K. Nichols, S. Blake, F. Baker, D. Black, "Definition of the Differentiated Services Field (DS Field) in the IPv4 and IPv6 Headers", RFC 2474, Dec. 1998.

[R2475] S. Blake, D. Black, M. Carlson, E. Davies, Z. Wang, W. Weiss, "An Architecture for Differentiated Services", RFC 2475, Dec. 1998.

[R2597] J. Heinanen, T. Finland, F. Baker, W. Weiss, J. Wroclawski: "Assured Forwarding PHB Group", IETF, RFC 2597, June 1999.

[SM01] A. Striegel, G. Manimaran: "A Scalable Protocol for Member Join/Leave in DiffServ Multicast", in Proc. of Local Computer Networks (LCN) 2001, Tampa, Florida, Nov. 2001.

[YFB02] Y. Shuqian Yan, M. Faloutsos, A. Banerjea, "QoS-aware multicast routing for the Internet: the design and evaluation of QoSMIC", IEEE/ACM Transactions on Networking, Vol. 10, No. 1, pp. 54-66, Feb. 2002.

[YM02] B. Yang, P. Mohapatra: "Multicasting in Differentiated Service Domains", IEEE Globecom 2002.

Author Index

Lecture Notes in Computer Science

For information about Vols. 1–2501

please contact your bookseller or Springer-Verlag

Vol. 2357: M. Blaze (Ed.), Financial Cryptography. Proceedings, 2002. VIII, 301 pages. 2003.

Vol. 2502: D. Gollmann, G. Karjoth, M. Waidner (Eds.), Computer Security – ESORICS 2002. Proceedings, 2002. X, 281 pages. 2002.

Vol. 2503: S. Spaccapietra, S.T. March, Y. Kambayashi (Eds.), Conceptual Modeling – ER 2002. Proceedings, 2002. XX, 480 pages. 2002.

Vol. 2504: M.T. Escrig, F. Toledo, E. Golobardes (Eds.), Topics in Artificial Intelligence. Proceedings, 2002. XI, 432 pages. 2002. (Subseries LNAI).

Vol. 2506: M. Feridun, P. Kropf, G. Babin (Eds.), Management Technologies for E-Commerce and E-Business Applications. Proceedings, 2002. IX, 209 pages. 2002.

Vol. 2507: G. Bittencourt, G.L. Ramalho (Eds.), Advances in Artificial Intelligence. Proceedings, 2002. XIII, 418 pages. 2002. (Subseries LNAI).

Vol. 2508: D. Malkhi (Ed.), Distributed Computing. Proceedings, 2002. X, 371 pages. 2002.

Vol. 2509: C.S. Calude, M.J. Dinneen, F. Peper (Eds.), Unconventional Models in Computation. Proceedings, 2002. VIII, 331 pages. 2002.

Vol. 2510: H. Shafazand, A Min Tjoa (Eds.), EurAsia-ICT 2002: Information and Communication Technology. Proceedings, 2002. XXIII, 1020 pages. 2002.

Vol. 2511: B. Stiller, M. Smirnow, M. Karsten, P. Reichl (Eds.), From QoS Provisioning to QoS Charging. Proceedings, 2002. XIV, 348 pages. 2002.

Vol. 2512: C. Bussler, R. Hull, S. McIlraith, M.E. Orlowska, B. Pernici, J. Yang (Eds.), Web Services, E-Business, and the Semantic Web. Proceedings, 2002. XI, 277 pages. 2002.

Vol. 2513: R. Deng, S. Qing, F. Bao, J. Zhou (Eds.), Information and Communications Security. Proceedings, 2002. XII, 496 pages. 2002.

Vol. 2514: M. Baaz, A. Voronkov (Eds.), Logic for Programming, Artificial Intelligence, and Reasoning. Proceedings, 2002. XIII, 465 pages. 2002. (Subseries LNAI).

Vol. 2515: F. Boavida, E. Monteiro, J. Orvalho (Eds.), Protocols and Systems for Interactive Distributed Multimedia. Proceedings, 2002. XIV, 372 pages. 2002.

Vol. 2516: A. Wespi, G. Vigna, L. Deri (Eds.), Recent Advances in Intrusion Detection. Proceedings, 2002. X, 327 pages. 2002.

Vol. 2517: M.D. Aagaard, J.W. O'Leary (Eds.), Formal Methods in Computer-Aided Design. Proceedings, 2002. XI, 399 pages. 2002.

Vol. 2518: P. Bose, P. Morin (Eds.), Algorithms and Computation. Proceedings, 2002. XIII, 656 pages. 2002.

Vol. 2519: R. Meersman, Z. Tari, et al. (Eds.), On the Move to Meaningful Internet Systems 2002: CoopIS, DOA, and ODBASE. Proceedings, 2002. XXIII, 1367 pages. 2002.

Vol. 2521: A. Karmouch, T. Magedanz, J. Delgado (Eds.), Mobile Agents for Telecommunication Applications. Proceedings, 2002. XII, 317 pages. 2002.

Vol. 2522: T. Andreasen, A. Motro, H. Christiansen, H. Legind Larsen (Eds.), Flexible Query Answering. Proceedings, 2002. XI, 386 pages. 2002. (Subseries LNAI).

Vol. 2523: B.S. Kaliski Jr., Ç.K. Koç, C. Paar (Eds.), Cryptographic Hardware and Embedded Systems – CHES 2002. Proceedings, 2002. XIV, 612 pages. 2002.

Vol. 2525: H.H. Bülthoff, S.-Whan Lee, T.A. Poggio, C. Wallraven (Eds.), Biologically Motivated Computer Vision. Proceedings, 2002. XIV, 662 pages. 2002.

Vol. 2526: A. Colosimo, A. Giuliani, P. Sirabella (Eds.), Medical Data Analysis. Proceedings, 2002. IX, 222 pages. 2002.

Vol. 2527: F.J. Garijo, J.C. Riquelme, M. Toro (Eds.), Advances in Artificial Intelligence – IBERAMIA 2002. Proceedings, 2002. XVIII, 955 pages. 2002. (Subseries LNAI).

Vol. 2528: M.T. Goodrich, S.G. Kobourov (Eds.), Graph Drawing. Proceedings, 2002. XIII, 384 pages. 2002.

Vol. 2529: D.A. Peled, M.Y. Vardi (Eds.), Formal Techniques for Networked and Distributed Sytems – FORTE 2002. Proceedings, 2002. XI, 371 pages. 2002.

Vol. 2531: J. Padget, O. Shehory, D. Parkes, N. Sadeh, W.E. Walsh (Eds.), Agent-Mediated Electronic Commerce IV. Proceedings, 2002. XVII, 341 pages. 2002. (Subseries LNAI).

Vol. 2532: Y.-C. Chen, L.-W. Chang, C.-T. Hsu (Eds.), Advances in Multimedia Information Processing – PCM 2002. Proceedings, 2002. XXI, 1255 pages. 2002.

Vol. 2533: N. Cesa-Bianchi, M. Numao, R. Reischuk (Eds.), Algorithmic Learning Theory. Proceedings, 2002. XI, 415 pages. 2002. (Subseries LNAI).

Vol. 2534: S. Lange, K. Satoh, C.H. Smith (Ed.), Discovery Science. Proceedings, 2002. XIII, 464 pages. 2002.

Vol. 2535: N. Suri (Ed.), Mobile Agents. Proceedings, 2002. X, 203 pages. 2002.

Vol. 2536: M. Parashar (Ed.), Grid Computing – GRID 2002. Proceedings, 2002. XI, 318 pages. 2002.

Vol. 2537: D.G. Feitelson, L. Rudolph, U. Schwiegelshohn (Eds.), Job Scheduling Strategies for Parallel Processing. Proceedings, 2002. VII, 237 pages. 2002.

Vol. 2538: B. König-Ries, K. Makki, S.A.M. Makki, N. Pissinou, P. Scheuermann (Eds.), Developing an Infrastructure for Mobile and Wireless Systems. Proceedings 2001. X, 183 pages. 2002.

Vol. 2539: K. Börner, C. Chen (Eds.), Visual Interfaces to Digital Libraries. X, 233 pages. 2002.

Vol. 2540: W.I. Grosky, F. Plášil (Eds.), SOFSEM 2002: Theory and Practice of Informatics. Proceedings, 2002. X, 289 pages. 2002.

Vol. 2541: T. Barkowsky, Mental Representation and Processing of Geographic Knowledge. X, 174 pages. 2002. (Subseries LNAI).

Vol. 2542: I. Dimov, I. Lirkov, S. Margenov, Z. Zlatev (Eds.), Numerical Methods and Applications. Proceedings, 2002. X, 174 pages. 2003.

Vol. 2544: S. Bhalla (Ed.), Databases in Networked Information Systems. Proceedings 2002. X, 285 pages. 2002.

Vol. 2545: P. Forbrig, Q. Limbourg, B. Urban, J. Vanderdonckt (Eds.), Interactive Systems. Proceedings 2002. XII, 574 pages. 2002.

Vol. 2546: J. Sterbenz, O. Takada, C. Tschudin, B. Plattner (Eds.), Active Networks. Proceedings, 2002. XIV, 267 pages. 2002.

Vol. 2547: R. Fleischer, B. Moret, E. Meineche Schmidt (Eds.), Experimental Algorithmics. XVII, 279 pages. 2002.

Vol. 2548: J. Hernández, Ana Moreira (Eds.), Object-Oriented Technology. Proceedings, 2002. VIII, 223 pages. 2002.

Vol. 2549: J. Cortadella, A. Yakovlev, G. Rozenberg (Eds.), Concurrency and Hardware Design. XI, 345 pages. 2002.

Vol. 2550: A. Jean-Marie (Ed.), Advances in Computing Science – ASIAN 2002. Proceedings, 2002. X, 233 pages. 2002.

Vol. 2551: A. Menezes, P. Sarkar (Eds.), Progress in Cryptology – INDOCRYPT 2002. Proceedings, 2002. XI, 437 pages. 2002.

Vol. 2552: S. Sahni, V.K. Prasanna, U. Shukla (Eds.), High Performance Computing – HiPC 2002. Proceedings, 2002. XXI, 735 pages. 2002.

Vol. 2553: B. Andersson, M. Bergholtz, P. Johannesson (Eds.), Natural Language Processing and Information Systems. Proceedings, 2002. X, 241 pages. 2002.

Vol. 2554: M. Beetz, Plan-Based Control of Robotic Agents. XI, 191 pages. 2002. (Subseries LNAI).

Vol. 2555: E.-P. Lim, S. Foo, C. Khoo, H. Chen, E. Fox, S. Urs, T. Costantino (Eds.), Digital Libraries: People, Knowledge, and Technology. Proceedings, 2002. XVII, 535 pages. 2002.

Vol. 2556: M. Agrawal, A. Seth (Eds.), FST TCS 2002: Foundations of Software Technology and Theoretical Computer Science. Proceedings, 2002. XI, 361 pages. 2002.

Vol. 2557: B. McKay, J. Slaney (Eds.), AI 2002: Advances in Artificial Intelligence. Proceedings, 2002. XV, 730 pages. 2002. (Subseries LNAI).

Vol. 2558: P. Perner, Data Mining on Multimedia Data. X, 131 pages. 2002.

Vol. 2559: M. Oivo, S. Komi-Sirviö (Eds.), Product Focused Software Process Improvement. Proceedings, 2002. XV, 646 pages. 2002.

Vol. 2560: S. Goronzy, Robust Adaptation to Non-Native Accents in Automatic Speech Recognition. Proceedings, 2002. XI, 144 pages. 2002. (Subseries LNAI).

Vol. 2561: H.C.M. de Swart (Ed.), Relational Methods in Computer Science. Proceedings, 2001. X, 315 pages. 2002.

Vol. 2562: V. Dahl, P. Wadler (Eds.), Practical Aspects of Declarative Languages. Proceedings, 2003. X, 315 pages. 2002.

Vol. 2566: T.Æ. Mogensen, D.A. Schmidt, I.H. Sudborough (Eds.), The Essence of Computation. XIV, 473 pages. 2002.

Vol. 2567: Y.G. Desmedt (Ed.), Public Key Cryptography – PKC 2003. Proceedings, 2003. XI, 365 pages. 2002.

Vol. 2568: M. Hagiya, A. Ohuchi (Eds.), DNA Computing. Proceedings, 2002. XI, 338 pages. 2003.

Vol. 2569: D. Gollmann, G. Karjoth, M. Waidner (Eds.), Computer Security – ESORICS 2002. Proceedings, 2002. XIII, 648 pages. 2002. (Subseries LNAI).

Vol. 2570: M. Jünger, G. Reinelt, G. Rinaldi (Eds.), Combinatorial Optimization – Eureka, You Shrink!. Proceedings, 2001. X, 209 pages. 2003.

Vol. 2571: S.K. Das, S. Bhattacharya (Eds.), Distributed Computing. Proceedings, 2002. XIV, 354 pages. 2002.

Vol. 2572: D. Calvanese, M. Lenzerini, R. Motwani (Eds.), Database Theory – ICDT 2003. Proceedings, 2003. XI, 455 pages. 2002.

Vol. 2574: M.-S. Chen, P.K. Chrysanthis, M. Sloman, A. Zaslavsky (Eds.), Mobile Data Management. Proceedings, 2003. XII, 414 pages. 2003.

Vol. 2575: L.D. Zuck, P.C. Attie, A. Cortesi, S. Mukhopadhyay (Eds.), Verification, Model Checking, and Abstract Interpretation. Proceedings, 2003. XI, 325 pages. 2003.

Vol. 2576: S. Cimato, C. Galdi, G. Persiano (Eds.), Security in Communication Networks. Proceedings, 2002. IX, 365 pages. 2003.

Vol. 2578: F.A.P. Petitcolas (Ed.), Information Hiding. Proceedings, 2002. IX, 427 pages. 2003.

Vol. 2580: H. Erdogmus, T. Weng (Eds.), COTS-Based Software Systems. Proceedings, 2003. XVIII, 261 pages. 2003.

Vol. 2581: J.S. Sichman, F. Bousquet, P. Davidsson (Eds.), Multi-Agent-Based Simulation II. Proceedings, 2002. X, 195 pages. 2003. (Subseries LNAI).

Vol. 2583: S. Matwin, C. Sammut (Eds.), Inductive Logic Programming. Proceedings, 2002. X, 351 pages. 2003. (Subseries LNAI).

Vol. 2588: A. Gelbukh (Ed.), Computational Linguistics and Intelligent Text Processing. Proceedings, 2003. XV, 648 pages. 2003.

Vol. 2589: E. Börger, A. Gargantini, E. Riccobene (Eds.), Abstract State Machines 2003. Proceedings, 2003. XI, 427 pages. 2003.

Vol. 2594: A. Asperti, B. Buchberger, J.H. Davenport (Eds.), Mathematical Knowledge Management. Proceedings, 2003. X, 225 pages. 2003.

Vol. 2598: R. Klein, H.-W. Six, L. Wegner (Eds.), Computer Science in Perspective. X, 357 pages. 2003.

Vol. 2600: S. Mendelson, A.J. Smola, Advanced Lectures on Machine Learning. Proceedings, 2002. IX, 259 pages. 2003. (Subseries LNAI).

Vol. 2601: M. Ajmone Marsan, G. Corazza, M. Listanti, A. Roveri (Eds.) Quality of Service in Multiservice IP Networks. Proceedings, 2003. XV, 759 pages. 2003.

Vol. 2602: C. Priami (Ed.), Computational Methods in Systems Biology. Proceedings, 2003. IX, 214 pages. 2003.

Vol. 2607: H. Alt, M. Habib (Eds.), STACS 2003. Proceedings, 2003. XVII, 700 pages. 2003.